Seaweeds and Seaweed-Derived Compounds

Fatih Ozogul • Monica Trif • Alexandru Rusu
Editors

Seaweeds and Seaweed-Derived Compounds

Meeting the Growing Need for Healthy Biologically Active Compounds

Springer

Editors
Fatih Ozogul
Department of Seafood Processing
Technology, Faculty of Fisheries
Cukurova University
Adana, Türkiye

Monica Trif
Food Research Department
Centre for Innovative Process
Engineering GmbH
Stuhr, Germany

Alexandru Rusu
Strategic Research Department
Biozoon Food Innovations GmbH
Bremerhaven, Germany

ISBN 978-3-031-65528-9 ISBN 978-3-031-65529-6 (eBook)
https://doi.org/10.1007/978-3-031-65529-6

This Springer imprint is published by the registered company Springer Nature Switzerland AG
The registered company address is: Gewerbestrasse 11, 6330 Cham, Switzerland

Preface

Seaweeds, also known as marine macroalgae, are large, multicellular algae that grow in marine environments such as oceans and seas. They are considered some of the most productive and diverse organisms on Earth and play a crucial role in marine ecosystems. Seaweed has been used for various purposes for centuries in different cultures. Today, seaweeds are gaining increasing attention for their various applications in various industries, including food, agriculture, pharmaceuticals, cosmetics, and environmental management.

These incredible organisms have been used for various purposes throughout human history, such as food, fertilizer, and even medicine. In recent years, there has been growing interest in seaweed and seaweed-derived compounds due to their potential health benefits and various biologically active properties. Meeting the growing need for healthy biologically active compounds is a crucial aspect of addressing various health challenges faced by individuals and populations worldwide. Biologically active compounds are natural or synthetic substances that have a positive impact on biological processes and can contribute to overall health and well-being.

While the potential health benefits of seaweed and seaweed-derived compounds are promising, more research is needed to fully understand their mechanisms of action and potential applications. Seaweeds offer many potential benefits, their sustainable harvest and cultivation are essential to protect marine ecosystems and ensure a stable supply for various applications. Proper management practices are necessary to maintain the balance between utilization and conservation of these valuable marine resources. Additionally, as with any natural product, it is crucial to ensure proper harvesting, processing, and quality control to maximize their benefits and minimize potential risks. As the demand for natural and sustainable products grows, seaweeds and seaweed-derived compounds have the potential to become essential components in various industries, including functional foods, nutraceuticals, cosmeceuticals, and pharmaceuticals.

Nutritional Value: Seaweeds are rich in essential nutrients, including vitamins (such as vitamins A, C, E, and K), minerals (such as iodine, iron, calcium, and magnesium), and trace elements. They also contain a good amount of dietary fiber and are relatively low in calories, making them a valuable addition to a healthy diet.

Antioxidant Properties: Many seaweed species contain antioxidants like polyphenols, flavonoids, and carotenoids. These compounds help combat oxidative stress in the body, reducing the risk of chronic diseases and promoting overall well-being.

Anti-inflammatory Effects: Seaweeds are known to possess anti-inflammatory properties, which can be attributed to various bioactive compounds they contain. These compounds may help in alleviating inflammation-related conditions and diseases.

Immunomodulatory Effects: Some seaweed-derived compounds have been found to modulate the immune system, potentially enhancing immune responses and promoting a balanced immune function.

Antimicrobial Activity: Seaweeds produce natural antimicrobial compounds as part of their defense mechanisms against pathogens. These antimicrobial properties have drawn attention to their potential application in various industries, including pharmaceuticals and food preservation.

Anticancer Potential: Certain seaweed-derived compounds have demonstrated promising anticancer properties in preclinical studies. These compounds may have the potential to inhibit tumor growth and induce apoptosis in cancer cells.

Neuroprotective Effects: Some research suggests that certain seaweed compounds could have neuroprotective properties, offering potential benefits in neurological disorders and brain health.

Weight Management: As a low-calorie and nutrient-dense food source, seaweeds may aid in weight management and could be beneficial in obesity-related interventions.

Wound Healing: Seaweeds contain compounds that may promote wound healing by accelerating cell proliferation and collagen synthesis.

Environmental Sustainability: Cultivating and harvesting seaweeds can be environmentally sustainable since they require no land, freshwater, or fertilizers, and they absorb carbon dioxide while releasing oxygen during photosynthesis.

Adana, Turkey
Stuhr, Germany
Bremerhaven, Germany

Fatih Ozogul
Monica Trif
Alexandru Rusu

Contents

Contributors

Lubna Ahmed Department of Agriculture, Food and Animal Health, School of Health and Science, Dundalk Institute of Technology, Dunkdal, Ireland

Shivananthini Balasundaram Centre for Plant Biotechnology, Department of Botany, St. Xavier's College (Autonomous), Palayamkottai, Tamil Nadu, India

Catherine Barry-Ryan Food Innovation Lab, School of Food Science and Environmental Health, Technological University Dublin, Dublin, Ireland

Darshee Baxi Department of Biomedical Sciences and Life Sciences, School of Science, Navrachana University, Vadodara, Gujarat, India

Touria Ould Bellahcen Health and Environment Laboratory, Faculty of Sciences Ain Chock, Hassan II University of Casablanca, Casablanca, Morocco

Sami Ben Hadj Ahmed Laboratory of Protein Engineering and Bioactive Molecules (LIP-MB), National Institute of Applied Sciences and Technology (INSAT), University of Carthage, Carthage, Tunisia

Medical Laboratory Sciences Department, College of Applied Medical Sciences, King Khalid University, Abha, Kingdom of Saudi Arabia

Geert Bruggeman Nutrition Sciences N.V., Drongen, Belgium

Martina Čagalj Department of Marine Studies, University of Split, Split, Croatia

Alejandra Cardelle-Cobas Laboratorio de Higiene, Inspección y Control de Alimentos (LHICA), Departamento de Química Analítica, Nutrición y Bromatología, Universidade de Santiago de Compostela, Lugo, Spain

Hannes Carmans Nutrition Sciences N.V., Drongen, Belgium

Alberto Cepeda Laboratorio de Higiene, Inspección y Control de Alimentos (LHICA), Departamento de Química Analítica, Nutrición y Bromatología, Universidade de Santiago de Compostela, Lugo, Spain

Ouafa Cherifi Laboratory of Water Biodiversity and Climate Changes, Cadi Ayyad University, Marrakech, Morocco

National Center for Studies and Research on Water and Energy, Cadi Ayyad University, Marrakech, Morocco

Eleonora Curcuraci Department of Earth and Marine Sciences DiSTeM, University of Palermo, Trapani, Italy

Nariman El Abed Laboratory of Protein Engineering and Bioactive Molecules (LIP-MB), National Institute of Applied Sciences and Technology (INSAT), University of Carthage, Carthage, Tunisia

Maryam El Bakali BV2MAP, FPL, Adelmalek Essaadi University, Tetouan, Morocco

Tejal K. Gajaria Department of Biomedical Sciences and Life Sciences, School of Science, Navrachana University, Vadodara, Gujarat, India

Marco Garcia-Vaquero School Agriculture and Food Science, University College Dublin, Dublin, Ireland

Vidyarani George Centre for Plant Biotechnology, Department of Botany, St. Xavier's College (Autonomous), Palayamkottai, Tamil Nadu, India

Cengiz Gokbulut Department of Pharmacology, Faculty of Medicine, Balikesir University, Balikesir, Turkiye

Wiktoria Grzebieniarz Department of Chemistry, University of Agriculture, Cracow, Poland

Claire Hellio LEMAR, IRD, CNRS, Ifremer, Université de Brest, Plouzane, France

Silvia Juliet Iruthayamani Centre for Plant Biotechnology, Department of Botany, St. Xavier's College (Autonomous), Palayamkottai, Tamil Nadu, India

Ewelina Jamróz Department of Chemistry, University of Agriculture, Cracow, Poland

G. D. T. M. Jayasinghe Analytical Chemistry Laboratory, National Aquatic Resources Research & Development Agency (NARA), Colombo, Sri Lanka

Universite de Pau et des Pays de l'Adour, E2S UPPA, CNRS, IPREM UMR 5254, Hélioparc, Pau, France

B. K. K. K. Jinadasa Analytical Chemistry Laboratory, National Aquatic Resources Research & Development Agency (NARA), Colombo, Sri Lanka

Department of Food Science and Technology (DFST), Faculty of Livestock, Fisheries & Nutrition (FLFN), Wayamba University of Sri Lanka, Makandura, Gonawila, Sri Lanka

Se-Kwon Kim Department of Marine Sciences and Convergence Engineering, College of Science and Technology, Hanyang University, Seoul, Gyeonggi-do, Republic of Korea

Aroa Lopez-Santamarina Laboratorio de Higiene, Inspección y Control de Alimentos (LHICA), Departamento de Química Analítica, Nutrición y Bromatología, Universidade de Santiago de Compostela, Lugo, Spain

Johnson Marimuthu Alias Antonysamy Centre for Plant Biotechnology, Department of Botany, St. Xavier's College (Autonomous), Palayamkottai, Tamil Nadu, India

Concetta Maria Messina Department of Earth and Marine Sciences DiSTeM, University of Palermo, Trapani, Italy

Jose Manuel Miranda Laboratorio de Higiene, Inspección y Control de Alimentos (LHICA), Departamento de Química Analítica, Nutrición y Bromatología, Universidade de Santiago de Compostela, Lugo, Spain

Armin Mirzapour-Kouhdasht School Agriculture and Food Science, University College Dublin, Dublin, Ireland

Liv Torunn Mydland Department of Animal and Aquacultural Sciences, Faculty of Biosciences, Norwegian University of Life Sciences, Ås, Norway

Pranav Nakhate Aachen-Maastricht Institute for Bio-based Materials (AMIBM), Maastricht University, Maastricht, the Netherlands

Ali Niazi Institute of Biotechnology, Shiraz University, Shiraz, Iran

Nikola Nowak Department of Chemistry, University of Agriculture, Cracow, Poland

Margareth Øverland Department of Animal and Aquacultural Sciences, Faculty of Biosciences, Norwegian University of Life Sciences, Ås, Norway

Fatih Ozogul Department of Seafood Processing Technology, Faculty of Fisheries, University of Cukurova, Adana, Turkey

Yeşim Ozogul Department of Seafood Processing Technology, Faculty of Fisheries, Cukurova University, Adana, Turkey

Parth Pandya Department of Biomedical Sciences and Life Sciences, School of Science, Navrachana University, Vadodara, Gujarat, India

Ratih Pangestuti Research Center for Food Processing and Technology, National Research and Innovation Agency (BRIN), Yogyakarta, Indonesia

Sevim Polat Department of Marine Biology, Faculty of Fisheries, Cukurova University, Adana, Turkey

Yanuariska Putra Research Center for Marine and Land Bioindustry, National Research and Innovation Agency (BRIN), Mataram, West Nusa Tenggara, Indonesia

Puji Rahmadi Research Center for Oceanography, National Research and Innovation Agency (BRIN), Jakarta Utara, Indonesia

A. V. Ramachandran Department of Biomedical Sciences and Life Sciences, School of Science, Navrachana University, Vadodara, Gujarat, India

Elizabeth Robin Department of Biomedical Sciences and Life Sciences, School of Science, Navrachana University, Vadodara, Gujarat, India

Alexandru Rusu Strategic Research Department, Biozoon Food Innovations GmbH, Bremerhaven, Germany

Andrea Santulli Department of Earth and Marine Sciences DiSTeM, University of Palermo, Trapani, Italy

Istituto di Biologia Marina, Consorzio Universitario della Provincia di Trapani, Trapani, Italy

Evi Amelia Siahaan Research Center for Marine and Land Bioindustry, National Research and Innovation Agency (BRIN), Mataram, West Nusa Tenggara, Indonesia

Vida Šimat Department of Marine Studies, University of Split, Split, Croatia

Laura I. Sinisterra-Loaiza Laboratorio de Higiene, Inspección y Control de Alimentos (LHICA), Departamento de Química Analítica, Nutrición y Bromatología, Universidade de Santiago de Compostela, Lugo, Spain

Mohammad Sadegh Taghizadeh Institute of Biotechnology, Shiraz University, Shiraz, Iran

Monica Trif Food Research Department, Centre for Innovative Process Engineering GmbH, Stuhr, Germany

Robin Vandercruyssen Nutrition Sciences N.V., Drongen, Belgium

Yvonne Van der Meer Aachen-Maastricht Institute for Bio-based Materials (AMIBM), Maastricht University, Maastricht, the Netherlands

Stefanie Verstringe Nutrition Sciences N.V., Drongen, Belgium

Pinar Yerlikaya Department of Seafood Processing Technology, Faculty of Fisheries, Akdeniz University, Antalya, Turkey

Chapter 1
Why Global Interest in Seaweed? Can Seaweed Conquer the World?

Martina Čagalj and Vida Šimat

1 Seaweeds: Some General Points

The seaweeds, also known as macroalgae, are a broad group of multi-cellular, macroscopic marine organisms that have photosynthetic pigments and perform photosynthesis. They convert inorganic CO_2 dissolved in seawater to organic forms using solar energy. Seaweeds can change their photosynthetic energy investment during the day or in different seasons from C to N and P. They have a simple structure with no or little complex tissues and cellular differentiation, making them talophytes (Gómez and Huovinen 2012; Gordillo 2012; Peñalver et al. 2020). This group of organisms is represented by over several thousand species and the final number of species remains unknown. Seaweeds grow at different depths, from the surface, intertidal shallow waters all the way to the maximal depth where a limiting factor is light with not less than 0.05% or 0.002% of the surface irradiance. Seaweeds can adjust their growth and photosynthesis to daily or seasonal temperature changes. Generally, their growth and photosynthetic rates increase with sea temperature, however, the maximum growth and photosynthetic rates occur at optimal sea temperature which is highly dependent on the species and geographical location. As primary producers, seaweeds are an essential part of aquatic ecosystems, providing them with oxygen and organic compounds. They are the essential for certain benthic communities being food to sea urchins, gastropods, and herbivorous fish. Besides, seaweeds provide habitats, hatcheries, and shelter for fish and other marine organisms, they reduce overfishing by being the alternative livelihoods to fish and protect the shoreline by acting as a buffer against strong waves. Some species prefer sheltered bays, while others grow in exposed open sea areas (Eggert 2012; Hanelt and

M. Čagalj (✉) · V. Šimat
University of Split, Department of Marine Studies, Split, Croatia
e-mail: mcagalj@unist.hr

F. Ozogul et al. (eds.), *Seaweeds and Seaweed-Derived Compounds*,
https://doi.org/10.1007/978-3-031-65529-6_1

Figueroa 2012; Peng et al. 2015; Pereira 2016; Lähteenmäki-Uutela et al. 2021; Cai et al. 2021).

Seaweeds are classified as brown (Phaeophyta), red (Rhodophyta) or green (Chlorophyta) based on the different pigmentation. Brown seaweeds contain chlorophyll a and c, and fucoxanthin, red seaweeds chlorophyll, phycocyanin and phycoerythrin, and green seaweeds chlorophyll a and b, and xanthophylls (Peng et al. 2015; Pereira 2016). Brown seaweeds are represented by yellow to dark brown coloured species, divided into kelps or fucales. Red seaweeds are the most primitive among the three, and have the highest number of representatives, while the green seaweeds are the smallest group represented by greenish-yellow to dark green coloured algae (Peñalver et al. 2020). Brown seaweeds can grow from small species of 30 cm to giant kelps that are over 50 m long. Usually smaller red seaweeds grow in a range from only a few centimetres to around 1 m. Furthermore, they are not always red, they can also be purple or brownish-red. Similar to red seaweeds, the length range of green seaweeds is from a few centimetres to less than 1 m (McHugh 2003).

Seaweeds are able to adapt and become invulnerable to the harsh living conditions in seas and oceans. They produce primary metabolites (polysaccharides, fatty acids, proteins and amino acids) involved in their development, growth, and reproduction, and secondary metabolites (phenolic compounds, sterols, pigments, vitamins and other) as a response to stressful environmental conditions they live and grow in. Some of their stressors are salinity and temperature oscillations, tides, UV radiation, predators, and environmental toxins (Pereira 2018; Generalić Mekinić et al. 2019). During their lifecycle, the seaweeds adapt to the environmental changes by specific changes in thallus development and in the chemical profile of primary and secondary metabolites.

2 Seaweed Production

Global seaweed production is increasing every year and it has reached the recording 35.762.504 tonnes in wet weight in 2019. valued at 14.7 billion USD (Cai et al. 2021). Seaweed cultivation is responsible for almost 97% of the total world seaweed production. Asian countries, namely China, Republic of Korea, Indonesia, Democratic People's Republic of Korea, Philippines and Japan together cover 96.73% of the world seaweed production, almost completely attained through seaweed cultivation. Only 2.16% of the world production is contributed by the Americas and Europe. However, seaweed production in these continents is mostly from wild harvesting (FAO 2021a; Bizzaro et al. 2022). In Europe, the largest collector of seaweeds (mainly kelp collected by dredging) is Norway with around 150.000 tonnes annually. Seaweed species that are being cultivated or harvested in Norway, Sweden, Denmark, Scotland and France are brown seaweeds *Ascophyllum nodosum*, *Laminaria hyperborean*, *Laminaria digitata*, *Saccharina latissima* and *Alaria esculenta* (Lähteenmäki-Uutela et al. 2021). For some red seaweeds such as dulse

(*Palmaria palmata*), used in both food and feed formulations, recent studies investigated the new hatchery methods (Schmedes and Nielsen 2020) and the possibilities of their cultivation in aerated tanks (Evans et al. 2021).

As the world population increases, seaweed have become an interesting source of nutrients, especially proteins. They also have high productivity and production sustainability compared to terrestrial plants (Felaco et al. 2020). Cultivation of seaweeds removes large amounts of phosphorus and nitrogen from coastal ecosystems, thus it is often associated with aquaculture production, and most recently even with food production process waters (Stedt et al. 2022). As marine nutrient pollution increases around the world and contributes to eutrophication, seaweeds could potentially have a role in targeted nutrient assimilation as a natural bio-extractant. Seaweeds can alleviate eutrophication, reduce ocean acidification, treat wastewater and capture/sequester carbon through the photosynthetic process (Racine et al. 2021). This opens new opportunities for seaweed exploitation since only a small number of species is cultivated. According to FAO statistics, only 27 species were cultivated in 2019. However, it is possible that not all data is recorded in their statistics due to underreporting, confidentiality or other reasons (Cai et al. 2021). Over 95% of global seaweed cultivation in 2019 was covered by five genera belonging to brown (*Laminaria/Saccharina* and *Undaria*) and red (*Kappaphycus/Eucheuma*, *Gracilaria*, and *Porphyra/Pyropia*) seaweeds showing the annual growth over 10%. Nevertheless, nearly 30% of world aquaculture production is attained from seaweeds (FAO 2021b; Cai et al. 2021). Even in wild harvesting, only a small number of species is collected, only 36 species in 2019, from which 15 species are also being cultivated and the other 21 are mostly kelps. Some of the issues and challenges in seaweed cultivation are limited or uncertain demand, reduced or limited availability of suitable nearshore farm sites, shortage of labour, low or declined seedling quality, and environmental and ecosystem impacts or risks (Cai et al. 2021). Seaweed cultivation offers development opportunities in coastal areas throughout the world, however, preventive measures have to be taken since intensive farming can have irreversible consequences on biodiversity (Loureiro et al. 2015).

On the other hand, the ecological impact of seaweed cultivation is positive and has a neutral carbon footprint as they use only sunlight, nutrients and CO_2 from the sea (Bizzaro et al. 2022). Seaweeds are commonly cultivated in open water. Their cultivation doesn't use freshwater, terrestrial land or feed directly. Besides, they have high growth rates. Seaweeds can also be used in the integrated aquaculture system as a way to treat wastes and as a food supply for produced marine animals at the same site. The environmental conditions for algal growth (temperature, salinity, light, nutrients, water movement and quality) depend on the cultivated species. Control of these conditions is rather difficult in open waters. Environmental parameters aside, algal growth is also under influence of wild animals and marine microbiota. To cope with the problems of extensive cultivation, seaweeds can be cultivated in intensive systems in ponds or tanks. There is a possibility to cultivate one or several seaweed species at once in tanks under artificial light and with added nutrients and phytohormones or in lakes, ponds and lagoons with the use of organic and inorganic fertilizers. However, intensive systems are expensive and can cause stress to seaweeds (McHugh 2003; Titlyanov and Titlyanova 2010; Cai et al. 2021; Zhu et al. 2022). Another method of intensive cultivation is the integrated multitrophic aquaculture (IMTA) system. So far, three methods of integrated seaweed aquaculture can be distinguished: (i) cultivation of seaweeds in tanks, ponds or pools connected to reservoirs with fish or invertebrates to receive their effluents; (ii) cultivation of seaweeds with crustaceans, fish or mollusks together in ponds; and (iii) cultivation of seaweeds near the cages with animals in the open sea (Titlyanov and Titlyanova 2010). The most suitable seaweed species for their inclusion in IMTA systems are kelps (Bizzaro et al. 2022). Although the IMTA system has many advantages like reduction of environmental pollution, high nutrient uptake and production of higher biomass, it is a complex system with many challenges. Farmers can lack the expertise to maintain a well-functioning ecosystem for all cultivated species, infrastructures and operations needed for the cultivation of a few species can interfere with each other, and production planning based on market demand is less flexible (Titlyanov and Titlyanova 2010; Cai et al. 2021).

Seaweeds can be cultivated vegetatively or involving a reproductive cycle, depending on the species. When seaweeds are being cultivated vegetatively, small seaweed pieces are taken and placed in the environment to grow. When seaweeds are fully grown, they can be harvested in two ways. Either whole seaweed is being removed with small pieces being cut for re-cultivation or most of the plant is removed and small pieces are left to regrow (McHugh 2003). Placing the seaweeds in the environment can be done differently: (i) the pieces can be tied on the long ropes between wooden stakes on the water surface or several meters below; (ii) the pieces can be tied to ropes on a raft that is anchored to the bottom; (iii) ropes can be exchanged for nets in (i) and (ii); (iv) the pieces can be placed in the pond at the bottom without any fixation; (v) the pieces can be lodged in the soft sediment in more open waters; and (vi) the pieces can be attached to the sand-filled plastic tubes on sandy bottoms (McHugh 2003; Titlyanov and Titlyanova 2010).

Some seaweed species can't be grown vegetatively from pieces of mature individuals and it is necessary to involve a reproductive cycle in cultivation by generations' alternation. The most cultivated brown seaweeds belonging to the *Laminaria* order are cultivated in this way. The life cycle of this seaweed includes microscopic gametophyte and a large sporophyte alternation. The gametophytes are germinated from spores. As they grow, they become fertile and release sperm and eggs needed to form the embryonic sporophytes. These sporophytes grow and develop into large seaweed that releases spores and continues the life cycle. The transition from spore to gametophyte to embryonic sporophyte is carried out with cautious control of water temperature, light and nutrients in tanks of land-based facilities. Cultivation techniques include placing the ropes in tanks with spores for them to settle or manually inserting sporophytes into ropes (McHugh 2003; Titlyanov and Titlyanova 2010).

3 The Uses and Applications of the Seaweeds

3.1 Nutrient and Chemical Composition of Seaweeds

Seaweed macro and micronutrients have a broad range of confirmed biological activities which has, among other reasons, brought seaweeds to the focus of the research community recently. The number of research publications is increasing exponentially, reaching over 2800 documents with the keyword "seaweed" or "macroalgae" in 2021 in the Scopus database (Fig. 1.1).

The chemical composition of seaweeds is related to their habitat, environmental factors, and seasonal growth. They are known as highly nutritious food and a good source of micro- and macronutrients such as important minerals and vitamins, proteins and essential amino acids, polyunsaturated fatty acids, and soluble and insoluble dietary fibers. As such, they are considered a functional food. Seaweed have the ability to benefit human health by increasing welfare, decreasing the risk of some diseases and having a beneficial effect on physiological functions (Pereira 2016). Besides, they are found to be beneficial for the health and wellness of clinically unwell and elderly people (Cai et al. 2021).

Seaweeds contain essential minerals and trace elements bound to the polysaccharides making them easily bioavailable which are critical for many body functions in humans. They contain calcium, magnesium, potassium, iron, copper, iodine, zinc, selenium, strontium, molybdenum, manganese, and germanium, which make them important for the production of food supplements. Among mentioned, iodine has a fundamental role in thyroid gland function and thyroid hormones production. In the last century, iodine deficiency has been well documented in 1/3 of the world population. According to World Health Organization, recommended daily intake of

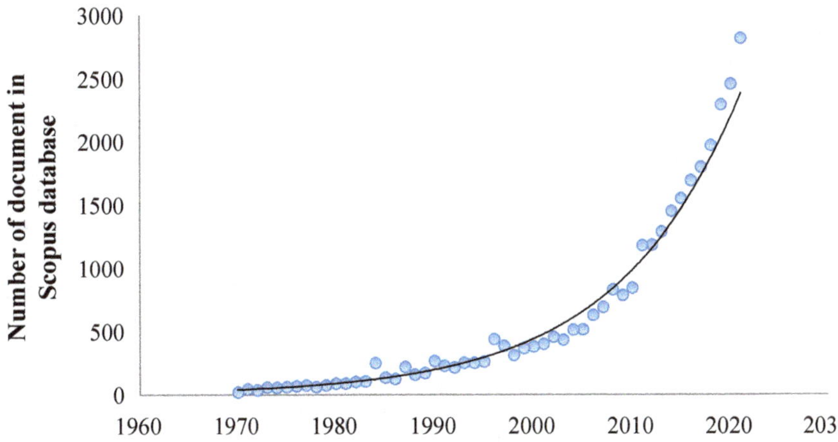

Fig. 1.1 The number of research publications with keyword "seaweed" or "macroalgae" from the Scopus database (data accessed in April 2022)

iodine for adults is 150 µg. Iodine levels in different seaweed species vary from 4.3 to 2660 mg/kg wet weight and their consumption may help to mitigate this problem (WHO 2007; Roohinejad et al. 2017; Correia et al. 2021). However, excessive iodine consumption over a long period of time can cause damage to human health through thyroid dysfunction. Consumers should follow producers' recommendations on seaweed daily intake limits (Darias-Rosales et al. 2020). Edible seaweeds might also be used for replacement of table salt which is high in sodium since they contain much less sodium (9–12%) but are rich in potassium.

Seaweeds are also an excellent source of vitamins; they contain both fat-soluble and water-soluble vitamins. Vitamins A (retinol), C (L-ascorbic acid), D (calciferol), E (tocopherols), K (phylloquinone) and B complex (riboflavin, cobalamin, folic acid) can be found in seaweeds. In comparison to terrestrial plants, vitamin content of seaweeds is superior as they contain all essential elements. Vitamin content varies depending on species, seasons, environmental conditions, time of harvesting, and the processing of seaweed (Peng et al. 2015; Peñalver et al. 2020; Polat et al. 2023). Commonly, seaweeds are marketed dry in order to stabilize them for long-term storage. Vitamins loss may occur during processing and storage of seaweeds, especially when drying is performed at high temperatures (Stévant et al. 2018; Rodríguez-Bernaldo de Quirós and López-Hernández 2021).

Recently, seaweeds became an alternative cheaper source of protein (Peng et al. 2015). Although protein content in seaweeds commonly ranges from 10% to 30%, red seaweeds contain up to 47% of protein which makes them a high-protein food. All essential amino acids are contained in the majority of seaweeds, and they represent almost half of the amino acids in seaweeds. Aspartic and glutamic acids, responsible for the "umami" taste, are the most abundant amino acids. However, serine, histidine, asparagine, lysine, valine, arginine, leucine, and isoleucine can also be found. The composition of amino acids in seaweed proteins varies seasonally (Tanna and Mishra 2018; Bizzaro et al. 2022; Polat et al. 2023). Besides, the protein content also varies seasonally, with the highest protein content found during winter and/or spring and the lowest during the summer and/or autumn for brown seaweeds (Garcia-Vaquero et al. 2021). Seaweed proteins can be hydrolysed chemically with acids and/or alkali, using high temperatures or enzymatically using proteases to produce protein hydrolysates and after purification, bioactive peptides. Bioactive peptides are short amino acid sequences, consisting from 2 to 20 amino acids encrypted in a protein. After being released through the extraction, food processing or digestion they exert beneficial effects on human health. Smaller peptide sequences have higher bioactive potential than proteins. They show antioxidant, antimicrobial, anti-inflammatory, antitumor, and antihypertensive activities mostly *in vitro*, but few studies also confirmed their bioactivity *in vivo* as well (Echave et al. 2021; Bizzaro et al. 2022).

Generally, the seaweeds contain low content of lipids, only 1–5% of the dry matter, making them a low-fat food. Polyunsaturated omega-3 and omega-6 fatty acids take up from 34% to 74% of total lipids in brown and red seaweeds (Polat et al. 2023). Brown seaweeds contain high amounts of oleic acid, α-linolenic acid, and linoleic acid and a range from 0.5 to 42 relative percentage of eicosapentaenoic acid (EPA). Red seaweeds may contain higher amounts of EPA, oleic acid, arachidonic acid, and palmitic acid when compared to brown seaweeds. Green seaweeds contain high amounts of docosahexaenoic acid (DHA), oleic acid, palmitic acid, linoleic acid and α-linolenic acid (Peñalver et al. 2020). The consummation of seaweeds with high content of polyunsaturated fatty acids, like arachidonic acid and especially omega-3 EPA and DHA, is beneficial for the decrease of blood pressure, heart rate, clot formation, and arrhythmia, reduction of cardiovascular diseases risk, cancer risk, inflammation, diabetes and osteoporosis, and protection against dementia, Parkinson's and Alzheimer's disease (Peng et al. 2015; Lange 2020; Ozogul et al. 2021). Seaweeds contain higher amounts of essential fatty acids like EPA and DHA than terrestrial plants (Peñalver et al. 2020).

The pigments that distinguish seaweed groups can be allocated into three categories: chlorophylls (a, b, c1, and c2), carotenoids, and phycobiliproteins (phycoerythrin and phycocyanin) (Wang et al. 2017; Polat et al. 2023). Pigments capture the energy of sunlight for photosynthesis. Chlorophylls are greenish tetrapyrrole lipid soluble pigments that play a key role in photosynthesis. In most seaweeds, chlorophylls are major pigments. Other lipid soluble pigments, carotenoids, belong to the tetraterpenoids group. They occur as yellow, orange and green pigments. Carotenoids are accessory pigments and their role is to absorb and pass the light energy to chlorophyll and to provide aid in photoprotection. They can be classified as carotenes (α-carotene, β-carotene, and lycopene) and xanthophylls (lutein, astaxanthin, zeaxanthin, and fucoxanthin). Water soluble pigment-protein complexes found only in red seaweeds are phycobiliproteins (red, blue and light-blue pigments). They are accessory pigments, just like carotenoids (Aryee et al. 2018; Šimat et al. 2020; Polat et al. 2023). Due to pigments' antibacterial, antiviral, anticancer, antioxidant, antimalarial, anti-inflammatory, antidiabetic, and anti-obesity activities, they have a wide application in nutraceuticals, cosmetics and food industries (Chakdar and Pabbi 2017; Wang et al. 2017; Zhao et al. 2018; Rengasamy et al. 2020). Natural pigments found in seaweeds are interesting as food colorants, especially because they act beneficially on human health (Aryee et al. 2018).

Seaweeds are the most important source of non-animal sulfated polysaccharides; however, polysaccharides' composition in seaweed varies with regard to the species, season, location of growth, and other environmental conditions. Seaweed polysaccharides are usually divided according to their place of origin into storage and parietal polysaccharides. Laminarin, fucoidan and alginate are the major sulfated polysaccharides found in brown seaweeds, carrageenan in red seaweeds and ulvan in green seaweeds, respectively (Šimat et al. 2020; Udayangani et al. 2020; Polat et al. 2023). They are a valuable ingredient for food industry, used for thickening and gelling but they also exhibit numerous biological activities including antibacterial, antivirus, antitumor, anti-inflammatory, antioxidant, anti-allergic, antidiabetic, anticoagulant, and cardioprotective activities (Sanjeewa et al. 2018; Sharanagat et al. 2020; Thakur 2020). Most algal polysaccharides are not digested by humans so they can be considered fiber. The soluble/insoluble fiber ratio in seaweeds is higher than the ratio found for terrestrial vegetables. It ranges in range from 10% to

65% of dray seaweed weight. Dietary fiber has been recognized as an important part of a healthy diet. It consists of two fractions, soluble fiber which has the ability to form viscous gels when in contact with water, and insoluble fiber which can retain water in its structural matrix, increasing the fecal mass and transit of food through the stomach (Peñalver et al. 2020).

Seaweeds phenolic are a group of compounds from relatively simple to highly complex molecules with an aromatic benzene ring. This ring has one or more hydroxyl groups. Phenolics protect the seaweeds from grazers, bacteria, fouling organisms, UV radiation, oxidative damage, etc. Brown seaweeds' polyphenols called phlorotannins (1, 3, 5-benzenetriol (phloroglucinol) polymers) became the most interesting group due to their bioactivity. They are known to be free radical scavengers and exceptionally good reducing agents (Generalić Mekinić et al. 2019; Šimat et al. 2020; Meng et al. 2021). Besides antioxidant activity, they exhibit various other biological activities such as antimicrobial, anti-inflammatory, cytotoxic and antitumor (Stiger-Pouvreau et al. 2014; Abdelhamid et al. 2018).

3.2 Application of Seaweeds in the Food and Feed Industry

The use of seaweeds is versatile (Fig. 1.2), they are used as human food or food ingredient (soup ingredients, snacks, salad, pickles), hydrocolloids for the food industry (carrageenan, alginate and agar), abalone feed, livestock feed, biofertilizers, biofuel, cosmetics, therapeutics, nutraceuticals, pharmaceuticals, biopackaging, etc. (FAO 2021b).

Seaweeds belonging to genera *Laminaria/Saccharina* (kelp), *Porphyra* (nori), *Undaria* (wakame), *Caulerpa*, *Gracilaria* and *Kappaphycus/Eucheuma* are consumed as food. They are marketed dried and fresh. Traditional use of seaweeds as food began in fourth century in Japan, followed by China in the sixth century. Nowadays, seaweeds are consumed daily in East Asia, especially in China, Japan and Korea, by the Pacific's island nations and Europe's Celtic nations. As the people migrated from countries that traditionally eat seaweeds and the Asian cuisine spread around the globe, seaweeds have gained popularity and are not that unusual to consume in many more countries (Pereira 2016; Peñalver et al. 2020). Recently, seaweed consumption is spreading as they are found in restaurant dishes and on supermarket shelves around the world. New products are being developed to bring seaweeds closer to consumers, like algenbrot, a bread with up to 3% of seaweeds in its composition sold in Germany and Austria, bara mor, a "bread of the sea" produced in Brittany from dulse and kombu, or beurre des algues, a butter with minced seaweeds that can be used for cooking or spreading on the bread. Another popular product in Europe is sea sauerkraut, fermented *Himanthalia elongata* (sea spaghetti). Furthermore, researchers are developing new low-fat meat products by adding seaweeds to burgers, sausages, steaks, pâtés, and frankfurters. Seaweeds are also used in the formulation of fish products to prolong their shelf-life because of their antioxidant potential. They can be added to flour, pasta or dairy products (milk,

Fig. 1.2 The different uses of seaweeds show their potential to "conquer" the world

yogurt, cheese, cream) to increase the product's nutritional value (Fleurence 2016; Peñalver et al. 2020). Besides the nutritional benefits, edible seaweeds have flavour-enhancing properties as well as a positive effect on physico-chemical properties such as texture, water- and fat-binding properties, and colour when applied to foods (Mouritsen 2017).

Water soluble seaweed polysaccharides have high viscosity, high molecular weights, and exceptional stabilizing, gelling, and emulsifying characteristics. Carrageenan, agar and alginate are some of the hydrocolloids produced from seaweeds. They are used in food industries as natural additives and have high economic significance (Pereira 2016). Carrageenan is obtained from *Kappaphycus/Eucheuma*, *Sarcothalia crispate*, *Mazzaella laminarioides*, *Gigartina skottsbergii*, *Chondracanthus chamissoi*, and *Gymnogongrus furcellatus*. On the other hand, agar is obtained from several genera of red seaweeds *Gelidium*, *Gracilaria*, and *Pterocladia*, while brown seaweeds are used to obtain alginate (FAO 2021b). In the search for alternatives to plastic in food contact materials in the food industry, seaweed polysaccharides (carrageenan, agar and alginate) have gained interest due to their properties. They are biodegradable and transparent, have antioxidant potential and excellent ability to form films, and they are not toxic which makes them ideal for sustainable "green" packaging production. Besides these advantages, they have some disadvantages like water solubility and low tensile strength. However, their shortcomings are managed by the addition of nanoparticles, hydrophobic components, natural active agents or other biopolymers. Seaweed polysaccharides have been used as coatings, edible films, and active and intelligent packaging. The results are promising, however, there is a need to improve and optimize processing technologies to reduce high production costs and stimulate industrial investment (Perera et al. 2021; Lomartire et al. 2022).

The nutritional value of seaweeds makes them suitable for feed production. Seaweeds or seaweed extracts applied in animal feed can improve the quality, ensure sustainability and reduce the use of synthetic substances contributing to product safety. Furthermore, phenolic compounds with antioxidant activity found in seaweeds can reduce oxidative stress that leads to the production of rancid flavors and odors that affect sensory attributes, nutritional quality and safety in feed. Besides, they have shown antimicrobial activity against major spoilage and pathogenic microorganisms found in feed (Fleurence 2016; Morais et al. 2020; Lomartire et al. 2021). This bioactivity also leads to the reduction of antibiotic use.

Farm animals, sheep and cattle, traditionally graze on seaweeds in Scotland. Seaweeds are a part of the meal in the farm animal diets in some European countries. For example, seaweeds are a part of diet for cattle and hens in France, pigs in Belgium, sheep and cattle in Finland and Norway, and racing horses and pigs in other countries. Outside of Europe, the United States is developing meals with added seaweeds to feed poultry, goats, sheep, horses, cats and dogs (Fleurence 2016; Lomartire et al. 2021). The addition of seaweeds or seaweed extracts to monogastric farm animals' diets increased their gut health and immune status and showed growth promoting effects (Øverland et al. 2019). A feed with added seaweeds has a beneficial effect even on poultry meat and their eggs, it improves their immune status and decreases microbial load in their digestive tracts (Morais et al. 2020).

Seaweeds, *Ascophyllum nodosum*, *Ulva* sp. and *Porphyra* sp., are being incorporated as meals or extracts in fish aquaculture for Atlantic salmon (*Salmo salar*) and sea bream (*Pagrus major*). Other seaweed species, *Gracilaria* sp., *Sargassum* sp., *Padina* sp., and *Ascophyllum nodosum* were also added to fish feed and showed promising results, enhancing the fish growth, stress response, physiological activity, lipid metabolism, and disease resistance of various farmed species. Research data support the use of seaweed as a part of fish feed to gain higher productivity. The inclusion of seaweed in aquaculture feed is not limited to fish, red seaweeds were used in the European abalone (*Haliotis tuberculata*) diet (Fleurence 2016; Morais et al. 2020). For example, for the production of abalone feed, kelp, *Gracilaria*, *Ulva*, and other seaweeds are being used (FAO 2021b).

Another great aspect of seaweed application in livestock feed lies in anti-methanogenic activity. Enteric methane produced by microbial fermentation of nutrients in animals is causing a problem on livestock farms with high emissions rates. It was found that the red seaweed *Asparagopsis armata* added to animal feed in low concentrations can reduce methane production without negative impacts (Morais et al. 2020).

3.3 Seaweeds as a Source of New Molecules and Bioactive Molecules with Health Benefits

Bioactive compounds are biologically active compounds naturally found in various foods in small quantities. They provide health benefits apart from the basic nutritional value of the food inducing the behavioral, physiological, and immunological effects. Bioactive compounds are grouped according to differences in their chemical structure and function (Kitts 1994; Hamzalioğlu and Gökmen 2016). Seaweeds are used to produce bioactive molecules or marine natural products. There are over 1200 new compounds derived from marine organisms, including green, red and brown seaweeds, reported in scientific publications every year. From 5% to 6% of new compounds are found in seaweeds each year, except for 2020 when a slight decrease was observed with only 3% of new compounds found (Fig. 1.3) (Blunt et al. 2018; Carroll et al. 2019, 2020, 2021).

There are two ways in which seaweeds can benefit human health, directly and indirectly. A direct way would be consuming the whole seaweed or seaweed food supplements and natural drugs which contribute to a healthier lifestyle. For example, iodine, fucoxanthin, fucoidan and phlorotannins from seaweeds are commercially used for food supplement production (Cai et al. 2021). On the other side, indirectly, the use of seaweeds as biofertilizers in agricultural production to provide healthy cultivation and enhance plant and soil conditions indirectly benefits human health through the reduction of chemicals from traditional fertilizers and enrichment of the agricultural product (Lomartire et al. 2021). Seaweed fertilizers are biodegradable, they are not toxic or polluting, and they present no hazard to humans, animals or birds. Some of the advantages of seaweed fertilizers application are the improved crop yield and performance, stimulation of seed germination, and an

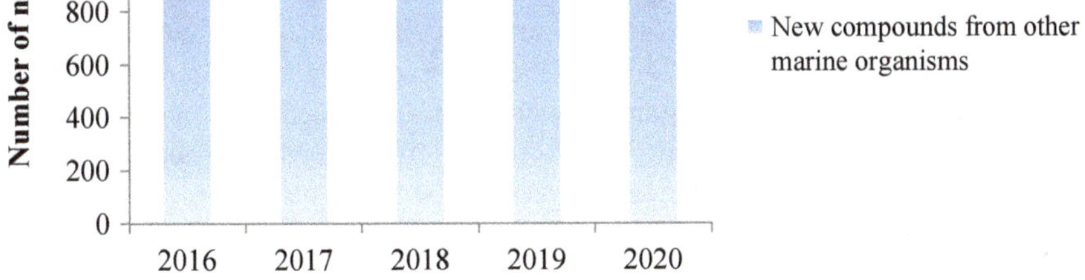

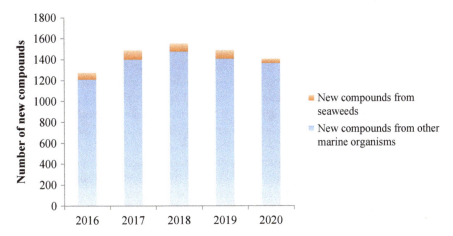

Fig. 1.3 The number of new marine derived compounds reported from 2016 to 2020. (Data adapted from reports on annual new marine components, Blunt et al. 2018; Carroll et al. 2019, 2020, 2021)

increase in abiotic and biotic stress resistance in plants (Raghunandan et al. 2019; Lomartire et al. 2021). A sustainable approach for the exploitation of seaweeds for the production of bioactive compounds, chemicals and biofuels is a biorefinery system. Biorefinery combines processes and technologies of biomass conversion to feed, food, energy and other value-added products with minimal, nearly zero waste. The goal is to extract polysaccharides, minerals, proteins, lipids and antioxidants for use as pharmaceuticals, nutraceuticals, cosmetics, food, feed and fertilizers, and to produce biogas, bioethanol or biodiesel from the leftover waste. It is important to adapt this process to achieve its efficiency by developing a sustainable industrial economy and reducing the impact on climate change (Balina et al. 2017; Torres et al. 2019).

Besides being considered functional foods, seaweeds are the source of compounds with a proven positive effect on human health and thus can be used for the production of nutraceuticals. Nutraceuticals can be used to enrich food products with health benefits beyond their nutritive value. Their regular consumption can have major long-term health benefits (Tanna and Mishra 2018). For example, fucoidans are sold in USA and Australia, and phlorotannins, agar and carrageenan are already available on the market as nutraceuticals (Lomartire et al. 2021).

Bioactives from seaweeds are attractive for biomedical applications. In Asian countries, seaweeds have been used in traditional and folk medicine to treat colds, Chinese influence, wounds, diarrhea, high blood pressure, goitre, obesity, nephritic diseases, gastric ulcers, and gout (Lomartire et al. 2021). Nowadays, with knowledge provided from scientific research, many other biomedical applications are discovered. Novel antitumor therapies could be developed using seaweed bioactives (polyphenols, terpenoids, including carotenoids, and alkaloids) as they show *in vitro* and *in vivo* antitumor activities against various human tumor cell lines. The most promising compounds to develop new anticancer agents are halogenated terpenes and bromophenols, seaweeds' secondary metabolites that are only biosynthesized in marine species (Rocha et al. 2018).

Neurodegenerative diseases (Alzheimer's, Parkinson's, Multiple Sclerosis, and other chronic diseases) could be prevented or treated by seaweeds due to their favorable therapeutic characteristics. There is a need for natural bioactive compounds with neuroprotective potential because synthetic neuroprotective agents can have various adverse side effects. Seaweed's potential as a neuroprotective agent was observed *in vitro*, in animal models, and a few human studies. There is a need for more clinical studies to confirm these observations (Pereira and Valado 2021).

Cardiovascular diseases (CVD) are associated to obesity, hyperlipidemia, type 2 diabetes mellitus and hypertension. Seaweed-isolated bioactive compounds (phlorotannins, polysaccharides, peptides, carotenoids) have shown anti-obesity, antihyperlipidemia, anti-diabetic, and anti-hypertensive effects *in vitro* and *in vivo*, however, for their application more clinical studies are needed to determine their effect on inhibition and alleviation of symptoms related to listed diseases (Cho et al. 2022).

Fucoidans with high molecular weights and fucose contents extracted from *L. hyperborea* and *S. latissima* were recently found to inhibit vascular endothelial growth factor and protect against oxidative stress *in vitro*. These findings make fucoidan a potential therapeutic for age-related macular degeneration (AMD) treatment. It could be used to prevent disease progression and vision loss in the early stages (Dörschmann and Klettner 2020).

The seaweeds are "green", natural and eco-friendly sources of fucoidan, carrageenan, laminarin, fucoxanthin, and mycosporine-like amino acids that have shown *in vitro* and *in vivo* anti-photoaging activity. This bioactivity has been useful for seaweed utilization in the skincare and cosmetic industries. Besides, these compounds were found to inhibit scar formation and accelerate wound healing and hair growth (Pereira 2018). Some of these compounds were clinically tested and are already available on the market as cosmeceuticals. Extracts or compounds from seaweeds available on the market offer a wide range of benefits related to anti-photoaging, such as photoptotection (UV-A), DNA protection, prevention of sunburn, modulation of skin immunity, increase of brightness and soothing, and antioxidant and anti-inflammation properties (Pangestuti et al. 2021). Seaweed extracts and bioactive compounds are also combined with other medicinal or therapeutic molecules from different sources in algotherapy and thalassotherapy (a special branch of herbal medicine) that uses the synergic effect of these combinations to attenuate, combat and prevent different (non-identified) diseases and alleviate their symptoms (Pereira 2018).

Seaweed compounds can have beneficial effects on human gut microbiota. They can enhance the abundance and diversity of gut bacteria. Researchers are investigating the potential therapeutic application of polyphenols, polysaccharides and peptides from seaweeds for modulation of the gut microbiota *in vitro* and *in vivo*. This is rather important as the gut bacteria imbalance was linked to the development of obesity, type-2-diabetes, hypertension, immunodeficiency, inflammatory bowel disease, and cancer. Seaweed-derived compounds (especially laminarin, fucoidan, alginate, porphyrin, and ulvan) have shown the potential to positively modulate the gut microbiota as prebiotics through the enhancement of bacterial populations. These bacteria produce the short-chain fatty acids that protect against pathogens, provide gastrointestinal epithelial cells with energy, influence immunomodulation, and induce colon cancer cells apoptosis. Large-scale clinical trials are necessary to validate seaweed-derived compounds' health effects on gut microbiota (Shannon et al. 2021).

Seaweed bioactives are used for the formulation of new and innovative drugs for the substitution of synthetic compounds in pharmaceutical applications. Seaweed extracts (containing sulfated polysaccharides, phenolic compounds or alkaloids) tested on cell lines or administered to animals showed antioxidant, antibacterial, antiviral, anti-inflammatory, anticancer, and anticoagulant activities (Lomartire et al. 2021).

Despite the bioactivity and health benefits of seaweeds and their extracted compounds, there are some risks related to their daily consumption. As seaweeds grow in coastal waters, they can accumulate iodine, heavy metals (lead, cadmium and mercury), arsenic, pesticides washed from farms, antibiotics and other toxic compounds. Food and Drug Administration (FDA) and the European Food Safety

Authority (EFSA) adopted recommendations and limits for the consumption of some seaweed species and their components in order to cope with this health risk (Cotas et al. 2021; Lähteenmäki-Uutela et al. 2021).

The great interest of the research community in seaweeds is surely justified. Seaweeds have high nutritive value, and are rich source of many essential minerals, vitamins, dietary fiber, and high-quality proteins. Besides, polysaccharides and bioactive compounds from seaweeds show great potential for exploitation and application in food, feed, cosmetic and pharmaceutical industries as they exert numerous health benefits and could be used for prevention of cancer, neurodegenerative, skin-related and cardiovascular diseases, and obesity, and for modulation of the gut microbiota. Seaweeds are also a good source of natural pigments, such as carotenoids, chlorophylls, and xanthophylls that show antioxidant, anti-inflammatory, anticancer, and antidiabetic properties and could be beneficial in the prevention of some health conditions such as neuro-degenerative diseases, cancers and gastric ulcers. These have proven *in vitro* but *in vivo* only in some cases. To contribute to full exploitation of algal bioactives and other valuable components more clinical studies are needed. Furthermore, there is room for the expansion of the seaweed cultivation, both in quantities and in introduction of new species to the process. Seaweed biorefinery processes should be further developed and adapted to extract the most from seaweed biomass leaving nearly zero waste. It will contribute to the more sustainable production of food, feed and other value-added products along with biofuel and reduce the impact on climate change.

References

Abdelhamid A, Jouini M, Bel Haj Amor H et al (2018) Phytochemical analysis and evaluation of the antioxidant, anti-inflammatory, and antinociceptive potential of phlorotannin-rich fractions from three mediterranean brown seaweeds. Mar Biotechnol 20:60–74. https://doi.org/10.1007/s10126-017-9787-z

Aryee AN, Agyei D, Akanbi TO (2018) Recovery and utilization of seaweed pigments in food processing. Curr Opin Food Sci 19:113–119. https://doi.org/10.1016/j.cofs.2018.03.013

Balina K, Romagnoli F, Blumberga D (2017) Seaweed biorefinery concept for sustainable use of marine resources. Energy Procedia 128:504–511. https://doi.org/10.1016/j.egypro.2017.09.067

Bizzaro G, Vatland AK, Pampanin DM (2022) The One-Health approach in seaweed food production. Environ Int 158:106948. https://doi.org/10.1016/j.envint.2021.106948

Blunt JW, Carroll AR, Copp BR et al (2018) Marine natural products. Nat Prod Rep 35:8–53. https://doi.org/10.1039/c7np00052a

Cai J, Lovatelli A, Aguilar-Manjarrez J et al (2021) Seaweeds and microalgae: an overview for unlocking their potential in global aquaculture development. FAO, Rome

Carroll AR, Copp BR, Davis RA et al (2019) Marine natural products. Nat Prod Rep 36:122–173. https://doi.org/10.1039/c8np00092a

Carroll AR, Copp BR, Davis RA et al (2020) Marine natural products. Nat Prod Rep 37:175–223. https://doi.org/10.1039/c9np00069k

Carroll AR, Copp BR, Davis RA et al (2021) Marine natural products. Nat Prod Rep 38:362–413. https://doi.org/10.1039/d0np00089b

Chakdar H, Pabbi S (2017) Algal pigments for human health and cosmeceuticals. In: Algal green chemistry. Elsevier B.V., Amsterdam, pp 171–188
Cho CH, Lu YA, Kim MY et al (2022) Therapeutic potential of seaweed-derived bioactive compounds for cardiovascular disease treatment. Appl Sci 12:1025. https://doi.org/10.3390/app12031025
Correia H, Soares C, Morais S et al (2021) Seaweeds rehydration and boiling: Impact on iodine, sodium, potassium, selenium, and total arsenic contents and health benefits for consumption. Food Chem Toxicol 155:112385. https://doi.org/10.1016/j.fct.2021.112385
Cotas J, Pacheco D, Araujo GS et al (2021) On the health benefits vs. risks of seaweeds and their constituents: the curious case of the polymer paradigm. Mar Drugs 19:1–19. https://doi.org/10.3390/md19030164
Darias-Rosales J, Rubio C, Gutiérrez ÁJ et al (2020) Risk assessment of iodine intake from the consumption of red seaweeds (Palmaria palmata and Chondrus crispus). Environ Sci Pollut Res 27:45737–45741. https://doi.org/10.1007/s11356-020-10478-9
Dörschmann P, Klettner A (2020) Fucoidans as potential therapeutics for age-related macular degeneration—current evidence from in vitro research. Int J Mol Sci 21:1–19. https://doi.org/10.3390/ijms21239272
Echave J, Fraga-Corral M, Garcia-Perez P et al (2021) Seaweed protein hydrolysates and bioactive peptides: extraction, purification, and applications. Mar Drugs 19:1–22. https://doi.org/10.3390/md19090500
Eggert A (2012) Seaweed responses to temperature. In: Seaweed biology novel insights into ecophysiology, ecology and utilization. Springer, Berlin Heidelberg, pp 47–66
Evans S, Rorrer GL, Langdon CJ (2021) Cultivation of the macrophytic red alga Palmaria mollis (Pacific dulse) on vertical arrays of mesh panels in aerated tanks. J Appl Phycol 33:3915–3926. https://doi.org/10.1007/s10811-021-02582-1
FAO (2021a) FAO Yearbook. Fishery and aquaculture statistics 2019/FAO annuaire. Statistiques des pêches et de l'aquaculture 2019/FAO anuario. Estadísticas de pesca y acuicultura 2019. FAO, Rome
FAO (2021b) Global seaweed and microalgae production, 1950–2019. World Aquaculture Performance Indicators (WAPI) factsheet. www.fao.org/3/cb4579en/cb4579en.pdf
Felaco L, Olvera-Novoa MA, Robledo D (2020) Multitrophic integration of the tropical red seaweed Solieria filiformis with sea cucumbers and fish. Aquaculture 527:735475. https://doi.org/10.1016/j.aquaculture.2020.735475
Fleurence J (2016) Seaweeds as food. In: Seaweed in health and disease prevention. Elsevier Inc, Amsterdam, pp 149–167
Garcia-Vaquero M, Rajauria G, Miranda M et al (2021) Seasonal variation of the proximate composition, mineral content, fatty acid profiles and other phytochemical constituents of selected brown macroalgae. Mar Drugs 19:204. https://doi.org/10.3390/md19040204
Generalić Mekinić I, Skroza D, Šimat V et al (2019) Phenolic content of brown algae (Pheophyceae) species: extraction, identification, and quantification. Biomolecules 9:244. https://doi.org/10.3390/biom9060244
Gómez I, Huovinen P (2012) Morpho-functionality of carbon metabolism in seaweeds. In: Seaweed biology novel insights into ecophysiology, ecology and utilization. Springer, Berlin and Heidelberg, pp 25–46
Gordillo FJL (2012) Environment and algal nutrition. In: Seaweed biology novel insights into ecophysiology, ecology and utilization. Springer, Berlin Heidelberg, pp 67–86
Hamzalioğlu A, Gökmen V (2016) Interaction between bioactive carbonyl compounds and asparagine and impact on acrylamide. In: Acrylamide in food: analysis, content and potential health effects. Elsevier, Amsterdam, pp 355–376
Hanelt D, Figueroa FL (2012) Physiological and photomorphogenic effects of light on marine macrophytes. In: Seaweed biology novel insights into ecophysiology, ecology and utilization. Springer, Berlin Heidelberg, pp 3–23

Kitts DD (1994) Bioactive substances in food: identification and potential uses. Can J Physiol Pharmacol 72:423–434. https://doi.org/10.1139/y94-062

Lähteenmäki-Uutela A, Rahikainen M, Camarena-Gómez MT et al (2021) European Union legislation on macroalgae products. Aquac Int 29:487–509. https://doi.org/10.1007/s10499-020-00633-x

Lange KW (2020) Lipids in the treatment of neurodegenerative diseases. In: Bailey's industrial oil and fat products. Wiley, Hoboken, pp 1–17. https://doi.org/10.1002/047167849x.bio118

Lomartire S, Marques JC, Gonçalves AMM (2021) An overview to the health benefits of seaweeds consumption. Mar Drugs 19:341. https://doi.org/10.3390/md19060341

Lomartire S, Marques JC, Gonçalves AMM (2022) An overview of the alternative use of seaweeds to produce safe and sustainable bio-packaging. Appl Sci 12:3123. https://doi.org/10.3390/app12063123

Loureiro R, Gachon CMM, Rebours C (2015) Seaweed cultivation: potential and challenges of crop domestication at an unprecedented pace. New Phytol 206:489–492. https://doi.org/10.1111/nph.13278

McHugh D (2003) A guide to the seaweed industry, vol 441. Food and Agriculture Organization of the United Nations, Rome

Meng W, Mu T, Sun H, Garcia-Vaquero M (2021) Phlorotannins: a review of extraction methods, structural characteristics, bioactivities, bioavailability, and future trends. Algal Res 60:102484. https://doi.org/10.1016/j.algal.2021.102484

Morais T, Inácio A, Coutinho T et al (2020) Seaweed potential in the animal feed: a review. J Mar Sci Eng 8:1–24. https://doi.org/10.3390/JMSE8080559

Mouritsen OG (2017) Those tasty weeds. J Appl Phycol 29:2159–2164. https://doi.org/10.1007/s10811-016-0986-1

Øverland M, Mydland LT, Skrede A (2019) Marine macroalgae as sources of protein and bioactive compounds in feed for monogastric animals. J Sci Food Agric 99:13–24. https://doi.org/10.1002/jsfa.9143

Ozogul F, Cagalj M, Šimat V et al (2021) Recent developments in valorisation of bioactive ingredients in discard/seafood processing by-products. Trends Food Sci Technol 116:559–582. https://doi.org/10.1016/j.tifs.2021.08.007

Pangestuti R, Shin KH, Kim SK (2021) Anti-photoaging and potential skin health benefits of seaweeds. Mar Drugs 19:172. https://doi.org/10.3390/MD19030172

Peñalver R, Lorenzo JM, Ros G et al (2020) Seaweeds as a functional ingredient for a healthy diet. Mar Drugs 18:1–27. https://doi.org/10.3390/md18060301

Peng Y, Hu J, Yang B et al (2015) Chemical composition of seaweeds. Elsevier Inc., Amsterdam

Pereira L (2016) Edible seaweeds of the world. CRC Press, Boca Raton

Pereira L (2018) Seaweeds as source of bioactive substances and skin care therapy—cosmeceuticals, algotheraphy, and thalassotherapy. Cosmetics 5:68. https://doi.org/10.3390/cosmetics5040068

Pereira L, Valado A (2021) The seaweed diet in prevention and treatment of the neurodegenerative diseases. Mar Drugs 19:1–25

Perera KY, Sharma S, Pradhan D et al (2021) Seaweed polysaccharide in food contact materials (Active packaging, intelligent packaging, edible films, and coatings). Foods 10:1–22. https://doi.org/10.3390/foods10092088

Polat S, Trif M, Rusu A et al (2023) Recent advances in industrial applications of seaweeds. Crit Rev Food Sci Nutr 63:4979–5008. https://doi.org/10.1080/10408398.2021.2010646

Racine P, Marley A, Froehlich HE et al (2021) A case for seaweed aquaculture inclusion in U.S. nutrient pollution management. Mar Policy 129:104506. https://doi.org/10.1016/j.marpol.2021.104506

Raghunandan BL, Vyas RV, Patel HK, Jhala YK (2019) Perspectives of seaweed as organic fertilizer in agriculture. In: Soil fertility management for sustainable development. Springer, Berlin, pp 267–289

Rengasamy KR, Mahomoodally MF, Aumeeruddy MZ et al (2020) Bioactive compounds in seaweeds: an overview of their biological properties and safety. Food Chem Toxicol 135:111013. https://doi.org/10.1016/j.fct.2019.111013

Rocha DHA, Seca AML, Pinto DCGA (2018) Seaweed secondary metabolites in vitro and in vivo anticancer activity. Mar Drugs 16:1–27. https://doi.org/10.3390/md16110410

Rodríguez-Bernaldo de Quirós A, López-Hernández J (2021) An overview on effects of processing on the nutritional content and bioactive compounds in seaweeds. Foods 10:2168. https://doi.org/10.3390/foods10092168

Roohinejad S, Koubaa M, Barba FJ et al (2017) Application of seaweeds to develop new food products with enhanced shelf-life, quality and health-related beneficial properties. Food Res Int 99:1066–1083. https://doi.org/10.1016/j.foodres.2016.08.016

Sanjeewa KKA, Kang N, Ahn G et al (2018) Bioactive potentials of sulfated polysaccharides isolated from brown seaweed Sargassum spp in related to human health applications: a review. Food Hydrocoll 81:200–208. https://doi.org/10.1016/j.foodhyd.2018.02.040

Schmedes PS, Nielsen MM (2020) New hatchery methods for efficient spore use and seedling production of Palmaria palmata (dulse). J Appl Phycol 32:2183–2193. https://doi.org/10.1007/s10811-019-01998-0

Shannon E, Conlon M, Hayes M (2021) Seaweed components as potential modulators of the gut microbiota. Mar Drugs 19:1–50. https://doi.org/10.3390/md19070358

Sharanagat VS, Singla V, Singh L (2020) Bioactive compounds from marine sources. In: Goyal MR, Rasul Suleria HA, Kirubanandan S (eds) Technological processes for marine foods-from water to fork: bioactive compounds, industrial applications and genomics. Apple Academic Press, Inc., Oakville

Šimat V, Elabed N, Kulawik P et al (2020) Recent advances in marine-based nutraceuticals and their health benefits. Mar Drugs 18:1–40. https://doi.org/10.3390/md18120627

Stedt K, Trigo JP, Steinhagen S et al (2022) Cultivation of seaweeds in food production process waters: evaluation of growth and crude protein content. Algal Res 63:102647. https://doi.org/10.1016/j.algal.2022.102647

Stévant P, Indergård E, Ólafsdóttir A et al (2018) Effects of drying on the nutrient content and physico-chemical and sensory characteristics of the edible kelp Saccharina latissima. J Appl Phycol 30:2587–2599. https://doi.org/10.1007/s10811-018-1451-0

Stiger-Pouvreau V, Jégou C, Cérantola S et al (2014) Phlorotannins in Sargassaceae species from Brittany (France). In: Bourgougnon N (ed) Advances in botanical research. Academic Press – Elsevier, Amsterdam, pp 379–411

Tanna B, Mishra A (2018) Metabolites unravel nutraceutical potential of edible seaweeds: an emerging source of functional food. Compr Rev Food Sci Food Saf 17:1613–1624. https://doi.org/10.1111/1541-4337.12396

Thakur M (2020) Marine bioactive components: sources, health benefits, and future prospects. In: Technological processes for marine foods-from water to fork: bioactive compounds, industrial applications and genomics. Apple Academic Press Inc., Oakville, pp 61–72

Titlyanov EA, Titlyanova TV (2010) Seaweed cultivation: methods and problems. Russ J Mar Biol 36:227–242. https://doi.org/10.1134/S1063074010040012

Torres MD, Kraan S, Domínguez H (2019) Seaweed biorefinery. Rev Environ Sci Bio/Technol 18(2):1–54

Udayangani RMAC, Somasiri GDP, Wickramasinghe I, Kim S (2020) Potential health benefits of sulfated polysaccharides from marine algae. In: Encyclopedia of marine biotechnology. Wiley-Blackwell, Hoboken, pp 629–635. https://doi.org/10.1002/9781119143802.ch22

Wang H, Li XC, Lee DJ, Chang JS (2017) Potential biomedical applications of marine algae. Bioresour Technol 244:1407–1415. https://doi.org/10.1016/j.biortech.2017.05.198

WHO (ed) (2007) Assessment of iodine deficiency disorders and monitoring their elimination: a guide for programme managers, 3rd edn. WHO, Geneva

Zhao C, Yang C, Liu B et al (2018) Bioactive compounds from marine macroalgae and their hypoglycemic benefits. Trends Food Sci Technol 72:1–12. https://doi.org/10.1016/j.tifs.2017.12.001

Zhu X, Soro AB, Tiwari BK, Garcia-Vaquero M (2022) Seaweeds in Ireland: main components, applications, and industrial prospects. In: Sustainable global resources of seaweeds, vol 1. Springer, Cham, pp 163–183

Chapter 2
Green and Environmentally Friendly Technologies to Develop Macroalgal Biorefineries

Armin Mirzapour-Kouhdasht, Mohammad Sadegh Taghizadeh, Ali Niazi, and Marco Garcia-Vaquero

1 Introduction

Macroalgae comprises a wide variety of widely spread marine organisms, expanding from coastal areas around the world including Europe (mainly Ireland and Iceland), Asia (China, Japan, India, and Iran), and South America which are currently considered as the main regions exploiting this biomass for their potential applications as foods and nutraceuticals. Nowadays, most macroalgae are currently farmed to account for the multiple ingredients demanded by several industrial applications, although wild harvesting practices still persist in most countries around the globe (Akbary et al. 2021; Buschmann et al. 2017; Mohammed et al. 2021).

Macroalgal composition is really variable, although it is generally described as favourable from a nutritional point of view, containing a wide variety of proteins, carbohydrates, minerals, fibres, and low levels of lipids that include a high proportions of polyunsaturated fatty acids (Ganesan et al. 2019; Schmid et al. 2018). The composition of macroalgae has been described as extremely variable, as it is influenced by the macroalgal class (red macroalgae or Rhodophyta, green macroalgae or Chlorophyta and brown macroalgae or Phaeophyta), as well as multiple biotic and abiotic factors affecting the biological development of the biomass, such as place of collection, season, temperature, solar irradiation and nutrients in the water amongst other multiple factors described in the literature (Garcia-Vaquero et al. 2021a; Schmid et al. 2018). The composition of selected macroalgal species belonging to red, green, and brown macroalgal classes are summarised in Table 2.1.

A. Mirzapour-Kouhdasht · M. Garcia-Vaquero (✉)
School Agriculture and Food Science, University College Dublin, Dublin, Ireland
e-mail: marco.garciavaquero@ucd.ie

M. S. Taghizadeh · A. Niazi
Institute of Biotechnology, Shiraz University, Shiraz, Iran

F. Ozogul et al. (eds.), *Seaweeds and Seaweed-Derived Compounds*,
https://doi.org/10.1007/978-3-031-65529-6_2

Table 2.1 Main compounds of selected red, green, and brown macroalgal species

Macroalgal species	Protein contents (%)	Carbohydrate contents (%)	Lipid contents (%)	Ash contents (%)	Fibre contents (%)	References
Red macroalgae (Rhodophyta)						
Gracilaria edulis	14.26–25.29	32.39–45.50	0.83–4.76	8.70	63.18	Debbarma et al. (2016) and Rosemary et al. (2019)
Hypnea musciformis	17.10–18.64	20.60	0.30–1.27	14.10–21.57	37.92	Carneiro et al. (2014) and Siddique et al. (2013)
Gracilaria changii	12.57	41.52	0.30	40.30	29.44	Chan and Matanjun (2017)
Solieria filiformis	18.80–20.30	64.41	0.30–2.17	8.81–15.10	5.66	Carneiro et al. (2014), Mæhre et al. (2014), and Martínez-Milián and Olvera-Novoa (2016)
Plocamium brasiliense	15.72–19.04	52.03	1.52–3.63	25.63	–	Cherry et al. (2019) and Gressler et al. (2011)
Gelidium pusillum	11.31	40.64	2.16	21.15	24.74	Siddique et al. (2013)
Green macroalgae (Chlorophyta)						
Ulva lactuca	10.69–17.10	43.19–59.71	0.86–3.60	12.41–23.67	5.6–13.57	Debbarma et al. (2016), Mwalugha et al. (2015), Pádua et al. (2004), and Rohani-Ghadikolaei et al. (2012)

(continued)

Table 2.1 (continued)

Macroalgal species	Protein contents (%)	Carbohydrate contents (%)	Lipid contents (%)	Ash contents (%)	Fibre contents (%)	References
Caulerpa racemosa	17.36–19.72	48.97–52.81	2.21–7.65	12.15–23.81	3.11–11.51	Bhuiyan et al. (2016) and Nagappan and Vairappan (2014)
Halimeda macroloba	1.60–5.28	26.70	1.95	22.50–66.07	9.88–14.70	Gunji et al. (2007) and Mwalugha et al. (2015)
Codium tomentosum	6.13–25.30	20.47–28.50	2.20–2.53	24.00–35.99	–	Manivannan et al. (2008), Rodrigues et al. (2015), and Soares et al. (2021)
Brown macroalgae (Phaeophyta)						
Ascophyllum nodosum	7.90–8.70	53.60	2.70–3.62	21.20–30.89	3.50	Cruz-Suárez et al. (2009) and Lorenzo et al. (2017)
Sargassum polycystum	0.90–5.40	33.49	0.29	3.80–42.40	39.67–65.70	Gunji et al. (2007) and Matanjun et al. (2009)
Laminaria digitata	1.10–8.20	33.9–76.0	0.85	11.10–31.60	–	Mæhre et al. (2014), Nielsen et al. (2016), and Schiener et al. (2015)
Sargassum muticum	7.00–16.90	–	1.60–3.20	13.20–30.50	–	Balboa et al. (2016) and Rodrigues et al. (2015)

In 2018 the global macroalgal production market was valued in approximately 10.31 billion USD with predictions of an increased annual rate of 8.9%, reaching an estimated value of 22.13 billion USD by the year 2024 (Smith et al. 2018). Approximately 85% of this valuable biomass is currently used by the food industry, while the remaining 15% is utilized in other applications including pharmaceuticals, nutraceuticals, and cosmeceuticals for their high-value compounds (Zhu et al. 2021).

An optimum exploitation of macroalgal molecules and their beneficial health effects requires their extraction from the biomass. Traditional extraction techniques or solid-liquid extraction protocols, such as maceration and steam distillation, have been applied to extract these compounds from macroalgae (Garcia-Vaquero et al. 2020a). Nevertheless, these traditional methods have some drawbacks as they are laborious and time-consuming and they show low selectivity, and low yields of extraction (Garcia-Vaquero et al. 2020a). Therefore, innovative technologies are currently gaining momentum for being fast, selective, low cost and highly efficient extracting high yields of compounds, as well as for their sustainability (Herrero and Ibanez 2015) that will contribute to the generation of green extraction protocols for the exploitation of macroalgal molecules. Among others, ultrasound-assisted extraction (UAE), microwave-assisted extraction (MAE), high pressure-assisted extraction (HPAE), pulsed electric field-assisted extraction (PEF), and those based on the application of compressed fluids such as supercritical fluid extraction (SFE), pressurized liquid extraction (PLE) as well as subcritical water extraction (SWE), are some of the most promising methods for the extraction of multiple target compounds from macroalgae (Garcia-Vaquero et al. 2021b; Gomez et al. 2020; Marr 1993; Mendiola et al. 2007; Silva et al. 2020). Moreover, the sequential application of these technologies targeting different compounds from macroalgae has been gaining momentum for the development of macroalgal biorefineries, aiming to generate exploitation models from the macroalgal biomass that will generate minimum or no waste (Garcia-Vaquero et al. 2019).

This chapter aims to provide an overview of the overall composition of macroalgae as well as the principles and main parameters analysed by researchers when applying innovative technologies (UAE, MAE, HPAE, PEF, SFE, PLE and SWE) to extract multiple compounds from macroalgae with promising biological activities. The critical factors to consider working with these technologies including their effects reported on product quality, process efficiency, production costs and the development of environmentally friendly processes for the exploitation of macroalgae at industrial level will also be covered in this chapter together with the future prospects of the use of innovative technologies for the development of macroalgal biorefineries.

2 Innovative Technologies for the Extraction of High-Value Compounds

Innovative extraction technologies are currently gaining momentum for the development of greener and more sustainable protocols for the extraction of high-value compounds from macroalgae. The main innovative technologies currently explored aiming to establish green and sustainable macroalgal biorefinery models are UAE, MAE, HPAE, PEF, SFE, PLE and SWE. All these technologies can be used for multiple purposes in the food industry and an overview of the main advantages and limitation of UAE, MAE, HPAE, PEF and SFE is provided in Table 2.2.

Table 2.2 Main advantages and limitations of selected innovative technologies for food applications. Content of this table originally published by Garcia-Vaquero and Rajauria (2022) and reproduced with permission from Elsevier

Technologies	Advantages	Limitations
Ultrasound processing	Reduction of processing temperature and time. Increased heat transfer. Possible operation as batch or continuous modes. Can be used alone or combined with other processing technologies. High throughput and low energy consumption.	Complex modes of action. Depth of penetration affected by solids and air in the product. Possible damage by free radicals. Undesirable modifications of food properties (structure, texture, flavour, colour, or nutritional value). Potential problems scaling-up processes.
Microwave heating	Reduced carbon footprint. Can be used alone or combined with conventional heating and other processing technologies. Heat generated within the products. Short processing times. Safe food products for consumers.	Needs a high engineering intelligence. High energy costs. Possible uneven heating or thermal runaway.
High pressure processing	No evidence of toxicity. Preservation of food properties (colours, flavours and nutrients). Reduced processing time. Uniformity of treatments. Possible and desirable texture changes. Possibility to process in-package.	Little effects on food enzymatic activities. High cost of equipment. Foods should be 40% free water for an anti-microbial effect and some microbes can survive. Regulatory issues.
Pulsed electric fields processing	Colours, flavours and nutrients of food are preserved. No evidence of toxicity. Relatively short treatment times.	Low or no effect on enzymes/spores. Difficulties operating with conductive materials. Only suitable for liquid foods. Air (bubbles) may lead to non-uniform treatment of liquids. By-products of electrolysis may adversely affect foods. Uncertainty on energy efficiency during processing. Regulatory issues remain to be resolved. Potential operational and safety issues.
Supercritical fluids	Allows the extraction of non-polar compounds without solvents (green extraction technique). Product with maintained qualities due to low operational temperatures (40–80 °C). Free of heavy metals and inorganic salts. High effectivity due to the low viscosity and high diffusivity of the supercritical fluids. Short processing times and high yields of compounds extracted.	Expensive equipment. Complex operation involving high pressures. Not suitable for the extraction of polar compounds. High energy consumption.

When these technologies are used for the purposes of extraction of compounds or macroalgal ingredients, these innovative systems have been explored alone, simultaneously or sequentially to recover a wide variety of compounds. Understanding the mechanisms of action of each technology, as well as the main results when extracting several high-value compounds from the biomass remains as a critical point for the future scale-up and industrial exploitation of these technological aids, and thus, it is necessary to summarize the main achievements to date in the macroalgal sector using these technologies for the purposes of extraction.

2.1 Application of UAE

UAE is based on the application of ultrasound energy in a solvent containing the biomass (Freitas de Oliveira et al. 2016). Ultrasound energy are mechanical waves with a frequency above 20 kHz, which can pass through liquid, solid, and gas mediums. During the process of extraction, when the pressure exceeds the tensile strength of the liquid, vapor bubbles will form and lately collapse under a powerful ultrasonic field and explode, causing high-speed interparticle collisions and turbulence in the particles. All these effects generate a rapid movement or solvent flow, followed by the collision and displacement of the molecules from their original positions in the biological matrices, and finally the release of active compounds (Kadam et al. 2015c; Shirsath et al. 2012). The application of UAE may induce raises of the temperature of up to 5000 K and pressure of up to 1000 atm, expediting the biochemical reactions in their vicinity of their intended application surfaces (Kumar et al. 2021). Several parameters can affect the extraction efficiency when using UAE, including ultrasound power, type and ratio of solvent, time, temperature, and particle size of the sample undergoing the extraction processes (Getachew et al. 2020). UAE has been described as beneficial for the extraction of multiple compounds when establishing macroalgal biorefineries (Filote et al. 2021). The main scientific literature exploring UAE for the recovery of multiple biologically active compounds, as well as the UAE conditions detailed by the researchers are summarized in Table 2.3.

As seen in Table 2.3, variations in the solvents used facilitate a more selective extraction of compounds. Thereby the use of alcoholic solutions (ethanol, methanol and others at variable percentages) combined with multiple UAE devices and conditions have been described as efficient methods for the extraction of compounds of hydrophobic nature, such as phenolic compounds, carotenoids, chlorophylls, fucoxanthin and phlorotannins (Fabrowska et al. 2018; Oliyaei and Moosavi-Nasab 2021; Putra et al. 2022; Ummat et al. 2020; Zhu et al. 2022). UAE protocols aiming the recovery of hydrophilic compounds, such as polysaccharides and proteins, use water based solutions when optimizing the rest of the extraction conditions (ultrasound power, type and ratio of solvent, time and temperature) depending on the chemical nature of the targeted compounds and the characteristics of the biomass (Flórez-Fernández et al. 2017; Garcia-Vaquero et al. 2019, 2020b; Hmelkov et al. 2018).

Table 2.3 Summary of UAE conditions used for the extraction of multiple compounds and associated biological activities from macroalgae

Macroalgae sp.	UAE conditions	Compounds (maximum yields)	Biological activity	References
Sargassum angustifolium	Methanol solution and US probe (150 W, 100% amplitude, 15 min)	Fucoxanthin (0.79 ± 0.01 mg/g)	Antibacterial	Oliyaei and Moosavi-Nasab (2021)
Cystoseira indica		Fucoxanthin (0.81 ± 0.01 mg/g)		
Chondrus ocellatus	61% ethanol solution at 25:1 S/A and 60 °C and US water bath (100 W, 41 min)	TPC (24.8%) PC (17.1%) TSC (6.94%)	Antioxidant and enzyme inhibition	Zhu et al. (2022)
Kappaphycus alvarezii	50% ethanol solution at 52.5 °C and 30:1 S/A and US probe (200 W, 26 kHz)	TPC (845.28 ± 93.17 mg GAE/kg)	–	Putra et al. (2022)
Kappaphycus denticulatum		TPC (802.08 ± 21.93 mg GAE/kg)		
Kappaphycus striatum		TPC (940.21 ± 5.40 mg GAE/kg)		
Ascophyllum nodosum	US probe (500 W, 50% amplitude, 5 min)	FSP (195.36 ± 1.36 mg fucose/100 g) TSC (2572.97 ± 269.37 mg GE/100 g) TPC (2340.54 ± 65.55 mg GAE/100 g)	Antioxidant	Garcia-Vaquero et al. (2020b)
Laminaria hyperborea	Hydrothermal followed by US water bath (80 W, 0.1 M HCl, 15–30 min)	FSP (487.4 ± 10.3 mg/100 g DW) Total glucans (908.0 ± 51.4 mg/100 g DW)	Antioxidant	Garcia-Vaquero et al. (2019)
Laminaria digitata		FSP (155.1 ± 1.0 mg/100 g DW) Total glucans (134.8 ± 11.8 mg/100 g DW)		
Ascophyllum nodosum		FSP (2971.7 ± 61.9 mg/100 g DW) Total glucans (494.2 ± 26.9 mg/100 g DW)		

(continued)

Table 2.3 (continued)

Macroalgae sp.	UAE conditions	Compounds (maximum yields)	Biological activity	References
Cladophora glomerata	96% ethanol solution at 40 °C US water bath (40 kHz frequency, 60 min)	Chlorophyll a (10.8 ± 0.6 μg/ml) Chlorophyll b (5.1 ± 0.3 μg/ml) TC (0.5 ± 0.1 μg/ml)	–	Fabrowska et al. (2018)
Cladophora rivularis		Chlorophyll a (3.4 ± 0.2 μg/ml) Chlorophyll b (1.7 ± 0.1 μg/ml) TC (0.6 ± 0.1 μg/ml)		
Ulva flexuosa		Chlorophyll a (17.6 ± 0.9 μg/ml) Chlorophyll b (20.1 ± 1.0 μg/ml) TC (2.2 ± 0.1 μg/ml)		
Fucus vesiculosus	50% ethanol solution at 10:1 S/A and US water bath (35 kHz, 30 min)	TPC (572.3 ± 3.2 mg GAE/g) Phlorotannins (476.3 ± 2.2 mg PE/g) TFC (281.0 ± 1.7 mg QE/g)	Antioxidant	Ummat et al. (2020)
Fucus evanescens	H_2O at 23 °C, 20:1 S/A and US water bath (150 W, 35 kHz, 5–30 min)	Fucoidan (3.71% (w/w) defatted macroalgae)	Cytotoxicity against the HCT-116 and HT-29 cells	Hmelkov et al. (2018)
Laminaria digitata	0.1 M HCl at 10:1 S/A, 76 °C and US probe (500 W, 100% amplitude, 10 min)	FSP (1060.7 ± 70.6 mg fucose/100 g DW) Total glucans (968.6 ± 13.3 mg/100 g DW)	Antioxidant	Garcia-Vaquero et al. (2018)
Eucheuma cottonii	US probe (35 Hz, power intensity 30%, 20:1 S/A, 30 min) followed by autoclaving	Agar (56.49%)	–	Din et al. (2019)
Gelidium amansii		Agar (49.1%)		
Ascophyllum nodosum	Water as solvent and US probe (750 W, 60% amplitude, 15 min)	Laminarin (5.290 ± 0.480%) TPC (0.156 ± 0.014 mg PE/g DW)	Antioxidant and antibacterial	Kadam et al. (2015b)
Laminaria hyperborea		Laminarin (5.975 ± 0.467%) TPC (0.365 ± 0.039 mg PE/g DW)		

(continued)

Table 2.3 (continued)

Macroalgae sp.	UAE conditions	Compounds (maximum yields)	Biological activity	References
Sargassum muticum	H_2O, 20:1 S/A, 25 °C and US water bath (150 W, 40 Hz, 5–30 min)	Fucose (87.9 g/kg extract) TPC (2.5% PE) SC (39.54 ± 0.001-mg sulfate ion/g) Phlorotannins (25 mg/g)	Antioxidant and antitumor	Flórez-Fernández et al. (2017)
Laminaria japonica	H_2O, 80 °C, 50:1 S/A and US probe (1050 W, 54 min)	Polysaccharides (5.75 ± 0.3%)	Antioxidant and enzyme inhibition	Wan et al. (2015)
Ascophyllum nodosum	50% ethanol, 25 °C and US water bath (40 kHz, 30 min)	TPC (4.66 g PE/100 g DW)	Antioxidant	Agregán et al. (2018)
Fucus vesiculosus		TPC (1.94 g PE/100 g DW)		
Bifurcaria bifurcata		TPC (5.74 g PE/100 g DW)		
Ascophyllum nodosum	0.03 M HCl, US probe (750 W, 20 kHz, 114 μm of amplitude, 25 min)	TPC (143.12 mg GE/g DW) Fucose (87.06 mg/g DW) Uronic acid (128.54 mg/g)	–	Kadam et al. (2015d)
Sargassum binderi and *Turbinaria ornata*	2% NaOH, pH 12, 10:1 S/A, 90 °C and US probe (150 W, 25 kHz, 30 min)	Alginates (50–55%)	–	Youssouf et al. (2017)
Kappaphycus alvarezii and *Euchema denticulatum*	H_2O, pH 7, 10:1 S/A, 90 °C and US probe (150 W, 25 kHz, 15 min)	Carrageenan (50–55%)		
Gelidium pusillum	Maceration in combination with US probe (41.97 W, 120 μm of amplitude, 30 °C, 10 min)	R-phycoerythrin (76.8%) R-phycocyanin (93.13%)	–	Mittal et al. (2017)
Porphyra yezoensis	H_2O, 40.5 °C, 20:1 S/A and US (20 KHz, 300 W, 38.3 min)	Taurine (13 mg/g)	–	Wang et al. (2015)

(continued)

Table 2.3 (continued)

Macroalgae sp.	UAE conditions	Compounds (maximum yields)	Biological activity	References
Grateloupia turuturu	US flow-through reactor (200–400 W, amplitude of 100%, 22–40 °C, 60–120 min)	R-phycoerythrin (3.6 ± 0.3 mg/g DW) Carbohydrates (439 ± 16 mg/g DW) Carbon content (83%) Nitrogen content (92%)	–	Le Guillard et al. (2015)
Padina tetrastromatica	80% ethanol solution, 50 °C, 10:1 S/A and US bath (50 Hz, 230 V, 30 min)	Fucoxanthin (750 μg/g DW) TPC (10.9 mg GAE/g DW)	Antioxidant	Raguraman et al. (2018)
Undaria pinnatifida	H_2O, 50 °C, 30:1 S/A and US probe in a Uwave 1000 apparatus (300 W, 20 kHz, 30 min)	Carotenoids (30 mg/mL)	–	Zhu et al. (2017)
Gracilaria lemaneiformis	Ultrasonic-microwave-assisted extraction (50 W ultrasonic power, 87 °C, 31.7 min, 60.7:1 S/A	Polysaccharides (34.8%)	Antitumor against the MCF-7, HepG-2, and Hela cells	Shi et al. (2018)
Porphyra haitanensis	Ultrasonic-microwave-assisted extraction (microwave power of 500 W, ultrasonic power of 50 W, 79.94 °C, 29.64 min, 41.79:1 S/A)	Polysaccharides (20.98%)	–	Xu et al. (2020)
Ulva fasciata	H_2O, 25 °C, 60:1 S/A and US (480 W, 30 min)	TSC (38.63%) Protein (0.28%) Sulfate (19.41%) Uronic acid (35.06%)	Antioxidant and antitumor	Shao et al. (2013)
Sargassum henslouianum		TSC (35.28%) Protein (2.83%) Sulfate (13.15%) Uronic acid (4.36%)		
Gloiopeltis furcata		TSC (66.2%) Protein (2.3%) Sulfate (24.1%) Uronic acid (1.35%)		
Enteromorpha tubulosa	UAE during 36 min at 69 °C and 45:1 S/A	Polysaccharides (16.04%)	Cytotoxicity against MCF-7 ell	Hu et al. (2018)

(continued)

Table 2.3 (continued)

Macroalgae sp.	UAE conditions	Compounds (maximum yields)	Biological activity	References
Padina australis	H_2O, 60 °C, 100:1 S/A and US water bath (40 kHz)	TPC (11.04 mg GAE/g DW)	Antioxidant and tyrosinase inhibitory	Hassan et al. (2021)
Sargassum wightii	Methanol solution, 37 °C, 10:1 S/A and US water bath (40 kHz, 120 min)	TPC (19.27 ± 0.05 mg GAE/g) TFC (14.94 ± 0.65 mg QE/g) Fucoxanthin (1.45 ± 0.22 mg/kg DW) Phloroglucinol (69.86 ± 5.25 mg/kg DW) Gallic acid (0.04 ± 0.00 mg/kg DW) Quercetin (0.05 ± 0.00 mg/kg DW) Ferulic acid (0.08 ± 0.00 mg/kg DW) Vanillin (0.22 ± 0.07 mg/kg DW)	–	Kumar et al. (2020)
Ulva intestinalis	H_2O, 66 °C, 50:1 S/A and US water bath (180 W, 40 min)	Ulvan (8.3 ± 0.04%)	Antioxidant	Rahimi et al. (2016)

Abbreviations in the table are as follows: *US* ultrasonic, *S/A* solvent/algae ratio, *DW* dry weight, *GAE* gallic acid equivalents, *PE* phloroglucinol equivalent, *QE* quercetin equivalent, *SC* sulfate content, *TPC* total phenolic compounds, *PC* protein content, *TSC* total sugar content, *TFC* total flavonoids content, *TAC* total alkaloid compounds, *TTC* total tannic compounds, *FSP* fucose-sulphated polysaccharides, *GE* glucose equivalents, *TC* total carotenoids

Other uses of UAE have also been explored by combining the application of UAE sequentially with other technologies for the purposes of extraction. Din et al. (2019) used UAE combined with autoclaving for the extraction of agar from *Eucheuma cottonii* and *Gelidium amansii*. Garcia-Vaquero et al. (2019) explored the use of UAE followed by hydrothermal treatments for the recovery of fucose containing polysaccharides and total glucans from *Laminaria hyperborea*, *Laminaria digitata*, and *Ascophyllum nodosum*. Moreover, UAE is a technology that can be easily incorporated into other technological processes aiming to improve their efficiency. The simultaneous application of ultrasounds together with other conventional techniques such as enzyme extraction have also shown promising results. Soft liquefaction of *Grateloupia turuturu* using UAE, MAE, and ultrasound-assisted enzymatic hydrolysis (UAEH) was studied by Le Guillard et al. (2016). According to the authors, the experiment was performed using an enzymatic cocktail of 4 industrial carbohydrates at 40 °C for 6 h, which resulted in observed a similar profile between UAE and MAE with the recovery of 71–74% of the initial material into the soluble phase. UAEH improved the recovery up to 91% of the solubilized material while also improving the carbohydrate, nitrogen, carbon, and amino acids contents, with yields of 439.43 ± 16.39 mg/g dry weight, 73.95 ± 1.16%, 92.27 ± 0.34%, 54.06 ± 0.19%, respectively (Le Guillard et al. 2016).

The simultaneous application of ultrasounds and microwave have also shown advantages in macroalgal biorefinery, particularly for its efficiency when extracting polysaccharides (Garcia-Vaquero et al. 2020b; Shi et al. 2018; Xu et al. 2020). Few research is available on the simultaneous use of these technologies in macroalgae, focusing mainly on developing extraction protocols for the recovery of polysaccharides from *Gracilaria lemaneiformis* (Shi et al. 2018) and *Porphyra haitanensis* (Xu et al. 2020) and studying the biological properties of the extracted polysaccharides. However, the efficiency of the protocols developed compared to other innovative technologies or protocols was not established. Garcia-Vaquero et al. (2020b) was one of the first studies determining the effect of the simultaneous application of both technologies in the extraction of polysaccharides and phenolics compounds from macroalgae. The authors determined that the simultaneous application of microwaves and ultrasounds had a significant effect on the extraction yields of fucose containing polysaccharides and total soluble carbohydrates from *Ascophyllum nodosum* with recovery yields of these compounds several folds higher compared to the use of ultrasounds and microwave alone (Garcia-Vaquero et al. 2020b). The exploration of the multiple extraction parameters of these technological combinations and its effects on the extraction efficiency of several compounds, aiming a selective or targeted extraction can be useful to expand the application of both technologies together in macroalgal biorefineries (Garcia-Vaquero et al. 2020b).

2.2 Application of MAE

MAE is based on the extraction of compounds from algae due to disruption or changes in the structure of cells by using non-ionizing electromagnetic radiation with a frequency ranging from 300 MHz to 300 GHz and wavelengths from 1 cm to 1 m (Gomez et al. 2020). In MAE, the energy transfers via the mechanisms of dipole rotation and ionic conduction simultaneously. When applying electromagnetic waves, the ion conduction mechanism results in migrating ions followed by causing homogenous heat in the solution. Subsequently, when these waves are removed, the dipole rotation mechanism results in re-orientating the molecules in line with the electric field followed by causing thermal agitation (Chemat et al. 2017). These events cause the breakdown of hydrogen bonds and the migration of soluble ions, accelerating the penetration of the solvent into the tissue and the extraction of compounds. On the other hand, after generating thermal agitation, the moisture content of the sample matrix evaporates, resulting in an increase in the intracellular pressure and breaking the cell walls (Sosa-Hernández et al. 2018). Thus, the heat transfer and mass transfer of solutes follows the same direction when using MAE compared to conventional heating (Gomez et al. 2020) as seen in Fig. 2.1.

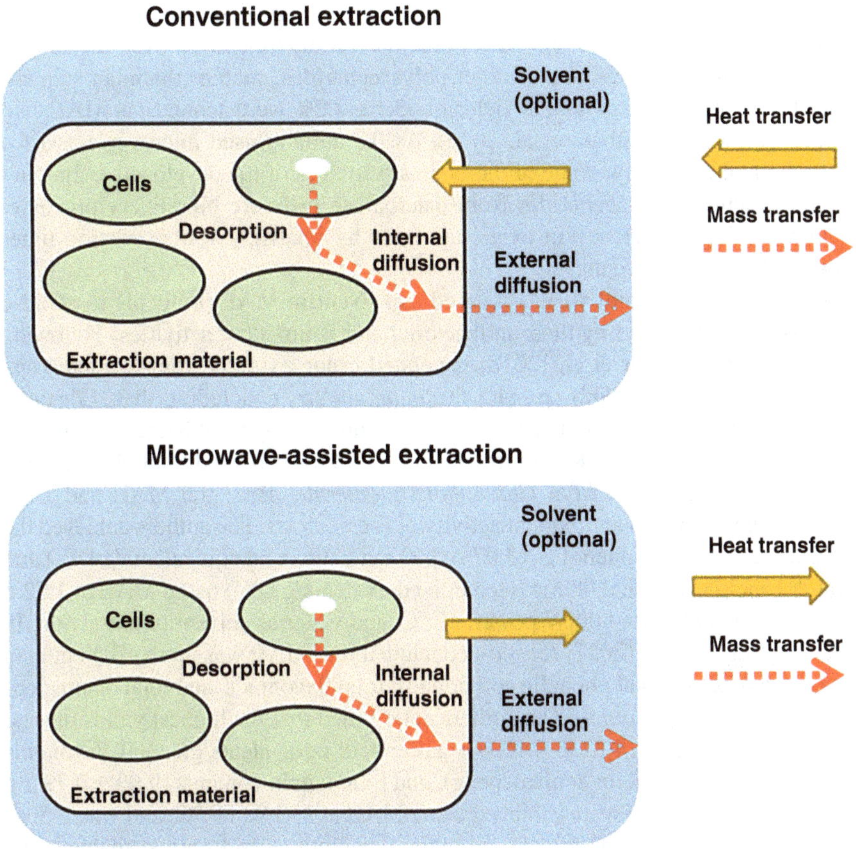

Fig. 2.1 Mass and heat transfer gradients in MAE and conventional heating methods. Image taken from Gomez et al. (2020) and reproduced with permission from Elsevier

The first use of microwave energy to extract target compounds dates back to 1986 (Ganzler et al. 1986). Similar to any technique, MAE also needs to be optimized to achieve high yields of compounds. Temperature, microwave power, matrix characteristic, extraction time, solvent type, solvent to material ratio, and target compound are parameters to control when optimizing the extraction of compounds using MAE. MAE is a suitable method for extracting lipids, carotenoids, polyphenolic compounds, polysaccharides, and proteins (Filote et al. 2021).

MAE has been used to extract multiple compounds from algae. Le et al. (2019) optimized the use of MAE for the improving the extraction yields of the polysaccharide ulvan from *Ulva pertusa* while monitoring the antioxidant activity of the extracts. The authors achieved yields of 41.91% and polysaccharides with high antioxidant power at optimized MAE conditions of 600 W power, solvent to material ratio of 55.45, pH of 6.57 and extraction time of 43.63 min (Le et al. 2019). Tsubaki et al. (2016) achieved maximum yields of the polysaccharide ulvan of 40.4 ± 3.2%

and 36.5 ± 3.1% from *Ulva meridionalis* and *Ulva ohnoi*, respectively using MAE conditions of 160 °C for 10 min with 4 min of come-up time. Moreover, the authors determined that the extraction of other polysaccharides, such as rhamnan sulphate from *Monostroma latissimum*, at yields of 53.1 ± 7.2% were achieved at MAE conditions of 140 °C (Tsubaki et al. 2016). As the authors used microwave-assisted hydrothermal extraction, some of the main advantages of the developed techniques when extracting polysaccharides from macroalgae in future biorefinery processes include the high efficiency in extraction rates by relatively low extraction times without the use of solvents.

Ethanol and methanol were also used as solvent in MAE protocols to achieve extracts to be evaluated by their antibacterial and antioxidant activities. Following these protocols, Munir et al. (2018) generated crude extracts from *Nitzschia* sp., *Stigeoclonium* sp., *Ulothrix* sp., and *Oedogonium* sp., concluding that. Ethanolic extracts from *Nitzschia* sp. and *Ulothrix* sp. exhibited high antibacterial activities (Munir et al. 2018). Also, Korzeniowska et al. (2020) recovered total phenolic and total flavonoid contents from *Cladophora glomerata* using the MAE and UAE methods and evaluated antioxidant activity of each extract. The authors achieved the highest yields of total phenolics (3.07 ± 0.09 mg gallic acid equivalents/g DW) and total flavonoids (1.46 ± 0.08 mg quercetin equivalents/g DW) using MAE at 1:22.5 sample to solvent ratio, 800 W power, 40 °C, and water as solvent of extraction. In addition, Čagalj et al. (2021) recently concluded that MAE was more efficient than UAE for extracting total phenolic compounds, total flavonoids, and total tannin contents from *Padina pavonica*. The authors determined that the highest yields of total phenolic compounds (4.32 ± 0.15 mg gallic acid equivalents/g), total flavonoids (2.87 ± 0.01 mg quercetin equivalents/g), and total tannin contents (9.03 ± 0.79 mg catechin equivalents/g) were achieved using MAE at 200 W, 60 °C, and water, 50% ethanol, or 70% ethanol as solvent for 5 min. Conflicting results are available in the literature concerning the efficiency of different methods. Garcia-Vaquero et al. (2021b) analysed the use of fixed extraction conditions (solvent: 50% ethanol; extraction time: 10 min; algae/solvent ratio: 1/10) for the recovery of total phenolics, total phlorotannins, total flavonoids and total tannins from *Fucus vesiculosus* and *Pelvetia canaliculate* using multiple innovative technologies including UAE, MAE, ultrasound–microwave-assisted extraction, hydrothermal-assisted extraction and HPAE. The authors concluded that UAE was the most effective technology for the recovery of all the phytochemicals analysed (Garcia-Vaquero et al. 2021b). However, further studies are needed evaluating multiple technologies as these results may change depending on the devices and models used to perform the extraction procedures, as well as the nature of the original biomass used during the extraction and other conditions, such as the solvent and time of extraction at which each technology may exert their effects.

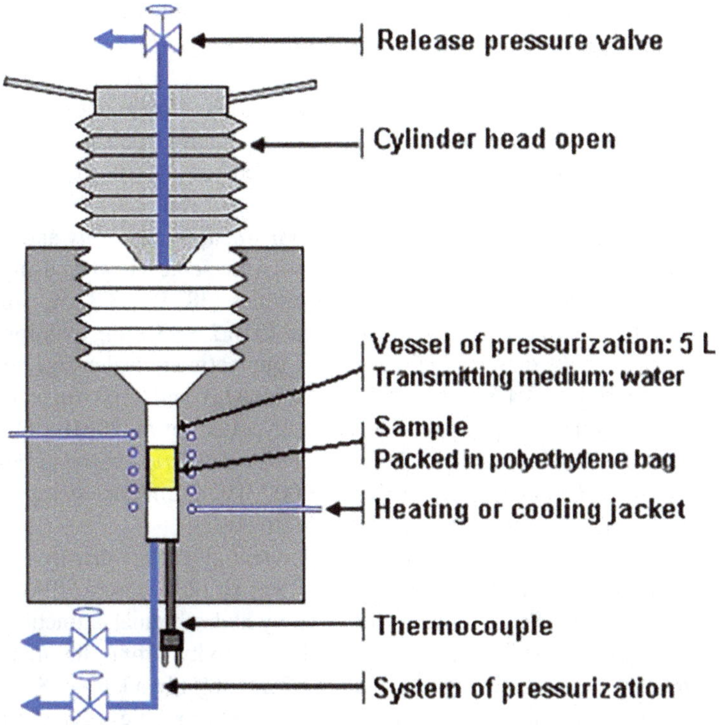

Fig. 2.2 Scheme of a HPAE system originally published by Chen et al. (2009) and reproduced with permission from Elsevier

2.3 *Application of HPAE*

HPAE is based on the application of pressure ranging from 100 to 1000 MPa. The functioning of HPAE is based on the isostatic and Le Chatelier principles. According to the isostatic principle, the pressure is applied uniformly on the biological matrix regardless of its shape or constitution and following the Le Chatelier's principle any conformational change accompanied by a decrease in volume is enhanced by pressure (Yordanov and Angelova 2010). From an extraction point of view HPAE alters the equilibrium of atoms position in the matrix, causing disruptions of the cell walls and facilitating the extraction of compounds (Moreira et al. 2020). A schematic representation of a HPAE system is provided in Fig. 2.2.

Some of the advantages of HPAE technology are the short extraction times, flexibility to choose the extract solutions composition, the enhancement of solvent permeability in cells, suitable for extraction of small molecules (Khan et al. 2019). However, HPAE is not suitable for extracting pressure-sensitive compounds and the costs of use are higher compared to other technologies. Using HPAE for the purposes of extraction require the control of certain parameters that include the extraction time, pressure, type and volume of solvent, number of cycles, and temperature, that could play a major role in the recovery yield of bioactives (Khan et al. 2019).

HPAE technology has been used to extract various compounds from algae. Mulchandani et al. (2015) recovered lipids from the microalgae *Chlorella saccharophila* using HPAE at an optimized pressure of 800 bar and 10 cycles to recover maximum yields of 89.91 ± 3.69%. Garcia-Vaquero et al. (2021b) used HPAE (600 MPa, 10 min) to recover compounds from *Fucus vesiculosus* and *Pelvetia canaliculata*. The maximum yields of compounds were achieved when using HPAE and fucus vesiculosus, with yields of total phenolics (387.9 ± 3.8 mg gallic acid equivalents/g DW extract), total phlorotannins (316.2 ± 3.1 mg phloroglucinol equivalents/g), total flavonoids (231.5 ± 1.3 mg quercetin equivalents/g), total tannin contents (148.4 ± 3.4 mg catechin equivalents/g) and total sugar contents (123.7 ± 5.6 mg glucose equivalents/g) (Garcia-Vaquero et al. 2021b). O'Connor et al. (2020) recovered protein from *Fucus vesiculosus*, *Alaria esculenta*, *Palmaria palmata*, and *Chondrus crispus* using HPAE (600 MPa, 4 min), achieving recovery yields of 23.7%, 15%, 14.9%, and 16.1%, respectively.

Furthermore, HPAE can also be combined sequentially with other techniques to increase the yields of extraction. For instance, Li et al. (2017) used HPAE at optimized pressure of 70 MPa for 2 cycles followed by hydrothermal extraction for the recovery of fucoidans with strong antioxidant activity from the macroalgae *Nemacystus decipients*, achieving yields of 16.67 ± 0.36%. In addition, Suwal et al. (2019) extracted and analysed the potential antioxidant properties of multiple compounds from *Palmaria palmata* and *Solieria chordalis* using HPAE (400 MPa, 20 min) combined with polysaccharidases to extract polysaccharides, proteins and polyphenols.

2.4 Application of PEF

PEF is a nonthermal technology that can be used to disrupt cell walls from algae, extracting target compounds. In this method, algae suspensions are placed between two electrodes and they receive direct-current high-voltage pulses (0.5–30 kV/cm) for very short time intervals (5–10,000 μs), causing an increase in permeability of the cell membranes by electroporation (Carullo et al. 2020). This allows the solvent to penetrate into the cell and leads to the selective release of compounds from the cell without the production of cell debris (Pataroa et al. 2019; Poojary et al. 2016). A representation of a PEF treatment chamber is presented in Fig. 2.3.

Since PEF is regularly conducted at ambient temperature, this method is suitable for the extraction of various compounds, such as carotenoids, sucrose, inulin, vitamins, and proteins (Puértolas et al. 2012). Various parameters can influence the efficiency of PEF, including field strength, pulse shape, conductivity of intact, pulse time, disintegrated cells, number of pulses, and specific treatment energy (Bāliņa

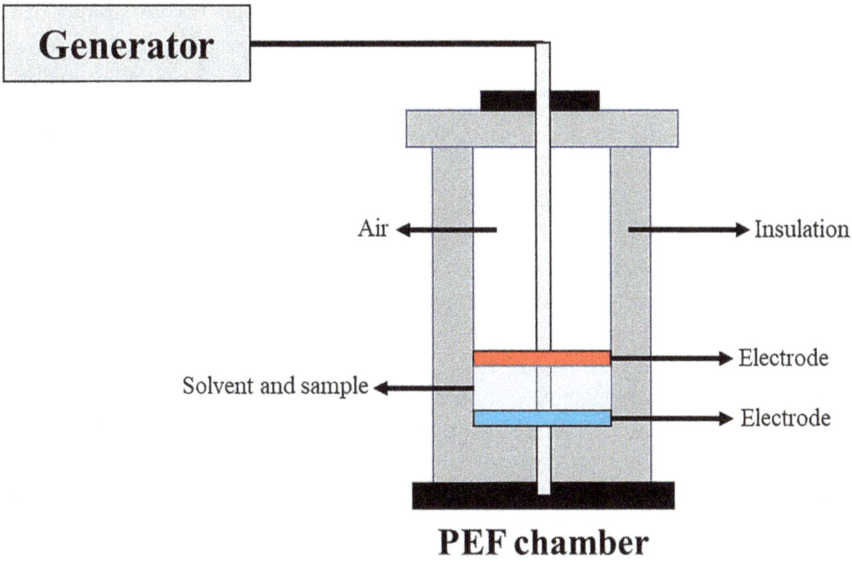

Fig. 2.3 Schematic representation of a PEF treatment chamber

et al. 2020; Goettel et al. 2013), which need to be optimized to achieve maximum yields of different compounds.

PEF is a non-destructive technology that have several advantages. PEF is energetically efficient as well as having processing advantages such as short processing times, high selectivity, no thermal effect, easy to scale and low cost. However, the variable efficiency of this technology based on the medium composition (conductivity) and the high initial cost of purchasing the equipment are among the main disadvantages of this method (Goettel et al. 2013). PEF is suitable for extracting lipids, carotenoids, polyphenolic compounds, polysaccharides, and proteins (Filote et al. 2021). Recently, the use of PEF for extracting target compounds from macroalgae was highlighted due to its multiple advantages over other extraction technologies. Einarsdóttir et al. (2022) used PEF to extract total phenolic compounds, total carbohydrates, and water-soluble proteins from *Alaria esculenta*, which obtained a yield of 10.4 ± 3.7 mg gallic acid equivalent/g DW, 21.4 ± 1.8 mg glucose equivalent/g DW, and 0.8 ± 0.4 mg/g DE, respectively. Castejón et al. (2021) compared aqueous extracts obtained from *Ulva lactuca*, *Alaria esculenta* and *Palmaria palmata* by PEF using exponential decay pulses with a width of 0.96 μs and 8 kV/cm electric field at 1.2 Hz for 10 min, and by conventional extraction (95 °C, 45 min). The authors reported similar levels of extraction between PEF and conventional technologies with the advantage of PEF being of non-thermal nature and its shorter extraction times compared to conventional methods. In addition, the highest content of phenolics (8.86 mg gallic acid/g DW) and flavonoids (12.09 mg quercetin/g DW) were obtained from *Alaria esculenta*, also exhibiting excellent antioxidant and anti-enzymatic activities (Castejón et al. 2021). In another study, Polikovsky et al. (2016)

extracted low levels of protein, achieving yields of less than 1% (59 μg/ml) from *Ulva* sp. using PEF with application of 75 pulses with duration of 7.5 μs and an average electric field strength of 2.964 ± 0.007 kV/cm for 5 min. Similar levels of 53.8 ± 0.69 μg/ml were also extracted from *Ulva* sp. using PEF with 50 pulses with 2.3 μs duration, applied at 26 kV, 7.26 kV/cm filed strength (Polikovsky et al. 2019). Postma et al. (2018) achieved a yield of extraction of 15.1 ± 0.7% proteins and 14.8 ± 3.3% carbohydrates from *Ulva lactuca* using PEF at optimized conditions of 7.5 kV/cm electric field strength with 2 pulses of 0.05 ms. Interestingly, Robin et al. (2018) showed that PEF technology at 247 kJ/kg fresh *Ulva spp.* delivered through 50 pulses of 50 kV/cm was produced about seven-fold increase in the protein yield in comparison with osmotic shock extraction (33 ± 20 mg). However, Prabhu et al. (2019) showed that PEF with 200 pulses with a field strength of 1 kV cm–1, pulse duration of 50 μs, and pulse repetition rate of 3 Hz was able to extract starch and protein with yields of 59.54% (1.14 times higher than control) and 14.94% (4.73 times higher than control) from *Ulva ohnoi*, respectively. Based on the variable results achieved with this technology, PEF seems to be promising when recovering protein from macroalgae, although further optimization will be needed when using different PEF systems and macroalgal biomass aiming to achieve the maximum yields of compounds from the biomass.

2.5 Application of SFE

SFE is another novel and environmentally friendly extraction method that uses an extraction medium, and the extraction is conducted based on differences in solubility. For this purpose, carbon dioxide, water, methane, ethane, ethylene, propylene, methanol, ethanol, and acetone can be used as extraction medium in this technique (Sapkale et al. 2010), at a critical temperature of up to 40 °C and a critical pressure of up to 7 MPa (Machmudah et al. 2020). Supercritical fluids have lower viscosities and better diffusion properties in comparison with those of certain liquids (Taylor 2009). A schematic representation of a SFE system as well as the mechanism of action for the extraction of bioactive compounds is presented in Fig. 2.4.

In particular, the use of supercritical carbon dioxide as an extraction medium presents numerous advantages, including the inert nature of the solvent that is also environment-friendly, inexpensive, tasteless and odourless, relatively non-toxic, easily available and non-flammable, it also requires short extraction times, its operation is simple, and it can improve the yields of extraction compared to conventional extraction using organic solvents. Also, solvent removal following extraction is much easier, allowing more accurate component concentrations to be determined (Cikoš et al. 2018; Machmudah et al. 2020). Carbon dioxide is widely used in this system because it becomes a supercritical fluid under low pressure (7.38 MPa) and temperature (31.06 °C) (Kadam et al. 2015a). When carbon dioxide is compressed and heated, it is able to act well as a processing fluid to extract a wide range of compounds, as under these conditions it has the solvating power of a liquid and the

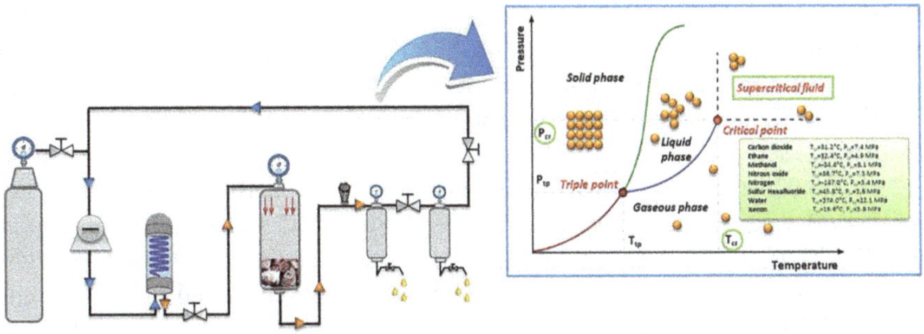

Fig. 2.4 Schematic representation of SFE and the mechanism of extraction. Image originally published by Al Khawli et al. (2019) and reproduced from MDPI

diffusivity of a gas. Therefore, this method can be used for compounds that are temperature-sensitive or easily oxidized. Supercritical carbon dioxide is effective at dissolving lipid-soluble or low-polar compounds, but it can be adapted to the extraction of highly polar and high molecular compounds by adding alcohols, such as methanol or organic solvents (Monteagudo-Olivan et al. 2019). However, the use of organic solvents may need higher temperature to become supercritical fluid (Wang and Weller 2006). Generally, SFE parameters also need to be optimized to achieve high yields of compounds during the extraction processes. These parameters include temperature, pressure, time, solvent flow rate, type of solvent, diffusion of solvent, solubility of solvent, and density of fluid (Sapkale et al. 2010). In addition, SFE technique may be used to enrich products with proteins as it has a really high selectivity for low polarity impurities. However, SFE systems normally need high energy inputs, the equipment is costly and the final products extracted may be contaminated by co-solvents when they are needed (Gordalina et al. 2021). It should be noted that this method is suitable for extracting lipids, carotenoids and polyphenolic compounds (Filote et al. 2021).

Many studies in recent years have used this technology to extract various compounds from macroalgae. Messyasz et al. (2018) used SFE at a pressure of 500 bar, temperature of 40 °C, and CO_2/ethanol as solvent to extract fatty acids from multiple macroalgae (*Ulva clathrata*, *Cladophora glomerata*, *Polysiphonia fucoides*, and multi-species mixtures). The authors reported that all extracts contained high concentrations of unsaturated fatty acids, mainly C16:1 and C18:1, and the fatty acids C16:0 and C18:1 were the predominant compounds achieved with SFE (Messyasz et al. 2018). Also, the authors measured the content of chlorophyll a and carotenoids extracted under the above conditions with a yield of 0.32 and 0.09 mg/g, respectively. In addition, the extracts also contained variable levels of total phenolics and total flavonoids, being 20.32 mg gallic acid/g and 1.51 mg quercetin acid/g from Baltic multi-species biomass, and 25.22 mg gallic acid/g and 1.08 mg quercetin/g from *Cladophora glomerata*, respectively, that showed good antioxidant activities (Messyasz et al. 2018). Roh et al. (2008) obtained the highest yields of fucoxanthin (0.00753 μg/g) and polyphenols (approximately 800 μg/g) from *Undaria pinnatifida* using SFE with CO_2 and ethanol as co-solvent at the pressure and temperature of 200 bar, 323 K, and 250 bar, 333 K, respectively. Ospina et al. (2017) showed that the extracts obtained from *Gracilaria mammillaris* using SFE at 30 MPa, 60 °C and 8% ethanol as co-solvent with a yield of 3.03% had the highest antioxidant activity (42.1%). However, the maximum yields of total phenolic (3.791 ± 0.035 mg gallic acid/g) and carotenoid compounds (5.038 ± 0.087 mg carotene/g) were extracted using SFE at 30 MPa, 60 °C and 8% ethanol, and at 30 MPa, 50 °C and 5% ethanol, respectively. Conde et al. (2015) achieved the highest yields of extraction of fucoxanthin (1.5 mg/100 g algae) and phenolic compounds (4% DW) from *Sargassum muticum* using SFE 10 MPa, 50 °C, and CO_2 as pure supercritical solvent. Interestingly, the addition of ethanol as co-solvent improved total yield (up to 3 times higher), radical scavenging property (up to 2.5 times higher), and fucoxanthin yields (up to 90 times higher) (Conde et al. 2015). In addition, extracts obtained from pure CO_2 SFE were contained the higher content of unsaturated fatty acids. Also, the selected extracts exhibited anti-browning activity on B16F10 murine cells and inhibition of lipogenesis in SW872 liposarcoma cells (Conde et al. 2015). De Corato et al. (2017) prepared crude extracts from *Laminaria digitata*, *Undaria pinnatifida*, *Porphyra umbilicalis*, *Eucheuma denticulatum* and *Gelidium pusillum* using SFE at 37.9 MPa and 50 °C to evaluate their antifungal activity against 3 postharvest pathogens of *Botrytis cinerea*, *Monilinia laxa* and *Penicillium digitatum*. The yield of total lipids was 48.6 ± 0.3%, 42.3 ± 0.2%, 42.8 ± 0.4%, 20.5 ± 0.5%, and 25.6 ± 0.01% for *Laminaria digitata*, *Undaria pinnatifida*, *Porphyra umbilicalis*, *Eucheuma denticulatum* and *Gelidium pusillum*, respectively (De Corato et al. 2017). Also, the content of total water-soluble polysaccharides these extracts ranged between 14.5% and 23% while the range of total phenolics varied between 0.15% and 2.3% depending on the macroalgal species. The authors showed that *Laminaria digitata* and *Undaria pinnatifida* extracts at 30 g/L completely inhibit and suppress conidial germination of *Botrytis cinerea* and *Monilinia laxa* (De Corato et al. 2017). In a study, total phenolic compounds of 3.45 ± 0.31% was obtained from *Sargassum muticum* using CO_2 modified with 12% ethanol at 15.2 MPa and 60 °C during 90 min, showed an antioxidant activity (β-carotene bleaching test) with an IC_{50} of 0.72 ± 0.22 mg/ml (Anaëlle et al. 2013). Park et al. (2012) were able to extract oil from *Laminaria japonica* using SFE at 25 MPa, 55 °C, and CO_2 as supercritical solvent with a yield of 2.59 g/250 g. The authors state that at constant temperature, the amount of oil extracted increases with the pressure (Park et al. 2012). Also, Goto et al. (2015) showed that at constant pressure (40 MPa), the amount of fucoxanthin extracted from *Undaria pinnatifida* increases with increased extraction time from 30 min to 150 min and temperature of up to 60 °C, achieving yields ranging from 47 to 58 μg/g (Goto et al. 2015). Kang et al. (2016) showed anti-inflammatory activity of essential oils extracted with SFE at 20 MPa and 45 °C from *Undaria pinnatifida* showing IC_{50} of 87, 134, and 158 μg per ear for edema, erythema, and blood flow, respectively. In another study, Saravana et al. (2017) optimized extraction conditions of total carotenoid, fucoxanthin and phlorotannin from *Saccharina japonica* using SFE. The extraction conditions achieving the maximum yield of total carotenoid (2.391 ± 0.419 mg/g) and fucoxanthin (1.421 ± 0.181 mg/g) were 50.62 °C, 300 bar, and 2% of sunflower oil as co-solvent, while the maximum yields of phlorotannins (0.927 ± 0.026 mg/g) required SFE at 48.98 °C, 300 bar, and 2% of water as co-solvent (Saravana et al. 2017). Similarly, Getachew et al. (2018) extracted fucoxanthin and total phenolic compounds from *Saccharina japonica* using SFE at 300 bar, 40 °C, and coffee oil as co-solvent with a yield of 2.08 ± 0.06 mg/g and 287.71 ± 13 mg/100 g, respectively.

Klejdus et al. (2010) optimized the extraction of isoflavones from *Chondrus crispus* using SFE at 35 MPa, 40 °C, and 3% MeOH/water mixture (9:1) as co-solvent for 60 min. Sivagnanam et al. (2015) determined yields of fucoxanthin in oils extracted by SFE at a temperature of 45 °C, pressure of 250 bar, and ethanol as co-solvent from *Saccharina japonica* and *Sargassum horneri* achieving levels of 0.41 ± 0.05 and 0.77 ± 0.07 mg fucoxanthin/g, respectively. Moreover, the authors also reported that the extracted fucoxanthin had powerful biological activities including antioxidant, antimicrobial, and antihypertension effects (Sivagnanam et al. 2015). Becerra et al. (2015) extracted fucosterol from *Lessonia vadosa* by SFE using CO_2 at 180 bar and 50 °C with 20 to 30% of cellulose as modifier with a yield of 16–20.3% (w/w of extract), that showed antileishmanial activity ($IC_{50} < 10$ μM) against intracellular amastigotes. SFE has also been used to extract carotenoids and phenolic compounds from *Cladophora glomerata*, *Ulva flexuosa* and *Chara fragilis* using an optimized conditions of 40 °C, 300 bar and 11.4% ethanol as co-solvent (Fabrowska et al. 2016). According to the authors, the extracts of *Chara fragilis* were the richest in total carotenoids and total phenols (24.90 mg fucoxanthin equivalents/g extract and 30.20 mg gallic acid equivalents/g extract, respectively), while the extracts of *Ulva flexuosa* showed the highest antioxidant activity (0.944 mmol trolox equivalents/g extract) (Fabrowska et al. 2016).

For agricultural applications, Michalak et al. (2016) evaluated extracts containing auxins, cytokinins, polyphenols, micro- and macro-elements obtained from *Polysiphonia*, *Ulva* and *Cladophora* spp. using SFE at a pressure of 500 bar and a temperature of 40 °C. The application of the extracts on the garden cress and wheat resulted in enhanced chlorophyll and carotenoid content in plant shoots, as well as root thickness and above-ground biomass (Michalak et al. 2016). In addition, Górka and Wieczorek (2017) measured the content of phytohormones from *Cladophora glomerata* extract obtained using SFE at 500 bar and 40 °C, resulting in contained phenylacetic acid and indoleacetic acid in the concentration of 229.30 ± 7.90 μg/g and 23.91 ± 0.80 μg/g, respectively. Overall, following the most commonly described SFE protocols in the literature SFE is regularly conducted at pressure from 7.8 to 70 MPa, the temperature from 313.15 to 349.15 K, and CO_2/algae mass ratio from 6 to 500 (Crampon et al. 2011).

2.6 Application of SWE

SWE is a novel technology for recovering target compounds from macroalgae. A schematic representation of SWE is provided in Fig. 2.5.

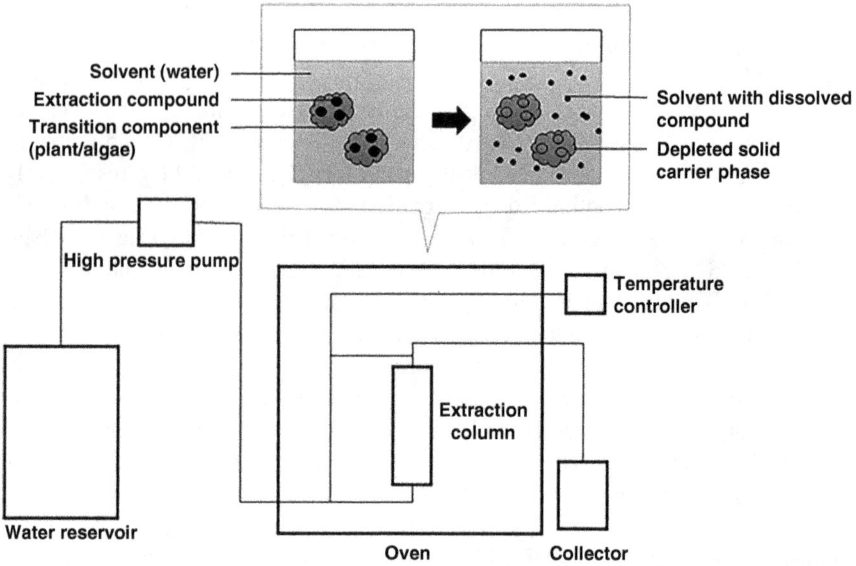

Fig. 2.5 Schematic representation of the SWE process. Originally published by Zakaria and Kamal (2016) and reproduced with permission from Springer Nature

SWE operates under high pressure (50–300 psi) and temperature (50–200 °C) for a time of 5–10 min. Since the critical pressure and temperature of water are 22.1 MPa and 374 °C (Franzese et al. 2003), liquid water below these critical points is known as subcritical water. When the temperature is between the boiling point and the critical point of water, i.e., between 100 and 374 °C, the pressure causes the water to remain liquid (Machmudah et al. 2020). It should be noted that applying high temperature and pressure induces a change in the physical and chemical properties of water during the extraction (Kalinichev and Churakov 1999). For instance, the dielectric constant of water is 80 at 25 °C, while it significantly decreases to 33 at 200 °C, being similar to the dielectric constant of methanol or ethanol (Turner and Ibanez 2011). Therefore, it is the properties of water at subcritical conditions the one generating the benefits of SWE for extracting different bioactives from macroalgae. The use of water as solvent in SWE technique has several advantages, as it is commonly available in nature, ecologically safe, non-toxic, non-flammable, inexpensive, and environmentally friendly extractant (Machmudah et al. 2020). Therefore, the use of water in replacement of organic solvents by the means of SWE represents a competitive advantage together with an increased extraction efficiency of non-polar compounds due to the increase of its penetration in the matrix by the decrease of the waters' viscosity and surface tension at its subcritical stage (Cikoš et al. 2018).

Several parameters need to be controlled when using SWE to ensure high extraction efficiency, including pressure, time, and temperature, together with selecting an appropriate solvent (Azmir et al. 2013). Generally, SWE is suitable for the extraction of lipids, carotenoids and polyphenolic compounds from macroalgae (Filote et al. 2021). There is some advantage of SWE over other techniques, including a faster process of the sample, the higher quality of the extracts, a low amount of solvents used, and its cost (Zakaria and Kamal 2016).

Daneshvar et al. (2012) used SWE to obtain aqueous and hexane extracts at 100–240 °C for 10 min from *Codium fragile*. The aqueous extract contained high levels of total organic carbon (14% at 210 °C) and the main product in this extract was hydrolysis derived-soluble sugars (50% at 210 °C). In addition, the yield of hexane-soluble materials increased up to 2% at 210 °C followed (Daneshvar et al. 2012). In another study, Rosdi et al. (2017) were able to extract carbohydrates and proteins using SWE from *Padina* sp., *Sargassum* sp., and *Enteromorpha* sp. The highest yields of both compounds were obtained from *Enteromorpha* sp. SWE at 140 °C, 15 min, and 10% biomass loading achieved carbohydrate yields of 31.8% w/w; while SFE of 220 °C, 25 min, and 10% biomass achieved the maximum yields of proteins (11.22% w/w). Bordoloi and Goosen (2020) optimized SWE targeting several outputs of extraction from *Ecklonia maxima*. The optimum condition for extraction of total phenolic compounds (41.20 ± 0.42 mg gallic acid/g), extract yield (76.02 ± 0.51%), and its antioxidant activity (1.81 ± 0.01 mmol Trolox equivalent/g) was SWE at 180 °C and solvent to algae ratio of 30 ml/g for 23.75 min. However, the maximum yield of polysaccharides (58.25 ± 0.93%), sulphate contents (48.53 ± 1.19 mg SO_4^{2-}/g), and alginates (15.65 ± 0.42%) were achieved at SWE at 120 °C and solvent to algae ratio of 30 ml/g for 5 min (Bordoloi and Goosen 2020). Recently, Pangestuti et al. (2021) optimised SWE for the recovery of bioactive compounds from *Caulerpa racemosa* and *Ulva lactuca*. The highest total phenolic compounds (39.82 ± 0.32 mg gallic acid/g), total saponins (13.22 ± 0.33 mg diosgenin/g), and total flavonoids (6.5 ± 0.47 mg quercetin/g) obtained from *U. lactuca* were achieved using SWE at 230 °C, 1:40 (w/v), and 10 min (Pangestuti et al. 2021). Moreover, these extracts also exhibited ABTS radical scavenging and total antioxidant activities with vales of 5.45 ± 0.11 mg ascorbic acid equivalents/g and 8.03 ± 0.06 mg trolox equivalents/g, respectively (Pangestuti et al. 2021).

The SWE technology can also be used to re-value macroalgal biomass as a biofertilizer. Soares et al. (2020) used SWE at 30 bar, 3 Hz, 180 °C, 30 min, solvent/algae ratio of 10:1 to treat *Saccorhiza polyschides*. The authors achieved an extract with total organic matter (34.1 ± 0.3 g/l), macro-elements (range from 10.2 ± 0.8 mg/l to 14.4 ± 0.2 g/l), micro-elements (range from 102 ± 3 μg/g DW to 360 ± 2 μg/g DW), and other minerals including Rb (67 ± 2.9 μg/g DW) and Sb (0.094 ± 0.005 μg/g DW) with low phytotoxicity (Soares et al. 2020).

Furthermore, SWE can also be used sequentially with other novel extraction techniques to improve the extraction of various compounds. For instance, *Eucheuma cottonii* and *Gracilaria* sp. extracts obtained using SFE (CO_2, 25 MPa, 60 °C, CO2 flow rate of 15 ml/min, and ethanol flow rate of 0.25 ml/min) followed by SWE (optimized at 180 °C and 7 MPa), contained 209.91 μg/g of β-carotene and 321.025 μg/g of linoleic acid content in *Eucheuma cottonii*, and 219.99 μg/g of β-carotene and 286.52 μg/g of linoleic acid content in *Gracilaria sp.* Under these conditions, the highest total phenolic contents in extracts generated from *Eucheuma cottonii* and *Gracilaria* sp. were 18.51 and 22.47 mg gallic acid/g, respectively (Setyorini et al. 2018).

3 Future Prospects and Challenges of Macroalgal Biorefineries

As the worlds' population increases, there will be a need of increase the production of food and thus, the agri-food sector is currently aiming to incorporate further improvements that could aid providing food supplies and access to other commodities to future generations; as well as improvements of processes aiming to fully utilise the currently available resources for food and other related health applications. Thus, the use of macroalgal compounds as food and other applications represents a huge opportunity both in terms of market growth as well as health, exploring the biological activities of these relatively unknown marine products for their use as nutraceuticals or functional foods.

The extraction and sequential re-use of the residual algal biomass for further and selective extraction of bioactive compounds targeting food and feed applications, cosmeceuticals, nutraceuticals and pharmaceuticals and finally low value applications, such as bioenergy, will ensure the full and sustainable use of the macroalgal biomass. Currently multiple researchers aim to establish these biorefinery approaches for the exploitation of macroalgae and the development of these models can be improved by the use of innovative technologies due to their sustainability and processing advantages.

Current technological developments including UAE, MAE, HPAE, PEF, SFE and SWE amongst others, have shown promising results being explored in the development of greener and more efficient extraction protocols of multiple compounds from macroalgae when compared with conventional extraction methods as summarised in this chapter. Optimization studies, aiming to increase the yields of compounds and reduce the time, energy inputs, costs and organic solvents used for the generation of macroalgal ingredients; as well as studies comparing the performance and sustainability claims between multiple innovative approaches and with other conventional methods will be needed in the future to continue to expand the knowledge of the effect of different extraction forces on the yields and also chemical structures and biological activities of the extracted compounds.

Innovative extraction technologies are also evolving, including more powerful apparatus and with higher capacity to reach and adapt to the needs of industry. Moreover, these technologies can also be applied simultaneously, such as the case of ultrasounds and microwaves, improving the recovery of several compounds. The simultaneous application of different extraction forces represents a new challenge for future researchers aiming to establish sustainable macroalgal biorefineries. Moreover, the effect of these new combined extraction forces and technologies on the chemical structure of the natural molecules and their biological properties for their future use represents a massive challenge and field for exploration by future generations of researchers.

Acknowledgments Dr. Armin Mirzapour-Kouhdasht works within the project AMBROSIA funded by the Department of Agriculture Food and the Marine (DAFM) under the umbrella of the European Joint Programming Initiative "A Healthy Diet for a Healthy Life" (JPI-HDHL) and of the ERA-NET Cofund ERA HDHL (GA No 696295 of the EU Horizon 2020 Research and Innovation Programme).

References

Agregán R, Munekata PES, Franco D, Carballo J, Barba FJ, Lorenzo JM (2018) Antioxidant potential of extracts obtained from macro- (Ascophyllum nodosum, Fucus vesiculosus and Bifurcaria bifurcata) and micro-algae (Chlorella vulgaris and Spirulina platensis) assisted by ultrasound. Medicines 5(2):33

Akbary P, Liao LM, Aminikhoei Z, Tavabe KR, Hobbi M, Erfanifar E (2021) Sterol and fatty acid profiles of three macroalgal species collected from the Chabahar coasts, southeastern Iran. Aquac Int 29(1):155–165

Al Khawli F, Pateiro M, Domínguez R, Lorenzo JM, Gullón P, Kousoulaki K, Barba FJ (2019) Innovative green technologies of intensification for valorization of seafood and their by-products. Mar Drugs 17(12):689

Anaëlle T, Serrano Leon E, Laurent V, Elena I, Mendiola JA, Stéphane C, Valérie S-P (2013) Green improved processes to extract bioactive phenolic compounds from brown macroalgae using Sargassum muticum as model. Talanta 104:44–52. https://doi.org/10.1016/j.talanta.2012.10.088

Azmir J, Zaidul ISM, Rahman MM, Sharif K, Mohamed A, Sahena F, Omar A (2013) Techniques for extraction of bioactive compounds from plant materials: a review. J Food Eng 117(4):426–436

Balboa EM, Gallego-Fábrega C, Moure A, Domínguez H (2016) Study of the seasonal variation on proximate composition of oven-dried Sargassum muticum biomass collected in Vigo Ria, Spain. J Appl Phycol 28(3):1943–1953. https://doi.org/10.1007/s10811-015-0727-x

Bāliņa K, Ivanovs K, Romagnoli F, Blumberga D (2020) Comprehensive literature review on valuable compounds and extraction technologies: the Eastern Baltic Sea seaweeds. Environmental and Climate Technologies 24(2):178–195

Becerra M, Boutefnouchet S, Córdoba O, Vitorino GP, Brehu L, Lamour I, Michel S (2015) Antileishmanial activity of fucosterol recovered from Lessonia vadosa Searles (Lessoniaceae) by SFE, PSE and CPC. Phytochem Lett 11:418–423

Bhuiyan K, Qureshi S, Mustafa Kamal A, AftabUddin S, Siddique A (2016) Proximate chemical composition of sea grapes Caulerpa racemosa (J. Agardh, 1873) collected from a sub-tropical coast. Virol Mycol 5(158):2161–0517

Bordoloi A, Goosen NJ (2020) A greener alternative using subcritical water extraction to valorize the brown macroalgae Ecklonia maxima for bioactive compounds. J Appl Phycol 32(4):2307–2319. https://doi.org/10.1007/s10811-020-02043-1

Buschmann AH, Camus C, Infante J, Neori A, Israel Á, Hernández-González MC, Tadmor-Shalev N (2017) Seaweed production: overview of the global state of exploitation, farming and emerging research activity. Eur J Phycol 52(4):391–406

Čagalj M, Skroza D, Tabanelli G, Özogul F, Šimat V (2021) Maximizing the antioxidant capacity of Padina pavonica by choosing the right drying and extraction methods. Processes 9(4):587

Carneiro JG, Rodrigues JAG, Teles FB, Cavalcante ABD, Benevides NMB (2014) Analysis of some chemical nutrients in four Brazilian tropical seaweeds. Acta Sci Biol Sci 36(2):137–145

Carullo D, Pataro G, Donsì F, Ferrari G (2020) Pulsed electric fields-assisted extraction of valuable compounds from arthrospira platensis: effect of pulse polarity and mild heating. Front Bioeng Biotechnol 8. https://doi.org/10.3389/fbioe.2020.551272

Castejón N, Thorarinsdottir KA, Einarsdóttir R, Kristbergsson K, Marteinsdóttir G (2021) Exploring the potential of icelandic seaweeds extracts produced by aqueous pulsed electric

fields-assisted extraction for cosmetic applications. Mar Drugs 19(12). https://doi.org/10.3390/md19120662
Chan PT, Matanjun P (2017) Chemical composition and physicochemical properties of tropical red seaweed, Gracilaria changii. Food Chem 221:302–310
Chemat F, Rombaut N, Meullemiestre A, Turk M, Perino S, Fabiano-Tixier A-S, Abert-Vian M (2017) Review of green food processing techniques. Preservation, transformation, and extraction. Innovative Food Sci Emerg Technol 41:357–377. https://doi.org/10.1016/j.ifset.2017.04.016
Chen R, Meng F, Zhang S, Liu Z (2009) Effects of ultrahigh pressure extraction conditions on yields and antioxidant activity of ginsenoside from ginseng. Sep Purif Technol 66(2):340–346. https://doi.org/10.1016/j.seppur.2008.12.026
Cherry P, O'Hara C, Magee PJ, McSorley EM, Allsopp PJ (2019) Risks and benefits of consuming edible seaweeds. Nutr Rev 77(5):307–329
Cikoš A-M, Jokić S, Šubarić D, Jerković I (2018) Overview on the application of modern methods for the extraction of bioactive compounds from marine macroalgae. Mar Drugs 16(10):348
Conde E, Moure A, Domínguez H (2015) Supercritical CO2 extraction of fatty acids, phenolics and fucoxanthin from freeze-dried Sargassum muticum. J Appl Phycol 27(2):957–964. https://doi.org/10.1007/s10811-014-0389-0
Crampon C, Boutin O, Badens E (2011) Supercritical carbon dioxide extraction of molecules of interest from microalgae and seaweeds. Ind Eng Chem Res 50(15):8941–8953
Cruz-Suárez LE, Tapia-Salazar M, Nieto-López MG, Guajardo-Barbosa C, Ricque-Marie D (2009) Comparison of Ulva clathrata and the kelps Macrocystis pyrifera and Ascophyllum nodosum as ingredients in shrimp feeds. Aquac Nutr 15(4):421–430. https://doi.org/10.1111/j.1365-2095.2008.00607.x
Daneshvar S, Salak F, Ishii T, Otsuka K (2012) Application of subcritical water for conversion of macroalgae to value-added materials. Ind Eng Chem Res 51(1):77–84
De Corato U, Salimbeni R, De Pretis A, Avella N, Patruno G (2017) Antifungal activity of crude extracts from brown and red seaweeds by a supercritical carbon dioxide technique against fruit postharvest fungal diseases. Postharvest Biol Technol 131:16–30. https://doi.org/10.1016/j.postharvbio.2017.04.011
Debbarma J, Rao M, Murthy LN, Mathew S, Gudipati V, Ravishankar C (2016) Nutritional profiling of the edible seaweeds Gracilaria edulis, Ulva lactuca and Sargassum sp. Indian J Fish 63:81–87. https://doi.org/10.21077/ijf.2016.63.3.60073-11
Din SS, Chew KW, Chang Y-K, Show PL, Phang SM, Juan JC (2019) Extraction of agar from Eucheuma cottonii and Gelidium amansii seaweeds with sonication pretreatment using autoclaving method. J Oceanol Limnol 37(3):871–880. https://doi.org/10.1007/s00343-019-8145-6
Einarsdóttir R, Þórarinsdóttir KA, Aðalbjörnsson BV, Guðmundsson M, Marteinsdóttir G, Kristbergsson K (2022) Extraction of bioactive compounds from Alaria esculenta with pulsed electric field. J Appl Phycol 34(1):597–608. https://doi.org/10.1007/s10811-021-02624-8
Fabrowska J, Ibañez E, Łęska B, Herrero M (2016) Supercritical fluid extraction as a tool to valorize underexploited freshwater green algae. Algal Res 19:237–245
Fabrowska J, Messyasz B, Szyling J, Walkowiak J, Łęska B (2018) Isolation of chlorophylls and carotenoids from freshwater algae using different extraction methods. Phycol Res 66(1):52–57. https://doi.org/10.1111/pre.12191
Filote C, Santos SCR, Popa VI, Botelho CMS, Volf I (2021) Biorefinery of marine macroalgae into high-tech bioproducts: a review. Environ Chem Lett 19(2):969–1000. https://doi.org/10.1007/s10311-020-01124-4
Flórez-Fernández N, López-García M, González-Muñoz MJ, Vilariño JML, Domínguez H (2017) Ultrasound-assisted extraction of fucoidan from Sargassum muticum. J Appl Phycol 29(3):1553–1561. https://doi.org/10.1007/s10811-016-1043-9
Franzese G, Marqués MI, Stanley HE (2003) Intramolecular coupling as a mechanism for a liquid-liquid phase transition. Phys Rev E 67(1):011103

Freitas de Oliveira C, Giordani D, Lutckemier R, Gurak PD, Cladera-Olivera F, Ferreira Marczak LD (2016) Extraction of pectin from passion fruit peel assisted by ultrasound. LWT Food Sci Technol 71:110–115. https://doi.org/10.1016/j.lwt.2016.03.027

Ganesan AR, Tiwari U, Rajauria G (2019) Seaweed nutraceuticals and their therapeutic role in disease prevention. Food Sci Human Wellness 8(3):252–263

Ganzler K, Salgó A, Valkó K (1986) Microwave extraction: A novel sample preparation method for chromatography. J Chromatogr A 371:299–306. https://doi.org/10.1016/S0021-9673(01)94714-4

Garcia-Vaquero M, Rajauria G (2022) Chapter 1 – Overview of the application of innovative and emerging technologies in the bio-marine food sector. In: Garcia-Vaquero M, Rajauria G (eds) Innovative and emerging technologies in the bio-marine food sector. Academic Press, pp 1–12

Garcia-Vaquero M, Rajauria G, Tiwari B, Sweeney T, O'Doherty J (2018) Extraction and yield optimisation of fucose, glucans and associated antioxidant activities from Laminaria digitata by applying response surface methodology to high intensity ultrasound-assisted extraction. Mar Drugs 16(8):257

Garcia-Vaquero M, O'Doherty JV, Tiwari BK, Sweeney T, Rajauria G (2019) Enhancing the extraction of polysaccharides and antioxidants from macroalgae using sequential hydrothermal-assisted extraction followed by ultrasound and thermal technologies. Mar Drugs 17(8):457

Garcia-Vaquero M, Rajauria G, Tiwari B (2020a) Chapter 7 – Conventional extraction techniques: solvent extraction. In: Torres MD, Kraan S, Dominguez H (eds) Sustainable seaweed technologies. Elsevier, pp 171–189

Garcia-Vaquero M, Ummat V, Tiwari B, Rajauria G (2020b) Exploring ultrasound, microwave and ultrasound–microwave assisted extraction technologies to increase the extraction of bioactive compounds and antioxidants from brown macroalgae. Mar Drugs 18(3):172

Garcia-Vaquero M, Rajauria G, Miranda M, Sweeney T, Lopez-Alonso M, O'Doherty J (2021a) Seasonal variation of the proximate composition, mineral content, fatty acid profiles and other phytochemical constituents of selected brown macroalgae. Mar Drugs 19(4):204

Garcia-Vaquero M, Ravindran R, Walsh O, O'Doherty J, Jaiswal AK, Tiwari BK, Rajauria G (2021b) Evaluation of ultrasound, microwave, ultrasound–microwave, hydrothermal and high pressure assisted extraction technologies for the recovery of phytochemicals and antioxidants from brown macroalgae. Mar Drugs 19(6):309

Getachew AT, Saravana PS, Cho YJ, Woo HC, Chun BS (2018) Concurrent extraction of oil from roasted coffee (Coffea arabica) and fucoxanthin from brown seaweed (Saccharina japonica) using supercritical carbon dioxide. J CO2 Util 25:137–146

Getachew AT, Jacobsen C, Holdt SL (2020) Emerging technologies for the extraction of marine phenolics: opportunities and challenges. Mar Drugs 18(8):389

Goettel M, Eing C, Gusbeth C, Straessner R, Frey W (2013) Pulsed electric field assisted extraction of intracellular valuables from microalgae. Algal Res 2(4):401–408. https://doi.org/10.1016/j.algal.2013.07.004

Gomez L, Tiwari B, Garcia-Vaquero M (2020) Emerging extraction techniques: microwave-assisted extraction. In: Sustainable seaweed technologies. Elsevier, pp 207–224

Gordalina M, Pinheiro HM, Mateus M, da Fonseca MMR, Cesário MT (2021) Macroalgae as protein sources—a review on protein bioactivity, extraction, purification and characterization. Appl Sci 11(17):7969

Górka B, Wieczorek PP (2017) Simultaneous determination of nine phytohormones in seaweed and algae extracts by HPLC-PDA. J Chromatogr B 1057:32–39

Goto M, Kanda H, Machmudah S (2015) Extraction of carotenoids and lipids from algae by supercritical CO2 and subcritical dimethyl ether. J Supercrit Fluids 96:245–251

Gressler V, Fujii MT, Martins AP, Colepicolo P, Mancini-Filho J, Pinto E (2011) Biochemical composition of two red seaweed species grown on the Brazilian coast. J Sci Food Agric 91(9):1687–1692

Gunji S, Santoso J, Yoshie-Stark Y, Suzuki T (2007) Effects of extracts from tropical seaweeds on DPPH radicals and Caco-2 cells treated with hydrogen peroxide. Food Sci Technol Res 13(3):275–279

Hassan IH, Pham HNT, Nguyen TH (2021) Optimization of ultrasound-assisted extraction conditions for phenolics, antioxidant, and tyrosinase inhibitory activities of Vietnamese brown seaweed (Padina australis). J Food Process Preserv 45(5):e15386. https://doi.org/10.1111/jfpp.15386

Herrero M, Ibanez E (2015) Green processes and sustainability: an overview on the extraction of high added-value products from seaweeds and microalgae. J Supercrit Fluids 96:211–216

Hmelkov AB, Zvyagintseva TN, Shevchenko NM, Rasin AB, Ermakova SP (2018) Ultrasound-assisted extraction of polysaccharides from brown alga Fucus evanescens. Structure and biological activity of the new fucoidan fractions. J Appl Phycol 30(3):2039–2046. https://doi.org/10.1007/s10811-017-1342-9

Hu Z, Hong P, Cheng Y, Liao M, Li S (2018) Polysaccharides from Enteromorpha tubulosa: optimization of extraction and cytotoxicity. J Food Process Preserv 42(1):e13373. https://doi.org/10.1111/jfpp.13373

Kadam SU, Álvarez C, Tiwari BK, O'Donnell CP (2015a) Chapter 9 – Extraction of biomolecules from seaweeds. In: Tiwari BK, Troy DJ (eds) Seaweed sustainability. Academic Press, San Diego, pp 243–269

Kadam SU, Donnell CP, Rai DK, Hossain MB, Burgess CM, Walsh D, Tiwari BK (2015b) Laminarin from Irish brown seaweeds Ascophyllum nodosum and Laminaria hyperborea: ultrasound assisted extraction, characterization and bioactivity. Mar Drugs 13(7):4270–4280

Kadam SU, Tiwari BK, O'Donnell CP (2015c) Effect of ultrasound pre-treatment on the drying kinetics of brown seaweed Ascophyllum nodosum. Ultrason Sonochem 23:302–307. https://doi.org/10.1016/j.ultsonch.2014.10.001

Kadam SU, Tiwari BK, Smyth TJ, O'Donnell CP (2015d) Optimization of ultrasound assisted extraction of bioactive components from brown seaweed Ascophyllum nodosum using response surface methodology. Ultrason Sonochem 23:308–316. https://doi.org/10.1016/j.ultsonch.2014.10.007

Kalinichev A, Churakov S (1999) Size and topology of molecular clusters in supercritical water: a molecular dynamics simulation. Chem Phys Lett 302(5–6):411–417

Kang J-Y, Chun B-S, Lee M-C, Choi J-S, Choi IS, Hong Y-K (2016) Anti-inflammatory activity and chemical composition of essential oil extracted with supercritical CO2 from the brown seaweed Undaria pinnatifida. J Essent Oil Bear Plants 19(1):46–51

Khan SA, Aslam R, Makroo HA (2019) High pressure extraction and its application in the extraction of bio-active compounds: a review. J Food Process Eng 42(1):e12896. https://doi.org/10.1111/jfpe.12896

Klejdus B, Lojková L, Plaza M, Šnóblová M, Štěrbová D (2010) Hyphenated technique for the extraction and determination of isoflavones in algae: ultrasound-assisted supercritical fluid extraction followed by fast chromatography with tandem mass spectrometry. J Chromatogr A 1217(51):7956–7965

Korzeniowska K, Łęska B, Wieczorek PP (2020) Isolation and determination of phenolic compounds from freshwater Cladophora glomerata. Algal Res 48:101912. https://doi.org/10.1016/j.algal.2020.101912

Kumar Y, Singhal S, Tarafdar A, Pharande A, Ganesan M, Badgujar PC (2020) Ultrasound assisted extraction of selected edible macroalgae: effect on antioxidant activity and quantitative assessment of polyphenols by liquid chromatography with tandem mass spectrometry (LC-MS/MS). Algal Res 52:102114. https://doi.org/10.1016/j.algal.2020.102114

Kumar K, Srivastav S, Sharanagat VS (2021) Ultrasound assisted extraction (UAE) of bioactive compounds from fruit and vegetable processing by-products: A review. Ultrason Sonochem 70:105325. https://doi.org/10.1016/j.ultsonch.2020.105325

Le Guillard C, Dumay J, Donnay-Moreno C, Bruzac S, Ragon J-Y, Fleurence J, Bergé J-P (2015) Ultrasound-assisted extraction of R-phycoerythrin from Grateloupia turuturu with and without enzyme addition. Algal Res 12:522–528. https://doi.org/10.1016/j.algal.2015.11.002

Le Guillard C, Bergé J-P, Donnay-Moreno C, Bruzac S, Ragon J-Y, Baron R, Dumay J (2016) Soft liquefaction of the red seaweed Grateloupia turuturu Yamada by ultrasound-assisted enzymatic hydrolysis process. J Appl Phycol 28(4):2575–2585. https://doi.org/10.1007/s10811-015-0788-x

Le B, Golokhvast KS, Yang SH, Sun S (2019) Optimization of microwave-assisted extraction of polysaccharides from Ulva pertusa and evaluation of their antioxidant activity. Antioxidants 8(5):129

Li G-Y, Luo Z-C, Yuan F, Yu X-B (2017) Combined process of high-pressure homogenization and hydrothermal extraction for the extraction of fucoidan with good antioxidant properties from Nemacystus decipients. Food Bioprod Process 106:35–42. https://doi.org/10.1016/j.fbp.2017.08.002

Lorenzo JM, Agregán R, Munekata PES, Franco D, Carballo J, Şahin S, Barba FJ (2017) Proximate composition and nutritional value of three macroalgae: Ascophyllum nodosum, Fucus vesiculosus and Bifurcaria bifurcata. Mar Drugs 15(11):360

Machmudah S, Kanda H, Goto M (2020) Emerging seaweed extraction techniques: supercritical fluid extraction. In: Sustainable seaweed technologies. Elsevier, pp 257–286

Mæhre HK, Malde MK, Eilertsen KE, Elvevoll EO (2014) Characterization of protein, lipid and mineral contents in common Norwegian seaweeds and evaluation of their potential as food and feed. J Sci Food Agric 94(15):3281–3290

Manivannan K, Thirumaran G, Devi GK, Hemalatha A, Anantharaman P (2008) Biochemical composition of seaweeds from Mandapam coastal regions along Southeast Coast of India. Am Eurasian J Bot 1(2):32–37

Marr R (1993) Extraction of natural products using near-critical solvents. Von King MB, Bott TR (Eds). Chapman & Hall, London, 325 S., zahlr. Abb. und Tab., £ 65.00: Wiley Online Library

Martínez-Milián G, Olvera-Novoa MA (2016) Evaluation of potential feed ingredients for the juvenile four-sided sea cucumber, Isostichopus badionotus. J World Aquacult Soc 47(5):712–719

Matanjun P, Mohamed S, Mustapha NM, Muhammad K (2009) Nutrient content of tropical edible seaweeds, Eucheuma cottonii, Caulerpa lentillifera and Sargassum polycystum. J Appl Phycol 21(1):75–80. https://doi.org/10.1007/s10811-008-9326-4

Mendiola JA, Herrero M, Cifuentes A, Ibañez E (2007) Use of compressed fluids for sample preparation: food applications. J Chromatogr A 1152(1–2):234–246

Messyasz B, Michalak I, Łęska B, Schroeder G, Górka B, Korzeniowska K, Chojnacka K (2018) Valuable natural products from marine and freshwater macroalgae obtained from supercritical fluid extracts. J Appl Phycol 30(1):591–603. https://doi.org/10.1007/s10811-017-1257-5

Michalak I, Górka B, Wieczorek PP, Rój E, Lipok J, Łęska B, Dobrzyńska-Inger A (2016) Supercritical fluid extraction of algae enhances levels of biologically active compounds promoting plant growth. Eur J Phycol 51(3):243–252

Mittal R, Tavanandi HA, Mantri VA, Raghavarao KSMS (2017) Ultrasound assisted methods for enhanced extraction of phycobiliproteins from marine macro-algae, Gelidium pusillum (Rhodophyta). Ultrason Sonochem 38:92–103. https://doi.org/10.1016/j.ultsonch.2017.02.030

Mohammed HO, O'Grady MN, O'Sullivan MG, Hamill RM, Kilcawley KN, Kerry JP (2021) An assessment of selected nutritional, bioactive, thermal and technological properties of brown and red Irish seaweed species. Food Secur 10(11):2784

Monteagudo-Olivan R, Cocero MJ, Coronas J, Rodríguez-Rojo S (2019) Supercritical CO2 encapsulation of bioactive molecules in carboxylate based MOFs. J CO2 Util 30:38–47. https://doi.org/10.1016/j.jcou.2018.12.022

Moreira SA, Pintado M, Saraiva JA (2020) Chapter 14 – High hydrostatic pressure-assisted extraction: a review on its effects on bioactive profile and biological activities of extracts. In: Barba FJ, Tonello-Samson C, Puértolas E, Lavilla M (eds) Present and future of high pressure processing. Elsevier, pp 317–328

Mulchandani K, Kar JR, Singhal RS (2015) Extraction of lipids from Chlorella saccharophila using high-pressure homogenization followed by three phase partitioning. Appl Biochem Biotechnol 176(6):1613–1626. https://doi.org/10.1007/s12010-015-1665-4

Munir N, Rafique M, Altaf I, Sharif N, Naz S (2018) Antioxidant and antimicrobial activities of extracts from selected algal species. Bangladesh J Bot 47:53–61

Mwalugha HM, Wakibia JG, Kenji GM, Mwasaru MA (2015) Chemical composition of common seaweeds from the Kenya Coast. J Food Res 4(6):28–38

Nagappan T, Vairappan CS (2014) Nutritional and bioactive properties of three edible species of green algae, genus Caulerpa (Caulerpaceae). J Appl Phycol 26(2):1019–1027

Nielsen MM, Manns D, D'Este M, Krause-Jensen D, Rasmussen MB, Larsen MM, Bruhn A (2016) Variation in biochemical composition of Saccharina latissima and Laminaria digitata along an estuarine salinity gradient in inner Danish waters. Algal Res 13:235–245

O'Connor J, Meaney S, Williams GA, Hayes M (2020) Extraction of protein from four different seaweeds using three different physical pre-treatment strategies. Molecules 25(8):2005

Oliyaei N, Moosavi-Nasab M (2021) Ultrasound-assisted extraction of fucoxanthin from Sargassum angustifolium and Cystoseira indica brown algae. J Food Process Preserv 45(11):e15929. https://doi.org/10.1111/jfpp.15929

Ospina M, Castro-Vargas HI, Parada-Alfonso F (2017) Antioxidant capacity of Colombian seaweeds: 1. Extracts obtained from Gracilaria mammillaris by means of supercritical fluid extraction. J Supercrit Fluids 128:314–322. https://doi.org/10.1016/j.supflu.2017.02.023

Pádua MD, Fontoura PSG, Mathias AL (2004) Chemical composition of Ulvaria oxysperma (Kützing) bliding, Ulva lactuca (Linnaeus) and Ulva fascita (Delile). Braz Arch Biol Technol 47(1):49–55

Pangestuti R, Haq M, Rahmadi P, Chun B-S (2021) Nutritional value and biofunctionalities of two edible green seaweeds (Ulva lactuca and Caulerpa racemosa) from Indonesia by subcritical water hydrolysis. Mar Drugs 19(10):578

Park J-N, Shin T-S, Lee J-H, Chun B-S (2012) Production of reducing sugars from Laminaria japonica by subcritical water hydrolysis. APCBEE Proc 2:17–21. https://doi.org/10.1016/j.apcbee.2012.06.004

Pataroa G, Carulloa D, Ferraria G (2019) PEF-assisted supercritical CO2 extraction of pigments from microalgae nannochloropsis oceanica in a continuous flow system. Chem Eng 74

Polikovsky M, Fernand F, Sack M, Frey W, Müller G, Golberg A (2016) Towards marine biorefineries: selective proteins extractions from marine macroalgae Ulva with pulsed electric fields. Innovative Food Sci Emerg Technol 37:194–200. https://doi.org/10.1016/j.ifset.2016.03.013

Polikovsky M, Fernand F, Sack M, Frey W, Müller G, Golberg A (2019) In silico food allergenic risk evaluation of proteins extracted from macroalgae Ulva sp. with pulsed electric fields. Food Chem 276:735–744. https://doi.org/10.1016/j.foodchem.2018.09.134

Poojary MM, Barba FJ, Aliakbarian B, Donsì F, Pataro G, Dias DA, Juliano P (2016) Innovative alternative technologies to extract carotenoids from microalgae and seaweeds. Mar Drugs 14(11):214

Postma PR, Cerezo-Chinarro O, Akkerman RJ, Olivieri G, Wijffels RH, Brandenburg WA, Eppink MHM (2018) Biorefinery of the macroalgae Ulva lactuca: extraction of proteins and carbohydrates by mild disintegration. J Appl Phycol 30(2):1281–1293. https://doi.org/10.1007/s10811-017-1319-8

Prabhu MS, Levkov K, Livney YD, Israel A, Golberg A (2019) High-voltage pulsed electric field preprocessing enhances extraction of starch, proteins, and ash from marine macroalgae Ulva ohnoi. ACS Sustain Chem Eng 7(20):17453–17463. https://doi.org/10.1021/acssuschemeng.9b04669

Puértolas E, Luengo E, Álvarez I, Raso J (2012) Improving mass transfer to soften tissues by pulsed electric fields: fundamentals and applications. Annu Rev Food Sci Technol 3:263–282. https://doi.org/10.1146/annurev-food-022811-101208

Putra VGP, Mutiarahma S, Chaniago W, Rahmadi P, Kurnianto D, Hidayat C, Setyaningsih W (2022) An ultrasound-based technique for the analytical extraction of phenolic compounds in red algae. Arab J Chem 15(2):103597. https://doi.org/10.1016/j.arabjc.2021.103597

Raguraman V, MubarakAli D, Narendrakumar G, Thirugnanasambandam R, Kirubagaran R, Thajuddin N (2018) Unraveling rapid extraction of fucoxanthin from Padina tetrastromatica: purification, characterization and biomedical application. Process Biochem 73:211–219. https://doi.org/10.1016/j.procbio.2018.08.006

Rahimi F, Tabarsa M, Rezaei M (2016) Ulvan from green algae Ulva intestinalis: optimization of ultrasound-assisted extraction and antioxidant activity. J Appl Phycol 28(5):2979–2990. https://doi.org/10.1007/s10811-016-0824-5

Robin A, Kazir M, Sack M, Israel A, Frey W, Mueller G, Golberg A (2018) Functional protein concentrates extracted from the green marine macroalga Ulva sp., by high voltage pulsed electric fields and mechanical press. ACS Sustain Chem Eng 6(11):13696–13705. https://doi.org/10.1021/acssuschemeng.8b01089

Rodrigues D, Freitas AC, Pereira L, Rocha-Santos TAP, Vasconcelos MW, Roriz M, Duarte AC (2015) Chemical composition of red, brown and green macroalgae from Buarcos bay in Central West Coast of Portugal. Food Chem 183:197–207. https://doi.org/10.1016/j.foodchem.2015.03.057

Roh M-K, Uddin MS, Chun B-S (2008) Extraction of fucoxanthin and polyphenol from Undaria pinnatifida using supercritical carbon dioxide with co-solvent. Biotechnol Bioprocess Eng 13(6):724–729. https://doi.org/10.1007/s12257-008-0104-6

Rohani-Ghadikolaei K, Abdulalian E, Ng W-K (2012) Evaluation of the proximate, fatty acid and mineral composition of representative green, brown and red seaweeds from the Persian Gulf of Iran as potential food and feed resources. J Food Sci Technol 49(6):774–780

Rosdi, A. N., Phang, S. M., & Harun, M. R. (2017). Optimization of bioactive compound production from seaweed via subcritical water extraction. Paper presented at the 5th international symposium on applied engineering and sciences (SAES2017), Universiti Putra Malaysia

Rosemary T, Arulkumar A, Paramasivam S, Mondragon-Portocarrero A, Miranda JM (2019) Biochemical, micronutrient and physicochemical properties of the dried red seaweeds Gracilaria edulis and Gracilaria corticata. Molecules 24(12):2225

Sapkale G, Patil S, Surwase U, Bhatbhage P (2010) Supercritical fluid extraction. Int J Chem Sci 8(2):729–743

Saravana PS, Getachew AT, Cho Y-J, Choi JH, Park YB, Woo HC, Chun BS (2017) Influence of co-solvents on fucoxanthin and phlorotannin recovery from brown seaweed using supercritical CO2. J Supercrit Fluids 120:295–303

Schiener P, Black KD, Stanley MS, Green DH (2015) The seasonal variation in the chemical composition of the kelp species Laminaria digitata, Laminaria hyperborea, Saccharina latissima and Alaria esculenta. J Appl Phycol 27(1):363–373. https://doi.org/10.1007/s10811-014-0327-1

Schmid M, Kraft LG, van der Loos LM, Kraft GT, Virtue P, Nichols PD, Hurd CL (2018) Southern Australian seaweeds: a promising resource for omega-3 fatty acids. Food Chem 265:70–77

Setyorini, D., Aanisah, R., Machmudah, S., Winardi, S., Kanda, H., & Goto, M. (2018). Extraction of phytochemical compounds from Eucheuma cottonii and Gracilaria sp using supercritical CO2 followed by subcritical water. Paper presented at the MATEC web of conferences

Shao P, Chen X, Sun P (2013) In vitro antioxidant and antitumor activities of different sulfated polysaccharides isolated from three algae. Int J Biol Macromol 62:155–161. https://doi.org/10.1016/j.ijbiomac.2013.08.023

Shi F, Yan X, Cheong K-L, Liu Y (2018) Extraction, purification, and characterization of polysaccharides from marine algae Gracilaria lemaneiformis with anti-tumor activity. Process Biochem 73:197–203. https://doi.org/10.1016/j.procbio.2018.08.011

Shirsath SR, Sonawane SH, Gogate PR (2012) Intensification of extraction of natural products using ultrasonic irradiations—a review of current status. Chem Eng Process Process Intensif 53:10–23. https://doi.org/10.1016/j.cep.2012.01.003

Siddique M, Khan M, Bhuiyan M (2013) Nutritional composition and amino acid profile of a sub-tropical red seaweed Gelidium pusillum collected from St. Martin's Island, Bangladesh. Int Food Res J 20(5):2287

Silva A, Silva SA, Carpena M, Garcia-Oliveira P, Gullón P, Barroso MF, Simal-Gandara J (2020) Macroalgae as a source of valuable antimicrobial compounds: extraction and applications. Antibiotics 9(10):642

Sivagnanam SP, Yin S, Choi JH, Park YB, Woo HC, Chun BS (2015) Biological properties of fucoxanthin in oil recovered from two brown seaweeds using supercritical CO2 extraction. Mar Drugs 13(6):3422–3442

Smith R, Ferdouse F, Løvstad Holdt S, Murúa P, Yang Z (2018) The global status of seaweed production, trade and utilization, vol 124. Food and Agriculture Organization of the United Nations
Soares C, Švarc-Gajić J, Oliva-Teles MT, Pinto E, Nastić N, Savić S, Delerue-Matos C (2020) Mineral composition of subcritical water extracts of Saccorhiza polyschides, a brown seaweed used as fertilizer in the North of Portugal. J Mar Sci Eng 8(4):244
Soares C, Paíga P, Marques M, Neto T, Carvalho AP, Paiva A, Grosso C (2021) Multi-step subcritical water extracts of Fucus vesiculosus L. and Codium tomentosum stackhouse: composition, health-benefits and safety. Processes 9(5):893
Sosa-Hernández JE, Escobedo-Avellaneda Z, Iqbal HMN, Welti-Chanes J (2018) State-of-the-art extraction methodologies for bioactive compounds from algal biome to meet bio-economy challenges and opportunities. Molecules 23(11):2953
Suwal S, Perreault V, Marciniak A, Tamigneaux É, Deslandes É, Bazinet L, Doyen A (2019) Effects of high hydrostatic pressure and polysaccharidases on the extraction of antioxidant compounds from red macroalgae, Palmaria palmata and Solieria chordalis. J Food Eng 252:53–59. https://doi.org/10.1016/j.jfoodeng.2019.02.014
Taylor LT (2009) Supercritical fluid chromatography for the 21st century. J Supercrit Fluids 47(3):566–573
Tsubaki S, Oono K, Hiraoka M, Onda A, Mitani T (2016) Microwave-assisted hydrothermal extraction of sulfated polysaccharides from Ulva spp. and Monostroma latissimum. Food Chem 210:311–316. https://doi.org/10.1016/j.foodchem.2016.04.121
Turner C, Ibanez E (2011) Pressurized hot water extraction and processing. In: Lebovka N, Vorobiev E, Chemat F (eds) Enhancing extraction processes in the food industry. CRC Press, Boca Raton, pp 223–254
Ummat V, Tiwari BK, Jaiswal AK, Condon K, Garcia-Vaquero M, O'Doherty J, Rajauria G (2020) Optimisation of ultrasound frequency, extraction time and solvent for the recovery of polyphenols, phlorotannins and associated antioxidant activity from brown seaweeds. Mar Drugs 18(5):250
Wan P, Yang X, Cai B, Chen H, Sun H, Chen D, Pan J (2015) Ultrasonic extraction of polysaccharides from Laminaria japonica and their antioxidative and glycosidase inhibitory activities. J Ocean Univ China 14(4):651–662. https://doi.org/10.1007/s11802-015-2648-3
Wang L, Weller CL (2006) Recent advances in extraction of nutraceuticals from plants. Trends Food Sci Technol 17(6):300–312. https://doi.org/10.1016/j.tifs.2005.12.004
Wang F, Guo X-Y, Zhang D-N, Wu Y, Wu T, Chen Z-G (2015) Ultrasound-assisted extraction and purification of taurine from the red algae Porphyra yezoensis. Ultrason Sonochem 24:36–42. https://doi.org/10.1016/j.ultsonch.2014.12.009
Xu S-Y, Chen X-Q, Liu Y, Cheong K-L (2020) Ultrasonic/microwave-assisted extraction, simulated digestion, and fermentation in vitro by human intestinal flora of polysaccharides from Porphyra haitanensis. Int J Biol Macromol 152:748–756. https://doi.org/10.1016/j.ijbiomac.2020.02.305
Yordanov DG, Angelova GV (2010) High pressure processing for foods preserving. Biotechnol Biotechnol Equip 24(3):1940–1945. https://doi.org/10.2478/V10133-010-0057-8
Youssouf L, Lallemand L, Giraud P, Soulé F, Bhaw-Luximon A, Meilhac O, Couprie J (2017) Ultrasound-assisted extraction and structural characterization by NMR of alginates and carrageenans from seaweeds. Carbohydr Polym 166:55–63. https://doi.org/10.1016/j.carbpol.2017.01.041
Zakaria SM, Kamal SMM (2016) Subcritical water extraction of bioactive compounds from plants and algae: applications in pharmaceutical and food ingredients. Food Eng Rev 8(1):23–34
Zhu Z, Wu Q, Di X, Li S, Barba FJ, Koubaa M, He J (2017) Multistage recovery process of seaweed pigments: investigation of ultrasound assisted extraction and ultra-filtration performances. Food Bioprod Process 104:40–47. https://doi.org/10.1016/j.fbp.2017.04.008
Zhu X, Healy L, Zhang Z, Maguire J, Sun DW, Tiwari BK (2021) Novel postharvest processing strategies for value-added applications of marine algae. J Sci Food Agric 101(11):4444–4455
Zhu Y, Chen W, Kong L, Zhou B, Hua Y, Han Y, Shen J (2022) Optimum conditions of ultrasound-assisted extraction and pharmacological activity study for phenolic compounds of the alga Chondrus ocellatus. J Food Process Preserv 46(3):e16400. https://doi.org/10.1111/jfpp.16400

Chapter 3
Functional Properties of Seaweed on Gut Microbiota

Aroa Lopez-Santamarina, Alejandra Cardelle-Cobas, Laura I. Sinisterra-Loaiza, Alberto Cepeda, and Jose Manuel Miranda

1 Introduction

There are roughly 100 trillion bacteria living in the human gastrointestinal system, which form a community. This class of microorganisms consists of bacteria, viruses, helminths, fungus, protozoa, and archaea. Each of these creates a highly intricate web of interactions with the host, becoming important to human health (Savage 1977; Suau et al. 1999; Rowan-Nash et al. 2019). The number of eukaryotic cells in the human body are 10–100 times more in the gastrointestinal tract, particularly in the distal section of the colon, where up to 10^{11}–10^{12} bacteria per gram are present (Lopez-Santamarina et al. 2020). Bacteria have been the category of microbes that have been studied the most, therefore when the term "microbiota" is used in scientific works, it usually refers to bacteria.

Firmicutes and Bacteroidetes make up 90% of the total gut microbiota (GM), while Actinobacteria, Proteobacteria, Fusobacteria, and Verrucomicrobia are the other major bacterial phyla found in the human gut (Roca-Saavedra et al. 2018). The *phylum* Firmicutes includes important genera such as *Clostridium*, which is the most abundant, *Lactobacillus*, *Bacillus*, *Enterococcus* and *Ruminococcus* (Hehemann et al. 2010). Within the phyla Bacteroidetes, the genus *Bacteroides* predominates together with *Prevotella* (Lopez-Santamarina et al. 2020). The *phylum* Actinobacteria is the least predominant but includes the important genus *Bifidobacterium*, known by its health benefits and probiotic properties. The *phylum* Proteobacteria includes genera that can be harmful to humans, such as *Salmonella*,

A. Lopez-Santamarina · A. Cardelle-Cobas (✉) · L. I. Sinisterra-Loaiza · A. Cepeda
J. M. Miranda
Laboratorio de Higiene, Inspección y Control de Alimentos (LHICA), Departamento de Quimica Analítica, Nutrición y Bromatología, Universidade de Santiago de Compostela, Campus Terra, Lugo, Spain
e-mail: alejandra.cardelle@usc.es

F. Ozogul et al. (eds.), *Seaweeds and Seaweed-Derived Compounds*,
https://doi.org/10.1007/978-3-031-65529-6_3

Klebsiella, *Yersinia* or *Escherichia* (Geerlings et al. 2018). The *phylum* Verrucomicrobia is important as it includes the genus *Akkermansia*, which is associated with good gut health (Roca-Saavedra et al. 2018).

The GM is established during infancy and the variance among individuals is conditioned by different factors such as the gestational period, the delivery method, the use of antibiotics during gestation or during the delivery, the feeding mode (breastfeeding or milk formula), etc. (Rinniella et al. 2019). During the 3–5 first years of life, the human intestinal microbiota is unstable and easily modifiable, being an interesting period of intervention. Over the course of the first 3–5 years, the microbiota stabilizes and develops a predictable composition that is challenging to change over time (Rodríguez et al. 2015). The microbiota composition changes once more in the elderly, decreasing both variety and abundance, albeit it has been noted to exhibit more intraindividual variation than in adults. Due to all of this, there is no agreement on the ideal GM, but it is known that a balanced and healthy GM is one that has a diverse microbial community rich in different bacterial species that shapes all the functional functions in symbiosis with the host to optimally perform metabolic and immune functions and prevent disease development (Rinniella et al. 2019; Imathiu 2020). An increasing number of studies link the composition of the GM to different non-communicable diseases such as cardiovascular disease, obesity, diabetes, cancer, gastrointestinal diseases, and even neurological disorders (Collins et al. 2012; Bhattarai et al. 2017; Tang et al. 2019; Pascale et al. 2019; Raza et al. 2019).

Numerous crucial aspects of human health are carried out by the GM, including the synthesis of vitamins, the control of the immune system, and the avoidance of illnesses caused by gut pathogens (The Human Microbiome Project 2012). Among all the variables that affect GM, food is a significant modifiable element that has a significant impact on microbial diversity, composition, and stability. Nutrients can directly interact with the GM, either encouraging or limiting their growth. In this way, the individuals who can get their energy from a certain type of food will be able to reproduce at the expense of everyone else. Indigestible carbohydrates are the most crucial dietary components and can be obtained from both terrestrial and marine sources (Kong et al. 2016; Zmora et al. 2019) and its consumption has been linked to prevention of carcinogenesis, cardiovascular disease, metabolic syndrome, type 2 diabetes, and obesity. The human genome only encodes a small number of hydrolases capable of hydrolyzing the glycosidic bonds of indigestible carbohydrates, collectively known as CAZymes). As a result, many polysaccharides, such as resistant starch, inulin, lignin, pectin, cellulose, and FOS, reach the large intestine undigested. Contrarily, the intestinal microbiome contains tens of thousands of CAZymes, which enable the bacteria to breakdown and digest these sugars. Primary degraders of these non-digested polysaccharides include species from de genus *Bacteroides*, *Bifidobacterium*, *Ruminococcus*, *Roseburia*, *Facealibacterium*, *Anaerostides*, and *Coprococcus* (Clemente et al. 2015).

The intestinal bacteria's metabolization of these polysaccharides produce beneficial metabolites such short-chain fatty acids (SCFAs) (Chater et al. 2016; Gentile and Weir 2018). SCFAs, such as acetate, propionate and butyrate are a source of energy for gastrointestinal epithelial cells, provide protection against pathogens, influence intestinal mucosal immunity and barrier integrity, and induce apoptosis of colon cancer cells (Markowiak-Kopec and Slizewska 2020; Parada Venegas et al. 2019). When undigested polysaccharides can promote a beneficial effect on the human health, there are called prebiotics. A prebiotic is a "selectively fermented ingredient that allows specific changes, both in the composition and/or activity in the GM that confer benefits upon host wellbeing and health" (Gibson et al. 2010).

Even though dietary fiber is one of the main food components and consumption of 25–35 g daily is recommended in most of dietary guidelines around the world, in Western societies the recommended amount is not achieved (Castro-Penalonga et al. 2018), so its consumption must be promoted, and population informed about the different food sources of fiber. Seaweeds are a source of fiber and a potential prebiotic food. Numerous research works have studied, *in vitro* and *in vivo*, in mice, the potential beneficial effect of seaweed polysaccharides on the human GM (Shannon et al. 2021) with promissory results and even though the assays with the whole seaweed are scarce, seaweeds arise as a promissory alternative to foods of terrestrial origin as a source of fiber (Lopez-Santamarina et al. 2020). In addition, it could be also used as nutraceuticals for microbiota modulation in pathological situations.

In this chapter a revision of the scientific works about the potential functional properties of seaweed on GM have been carried out. Among the seaweed components that can reach the colon and modulate the abundance and diversity of GM, arising as potential prebiotic ingredients, non-digestible polysaccharides, polyphenols, and peptides are included.

2 Seaweed Components as Potential Modulators of Gut Microbiota

2.1 *Polysaccharides*

The most prevalent type of carbohydrates in food are polysaccharides. They are made up of monosaccharide building blocks connected by glycosidic linkages. These monosaccharides can be made up of just one kind of monosaccharide and/or its derivatives, resulting in a homopolysaccharide, or they can be made up of several kinds of sugar molecules and/or their derivatives, resulting in a heteropolysaccharide. Cellulose is an illustration of a homopolysaccharide and is a crucial structural element of the cell wall of plants and the majority of seaweeds (Stiger-Pouvreau et al. 2016) and, for example, carrageenan (*Rhodophyta*) and ulvan (*Cholophyta*) are examples of heteropolysaccharides (Stiger-Pouvreau et al. 2016). They are present in all plants, fungi, and algae and can be linear or extremely branching in structure. Examples include structural polysaccharides like cellulose and chitin as well as intercellular mucilage, as well as storage polysaccharides like starch, glycogen, and galactogen. Seaweeds' cell walls include polysaccharides along with other substances like proteins, proteoglycans, polyphenols, and some mineral elements like calcium and potassium (Charoensiddhi et al. 2017). The most abundant polysaccharides are structural ones, and the type of seaweed as well as environmental elements like salt, water temperature, and sunlight intensity can affect how they are composed (Collins et al. 2016; Torres et al. 2019). These polysaccharides are species-specific (Chang 2012). The differential characteristics of the seaweed polysaccharides is that in the case of seaweeds, they are sulfated polysaccharides whereas in land plants they are not sulfated. These sulfated polysaccharides participate in ionic regulation (Ruperez et al. 2013).

It has been suggested that most of the polysaccharides in seaweed can be considered as dietary fiber. According to reports, the total fiber content varies by species and ranges from 35% to 62% for brown, 10% to 57% for red, and 29% to 67% for green (Shannon et al. 2021). The physiological characteristics of these fibers will alter depending on whether they are water-soluble or not (Gómez-Ordóñez et al. 2010). Thus, it has been shown that water-soluble polysaccharides like guar gum or pectins have hypoglycemic and hypocholesterolemic effects, whereas water-insoluble polysaccharides like cellulose are associated with a slowed transit time through the digestive system (Burtin 2003).

Depending on the seaweed, the polysaccharides also vary in type. Fucoidan, laminarin, and alginate are the primary polysaccharides in brown seaweeds; porphyran, carrageenan, hypnean, and floridean starch are the predominant polysaccharides in red seaweeds; and ulvan, sulphated-rhamnans, -arabinogalactans, and -mannans are the predominant polysaccharides in green seaweeds (Fernando et al. 2019; Fig. 3.1).

Since the selective modulation of microbiota strongly depends on carbohydrate structure, including, degree of polymerization and type of linkage, the effect of the different types of seaweeds on the GM may be different and will also depend on external factors such as the location and the season of collection.

In the literature, there are numerous studies showing the modulator effect of polysaccharides, *in vitro* and *in vivo*. These polysaccharides have shown to reach the colon in a higher proportion than polysaccharides from plants since they are less degradable by the gastrointestinal tract than land plants (Shannon et al. 2021). In fact, these polysaccharides have been shown to exhibit greater prebiotic activity than land plant polysaccharides in *in vitro* models (Seong et al. 2019; Han et al. 2019). They have been demonstrated to have the ability to alter gut metabolism, including fermentation, to prevent pathogen adherence and evasion, and to possibly treat inflammatory bowel illnesses (Charoensiddhi et al. 2017; Lean et al. 2015). Additionally, research has concentrated on their usage as prebiotics to aid in the prevention of prevalent chronic non-communicable diseases as obesity, diabetes, cardiovascular disease, or some forms of cancer (Gurpilhares et al. 2019).

Due to the inclusion of various non-digestible polysaccharides, such as laminarin, fucoidan, and alginate, brown seaweeds are recognized to be a rich source of fiber. These particular polysaccharides have demonstrated to encourage the diversity of gut bacteria (Li et al. 2016; Vázquez-Rodríguez et al. 2021). The primary

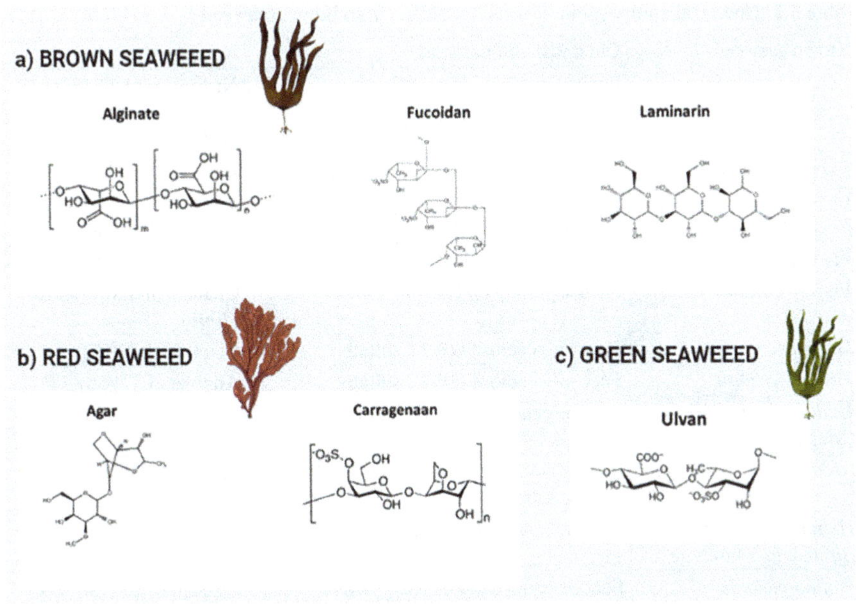

Fig. 3.1 Main polysaccharides in (**a**) brown, (**b**) red, and (**c**) green seaweed. (Created by Biorender.com)

structural polysaccharides in red seaweeds include agar, carrageenans, xylans, sulphated galactans, and porphyrins, while starch serves as the primary storage carbohydrate (You et al. 2019; Usov 2011). The primary structural polysaccharides found in green seaweeds are sulfated ulvans, galactans, xylans, and mannans. The gut bacteria partially ferments these polymers, which are mostly made of the sugars rhamnose, xylose, glucose, and glucuronic acid (Jiao et al. 2011; Wells et al. 2017; Cherry et al. 2019). Work to date on seaweed consumption and its effect on GM is based on the extraction of bioactive compounds from seaweed, mostly polysaccharides, except for two studies using whole seaweed as substrate (Bajury et al. 2017; Lopez-Santamarina et al. 2022).

Tables 3.2 and 3.3 list the different studies (both *in vitro* and *in vivo,* in animals) conducted about the prebiotic properties of seaweed compounds. *In vitro* batch fermentation models frequently use 24-hour anaerobic human fecal fermentation as a model for gut fermentation procedures. For instance, Devillé et al. (2007) shown that laminarin enhanced SCFA synthesis, particularly butyrate and propionate. *Bifidobacterium* populations significantly increased, as did acetate and propionate levels, in response to low molecular weight (MW) polysaccharides from the red alga *Gelidium sesquipidale* (Ramnani et al. 2012). Rodrigues et al. (2016) obtained similar results using an extract of the brown alga *Osmundea pinnatifida* (Charoensiddhi et al. 2016, 2017). According to Charoensiddhi et al. (2017), several *Ecklonia radiata* extracts encouraged the growth of SCFA-producing bacteria like *Bifidobacterium* and *Lactobacillus*, increasing the amount of SCFAs as a result.

Table 3.1 Chemical composition of some fucoidans from brown seaweed

Brown seaweed	Chemical composition	Reference
Fucus vesiculosus	Fucose, sulfate	Nishino et al. (1994) and Conchie and Percival (1950)
Fucus evanescens C. Ag.	Fucose/sulfate/acetate (1/1.23/0.36)	Bilan et al. (2002)
Fucus distichus	Fucose/sulfate/acetate (1/1.21/0.08)	Bilan et al. (2004)
Fucus serratus L	Fucose/sulfate/acetate (1/1/0.1)	Bilan et al. (2006)
Lessonia vadosa	Fucose/sulfate (1/1.12)	Chandía and Matsuhiro (2008)
Macrocystis pyrifera	Fucose/galactose (18/1), sulfate	Black et al. (1952)
Pevetia wrightii	Fucose/galactose (10/1), sulfate	Anno et al. (1966)
Undaria pinnatifida (Mekabu)	Fucose/galactose (1/1.1), sulfate	Lee et al. (2004)
Ascophyllum nodosum	Fucose (49%), xylose (10%), GlcA (11%), sulfate	Percival (1968)
Himanthalia lorea and Bifurcaria bifurcante	Fucose, xylose, GlcA, sulfate	Mian and Percival (1973)
Padina pavonia	Fucose, xylose, mannose, glucose, galactose, sulfate	Hussein et al. (1980a, b)
Laminaria angustata	Fucose/galactose/sulfate (9/1.9)	Kitamura et al. (1991)
Ecklonia kurome	Fucose/galactose, mannose, xylose, GlcA, sulfate	Nishino et al. (1989)
Sargassum stenopjyllum	Fucose/ galactose, mannose, GlcA, Glucose, xylose, sulfate	Duarte et al. (2001)
Adenocytis utricularis	Fucose, galactose, mannose, sulfate	Ponce et al. (2003)
Hizika fusiforme	Fucose, galactose, mannose, mqnnose, xylose, GlcA, sulfate	Li et al. (2006)
Dictyota mesntrualis	Fucose/xylose/uronic acid/galactose/sulfate (1/0.8/0.7/0.8/0.4) and (1/0.3/0.4/1.5/1.3)	Albuquerque et al. (2004)
Spatoglossum schroederi	Fucose/xylose/galactose/sulfate (1/0.5/2/2)	Rocha et al. (2005)

Reproduced from Li et al. (2008)

Additionally, Kuda et al. (2005) found that the pathogens *Salmonella Typhimurium*, *Listeria monocytogenes*, and *Vibrio parahaemolyticus* were prevented from adhering to and invading enterocyte cells by sodium alginate and laminarin from brown seaweeds. In an assay simulating small intestinal and caecal conditions with pig fecal samples, showed as the use of the whole brown seaweed *Ascophyllum nodosum* on the microbiota decrease its fermentative activity (Dierick et al. 2010).

In many of these studies, before simulating colonic fermentation, the substrate/s are submitted to the digestion in the upper gastrointestinal part (oral, gastric, and small intestine digestion) using enzymatic solutions, acidic conditions, etc. as the physiological. Thus, it is possible, by taking samples at each part (or only after digestion in the small intestine) determine, the digestibility and absorption of

Table 3.2 Microbiota modulation by polysaccharides or polysaccharide-extracts obtained from different species of brown seaweeds

Substrate	Type of study	Structure	Study conditions (SC) and Microbiota analysis (MA)	Source	Effect on gut microbiota	Reference
Fucoidans	*In vitro*	Two fractions MWCOL, MW < 30 kDa and MWCOH, MW > 30 kD)	**SC:** Simulated upper gastrointestinal digestion and colon fermentation. Fecal samples from three healthy donors, anaerobiosis, temperature controlled (37 °C), no pH controlled. Substrate: 8%, 0–48 h **MA**: Cultivation in different media by the spread plate method	*Laminaria japonica*	↓ *Enterobacter* spp. and ↑ *Lactobacillus* and *Bifidobacterium*	Kong et al. (2016)
	In vivo	*High purity fucoidan* (Marinova Pty Ltd., Tasmania, Australia). Neutral carbohydrates (fucose), 59.5%, sulfate content 26.6%; MW 61.7 kDa	**SC:** Male C57BL/6 mice (aged 8–10 weeks; 21–27 g, average weight ~25 g)- Colitis induced. Substrate concentration-not provided. Control group (n = 10) vs treatment group (n = 10) for 7 days. **MA:** microbiota was not evaluated	*Fucus vesiculosus*	Diarrhoea reduction and faecal blood loss. ↓ of at least 15 proinflammatory cytokines by the colon tissue	Lean et al. (2015)
		Type I, sulphate content 18.4%; MW 310 kDa	**SC:** Healthy C57BL/6 male mice (n = 12, 6 control vs 6 test) oral gavage, 100 mg/kg/day for 6 weeks **MA:** 16S rRNA high throughput sequencing (V3–V4 hypervariable regions)	*Laminaria japonica*	↑ *Lactobacillus* and *Ruminococcaceae*, ↓ the pathogenic bacteria-*Peptococcus*	Shang et al. (2016)
		Type II, sulphate content 21.0%; MW 1330 kDa		*Ascophyllum nodosum*	↑ *Ruminococcaceae* but ↓ *Akkermansia*, Alistipes and Clostridiales	

(continued)

Table 3.2 (continued)

Substrate	Type of study	Structure	Study conditions (SC) and Microbiota analysis (MA)	Source	Effect on gut microbiota	Reference
		Type I, sulfate content 18.4%; MW 310 kDa	**SC:** C57BL/6J mie with high-fat diet-induced metabolic syndrome oral gavage, 200 mg/kg/day for 6 weeks. **MA:** 16S rRNA high throughput sequencing (V4 hypervariable region)	*Laminaria japonica*	Improvement of metabolic syndrome. Attenuation of gut dysbiosis by ↑ *Bacteroides* and *Akkermansia* and ↓ *Rikenellaceae* and *Alistipes*	Shang et al. (2017a)
		Type II, sulphate content 21.0%; MW 1330 kDa		*Ascophyllum nodosum*	Improvement of metabolic syndrome. Attenuation of gut dysbiosis by ↑ *Akkermansia*, clostridiales vadinBB60, Alloprevotella and Blautia and ↓ Rikenellaceae and *Alistipes*	
		Type II	**SC:** BALB/c mice with high-fat diet-induced obesity, oral gavage, 100 mg/kg/día for 8 weeks	*Undaria pinnatifida*	↑ in *Alloprevotella* and ↓ *Staphylococcus* and *Streptococcus*	Liu et al. (2018)
		Data not provided	**SC:** 60 female Spraque-Dawley (SD) 6–8 weeks old mice divided in control group (C), model group (M), low dose (200 mg/kg bw) of fucoidan group (F1) and high dose (400 mg/kg bw) of fucoidan group (F2) based on body weight (bw). Breast cancer induced in M, F1 and F2.) Oral gavage. **MA:** 16S rDNA high throughput sequencing	*Sargassum* spp.	↑ Firmicutes and *Prevotella* and ↓ Bacteroidetes	Xue et al. (2018)
		Type II; sulphate content 14.55%; MW 205.8 kDa	**SC:** ICR mice with diabetes, oral gavage, 100 mg/kg/day for 6 weeks **MA:** 16S rRNA high throughput sequencing (V4 hypervariable region)	*Sargassum fusiforme*	↑ *Alloprevotella*, *Alistipes*, *Odoribacter*, *Millionella* and *Roseburia*. ↓ *Desulfovibrio*, *Parabacteroides* and *Muscipirillum* in diabetic mice	Cheng et al. (2019)

		Type II; sulphate content 27.8%; MW 309 kda	**SC:** SD rats with high-fat diet-induced obesity, oral gavage, 100 mg/kg 7 day and for 8 weeks **MA**: 16S rRNA high throughput sequencing (V4 hypervariable region)	*Undaria pinnatifida*	Attenuation of gut dysbiosis by ↑ *Bacillus*, *Ruminococcus*, *Adlercreutzia*, *Prevotella*, *Oscillospira* and *Desulfovibrio* and ↓ *Clostridium*, *Corynebacterium*, *Staphylococcus* and *Streptococcus*	Chen et al. (2019)
Alginate	*In vitro*	Alginate 100 kDa Sangon Biotech Company (Shanghai, China)	**SC:** Fecal samples from six healthy donors (22–35 years), anaerobiosis, temperature controlled (37 °C), pH not controlled. Substrate: 0.5%: time; 0–48 h **MA:** the V3 region of the 16S rRNA gene (positions 341 to 534 of the Escherichia coli gene) of all fermentations at 0 and 48 h was analyzed using PCR-DGGE	Not provided	alginate was completely degraded and utilized at various rates by fecal microbiota. The bacteria having a pronounced ability to degrade Alg isolated from human fecal samples were identified as *Bacteroides ovatus*, *Bacteroides xylanisolvens*, and *Bacteroides thetaiotaomicron*. Alg also the production of SCFA.	Li et al. (2016)
		Alginate produced by degradation with acid; MW 55.97 kDa) and with hydrogen peroxide; MW 31.04 kDa	**SC:** Fecal samples from three healthy donors (21–31 years old), anaerobiosis, temperature controlled (37 °C), pH controlled. **MA:** Fluorescent in-situ hybridization (FISH)	*Ascophylum nodosum*	↑ *Bifidobacterium* spp. and *Lactobacillus* spp. ($p > 0.05$); ↑ acetic acid and total SCFas with both substrates ($p < 0.05$) and ↑ propionic acid for the alginate with lower MW ($p < 0.05$)	Ramnani et al. (2012)

(continued)

Table 3.2 (continued)

Substrate	Type of study	Structure	Study conditions (SC) and Microbiota analysis (MA)	Source	Effect on gut microbiota	Reference
		Extracted alginate; >92% MW: 600 kDa; ratio mannuronic/glucuronic acid = 0.5–0.8	**SC:** Fecal samples from three healthy donors (22–25 years), anaerobiosis, temperature controlled (37 °C), pH not-controlled. Substrate: 0.5%: time; 0–72 h **MA:** High-throughput sequencing analysis (V3–V4 region of the 16S rRNA gene)	*Laminaria japonica*	↑ *Bacteroides*, specifically *B. finegoldii* and total SCFA	Ai et al. (2019)
	In vivo	Sodium alginate (SA) from a brown seaweed-not specifications provided	**SC:** Wild-type (WT) C57BL/6 mice Mice were fed a HFD from 9 weeks of age for 4 or 12 weeks. 2 different High Fat Diets (60 kcal% fat) (i) a high fat purified diet with 50.0 g cellulose and (ii) 11.3 g cellulose and 38.7 g SA (D18102802, 5% SA + HFD).	*Not specified*	SA changed the gut microbiota (GM) composition and significantly ↑ the abundance of *Bacteroides*. SA ↓ the number of colonic inflammatory monocytes, which promote MetS development	Ejima et al. (2021)
		Extracted alginate; MW: 108 kDa.	**SC:** Female BALB/c mice; 3 groups (n = 15); (i) control, (ii) OVA induced diarrhoea, (iii) OVA induced diarrhea treated with alginate (oral administration, 2 mg) **MA:** 16S rRNA gene high-throughput Sequencing (V3–V4 hypervariable regions)	*Laminaria japonica*	Alginate treatment attenuated Ovoalbumin (OVA) induced GM disorder in mice. ↓ morphological damage in the duodenum. ↑ microbiota richness and diversity. The relative abundances of *Alloprevotella*, *Bacteroides*, *Parabacteroides* and *Rikenellaceae* RC9 displayed a trend of recovery to normal values in the alginate treated group	Yao et al. (2021)

		Extracted sodium alginate (SA); 100 kDa, 99.6%	**SC**: 40 male BALB/c mice (6–7 weeks old, 22 ± 2 g, specific-pathogen-free grade); 4 groups: (i) normal control group (N), (ii) CTX-induced immunosuppression model group (M), (iii) SA low-dose (50 mg/kg/day) group (ALM), and (iv) SA high-dose (250 mgkg/day) group (AHM), 28 days. **MA:** 16S rRNA gene high-throughput Sequencing (V3–V4 hypervariable regions)	*Laminaria*	↑ *Lactobacillus*, *Roseburia*, and *Lachnospiraceae* NK4A136 and ↓ Helicobacter, Peptococcus, and Tyzzerella SA also ↓ gut inflammation by ↓ serum D-lactic acid (D-LA) and lipopolysaccharide (LPS) concentrations.	Huang et al. (2021)
	Clinical studies	Alginate produced by depolymerization by alkaline extraction. MW: 30 kDa, ratio of p-1, 4-D-mannuranate to a-1, 4-L-guluronate of alginate was 0.6	**SC**: 8 healthy male volunteer students (aged 18–23 year) received an experimental drink containing 10 g of alginate/once a day in addition to their normal diet, for 2 weeks. **MA:** not provided and original reference not found	*Lessonia fravicans and L. nigrescens*	↑ *Bifidobacterium* spp., ↑ SCFAs and ↓ the putrefactive activity of intestinal bacteria.	Terada et al. (1995)
Laminarin	*In vitro*	Laboratory extracted laminarin and commercial laminarin (Sigma L-9634)	**SC**: Fecal samples from six Chinese healthy donors (25–35 years), anaerobiosis, temperature controlled (37 °C), pH not-controlled. Substrate: 1%: time; 0–48 h. **MA:** biochemical and microbiological parameters	*Laminaria digitata*	Laminarin was more than 90% used after 24 h. It was not selectively used by *bifidobacterium* and *lactobacillus*, but ↑ the production of propionate and butyrate.	Devillé et al. (2007)

(continued)

Table 3.2 (continued)

Substrate	Type of study	Structure	Study conditions (SC) and Microbiota analysis (MA)	Source	Effect on gut microbiota	Reference
		Glucose residues linked by β-d-1,6-branched β-d-1,3-glycosidic bonds. Ratio 1:2; DP: 22–25; MW: 5 kDa	**SC:** in this work HT-29-Luc cell adhesion and invasion assay with *Salmonella* Typhimurium, *Listeria monocytogenes* and *Vibrio parahaemolyticus* **MA:** *viable counts*	*Eisenia bicyclis*	Inhibition of *S. Typhimurium*, *L. monocytogenes* and *V. parahemolyticus* adhesion and invasion.	Kuda et al. (2005)
		β-(1–3) linked linear chain; MW 6 kDa	**SC**: batch fermentation system under strict anaerobic conditions with individual bifidobacteria including *B. infantis* (JCM 1222), *B. longum* (JCM 1217), and *B. adolescentis* (JCM 1275), temperature controlled (37 °C), pH not controlled. Substrate: 0.5%: time; 0–24 h. **MA**: viable counts	*L. digitata*	*B. infantis* showed the highest increase and produced almost double the amount of total short-chain fatty acids (SCFAs) than the other two bifidobacteria.	Zhao and Cheung (2011)
	In vivo	Tokyo Kasei, Tokyo, Japan	**SC:** Four-week-old male Wistar rats three groups (n = 6) and given a control diet or experimental diet containing 2% (wt/wt) of laminaran (Tokyo Kasei, Tokyo, Japan) for 2 weeks **MA**: Pyrosequencing of bacterial 16S rRNA fragments	*Not provided*	*Clostridium ramosum* (3.14%) and *Parabacteroides distasonis* (1.36%) were only detected in the laminarin group. Laminarin ↑ cecal organic acid levels, particularly propionic acid. ↓ H2S, and phenol	An et al. (2013)

		BioAtlantis Limited (Kerry Technology Park).	**SC:** 28 pigs (progeny of large white × (large white × landrace sows) The dietary treatments were as follows: (1) basal diet; (2) basal diet + 240 parts per million (ppm) FUC; (3) basal diet + 300 ppm LAM; (4) basal diet + 300 ppm LAM and 240 ppm FUC, 8 d *ad libitum* **MA**: 16S rRNA gene high-throughput Sequencing (V3–V4 hypervariable regions)	*Laminaria spp.*	FUC diet ↓ *Enterobacteriaceae*, this effect was not observed when combined with LAM. LAM diet ↓ *E. coli*, but this effect was not observed when combined with FUC	Walsh et al. (2013)
		Laminarin (Sigma)- more data not provided	**SC:** 18 4-week-old female BALB/c mice. Three dietary groups (3 mice per cage, 2 cages per group) fed: (i) a normal diet (CTL), (ii) a high-fat diet (HFD) (45% kcal from fat), or (iii) a high-fat diet and 1% laminarin-supplemented water (HFL). **MA:** 16S rRNA gene high-throughput Sequencing (V3–V4 hypervariable regions)	*Laminaria digitata*	↓ Firmicutes and ↑ Bacteroidetes phylum, especially the genus *Bacteroides*. Laminarin could reduce the adverse effects of a high-fat diet by shifting GM towards higher energy metabolism	Nguyen et al. (2016)

Table 3.3 Microbiota modulation by polysaccharides or polysaccharide-extracts obtained from different species of red and green seaweeds

Substrate	Type of study	Structure	Study conditions (SC) and microbiota analysis (MA)	Source	Effect on gut microbiota (GM)	Reference
Red seaweeds						
Porphyran	*In vitro*	Purity ≥ 80%, Carbosynth, UK	**SC:** Fresh fecal samples were collected from 5 healthy adults in water-jacketed fermenter vessels were filled with pre-sterilized basal growth medium. 1% of seaweed polysaccharides, 37 °C, pH 6.8, anaerobic conditions, 12-24 h. **MA:** quantitative real-time polymerase chain.	*Porphyra* spp.	↑ *Bifidobacterium* spp. and *Lactobacillus* spp. ↑ *L. rhamnosus*, *L. plantarum*, *B. bifidum*, *B. ovatus*, and *B. dorei* No significant changes in SCFAs	Seong et al. (2019)
Carrageenan	*In vitro*	Carrageenans oligosaccharides with a MW of about 2.0 kDa of Qingdao International Oligose Preparation Center (Qingdao, China)	**SC:** Batch fermentations of carrageenans were performed in anaerobic conditions, 37 °C using Hungate tubes. 8 g/L of carrageenans, 1 mL of fecal suspensions (20% w/v), 48 h. **MA**: rRNA gene amplicon sequencing.	*Not provided.*	Carrageenans were fermented primarily by *Bacteroides* spp. *Bacteroides* enterotype microbiota and *Prevotella* enterotype microbiota produced a higher amount of acetate and butyrate during fermentation of carrageenans.	Fu et al. (2022)
	In vivo	κ-, ι- and λ-carrageenan from Sigma (Shanghai, China)	**SC:** 24 specific pathogen-free C57BL/6J mice (8-week-old). Three treatment groups and one control group with six animals each. Supplemented in drinking water at a concentration of 20 mg/L of different carrageenans. 6 weeks. **MA:** high-throughput sequencing	*Not provided*	↑ *Bacteroidales*, *Bacteroides*, *Lachnospiraceae* and *Alistipes* spp. ↓ *Akkermansia muciniphila*	Shang et al. (2017b)

		i-carrageenans	**SC:** Male Wistar rats (8–9 weeks old; 338 ± 1 g, n = 48). Supplemented with 5% of *Sarconema filiforme* for 16 weeks. **MA:** amplifying and sequencing the 16S rRNA gene.	*Sarconema filiforme*	*S. filiforme* supplementation modulated GM without changing the Firmicutes to Bacteroidetes ratio.	Du Preez et al. (2020)
		k-carrageenans	**SC:** fifty-four male, specific-pathogen free (SPF) C57BL/6J mice (14–18 g, 4-weeks old). Supplemented with 5% of k-carrageenans for 16 weeks. **MA:** 16S rRNA analysis	*Kappaphycus alvareziii*	↓ Bacteroidetes/Firmicutes ↑ *Prevotellaceae* and *Alistipes* ↑ Acetate and butyrate	Chin et al. (2019)
Green seaweeds						
Ulvans	*In vitro*	Purity ≥ 80% (Elicityl, France).	**SC:** Fresh fecal samples were collected from 5 healthy adults in water-jacketed fermenter vessels were filled with pre-sterilized basal growth medium. 1% of seaweed polysaccharides, 37 °C, pH 6.8, anaerobic conditions, 12–24 h. **MA:** quantitative real-time polymerase chain.	*Ulvan* spp.	↑ *Lactobacillus* and *Bifidobacterium*	Seong et al. (2019)

nutrients to know the components reaching the large bowel and, in consequence, its resident microbiota. Neoagaro-oligosaccharides and glycerol galactosides, for example, from red seaweeds have demonstrated that they reach the colon intact because they are not digested by the enzymes ordinarily present in the small intestine (Charoensiddhi et al. 2016).

Recent research has used additional models, such as multi-stage, dynamic *in vitro* models or *in vitro* dynamic continuous culture systems, to simulate the continuous function of the large bowel, small intestine, and even the stomach. An example of these models is the simulator of the TNO *in vitro* model of the colon (TIM-2) (Venema and The 2015) and the human intestinal microbial ecosystem (SHIME®) (Van de Wiele et al. 2015). Most of the dynamic fermentation models simulate the conditions of stomach, small intestine, and all parts of the colon (Wang et al. 2019).

These *in vitro* models have a number of benefits over in vivo tests. Although *in vivo* research can provide immediate findings, because of their high cost, the need for ethical review, and the fact that everyone's digestive system is different, *in vitro* models are typically favoured. Usually, they are used to test a battery of substrates and then to select the best depending on their effects, they can be also proved in terms of their efficacy or to test their bioavailability, etc. In brief, the *in vitro* studies are an useful tool to develop pre-clinical assays.

Thus, and focusing the theme on seaweed, the *in vitro* assays are a powerful tool to select the best polysaccharide extract, the best seaweed, saving time and costs in the *in vivo* assays, etc.

Most of the studies developed with seaweed, *in vivo* with rats, have used a seaweed-supplemented diet in a group comparing the results against a group fed with a non-supplemented diet. The germ-free mice or the gnobiotic animals (colonized by one or a few kinds of known microorganisms) are other animal models used to study the effect of functional foods as the seaweed. However, these studies require specialized facilities, high costs, and skilled volunteers leading to the limited use of this mice model in much research (Kennedy et al. 2018).

The results obtained with seaweed from these *in vivo* studies showed positive and promising results (Tables 3.2 and 3.3). Thus, the findings of Liu et al. (2015) demonstrated that rats supplemented with the red seaweed *Chondrus crispus* had increased abundance of good gut bacteria like *Bifidobacterium* breve and decreased abundance of dangerous bacteria like Streptococcus pneumonia and *Clostridium septicum*. Additionally, there was an improvement in the modulation of host immunity as seen by the SCFA synthesis and plasma immunoglobulin levels. Kim et al. (2016) discovered that supplementing rats with the brown seaweeds *Undaria pinnatifida* and *Laminaria japonica* decreased the ratio Firmicutes/Bacteroidetes, which is linked to obesity when it is >1, as well as suppressed rat weight increase. The number of pathogenic bacteria, such as the *Clostridium*, *Escherichia*, and *Enterobacter* genera, was decreased.

The use of extracts of fucoidans has shown their potential effect in inflammatory markers when administered orally in mice with induced colitis (Lean et al. 2015). Additionally, rats fed a diet containing laminarin and low MW alginate showed decreased synthesis of indole and sulfide, which have been identified as risk factors for colon cancer, according to Kuda et al. (2005). Neoagaro-oligosaccharides produced by the hydrolysis of agarose by -agarase led to a rise in *Lactobacillus* and *Bifidobacterium* and a decrease in putrefactive bacteria in the feces or cecal content of mice (Charoensiddhi et al. 2016).

2.1.1 Brown Seaweed Polysaccharides

The brown seaweed species most studied in terms of their relationship with the GM are *Ecklonia, Sargassum, Laminaria, Ascophyllum, Fucus, Undaria, Saccorhiza*, or *Porphyra* (Lopez-Santamarina et al. 2020). Fucoidans, alginate and laminarin are the main polysaccharides of this group. Table 3.2 shows *in vitro* and *in vivo* studies of the effect of the main brown seaweed polysaccharides on the GM.

Fucoidans Alginate, fucoidans, and laminarin are the three primary components of brown seaweed polysaccharides. Sulfate L-fucose serves as the primary sugar unit in the sulfated polysaccharide known as fucoidan, in which sulfate ester groups are connected to fucose by -1,3 connections. Fucoidans have a complicated structure, as indicated by several research studies, and they may have a linear backbone made of (13)-L-Fuc (type I fucoidan) or alternate (13)-L-Fuc and (14)-L-Fuc (type II fucoidan), with (12)-L-Fuc occasionally present in the backbone branching. Sulfate groups frequently occupy the fucose C-2 or/and C-3, C-4. Due to the stability of this structure and the presence of additional sugars in the branching of the structural core, fucoidans with high uronic acid (UA) and hexose content may have their structural core built of alternating UA-hexose. In addition to other monosaccharides like galactose, glucose, mannose, or xylose, some fucoidans also contain acetate in their structural makeup (Table 3.1).

These fucoidans present numerous biological activities which make them very attractive for research in different fields. One of them is, of course their use as microbiota modulators. In this sense different works have been carried out as it can be observed in Table 3.2.

The *in vitro* assays carried out to date with faecal samples from healthy volunteers indicate a growth stimulation of *Bifidobacterium* and *Lactobacillus* spp. (Kong et al. 2016) whereas the *in vivo* studies also shown a decrease in proinflammatory cytokines (Lean et al. 2015), an improvement of the metabolic syndrome (Shang et al. 2017a) and a decrease of pathogenic bacteria such as *Peptococcus, Desulfovibrio, Alistipes* and *Clostridiales* (Shang et al. 2016, 2017a; Cheng et al. 2019). However, as can it observed in Table 3.2 these results shortly depend on the fucoidan structure, the dose, the time of assay, etc. In addition, these fucoidans were used in a specific mice model of colitis, obesity, or diabetes, so, the results also must be regarded carefully.

Alginate Brown seaweeds' cell walls contain a linear polymer known as alginate. It comprises of D-mannuronic acid and l-glucuronic acid units that are -(1–4) connected. M-blocks (homopolymer of -D-mannuronic acid), G-blocks (homopolymer of poly a-L-guluronic acid), and alternating GM-blocks (heteropolymer of randomly placed residues of M and G) are the three residue sequences that make up this compound (Draget and Taylor 2011; Kabisch et al. 2014). MG block's gel-forming capabilities are distinct from those of M and G blocks (Morris et al. 1973). Figure 3.2 depicts the different blocks in detail.

Brown seaweed is used to extract alginates, which are traditionally utilized as gelling agents in food, medicine, cosmetics, feeds, and other industrial purposes.

According to many publications, oligosaccharides from alginate contain a variety of intriguing properties, including antifungal, anti-inflammatory, and immunomodulatory effects (Andriamanantoanina and Rinaudo 2010). The medication propylene glycol alginate sodium sulfate, an alginate derivative, is used to treat cardiovascular disease.

In addition, alginate and its oligomer derivatives, as other edible dietary fibers, are resistant to the human digestion and can reach the colon and utilized by the GM. In Table 3.2 different assays, *in vitro* and *in vivo*, carried out with alginates and derivatives from brown seaweed are presented. Alginates can exist in the form of a salt or in its acidic state, alginic acid, which compresses two different types of hexuronic acid monomers: -D-mannuronic acid and -L-guluronic acid, coupled by 1–4 bonds (Andriamanantoanina and Rinaudo 2010). Sodium alginate is often used as the salt form. The *in vitro* assays with alginates with different degree of polymerization and MW, obtained from different methods showed as alginate can be completely degraded by fecal microbiota being *Bacteroides* spp. the main bacterial genera using the carbohydrate in studies with alginate MW ranging: 100–600 (Li et al. 2016; Ai et al. 2019). Alginate with lower MW shown an increase in *Bifidobacterium* and *Lactobacillus* (Ramnani et al. 2012). The *in vivo* assays in mice confirm these results, also reporting an increase in *Bacteroides*, in microbiota richness and diversity and decreasing inflammatory markers (Ejima et al. 2021; Huang et al. 2022; Yao et al. 2021).

A clinical study developed in 1995 with 8 healthy male volunteers showed that the intake of 10 g of alginate (30 kDa) from *Lessonia favricans* and *Lessonia*

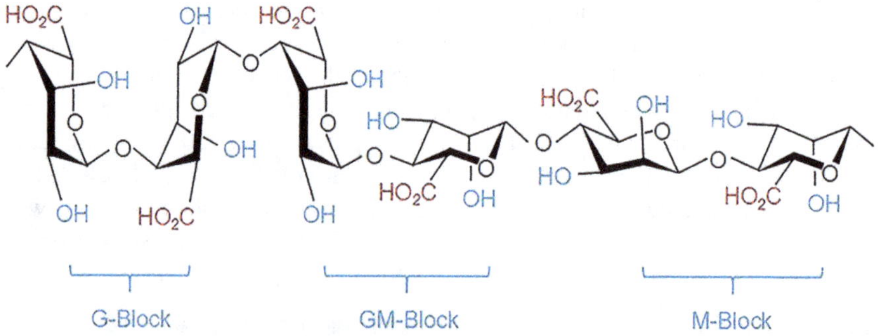

Fig. 3.2 Alginate residues sequences: G, GM, and M Blocks. (Reprinted from Guo et al. (2020), Copyright (2020), with permission from Elsevier)

nigrescens once a day for 2 weeks produced an increase in the bifidobaceria population and SCFAs together with a decrease of the putrefactive activity on intestinal bacteria (Terada et al. 1995).

Laminarin Laminarin is a polysaccharide of low MW. It contains β-glucans, which are in the form of β(1–3) link accompanying β(1,6)-linked side chains of different distributions and lengths (approx. 20–25 glucose moieties).

Among the different properties attributed to laminarin, it seems, as it has been shown in different studies, to improve immunity to *Escherichia coli* lipopolysaccharides (Neyrinck et al. 2007), it seems to induce TNF-alpha production from human peripheral blood monocytes (Miyanishi et al. 2003) and increase production of SCFAs, especially butyrate, in the gut (Devillé et al. 2007), which could provide a protective intestinal barrier against metabolic diseases (Brahe et al. 2013). Although these effects may not be directly linked to obesity prevention, laminarin may improve gut health and shift GM towards higher energy metabolism in individuals consuming high fat-diets. Laminarin's anti-obesity benefits are relatively rare. The *in vitro* studies carried out with these polysaccharides (Table 3.2) showed an increase in *Bifidobacterium* and *Lactobacillus* together an increase of SCFAs (Devillé et al. 2007; Zhao and Cheung 2011). Also, inhibition of pathogens such as *S. Typhimurium*, *L. monocytogenes,* have been showed in *in vivo* studies as well as an increase is Bacteroidetes (Nguyen et al. 2016).

In addition to the studies shown in Table 3.2, there are further *in vitro* studies on polysaccharide extracts and their effect on microbiota although the polysaccharide is not specified. For example, using polysaccharide extracts from different brown seaweeds, an increase in bacterial diversity has been observed (Charoensiddhi et al. 2016; Yang et al. 2019) as well as an increase in *Bifidobacterium*, *Lactobacillus*, *Parabacteroides*, *Lachnospiraceae* and *Prevotella* (Charoensiddhi et al. 2016, 2017; Strain et al. 2019; Praveen et al. 2019). In addition, a reduction in Firmicutes to Bacteroidetes ratio was also reported by Fu et al. (2018). There is only one study in which a whole seaweed (*Himanthalia elongata*) was used in an *in vitro* assay to test its possible prebiotic effect (Lopez-Santamarina et al. 2022). In this study, in contrast to inulin, which significantly increased the relative abundance of *Parabacteroides distasonis*, *H. elongata* was selectively exploited by some GM members, leading to an increase in *Bacteroides* species. Even though there were taxonomic differences between the two substrates used, the metabolic pathways linked to the GM function were not significantly different, suggesting the possibility of using the complete seaweed as a prebiotic.

2.1.2 Red Seaweed Polysaccharides

The most studied red seaweed in relation to their effect on GM has been the genera *Acanthopora*, *Gracilaria*, *Kappaphycus*, *Euchema* and *Grateloupia*, *Chondrus*, *Gelidium*, or *Osmundea* (Lopez-Santamarina et al. 2020). As indicated above in red seaweed, porphyrin and carrageenans are the prevalent polysaccharides. Table 3.3

shows *in vitro* and *in vivo* studies of the effect of the main red seaweed polysaccharides on the GM.

Porphyran Porphyran, a sulphated polysaccharide that is unique to red seaweed in the genus *Porphyra*, makes up roughly 11–21% of the dry mass of the seaweed. It is made up of alternating units of galactose-6-sulphate and 6-O-methyl-galactose, as well as repeating units of galactose and 3,6-anhydrogalactose. Its MW can range from 14 to 201 kDa on average. Although porphyran is soluble in hot water, its viscosity prevents it from being used in pharmaceuticals (Shannon et al. 2021). Regarding their potential prebiotic function, investigations using entire red seaweed and *in vitro* digestion simulations showed that porphyran increased the production of SCFA and good gut bacteria (Seong et al. 2019) (Table 3.3).

Carragenan As for linear sulphated polysaccharides, carrageenans are structural components of the extracellular matrix. There are 15 different types of carrageenan, with the most popular and approved for usage in the food industry being iota I kappa (k), and lambda (l) (Fig. 3.3). K and i-carrageenan are composed of D-galactose and 3,6-anhydro-galactose units with differing proportions of alternately positioned sulfate groups. Lambda-carrageenan lacks 3,6-anhydro-galactose and alternates between 1,3 and 1,4 inter-galactose links. Their MW is roughly 500 kDa on average. The melting point of carrageenan gel (40–70 °C) is above which all forms of carrageenan are soluble in water (Shannon et al. 2021). The *in vitro* gastrointestinal investigations conducted to date (Table 3.3) have discovered that k-carrageenan can have both positive and negative effects because occasionally inflammation indicators were found to increase and other times to decrease along with an increase in SCFA and beneficial gut flora (Sun et al. 2019).

In Table 3.3, the studies conducted with characterized carrageenans extracted from red seaweeds are include.

Agar Agar is a linear polysaccharide made up of 1,3-linked D-galactose and 1,4-linked 3,6-anhydro-L galactose that are alternated and to some extent substituted with sulfate, methyl, or pyruvate groups. These structures reveal that agar is made up of agarose and agaropectin. These polysaccharides' molecular structures vary depending on the species. Agarose is a lineal sugar made up of four connected 3,6-anhydro-a-L-galactose and three linked b-D-galactose. Agaropectin is an acid polysaccharide with pyruvic acid and D-glucuronic acid attached to agarobiose as sulfate groups. Agarose, which makes up 70% of the mixture, causes the mixture to gel, whereas agaropectin causes the mixture to thicken. It is possible to degrade agarose chemically or enzymatically to produce several derived agarose molecules. The majority of the hydrolysis products, including agarooligosaccharides (AOS), neoagaroligosaccharides (NAOSs), neoagarobiose (NAB), and 3,6-anhydro-L-galactose (L-AHG), have biological properties (Torres et al. 2019).

Agar is an approved food additive in Europe and a generally regarded as safe (GRAS) food additive in the United States (E406). Agar cannot be digested in the digestive tract because humans lack the -agarases, the enzymes necessary for its

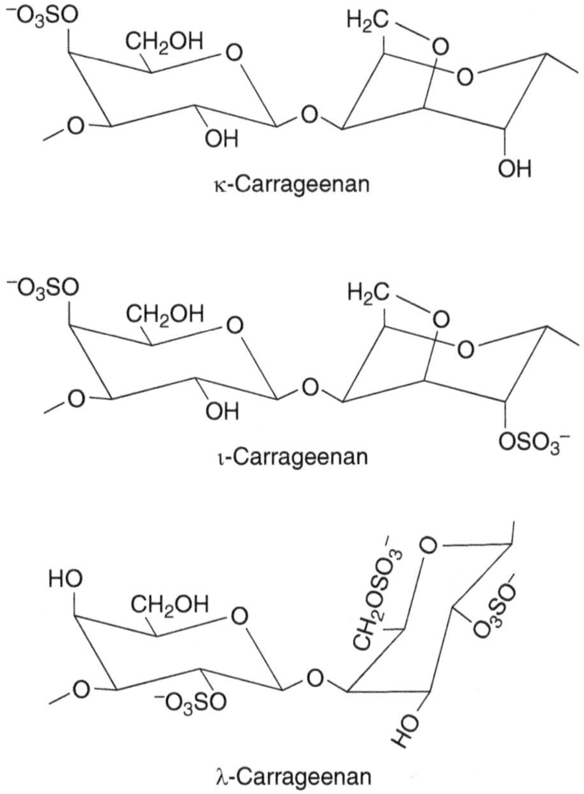

Fig. 3.3 Forms of carrageenan used in food industry. (Reprinted from BeMiller (2019), Copyright (2019), with permission from Elsevier)

hydrolysis, but it can be converted to D-galactose by intestinal bacteria, and for this reason, numerous research have been conducted to assess its possible prebiotic effect. One of the notable polysaccharides in red seaweed is agarose, which reaches the distal parts of the colon, where it is fermented and metabolized by the GM (Ramnani et al. 2012; Yun et al. 2021).

Agarose oligosaccharides have also been demonstrated to encourage the growth of advantageous bacterial strains, including *B. adolescentis* ATCC 15703 and *B. infantis* ATCC 15697 (Hu et al. 2006). After 24 hours of *in vitro* fermentation with human feces, Low MW agar had a bifidogenic effect along with an increase in the amounts of SCFA and propionate acetate (Ramnani et al. 2012).

It is noteworthy thar most of the studies carried out with red seaweed polysaccharides have been without purification or characterization of these polysaccharides, so in many cases a mixture of polysaccharides, or polysaccharides with other secondary compounds, are tested (*in vitro* and/or *in vivo*).

In general, administration of extracted polysaccharides from red seaweed increased the production of SCFA, stimulating the growth of beneficial bacteria such as *Lactobacillus* (Praveen et al. 2019) or *Bifidobacterium* (Liu et al. 2015;

Ramnani et al. 2012; Bajury et al. 2017), while inhibiting the growth of potentially pathogenic bacteria (Liu et al. 2015; Praveen et al. 2019).

The pathogens *Clostridium septicum* and *Streptococcus pneumonia* significantly decreased in rats fed with the whole seaweed *C. crispus*, whereas *Bifidobacterium brevis* and SCFA synthesis increased (Liu et al. 2015). In other study conducted with the whole seaweed, *Kappaphycus alvarezii* an increase in acetate and propionate production was observed as well as *Bifidobacteria*, decreasing *Clostridium coccoides* and *Eubacterium rectale* (Bajury et al. 2017).

Polysaccharides isolated from the red seaweed *Grateloupia filicina* and *Eucheuma spinosum* stimulated the growth of *Bifidobacteria* (Chen et al. 2018). *Gracilaria rubra* polysaccharides increase the relative abundances of beneficial genera such as *Bacteroides*, *Prevotella* and *Phascolarctobacterium in vitro* (Di et al. 2018).

Another study in chickens described an increase in the abundance of beneficial bacteria such as *Bifidobacterium longum* and *Streptococcus salivarius* when the seaweed *Sarcodiotheca gaudichaudii* and *C. crispus* were fed (Kulshreshtha et al. 2014). Extracts of red seaweed (*Osmundea pinnatifida* and *Sargassum muticum*) showed *in vitro* an increase in acetate and propionate production and *Bifidobacterium* population (Rodrigues et al. 2016). The seaweed *Acanthophora spicifera* has been shown to demonstrate prebiotic activity, promoting the growth of *Lactobacillus plantarum* and suppressing the growth of the pathogen *Salmonella typhimurium* (Praveen et al. 2019).

2.1.3 Green Seaweed Polysaccharides

The main genera studied of green seaweeds have been *Enteromorpha* and *Ulva* (Lopez-Santamarina et al. 2020). The main polysaccharides of the green seaweed are the ulvans. Table 3.3 shows *in vitro* and *in vivo* studies of the effect of the main red seaweed polysaccharides on the GM.

Ulvans Green algae's primary polysaccharides are ulvans, which make up 38–54% of the thallus' dry mass. Ulvans are repeating units of sulphated L-rhamnose, D-xylose, D-glucuronic acid, and its epimer Liduronic acid that are water-soluble, gelling polysaccharides. Molecules can weigh from between 1 and 2000 kDa. In vitro and animal research have demonstrated that ulvans have the potential to be used as prebiotics (Shannon et al. 2021). Studies on the prebiotic potential of green seaweed are scarce, in part because their fermentation necessitates uncommon and particular l-rhamnosidase activity in the gastrointestinal tract (Munoz-Munoz et al. 2017).

Ulvans are one of the most prevalent polysaccharides in green seaweed, and their potential prebiotic impact has also been investigated. In fact, it has been discovered that they promote the development of populations of *Bifidobacterium* and *Lactobacillus* as well as boost the production of SCFA (lactic and acetic) (Seong et al. 2019). They also seem to inhibit of the growth of potentially pathogenic bacteria (Zmora et al. 2019; Ren et al. 2017). An increase in Firmicutes and Actinobacteria has also been seen in addition to a decrease in Bacteroidetes and Proteobacteria (Ren et al. 2017). The number of *Akkermansia muciniphila*, Bacteroides, *Alloprevotella*, *Ruminococcaceae*, and *Blautia* in the digestive tract of mice increased after *Enteromorpha clathrata* administration (Huebbe et al. 2017; Shang et al. 2017b).

E. clathrata polysaccharides were found to exert prebiotic effects by increasing *A. muciniphila*, *Bifidobacterium* and *Lactobacillus* in male and female mice, decreasing the proportion of *Helicobacter* (Shang et al. 2018). *Enteromorpha compressa* showed the highest *in vitro* prebiotic activity score promoting the growth of *L. plantarum* and suppressing the growth of the pathogen *S. Typhimurium* (Praveen et al. 2019).

Other bioactive substances found in seaweeds offer many health benefits, including anti-inflammatory, antiviral, antimicrobial, antidiabetic, antitumor, antihypertensive, antimicrobial, antiallergic, and immunomodulatory properties (Gomez-Zavaglia et al. 2019; Okolie et al. 2017; Freitas et al. 2015; Rosa et al. 2019). Alkaloids, flavonoids, polyoletides (such as plorotannins), isoprenoids (such as terpenes, carotenoids, and steroids), and polyoletides are the major categories of secondary metabolites identified in seaweed (Okolie et al. 2017; Gomez-Zavaglia et al. 2019).

2.2 *Phenolic Compounds*

Complexly structured ingested polyphenols can also reach it to the large intestine, where they may be transformed into advantageous bioactive metabolites by the microorganims (Charoensiddhi et al. 2016). The existence, metabolism, and functions of various kinds of polyphenols are diverse. Flavonoids, stilbenes, phenolic acids, and lignans are the main types. Most polyphenols enter the colon without being absorbed in the small intestine and can be linked to organic acids and carbohydrates (Charoensiddhi et al. 2016). Brown seaweed's phlorotannins have been demonstrated to do this (Corona et al. 2016). According to Charoensiddhi et al. (2017), fermentation of the phlorotannin-enriched fraction of the brown seaweed Ecklonia radiata led to a decline in *Enterococcus*, which is frequently associated with unfavorable results for gut health. However, compared to other seaweed fractions and controls, only small amounts of SCFA were generated. Although the growth of certain populations of gut bacteria is inhibited, phlorotannins in brown seaweed appear to have some antibacterial properties (Charoensiddhi et al. 2016), which may explain the reduced SCFA generation. The most common bacteria that break down polyphenols are *Lachnospiraceae* CG191, *Eubacterium ramulus, Eubacterium cellulosolvens*, *Bacteroides distasonis*, *Bacteroides ovatus*, and *Bacteroides uniformis*. The aglycones and monomers are subsequently broken down into simpler compounds like hydroxyphenyl propionic acid and hydroxyphenyl acetic acid through ring-cleavage and decarboxylation. Flavonoids, lignans, and phenolic acids have all been linked to these processes (Braune and Blaut 2016).

Several bromophenols found in seaweeds have been shown to be cytotoxic to some cancer cells and antibacterial to several strains of both Gram-positive and Gram-negative bacteria (Nwosu et al. 2011; Liu et al. 2015). Additionally, seaweeds' flavonoids and their glycosides have been found to have antioxidant qualities and to be useful in the treatment of cancer and arteriosclerosis. Some show promise against obesity, such fucoxanthin (Gomez-Zavaglia et al. 2019). In diabetic mice, flavonoids also affect the balance of the GM by boosting the populations of the taxa Alistipes, Lachnospiraceae, and Odoribacter (Tang et al. 2013). Lectins, a type of seaweed peptide, also have advantageous properties like antiviral, antibacterial, and antifungal action (Rosa et al. 2019).

2.3 *Proteins and Peptides*

Depending on heredity and other dietary circumstances, around 66–95% of proteins are typically absorbed before entering the large intestine (Shannon et al. 2021). The principal elements that adversely impact the digestion of algal proteins are phlorotannins and seaweeds' high polysaccharide content (Bleakley and Hayes 2017). While there has not been much research on the association between gut health and protein fermentation, dietary fiber fermentation shows various benefits for gut health, including the production of SCFA and the modulation of gut microbiomes. In vitro research has dominated most investigations. For the proteolysis, *Bacteroides*, *Clostridium*, *Propionibacterium*, *Fusobacterium*, *Lactobacillus*, and *Streptococcus* are to blame (Charoensiddhi et al. 2016). The fermentation of aspartate, alanine, threonine, and methionine generates propionate, whereas the fermentation of glutamate, lysine, histidine, cysteine, serine, and methionine leads to the formation of butyrate (Yadav et al. 2018). Bacteroidetes and Firmicutes ferment amino acids and peptides to create propionate and butyrate (Scott et al. 2006). The distal colon is where the fermentation process primarily takes place. Bacterial exo- and endopeptidases initially break down the complex proteins to release free amino acids and small peptides (Macfarlane and Macfarlane 2006). Branched chain fatty acids (2-methyl butyrate, isobutyrate, and isovalerate), organic acids, gases (H_2 and CO_2), and potentially hazardous metabolites such phenols, amines, indoles, sulfides, and ammonia are produced after fermentation of amino acids and short peptides (Windey et al. 2012). This is linked to a large intake of processed and red meat, which may raise the risk of colon cancer (Chao et al. 2005). However, Cian et al. (2015) showed that proteins and peptides derived from them, along with polysaccharides and minerals, could balance the function of the intestinal mucosal barrier, acting as prebiotics, regulating intestinal epithelial cell, macrophage, and lymphocyte proliferation and differentiation, and modulating the immune response.

Seaweed also contains a few minor nutrients that are good for human health. As an illustration, phycobiliproteins, the pigments that give seaweed its vivid color and are used commercially in foods and nutraceuticals because they have antimicrobial, antioxidant, anti-inflammatory, neuroprotective, hepatoprotective, immunomodulatory, and anticarcinogenic properties, are used to color seaweed (Aryee et al. 2018; Fernández-Rojas et al. 2014; Hao et al. 2018; Jiang et al. 2017; Manirafasha et al. 2016).

3 Conclusions

Seaweeds are important sources of bioactive substances, such as polysaccharides, polyphenols, and proteins that may be beneficial for gastrointestinal health. It is becoming more widely accepted to use these seaweed-derived substances or even the entire seaweed in functional foods and nutraceuticals, especially prebiotic supplements. The majority of seaweed polysaccharides can be thought of as dietary fiber that resists digestion by digestive enzymes found in the human gastrointestinal system and specifically promotes the growth of good bacteria in the gut. The use of seaweeds and chemicals derived from them to modify the synthesis of SCFA and the gut microbiome is encouraged by growing scientific evidence from both in vitro and in vivo research.

Based on the research accomplishments of the structure-function correlations and molecular mechanisms, seaweed polysaccharides, especially those with sulfated groups, have shown to possess a variety of biological activities including metabolic modulation, immune regulation, antioxidation, anticoagulation, and anti-tumorigenesis. These biological activities are promising for in-depth product development in the health industry. Further research is therefore required, particularly for polysaccharides like carrageenan with variable biological activities, to determine the precise mechanism by which the GM catabolizes seaweed-derived polysaccharides and the systematic pathways activated by the resulting microbial metabolites. However, for better quality and higher cost performance in industry, seaweed polysaccharide production processes also need to be updated.

Despite the evidence, in vivo clinical investigations employing human models are still lacking in order to verify the outcomes of *in vitro* and in vivo animal studies.

References

Ai C, Jiang P, Liu Y, Duan M, Sun X, Luo T, Guoping J, Song S (2019) The specific use of alginate from Laminaria japonica by Bacteroides species determined its modulation of the Bacteroides community. Food Funct 10:4304–4314

Albuquerque IRL, Queiroz KCS, Alves LG, Santos EA, Leite EL, Rocha HAO (2004) Heterofucans from Dictyota menstrualis have anticoagulant activity. Braz J Med Biol Res 37:167–171

An C, Kuda T, Yazaki T, Takahashi H, Kimura B (2013) FLX pyrosequencing analysis of the effects of the brown-algal fermentable polysaccharides alginate and laminaran on rat cecal microbiotas. Appl Environ Microbiol 79:860–866

Andriamanantoanina H, Rinaudo M (2010) Relationship between the molecular structure of alginates and their gelation in acidic conditions. Polym Int 59(11):1531–1541

Anno K, Terahata H, Hayashi Y (1966) Isolation and purification of fucoidin from brown seaweed Pelvetia wrightii. Agric Biol Chem 30:495–499

Aryee AN, Agyei D, Akanbi TO (2018) Recovery and utilization of seaweed pigments in food processing. Curr Opin Food Sci 19:113–119

Bajury DM, Rawi MH, Sazali IH, Abdullah A, Sarbini SR (2017) Prebiotic evaluation of red seaweed (*Kappaphycus alvarezii*) using *in vitro* colon model. Int J Food Sci Nutr 68:821–828

BeMiller JN (2019) 13- Carrageenas. In: Carbohydrate chemistry for food scientists, 3rd edn. Elsevier, pp 279–291

Bhattarai Y, Muniz Pedrogo DA, Kashyap PC (2017) Irritable bowel syndrome: a gut microbiota-related disorder? Am J Physiol Gastrointest Liver Physiol 312:52–62

Bilan MI, Grachev AA, Ustuzhanina NE (2002) Structure of a fucoidan from the brown seaweed Fucus evanescens C. Ag. Carbohydr Res 337:719–730

Bilan MI, Grachev AA, Ustuzhanina NE, Shashkov AS, Nifantiev NE, Usov AI (2004) A highly regular fraction of a fucoidan from the brown seaweed Fucus distichus L. Carbohydr Res 339:511–517

Bilan MI, Grachev AA, Shashkov AS, Nifantiev NE, Usov AI (2006) Structure of a fucoidan from the brown seaweed Fucus serratus L. Carbohydr Res 341:238–245

Black WAP, Dewar ET, Woodward FN (1952) Manufacture of algal chemicals. IV. -Laboratory-scale isolation of fucoidin from brown marine algae. J Sci Food Agric 3:122–129

Bleakley S, Hayes M (2017) Algal proteins: extraction, application, and challenges concerning production. Foods 6(5):1–33

Brahe LK, Astrup A, Larsen LH (2013) Is butyrate the link between diet, intestinal microbiota and obesity-related metabolic diseases? Obes Rev 14:950–959

Braune A, Blaut M (2016) Bacterial species involved in the conversion of dietary flavonoids in the human gut. Gut Microbes 7(3):216–234

Burtin P (2003) Nutritional value of seaweeds. Elec J Environ Agric Food Chem 2:498–503

Castro-Penalonga M, Roca-Saavedra P, Miranda JM, PortoArias JJ, Nebot C, Cardelle-Cobas A, Franco CA, Cepeda A (2018) Influence of food consumption patterns and Galician lifestyle on human gut microbiota. J Physiol Biochem 74(1):85–92

Chandía NP, Matsuhiro B (2008) Characterization of a fucoidan from Lessonia vadosa (Phaeophyta) and its anticoagulant and elicitor properties. Int J Biol Macromol 42:235–240

Chang C (2012) Carbohydrates – comprehensive studies on glycobiology and glycotechnology [Internet]. IntechOpen, London, 572 p [cited 2022 Nov 16]. Available from: https://www.intechopen.com/books/2323

Chao A, Thun MJ, Connell CJ, McCullough ML, Jacobs EJ, Flanders WD, Rodriguez C, Sinha R, Calle EE (2005) Meat consumption and risk of colorectal cancer. JAMA 293(2):172–182

Charoensiddhi S, Conlon MA, Vuaran MS, Franco CM, Zhang W (2016) Impact of extraction processes on prebiotic potential of the brown seaweed Ecklonia radiata by *in vitro* human gut bacteria fermentation. J Funct Foods 24:221–230

Charoensiddhi S, Conlon MA, Vuaran MS, Franco CM, Zhang W (2017) Polysaccharide, and phlorotannin-enriched extracts of the brown seaweed *Ecklonia radiata* influence human gut microbiota and fermentation *in vitro*. J Appl Phycol 29:2407–2416

Chater PI, Wilcox M, Cherry P, Herford A, Mustar S, Wheater H, Brownlee I, Seal C, Pearson J (2016) Inhibitory activity of extracts of Hebridean brown seaweeds on lipase activity. J Appl Phycol 28:1303–1313

Chen X, Sun Y, Hu L et al (2018) In vitro prebiotic effects of seaweed polysaccharides. J Oceanol Limnol 36:926–932

Chen Q, Liu M, Zhang P, Fan S, Huang J, Yu S, Zhang C, Li H (2019) Fucoidan and galactooligosaccharides ameliorate high-fat diet–induced dyslipidemia in rats by modulating the gut microbiota and bile acid metabolism. Nutrition 65:50–59

Cheng Y, Sibusiso L, Hou L, Jiang H, Chen P, Zhang X, Wu M, Tong H (2019) Sargassum fusiforme fucoidan modifies the gut microbiota during alleviation of streptozotocin-induced hyperglycemia in mice. Int J Biol Macromol 131:1162–1170

Cherry P, Yadav S, Strain CR, Allsopp PJ, McSorley EM, Ross RP, Stanton C (2019) Prebiotics from seaweeds: an ocean of opportunity? Mar Drugs 17:327

Chin YX, Mi Y, Cao WX, Lim PE, Xue CH, Tang QJ (2019) A pilot study on anti-obesity mechanisms of *Kappaphycus Alvarezii*: the role of native κ-carrageenan and the leftover sans-carrageenan fraction. Nutrients 11(5):1133

Cian RE, Drago SR, De Medina FS, Martínez-Augustin O (2015) Proteins, and carbohydrates from red seaweeds: evidence for beneficial effects on gut function and microbiota. Mar Drugs 13:5358–5383

Clemente JC, Pehrsson EC, Blaser MJ, Sandhu K, Gao Z, Wang B, Magris M, Hidalgo G, Contreras M, Noya-Alarcón Ó (2015) The microbiome of uncontacted Amerindians. Sci Adv 1:e1500183

Collins SM, Surette M, Bercik P (2012) The interplay between the intestinal microbiota and the brain. Nat Rev Microbiol 10:735–742

Collins KG, Fitzgerald GF, Stanton C, Ross RP (2016) Looking beyond the terrestrial: the potential of seaweed derived bioactives to treat non-communicable diseases. Mar Drugs 14(3):60

Conchie J, Percival EGV (1950) Fucoidin. Part II. The hydrolysis of a methylated fucoidin prepared from Fucus vesiculosus. J Chem Soc:827–833

Corona G, Ji Y, Anegboonlap P, Hotchkiss S, Gill C, Yaqoob P et al (2016) Gastrointestinal modifications and bioavailability of brown seaweed phlorotannins and effects on inflammatory markers. Br J Nutr 2016(115):1240–1253

Devillé C, Gharbi M, Dandrifosse G, Peulen O (2007) Study on the effects of laminarin, a polysaccharide from seaweed, on gut characteristics. J Sci Food Agric 87:1717–1725

Di T, Chen G, Sun Y, Ou S, Zeng X, Ye H (2018) In vitro digestion by saliva, simulated gastric and small intestinal juices and fermentation by human fecal microbiota of sulfated polysaccharides from *Gracilaria rubra*. J Funct Foods 40:18–27

Dierick N, Ovyn A, De Smet S (2010) In vitro assessment of the effect of intact marine brown macro-algae Ascophyllum nodosum on the gut flora of piglets. Livest Sci 133(1–3):154–156

Draget KI, Taylor C (2011) Chemical, physical and biological properties of alginates and their biomedical implications. Food Hydrocoll 25(2):251–256

Du Preez R, Paul N, Mouatt P, Majzoub ME, Thomas T, Panchal SK, Brown L (2020) Carrageenans from the red seaweed Sarconema filiforme attenuate symptoms of diet-induced metabolic syndrome in rats. Mar Drugs 18(2):97

Duarte MER, Cardoso MA, Noseda MD, Cerezo AS (2001) Structural studies on fucoidans from the brown seaweed Sargassum stenophyllum. Carbohyd Res 333(4):281–293

Ejima R, Akiyama M, Sato H, Tomioka S, Yakabe K, Kimizuka T, Seki N, Fujimura Y, Hirayama A, Fukuda S, Hase K, Kim YG (2021) Seaweed dietary fiber sodium alginate suppresses the migration of colonic inflammatory monocytes and diet-induced metabolic syndrome via the gut microbiota. Nutrients 13:2812

Fernández-Rojas B, Hernández-Juárez J, Pedraza-Chaverri J (2014) Nutraceutical properties of phycocyanin. J Funct Foods 11:375–392

Fernando IPS, Kim K-N, Kim D, Jeon Y-J (2019) Algal polysaccharides: Potential bioactive substances for cosmeceutical applications. Crit Rev Biotechnol 39:99–113

Freitas AC, Pereira L, Rodrigues D, Carvalho AP, Panteleitchouk T, Gomes AM, Duarte AC (2015) Marine functional foods. In: Springer handbook of marine biotechnology. Springer, Heidelberg, pp 969–994

Fu X, Cao C, Ren B, Zhang B, Huang Q, Li C (2018) Structural characterization and *in vitro* fermentation of a novel polysaccharide from *Sargassum thunbergii* and its impact on gut microbiota. Carbohydr Polym 183:230–239

Fu T, Zhou L, Fu Z, Zhang B, Li Q, Pan L, Zhou C, Zhao Q, Shang Q, Yu G (2022) Enterotype-specific effect of human gut microbiota on the fermentation of marine algae oligosaccharides: a preliminary proof-of-concept *in vitro* study. Polymers 14:770

Geerlings SY, Kostopoulos I, de Vos WM, Belzer C (2018) *Akkermansia muciniphila* in the human gastrointestinal tract: when, where, and how? Microoorganisms 6(3):75

Gentile CL, Weir TL (2018) The gut microbiota at the intersection of diet and human health. Science 362:776–780

Gibson GR, Scott KP, Rastall RA, Tuohy KM, Hotchkiss A, Dubert-Ferrandon A, Gareau M, Murphy EF, Saulnier D, Loh G, Macfarlane S, Delzenne N, Ringel Y, Kozianowski G, Dickmann R, Lenoir-Wijnkoop I, Walker C, Buddington R (2010) Dietary prebiotics: current status and new definition. IFIS Funct Foods Bull 7(1):1–19

Gómez-Ordóñez E, Jiménez-Escrig A, Rupérez P (2010) Dietary fibre and physicochemical properties of several edible seaweeds from the northwestern Spanish coast. Food Res Int 3(9):2289–2294

Gomez-Zavaglia A, Prieto Lage MA, Jimenez-Lopez C, Mejuto JC, Simal-Gandara J (2019) The potential of seaweeds as a source of functional ingredients of prebiotic and antioxidant value. Antioxidants 8:406

Guo X, Wang Y, Qin Y, Shen P, Peng Q (2020) Structures, properties and application of alginic acid: a review. Int J Biol Macromol 162:618–628

Gurpilhares DB, Cinelli LP, Simas NK, Pessoa A Jr, Sette LD (2019) Marine prebiotics: polysaccharides and oligosaccharides obtained by using microbial enzymes. Food Chem 280:175–186

Han Z, Yang M, Fu X, Chen M, Su Q, Zhao Y, Mou H (2019) Evaluation of prebiotic potential of three marine seaweed oligosaccharides from enzymatic hydrolysis. Mar Drugs 17:173

Hao S, Yan Y, Li S, Zhao L, Zhang C, Liu L, Wang C (2018) The *in vitro* anti-tumor activity of phycocyanin against non-small cell lung cancer cells. Mar Drugs 16:178

Hehemann J, Correc G, Barbeyron T, Helbert W, Czjzek M, Michel G (2010) Transfer of carbohydrate-active enzymes from marine bacteria to Japanese gut microbiota. Nature 464:908–912

Hu B, Gong Q, Wang Y, Ma Y, Li J, Yu W (2006) Prebiotic effects of neoagaro-oligosaccharides prepared by enzymatic hydrolysis of agarose. Anaerobe 12(5–6):260–266

Huang J, Huang J, Li Y, Wang Y, Wang F, Qiu X, Liu X, Li H (2021) Sodium alginate modulates immunity, intestinal mucosal barrier function, and gut microbiota in cyclophosphamide-induced immunosuppressed BALB/c mice. J Agric Food Chem 69(25):7064–7073

Huang W, Tan H, Nie S (2022) Beneficial effects of seaweed-derived dietary fiber: highlights of the sulfated polysaccharides. Food Chem 373:131608

Huebbe P, Nikolai S, Schloesser A, Herebian D, Campbell G, Gluer CC, Zeyner A, Demetrowitsch T, Schwarz K, Metges CC et al (2017) An extract from the Atlantic brown seaweed *Saccorhiza polyschides* counteracts diet-induced obesity in mice via a gut related multi-factorial mechanisms. Oncotarget 8:73501–73515

Hussein MM, Abdel A, Salem HM (1980a) Sulfated heteropolysaccharides from Padina pavoia. Phytochemistry 19:2131–2213

Hussein MM, Abdel A, Salem HM (1980b) Some structural features of a new sulfated heteropolyssaride from Padina pavoia. Phytochemistry 19:2133–2135

Imathiu S (2020) Benefits and food safety concerns associated with consumption of edible insects. NFS J 18:1–11

Jiang L, Wang Y, Yin Q, Liu G, Liu H, Huang Y, Li B (2017) Phycocyanin: a potential drug for cancer treatment. J Cancer 8:3416–3429

Jiao G, Yu G, Zhang J, Ewart HS (2011) Chemical structures and bioactivities of sulfated polysaccharides from marine seaweed. Mar Drugs 9:196–223

Kabisch A, Otto A, König S et al (2014) Functional characterization of polysaccharide utilization loci in the marine *Bacteroidetes* '*Gramella forsetii*' KT0803. ISME J 8:1492–1502
Kennedy EA, King KY, Baldridge MT (2018) Mouse microbiota models Comparing germ-free mice and antibiotics treatment as tools for modifying gut bacteria. Front Physiol 9:1–16
Kim J, Yu D, Kim J, Choi E, Lee C, Hong Y, Kim C, Lee S, Choi I, Cho K (2016) Effects of *Undaria linnatifida* and *Laminaria japonica* on rat's intestinal microbiota and metabolite. J Nutr Food Sci 6(3):1–7
Kitamura K, Matsuo M, Yasui T (1991) Fucoidan from brown seaweed Laminaria angustata var. longissima. Agric Biol Chem 55(2):615–616
Kong Q, Dong S, Gao J, Jiang C (2016) *In vitro* fermentation of sulfated polysaccharides from *E. prolifera* and *L. japonica* by human fecal microbiota. Int J Biol Macromol 91:867–871
Kuda T, Yano T, Matsuda N, Nishizawa M (2005) Inhibitory effects of laminaran and low molecular alginate against the putrefactive compounds produced by intestinal microflora in vitro and in rats. Food Chem 91(4):745–749
Kulshreshtha G, Rathgeber B, Stratton G, Thomas N, Evans F, Critchley A, Hafting J, Prithiviraj B (2014) Feed supplementation with red seaweeds, *Chondrus crispus* and *Sarcodiotheca gaudichaudii*, affects performance, egg quality, and gut microbiota of layer hens. Poult Sci 93:2991–3001
Lean QY, Eri RD, Fitton JH, Patel RP, Gueven N (2015) Fucoidan extracts ameliorate acute colitis. PLoS One 10:e0128453
Lee JB, Hayashi K, Hashimoto M, Nakano T, Hayashi T (2004) Novel antiviral fucoidan from sporophyll of Undaria pinnatifida (Mekabu). Chem Pharm Bull 52:1091–1094
Li B, Xin JW, Sun JL, Xu SY (2006) Structural investigation of a fucoidan containing a fucose-free core from the brown seaweed Hizikia fusiforme. Carbohydr Res 341:1135–1146
Li B, Lu F, Wei X, Zhao R (2008) Fucoidan: structure and bioactivity. Molecules 13:1671–1695
Li M, Li G, Shang Q, Chen X, Liu W, Zhu L, Yin Y, Yu G, Wang X (2016) *In vitro* fermentation of alginate and its derivatives by human gut microbiota. Anaerobe 39:19–25
Liu J, Kandasamy S, Zhang J, Kirby CW, Karakach T, Hafting J, Critchley AT, Evans F, Prithiviraj B (2015) Prebiotic effects of diet supplemented with the cultivated red seaweed *Chondrus crispus* or with fructo-oligosaccharide on host immunity, colonic microbiota and gut microbial metabolites. BMC Complement Altern Med 15:279
Liu M, Ma L, Chen Q, Zhang P, Chen C, Jia L, Li H (2018) Fucoidan alleviates dyslipidemia and modulates gut microbiota in high-fat diet-induced mice. J Funct Foods 48:220–227
Lopez-Santamarina A, Miranda JM, Mondragon AC, Lamas A, Cardelle-Cobas A, Franco CM, Cepeda A (2020) Potential use of marine seaweeds as prebiotics: a review. Molecules 25:1004
Lopez-Santamarina A, Cardelle-Cobas A, Mondragon AC, Sinisterra-Loaiza L, Manuel Miranda J, Cepeda A (2022) Evaluation of the potential prebiotic effect of *Himanthalia elongata*, an Atlantic brown seaweed, in an *in vitro* model of the human distal colon. Food Res Int 2022(156):111156
Macfarlane S, Macfarlane GT (2006) Composition and metabolic activities of bacterial biofilms colonizing food residues in the human gut. Appl Environ Microbiol 72(9):6204–6211
Manirafasha E, Ndikubwimana T, Zeng X, Lu Y, Jing K (2016) Phycobiliprotein: Potential microseaweed derived pharmaceutical and biological reagent. Biochem Eng J 109:282–296
Markowiak-Kopec P, Slizewska K (2020) The effect of probiotics on the production of short–chain fatty acids by human intestinal microbiome. Nutrients 12:1107
Mian J, Percival E (1973) Carbohydrates of the brown seaweeds Himanthalia lorea and Bifurcaria bifurcata Part II. structural studies of the "fucans". Carbohydr Res 26:147–161
Miyanishi Y, Iwamoto EW, Odaz T (2003) Induction of TNF-alpha production from human peripheral blood monocytes with beta-1,3-glucan oligomer prepared from laminarin with beta-1,3-glucanase from *Bacillus clausii* NM-1. J Biosci Bioeng 363(95):192–195
Morris R, Rees DA, Thom D (1973) Characterization of polysaccharide structure and interactions by circular dichroism: order-disorder transition in the calcium alginate system. J Chem Soc Chem Commun:245–246

Munoz-Munoz J, Cartmell A, Terrapon N, Henrissat B, Gilbert HJ (2017) Unusual active site location and catalytic apparatus in a glycoside hydrolase family. Proc Natl Acad Sci 114:4936–4941

Neyrinck AM, Mouson A, Delzenne NM (2007) Dietary supplementation with laminarin, a fermentable marine beta (1–3) glucan, protects against hepatotoxicity induced by LPS in rat by modulating immune response in the hepatic tissue. Int Immunopharmacol 7(12):1497–1506

Nguyen SG, Guevara R, Lee J-H, Kim J (2016) Laminarin favorably modulates gut microbiota in mice fed a high-fat diet. Food Funct 7(10):4193–4201

Nishino T, Yokoyama G, Dobahi K (1989) Isolation, purification and characterization of fucose-containing sulfated polysaccharides from the brown seaweed Ecklonia kurome and their blood-anticoagulant activities. Carbohydr Res 186:119–129

Nishino T, Nishioka C, Ura H (1994) Isolation and partial characterization of a novel amino sugar-containing fucan sulfate from commercial Fucus vesiculosus fucoidan. Carbohydr Res 255:213–224

Nwosu F, Morris J, Lund VA, Stewart D, Ross HA, McDougall GJ (2011) Anti-proliferative and potential anti-diabetic effects of phenolic-rich extracts from edible marine seaweed. Food Chem 126:1006–1012

Okolie CL, CK Rajendran SR, Udenigwe CC, Aryee AN, Mason B (2017) Prospects of brown seaweed polysaccharides (BSP) as prebiotics and potential immunomodulators. J Food Biochem 41:e12392

Parada Venegas D, De la Fuente MK, Landskron G, González MJ, Quera R, Dijkstra G, Harmsen HJM, Faber KN, Hermoso MA (2019) Short chain fatty acids (SCFAs)-mediated gut epithelial and immune regulation and its relevance for inflammatory bowel diseases. Front Immunol 10:277

Pascale A, Marchesi N, Govoni S, Coppola A, Gazzaruso C (2019) The role of gut microbiota in obesity, diabetes mellitus, and effect of metformin: new insights into old diseases. Curr Opin Pharmacol 49:1–5

Percival E (1968) Glucoroxylofucan, a cell-wall component of Ascophyllum nodosum. Carbohydr Res 7:272–283

Ponce NMA, Pujol CA, Damonte EB (2003) Fucoidans from the brown seaweed Adenocystis utricularis: extraction methods, antiviral activity and structural studies. Carbohydr Res 338:153–165

Praveen MA, Parvathy KK, Jayabalan R, Balasubramanian P (2019) Dietary fiber from Indian edible seaweeds and its in-vitro prebiotic effect on the gut microbiota. Food Hydrocoll 96:343–353

Ramnani P, Chitarrari R, Tuohy K, Grant J, Hotchkiss S, Philp K, Campbell R, Gill C, Rowland I (2012) *In vitro* fermentation and prebiotic potential of novel low molecular weight polysaccharides derived from agar and alginate seaweeds. Anaerobe 18:1–6

Raza MH, Gul K, Arshad A, Riaz N, Waheed U, Rauf A, Aldakheel F, Alduraywish S, Rehman MU, Abdullah M et al (2019) Microbiota in cancer development and treatment. J Cancer Res Clin Oncol 145:49–63

Ren X, Liu L, Gamallat Y, Zhang B, Xin Y (2017) Enteromorpha and polysaccharides from *Enteromorpha* ameliorate loperamide-induced constipation in mice. Biomed Pharmacother 96:1075–1081

Rinniella E, Raoul P, Cintoni M, Francesci F, Miggiano GAD, Gasbarrini A, Mele MC (2019) What is the healthy gut microbiota composition? A changing ecosystem across age, environment, diet, and diseases. Microorganisms 7:14

Roca-Saavedra P, Mendez-Vilabrille V, Miranda JM, Nebot C, Cardelle-Cobas A, Franco CM, Cepeda A (2018) Food additives, contaminants and other minor components: effects on human gut microbiota—a review. J Physiol Biochem 74:69–83

Rocha HAO, Moraes FA, Trindade ES, Franco CRC, Torquato RJS, Veiga SS, Valente AP, Mourão PAS, Leite EL, Nader HB, Dietrich CP (2005) Structural and hemostatic activities of a sulfated galactofucan from the brown alga Spatoglossum schroederi. J Biol Chem 280:1278–41288

Rodrigues D, Walton G, Sousa S, Rocha-Santos TA, Duarte AC, Freitas AC, Gomes AM (2016) *In vitro* fermentation and prebiotic potential of selected extracts from seaweeds and mushrooms. LWT-Food Sci Technol 73:131–139

Rodríguez JM, Murphy K, Stanton C, Ross RP, Kober OI, Juge N, Avershina E, Rudi K, Narbad A, Jenmalm MC, Marchesi JR, Collado MC (2015) The composition of the gut microbiota throughout life, with an emphasis on early life. Microb Ecol Health Dis 26:26050

Rosa GP, Tavares WR, Sousa PMC, Pagès AK, Seca AML, Pinto DCGA (2019) Seaweed secondary metabolites with beneficial health effects: an overview of successes in in vivo studies and clinical trials. Mar Drugs 18:8

Rowan-Nash AD, Korry BJ, Mylonakis E, Belenky P (2019) Cross-domain and viral interactions in the microbiome. Microbiol Mol Biol Rev 83(1):e00044-18

Ruperez P, Gómez-Ordoñez E, Jiménez-Escrig A (2013) Chapter 11: Biological activity of algal sulfated and nonsulfated polysaccharides. In: Hernández-Ledesma B, Herrero M (eds) Bioactive compounds from marine foods: plant and animal sources. Wiley

Savage DC (1977) Microbial ecology of the gastrointestinal tract. Ann Rev Microbiol 31:107–133

Scott KP, Martin JC, Campbell G, Mayer CD, Flint HJ (2006) Whole-genome transcription profiling reveals genes up-regulated by growth on fucose in the human gut bacterium "*Roseburia inulinivorans*". J Bacteriol 188(12):4340–4349

Seong H, Bae J, Seo JS, Kim S, Kim T, Han NS (2019) Comparative analysis of prebiotic effects of seaweed polysaccharides laminaran, porphyran, and ulvan using in vitro human fecal fermentation. J Funct Foods 57:408–416

Shang Q, Shan X, Cai C, Hao J, Li G, Yu G (2016) Dietary fucoidan modulates the gut microbiota in mice by increasing the abundance of *Lactobacillus* and *Ruminococcaceae*. Food Funct 7:3224–3232

Shang Q, Guanrui S, Zhang M, Shi J, Xu C, Hao J, Li G, Yu G (2017a) Dietary fucoidan improves metabolic syndrome in association with increased *Akkermansia* population in the gut microbiota of high-fat diet-fed mice. J Funct Foods 28:138–146

Shang QS, Sun WX, Shan XD, Jiang H, Cai C, Hao JJ et al (2017b) Carrageenan-induced colitis is associated with decreased population of anti- inflammatory bacterium, Akkermansia muciniphila, in the gut microbiota of C57BL/6J mice. Toxicol Lett 279:87–95

Shang QS, Jiang H, Cai C, Hao JJ, Li GY, Yu JL (2018) Gut microbiota fermentation of marine polysaccharides and its effects on intestinal ecology: an overview. Carbohyr Polym 179:172–185

Shannon E, Conlon M, Hayes M (2021) Seaweed components as potential modulators of the gut microbiota. Mar Drugs 19:358

Stiger-Pouvreau V, Bourgougnon N, Deslandes É (2016) Chapter 8: Carbohydrates from seaweeds. In: Fleurence J, Levine I (eds) Seaweed in health and disease prevention. Academic Press, pp 223–274. ISBN 9780128027721

Strain CR, Collins KC, Naughton V, McSorley EM, Stanton C, Smyth TJ, Soler-Vila A, Rea MC, Ross PR, Cherry P (2019) Effects of a polysaccharide-rich extract derived from Irish-sourced *Laminaria digitata* on the composition and metabolic activity of the human gut microbiota using an *in vitro* colonic model. Eur J Nutr 1–17:309–325

Suau A, Bonnet R, Sutren M, Godon JJ, Gibson GR, Collins MD, Doré J (1999) Direct analysis of genes encoding 16S rRNA from complex communities reveals many novel molecular species within the human gut. Appl Environ Microbiol 65(11):4799–4807

Sun Y, Cui X, Duan M, Ai C, Song S, Chen X (2019) In vitro fermentation of κ-carrageenan oligosaccharides by human gut microbiota and its inflammatory effect on HT29 cells. J Func Foods 59:80–91

Tang Z, Gao H, Wang S, Wen S, Qin S (2013) Hypolipidemic and antioxidant properties of a polysaccharide fraction from *Enteromorpha prolifera*. Int J Biol Macromol 58:186–189

Tang WHW, Bäckhed F, Landmesser U, Hazen SL (2019) Intestinal microbiota in cardiovascular health and disease. J Am Coll Cardiol 73:2089–2105

Terada A, Hara H, Mitsuoka T (1995) Effect of dietary alginate on the faecal microbiota and faecal metabolite activity in humans. Microb Ecol Health Dis 8:259–266

The Human Microbiome Project Consortium (2012) Structure, function and diversity of the healthy human microbiome. Nature 486:207–214

Torres MD, Flórez-Fernández N, Domínguez H (2019) Integral utilization of red seaweed for bioactive production. Mar Drugs 17:314

Usov AI (2011) Polysaccharides of the red seaweed. In: Advances in carbohydrate chemistry and biochemistry, vol 65. Elsevier, San Diego, pp 115–217

Van de Wiele T, Van den Abbeele P, Ossieur W, Possemiers S, Marzorati M (2015) Chapter 27: The simulator of the human intestinal microbial ecosystem (SHIME®). In: Verhoeckx K, Cotter P, López-Expósito I, Kleiveland C, Lea T, Mackie A, Requena T, Swiatecka D, Wichers H (eds) The impact of food bioactives on health: in vitro and ex vivo models [Internet]. Springer, Cham

Vázquez-Rodríguez B, Santos-Zea L, Heredia-Olea E, Acevedo-Pacheco L, Santacruz A, Gutiérrez-Uribe JA, Cruz-Suárez LE (2021) Effects of phlorotanin and polysaccharide fractions of brown seaweed Silvetia compressa on human gut microbiota composition using an in vitro colonic model. J Funct Foods 84:104596

Venema K, The TNO (2015) Chapter 26: In vitro model of the colon (TIM-2). In: Verhoeckx K, Cotter P, López-Expósito I, Kleiveland C, Lea T, Mackie A, Requena T, Swiatecka D, Wichers H (eds) The impact of food bioactives on health: in vitro and ex vivo models [Internet]. Springer, Cham. PMID: 29787064

Walsh AM, Sweeney T, O'Shea CJ, Doyle DN, Doherty JVO (2013) Effect of supplementing varying inclusion levels of laminarin and fucoidan on growth performance, digestibility of diet components, selected fecal microbial populations and volatile fatty acid concentrations in weaned pigs. Anim Feed Sci Technol 183:151–159

Wang M, Wichienchot S, He X, Fu X, Huang Q, Zhang B (2019) *In vitro* colonic fermentation of dietary fibers: fermentation rate, short-chain fatty acid production and changes in microbiota. Trends Food Sci Technol 88:1–9

Wells ML, Potin P, Craigie JS, Raven JA, Merchant SS, Helliwell KE, Smith AG, Camire ME, Brawley SH (2017) Seaweed as nutritional and functional food sources: revisiting our understanding. J Appl Phycol 29:949–982

Windey K, De Preter V, Verbeke K (2012) Relevance of protein fermentation to gut health. Mol Nutr Food Res 56(1):184–196

Xue M, Ji X, Liang H, Liu Y, Wang B, Sun L, Li W (2018) The effect of fucoidan on intestinal flora and intestinal barrier function in rats with breast cancer. Food Funct 9:1214–1223

Yadav M, Verma MK, Chauhan NS (2018) A review of metabolic potential of human gut microbiome in human nutrition. Arch Microbiol 200(2):203–217

Yang CF, Lai SS, Chen YH, Liu D, Liu B, Ai C, Wan WZ, Gao LY, Chen XH, Zhao C (2019) Antidiabetic effect of oligosaccharides from seaweed *Sargassum confusum* via JNK-IRS1/PI3K signaling pathways and regulation of gut microbiota. Food Chem Toxicol 131:110562

Yao L, Yang P, Lin Y, Bi D, Yu B, Lin Z, Wu Y, Xu H, Hu Z, Xu X (2021) The regulatory effect of alginate on ovalbumin-induced gut microbiota disorders. J Funct Foods 86:104727

You L, Gong Y, Li L, Hu X, Brennan C, Kulikouskaya V (2019) Beneficial effects of three brown seaweed polysaccharides on gut microbiota and their structural characteristics: an overview. Int J Food Sci Technol 55:1199–1206

Yun EJ, Yu S, Park NJ et al (2021) Metabolic and enzymatic elucidation of cooperative degradation of red seaweed agarose by two human gut bacteria. Sci Rep 11:13955

Zhao J, Cheung PC (2011) Fermentation of b-glucans derived from different sources by bifidobacteria: evaluation of their bifidogenic effect. J Agric Food Chem 59:5986–5992

Zmora N, Suez J, Elinav E (2019) You are what you eat: diet, health and the gut microbiota. Nat Rev Gastroenterol Hepatol 16:35–56

Chapter 4
Seaweeds: A Holistic Approach to Heathy Diets and to an Ideal Food

Pınar Yerlikaya

1 Introduction

Seaweeds are multicellular macroscopic marine algae, visible to the naked eye and inhabit the coastal regions of the ocean. Seaweeds are classified into three groups depending on their pigmentation (Gamero-Vega et al. 2020; Peñalver et al. 2020; Polat et al. 2021):

Rhodophyta: red algae; sometimes purple, even brownish red; the largest and the most primitive group of algae. Many of red algae are found at greater depths than brown and green algae, in tropical and subtropical areas. The main reserve polysaccharide is floridean starch, and cell wall components are cellulose, carrageenan, and agar which is used in many industrial applications.

Ochrophyta, *Phaeophyceae*: brown algae; second most abundant group presents in rocky shores of temperate zones. They are large in size and provide ease of harvesting. Brown seaweeds are used in food industries, pharmaceutical, cosmetic, agriculture and animal feeding. Commonly, phycocolloids as alginates are obtained from brown seaweed.

Chlorophyta: blue-green algae; prokaryotes containing chlorophylls. It is the less presence and less consumed among other algae species.

The changes in ecological conditions can stimulate or inhibit the biosynthesis of several bioactive compounds (Marinho-Soriano et al. 2006; Li et al. 2011). The nutritional value of seaweeds is attributed to their high content of dietary fiber, amino acids, bioactive peptides, certain polyunsaturated fatty acids, vitamins, mineral and possess trace amounts of secondary compounds in cell walls (Bocanegra

P. Yerlikaya (✉)
Department of Seafood Processing Technology, Faculty of Fisheries, Akdeniz University, Antalya, Türkiye
e-mail: pyerlikaya@akdeniz.edu.tr

F. Ozogul et al. (eds.), *Seaweeds and Seaweed-Derived Compounds*,
https://doi.org/10.1007/978-3-031-65529-6_4

et al. 2009; Astorga-Espana et al. 2015; Pirian et al. 2017; Gaillard et al. 2018; Salma et al. 2021; Pradhan et al. 2022). Several environmental factors pose great variation in the composition of seaweeds. Chemical composition of seaweeds differs depending on species, developmental stage, environmental factors, geographic locations (Balboa et al. 2013; Astorga-Espana and Mansilla 2014).

Several researches have been revealed that seaweed is a healthy diet and an ideal food so far. Although being respected to all these studies, limited references could be cited in this chapter.

2 Nutritional Compounds

2.1 Polysaccharides

Energy need of metabolic activities are supplied by carbohydrates. Algal carbohydrates are composed of mainly polysaccharides and followed by few amounts of disaccharides and monosaccharides. *Gracilaria cervicornis* and *Sargassum vulgare* collected monthly from July 2000 to June 2001 from Brazil had carbohydrate concentrations 57.7–68.2% and 52.6–68.5% dw, respectively (Marinho-Soriano et al. 2006). Astorga-Espana and Mansilla (2014) reported that the highest carbohydrate content was 9.5% in *P. columbina*, while the lowest was 2.3% in *C. variegata* seaweeds collected from Chile. *Padina boryana* had carbohydrate content of 74.78% dw with 1.59% dw fucoidan yield, while *Turbinaria ornata* contained 70.30% dw carbohydrate with 105.19% dw alginate yield. 0.76% dw laminarin yield was determined in *Sargassum polycystum* obtained from Malaysia, having 71.65% dw carbohydrate content (Fauziee et al. 2021).

Seaweeds, particularly the red and brown algae, are a source of complex polysaccharides not present in land plants. The cell walls of seaweed, mainly brown algae, is strength, flexible, ionically balanced and not desiccate, due to the presence of fucoidan, alginate and laminarin (3:1:1) (Balboa et al. 2013; Vazquez-Rodriguez et al. 2021). Extractible polysaccharide content in seaweeds range from 6.5% to 38% dw (Angell et al. 2016a). Holdt and Kraan (2011) reported a wider polysaccharide content ranging from 4% (*Sargassum*—brown seaweed) to 76% (*Porphyra*—red seaweed) of dw. Seaweeds are considered the main source of polysaccharides, which may be sulfated (fucoidans, carrageenans, galactans, and agars) and non-sulfated (alginates, laminarin) (Hentati et al. 2020). Brown seaweeds are rich in polysaccharides of the laminarin, alginate, and fucoidan (sulphated fucose) while major component in cell wall of red seaweeds are carrageenans, agars, sulphated galactans, xylan and floridean. The basic polysaccharides present in green seaweeds are cellulose, mannan, sulfated rhamnan and water soluble ulvans (Hentati et al. 2020; Zheng et al. 2020). These complex polysaccharides are metabolized by the fecal microbiota and thus, stimulating the growth of certain beneficial populations (namely, *Bifidobacterium*, *Bacteroides*, *Lactobacillus*, etc) (Cassani et al. 2020).

Hydrocolloids are dissolve in water and give a viscous texture. This feature is used in food industry as gelling agents, stabilizers, thickeners and emulsifiers. Mainly, red and brown seaweeds are good sources of hydrocolloid structured alginate, agar-agar and carrageenan. These water-soluble carbohydrates are used to increase the viscosity of aqueous solutions, to form gels of varying degrees of firmness, to form water soluble films, and to stabilize some products, also used in the preparation of culture media (Hugh 2003).

Alginate occurs in the cell walls of seaweeds with the major cations being Na, Ca, Mg, and K together with a number of minor metal counterions and partly responsible for the flexibility of the seaweed. The gelling and chelating abilities of alginates and fucoidans present especially, in brown seaweeds (Phaeophyta). The charged, biocompatible, non-toxic and biodegradable and hydrophilic characteristics of these polysaccharides makes these seaweeds utilizable in food, agricultural and pharmaceutical industries as hydrogels, thickeners, stabilizers and additives (Cardozo et al. 2007; Hentati et al. 2018; Khajoei et al. 2021). High quality alginates can be obtained from brown seaweeds; *Ascophyllum*, *Durvillaea*, *Ecklonia*, *Laminaria*, *Lessonia*, *Macrocystis* and *Sargassum.* The latter genera have usually low alginate content compared to the others, and the quality of the alginate poor, although there are exceptions (Hugh 2003). Similarly, Mohammed et al. (2020) reported the same commercial brown algal species such as *Laminaria*, *Eclklonia*, and *Macrocystis*, and to a lesser extent *Sargassum*, that those contain sodium alginate estimated up to 40% dw. The alginate content of seaweeds depends on species, harvesting season, age of algae, extraction procedures (Holdt and Kraan 2011; Khajoei et al. 2021; Cebrian-Lloret et al. 2022). Alginate was extracted from four brown seaweeds (*Sargassum baccularia*, *Sargassum binderi*, *Sargassum siliquosum* and *Turbinaria conoides*) by hot (50 °C) and cold (room temperature) methods (Chee et al. 2011). The highest yield 49.9% of alginate was extracted from *S. siliquosum*, followed by *T. conoides* (41.4%) with using hot method. However, alginate extracted by the cold method gave higher molecular weight. The sensitivity of alginate against temperature, pH and the presence of other molecules are effective on their stability. Alginates, which are not found in any land plants, decrease the level of cholesterol, protects against carcinogens, have antihypertension effect, and have ability to bind divalent metallic ions that chelated heavy metals cannot be absorbed into the body tissue (Holdt and Kraan 2011).

Agar is a gelling hydrocolloid composed of agarose and agaropectin. These polysaccharides are the structural features of the cell wall of red seaweeds. Most of the raw material of *Gelidium* and *Gracilaria* (red seaweed) used for the extraction of agar. The agar content of *Gelidium latifolium* was 26% dw in March–April, while 42.5% in November (Mouradi-Givernaud et al. 1992). *Gracilaria* species produce agars with low quality due to high sulfate content however, the real agar can be obtained by alkali treatment. The untreated agar mass yield of *G. corticata* and *G. salicornia* were 26.2% and 15.8%, whereas treated yields were 27% and 21.9%, respectively. *G. edulis* specie show smallest mass yield 17.2% even alkali treatment (Vuai 2022). Meena et al. (2008) found out that the yield of agar decreases with the increase in alkali concentration for *Gracilaria* species (*G. edulis*, *G. crassa*, *G.*

foliifera, and *G. corticata*). The yield and quality of agar varies with species, environmental condition, seasonal variations, physiological factors and extraction methods. Sulphated residues strongly decrease the gel strength and methyl-ether groups modify the elasticity and gelling temperature (Mouradi-Givernaud et al. 1992). Gel forming abilities are depending on the amount and position of sulfate groups and the amount of other fractions. Each agar sources have different uses. The agar quality of *Gelidium* species is higher than *Gracilaria* considering gel strength. Only *Gelidium* can be used to prepare culture media due to low gelling temperature (34–36 °C) of resulting agar. While, *Gracilaria* can be used in confectionary with high sugar content. Because the presence of sugar (sucrose) increases the strength of the gel (Hugh 2003). Agar is also used in beverages, bakeries, meat, fish and dairy products.

Carrageenan is one of the main hydrocolloids in the food industry, after gelatin and starch. Red seaweeds such as *Eucheuma*, *Gigartina*, *Chondrus*, and *Hypnea* are the major sources of carrageenan (Ruiter and Rudolph 1997). These sulfated galactans obtained from certain species of marine red seaweed are used as gelling and thickening agents, and improve water holding capacities. Carrageenan remains stable over a wide pH range. The chemical structure of carrageenan varies in three forms, thus affecting their gelling properties and uses; (i) ι-iota; elastic gels (Ca salts) and clear gels, freeze/thaw stable, (ii) κ-kappa; strong and rigid gels (K salts), brittle forms (Ca salts), Slightly opaque gel, becomes clear with sugar addition, some synaeresis, (iii) λ-lambda; no gel formation, forms high viscosity solutions. Most carrageenan is extracted from *Kappaphycus alvarezii* (mainly kappa) and *Eucheuma denticulatum* (mainly iota). The original source of carrageenan is the red seaweed *Chondrus crispus* (mixture of kappa and iota) (Hugh 2003). De Souza et al. (2007) reported that λ-carrageenan (*Gigartina acicularis* and *G. pistillata*) have the best antioxidant potential compared to ι-carrageenan (*Eucheuma spinosum*) and κ-carrageenan (*Eucheuma cottonii*). Biological activities of carrageenan can be summarized as antioxidant, antitumor, immunomodulation, anti-inflammatory, antivirus and antibacterial activities. Also, carrageenan can be utilized as cryoprotectans due to ability to suppress the nucleation and the growth of ice crystals, thus protect the products from freeze-induced damage (Guo et al. 2022).

Complex and heterogeneous structured fucoidan (fucose-rich sulfated polysaccharides), are only present in the cell walls of brown seaweeds. They are not found in other parts of algae and not in land plants. Fucoidan is considered to protect seaweeds against the effects of dehydration when exposed at low tide (Holdt and Kraan 2011). Fucoidans are sulphated, α-1,4-bonded l-fucose-4-O-sulfate units, and sometimes acetylated homo- and heteropolysaccharides (Imbs et al. 2018). Fucoidans have complicated composition because, in addition to fucose, they contain xylose, galactose, mannose, and glucuronic acid (Bilan et al. 2007). The low molecular weight fucoidans are more biocompatible and active. The health effects of fucoidans are summarized as antithrombotic, anticoagulant, antitumor, antiviral, anti-inflammatory, protection against radiation and oxidative damage, osteoarthritis, gastric ulcer, Alzheimer and other neurodegenerative diseases (Balboa et al. 2013; Bauer et al. 2021). Fucoidan content was 5.2% w/w in *Cystoseira compressa*, while higher yields (up to 10%) for fucoidans isolated from *Agarum cribrosum* and *Saccharina japonica* (Hentati et al. 2018). Abdel-Latif et al. (2022) reported that fucoidan content of brown seaweeds ranged from 19% in *Ecklonia radiata* to 51.2% in *Cladosiphon* sp. The presence of fucoidans varies with the algal sources, weather of harvest, local climatic conditions, harvest site and extraction processes (Hentati et al. 2018).

Laminarin, having a molecular weight in the range of 1–10 kDa, can be isolated from brown seaweeds. The ratio of laminarin types (terminated by D-mannitol residues and terminated by D-glucose residues) which can vary depending on seaweed species, environmental factors and frond age, are related with their antioxidant activities (Sellimi et al. 2018). Although laminarin has less antioxidant activity than other compounds, it has a unique role in intestinal metabolism and immune system (Deville et al. 2007). Non-toxic, hydrophilic and biodegradable laminarins are reported to have antiapoptotic, anti-inflammatory, immunoregulatory, antitumor, anticoagulant and antioxidant activities (Moroney et al. 2015). Laminarin present in *Laminaria/Saccharina* and, to a lesser extent, in *Ascophyllum* and *Fucus* and *Undaria* (Holdt and Kraan 2011). The laminarin content in *Laminaria digitate* and *Ascophyllum nodosum* were 14.4 and 4.5 g/100 g (MacArtain et al. 2007). Sellimi et al. (2018) obtained 7.27% laminarin extraction yield in *Cystoseira barbata* collected from Tunisia in November. It was reported that *Laminaria* sp. accumulate carbohydrates mannitol and laminarin during summer and autumn to utilize as an energy source for new tissue growth during winter (Lafarga et al. 2020).

Sulfated polysaccharides, including fucoidan, sulfated galactan, carrageenan, agar and porphyran isolated from marine alga have been shown to exert radical scavenging activities due to the ease of abstraction of the anomeric hydrogen from the internal monosaccharide unit (Hu et al. 2001; Athukorala et al. 2006). The sulfated polysaccharides obtained from *Sargassum fulvellum* (brown seaweed) had promising DPPH radical scavenging activity and higher nitric oxide scavenging activity than commercial antioxidants such as α-tocopherol and BHA (Choi et al. 2007). Fucoidan and sodium alginate extracted and purified from *Cystoseira compressa* (brown seaweed) had shown the high antioxidant activity according to DPPH and FRAP assays (Hentati et al. 2018). Not only brown seaweed (*Sargassum stenophyllum*) but also red seaweed (*Spyridia hypnoides*, *Mastocarpus stellatus*, *Pyropia spiralis*) and green seaweed (*Ulva pertusa*, *Halimeda opuntia*, *Caulerpa* spp.) are reported to have antioxidant properties related to sulfate content by many researchers (Gomez-Ordonez et al. 2014; Li et al. 2018; Sudharsan et al. 2018; Tanna et al. 2018; Nazarudin et al. 2022; Urrea-Victoria et al. 2022).

Algal polysaccharides that are varying among species, are commonly used as texturing agents in food, cosmetic and pharmaceutical industries. Moreover, health claims of algal polysaccharides are summarized as antioxidant activity, anticoagulant and antithrombotic activities, anticancer and antitumor activities, neuroprotective activity and immunomodulatory property (Hentati et al. 2020; Fauziee et al. 2021). High content of polysaccharides and phlorotannins in seaweeds encourage the activity of beneficial microbiota in human and stimulate the synthesis of short-chain fatty acids (Raposo et al. 2016; Zheng et al. 2020).

The use of polysaccharides in food industry is wide. The addition of seaweed into bakery products allows the dough to absorb more water, less sticky properties and increase the firmness. The concentrations of seaweeds, most of which are brown and red seaweeds, as food ingredients in bread ranges of 0.5–8%; in noodles 3–30%; in pasta 5–20%; in cake 2.5–20%; in biscuit 5–60%; in cookies 3–9%; in extrudes maize 3.5% (Quitral et al. 2022).

Polysaccharides derived from seaweeds can be used as edible film-forming material precursors and play an essential role in bioplastics manufacturing due to being easily accessible, non-toxic, biocompatible and economic features. Seaweeds can be utilized in developing sustainable packaging, biodegradable plastics, active packaging, edible packaging and sachets (Carina et al. 2021). The main disadvantage of seaweed films is mechanically weakness or less water-resistant when they are the lone ingredient. Combination of two biodegradable polymers in a biocomposite structure makes more durable, better mechanical and thermal performance bioplastics suitable for food coatings or wrapping (Gade et al. 2013; Tavassoli-Kafrani et al. 2016; Lim et al. 2021; Mouritsen et al. 2021; Yong et al. 2022).

2.1.1 Fiber

Dietary fibers are generally regarded as beneficial for human gut health. Soluble fibers in seaweeds are; alginic acid, fucoidan and laminarin in brown seaweeds, carrageenan, agar and agarose in red seaweeds, as well as ulvan in green seaweeds (Huang et al. 2022). These highly fermented fibers are in contact with water and encourages the development of intestinal microbiota. They are non-digested in the upper gastrointestinal tract and are considered as dietary fiber (Lafarga et al. 2020). Moreover, soluble fibers help to decrease of cholesterol and glucose in blood. While insoluble fibers such as cellulose, mannans and xylan are capable of retaining water in its structural matrix, producing an increase in fecal mass that accelerates intestinal transit leading to laxative and intestinal regulating effect (Peñalver et al. 2020). Some of these fibers are not digested by human enzyme metabolism, although some are degradable by colonic microorganisms.

While soluble fibers have hypocholesterolemic and hypoglycemic effects, insoluble fibers are mainly associated with a decrease in digestive tract transit time. Seaweeds also have very high amounts of dietary fiber-higher than in terrestrial foodstuffs and are especially rich in soluble fiber (Ruperez and Saura-Calixto 2001; Gamero-Vega et al. 2020). Non-digestible polysaccharides including laminarin, fucoidan, and alginate favorably alter human gut microbiota composition and activity (Strain et al. 2020).

Dietary fiber contents of red seaweed species were summarized by Gamero-Vega et al. (2020) and reported that the minimum fiber content was 5.7% in *Gracilaria cervicornis*, while the maximum content was 64.7% in *Gracilaria changii* which were higher than those of fruits and vegetables are. Total dietary fiber content in red seaweed investigated through 8-month period averaged 28.17 ± 1.82% dw, with little change (Afonso et al. 2021). The same researcher reported that dietary fiber in red seaweeds varies between 5.7% and 64.7%, in green seaweeds from 29% to 66%, and in brown seaweeds from 10% to 64% in a dry weight basis. The proportion of soluble fiber to insoluble fiber is 3.91 in *Garateloupia turuturu* (red seaweed) (Denis et al. 2010), 1.09 in *Caulerpa lentillifera* (green seaweed) (Matanjun et al. 2009) and 1.74 and 1.30 in *Himanthalia elongate* and *Saccharina latissimi* (brown seaweed), respectively (Gómez-Ordóñez et al. 2010). The highest insoluble fiber (9.7% dw) and soluble fiber (40.3% dw) content were recorded in brown alga *D. antarctica* (Astorga-Espana and Mansilla 2014).

2.2 *Proteins*

Proteins are unique substances in biological systems and enzymatic reactions. The protein content of a food determines how that food is supplemental for human. The main factors in determining the protein quality are the amino acids profile, protein digestibility and functional properties. Proteins are present in intracellular components and cell walls of seaweeds. Some forms of proteins in seaweeds are bound to pigments (hycobiliproteins) and polysaccharides.

One of the major factors affecting protein content of seaweeds is species. Nine edible seaweed species were evaluated, and the protein content was the highest in red species (19.1–28.2 g/100 g dw), followed by the green seaweed *Ulva* spp. (20.5–23.3 g/100 g dw), the lowest content found in brown seaweeds (6.90–19.5 g/100 g dw) (Vieira et al. 2018). Similarly, protein fraction of green and red seaweeds is high (4–50% dw) compared with that of brown seaweeds (1–29% dw) (Fleurence 1999; Rioux et al. 2017; Ganesan et al. 2019). Although brown seaweed has a lower protein content than red and green seaweed, its rapid growth, distinctive compounds and possibility to co-extract other compounds make it economically attractive. Protein content of seaweeds ranged from 25.0 ± 0.6% in *C. variegata* (red algae) to 8.2 ± 1.3% in *D. antartica* (brown algae) collected from Santa Ana area, Chile (Astorga-Espana and Mansilla 2014). Similarly, Mansilla and Avila (2011) reported that the protein content of red algae *P. columbina* and *C. variegata* were higher (25.0% and 21.2% dw, respectively) than those for the brown algae *M. pyrifera* and *D. antarctica* (10.9% and 8.2% dw, respectively). Angell et al. (2014) found that the highest protein content (as determined by total amino acids) was 32.2% dw for green seaweeds, 28.7% dw for red seaweeds and 15.9% dw for brown seaweeds. The protein content of *Halymenia floresii* (red seaweed) (3.05% dw) was higher than (brown seaweed) (1.12% dw) (Polat and Ozogul 2008). Generally, Chlorophyceae and Rhodophyceae have higher protein contents compared to Pheophyceae (Hentati et al. 2020; Yong et al. 2022). However, Gómez-Ordóñez et al. (2010) found that protein content of brown seaweeds ranged from 10.9% to 25.7%, while being much higher for *Laminaria* (25.7%), which was cultured under sea natural conditions, followed by the red ones (15.5–21.3%). Even in the same genus, the protein content may differ. The highest protein concentration for red seaweed was reported for *Porphyra* genus reaching average 29.3 g/100 g, while red seaweed of *Gelidium* genus have the lowest content average of 11.6 g/100 g (Gamero-Vega et al. 2020). The protein content of *Gracilaria corticata* (red seaweed) was 42.38 g/100 g (Raja et al. 2020). Taxonomic groupings do not provide certainty in selecting species with a high quality of protein, however, Rawiwan et al. (2022) claimed that red seaweeds contain higher proteins than the green and brown species and summarized their protein contents ranging from 2.7% to 47.0% dw.

Protein content of a red seaweed, *A. vermiculophyllum*, ranged between 21.6% dw (September) and 6.2% dw (August) showing notorious variations along the seasons (Afonso et al. 2021). The highest protein content (21.9%) of *Palmaria palmate* (dulse) was found in the winter-spring period and the lowest (11.9%) in the summer-early autumn period (Galland-Irmouli et al. 1999). The protein content of *Macrocystis pyrifera* (brown seaweed) flour was 10.24% in summer, while 17.48% in winter (Mansilla and Avila 2011). Species and season interactively affected the content of total amino acids in crude protein in three red (*Mastocarpus stellatus*, *Palmaria palmata*, and *Porphyra* sp.), four brown (*Alaria esculenta*, *Laminaria digitata*, *Pelvetia canaliculata*, and *Saccharina latissima*), and two green (*Cladophora rupestris*. and *Ulva* sp.) seaweed species with value ranging from 67.2 for *Laminaria* in Spring to 90.2 g AA/16 g N for *Ulva* in Autumn. Seaweeds harvested in Spring compared to Autumn were richer in crude protein (Gaillard et al. 2018). Nutrient limitation during summer stratification of coastal waters lessens the macroalgal protein content. An inverse relationship between the protein content of seaweeds (higher in winter) and the polysaccharide content (higher in summer) was introduced basing on a sesonal study (Wells et al. 2017). This can also be attributed to the high nitrogen level in the upwelling in northern hemisphere (Rawiwan et al. 2022). Holdt and Kraan (2011) reported that protein contents of *Saccharina* and *Laminaria* were high during the period from February to May, and young parts considerably richer than old parts. Not only sampling season and species, but also sampling parts, salinity, solar radiation, geographic distribution, nutrient supply affect the protein concentration of seaweeds. The protein levels of *Gracilaria cervicornis* (14.29–22.70 g/100 g dw) and *Sargassum vulgare* (9.18–19.93 g/100 g dw), which were collected monthly from Northwest of Brazil, were negatively correlated with water temperature and salinity (Marinho-Soriano et al. 2006).

The essential amino acids (EAA), eight of which cannot be synthesized by animals, nor can they be replaced by other components are the building blocks. Since the amino acids present in eggs matches very closely the pattern the human body needs, the nutritional value of food proteins is compared with that of egg. Raja et al. (2022) reported that isoleucine and threonine content of seaweeds is similar to that of leguminous protein, while histidine, an essential amino acid, is equivalent to leguminous and egg proteins. Red seaweeds reported to have higher average proportion of total EAA, as well as lysine, followed by methionine, when compared to other seaweeds (Angell et al. 2016a). Essential amino acids of *P. palmata* (red seaweed) represented as much 26–50% of total amino acids which has amino acid profile close to that of egg protein (Galland-Irmouli et al. 1999) and leucine, valine and methionine are well represented in the essential amino acid fraction (Fleurence 1999). The same researcher reported that valine, leucine, and lysine are the main

EAA present in *Ulva persuta*, while histidine present at a similar level to egg proteins. Leucine, lysine, and valine were the most abundant EAA, while glutamic and aspartic acids were the predominant non-essential amino acids (nEAA) in the protein concentrates from three *Sargassum* (brown seaweed). The total EAA was in the range of 364–397 g/mg protein concentration (Wong and Cheung 2001a). Meanwhile, cysteine, methionine, histidine, tryptophan, and tyrosine were reported to have lower levels in brown seaweed (Holdt and Kraan 2011). The EAA in *G. chilensis* (red seaweed), *C. fragile* (green seaweed) and *M. pyrifera* (brown seaweed) were 59.7%, 49.6% and 38.9% of total protein, respectively with prevalence of glutamic acid in all seaweeds (Ortiz et al. 2009). Red seaweeds (*Hypnea charoides* and *Hypnea japonica*) and one green seaweed (*Ulva lactuca*) were rich in leucine, valine, and threonine. The essential amino acids leucine, lysine, threonine, valine present in relatively high levels than FAO/WHO/UNU (1985) requirements in *Porphyra columbina*. The limiting amino acid was tryptophan with a chemical score of 57%. Meanwhile, the main free amino acids in *Porphyra columbina* were alanine (0.47 g/100 g protein), aspartic acid (0.29 g/100 g protein) and glutamic acid (0.19 g/100 g protein) (Cian et al. 2013). These acids contribute such features in flavor development. Glutamic acid is the main component in the taste sensations of umami (Rioux et al. 2017). Brown seaweeds are presented the lowest contents of EAA but higher concentrations of free amino acids, especially histidine content 4.49–7.40 g/100 g protein in *S. polyschides* is considerable. The fraction of free amino acid was higher in the brown seaweed species (6.47–24.0 g/100 g protein), followed by the analyzed green seaweed *Ulva* spp. (8.39–14.0 g/100 g protein) and red seaweed species (3.40–13.1 g/100 g protein) (Vieira et al. 2018).

The limiting amino acids of the seaweeds' protein concentrations were the sulphur-containing amino acids (EAA score ranged from 0.24 to 0.79) such as cystine and methionine and lysine (EAA score ranged from 0.68 to 0.80) even though their amount are generally lower than vegetables and cereals (Galland-Irmouli et al. 1999; Holdt and Kraan 2011; Wong and Cheung 2001b; Peñalver et al. 2020). Conversely, lysine was found in high concentrations, especially in red (2.71–3.85% protein) and green (2.84–4.24% protein) seaweeds. Tryptophan, methionine, and leucine were the limiting EAAs in nine edible seaweed species of red, brown, and green seaweeds collected from Portuguese North-Central coast (Vieira et al. 2018). Phenylalanine and methionine were also reported as limiting amino acids in some seaweeds like *C. crispus*, *Gracilariasp.*, *O. pinnatifida*, *Porphyra* spp. (Ganesan et al. 2019).

Angell et al. (2016b), who made a literature study, reported that red seaweeds had a higher mean N-protein factor (5.10) compared to green and brown seaweeds (4.49 and 4.59, respectively). The reason behind higher N-protein factors of red seaweeds attributed to lower non-total amino acids nitrogen in green and brown seaweeds. Average factors of red, green, and brown seaweeds were 4.59, 5.13 and 5.38, respectively for 19 tropic marine algae (Lourenço et al. 2002). The conversion factor for crude biomass of *Chlorella vulgaris* (walled) was 6.35, whereas it was 5.96 based on direct protein extracts (Safi et al. 2013). The accurate N-to-protein conversion factor should be determined for each seaweed considering the distribution of N in protein and other non-protein N compounds. Angell et al. (2016b) propose that the overall median value of 5 be used as the most accurate universal seaweed nitrogen-to-protein conversion factor after meta-analyses of 103 macroalgae. Meanwhile, many researchers determined protein content by using 6.25 N-to protein conversion factor and found in the range of 33.4% and 86.3% dw in protein concentrations of seaweeds (Wong and Cheung 2001a, b; Marinho-Soriano et al. 2006; Kandasamy et al. 2012; Cian et al. 2013; Astorga-Espana and Mansilla 2014). Misurcova et al. (2010) determined nitrogen content in blue-green algae (*Spirulina pacifica*, *S. platensis*), green algae (*Chlorella pyrenoidosa*), red algae (*Palmaria palmata*, *Porphyra tenera*), and brown algae (*Eisenia bicyclis*, *Hizikia fusiformis*, *Laminaria japonica*, *Undaria pinnatifida*), while using 6.25 N-to protein conversion factor on presumption that proteins contain 16% of nitrogen. All of the freshwater algal samples showed about 200% higher content of total nitrogen than the marine samples (the lowest contents of total nitrogen were found in brown seaweed).

Seaweed protein concentrations with high in vitro digestibility, have the ability of emulsifying and foaming properties and water holding capacities. An excellent amino acid profile alone is not sufficient in case of poor bioavailability. Bioavailability is defined as the combination of bioactivity (uptake into tissues, metabolism, physiological effects) and bioaccessibility (release from food matrix, transformation through digestive system) (Wells et al. 2017). Amino acid bioaccessibility of both raw and 30 min-boiled *Palmaria palmata* (red algae) were higher than from an equivalent dry weight of wheat, rice, or corn flour in a simulated in vitro gastrointestinal digestion model (Maehre et al. 2016). Wong and Cheung (2001b) determined that in vitro protein digestibility of green seaweed (*Ulva lactuca*) was slightly lower than red seaweeds (*Hypnea charoides* and *Hypnea japonica*). *Porphyra*, followed by *Palmaria* and the green seaweeds (*Ulva* and *Cladophora*) can be considered as relevant sources of protein basing on their amino acid content and degradability (Gaillard et al. 2018). While being reported to have lower protein content, brown seaweeds (*Eisenia bicyclis*, *Hizikia fusiformis*, *Undaria pinnatifida*) had poorest protein digestibility compared to blue-green, green and red species using pepsin with different in vitro digestion methodologies (Misurcova et al. 2010). The in vitro digestibility of algal proteins can differ according to the species and seasonal variations in antinutritional factor content such as phenolic molecules, polysaccharides, or dietary fiber (Fleurence 1999; Wong and Cheung 2000). In order to improve the accessibility, digestibility and nutritional value, this can be overcome by an enzymatic treatment.

Oil absorption and water holding capacities of protein are important factors in the formulations of emulsified foods such as sausages and mayonnaise and improve the viscous nature of food formulations. Moreover, protein as a good foaming agent can adsorb rapidly at the air-water interface during bubbling and the ability to undergo rapid conformational changes and re-arrangement at the interface (Kandasamy et al. 2012).

Lectins, carbohydrate-binging proteins, present in macroalgal species such as *Ulva* sp., *Eucheuma* spp. and *Gracilaria* sp. (Holdt and Kraan 2011). The interest in seaweed lectins is notably due to their molecular structure and glycosidic bonding different from those found in plants or other life forms. The low molecular weight, monomeric structure, thermostability (due to the presence of disulfide bridges), and high affinity for glycoproteins instead of monosaccharides make them unique structures. Lectins have substantial therapeutic potential, exhibiting antiviral, anti-inflammatory, and antitumor effects (Rioux et al. 2017; Fontenelle et al. 2018). Lectins from *Eucheuma serra* and *Galaxaura marginata* have the capacity to inhibit the growth of marine *Vibrio vulnificus* (Liao et al. 2003).

Taurine which contains a sulfonated acid group rather than the carboxylic acid moiety, is not regarded as amino acid and not used as building blocks. However, taurine plays role in the formation of bile to form emulsion with lipids. Thus, lessen the cholesterol levels in the blood. The taurine content is about 400 mg/100 g dw in *Laminaria saccharina* (konbu) and *Porphyra tenera* (Asakusanori), which is also the case in lobster, crab, shellfish, and squid (Murata and Nakazoe 2001). Taurine ranges 0.30–0.42 g/100 g protein in red seaweed (*C. crispus*, *Gracilaria* sp., *O. pinnatifida*), 0.37–0.86 in brown seaweed (*Porphyra* spp., *A. nodosum*, *F. spiralis*, *S. polyschides*, *U. pinnatifida*) and 0.33–0.39 in green seaweed (*Ulva* spp.) (Vieira et al. 2018).

The protein content of seaweeds which is similar to that of some terrestrial plants, vegetables, seeds, grains, eggs, makes them sources of biologically good proteins for human nutrition (Fleurence 2004; Ortiz et al. 2009). Especially, red seaweeds seem to be a potential source of food proteins with the highest average proportion of total EAA such as lysine, compared to brown and green seaweeds. Protein bioavailability of seaweed is promising in terms of nutritional value and food quality. Moreover, the functional properties of seaweed proteins and bioactive peptides are valuable components for food formulations.

2.3 *Lipids*

It is a well-known fact that the intake of polyunsaturated fatty acids (PUFA) and balanced intake of omega-3 to omega-6 PUFA, are crucial in nutrition, prevention of diseases and having numerous beneficial influence on human health. Seaweeds have the ability de novo synthesis of essential n-6 fatty acids such as linoleic acid (18:2n-6) (LA) and the n-3 PUFA α-linolenic acid (18:3n-3) (ALA) which cannot be synthesized by most heterotrophic organisms. Long chain-PUFA such as eicosapentaenoic (EPA, 20:5n-3), arachidonic (ARA, 20:4n-6) and DHA (22:6n-3) have significant role in biochemical and physiological changes in the body and these can be synthesized by seaweeds (Schmid et al. 2018). This synthesis is very important because EPA and DHA can be synthesized from ALA, even very slowly and in small amounts, and AA from LA by a series of metabolic steps with the help of desaturase and elongase enzymes. Humans only convert $<5\%$ of ALA to EPA and $<0.05\%$ of ALA to DHA (Narayan et al. 2006; Yerlikaya et al. 2022). Therefore, intake of PUFA in daily diet is unavoidable. Long chain PUFA is not the fish itself, but seaweeds and phytoplankton is the major dietary source of these precious compounds.

Generally, lipid content of seaweeds is higher in brown algae (Ochrophyta) than green algae (Phaeophyceae), following red algae (Rhodophyta). Total lipid concentrations of 61 seaweed species varied in the range of 0.6 and 7.8 in % of dw, in which the highest concentrations being in the brown, then the green, and with the red seaweeds recording the lowest average concentrations (Schmid et al. 2018). Total of 22 species belonging to Chlorophyta (6), Phaeophyta (5) and Rhodophyta (11) were analyzed for their chemical composition and lipid contents were determined as 1.83–3.03%, 1.23–2.50% and 0.57–1.97% dw, respectively (Kumar et al. 2011). The highest lipid content ever reported belongs to *Ulva lactuca* seaweed collected in July in Tunisia with 7.87% dw (Yaich et al. 2011). Lipid content of seaweeds collected from Chile varied from 0.3 ± 0.1% to 4.6 ± 0.5% which were dominated by polyunsaturated (PUFA) and saturated fatty acids (SFA) (Astorga-Espana and Mansilla 2014). Total lipid content of seven sea weed species from the North Sea and two from tropical seas were in the range of 0.7–4.5% dw (van Ginneken et al. 2011). Lipid content of *Halimeda opuntia* (green seaweed) was 1.60% dw (Nazarudin et al. 2022). Red seaweeds had lipid contents ranging from 0.2% to 12.9% dw (Rawiwan et al. 2022) in which the highest 12.9% was obtained from *Palmaria palmata* (wild type) by Tibbetts et al. (2016). Seaweeds have low levels of lipids thus makes them lower calorie foods.

Afonso et al. (2021) reported that different collection sites did not influence total lipids content of red seaweed (*A. vermiculophyllum*), however there was a statistically significant difference between the sampling months in which the highest lipid contents were found in April and December (2.0% dw) while the lowest were in January and August (1.2% dw). Total PUFAs of *A. vermiculophyllum* ranged from 25.82% in October to 57.57% in March. Aquatic species living in cold waters contain higher amounts of PUFA, including EPA/DHA because a decrease in environmental temperature causes an increase in PUFA content (Narayan et al. 2006).

The predominant PUFA of seaweeds (included two species in the division Ochrophyta (*M. pyrifera* and *D. antarctica*) and two in the division Rhodophyta (*P. columbina* and *C. variegata*)) of Chile coast was composed of ARA and EPA, while DHA was not detected in any of the samples. The highest values of EPA (37.59 ± 2.67%) and ARA (23.57 ± 2.36%) were found in *D. antartica* and *M. pyrifera*, respectively, both of which had 1.8% crude lipid content (Astorga-Espana and Mansilla 2014). Most common fatty acids were palmitic (16:0), oleic (OLE, 18:1 n-9), ALA, ARA and EPA in 16 seaweed species (nine Phaeophyceae, five Rhodophyta and two Chlorophyta) collected from Irish west coast. *Palmaria palmate*, *P. diocia* and *C. virgatum* was identified as most promising species as a source of EPA amongst all species investigated in both June and November samplings and this species assumes that sufficient biomass can be provided sustainably (Schmid et al. 2014). Similarly, van Ginneken et al. (2011) reported that the highest relative concentration of PUFA was observed for EPA accounting for 59% of the total fatty acid content in *Palmaria palmate.* ARA was the most abundant PUFA in *A. vermiculophyllum* and sampled ranging from 14.83% to 52.88% total fatty acids throughout the year, while EPA and DHA were only detected in low amounts in some months (Afonso et al. 2021). These findings can be summarized as green seaweeds have in greater quantity of LA, ALA, palmitic acid, oleic acid and characteristically some DHA content. Apart from containing high oleic and palmitic acids, red seaweeds consist of higher ARA and EPA compared to brown seaweed (Wells et al. 2017; Peñalver et al. 2020). Red and brown seaweeds were generally richer in LC-PUFA, while high levels of SFA were observed in green species (van Ginneken et al. 2011; Schmid et al. 2014).

Both n-3 and n-6 fatty acids are competing for the same enzymes; desaturase and elongase (Yerlikaya et al. 2013; Tallima and el Ridi 2018). δ-6-desaturase enzyme has greater affinity to ALA than to LA. However, high LA intake interferes with the desaturation and elongation. Transformation of ALA to EPA needs energy supply derived from LA (Simopoulos 2008; Gomez Candela et al. 2011). In this context, the balance between n-3 and n-6 is an important issue. The ratio of n-3/n-6 is an index to assess the impact of lipid consumption on nutritional value and healthy metabolic profile of cardiovascular and obesity-related issues. The World Health Organization (WHO) recommends the ratio of n-3/n-6 in lipids ranging from 5:1 to 10:1 (WHO 2003). Seaweeds generally have favorable ratio of n-3 PUFA/n-6 PUFA of around 1 (Schmid et al. 2014, 2018). Algae obtained from cold water (Canada) are generally higher in n3/n6 fatty acids ratio and higher degree of unsaturation than warm water (China) samples (Colombo et al. 2006). Both brown and red seaweeds are reported to have balanced sources of n-3 and n-6 acids (Holdt and Kraan 2011; Balboa et al. 2013). Ganesan et al. (2019) reported that the ratio between n-3 and n-6 can be affected by the amount of nitrogen loaded in the cultivation site of seaweed. The presence of high nitrogen on the site had an attractive combination of n-6/n-3 (0.3) and 18:2n-6/18:3n-3 (0.5) ratios in *Ulva* species.

EPA and DHA are considered the two most important PUFA of marine lipids. Many national and international organizations recommend levels for EPA + DHA; World Health Organization (WHO) recommends 200–1000 mg/day. Fatty acid profile of 11 Chlorophyta, 17 Phaeophyceae (Ochrophyta) and 33 Rhodophyta species were determined by Schmid et al. (2018). EPA + DHA ranged between 30.3 mg/10 g dw in *Scytosiphon lomentaria* and 1.1 mg/10 g dw in *Carpoglossum confluens*. Although EPA has been detected in many studies, the presence of DHA is not very common in seaweed. *Sargassum natans* obtained from tropical seas reported to contain DHA 13% of total fatty acids (van Ginneken et al. 2011). Wells et al. (2017) reported that the bioaccessibility of DHA and EPA derived from seaweed is well quantified for humans, ranging from ~50% to 100% depending on the matrix.

The main source of lipids in marine seaweeds are phospholipids and glycolipids, followed by neutral lipids. Lipids of seaweed contains fatty acids, tocopherols and sterols. There are also carotenoids, the unsaponifiable fraction of seaweed, such as β-carotene, lutein, and violaxanthin found in red and green algae, and fucoxanthin in brown algae (Holdt and Kraan 2011). Sterols such as fucosterol, clionasterol, isofucosterol and cholesterol are the main nutritional constituents of marine seaweeds and have biological properties; anticancer, antioxidant, antiobesity, antitumoral, antiviral, protection against cardiovascular diseases (Kendel et al. 2015; Hentati et al. 2020). Fucosterol exists in many seaweeds, especially red and brown macroalgae. However, little is known about the in vivo effects of this sterol when seaweeds are consumed by human (Pereira et al. 2016; Wells et al. 2017).

The proportion of essential fatty acids in seaweed is higher than in terrestrial plants due to their ability to synthesize LC-PUFA (Peñalver et al. 2020). Moreover, seaweeds may provide cleaner and more concentrated sources of PUFAs than fish oils. Lipid content and fatty acid profile of seaweeds may vary spatially and temporally, depending upon species, mineral contents of the growth medium, stress conditions (Colombo et al. 2006; Schmid et al. 2014; Afonso et al. 2021).

2.4 *Polyphenols*

Polyphenols are secondary metabolites of marine and terrestrial plants and ability to have radical scavenging activities. Phenolic compounds are the structural components of cell wall and defense the seaweed from environmental stress and absorb ultraviolet radiation. Living organisms exposed to extreme conditions, develop various systems to be more durable. Similarly, some microalgae generate free radicals and oxidizing agents. Brown seaweeds are reported to obtain higher amounts and more active antioxidants than green and red seaweeds (Balboa et al. 2013). Brown seaweeds are mainly characterized by significant levels of phlorotannins, whereas red and green seaweeds are rich in flavonoids, phenolic acids and bromophenols (Hentati et al. 2020). Total phenolic content of red and green seaweeds ranged from 8.48 to 8.99 mg/dw (Wong and Cheung 2001b).

The only significant class of polyphenolic compounds is phlorotannins or algal polyphenolic compounds present in Phaeophyta. These analogous to tannins from terrestrial plants, constitute up 2–30% of algal dry weight (Kaushalya and Gunathilake 2022). Sharifian et al. (2019) determined that the phlorotannins content of ten different brown seaweed was ranged between 1.09 and 13.21 mg phloroglucinol/g extract in ethyl-acetate fraction.

Phlorotannins (polymers of phloroglucinol) are found only in brown seaweeds and protect algae from stress conditions and herbivores. Seaweeds form free radicals and other strong oxidizing agents to resist adverse environmental conditions such as light and high oxygen concentrations. These substances have ability to lessen lipid oxidation and not found in terrestrial plants (Balboa et al. 2013; Maqsood et al. 2013; Wang et al. 2017). Many interconnected aromatic rings and hydroxyl groups (acting as hydrogen donors) lead to improve the free radical scavenging capacity of phlorotannins (Cassani et al. 2020). The redox potential of phlorotannins absorb and neutralize free radicals quenching singlet and triple oxygen or decomposing peroxide. Antioxidant activity of seaweed derives from phenolic compounds and result of specific scavenging of radicals formed during peroxidation, scavenging of oxygen-containing compounds, or metal-chelating ability (Li et al. 2011). However, higher molecular weights demonstrated decreased activity.

Bogolitsyn et al. (2019) reported that high radical scavenging activity (776 ± 36 mg of ascorbic acid/g extract) was observed for phlorotannin subfractions with average molecular weights from 8 to 18 kDa. Kang et al. (2003) reported that *Eisenia bicyclis*, *Ecklonia stolonifera*, *Ecklonia cava*, *Ecklonia kurome* and *Hizikia fusiformis* had excellent free-radical scavenging activity; better than that of well-known antioxidants such as catechin, tocopherol, t-butylhydroxy anisole (BHA) and t-butylhydroxytoluene (BHT). Similarly, Shibata et al. (2007) claimed that phlorotannins such as bieckol and dieckol obtained from the same marine brown seaweed could be potent free radical scavengers, even stronger than ascorbic acid or α-tocopherol. The maximum antioxidant activity of phlorotannin was recorded as 89.47% which is higher than the standard, ascorbic acid (88.62%) by Surendhiran et al. (2019).

Phorotannins are classified according to the criteria of inter-phloroglucinol linkages into four primary types: (i) fucols (with only phenyl linkages), (ii) phlorethols (phlorotannins with only arylether linkages) and (iii) fucophlorethols (with arylether and phenyl linkages) (iv) eckols (with dibenzodioxin linkages) (Li et al. 2011; Martínez and Castaneda 2013). Phloroglucinol, fucofuroeckol-A, eckol, phlorofucofuroeckol-A, 8,8′-bieckol, dioxinodehydroeckol, 7-phloroeckol, 6,6′-bieckol, 8,4″-dieckol, triphloroethol A and dieckol, compounds present in brown seaweeds were reported to have antimicrobial activities against food-borne pathogenic bacteria (Eom et al. 2012). Their antimicrobial effect on *Staphylcoccus aureus* (resistant to several antibiotics, most notably penicillin), *Salmonella*, *Escherichia coli*, *Vibrio parahaemolyticus*, *Camphylobacter jejuni*, *Klebsiella pneumonia*, *Shigella flexneri*, *Listeria monocytogenes* were proven by previous studies. Any remarkable difference in susceptibility to phlorotannins between Gram-positive and Gram-negative bacteria was not detected (Nagayama et al. 2002; Choi et al. 2010; Lee et al. 2008, 2010a, 2014). However, Lopes et al. (2012) found that gram-positive microorganisms are more susceptible to the phlorotannins, *Staphylococcus epidermidis* being the most susceptible species. This finding was supported by Jiménez et al. (2010) that acetone extract of *A. nodosum* showed considerable antibacterial activity being more effective against gram-positive bacteria (*Micrococcus luteus*, *Staphylococcus aureus*) than against gram-negative bacteria (*Escherichia coli*, *Enterococcus aerogenes*). Morphological and physical changes were reported in the cell membrane of the microorganism that were exposed to phlorotannins (Cassani et al. 2020). The main target of phlorotannins on bacteria was cell membrane, ATP, protein and DNA (Surendhiran et al. 2019). Eight interconnected rings present in phlorotannins makes seaweed enlarge antimicrobial activity and stronger agent against microorganism than those extracted polyphenols from terrestrial plants. Phenolic aromatic rings and hydroxyl groups of phlorotannins bind to bacterial proteins by hydrophobic and hydrogen bonding interactions, respectively (Abu-Ghannam and Rajauria 2013; Shannon and Abu-Ghannam 2016). Additionally, anti-viral activity of phlorotannins were summarized by Shrestha et al. (2021).

Phlorotannins, bioactive therapeutic agents belonging only brown seaweed, are metabolized and absorbed, predominantly in the large intestine and to a lesser extent in small intestine (Corona et al. 2016). However, their bioavailability is low due to low water solubility. Some encapsulation studies with phlorotannins were performed in order to extend shelf-life, control the delivery, mask the taste, and reduce the damaging effects through the gastrointestinal tract. Sodium alginate and poly (ethylene oxide) blended nanofibers encapsulated with phlorotannins via electrospinning process was used to perform antimicrobial, antioxidant effects on chicken meat (Surendhiran et al. 2019). The authors reported that chicken meat was preserved from Salmonella contaminations, moreover sensorial quality was enhanced. A slow and kinetic release was observed in phlorotannins encapsulated polyvinylpyrrolidone in simulated gastrointestinal fluid (Bai et al. 2020). The vitality of phlorotannins, isolated from *Sargassum ilicifolium* and encapsulated in the chitosan-tripolyphosphate carrier, were studied by Kaushalya and Gunathilake (2022).

Polyphenol oxidase, also called as phenoloxidase and tyrosinase, catalyses the production of quinones to melanin in the presence of oxygen. Thus, acceptability of crustaceans decreases due to discoloration and lipid oxidation. The researchers are looking for natural preservatives to replace synthetic substances in order to extend shelf-life of shrimps. The presence of phlorotannins suppressed the synthesis of tyrosinase and melanine. Phlorotannins originated from *S. tenerimum* and *P. yezoensis* inhibits the polyphenol oxidase activity and melanosis formation in refrigerated white shrimps (Li et al. 2017; Sharifian et al. 2019). Tyrosinase inhibitory activity of 43 marine seaweeds were determined by Cha et al. (2011) and found that aqueous extract from *Schizymenia dubyi* had 90.75% and *Endarachne binghamiae* had 81.26% tyrosinase inhibitory activity at 20 °C and 70 °C, respectively. Tyrosinase inhibitor activity of acetone extract of *Ascophyllum nodosum* presented 65.6% inhibition of tyrosinase activity at the IC50 value of 0.1 mg/ml (Jiménez et al. 2010).

Significant amount of protein is derived as by-products during fish processing, especially filleting. Prooxidants such as heme proteins and enzymes, make these by-products very susceptible to lipid oxidation and quality loss. Moreover, polyunsaturated fatty acids, that are prone to oxidation, makes these proteins difficult to utilize. The proteins can be selectively isolated from complex raw materials by precipitation at their isoelectric point. This goal was achieved by seaweed. Brown seaweeds are rich in phlorotannins which interact with proteins and encourage their precipitation (Stern et al. 1995). Seaweed provides new colors, higher protein concentrations of isolates and enables clean label in food industry (Abdollahi et al. 2020).

Phlorotannins known for their bioactivities and multiple health benefits such as prebiotic, antiphotocarcinogenic, anticancer, antiallergic, antihypertension effects, protection against oxidative stress, vascular diseases and neurodegenerative diseases were studied by many researchers (Kang et al. 2003; Athukorala et al. 2006; Hwang et al. 2006; Sugiura et al. 2006; Lee et al. 2010b; Li et al. 2011; Barbosa et al. 2020; Zheng et al. 2020; Shrestha et al. 2021) and most of which were attributed to the radical scavenging compounds of seaweeds. However, ensuring the high content of pholorotannins are challenging due to (i) intrinsic factors; age, reproductive status, plant size, (ii) environmental factors; salinity, nutrient and light availability, UV radiation, (iii) pre-processing; grinding, drying methods, (iv) extraction conditions; extraction method, time, temperature, solvent, particle size, (v) storage; temperature, time, oxygen, light, (vi) subsequent incorporation into foods, (vii) passage through the gastrointestinal tract; pH, enzymes, other nutrients (Cassani et al. 2020).

Rajan et al. (2021) reported that approximately 150 phlorotannins have been reported from brown seaweeds so far. Amongst, dieckol is primarily used as a food ingredient, also in pharmacological sectors and exhibit a wide range of therapeutic properties. As the other phlorotannins, the functional aspects of dieckol was technically validated and proven to exhibit anti-bacterial, anti-inflammatory, antiviral, anti-cancer, anti-oxidant, anti-hyperlipidemic, and other activities. 11 Hydrogen bond donor and 18 hydrogen bond acceptor prove how bioactive dieckol ($C_{36}H_{22}O_{18}$) is.

2.5 *Vitamins*

Vitamins, even being essential for metabolic functions, cannot be synthesized sufficiently and must be obtained from diet. Seaweeds consist of water-soluble vitamins B-group (particularly B1, B12), vitamin C, niacin, folic acid, pantothenic acid and riboflavin, as well as fat-soluble vitamins A (derived from β-carotene) and vitamin E (α-tocopherol, active form) (Gupta and Abu-Ghannam 2011; Wells et al. 2017; Hentati et al. 2020; Choudhary et al. 2021). Ortiz et al. (2009) reported that 100 g of seaweed provides more than the daily requirement of vitamins A, B2 and B12 and two thirds of the vitamin C requirements. Moreover, red and brown algae are rich sources of folic acid and folate derivatives (Lordan et al. 2011).

The antioxidant activity of seaweeds derives from vitamin C or ascorbic acid, vitamin E or α-tocopherol, as well as carotenoids, polyphenols, phycobiliproteins (de Quirós et al. 2010). Seaweeds are considered as rich sources of vitamin C range from 500 to 3000 ppm (Rajapakse and Kim 2011). The average vitamin C content in 92 seaweed species within 132 data entries was 0.773 mg/g dw with a 90% of 2.06 mg/g and no significant differences were found among seaweed species (Nielsen et al. 2021). Afonso et al. (2021) found that the ascorbic acid content of *A. vermiculophyllum* were 0.18 mg/g dw with a maximum of 0.28 mg/g dw in March. Vitamin C was 0.348 mg/g in *D. antarctica* and 0.011 mg/g *C. variegata* obtained from Chile between October and December (Astorga-Espana and Mansilla 2014). Ascorbic acid values of seaweed samples of each species; *Eucheuma cottonii* (Rhodophyta), *Caulerpa lentillifera* (Chlorophyta) and *Sargassum polycystum* (Phaeophyta) was in the range of 34.5–35.3 g/100 g and not statistically different from each other (Matanjun et al. 2009). Environmental and seasonal differences as well as drying methods may affect the presence of vitamin C in seaweeds (Chan et al. 1997; Hernandez-Carmona et al. 2009; Wells et al. 2017). Freeze-dried seaweed had the highest content of total amino acids, total polyunsaturated fatty acids, and total vitamin C when compared with sun-dried and oven-dried seaweed. While oven-dried seaweed had the greatest nutrient losses (Chan et al. 1997). The concentration of vitamin C in *Eisenia arborea* (brown algae) was 22.8 in October and 41.5 in June, while vitamin E was 6.2 in May and 9.6 in September. Vitamin D3, B2,

B1 and vitamin A determinations were also varied throughout the year (Hernandez-Carmona et al. 2009).

Many prokaryotes capable of synthesizing vitamin B12 interact with seaweeds and promotes vitamin levels in macroalgae (Cotas et al. 2020). Vitamin B12 (cobalamin) content in seaweeds ranges between 0.06 and 0.786 g/100 g dw for green seaweeds, 0.0961 and 1.34 g/100 g dw for red seaweeds and from 0.0164 to 0.0431 g/100 g dw for brown seaweeds (Hentati et al. 2020). *Neopyropia yezoensis* (formerly *Porphyra yezoensis*, red seaweed) contained 51.49 μg B12/100 g dw and 80% of the vitamin was bioaccessible in mammalian rat intestinal (Watanabe et al. 2000).

Vitamin A is derived from provitamin A carotenoid, especially β-carotene which has high provitamin A activity due to its two its β-ionone ring structures (Grande et al. 2016). The proposal of mean requirements of vit A is 300 μg RE/day for men and 270 μg RE/day for women (EFSA 2015). The content of β-carotene (provitamin A) in *Codium fragile* (green algae) (197.9 μg/g dw) and *Gracilaria chilensis* (red algae) (113.7 μg/g dw) exceed those found in fresh carrot (88.33 mg/g) (Ortiz et al. 2009). The same researchers found that total α-tocopherols in *C. fragile*, *G. chilensis* and *M. pyrifera* were 453.5, 86.5 and 1327.7 μg/g lipids. Despite their low lipid contents, the presence of vitamin E contributes to the stability of the PUFA as they prevent the formation of free radicals. α-tocopherol, which has the highest vit E activity among tocopherols and tocotrienols, is common in organelles and cell membranes (Faramarzi 2012). Tocopherol content is higher in brown algae than in red and green algae. α-tocopherol content of *Sargassum polycystum* (brown algae) was 11.29 mg/100 g dw, followed by *Caulerpa lentillifera* (green algae) 8.41 mg/100 g dw and to a lesser extend in *Eucheuma cottonii* (red algae) 5.85 mg/100 g dw (Matanjun et al. 2009).

Seaweeds are the sources of enzyme co-factor vitamins some of which have antioxidant activity. Metabolic pathway of higher organisms cannot synthesize vitamins. Thus, seaweeds are alternative vitamin sources for vegetarian consumers as well as for all consumers.

2.6 Minerals and Trace Elements

Seaweeds are reported to contain high concentrations of mineral, trace and ultra-trace elements, even higher than terrestrial plants due to their strong bioadsortive and bioaccumulative capacities (Rohani-Ghadikolaei et al. 2012; Circuncisao et al. 2018; Munoz and Díaz 2020). Seaweeds are excellent sources of macro minerals (Ca, P, Na, Mg, K) and trace elements (I, Zn, Mn). The presence of Se, Zn, and Cu in seaweeds participates in the structure of some antioxidant enzymes and contributes to their activities (Dhargalkar and Pereira 2005; Polat and Ozogul 2008; Mohamed et al. 2012).

The accumulation of minerals in seaweeds are related with their concentration in the surrounding water and the uptake capacity of seaweeds. Ash content in

Gracilaria cervicornis were in the range of 8.07% and 13.1% dw, while higher contents 13.0–30.3% dw were found in *Sargassum vulgare* (Marinho-Soriano et al. 2006). Wong and Cheung (2000) determine ash content in *Hypnea japonica, Hypnea charoides* (red seaweeds) and *Ulva lactuca* (green seaweed) in the range of 21.3–22.8% dw. The highest ash content of *Macrocystis pyrifera* (brown seaweed) was recorded 37.18% dw in spring, while the lowest score was 29.88 in winter (Mansilla and Avila 2011). Three other brown seaweeds *Turbinaria ornata*, *Sargassum polycystum*, and *Padina boryana* had ash contents 5.55%, 5.69% and 4.09% dw, respectively (Fauziee et al. 2021). The ash content of *Eucheuma cottonii* (Rhodophyta), *Caulerpa lentillifera* (Chlorophyta) and *Sargassum polycystum* (Phaeophyta) were 37.15–46.19% dw (Matanjun et al. 2009). The content of minerals in seaweeds differ depending on species, environmental features such as salinity, temperature, pH, season or harvesting time. The effect of geographic region or zone on the mineral and trace element concentrations in seaweed is related with different physical and oceanographic features such as depth, tidal range, hydrographic conditions (Marinho-Soriano et al. 2006; Circuncisao et al. 2018). Even under the same conditions, the families of seaweed, their genera and species affect the amount of minerals (Astorga-Espana et al. 2015). These variations in ash content can be attributed to biological uptake of minerals depending on the selective adsorption property of polysaccharides in their cell wall (Bocanegra et al. 2009). Algal polysaccharides are in association with the cations and absorb inorganic salt from seawater. Fast-growing seaweeds accumulates less minerals. Each seaweed has different affinities for different trace elements (Akcali and Kucuksezgin 2011).

Brown, red and green seaweeds are reported to have high concentrations of Na, K, Ca, Mg and P, meanwhile Fe was the trace element with high concentrations. The brown seaweed had the highest K and Ca concentrations while the green seaweed had the highest mean Fe, Mg, Cu and Mn. Also it was reported that the serving size of 8 g/dw of green seaweed contributes to the intake of Fe, Mg and Mn more than recommended daily intake (RDI) (Astorga-Espana et al. 2015). On the other hand, Ca and Zn content determined in 100 g dw of red seaweed were lower than RDI (Afonso et al. 2021). The Ca in seaweeds (4–7%) is in the form of calcium phosphate which is more bioavailable than other forms (Rajapakse and Kim 2011). Bocanegra et al. (2009) reported that Ca and P contents of seaweeds are higher than those of apples, oranges, carrots and potatoes. Moreover, the high Ca/P ratio in algae (3:5) can compensate for the deficit of Ca in several foods, such as cereals and meats. Ca/Mg ratio is also crucial with respect to the deficiencies in Mg intake result in excessive accumulation of Ca. Thus, kidney stones and appearance of arthritis may occur. While Mg content is higher in green seaweeds, Ca is higher in brown seaweeds (Circuncisao et al. 2018).

Fe, Ca and Zn contents in *A. vermiculophyllum* (red seaweed) averaged 96.5, 114.63, and 1.63 mg/100 g dw, respectively. Calcium reached a maximum in October 2017, but Zn and Fe were higher in January 2018 (Afonso et al. 2021). The content of Fe in seaweeds is quite high compared to other trace elements, both because of need for growth and biomagnifying from the contaminated surrounding environment (Storelli et al. 2001). The Fe content is more substantial in green seaweeds, while red and brown seaweeds are rich in Mn and I, respectively (Circuncisao et al. 2018).

Potassium was the most abundant element followed by Na in all species examined by Astorga-Espana and Mansilla (2014) and Schiener et al. (2015). The most concentrated mineral, K in *Eisenia arborea* ranged from 1.980 mg/100 g dw in September to 7.946 mg/100 g dw in June. Sodium presented the second highest concentration (1.987 in July and 3.101 mg/100 g dw in December). Significant monthly differences were also found in Mg, Ca, Fe, Zn and Cu during the year for all the seven minerals analyzed (Hernandez-Carmona et al. 2009). The accumulation of mainly K and Na ions doubled in concentration from summer to winter months in *Laminaria digitata*, *Laminaria hyperborea*, *Saccharina latissima* and *Alaria esculenta* over a 14-month period (Schiener et al. 2015). High ratios of Na/K are related to higher incidence of hypertension. Na/K ratios are high in green seaweeds with the exception of *Ulva clathrate*, while the lowest Na/K ratio present in red seaweed *P. palmata* (Munoz and Díaz 2020). Conversely, Na/K ratio of red seaweed (*Eucheuma cottonii* 9.42 μg/g) was higher than those brown (*Sargassum polycystum* 7.66 μg/g) and green (*Caulerpa lentillifera* 4.78 μg/g) seaweeds (Matanjun et al. 2009). Circuncisao et al. (2018) summarized the Na/K ratio as ranging 0.9–1 for green seaweeds, 0.1–1.8 for red seaweeds, and 0.3–1.5 for brown seaweeds.

Iodine is an essential element in ensuring thyroid function. Iodine deficiency is tried to be compensated by iodized salt consumption. Iodine content in seaweed is higher than that in the land plants and even exceeds its dietary minimum requirement (150 mg/day). Brown seaweed iodine content is reported to be higher than red and green seaweeds (Rajapakse and Kim 2011; Cardoso et al. 2014). According to the ranking indicated by Matanjun et al. (2009), the highest I content present in red, brown and green seaweeds, respectively. Dawczynski et al. (2007) determined macro, trace and ultra-trace elements in 34 edible dried seaweeds and found that a daily intake of 5 g algae contribution of the essential elements to the diet is low, with the exception of I.

Brown seaweeds can be used as salt replacers due to being a major Na and K contributors. *Ulva* spp. (green seaweed) are major sources of Mg, Fe and Mn, while *Chondrus crispus* and *Phymatolithon calcareum* (red seaweeds) can be highlighted with their Ca richness. Moreover, Cu, Zn, Mo, and Se were reported to occur in various seaweeds (Circuncisao et al. 2018). Good correlations between carbohydrate, polysaccharides and dietary fiber in terms of bioavailability of metals were exhibited, while negative correlation was observed between proteins, lipids and minerals (Demarco et al. 2022). The valuable mineral content of seaweeds allows them to be a candidate for nutritional and functional food applications.

Unfortunately, seaweeds accumulate undesirable minerals as well as desirable minerals from the environment, which can endanger their food safety. Seaweeds can only obtain metal ions from the dissolved phase depending on the nature of the suspended particles. Therefore, trace metal concentration in seaweeds are the indicators of the bioavailable levels of a metals (Zibikowski et al. 2006). The mineral content of seaweeds is correlated with the sediment, reflecting the contamination of water column through environment (Akcali and Kucuksezgin 2011). The presence of toxic metals such as As, Cd, Cu, Hg and Pb in varying concentrations can be higher than terrestrial plants (Circuncisao et al. 2018). Seaweeds can accumulate elevated concentrations of metals, such as Zn, Cd and Cu (Jarvis and Bielmyer-Fraser 2015). Chen et al. (2018) revealed the distribution of 10 metals and metalloids in 295 dried brown and red seaweeds and found that the elements sequenced in descending order by mean values: Al > Mn > As > Cu > Cr > Ni > Cd > Se > Pb > Hg. The levels of Cd, Cu, Mn and Ni in red seaweeds were reported to be higher than those in brown seaweeds s (*Laminaria* and *Undaria*). EU (Commission Regulation (EC) No 629/2008) claims that seaweed accumulates cadmium naturally. Thus, higher levels of Cd can be present in food supplements consisting exclusively or mainly of dried seaweed or of products derived from seaweed. The accumulating of heavy metals varies among seaweeds due to their morphology such as larger surface area or their fast growing patterns. Metal levels of seaweeds is less in spring period due to environmental factors and algal growth rates. It was reported that metal concentrations decrease during growing periods and increase in winter (Phillips 1994). The heavy metal contamination risk of seaweeds can be reduced by harvesting in an unpolluted aquaculture system (Raja et al. 2022).

2.7 Pigments

Pigment composition is the main factor in the classification of seaweeds as Chlorophyta (green seaweed), Phaeophyta (brown seaweed) and Rhodophyta (red seaweed). Chlorophyll a and b are responsible for green pigmentation in Chlorophyta. The dominance of fucoxanthin and xanthophyll are responsible for the brown color in Phaeophyceae. Phycoerythrin and phycocyanin, while masking the pigments Chlorophyll a and β-carotene, yield the red color in Rhodophyta (Gupta and Abu-Ghannam 2011; Baweja et al. 2016). Seaweed pigments are composed of many phytochemicals such as chlorophyll, carotenoids such as carotenes and xanthophylls (fucoxanthin, violaxanthin, antheraxanthin, zeaxanthin, lutein, neoxanthin) and phycobiliprotein.

Chlorophyll is non-covalently attached to protein to form pigment-protein complexes and has many biological functions. These green lipid-soluble pigments convert light into biological energy through photosynthesis. Chlorophyll a and b of *Halimeda opuntia* (green seaweed) collected from Malaysia were 148.73 mg/g and 290.83 mg/g, respectively (Nazarudin et al. 2022). The same researcher also determined chlorophyll contents of *Halimeda macroloba*, *Ulva intestinalis* and *Sargassum ilicifolium* and found that the highest chlorophyll a (313.09 μg/g dw) and b (292.52 μg/g dw) records in green seaweed *U. intestinalis* (Nazarudin et al. 2020). Chlorophyll contents of red (*Eucheuma denticulatum*, *Gracilaria tikvahiae*, and *Kappaphycus striatum*), green (*Caulerpa lentillifera*) and brown seaweed (*Padina pavonica*) species were investigated and found that the highest chlorophyll content 7.5 μg/g dw belonged to *P. pavonica* (Othman et al. 2018).

Yellow, green and orange carotenoid pigments are widely distributed in nature. Carotenoids containing beta-ionone ring (α-carotene, β-carotene, γ-carotene, and β-cryptoxanthin) can be converted by the body to vitamin A (provitamin A carotenoids) and act as antioxidants. Moreover, carotenoids have the ability of photoprotection and used as colorant (Aryee et al. 2018). These natural colorants offer safe and variety of hues. Pigments present in Phaeophyceae are constituted by chlorophylls "a" and "c" masked by carotenes and xanthophylls. The brown color derives from fucoxanthin (a xanthophyll) (Hentati et al. 2018). Similarly, Haugan and Liaaen-Jensen (1994) mentioned that all-trans fucoxanthin was the major carotenoid (primary xanthophyll) and β,β-carotene the only carotene present in brown seaweed species. β-carotene, lutein, violaxanthin, neoxanthin and zeaxanthin are present in green seaweed, while red seaweed contains mainly α-and β-carotene, lutein and zeaxanthin. Total carotenoid content of green seaweeds *Halimeda opuntia*, *Halimeda macroloba* and *Caulerpa lentillifera* were 115.57 μg/g, 117.36 μg/g and 63.5 μg/g dw, respectively (Othman et al. 2018; Nazarudin et al. 2020, 2022). Carotenoid content of *P. pavonica* (brown seaweed) was 100.9 μg/g dw (Othman et al. 2018).

The xanthophyll fucoxanthin is found in microalgae, brown macroalgae and diatomeae. It is the most characteristic pigment of brown seaweeds. Fucoxanthin, having antioxidative, protective and photosynthetic functions, present in the seaweeds ranging from 1–6 mg/100 g in *Laminaria saccharina* to 29.7 mg/100 g in *Undaria pinnatifida* (de Quirós et al. 2010). Fucoxanthin has role in prevention of cerebrovascular, bone, ocular and chronic diseases, such as cancer, obesity, diabetes and liver disease (Sivri and Yur 2020).

Phycobiliproteins, water soluble pigment protein complex, covalently bound via cysteine amino acids to pigmented phycobilins (phycoerythrin and phycocyanin). Phycobiliproteins, collect light and, through fluorescence resonance energy transfer, which is implied in the photosynthesis (Rioux et al. 2017). These pigments capture the blue light present in greater marine depths and mask chlorophyll and carotenoids. Thus, red color of algae, varying from red to violet occurs (Gamero-Vega et al. 2020). Light and UV exposure impact the composition of pigments.

Red seaweeds—exhibit phycoerythrin, phycocyanin, chlorophyll a and carotene (Afonso et al. 2021). A particular protein called phycoerythrin, red pigments, can be used as natural dye in food industry and has high economic value. Phycoerythrin, posing antiviral, antioxidant, anti-inflammatory, antidiabetic, antitumor, antihypertensive, immunosuppressive, and neuroprotective activities, unfortunately, released as wastewater during carrageenan processing (Uju et al. 2020). Instability of phycoerythrin to heat and pH variabilities limits its utilization. Its extraction is performed by freeze thawing, ultrasonication (Fleurence 1999; Uju et al. 2020).

Besides being natural food dye, pigments derived from seaweed are bioactive compounds such as antioxidants thus impeding oxidative damage and inhibition of cancer, immune modulatory, antidiabetic, antiangiogenic, antiviral, anti-inflammatory and neuroprotective properties (Zhu et al. 2017; Hentati et al. 2020; Nazarudin et al. 2022).

3 Conclusion

Seaweeds are summarized as part of the diet, exotic component of the menu or consumed fresh as salad components; nori or purple laver (Porphyra spp.), Aonori or green laver (Monostroma spp. and Enteromorpha spp.), kombu (dried seaweed) or haidai (Laminaria japonica), wakame, quandai-cai (Undaria pinnatifida), hiziki (Hizikia fusiforme), mozuku (Cladosiphon okamuranus), sea grapes or green caviar (Caulerpa lentillifera), dulse (Palmaria palmata), Irish moss or carrageenan moss (Chondrus crispus), winged kelp (Alaria esculenta), ogo, ogonori or sea moss (Gracilaria spp.) (Hugh 2003). The consumption of seaweeds with high nutritional value is increasing all over the world in addition to the Far East. It may take time to get used to consume seaweeds directly in our diets, but it can be recommended to add farina or flour of these sea vegetables into various foods. Govaertz and Olsen (2022) revealed in the light of consumer behavior that there is a positive relationship between awareness of more sustainable and healthy food source and intention to consume seaweed.

Some countries such as Germany, Austria, Switzerland prepare lists of seaweeds applicable for foods including food supplements and fortified foods. However, some others such as Norway, Iceland do not agree national lists of algae items authorized as food and food supplements. In the European Union, foods classified as novel are subject to the pre-market authorization requirements of the novel food regulation (EU) 2015/2283. A list was elaborated by European Commission's science and knowledge service about algae as food and food supplements in Europe. In that Technical Report of Joint Research Centre, 54 algae species from which 31 are not currently included in the novel food catalogue as food, food supplements or both were presented. Those identified but not present in any official list were excluded. The complexity of taxonomic understanding of algae and identification of some problematic genera and species were clarified by molecular techniques. This is also important in terms of safety, nutritional quality, and functional attributes of the commercial application of algal products as food (Araujo and Peteiro 2021).

Seaweeds with the variety of biological active properties that are excellent natural antioxidant and antimicrobial agent successfully employed in the development of novel and improved foods. It provides a natural contribution to the color, strong flavor, certain vitamins such as A, C and E, essential trace elements and protein content of the products in which it is used. Seaweeds are utilized as non-synthetic food colorant and antioxidant food additives in food applications. Natural plant-based ingredients also serve vegetarian and vegan consumers alternative foods. Extending shelf-life of foods without using synthetic additives is another contribution of seaweed to the food industry. In addition to being a part of cuisine and food supplement, seaweed plays an important role in reaching the ideal food, albeit indirectly, with its agricultural practices. Apart from extending shelf life of the products, seaweed improves preventive medicine with their well-balanced unique components. There is still need to screen bioactivity and bioavailability of seaweed active substances.

One of the biggest problems that humanity will face in the future is food security due to growth of population, climate change and its effect on environment, limited arable land for agriculture and fresh water sources, declining livestock protein and live fish stocks. Serving important alternative dietary, nutritive and functional substances makes seaweeds the foods of the future. Seaweed production requires low nutrient demand, no irrigation or arable land. Its easy cultivation derived from rapid growth and simple technology requirement. Thus, production of bioactive compounds can be controlled and standardized. Seaweed utilization promotes more sustainable food production.

References

Abdel-Latif HMR, Dawood MAO, Alagawany M, Faggio C, Nowosad J, Kucharczyk D (2022) Health benefits and potential applications of fucoidan (FCD) extracted from brown seaweeds in aquaculture: an updated review. Fish Shellfish Immunol 122:115–130. https://doi.org/10.1016/j.fsi.2022.01.039

Abdollahi M, Olofsson E, Zhang J, Alminger M, Undeland I (2020) Minimizing lipid oxidation during pH-shift processing of fish by-products by cross-processing with lingonberry press cake, shrimp shells or brown seaweed. Food Chem 327:127078. https://doi.org/10.1016/j.foodchem.2020.127078

Abu-Ghannam N, Rajauria G (2013) Antimicrobial activity of compounds isolated from algae. In: Functional ingredients from algae for foods and nutraceuticals. Woodhead Publishing Ltd./Elsevier, Sawston, pp 287–306

Afonso C, Correia AP, Freitas MV, Baptista T, Neves M, Mouga T (2021) Seasonal changes in the nutritional composition of Agarophyton vermiculophyllum (Rhodophyta, Gracilariales) from the Center of Portugal. Foods 10:1145. https://doi.org/10.3390/foods10051145

Akcali I, Kucuksezgin F (2011) A biomonitoring study: heavy metals in macroalgae from eastern Aegean coastal areas. Mar Pollut Bull 62:637–645. https://doi.org/10.1016/j.marpolbul.2010.12.021

Angell AR, Mata L, de Nys R, Paul NA (2014) Variation in amino acid content and its relationship to nitrogen content and growth rate in *Ulva ohnoi* (Chlorophyta). J Phycol 50:216–226. https://doi.org/10.1111/jpy.12154

Angell AR, Angell SF, de Nys R, Paul NA (2016a) Seaweed as a protein source for mono-gastric livestock. Trends Food Sci Technol 54:74–84. https://doi.org/10.1016/j.tifs.2016.05.014

Angell AR, Mata L, de Nys R, Paul NA (2016b) The protein content of seaweeds: a universal nitrogen-to-protein conversion factor of five. J Appl Phycol 28(1):511–524. https://doi.org/10.1007/s10811-015-0650-1

Araujo R, Peteiro C (2021) Algae as food and food supplements in Europe. JRC technical report. Publications Office of the European Union, Luxembourg. https://doi.org/10.2760/049515

Aryee AN, Agyei D, Akanbi TO (2018) Recovery and utilization of seaweed pigments in food processing. Curr Opin Food Sci 19:113–119. https://doi.org/10.1016/j.cofs.2018.03.013

Astorga-Espana MS, Mansilla A (2014) Sub-Antarctic macroalgae: opportunities for gastronomic tourism and local fisheries in the Region of Magallanes and Chilean Antarctic Territory. J Appl Phycol 26:973–978. https://doi.org/10.1007/s10811-013-0141-1

Astorga-Espana MS, Galdon BR, Rodriguez EMR, Romero CD (2015) Mineral and trace element concentrations in seaweeds from the sub-Antarctic ecoregion of Magallanes (Chile). J Food Compos Anal 39:69–76. https://doi.org/10.1016/j.jfca.2014.11.010

Athukorala Y, Kim KN, Jeon YJ (2006) Antiproliferative and antioxidant properties of an enzymatic hydrolysate from brown alga, Ecklonia cava. Food Chem Toxicol 44:1065–1074. https://doi.org/10.1016/j.fct.2006.01.011

Bai Y, Sun Y, Gu Y, Zheng J, Yu C, Qi H (2020) Preparation, characterization and antioxidant activities of kelp phlorotannin nanoparticles. Molecules 25(19):4550. https://doi.org/10.3390/molecules25194550

Balboa EM, Conde E, Moure A, Falque E, Dominguez H (2013) In vitro antioxidant properties of crude extracts and compounds from brown algae. Food Chem 138:1764–1785. https://doi.org/10.1016/j.foodchem.2012.11.026

Barbosa M, Valentao P, Ferreres F, Gil-Izquierdo A, Andrade PB (2020) In vitro multifunctionality of phlorotannin extracts from edible Fucus species on targets underpinning neurodegeneration. Food Chem 333:127456. https://doi.org/10.1016/j.foodchem.2020.127456

Bauer S, Jin W, Zhang F, Linhardt RJ (2021) The application of seaweed polysaccharides and their derived products with potential for the treatment of Alzheimer's disease. Mar Drugs 19(2):89. https://doi.org/10.3390/md19020089

Baweja P, Kumar S, Sahoo D, Levine I (2016) Biology of seaweeds. In: Seaweed in health and disease prevention. Elsevier, Amsterdam, pp 41–106

Bilan MI, Zakharova AN, Grachev AA, Shashkov AS, Nifantiev NE, Usov AI (2007) Polysaccharides of algae: 60. Fucoidan from the pacific brown alga *Analipus japonicus* (Harv.) winne (Ectocarpales, Scytosiphonaceae). Russ J Bioorg Chem 33(1):38–46. https://doi.org/10.1134/S1068162007010049

Bocanegra A, Bastida S, Benedí J, Ródenas S, Sánchez-Muniz FJ (2009) Characteristics and nutritional and cardiovascular-health properties of seaweeds. J Med Food 12:236–258. https://doi.org/10.1089/jmf.2008.0151

Bogolitsyn K, Druzhinina A, Kaplitsin P, Ovchinnikov D, Parshina A, Kuznetsova M (2019) Relationship between radical scavenging activity and polymolecular properties of brown algae polyphenols. Chem Pap 73:2377–2385. https://doi.org/10.1007/s11696-019-00760-7

Cardoso S, Carvalho L, Silva P, Rodrigues M, Pereira O, Pereira L (2014) Bioproducts from seaweeds: a review with special focus on the Iberian peninsula. Curr Org Chem 18:896–917. https://doi.org/10.2174/138527281807140515154116

Cardozo KHM, Guaratini T, Barros MP, Falcao VR, Tonon AP, Lopes NP, Campos S, Torres MA, Souza AO, Colepicolo P, Pinto E (2007) Review: metabolites from algae with economical impact. Comp Biochem Physiol Toxicol Pharmacol 146:60–78. https://doi.org/10.1016/j.cbpc.2006.05.007

Carina D, Sharma S, Jaiswal AK, Jaiswal S (2021) Seaweeds polysaccharides in active food packaging: a review of recent progress. Trends Food Sci Technol 110:559–572. https://doi.org/10.1016/j.tifs.2021.02.022

Cassani L, Gomez-Zavaglia A, Jimenez-Lopez C, Lourenço-Lopes C, Prieto MA, Simal-Gandara J (2020) Seaweed-based natural ingredients: stability of phlorotannins during extraction, storage, passage through the gastrointestinal tract, and potential incorporation into functional foods. Food Res Int 137:109676. https://doi.org/10.1016/j.foodres.2020.109676

Cebrian-Lloret V, Metz M, Martinez-Abad A, Knutsen SH, Balance S, Lopez-Rubio A, Martinez-Sanz M (2022) Valorization of alginate-extracted seaweed biomass for the development of cellulose-based packaging films. Algal Res 61:102576. https://doi.org/10.1016/j.algal.2021.102576

Cha SH, Ko SC, Kim D, Jeon YJ (2011) Screening of marine algae for potential tyrosinase inhibitor: those inhibitors reduced tyrosinase activity and melanin synthesis in zebrafish. J Dermatol 38(4):354–363. https://doi.org/10.1111/j.1346-8138.2010.00983.x

Chan JCC, Cheung PCK, Ang PO (1997) Comparative studies on the effect of three drying methods on the nutritional composition of seaweed *Sargassum hemiphyllum* (Turn) C Ag. J Agric Food Chem 45:3056–3059. https://doi.org/10.1021/jf9701749

Chee SY, Wong PK, Wong CL (2011) Extraction and characterisation of alginate from brown seaweeds (Fucales, Phaeophyceae) collected from Port Dickson, Peninsular Malaysia. J Appl Phycol 23(2):191–196. https://doi.org/10.1007/s10811-010-9533-7

Chen Q, Pan X, Huang B, Han J (2018) Distribution of metals and metalloids in dried seaweeds and health risk to population in southeastern China. Sci Rep 8:3578. https://doi.org/10.1038/s41598-018-21732-z

Choi DS, Athukorala Y, Jeon YJ, Senevirathne M, Cho KR, Kim SH (2007) Antioxidant activity of sulfated polysaccharides isolated from *Sargassum fulvellum*. Prev Nutr Food Sci 12:65–73. https://doi.org/10.3746/jfn.2007.12.2.065

Choi JG, Kang OH, Brice OO, Lee YS, Chae HS, Oh YC, Sohn DH, Park H, Choi HG, Kim SG (2010) Antibacterial activity of Ecklonia cava against methicillin-resistant *Staphylococcus aureus* and *Salmonella* spp. Foodborne Pathog Dis 7:435–441. https://doi.org/10.1089/fpd.2009.0434

Choudhary B, Chauhan OP, Mishra A (2021) Edible seaweeds: a potential novel source of bioactive metabolites and nutraceuticals with human health benefits. Front Mar Sci 8:740054. https://doi.org/10.3389/fmars.2021.740054

Cian RE, Fajardo MA, Alaiz M, Vioque J, Gonzalez RJ, Drago SR (2013) Chemical composition, nutritional and antioxidant properties of the red edible seaweed *Porphyra columbina*. Int J Food Sci Nutr 65(3):299–305. https://doi.org/10.3109/09637486.2013.854746

Circuncisao AR, Catarino MD, Cardoso SM, Silva AMS (2018) Minerals from macroalgae origin: health benefits and risks for consumers. Mar Drugs 16:400. https://doi.org/10.3390/md16110400

Colombo ML, Rise P, Giavarini F, De Angelis L, Galli C, Bolis CL (2006) Marine macroalgae as sources of polyunsaturated fatty acids. Plant Foods Hum Nutr 61:67–72. https://doi.org/10.1007/s11130-006-0015-7

Corona G, Ji Y, Anegboonlap P, Hotchkiss S, Gill C, Yaqoob P et al (2016) Gastrointestinal modifications and bioavailability of brown seaweed phlorotannins and effects on inflammatory markers. Br J Nutr 115:1240–1253. https://doi.org/10.1017/S0007114516000210

Cotas J, Leandro A, Pacheco D, Gonçalves AM, Pereira L (2020) A comprehensive review of the nutraceutical and therapeutic applications of red seaweeds (Rhodophyta). Life 10:19. https://doi.org/10.3390/life10030019

Dawczynski C, Schäfer U, Leiterer M, Jahreis G (2007) Nutritional and toxicological importance of macro, trace, and ultra-trace elements in algae food products. J Agric Food Chem 55:10470–10475. https://doi.org/10.1021/jf0721500

de Quirós ARB, Frecha-Ferreiro S, Vidal-Pérez AM, López-Hernández J (2010) Antioxidant compounds in edible brown seaweeds. Eur Food Res Technol 231:495–498. https://doi.org/10.1007/s00217-010-1295-6

De Souza MCR, Marques CT, Dore CMG, Da Silva FRF, Rocha HAO, Leite EL (2007) Antioxidant activities of sulfated polysaccharides from brown and red seaweeds. J Appl Phycol 19:153–160. https://doi.org/10.1007/s10811-006-9121-z

Demarco M, de Moraes JO, Matos AP, Derner RB, Neves FF, Tribuzi G (2022) Digestibility, bioaccessibility and bioactivity of compounds from algae. Trends Food Sci Technol 121:114–128. https://doi.org/10.1016/j.tifs.2022.02.004

Denis C, Morançais M, Li M, Deniaud E, Gaudin P, Wielgosz-Collin G, Barnathan G, Jaouen P, Fleurence J (2010) Study of the chemical composition of edible red macroalgae *Grateloupia turuturu* from Brittany (France). Food Chem 119:913–917. https://doi.org/10.1016/j.foodchem.2009.07.047

Deville C, Gharbi M, Dandrifosse G, Peulen O (2007) Study on the effects of laminarin, a polysaccharide from seaweed, on gut characteristics. J Sci Food Agric 87:1717–1725. https://doi.org/10.1002/jsfa.2901

Dhargalkar VK, Pereira N (2005) Seaweed: promising plant of the millennium. Sci Cult 71:60–66

EFSA (2015) Scientific opinion on dietary reference values for vitamin A. EFSA Journal 2015, 13(3):4028

Eom S, Kim Y, Kim S (2012) Antimicrobial effect of phlorotannins from marine brown algae. Food Chem Toxicol 50:3251–3255. https://doi.org/10.1016/j.fct.2012.06.028

EU (2008) Commission Regulation (EC) No 629/2008 of 2 July 2008 amending Regulation (EC) No 1881/2006 setting maximum levels for certain contaminants in foodstuffs. Off J Eur Union 3:7

FAO/WHO/UNU (United Nations University) (1985) Energy and protein requirements. Report of a joint FAO/WHO/UNU expert consultation. Technical report series no. 724. WHO, Geneva

Faramarzi M (2012) The influences of vitamins C and E on the growth factors and carcass composition of common carp. Glob Vet 8(5):498–502

Fauziee NAM, Chang LS, Mustapha WAW, Nor ARM, Lim SJ (2021) Functional polysaccharides of fucoidan, laminaran and alginate from Malaysian brown seaweeds (*Sargassum polycystum*, *Turbinaria ornate* and *Padina boryana*). Int J Biol Macromol 167:1135–1145. https://doi.org/10.1016/j.ijbiomac.2020.11.067

Fleurence J (1999) Seaweed proteins: biochemical, nutritional aspects and potential uses. Trends Food Sci Technol 10:25–28. https://doi.org/10.1016/S0924-2244(99)00015-1

Fleurence J (2004) Seaweed proteins. In: Yada R (ed) Proteins in food processing. Woodhead Publishing Limited, Cambridge, pp 197–213

Fontenelle TPC, Lima GC, Mesquita JX, Lopes JLS, de Brito TV, Vieira Junior FD, Sale AB, Aragao KS, Souza MHLP, Barbosa ALR, Freitas ALP (2018) Lectin obtained from the red seaweed Bryothamnion triquetrum: secondary structure and anti-inflammatory activity in mice. Int J Biol Macromol 112:1122–1130. https://doi.org/10.1016/j.ijbiomac.2018.02.058

Gade R, Tulasi MS, Bhai V (2013) Seaweeds: a novel biomaterial. Int J Pharm Pharm Sci 5(2):40–44. https://doi.org/10.1017/CBO9781107415324.004

Gaillard C, Bhatti HS, Garrido MN, Lind V, Roleda MY, Weisbjerg MR (2018) Amino acid profiles of nine seaweed species and their in situ degradability in dairy cows. Anim Feed Sci Technol 241:210–222

Galland-Irmouli AV, Fleurence J, Lamghari R, Lucon M, Rouxel C, Barbaroux O, Bronowicki JP, Vuillaume C, Gueant JL (1999) Nutritional value of proteins from edible seaweed *Palmaria palmate* (Dulse). J Nutr Biochem 10(6):353–359. https://doi.org/10.1016/s0955-2863(99)00014-5

Gamero-Vega G, Palacios-Palacios M, Quitral V (2020) Nutritional composition and bioactive compounds of red seaweed: a mini-review. J Food Nutr Res 8(8):431–440. https://doi.org/10.12691/jfnr-8-8-7

Ganesan AR, Tiwari U, Rajauria G (2019) Seaweed nutraceuticals and their therapeutic role in disease prevention. Food Sci Human Wellness 8:252–263. https://doi.org/10.1016/j.fshw.2019.08.001

Gomez Candela C, Bermejo Lopez LM, Loria Kohen V (2011) Importance of balanced omega6/omega3 ratio for the maintenance of health. Nutritional recommendations. Nutr Hosp 26(2):323–329. https://doi.org/10.1590/S0212-16112011000200013

Gómez-Ordóñez E, Jiménez-Escrig A, Rupérez P (2010) Dietary fibre and physicochemical properties of several edible seaweeds from the northwestern Spanish coast. Food Res Int 43:2289–2294. https://doi.org/10.1016/j.foodres.2010.08.005

Gomez-Ordonez E, Jiménez-Escrig A, Rupérez P (2014) Bioactivity of sulfated polysaccharides from the edible red seaweed *Mastocarpus stellatus*. Bioact Carbohydr Diet Fibre 3:29–40. https://doi.org/10.1016/j.bcdf.2014.01.002

Govaertz F, Olsen SO (2022) Exploration of seaweed consumption in Norway using the norm activation model: the moderator role of food innovativeness. Food Qual Prefer 99:104511. https://doi.org/10.1016/j.foodqual.2021.104511

Guo Z, Wei Y, Zhang Y, Xu Y, Zheng L, Zhu B, Yao Z (2022) Carrageenan oligosaccharides: a comprehensive review of preparation, isolation, purification, structure, biological activities and applications. Algal Res 61:102593. https://doi.org/10.1016/j.algal.2021.102593

Grande F, Giuntini EB, Lajolo FM, de Menezes EW (2016) How do calculation method and food data source affect estimates of vitamin A content in foods and dietary intake? J Food Compos Anal 46:60–69. https://doi.org/10.1016/j.jfca.2015.11.006
Gupta S, Abu-Ghannam N (2011) Bioactive potential and possible health effects of edible brown seaweeds. Trends Food Sci Technol 22(6):315–326. https://doi.org/10.1016/j.tifs.2011.03.011
Haugan JA, Liaaen-Jensen S (1994) Algal carotenoids. 54. Carotenoids of brown-algae (Phaeophyceae). Biochem Syst Ecol 22:31–41. https://doi.org/10.1016/0305-1978(94)90112-0
Hentati F, Delattre C, Ursu AV, Desbrières J, Le Cerf D, Gardarin C, Abdelkafi S, Michaud P, Pierre G (2018) Structural characterization and antioxidant activity of water-soluble polysaccharides from the Tunisian brown seaweed *Cystoseira compressa*. Carbohydr Polym 198:589–600. https://doi.org/10.1016/j.carbpol.2018.06.098
Hentati F, Tounsi L, Djomdi D, Pierre G, Delattre C, Ursu AV, Fendri I, Abdelkafi S, Michaud P (2020) Bioactive polysaccharides from seaweeds. Molecules 25(14):3152. https://doi.org/10.3390/molecules25143152
Hernandez-Carmona G, Carrillo-Dominguez S, Arvizu-Higuera DL, Rodriguez-Montesinos YE, Murillo-Alvarez JI, Munoz-Ochoa M, Castillo-Dominguez RM (2009) Monthly variation in the chemical composition of *Eisenia arborea* JE Areschoug. J Appl Phycol 21:607–616. https://doi.org/10.1007/s10811-009-9454-5
Holdt SL, Kraan S (2011) Bioactive compounds in seaweed: functional food applications and legislation. J Appl Phycol 23:543–597. https://doi.org/10.1007/s10811-010-9632-5
Hu JF, Geng MY, Zhang JT, Jiang HD (2001) An in vitro study of the structure-activity relations of sulfated polysaccharide from brown algae to its antioxidant effect. J Asian Nat Prod Rep 73:353–358. https://doi.org/10.1080/10286020108040376
Huang W, Tan H, Nie S (2022) Beneficial effects of seaweed-derived dietary fiber: highlights of the sulfated polysaccharides. Food Chem 373:131608. https://doi.org/10.1016/j.foodchem.2021.131608
Hugh D (2003) A guide to the seaweed industry. Food and Agriculture Organization of the United Nations, Rome, 105 pp
Hwang H, Chen T, Nines RG, Shin HC, Stoner GD (2006) Photochemoprevention of UVB-induced skin carcinogenesis in SKH-1 mice by brown algae polyphenols. Int J Cancer 119:2742–2749. https://doi.org/10.1002/ijc.22147
Imbs TI, Silchenko AS, Fedoreev SA, Isakov VV, Ermakova SP, Zvyagintseva TN (2018) Fucoidanase inhibitory activity of phlorotannins from brown algae. Algal Res 32:54–59. https://doi.org/10.1016/j.algal.2018.03.009
Jarvis TA, Bielmyer-Fraser GK (2015) Accumulation and effects of metal mixtures in two seaweed species. Comp Biochem Physiol Part C Toxicol Pharmacol 171:28–33. https://doi.org/10.1016/j.cbpc.2015.03.005
Jiménez J, Ó'Connell SH, Lyons H, Bradley B, Hall M (2010) Antioxidant, antimicrobial, and tyrosinase inhibition activities of acetone extract of *Ascophyllum nodosum*. Chem Pap 64(4):434–442. https://doi.org/10.2478/s11696-010-0024-8
Kandasamy G, Karuppiah SK, Rao PVS (2012) Salt- and pH-induced functional changes in protein concentrate of edible green seaweed Enteromorpha species. Fish Sci 78:169–176. https://doi.org/10.1007/s12562-011-0423-y
Kang K, Park Y, Hye JH, Seong HK, Jeong GL, Shin HC (2003) Antioxidative properties of brown algae polyphenolics and their perspectives as chemopreventive agents against vascular risk factors. Arch Pharm Res 26:286–293. https://doi.org/10.1007/BF02976957
Kaushalya KGD, Gunathilake KDPP (2022) Encapsulation of phlorotannins from edible brown seaweed in chitosan: effect of fortification on bioactivity and stability in functional foods. Food Chem 377:132012. https://doi.org/10.1016/j.foodchem.2021.132012
Kendel M, Wielgosz-Collin G, Bertrand S, Roussakis C, Bourgougnon N, Bedoux G (2015) Lipid composition, fatty acids and sterols in the seaweeds *Ulva armoricana*, and *Solieria chordalis* from Brittany (France): an analysis from nutritional, chemotaxonomic, and antiproliferative activity perspectives. Mar Drugs 13:5606–5628. https://doi.org/10.3390/md13095606

Khajoei RA, Keramat J, Hamdami N, Ursu A, Delattre C, Gardarin C, Lecerf D, Desbrieres J, Djelveh G, Michaul P (2021) Effect of high voltage electrode discharge on the physicochemical characteristics of alginate extracted from an Iranian brown seaweed (*Nizimuddinia zanardini*). Algal Res 56:102326. https://doi.org/10.1016/j.algal.2021.102326

Kumar M, Kumari P, Trivedi N, Shukla MK, Gupta V, Reddy, CRK, Jha B (2011) Minerals, PUFAs and antioxidant properties of some tropical seaweeds from Saurashtra coast of India. J Appl Phycol 23:797–810. https://doi.org/10.1007/s10811-010-9578-7

Lafarga T, Acien-Fernandez FG, Garcia-Vaquero M (2020) Bioactive peptides and carbohydrates from seaweed for food applications: natural occurrence, isolation, purification, and identification. Algal Res 48:101909. https://doi.org/10.1016/j.algal.2020.101909

Lee DS, Kang MS, Hwang HJ, Eom SH, Yang JY, Lee MS, Lee WJ, Jeon YJ, Choi JS, Kim YM (2008) Synergistic effect between dieckol from Ecklonia stolonifera and b-lactams against methicillin-resistant Staphylococcus aureus. Biotechnol Bioprocess Eng 13:758–764. https://doi.org/10.1007/s12257-008-0162-9

Lee SH, Park MH, Heo SJ, Kang SM, Ko SC, Han JS et al (2010a) Dieckol isolated from *Ecklonia cava* inhibits a-glucosidase and a-amylase in vitro and alleviates postprandial hyperglycemia in streptozotocin-induced diabetic mice. Food Chem Toxicol 48:2633–2637. https://doi.org/10.1016/j.fct.2010.06.032

Lee MH, Lee KB, Oh SM, Lee BH, Chee HY (2010b) Antifungal activities of dieckol isolated from the marine brown alga *Ecklonia cava* against *Trichophyton rubrum*. Food Sci Biotechnol 53:504–507. https://doi.org/10.3839/jksabc.2010.076

Lee JH, Eom SH, Lee EH, Jung YJ, Kim HJ, Jo MR et al (2014) In vitro antibacterial and synergistic effect of phlorotannins isolated from edible brown seaweed *Eisenia bicyclis* against acne-related bacteria. Algae 29(1):47–55. https://doi.org/10.4490/algae.2014.29.1.047

Li YX, Wijesekara I, Li Y, Kim SK (2011) Phlorotannins as bioactive agents from brown algae. Process Biochem 46:2219–2224. https://doi.org/10.1016/j.procbio.2011.09.015

Li Y, Yang Z, Li J (2017) Shelf-life extension of Pacific white shrimp using algae extracts during refrigerated storage. J Sci Food Agric 97(1):291–298. https://doi.org/10.1002/jsfa.7730

Li W, Jiang N, Li B, Wan M, Chang X, Liu H, Zhang L, Yin S, Qi H, Liu S (2018) Antioxidant activity of purified ulvan in hyperlipidemic mice. Int J Biol Macromol 113:971–975. https://doi.org/10.1016/j.ijbiomac.2018.02.104

Liao WR, Lin JY, Shieh WY, Jeng WL, Huang R (2003) Antibiotic activity of lectins from marine algae against marine vibrios. J Ind Microbiol Biotechnol 30:433–439. https://doi.org/10.1007/s10295-003-0068-7

Lim C, Yusoff S, Ng CG, Lim PE, Ching YC (2021) Bioplastic made from seaweed polysaccharides with green production methods. J Environ Chem Eng 9:105895. https://doi.org/10.1016/j.jece.2021.105895

Lopes G, Sousa C, Silva LR, Pinto E, Andrade PB, Bernardo J et al (2012) Can phlorotannins purified extracts constitute a novel pharmacological alternative for microbial infections with associated inflammatory conditions? PLoS One 7(2):e31145. https://doi.org/10.1371/journal.pone.0031145

Lordan S, Ross RP, Stanton C (2011) Marine bioactives as functional food ingredients: potential to reduce the incidence of chronic diseases. Mar Drugs 9:1056–1100. https://doi.org/10.3390/md9061056

Lourenço SO, Barbarino E, De-Paula JC, Pereira LODS, Marquez UML (2002) Amino acid composition, protein content and calculation of nitrogen-to-protein conversion factors for 19 tropical seaweeds. Phycol Res 50:233–241. https://doi.org/10.1046/j.1440-1835.2002.00278.x

MacArtain P, Gill CIR, Brooks M, Campbell R, Rowland IR (2007) Nutritional value of edible seaweeds. Nutr Rev 65:535–543. https://doi.org/10.1301/nr.2007.dec.535-543

Maehre HK, Edvinsen GK, Eilertsen KE, Elvevoll EO (2016) Heat treatment increases the protein bioaccessibility in the red seaweed dulse (*Palmaria palmata*), but not in the brown seaweed winged kelp (*Alaria esculenta*). J Appl Phycol 28:581–590. https://doi.org/10.1007/s10811-015-0587-4

Mansilla A, Avila M (2011) Using *Macrocystis pyrifera* (L.) C. Agardh from southern Chile as a source of applied biological compounds. Rev Bras 21:262–267. https://doi.org/10.1590/S0102-695X2011005000072

Maqsood S, Benjakul S, Shahidi F (2013) Emerging role of phenolic compounds as natural food additives in fish and fish products. Crit Rev Food Sci Nutr 53(2):162–179. https://doi.org/10.1080/10408398.2010.518775

Marinho-Soriano E, Fonseca PC, Carneiro MAA, Moreira WSC (2006) Seasonal variation in the chemical composition of two tropical seaweeds. Bioresour Technol 97:2402–2406. https://doi.org/10.1016/j.biortech.2005.10.014

Martínez IJH, Castaneda THG (2013) Preparation and chromatographic analysis of phlorotannins. J Chromatogr Sci 51(8):825–838. https://doi.org/10.1093/chromsci/bmt045

Matanjun P, Mohamed S, Mustapha NM, Muhammad K (2009) Nutrient content of tropical edible seaweeds, *Eucheuma cottonii*, *Caulerpa lentillifera* and *Sargassum polycystum*. J Appl Phycol 21:75–80. https://doi.org/10.1007/s10811-008-9326-4

Meena R, Prasad K, Ganesan M, Siddhanta AK (2008) Superior quality agar from *Gracilaria* species (Gracilariales, Rhodophyta) collected from the Gulf of Mannar, India. J Appl Phycol 20:397–402. https://doi.org/10.1007/s10811-007-9272-6

Misurcova L, Kracmar S, Klejdus B, Vacek J (2010) Nitrogen content, dietary fiber, and digestibility in algal food products. Czech J Food Sci 28:27–35. https://doi.org/10.17221/111/2009-cjfs

Mohamed S, Hashim SN, Rahman HA (2012) Seaweeds: a sustainable functional food for complementary and alternative therapy. Trends Food Sci Technol 23(2):83–96. https://doi.org/10.1016/j.tifs.2011.09.001

Mohammed A, Rivers A, Stuckey DC, Ward K (2020) Alginate extraction from *Sargassum* seaweed in the Caribbean region: optimization using response surface methodology. Carbohydr Polym 245:116419. https://doi.org/10.1016/j.carbpol.2020.116419

Moroney NC, O'Grady MN, Lordan S, Stanton C, Kerry JP (2015) Seaweed polysaccharides (Laminarin and Fucoidan) as functional ingredients in pork meat: an evaluation of anti-oxidative potential, thermal stability and bioaccessibility. Mar Drugs 13:2447–2464. https://doi.org/10.3390/md13042447

Mouradi-Givernaud A, Givernaud T, Morvan H, Cosson J (1992) Agar from *Gelidium latifolium* (Rhodophyceae, Gelidiales): biochemical composition and seasonal variations. Bot Mar 35:153–159. https://doi.org/10.1515/botm.1992.35.2.153

Mouritsen OG, Rhatigan P, Cornish ML, Critchley AT, Pérez-Lloréns JL (2021) Saved by seaweeds: phyconomic contributions in times of crises. J Appl Phycol 33:443–458. https://doi.org/10.1007/s10811-020-02256-4

Munoz IL, Díaz NF (2020) Minerals in edible seaweed: health benefits and food safety issues. Crit Rev Food Sci Nutr 62(6):1592–1607. https://doi.org/10.1080/10408398.2020.1844637

Murata M, Nakazoe J (2001) Production and use of marine algae in Japan. Japan Agricultural Research Quarterly 35(4):281–290. https://doi.org/10.6090/jarq.35.281

Nagayama K, Iwamura Y, Shibata T, Hirayama I, Nakamura T (2002) Bactericidal activity of phlorotannins from the brown alga *Ecklonia kurome*. Antimicrob Agents Chemother 50:889–893. https://doi.org/10.1093/jac/dkf222

Narayan B, Miyashita K, Hosakawa M (2006) Physiological effects of eicosapentaenoic acid (EPA) and docosahexaenoic acid (DHA): a review. Food Rev Int 22:291–307. https://doi.org/10.1080/87559120600694622

Nazarudin MF, Isha A, Mastuki SN, Ain NM, Mohd Ikhsan NF, Abidin AZ, Aliyu-Paiko M, Vilas-Boas M (2020) Chemical composition and evaluation of the a-glucosidase inhibitory and cytotoxic properties of marine algae *Ulva intestinalis*, *Halimeda macroloba*, and *Sargassum ilicifolium*. Evid Based Complement Alternat Med 2020:1–13. https://doi.org/10.1155/2020/2753945

Nazarudin MF, Yasin ISM, Mazli NAIN, Saadi AR, Nooraini MA, Saad N, Ferdous UT, Fakhrulddin IM (2022) Preliminary screening of antioxidant and cytotoxic potential of green seaweed, *Halimeda opuntia* (Linnaeus) Lamouroux. Saudi J Biol Sci 29:2698–2705. https://doi.org/10.1016/j.sjbs.2021.12.066

Nielsen CW, Rustad T, Holdt SL (2021) Vitamin C from seaweed: a review assessing seaweed as contributor to daily intake. Foods 10:198. https://doi.org/10.3390/foods10010198

Ortiz J, Uquiche E, Robert P, Romero N, Quitral V, Llanten C (2009) Functional and nutritional value of the Chilean seaweeds *Codium fragile*, *Gracilaria chilensis* and *Macrocystis pyrifera*. Eur J Lipid Sci Technol 111:320–327. https://doi.org/10.1002/ejlt.200800140

Othman R, Amin NA, Sani MSA, Fadzillah NA, Jamaludin MA (2018) Carotenoid and chlorophyll profiles in five species of Malaysian seaweed as potential halal active pharmaceutical ingredient (API). Int J Adv Sci Eng Inf Technol 8(4–2):1610–1616. https://doi.org/10.18517/ijaseit.8.4-2.7041

Peñalver R, Lorenzo JM, Ros G, Amarowicz R, Pateiro M, Nieto G (2020) Seaweeds as a functional ingredient for a healthy diet. Mar Drugs 18(6):301. https://doi.org/10.3390/md18060301

Pereira CMP, Nunes CFP, Zambotti-Villela L, Streit NM, Dias D, Pinto E, Gomes CB, Colepicolo P (2016) Extraction of sterols in Brown macroalgae from Antarctica and their identification by liquid chromatography coupled with tandem mass spectrometry. J Appl Phycol 29:751–757. https://doi.org/10.1007/s10811-016-0905-5

Phillips DJH (1994) Macrophytes as biomonitors of trace metals. In: Kramer KJM (ed) Biomonitoring of coastal waters and estuaries. CRC Press, Boca Raton, pp 85–103

Pirian K, Jeliani ZZ, Sohrabipour J, Arman M, Faghihi MM, Yousezadi M (2017) Nutritional and bioactivity evaluation of common seaweed species from the Persian Gulf. Iran J Sci Technol 42:1795–1804. https://doi.org/10.1007/s40995-017-0383-x

Polat S, Ozogul Y (2008) Biochemical composition of some red and brown macroalgae from the northeastern Mediterranean Sea. Int J Food Sci Nutr 59:566–572. https://doi.org/10.1080/09637480701446524

Polat S, Trif M, Rusu A, Šimat V, Čagalj M, Alak G, Meral R, Ozogul Y, Polat A, Özogul F (2021) Recent advances in industrial applications of seaweeds. Crit Rev Food Sci Nutr 63(21):4979–5008. https://doi.org/10.1080/10408398.2021.2010646

Pradhan B, Bhuyan PP, Patra S, Nayak R, Behera PK, Behera C, Behera AK, Ki J, Jena M (2022) Beneficial effects of seaweeds and seaweed-derived bioactive compounds: current evidence and future prospective. Biocatal Agric Biotechnol 39:102242. https://doi.org/10.1016/j.bcab.2021.102242

Quitral V, Sepulveda M, Gamero-Vega G, Jimenez P (2022) Seaweeds in bakery and farinaceous foods: a mini-review. Int J Gastron Food Sci 28:100403. https://doi.org/10.1016/j.ijgfs.2021.100403

Raja R, Hemaiswarya S, Sridhar S, Alagarsamy A, Ganesan V, Elumalai S et al (2020) Evaluation of proximate composition, antioxidant properties, and phylogenetic analysis of two edible seaweeds. Smart Sci 8(3):95–100. https://doi.org/10.1080/23080477.2020.1795338

Raja K, Kadirvel V, Subramaniyan T (2022) Seaweeds, an aquatic plant-based protein for sustainable nutrition – a review. Future Foods 5:100142. https://doi.org/10.1016/j.fufo.2022.100142

Rajan DK, Mohan K, Zhang S, Ganesan AR (2021) Dieckol: a brown algal phlorotannin with biological potential. Biomed Pharmacother 142:111988. https://doi.org/10.1016/j.biopha.2021.111988

Rajapakse N, Kim S (2011) Nutritional and digestive health benefits of seaweed. Adv Food Nutr Res 64:17–28. https://doi.org/10.1016/B978-0-12-387669-0.00002-8

Raposo MFJ, de Morais AMMB, de Morais RMSC (2016) Emergent sources of prebiotics: seaweeds and microalgae. Mar Drugs 14:27. https://doi.org/10.3390/md14020027

Rawiwan P, Peng Y, Paramayuda GPB, Quek SY (2022) Red seaweed: a promising alternative protein source for global food sustainability. Trends Food Sci Technol 123:37–56. https://doi.org/10.1016/j.tifs.2022.03.003

Rioux L, Beaulieu L, Turgeon SL (2017) Seaweeds: a traditional ingredients for new gastronomic sensation. Food Hydrocoll 68:255–265. https://doi.org/10.1016/j.foodhyd.2017.02.005

Rohani-Ghadikolaei K, Abdulalian E, Ng W (2012) Evaluation of the proximate, fatty acid and mineral composition of representative green, brown and red seaweeds from the Persian Gulf of Iran as potential food and feed resources. J Food Sci Technol 49(6): 774–780. https://doi.org/10.1007/s13197-010-0220-0

Ruiter GA, Rudolph B (1997) Carrageenan biotechnology. Trends Food Sci Technol 8(12):389–395. https://doi.org/10.1016/S0924-2244(97)01091-1

Ruperez P, Saura-Calixto F (2001) Dietary fibre and physicochemical properties of edible Spanish seaweeds. Eur Food Res Technol 212(3):349–354. https://doi.org/10.1007/s002170000264

Safi C, Charton M, Pignolet O, Silvestre F, Vaca-Garcia C, Pontalier PY (2013) Influence of microalgae cell wall characteristics on protein extractability and determination of nitrogen-to-protein conversion factors. J Appl Phycol 25:523–529. https://doi.org/10.1007/s10811-012-9886-1

Salma S, Elsanti, Nurida NL, Dariah A (2021) Bio-decomposer of seaweed composting. IOP Conf Ser Earth Environ Sci 637(1):012080. https://doi.org/10.1088/1755-1315/637/1/012080

Schiener P, Black KD, Stanley MS, Green DH (2015) The seasonal variation in the chemical composition of the kelp species *Laminaria digitata*, *Laminaria hyperborea*, *Saccharina latissima* and *Alaria esculenta*. J Appl Phycol 27:363–373. https://doi.org/10.1007/s10811-014-0327-1

Schmid M, Guihéneuf F, Stengel DB (2014) Fatty acid contents and profiles of 16 macroalgae collected from the Irish coast at two seasons. J Appl Phycol 26(1):451–463. https://doi.org/10.1007/s10811-013-0132-2

Schmid M, Kraft LGK, Van Der Loos LM, Kraft GT, Virtue P, Nichols PD, Hurd CL (2018) Southern Australian seaweeds: a promising resource for omega-3 fatty acids. Food Chem 265:70–77. https://doi.org/10.1016/j.foodchem.2018.05.060

Sellimi S, Maalej H, Rekik DM, Benslima A, Ksouda G, Hamdi M, Sahnoun Z, Li S, Nasri M, Hajji M (2018) Antioxidant, antibacterial and in vivo wound healing properties of laminaran purified from *Cystoseira barbata* seaweed. Int J Biol Macromol 119:633–644. https://doi.org/10.1016/j.ijbiomac.2018.07.171

Shannon E, Abu-Ghannam N (2016) Antibacterial derivatives of marine algae: an overview of pharmacological mechanisms and applications. Mar Drugs 14(4):81. https://doi.org/10.3390/md14040081

Sharifian S, Shabanpour B, Taheri A, Kordjazi M (2019) Effect of phlorotannins on melanosis and quality changes of Pacific white shrimp (*Litopenaeus vannamei*) during iced storage. Food Chem 298:124980. https://doi.org/10.1016/j.foodchem.2019.124980

Shibata T, Ishimaru K, Kawaguchi S, Yoshikawa H, Hama Y (2007) Antioxidant activities of phlorotannins isolated from Japanese Laminariaceae. J Appl Phycol 20(5):705–711. https://doi.org/10.1007/s10811-007-9254-8

Shrestha S, Zhang W, Smid SD (2021) Phlorotannins: a review on biosynthesis, chemistry and bioactivity. Food Biosci 39(2021):100832. https://doi.org/10.1016/j.fbio.2020.100832

Simopoulos AP (2008) The importance of the omega-6/omega-3 fatty acid ratio in cardiovascular disease and other chronic diseases. Exp Biol Med 233:674–688. https://doi.org/10.3181/0711-MR-311

Sivri S, Yur F (2020) Effects of fucoxanthin on health. Van Health Sci J 13(3):93–102

Stern JL, Hagerman AE, Steinberg PD, Mason PK (1995) Phlorotannin protein interactions. J Chem Ecol 22(10):1877–1899. https://doi.org/10.1007/BF02028510

Storelli MM, Storelli A, Marcotrigiano GO (2001) Heavy metals in the aquatic environment of the Southern Adriatic Sea, Italy: macroalgae, sediments and benthic species. Environ Int 26:505–509. https://doi.org/10.1016/s0160-4120(01)00034-4

Strain CR, Collins KC, Naughton V, Mcsorley EM, Stanton C, Smyth TJ et al (2020) Effects of a polysaccharide-rich extract derived from Irish sourced *Laminaria digitata* on the composition and metabolic activity of the human gut microbiota using an in vitro colonic model. Eur J Nutr 59:309–325. https://doi.org/10.1007/s00394-019-01909-6

Sudharsan S, Giji S, Seedevi P, Vairamani S, Shanmugam A (2018) Isolation, characterization and bioactive potential of sulfated galactans from *Spyridia hypnoides* (Bory) Papenfuss. Int J Biol Macromol 109:589–597. https://doi.org/10.1016/j.ijbiomac.2017.12.097

Sugiura Y, Matsuda K, Yamada Y, Nishikawa M, Shioya K, Katsuzaki H, Imai K, Amano H (2006) Isolation of a new anti-allergic phlorotannin, phlorofucofuroeckol-B, from an edible brown alga, *Eisenia arborea*. Biosci Biotechnol Biochem 70:2807–2811

Surendhiran D, Cui H, Lin L (2019) Encapsulation of phlorotannin in alginate/PEO blended nanofibers to preserve chicken meat from Salmonella contaminations. Food Packag Shelf Life 21:100346. https://doi.org/10.1016/J.FPSL.2019.100346

Tallima H, El Ridi R (2018) Arachidonic acid: physiological roles and potential health benefits: a review. J Adv Res 11:33–41. https://doi.org/10.1016/j.jare.2017.11.004

Tanna B, Choudhary B, Mishra A (2018) Metabolite profiling, antioxidant, scavenging and anti-proliferative activities of selected tropical green seaweeds reveal the nutraceutical potential of *Caulerpa* spp. Algal Res 36:69–105. https://doi.org/10.1016/j.algal.2018.10.019

Tavassoli-Kafrani E, Shekarchizadeh H, Masoudpour-Behabadi M (2016) Development of edible films and coatings from alginates and carrageenans. Carbohydr Polym 137:360–374. https://doi.org/10.1016/j.carbpol.2015.10.074

Tibbetts SM, Milley JE, Lall SP (2016) Nutritional quality of some wild and cultivated seaweeds: nutrient composition, total phenolic content and in vitro digestibility. J Appl Phycol 28(6):3575–3585. https://doi.org/10.1007/s10811-016-0863-y

Uju, Dewi NPSUK, Santoso J, Setyaningsih I, Hardingtyas SD, Yopi (2020) Extraction of phycoerythrin from *Kappaphycus alvarezii* seaweed using ultrasonication. Earth Environ Sci 414:012028. https://doi.org/10.1088/1755-1315/414/1/012028

Urrea-Victoria V, Furlan CM, dos Santos DYAC, Chow F (2022) Antioxidant potential of two Brazilian seaweeds in response to temperature: *Pyropia spiralis* (red alga) and *Sargassum stenophyllum* (brown alga). J Exp Mar Biol Ecol 549:151706

van Ginneken VJT, Helsper J, de Visser W, van Keulen H, Brandenburg WA (2011) Polyunsaturated fatty acids in various macroalgal species from north Atlantic and tropical seas. Lipids Health Dis 10:104. https://doi.org/10.1186/1476-511X-10-104

Vazquez-Rodriguez B, Santos-Zea L, Heredia-Olea E, Acevedo-Pacheco L, Santacruz A, Gutierrez-Uribe JA, Cruz-Suarez LE (2021) Effects of phlorotannin and polysaccharide fractions of brown seaweed *Silvetia compressa* on human gut microbiota composition using an in vitro colonic model. J Funct Foods 84:104596. https://doi.org/10.1016/j.jff.2021.104596

Vieira EF, Soares C, Machado S, Correia M, Ramalhosa MJ, Oliva-Teles MT, Paula Carvalho A, Domingues VF, Antunes F, Oliveira TAC et al (2018) Seaweeds from the Portuguese coast as a source of proteinaceous material: total and free amino acid composition profile. Food Chem 269:264–275. https://doi.org/10.1016/j.foodchem.2018.06.145

Vuai SAH (2022) Characterization of agar extracted from Gracilaria species collected along Tanzanian coast. Heliyon 8:e09002. https://doi.org/10.1016/j.heliyon.2022.e09002

Wang T, Li Z, Yuan F, Lin H, Pavase TR (2017) Effects of brown seaweed polyphenols, α-tocopherol, and ascorbic acid on protein oxidation and textural properties of fish mince (*Pagrosomus major*) during frozen storage. J Sci Food Agric 97:1102–1107. https://doi.org/10.1002/jsfa.7835

Watanabe F, Takenaka S, Katsura H, Miyamoto E, Abe K, Tamura Y et al (2000) Characterization of a vitamin B12 compound in the edible purple laver, *Porphyra yezoensis*. Biosci Biotechnol Biochem 64:2712–2715. https://doi.org/10.1271/bbb.64.2712

Wells ML, Potin P, Craigie JS et al. (2017) Algae as nutritional and functional food sources: revisiting our understanding. J Appl Psychol 29:949–982. https://doi.org/10.1007/s10811-016-0974-5

WHO (2003) Diet, nutrition, the prevention of chronic disease. Report of a joint WHO/FAO expert consultation. Technical reports series no. 916. WHO, Geneva

Wong K, Cheung PCK (2000) Nutritional evaluation of some subtropical red and green seaweeds Part I – proximate composition, amino acid proÆles and some physico-chemical properties. Food Chem 71:475–482. https://doi.org/10.1016/S0308-8146(00)00175-8

Wong K, Cheung PCK (2001a) Influence of drying treatment on three Sargassum species 2. Protein extractability, in vitro protein digestibility and amino acid profile of protein concentrates. J Appl Phycol 13:51–58. https://doi.org/10.1023/A:1008188830177

Wong KH, Cheung PCK (2001b) Nutritional evaluation of some subtropical red and green seaweeds Part II. In vitro protein digestibility and amino acid profiles of protein concentrates. Food Chem 72:11–17. https://doi.org/10.1016/S0308-8146(00)00176-X

Yaich H, Garna H, Besbes S, Paquot M, Blecker C, Attia H (2011) Chemical composition and functional properties of Ulva lactuca seaweed collected in Tunisia. Food Chem 128:895–901. https://doi.org/10.1016/j.foodchem.2011.03.114

Yerlikaya P, Topuz OK, Buyukbenli HA, Gokoglu N (2013) Fatty acid profiles of different shrimp species: effects of depth of catching. J Aquat Food Prod Technol 22:290–297. https://doi.org/10.1080/10498850.2011.646388

Yerlikaya P, Alp AC, Tokay FG, Aygun T, Kaya A, Topuz OK, Yatmaz HA (2022) Determination of fatty acids and vitamins A, D and E intake through fish consumption. Int J Food Sci Technol 57:653–661. https://doi.org/10.1111/ijfs.15435

Yong WTL, Thien VY, Rupert R, Rodrigues KF (2022) Seaweed: a potential climate change solution. Renew Sust Energ Rev 159:112222. https://doi.org/10.1016/j.rser.2022.112222

Zheng LX, Chen XQ, Cheong KL (2020) Current trends in marine algae polysaccharides: the digestive tract, microbial catabolism, and prebiotic potential. Int J Biol Macromol 151:344–354. https://doi.org/10.1016/j.ijbiomac.2020.02.168

Zhu Z, Wu Q, Di X, Li S, Barba FJ, Koubaa M, Roohinejad S, Xiong X, He J (2017) Multistage recovery process of seaweed pigments: investigation of ultrasound assisted extraction and ultra-filtration performances. Food Bioprod Process 104:40–47. https://doi.org/10.1016/j.fbp.2017.04.008

Zibikowski R, Szefer P, Latala A (2006) Distribution and relationships between selected chemical elements in green alga *Enteromorpha* sp. from the southern Baltic. Environ Pollut 143(3):435–448. https://doi.org/10.1016/j.envpol.2005.12.007

Chapter 5
Seaweed-Derived Polysaccharides and Peptides and Their Potential Benefits for Human Health and Food Applications

Nariman El Abed, Sami Ben Hadj Ahmed, and Fatih Ozogul

Abbreviations

AA	Amino acid
AA	Arachidonic Acid
ABTS	2,2′-azino-bis(3-ethylbenzothiazoline-6-sulfonic acid)
ACE	Angiotensin I-Converter Enzyme
ALA	α-linolenic Acid
DHA	Docosahexaenoic Acid
DPPH	1,1-diphenyl-2-picrylhydrazyl (DPPH)-2,2-diphenyl-1-picrylhydrazyl.
EAA	Essential Amino Acid
EPA	Eicosapentaenoic Acid
FAO	Food and Agriculture Organization
FRAP	Ferric Reducing Antioxidant Power
GRG	Carrageenan
HCMV	Human cytomegalovirus
HIV	Human Immunodeficiency Virus
HSV-1	Herpes simplex virus type 1
NEAA	Non-essential Amino Acid
ORAC	Oxygen Radical Absorbance Capacity

N. El Abed (✉)
Laboratory of Protein Engineering and Bioactive Molecules (LIP-MB), National Institute of Applied Sciences and Technology (INSAT), University of Carthage, Carthage, Tunisia

S. Ben Hadj Ahmed
Laboratory of Protein Engineering and Bioactive Molecules (LIP-MB), National Institute of Applied Sciences and Technology (INSAT), University of Carthage, Carthage, Tunisia

Medical Laboratory Sciences Department, College of Applied Medical Sciences, King Khalid University, Abha, Kingdom of Saudi Arabia

F. Ozogul
Department of Seafood Processing Technology, Faculty of Fisheries, Cukurova University, Adana, Turkey

F. Ozogul et al. (eds.), *Seaweeds and Seaweed-Derived Compounds*,
https://doi.org/10.1007/978-3-031-65529-6_5

PUFA	Polyunsaturated Fatty Acid
SP	Sulfated polysaccharides
WHO	World Health Organization

1 Introduction

In recent years, food security and safety are considered among the principal problems throughout the world, which have important implications for human health. Indeed, food security and human lifestyle are closely correlated. As a result, consumers have become increasingly aware of the relationship between their diet and health. Thus, they demand more and more nutritious and healthy foods with functional properties to prevent the occurrence of different diseases, such as cardiovascular, diabetes, and obesity. Generally, biological marine products, including seaweeds, are excellent sources to develop new functional food ingredients owing to their high bioactivity and low toxicity. Consequently, the consumption of marine resources has been worldwide attracted attention. So, these products can provide a wide range of benefits and thus enhanced human health by their potential to reduce the risk of diseases.

Seaweeds, which known also by macroalgae, are considered an extensive and vast group of marine macroscopic organisms that included several species (Peng et al. 2015). In this context, these algae were represented as vital and important compounds of marine resources (Pradhan et al. 2022). They grow in a wide area of habitats under extremely variable environmental factors. In this regard, these marine products are the submerged and floating plants of superficial marine meadows. In fact, seaweeds can develop as epiphytes on pebbles, dead corals, stones, rocks, and other substrates (Baweja et al. 2016). Therefore, more than 150,000 categories of seaweeds are developed in the ocean, tropical waters, and intertidal zone. Further, the length of brown macroalgae varied between 30 cm and 70 cm, and thus, they more notable compared to the green and red seaweeds (Pradhan et al. 2022).

Seaweeds consist of 6000 species and are considered as potentially renewable resource in the ocean. These macroscopic marine algae are taxonomically classified as green (Chlorophyta), brown (Phaeophyceae), or red (Rhodophyta), depending on the type of their pigments (Vieira et al. 2018). In this context, the chlorophyll a and b are responsible for the pigmentation in Chlorophyta, The carotenoids, fucoxanthin, and xanthophyll are responsible for pigmentation in Phaeophyceae (Pradhan et al. 2022).

Generally, the macroalgae are recognized for their richness in vitamins, and minerals as well as the existence of bioactive compounds, such as proteins, peptides, phenolic compounds, and lipids, which supposed to have beneficial effects and can treat or prevent several diseases (Vieira et al. 2018).

Polysaccharides and peptides from seaweeds have an excellent role in different fields of biotechnology, like food and medicine industries (Lafarga et al. 2020; Pradhan et al. 2022). Therefore, diets rich in these bioactive compounds can provide considerable health benefits, improve well-being, and decrease the disease risk (Qiu et al. 2022). Indeed, seaweed-derived polysaccharides and peptides posses special chemical structure and biological properties, which make them as functional and nutraceuticals sources for food products (Qiu et al. 2022). Thus, their biological activities can be represented by prebiotic, antioxidant, anticancer, immunomodulatory, antimicrobial activities. This chapter provides information on seaweed-derived bioactive polysaccharides and peptides, as bioactive compounds, and their beneficial impacts on human health with their potential applications in the food industry.

2 Seaweeds

Seaweeds are photosynthetic and eukaryotic organisms that grow principally in the oceans and seas (Wang et al. 2020). They are found mostly in shallow littoral waters and are attached to a substrate necessary for their growth. Generally, Seaweeds are considered multi-cellular species, commonly subdivided into different classes, for instance brown, green, blue, and red algae due to their pigment composition according to Phaeophyceae, Ocrophyta, Chlorophyte, and Rhodophyta (Šimat et al. 2020; Pradhan et al. 2022).

2.1 *Bio-chemistry of Seaweeds*

The macroalgae are considered a rich source of valuable constituents, which including micro- and macronutrients (Roohinejad et al. 2017; Lafarga et al. 2020; Bizzaro et al. 2022) (Fig. 5.1). Generally, the chemical composition of seaweeds varies depending on the species (Fleurence 2016). In fact, the chemical and biochemical composition of seaweeds depends not only on the species, but also on the period of harvest, location, habitat, and on the external and abiotic conditions, for example the salinity, the amount of nutrients in the water, temperature (Makkar et al. 2016; Bizzaro et al. 2022). It has been reported that the seaweeds are known as a potential source of bioactive substances, which characterized by their nutritional properties and beneficial impacts on human health (Mabeau and Fleurence 1993). Therefore, the macroalgae are recognized as functional ingredients to enhance the sensory, textural, and nutritional attributes of food products, such as dairy products, meat products, bakery... etc. (Roohinejad et al. 2017; Øverland et al. 2019; Lafarga et al. 2020).

In this context, the macroalgae are considered an abundant source of micronutrient components, for instance, minerals (cadmium, calcium, potassium, magnesium, nickel, zinc, copper, mercury, iron, sodium, selenium, and iodine); sterols, vitamins (vitamins A, B1, B2, B3, B6, B12, C, D, E, folic acid, pantothenic acid, niacin, riboflavin, and folate derivatives), and pigments (including three classes: carotenoid, chlorophyll, and phycobiliproteins) (Zimmermann 2008; Patarra et al. 2011; Peña-Rodríguez et al. 2011; Ferraces-Casais et al. 2012; Gullón et al. 2020).

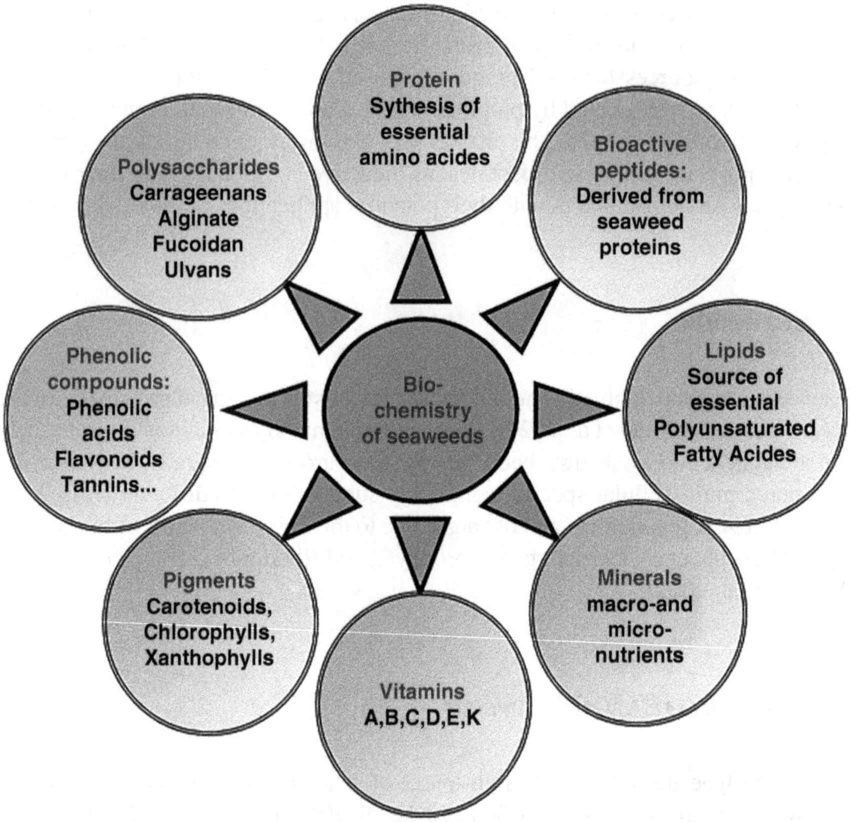

Fig. 5.1 Schematic representation of biochemical composition of seaweeds

Furthermore, the seaweeds are considered also a rich source of macro-nutrient ingredients, such as proteins, amino acids (e.g. non-essential and essential amino acids), carbohydrates, fats (e.g. Fatty acids), lipids, polyphenols, and fibers (Marsham et al. 2007; Peña-Rodríguez et al. 2011; Roohinejad et al. 2017; Bizzaro et al. 2022). In this context, it has been reported that the red seaweed presented an elevated proportion of essential amino acid and non-essential amino acid «EAA/NEAA» equivalent to 0.98 to 1.02, while the ratio of EAA/NEAA characterized the green seaweed equal to the values of 0.72 to 0.97 and the brown seaweeds contain aunt of 0.73 (Ganesan et al. 2019). In this regard, the most abundant amino acids in several macroalgae are the aspartic and glumatic acids, which known for their characteristic flavor (Lafarga et al. 2020; Bizzaro et al. 2022). Besides, the most of seaweeds have characterized by the existence of the essential amino acids, including isoleucine, lysine, methionine, leucine (Bleakley and Hayes 2017; Lafarga et al. 2020). Furthermore, the EAA are found with a few free amino acid portions in the range of 6.47% to 24%, particulary in *Ascophyllum nodosum* is equal to 34.4 g/100 g protein; *Ulva* spp. with an amount of 27 g/100 g protein; and the *Fucus vesiculosus* in the range of 25.1 g/100 g protein (Vieira et al. 2018; Ganesan et al. 2019). Therefore, the edible macroalgae can represent an important source for the application of essential amino acids in healthful human diets due to their significant role similar to legumes, animal proteins, and soybean (Ganesan et al. 2019; Bizzaro et al. 2022).

On the other hand, previous works have reported that the green seaweeds, for instance *Caulerpa racemosa, Caulerpa fastigiata, Codium spongiosum, Codium decorticatum, Codium taylorii, Ulva fasciata, Ulva pertusa*, contain high amounts of proline, alanine, and leucine (Lafarga et al. 2020). However, the red seaweed, such as *Palmaria palmate* has been characterized by its richness with EAA fractions, including valine, threonine, leucine, methionine, and isoleucine (Fleurence 1999; Lafarga et al. 2020). Whereas, the brown macroalgae, for example *Chnoospora minima, Dictyota menstrualis,* and *Padina gymnospora* have been recognized as rich sources of alanine, leucine, valine, thereonine, glycine, and lysine (Holdt and Kraan 2011).

Otherwise, the seaweeds have known by their high richness with protein contents, especially the green and red macroalgae, on the contrary, the brown algae have characterized by their lower ratio of protein (Lourenço et al. 2002). In fact, the proportion of this macronutrient varied between the different species of seaweeds. In this context, it has been reported that the green seaweeds, such as *Ulva lactuca,* contain an amount of proteins more than 26% of their total dry weight (Bizzaro et al. 2022). Therefore, this proportion may vary according to several factors, like the geographical area and seasonal variations (Fleurence, 2016; Biancarosa et al. 2017; Bizzaro et al. 2022). Besides, previous works have revealed that the red seaweeds, for example *Porphyra umbelicalis* and *Porphyra dioica* have an elevated amount of proteins in the range of 24.0 ± 0.2% and 31.0 ± 0.2 of their total dry weight (Garcia-Vaquero and Hayes 2016; Biancarosa et al. 2017).

On the other hand, seaweeds have also been characterized by the presence of other macronutrients in addition to proteins and amino acids. In this regard, these marine algae have known for their richness with carbohydrates and their amount in the range of 50% (Bizzaro et al. 2022). Previous studies have reported that the carbohydrates or the polysaccharides, as macronutrient compounds from seaweed, have been demonstrated several biological properties, and thus can enhance the sensory and nutritional attributes of food products (Roohinejad et al. 2017). In fact, the principal polysaccharides characterized the seaweeds are laminaran, fucoidan, alginate, ulvan, and carrageenan, which have revealed their beneficial effect on the enhancement of human health against many diseases, such as hypertension, obesity, diabetes, and dyslipidemia (Costa et al. 2010). Moreover, several seaweeds contain also an amount of fiber equivalent to 20–30% (Bizzaro et al. 2022). Indeed, the dietary fibers have been revealed many beneficial impacts, including anti-obesity properties due to their role in the decrease of the absorption of nutrients and gastric evacuation as well as their effect on the enhancement of the satiation (Brownlee et al. 2005). Also, the seaweeds have characterized by a low amount of lipids in a proportion does not surpass 5% of their total dry weight (Gullón et al. 2020; Bizzaro et al. 2022). Thereby, the fatty acid contents of seaweeds have gained a lot of attention owing to their richness with polyunsaturated fatty acids (PUFA), for instance docosahexaenoic (DHA, 22:6 n-3), eicosapentaenoic (EPA, 20:5 n-3), arachidonic (AA, 20:4 n-6), octadecatetraenoic (18:4 n-3), and α-linolenic (ALA, 18:3 n-3) acids (Gullón et al. 2020). Indeed, this category of acids has considerable impacts on human health due to to their nutritional attributes, and anti-obesity, antiviral, and anti-tumoral properties as well as their role in the prevention of cardiovascular diseases (Kendel et al. 2015).

2.2 *Bioavailability of Seaweed Micro- and Macro-elements*

The bioavailability of seaweed products has an important and beneficial role in human health and their consumption contributes to diverse biological benefits (Bizzaro et al. 2022). Generally, bioavailability can be identified as the proportion of ingested either food or seaweeds compounds into the digestive tract, including mouth, stomach, small and large intestines, and their absorption, distribution, metabolism, as well as their utilization for maintaining physiological functions (Guerra et al. 2012; Cassani et al. 2020; Bizzaro et al. 2022; Demarco et al. 2022). In addition to the digestion process of food or seaweed products; the bioavailability of these products can be influenced by ordinary processing techniques, for instance, freezing, and drying and their distribution in the human body (Bizzaro et al. 2022). Indeed, these processing techniques can decrease some compounds of seaweed and alter their properties. Thus, it can conduct to the occurrence of diverse metabolic functions in the human body (Bizzaro et al. 2022). In this context, the determination of bioaccessibility of macro- and micronutrients of seaweeds can contribute to the estimation of the effective and beneficial biomass of marine algae that can be consumed (Cassani et al. 2020). Therefore, the bioaccessibility of seaweeds needs to be investigated and controlled via the use of an appropriate method, such as the *in vitro* digestion model (Bonfanti et al. 2018). In this regard, several previous studies have reported that analysis of bioavailability and bioaccessibility of food products can determine through the most effective and standard assay and such one is the in vitro digestion, which attempts to stimulate the gastrointestinal process of food (Demarco et al. 2022). In this context, the human digestive process is a very complex setup where seaweed products ingested are subjected to biotransformation via enzymatic mechanisms for absorption in the small and large intestines and assimilation as nutrients into the bloodstream (Banwo et al. 2021; Demarco et al. 2022).

The bioaccessibility and bioavailability of micronutrients of seaweeds are affected by several parameters, such as the genres of processing done on marine algae and the micro-constituents (Cassani et al. 2020). It has been reported that several seaweeds are characterized by a high accumulation of heavy metals and also known by higher concentrations of macro-minerals, such as sodium (Na), potassium (K), magnesium (Mg), and calcium (Ca) and lower micro-minerals, including copper (Cu), iron (Fe), and Zinc (Zn) (Santoso et al. 2006; Cassani et al. 2020). Generally, the biological availability of macro-minerals and trace-minerals is based on the diet constitution and also is affected by the current nutrient levels and forms or non-nutritive constituents, and antagonistic or synergistic interactions of nutrients (Mišurcová et al. 2011). In this regard, the minerals can be incorporated into different metallo-proteins, for instance the iron (Fe) in hemoglobin (Mišurcová et al. 2011). Besides, other minerals, including the magnesium and calcium have a structural function in teeth and bones (Mišurcová et al. 2011).

Previous works have been reported that the phenolic compounds and dietary fiber could reduce the mineral availability due to the development of insoluble complexes leading to the decrease of the absorption of minerals (Mišurcová et al. 2011). It has been reported that the seaweed of *C. racemosa* collected from different locations of Spain are characterized by an elevated bioavailability of copper in the range of 44.3% to 56.3%, cadmium with 41.8% to 46.7%, the manganese equivalent to 71.8% to 85.3%, and zinc in the range of 37.7% to 47.4% (Domínguez-González et al. 2010). However, the bioaccessibility of iron, lead of this species of seaweed has a lower level in the range of 11.5% to16.5% and 22.3% to 32%, respectively (Domínguez-González et al. 2010). Moreover, the bioaccessibility of zinc, manganese, and copper obtained from marine algae of *Laminaria digitata* of the European regions is in the range of 77% to 31%, 22% to 75%, and 43%, while the bioavailability of iron does not identify due to the lower concentration in digestion (Cassani et al. 2020). Further, the investigation of the bioavailability of iodine in cooked seaweed of Gelidium and Sea spaghetti via the in vitro digestibility method has demonstrated that this mineral is highly dialyzed with a percentage of 17% (Domínguez-González et al. 2010).

Furthermore, the stability of chlorophyll pigment as micronutrient in three types of seaweeds has evaluated through the in *vitro* digestion assay and it has demonstrated that the bioavailability of this pigment in the nori seaweed is in the range of 8%, which followed by the two types of seaweeds sea lettuce and Kombu (Chen and Roca 2018). In fact, it has been reported that the bioavailability of the chlorophyll in seaweeds is identical to the land plants.

On the other hand, seaweeds are characterized by their richness with macronutrients, which have a several biological activities. Indeed, it has previously reported that that the polyphenols in seaweeds have known a great interest, but the information about their bioavailability remain rare, which limits the understanding of their bioactivity and mechanism of action *in vivo* (Cassani et al. 2020). Indeed, the evaluation of the absorption mechanism of the polyphenols, can apply to determine the bioavailability of theses compounds, such as the phlorotannins and their interaction with intestinal bacteria (Cassani et al. 2020).

Furthermore, the green marine algae of *Enteromorphoa* and *Ulva lactuca* have contain the proteins as macromolecules with an amount of 20% of dry weight and they are known by the presence of polyphenols, which can interact with hydrogen bonding of protein-phenol (Ganesan et al. 2019). In this context, the absorption of proteins can inhibit by the high concentrations of phenolic compounds, such as quercetin, catechol, and phlorotannins in several seaweeds, particulary the brown algae (Ganesan et al. 2019). Therefore, this reaction can provoke to the decrease of the bioavailability of protein *in vivo*.

In this context, the bioavailability, the digestibility and the amount of essential amino acids of proteins are considered essential parameters to define their nutritional attribute for the human health (Demarco et al. 2022). In fact, the determination of the digestibility of the protein is considered an essential factor to investigate their bioavailability. Furthermore, previous research study has demonstrated that the proteins derived from seaweed re excellently hydrolyzed (Kazir et al. 2019). Therefore, these findings have confirmed that the seaweed proteins are bioavailable for human absorption (Kazir et al. 2019). Nevertheless, the biological value of the proteins derived from the seaweed is not clear during their absorption in the gastrointestinal system.

On the other hand, the bioavailability and the amount of lipids and fatty acids from the edible brown macroalgae "*Fucus spiralis*" has been investigated before and after the stimulation *in vitro* of the human digestive system. Indeed, it has been reported that this species of seaweed characterized by lipid bioavailability in the range of 12.1% and the presence of lipid amount around 3.5% of dry weight (Francisco et al. 2020). Besides, this research study has revealed that the bioaccessibility of eicosapentaenoic acid as ω-3 fatty acid in the seaweed of *Fucus spiralis* is in the range of 13% (Francisco et al. 2020). In this context, the weak bioaccessibility of lipid can be correlated to the high amount of glycolipids and phospholipids as the two forms of intracellular seaweed lipid, which are related to cell walls (Demarco et al. 2022). Moreover, the seaweed lipids, such as docosahexaenoic acid (DHA), have characterized by their nutraceutical properties and their great role in the improvement of the human health even with a low amount (Arterburn et al. 2007). Indeed, the DHA is recognized as a safe compound for human health.

The carbohydrates and polysaccharides are considered among the essential and important macronutrients in the seaweeds. They are recognized as a source of dietary fibers and they are generally not totally digested by the human stomach and intestinal enzymes (Shofia et al. 2018; Zheng et al. 2020). Hence, they can interfere with the bioaccessibility of other dietary substances (Zheng et al. 2020). Therefore, the seaweed polysaccharides are considered as prebiotic foods for human health.

It has been reported that the carbohydrates and polysaccharides are available in macroalgae with higher amount and they are recognized as functional ingredients in food processing due to their ability of the retaining water and their great role in the enhance of the consistence and sensorial attributes of food products as well as their properties of stabilization, emulsification (Qin 2018).

2.3 Benefits and Industrial Production of Seaweeds

Several previous research studies have been revealed that diverse seaweed have characterized by their biochemical and nutritional values (Rizk 1997). In fact, their application for human food are due generally to their excellent nutritional value in fibers, bioactive peptides, proteins, carbohydrates, vitamin, unsaturated fatty acids, and minerals (Roohinejad et al. 2017; Salehi et al. 2019; Ferrara 2020; Peñalver et al. 2020; Harb and Chow 2022). Furthermore, these macroalgae are considered an important source of other phytochemical and bioactive compounds, including polyphenol, alginates, laminarin, carrageenan, and agar (Stiger-Pouvreau et al. 2016). Therefore, the seaweeds are characterized by their bioactive compounds, and thus, they can be applied to diverse food products, such as fish, dairy, meat, vegetable-based products, in order to make them more effective and functional (Roohinejad et al. 2017).

In this regards, the functional ingredients in seaweed showed numerous bioactive properties, for instance antioxidant, antimicrobial, antiviral, antifungal, anti-inflammatory, anti-tumor, anti-methanogenic, and several more properties. Hence, the macroalgae have attracted a great attention in diverse domains, such as for medicinal and pharmaceutical preparations, and in food and nutraceutical industries as well as in the cosmeceutical field (Wijesinghe and Jeon 2012; Roohinejad et al. 2017).

In recent years, the production of seaweeds has received a great attention due to its health-promoting properties and its identification as a sustainable and functional source of renewable biomass (Bizzaro et al. 2022). It has been reported that the wild harvesting of seaweed species can conduct to the reduction of those natural ressources and also the utilization of the mechanical harvesting can contribute to the increase of the preoccupation about the sustainability of these applications and their destructive impacts on the marine ecosystem (Bizzaro et al. 2022). Therefore, further research studies are necessary to develop novel and effective practices for the production of valuable products.

3 Seaweeds as a Potential Source of Bioactive Compounds

3.1 Seaweed-Derived Bioactive Polysaccharides

3.1.1 Polysaccharide Contents in Seaweeds

Seaweeds are considered an important source of several polysaccharide compounds, which known by their biological activities. In fact, the polysaccharides are important macromolecules excessively present and distributed in the nature and the marine environment (Thanigaivel et al. 2022).

Previous studies have revealed that the total polysaccharides amounts of marine macroalgae are in the range of 4 to 76% of dry weight with an important content in the species of *Palmaria, Ascophyllum*, and *Porphyra*, depending on the species (Lafarga et al. 2020; Gullón et al. 2020). So, their concentrations depended on the seaweed species (Gullón, et al. 2020). At this regard, the green macroalgae, such as the *Ulva* species are characterized by a polysaccharides amount in the range of 65% of dry weight (Holdt and Kraan 2011).

Furthermore, the polysaccharides extracted from marine seaweeds are present generally in non-sulfated and sulfated forms (Hentati et al. 2020; Gullón et al.

2020). In this context, previous studies have reported that the sulfated polysaccharides are the most principal compounds, which characterized the green, red, and brown seaweeds (Raposo et al. 2015). In this regard, the sulfated polysaccharides have structural and storage roles in macroalgae and then can exhibit several important biological properties (Jiménez-Escrig et al. 2011). Several works have revealed that the composition of seaweeds, such as sulfated polysaccharide and other compounds and their biological activities could relate to several environmental factors, for instance the seasonal variation, localization, geographical, and nutritional characteristics of sea water as well as the post harvest factors, for instance the age, the extraction procedures of these bioactive compounds and the seaweed drying (Rioux et al. 2007).

On the other hand, the red seaweeds are characterized by their richness on carrageenan, porphyran, xylogalactans, agar, xylan, sulfated galactans, and floridean starch; the brown macroalgae contain fucoidans, laminarin, alginic acid (alginate), and fucans; while the green seaweeds are rich with ulvans (Deniaud-Bouët et al. 2017; Roohinejad et al. 2017; Pangestuti and Kurnianto 2017; Ciancia et al. 2020; Thanigaivel et al. 2022; Harb and Chow, 2022). In this context, these different groups of polysaccharides are recognized as bioactive metabolites due to their functional properties, such as antimicrobial, antimutagenic, antitumor, immunomodulatory, anti-inflammatory, antithrombotic, and antiviral activities (Raposo et al. 2015; Hentati et al. 2020; Gullón et al. 2020).

Furthermore, the seaweed polysaccharides can be used in the food industries as stabilizers, emulsifiers, and thickeners (Lafarga et al. 2020). Besides, these bioactive compounds are recognized as dietary fibers, which are not completely digested in the gastrointestinal tract owing to the enzymatic activity of microorganisms in the gut (Ganesan et al. 2019). It has been demonstrated that the edible macroalgae, such as *Ulva, Gracilaria, Acanthopora,* and *Codium,* are characterized by their richness on dietary fiber with an amount varied between 23.5% and 64% of dry weight (Garcia-Vaquero et al. 2017).

Qiu et al. (2022) have reported that there are different extraction and purification procedures can be used to isolate and purify the polysaccharides from several seaweeds. In fact, the choice of each extraction and purification assay related to the properties of the polysaccharides.

Carrageenans

Carrageenans "GRGs" are recognized as among the main principal group of phycocolloids or marine algal polysaccharides (Venkatesan et al. 2017; Ozogul et al. 2021; Pradhan et al. 2022). In fact, this group is generally derived from red seaweeds, such as *Chondrus*, *Eucheuma*, *Gigartina*, *Hypnea*, and *Iridaea* (Pradhan et al. 2022). Moreover, previous studies have reported that the species of *Kappaphycus* sp. and *Chondrus crispus* are characterized by an amount of 88% and 71% of carrageenan, respectively (Holdt and Kraan 2011).

On the other hand, there are three significant industrially forms of GRGs exploited, known by the iota, lambda, and Kappa (Pradhan et al. 2022; Lafarga et al. 2020). These forms of this polysaccharide are distinguished by the existence of 1, 2, and 3 ester sulfate groups (Lafarga et al. 2020).

It was also revealed that GRG has several functional properties and can widely used in food industries for the stabilizing, gelling, and thickening the food products, such as coffee, and frozen poultry, meat products (Pereira et al. 2009; Saha and Bhattacharya, 2010; Prajapati et al. 2014).

Alginate

Alginates are among the principal group of seaweed polysaccharides that are extracted principally from the cell walls of brown macroalgae, such as *Ascophyllum, Cystoseira, Durvillaea, Ecklonia, Hormophysa, Laminaria, Lessonia, Macrocystis, Saccharina,* and *Turbinaria,* etc. (Lee and Mooney 2012; Delaney et al. 2016; Roohinejad et al. 2017; Pradhan et al. 2022). Previous studies have demonstrated that the species of *Macrocystis pyrifera* and *Durvillaea potatorum* are characterized by their richness on alginates with an amount more than 45% and 55% of dry weight, respectively (Westermeier et al. 2012; Lorbeer et al. 2017).

This type of algal polysaccharides is considered as anionic polymers and namely also by algin or alginic acid (Roohinejad et al. 2017; Venkatesan et al. 2017). This biopolymer have a great commercial value and can be also used in the food industries due to their functional activities, such as gelling properties as well as their several biological activities, for instance biodegradable, non-immunogenic, biocompatible, and bio-adhesive (Delaney et al. 2016). They are therefore applied to many food processing as gelling and thickening agents to restructure the vegetable, baked and meat products, as well as for cream and desserts (Bixler and Porse 2011; Upadhyay et al. 2012). It has been revealed that the alginates can apply in food industry either alone or in combination with other polymers, principally, on frozen desserts and creams (Roohinejad et al. 2017). This algal polysaccharide is also utilized in the pharmaceutical industry, for instance, as binding agents for tablets, gastric alkalis, drug delivery, dental impressions, and wound dressings (Delaney et al. 2016; Venkatesan et al. 2017). Further, alginates have other industrial applications, such as the production of electrodes, textiles, and water processing (Delaney et al. 2016).

Fucoidans

Fucoidans are sulfated polysaccharides and commonly present in in the cell wall matrix of several brown macroalgae (Wijesinghe and Jeon 2012; Venkatesan et al. 2017; Lafarga et al. 2020). The amounts of Fucoidans of marine algae varied on the basis of diverse factors, such as the seaweed species, age, and the seasons (Lafarga et al. 2020). Previous studies have reported that the *Undaria pinnatifida* had a fucoidan content equivalent to 1.5% and the species of *Fucus vesiculosus* are characterized by a fucoidan amount in the range of 16 to 20% (Holdt and Kraan 2011).

This polymer has attracted a great attention due to their biological activities as well as they have crucial roles in the protection of marine algae against several environmental factors (Garcia-Vaquero et al. 2017). Moreover, this algal polysaccharide is known by their nutraceutical properties (Venkatesan et al. 2017). Further, the fucoidans can be applied for the treatment of cancer diseases and also in the tissue engineering (Jin and Kim 2011; Lee et al. 2012; Jeong et al. 2013; Kwak 2014; Puvaneswary et al. 2015).

Ulvan

Ulvans are cell wall sulfated polysaccharides and generally isolated and extracted from the green seaweeds (Venkatesan et al. 2017; Kidgell et al. 2019). It has been reported that the amount of this algal polysaccharide varied between 9 and 36% dry weight of the total biomass of the seaweed species of *Ulva* sp. (Kidgell et al. 2019). This polymer is essentially composed of xylose, sulfated rhamnose, and uronic acids (iduronic acid and glucuronic acid) (Lahaye and Robic 2007). Ulvans as bioactive compound have characterized by several industrial applications in biomaterial science, such as tissue engineering, wound dressings, excipients, and biofilm prevention, drug delivery (Alves et al. 2012a, b; Dash et al. 2014; Cunha and Grenha 2016; Venkatesan et al. 2015; Venkatesan et al. 2017; Kidgell et al. 2019). Besides, this seaweed polysaccharide is known by several functional properties, such as the antihyperlipidemic, anticoagulant, antioxidant, immunostimulatory, anticancer, antiviral, and anticoagulant, … etc. (Lahaye and Robic 2007; Leiro et al. 2007; Qi et al. 2012; Venkatesan et al. 2017; Kidgell et al. 2019). In this regard, the biological properties of ulvans as polysaccharides are principally correlated to their chemical structure (Kidgell et al. 2019).

3.1.2 Chemical Structures of Seaweed Polysaccharides

The seaweed walls are characterized by the predominance of mucilage on their skeleton, the abundance of the intercellular matrix and the presence of high quantity of sulfated macromolecules contrary to the walls of terrestrial plants (Stiger-Pouvreau et al. 2016; Hentati et al. 2020). The cell walls of seaweeds are constituted by two elements; like the phase, namely crystalline, which has the role of the skeleton and the amorphous phase known by the matrix, which included the skeleton. In fact, the crystalline phase composed of cellulose (β-(1 → 4)-D-Glc*p*) in each of three classes of seaweeds has characterized by a few differences from the green to red macroalgae (Hentati et al. 2020). Previous works have revealed that the polysaccharides characterized the red and green are of the type α-(1 → 4) and α-(1 → 6)-D-glucans with similar structures to starch, which characterized the terrestrial plants; however, the β-(1 → 3)-D-glucans (laminarin) is present in the brown macroalgae (Stiger-Pouvreau et al. 2016). Moreover, the matrix as an amorphous phase of seaweed polysaccharides are widely diversified from one macroalgae group to another (Stiger-Pouvreau et al. 2016).

Generally, structural polysaccharides have several properties and are well recognized as the most exploited compounds among the chemicals isolated from seaweeds (Usov and Zelinsky 2013). Previous studies have reported that the seaweed polysaccharides consist of only carbohydrate monomer structures and present also uronic acids (alginate) or dehydrated galactopyranosyl moieties (agar and carrageenans) (Rhein-Knudsen and Meyer 2021). Therefore, the structural characteristics of seaweeds give to their cells several properties, such as the enhancement of the rigidity and elasticity of the system by the incorporation of the combination between the magnesium carbonate ($MgCO_3$) with Calcium carbonate ($CaCO_3$) and strontium (Sr); the mechanical strength of the thalli (maintained by alginates, celluloses, and galactans); and the augmentation of the ionic interchanges by taking cations, like Na^+; Ca^{2+}, Mg^{2+}; and also the effective adaptation against dehydration due to the existence of sulfated polysaccharides (Hentati et al. 2020). Furthermore, the basic configurations of the diverse seaweed polysaccharides are well determined, however, specific structural details, for instance, the extent of 3,6-anhydrogalactopyranosyls in agar and carrageenans and the equilibrium between the category of uronic acid in alginate differ with seaweed species, growth site, life-stage, and water depth (Hentati et al. 2020).

On the other hand, the capacity to produce viscous solutions or gels is related to the composition and chemical structure of polysaccharides (Rhein-Knudsen and Meyer 2021). Indeed, gelling seaweed polysaccharides (phycocolloids), for instance, alginates, agars, and carrageenans are generated a large scale and they have a several domain of applications in the pharmaceutical, food, and cosmetic industries (Usov and Zelinsky 2013). Further, several algal polysaccharides without gelling capacity are being evaluated as biologically active substances (Rinaudo 2007).

In this context, the functional and physico-chemical properties of seaweed polysaccharides essentially depend on the sequence of monomer units in their molecules, for example: Although the chemical structures of the algal polysaccharides can be masked by the different changes occurred in the final steps of biosynthesis, the algenic acids have characterized by linear backbones consisted by diverse blocks of two monomeric units as well as the ulvans of the green seaweeds or the sulfated galactans of the red macroalgae are characterized by linear backbones, which composed of disaccharide repeating units (Usov and Zelinsky 2013).

In this context, the seaweed polysaccharides of carrageenans are sulfated linear galactans, which characterized by the presence of β-1,3- and α-1,4-linked D-galactosyl residues «D- and G- units» (Lafarga et al. 2020). Besides, this type of polysaccharide depended on the basis of the degree of sulphation of the molecule as well as the structure of the disaccharide repeating unit by the existence of 3,6-anhydrogalactose (DA-unit) and the pattern of sulfation (Knutsen et al. 1994).

It has been demosnstrated also that the fucoidans are sulfated polysaccharides, constituted with a backbone of (1 → 3)-linked α-L-fucopyranosy or alternating

(1 → 3)- and (1 → 4)-linked α-L-fucopyranosyl residues with variable degrees of sulphation. Chemically, these seaweed polysaccharides are considered a specific class with molecular weights varied between 43 and 1600 kDa (Lafarga et al. 2020).

In addition, the algenic acids are a linear polymer and they are essentially characterized by the presence of α-L-guluronic acid and 1,4-β-D-mannuronic acid (Roohinejad et al. 2017). Besides, in its salt form can be blended with many cations, like Mg^{2+}, Na^{+}, Ca^{2+}, and K^{+} and then presented as compounds of the cell walls and the intercellular matrix in all brown seaweeds (Usov and Zelinsky 2013).

As well the laminarins are considered as a group of seaweed polysaccharides characterized by energy storage. They constituted by 1,3-linked β-D-glucose monosaccharides with varying ramifications at β-(1,6) (Garcia-Vaquero et al. 2017). Indeed, the chemical structures of laminarin can change in the degree of polymerization and the degree of ramifications depending on the season, age, seaweed species and other parameters, like the purification, isolation, and extraction methods (Garcia-Vaquero et al. 2017). In this context, the *Laminaria* sp. species have characterized by their richness on laminarin with an amount of 32% depending on the season. Therefore, the amount of polysaccharides on *Laminaria* sp. is higher during the autumn and summer and then can be used as an energy source during winter (Lafarga et al. 2020). Besides, the Laminarins are present in *Ascophyllum, Undaria* sp., *Saccharina* sp., and *Fucus* (Lafarga et al. 2020).

On the other hand, the chemical structures of ulvans as polysaccharides of green seaweeds demonstrated an important variability and complexity. Indeed, its structure has characterized by the presence of several oligosaccharide-repeating structural units, which specified either its native structure or chemically altered. Therefore, the essential repeating disaccharide units known are ulvanobiouronic acid 3-sulfate types containing iduronic and glucuronic acids (Lahaye and Robic 2007).

3.1.3 Biological Activity of Polysaccharides

The seaweed polysaccharides are considered good functional compounds characterized by several biological activities, such as antioxidant, antimicrobial, anti-inflammatory, antitumor, anti-proliferative, antiviral, anti-diabetic, anticoagulant, and immunostimulatory (Table 5.1).

Antioxidant Activity

The reactive oxygen species (ROS) and other free radical groups are generated by abnormal metabolic reactions (Inanli et al. 2020; Mohan et al. 2020). Consequently, they lead to oxidative stress, which can cause damage and deterioration to living organisms and cellular macro-molecules (Šimat et al. 2020; Mohan et al. 2020). Therefore, this oxidative damage can contribute to the generation of several

Table 5.1 Overview of some relevant research, providing the biological properties of seaweed-derived polysaccharides

Polysaccharides	Seaweeds	Biological activities	Main findings	References
Sulfated polysaccharides (SP)	*Boergeseniella thuyoides*; *Sphaerococcus coronopifolius*	Antiviral activity	This study demonstrated that the sulfated polysaccharides can inhibit the replication *in vitro*, both of the *Herpes simplex* virus type 1 (HSV-1) on Vero cells and the Human Immunodeficiency Virus (HIV)	Bouhlal et al. (2011)
	Fucus evanescens	Antioxidant	This study revealed that the polysaccharides characterized by an antioxidant activity via DPPH and ABTS radical-scavenging assays	Imbs et al. (2015)
	Ulva lactuca	Antioxidant	The findings of this work shown that these biopolymers can exhibit an excellent scavenging activity via DPPH radical (IC_{50} = 13.56 μg/mL)	Yaich et al. (2017)
	Gracilaria birdiae	Antioxidant anticoagulant	This work proven that the SP characterized by antioxidant and anticoagulant activities depending on the extraction conditions	Fidelis et al. (2014)
Fucoidan; Laminarin	*Eisenia bicyclis*	Antitumor activity	This work revealed that these polysaccharides isolated from *E. bicyclis* were non-cytoxic to human melanoma SK-MEL-28, and colon cancer DLD-1 cells. Besides, they have antitumor activity due to the inhibition of the development of those cells	Ermakova et al. (2013)
	Laminaria hyperborea	Antimicrobial and antioxidant activities	This study demonstrated that the these polysaccharides can inhibit the bacterial growth of *Escherichia coli, Listeria monocytogenes, Salmonella typhimurium*, and *Staphylcoccus aureus*	Kadam et al. (2015)
	Ascophyllum nodosum		The *L. hyperborean* and *A. nodosum* demonstrated antioxidant activities via DPPH with inhibition levels of 87.57% and 93.23%, respectively	
	Fucus evanescens	Anticancer activity	The findings revealed that the these polysaccharides extracted possess an anticancer activity in vitro via the cells of human colon carcinoma	Hmelkov et al. (2017)

(continued)

Table 5.1 (continued)

Polysaccharides	Seaweeds	Biological activities	Main findings	References
Fucoidan	*Cystoseira sedoides*	Anti-inflammatory antioxidant, and gastroprotective activities	Fucoidans extracted from three different seaweeds characterized by their anti-inflammatory activity. Therefore, the treatement with these bioactive compounds of *C. crinita*, *C. compressa*, and *C. sedoides*:	Hadj et al. (2015)
			Can inhibit the development of the oedema by 58.21%, 56.81%, and 51%, respectively	
	Cystoseira crinite		Can exhibit an antioxidant activity with EC_{50} values of 0.76 mg/mL, 0.84 mg/mL, and 0.96 mg/mL, respectively	
	Cystoseira compressa		Can provide an excellent protective impact against gastric mucosal lesion	
Ulvan	*Ulva armoricana*	Immunostimulatory activity	The ulvan demonstrated an immunostimulatory activity:	Berri et al. (2017)
			Can increase the protein expression of cytokines (CCL20, TNFα, and IL8) and the mRNA via the ELISA and RT-qPCR assays	
			Can stimulate the TLR4	
			Can increase the phosphorylation of Akt and the p65 subunit of nuclear factor-κB	

degenerative diseases, such as cancers, cardiovascular disorders, inflammation, atherosclerosis, neurodegenerative diseases (Šimat et al. 2020; Mohan et al. 2020; Inanli, et al. 2020; Qiu et al. 2022). Generally, the free radical species are eliminated by the intervention of antioxidant enzymes in human organisms. Besides, it has been reported by several research studies that several natural compounds can be used for the treatment of the harmful and negative effects of free radicals on biological systems (Mohan et al. 2020). In this context, several research studies have reported that the diverse seaweed polysaccharides can exhibit an excellent antioxidant activity.

Furthermore, it has been reported that the sulfated polysaccharides, such as carrageenan, porphyran, fucoidan, alginic acid, laminaran, and sulfated galactan extracted from red and brown macroalgae, have characterized by significant antioxidant abilities (Jiménez-Escrig et al. 2011; Hentati et al. 2020; Lafarga et al. 2020).

Moreover, several works have demonstrated that the red seaweed polysaccharides isolated from *Gracilaria* genus can provide a strong scavenging activity against ABTS and superoxide radicals with approximately 50% at the concentration of 2.5 mg/mL (Di et al. 2017). Indeed, the considerable antioxidant activity of sulfated polysaccharides can be due to their lower size and the existence of sulfate groups that lead to the feasibility of the interaction between the between polysaccharide and target focus of cationic proteins, while the existence of non-sulfated sugar units at polysaccharide can decrease the antioxidant activity of these biopolymer (Jiménez-Escrig et al. 2011; Roohinejad et al. 2017). Moreover, the anti-DPPH, chelating and FRAP activities of seaweed polysaccharides depend on the existence of a number of functional groups, for example the carboxylic groups (–COOH) of the L-GulpA units, and the D-ManpA for alginate and the sulfates groups ($SO^{-}{}_{4}$) in the ortho-position for fucoidans (Kadam et al. 2013).

Previous work of Kadam et al. (2013) investigated the antioxidant activity the polysaccharides of fucoidans and the sodium alginates extracted from the brown seaweed *Cystoseira compressa* through the DPPH and FRAP methods. It has revealed that these two seaweed polysaccharides have powerful antioxidant abilities. Further, it has been proved that the fucoidans can enhance the antioxidant defense and extremely decrease the oxidative damage (Wijesinghe and Jeon 2012). Besides, It has been demonstrated that the sulfated fucans isolated from the brown macroalgae *Sargassum fulvellum* has a good NO scavenger more than the commercial antioxidants, for instance, the BHA, and α-tocopherol (Choi et al. 2007). Further, it has been revealed that the ulvans and carrageenans extracted from the *Ulva pertusa* "Chlorophyceae" and Spyridia hypnoides "Rhodophyceae", respectively, have a good antioxidant capacity due to their sulfate contents (Li et al. 2018; Sudharsan et al. 2018; Kidgell et al. 2019).

Antimicrobial and Antiviral Activities

Microbial contaminations and infections are considered among the main risks for human health around the world and therefore there is considerable concern has been given to preclude these problems due to their effects on the high mortality and morbidity. Besides, the apparition of multi-resistant microbial strains and the absence of the invention of new antibiotic drugs have increased the effect of microbial infections in the human population (Sanjeewa et al. 2018). Consequently, the natural compounds with antimicrobial abilities can develop and use as antimicrobial drugs. Indeed, the seaweed polysaccharides have been considered as excellent natural compounds to substitute the chemical compounds. In this context, it has been revealed that the sulfated polysaccharides isolated from the *Sargassum wightii* have a higher antibacterial capacity *in vitro* against several human pathogenic strains (Vijayabaskar et al. 2012). Moreover, previous work has demonstrated that the fucoidan extracted from *Sargassum wightii* has a powerful antibacterial activity against the *Vibrio parahaemolyticus* in the shrimp *Penaeus monodon* post-larvae (Sivagnanavelmurugan et al. 2015). Besides, the investigation *in vitro* of the antibacterial activity of sulfated polysaccharides extracted from *Sargassum fusiforme* has proven its efficacy against *Vibrio harveyi* in shrimp *Fenneropenaeus chinensis* (Huang et al. 2006).

On the other hand, the viral curative target various phases of the viral replication cycle. Indeed, the seaweed polysaccharides have recognized as powerful antiviral compounds, such as the sulfated polysaccharides (ulvans, fucose, carrageenans… etc.). In this regard, it has been demonstrated that the carrageenans isolated from *Chondrus crispus* has characterized by an excellent antiviral activity against HIV virus, herpes simplex viruses, and other pathogen virus (Campo et al. 2009). Besides, Gupta and Abu-Ghannam (2011) has revealed that the fucoidan has a powerful antiviral capacity against viruses, for instance, human cytomegalovirus (HCMV), and HIV virus. In this context, the galactofucan sulfates isolated from *Ulva pinnatifida* have demonstrated their potential antiviral ability against herpes viruses HSV-1, HSV-2, and HCMV (Wijesinghe and Jeon 2012). Furthermore, other studies have demonstrated that the sulfated polysaccharides containing the fucoses have characterized by their antiviral properties (Hayashi et al. 2008). Moreover, the ulvans isolated from *Ulva clathrata*, *Ulva lactuca*, *Ulva armoricana*, *Ulva compressa*, *Ulva pertusa*, and *Ulva intestinalis* have demonstrated their antiviral activities (Kidgell et al. 2019).

Other Biological Activities of Seaweed Polysaccharides

Cardiovascular diseases are the principal causes of death around the globe at present. Indeed, it has been revealed that the sulfated polysaccharides isolated from seaweeds have a good anti-coagulant activity and can be used as powerful anticoagulant compounds in the prevention and treatment of thromboembolic disorders (Sanjeewa et al. 2018). Besides, the fucoidan extracted from *Cladosiphon okamuranus* has revealed its cardioprotective properties (Thomes et al. 2010). Hence, it can decrease the myocardial risk, which appeared by the enhancement in divers parameters, like lactate dehydrogenase, aspartate transaminase, alanine transaminase, and creatinine phosphokinase (Wijesinghe and Jeon 2012).

As reported by Cui et al. (2018), the fucoidan isolated from the *Nemacystus decipiens* (Pheophyceae) has an excellent fibrinolytic activity by the increase of plasma t-PA/PAI-1 amount. Thus, the fucoidan can be used as an antithrombotic compound. Besides, the sulfated fucoidans extracted from seaweeds have demonstrated in vivo their anticoagulant capacity in Wistar rats (Wijesinghe et al. 2011).

Furthermore, it has been proved that *Cladosiphon fucoidan* can considered safe compound with a potential role in the gastric protection (Sanjeewa et al. 2018). Besides, the sulfated polysaccharides can exhibited anticancer and antitumor activities, such one extracted from *Sargassum polycystum* that can provide an antiproliferative activity at a concentration of 50 μg/mL, and thus, can provoke apoptosis mediated cell death against the breast cancer cell line MCF-7 due to the activation of caspase-8 (Hentati et al. 2020). Moreover, it has been demonstrated that the fucoidans isolated from *Sargassum* spp. can induce *in vitro* an apoptosis in melanoma B16 cancer cells and Lewis lung carcinoma cells (Sanjeewa et al. 2018).

3.1.4 Commercial Importance of Seaweed Polysaccharides

Seaweeds are recognized as a food in several regions in the word. Indeed, the commercial values of macroalgae are rising and depending to their richness with several bioactive constituents, for example carrageenan, alginate, and agar, ... etc. Therefore, these polysaccharides are mainly used in food industries and for other industrial applications. It has been reported that the principal processes of seaweed polysaccharides are their cultivation, harvesting, processing, extraction/isolation and their diverse applications (Venkatesan et al. 2017). In this regard, the commercial importance of algal polysaccharide raised in the modern research, for instance, the application of carrageenans and alginate in drug-delivery products and tissue engineering (BeMiller and Whistler 2012; Renn 1997; Venkatesan et al. 2017). Hence, these modern applications and other industrial utilization can increase the commercial importance of seaweed polysaccharides (Venkatesan et al. 2017).

3.2 *Seaweed-Derived Bioactive Peptides*

3.2.1 Peptides Contents in Seaweeds

The diverse classes of seaweeds have characterized by their nutritional compound and the presence of bioactive peptides. Generally, the peptides are defined as a protein fraction with a large spectrum of bioactivities. These fractions characterized by the presence of 3 to 40 amino acids (AA). In this context, peptides and proteins are considered an excellent source of the major AA, such as alanine, glutamic, proline, arginine, glycine, and aspartic acids, while threonine, lysine, methionine, cysteine, and tryptophan are limited. Generally, they are cationic and amphipathic compounds (Cunha and Pintado 2022). The AA profiles of seaweeds are recognized as safe and bioactive compounds by the WHO and FAO (Raja et al. 2022) The importance of seaweed-derived peptides is related to the role of some AA sequences, which are not functional in seaweed proteins, and exhibited a large spectrum of biological activities after being liberated from the seaweed proteins. In fact, peptides are generated from the protein molecules via the digestion process in the gastrointestinal tract, as well as produced during fermentation, thermal, chemical, or by enzymatic hydrolysis (Pliego-Cortés et al. 2020). Indeed, the protein contents in the seaweeds varied from 5% to 47% of dry weights. These bioactive compounds are abundant in the green and red macroalgae compared to the brown seaweeds (Gullón et al. 2020). These variations on amounts may be due to the method of extraction used to extract these compounds. In this regard, the protein amount in divers seaweeds could be appropriate to integrate it as functional and supplementary foods.

On the other hand, the peptide contents in different seaweeds depended on diverse factors, such as the seaweed species, seasons of harvest, and age. Recently, there has been a great demand to extract bioactive peptides from marine macroalgae (Lafarga et al. 2020). Indeed, the development of seaweed-derived peptides are considered an important sector due to the different biological and physiological activities of these molecules (Echave et al. 2022). However, the contents of these functional compounds produced from seaweeds are still restricted. Indeed, diverse functional foods rich with seaweed-derived peptides are recently occupied the markets of some country, like Japon (Lafarga et al. 2020).

Further, owing to the seaweed variety, an elevated diversity of seaweed bioactive peptides can be detected and complement those already isolated from other natural sources, including cereals, fish, meat, and milk (Haque and Chand 2008; Pampanin et al. 2012; Malaguti et al. 2014; Bizzaro et al. 2022).

3.2.2 Biological Activity of Seaweed Bioactive Peptides

Generally, it has been demonstrated that the seaweed-derived peptides have diverse biological properties, including antioxidant, antimicrobial, anti-hypertensive, anti-inflammatory, anti-diabetic, anti-obesity, anticancer, and anti-tumoral (Lafarga et al. 2020; Bizzaro et al. 2022; Echave et al. 2022) (Table 5.2). Indeed, the principal bioactivities characterized the seaweed peptides are the antioxidant and anti-hypertensive activities.

Antioxidant Activity

In the human body, the existence of an imbalance between the antioxidant compounds and activated oxygen species, can induce an oxidative stress, which related to diverse degenerative diseases, such as cancer, inflammatory, and aging diseases (Bizzaro et al. 2022). In this context, among the several bioactive seaweed compounds, the peptide substances are excellent antioxidant agents, such as the glutathione and carnosine (Lafarga et al. 2020). Further, the seaweed peptides <1 kDa extracted from *S. chordalis* can exhibit a powerful antioxidant abilities via the following assays, FRAP, DPPH, and ORAC, compared to the parent seaweed proteins (Pliego-Cortés et al. 2020). In this regard, previous study has reported that the seaweed peptides SDIAPGGNM exhibited an antioxidant activity by the FRAP assay in the range of 21 nmoL TE/μmoL peptide as well as demonstrated an ORAC in the range of 152 nmoL TE/μmoL peptide (Harnedy et al. 2017).

Furthermore, seaweeds can present relatively elevated protein amount, and seaweed-derived proteins can be utilized for the production of antioxidant hydrolysates and peptides (Heo et al. 2005; Lafarga et al. 2020). Besides, the study of Heo et al. (2005) has revealed that a considerable content of antioxidants hydrolysates of proteins or peptides can extracted from *Ishige okamurae*, *Ecklonia cava*, *Sargassum coreanum*, *Scytosipon lomentaria*, *Sargassum horneri*, *Sargassum thunbergii*, and

Table 5.2 Overview of some relevant research, providing the biological properties of seaweed derived peptides

Seaweeds	Peptide sequences	Biological activities	Main findings	References
Palmaria palmata	SDITRPGGNM	Antioxidant	This research study has demonstrated that the peptide sequence isolated from *Palmaria palmate* can exhibit an excellent antioxidant activity via the FRAP and ORAC assays with values of 21.23 ± 0.90 and 152.43 ± 2.73 nmol TE/μmol peptide, respectively	Harnedy et al. (2017)
			This seaweed-derived peptides can use as food preservatives due to their antioxidant activity	
	IRLIIVLMPILMA	Renin inhibitory; Anti-hypertensive in spontaneously hypertensive rats (SHRs)	This work demonstrated that the peptides extracted from seaweed *Palmaria palmate* characterized by renin inhibitory activity and can use for the treatment of hypertension	Fitzgerald et al. (2014)
	NIGK	Anti-atherosclerosis	This work revealed that the macroalgae-derived protein hydrolyzate can employ for the inhibition of platelet activating factor acetylhydrolase (PAF-AH)	Fitzgerald et al. (2013)
	VYRT, AGGEY, LRY,YRD, IKGHY, LDY, VDHY, FEQDWAS, LKNPG	ACE-I inhibitory	The results have demonstrated that the Palmaria palmate derived peptides can inhibit the ACE	Furuta et al. (2016)
			The peptide sequence LRY has an excellent ACE inhibitory activity with IC_{50}: 0.044 μmoL	

(continued)

Table 5.2 (continued)

Seaweeds	Peptide sequences	Biological activities	Main findings	References
Pyropia haitanensis	QTDDNHSNVLWAGFSR	Anti-proliferative and anticancer	This work has demonstrated that the *Pyropia haitanensis* derived peptides characterized by excellent anti-proliferation impacts on human liver cancer cells (HepG-2), human breast cancer cells (MCF-7), and human lung cancer cells (A549), with IC_{50} values of 59.09–272.67 μg/mL	Mao et al. (2017)
Porphyra spp.	ELS and GGSK	α-Amylase inhibitory	This work proven that the peptides isolated from *Porphyra* spp. characterized α-amylase inhibitory capacity, and thus, it can decrease the blood sugar	Admassu et al. (2018)
Porphyra yezoensis	NMEKGSSSVVSSRM (+15.99) KQ	Anticoagulant	The findings have revealed that the peptides enzymatically extracted from the seaweed *Porphyra yezoensis* can demonstrate their anticoagulation effect with IC_{50} values of 0.3 μM	Indumathi and Mehta (2016)
			This study demonstrated a dose-dependent extension of stimulating thromboplastin time varied between 35 s to 320 s	
	TPDSEAL	Antimicrobial	The peptide sequence provokes damage of the cell membrane and cell wall of *Staphylococcus aureus*	Jiao et al. (2019)

Table 5.2 (continued)

Seaweeds	Peptide sequences	Biological activities	Main findings	References
Gracilariopsis lemaneiformis	QVEY	Anti-hypertensive	The data of this work demonstrated that the peptides derived from seaweed *Gracilariopsis lemaneiformis* have remarkably ACE inhibitory activity with 78.15 ± 1.56%	Cao et al. (2017)
	FQIN [M(O)] CILR	Anti-hypertensive	The findings revealed that the peptides extracted from seaweed *Gracilariopsis lemaneiformis* have ACE inhibitory activity with IC_{50} of 9.64 ± 0.36 μM	Deng et al. (2018)
Porphyra dioica	DYYKR	Antioxidant	This work demonstrated that the peptides extracted from seaweed *Porphyra dioica* characterized by an antioxidant activity via the ORAC activity with 4.27 μmol TE/μM	Cermeño et al. (2019)
	TYIA	Anti-hypertensive	The peptides extracted from seaweed *Porphyra dioica* have ACE inhibitory activity with IC_{50} of 89.7 μM	
	YLVA	Anti-diabetic	The peptide sequence characterized by anti-diabetic activity via the DPP-IV inhibitor ability with IC_{50} of 439 mM	
Spirulina maxima	LDAVNR	Anti-inflammatory	This peptide can inhibit the histamine release and generation of antigen-stimulated mast cells; Inhibition of intracellular ROS production in cells; prevent IL-8 production from endothelial cells	Vo et al. (2013)

Sargassum fullvelum, by the use of commercial enzymes, such as Alcalase, Kojizyme, Neutrase, Flavourzyme, and Protamex. Moreover, the antioxidant capacity of peptides isolated from *Pyropia columbina* enhanced after a simulated gastrointestinal digestion (Cian et al. 2015).

Antimicrobial Activity

Generally, several peptides have demonstrated their multifunctional properties depending on their sequences of AA. Indeed, specific peptide sequences can present two or more different functional and biological properties. In this context, several research works have revealed that diverse peptides extracted from different sources of food materials have characterized by their ability to inhibit several microbial and viral species (Agyei and Danquah 2012). The antimicrobial peptides are considered excellent agents against diverse pathogens either in plants or animals (Cunha and Pintado 2022). However, there are few studies have investigated the antimicrobial capacity from seaweed proteins and seaweed-derived peptides (Echave et al. 2022). Indeed, few works have investigated the antimicrobial activity of macroalgae peptides. For example, it has been demonstrated that the seaweed peptides can exhibit an excellent antibacterial activity against *S. aureus, E. coli,* and MRSA (Jiao et al. 2019; Guzmán et al. 2019). Hence, these recent research works prove that the seaweeds can be considered an interesting source of antimicrobial peptides. Therefore, it would be a considerable interest to investigate the antimicrobial capacity of seaweeds peptides and thus can be examined for their industrial applications.

On the other hand, it has been reported in previous work that the seaweed proteins "phycobiliproteins" isolated from *Hydropuntia cornea* can exhibit an antifungal activity against the fungi species of *Botrytis cinerea* (Righini et al. 2020). This study has demonstrated that the phycobiliproteins at a concentration of 0.3 mg/mL can decrease the germination of spores and fungal growth by 80% and 33%, respectively.

Anti-hypertensive Activities

Previous works have revealed that the marine species, principally the seaweeds, have characterized by their excellent anti-hypertensive activity. Indeed, several studies have proven the hypotensive impact of seaweed peptides from all macroalgae groups. In this context, these research studies have suggested that this activity may be due to the inhibition of angiotensin I-converter enzyme (ACE) or the renin inhibition (Echave et al. 2022). In fact, the main of the antihy-pertensive peptides are characterized by MW under 1.5 KDa and their role represented by the inhibition of ACE in a dose-dependent manner (Cunha and Pintado 2022). However, it has been demonstrated *in vitro* that one bioactive peptide extracted from the macroalgae *Palmaria palmata* has a renin inhibitory property (Cunha and Pintado 2022). Besides, the peptides FY, IW IY, KY, YNKL, YH, and IW isolated from *Undaria pinnatifida* and the peptides TGAPCR and FQIN [M (O)] CILR isolated from *Gracilariopsis lemaneiformis* have revealed *in vitro* their effective ACE inhibitory activities (Deng et al. 2018; Suetsuna and Nakano 2000). Therefore, the seaweed peptides IY, FY, KY, and YH were produced by hot water extraction, while all the others were extracted by enzymatic hydrolysis assay with trypsin or pepsin (Cunha and Pintado 2022).

4 The Potential Benefits of Seaweed-Derived Peptides and Polysaccharides for Human Health

Several previous works have revealed the important roles of the seaweed-derived bioactive compounds for human health (Fig. 5.2). In this context, bioactive peptides and polysaccharides extracted from diverse seaweed species are beneficial to human health via different modalities, including the enhancement of digestive efficacy (Bizzaro et al. 2022).

Generally, the human gastrointestinal microbiome characterized by the presence of a variety of microbial communities. Indeed, these microbe species have important and effective roles in the human body, including immunologic, protective, and also metabolic functions. Indeed, the diseases related to dysbiosis of the gut microbiota can be enhanced by curative methods (Qiu et al. 2022). However, the application of natural compounds, such as seaweed-derived bioactive compounds, both polysaccharides and peptides can also induce prebiotic activity through the regulation of the homeostasis of gut microbiota and the increase of their energy metabolism (Qiu et al. 2022).

In this regard, the seaweed-derived polysaccharides, such as laminarin, carrageenan, fucoidan, alginates, floridoside, laminaran, isofloridoside, and agar are recognized as dietary fibres because they are undigested by the human gastrointestinal track. Thereby, they are completely or partially fermented by the gut microbiota

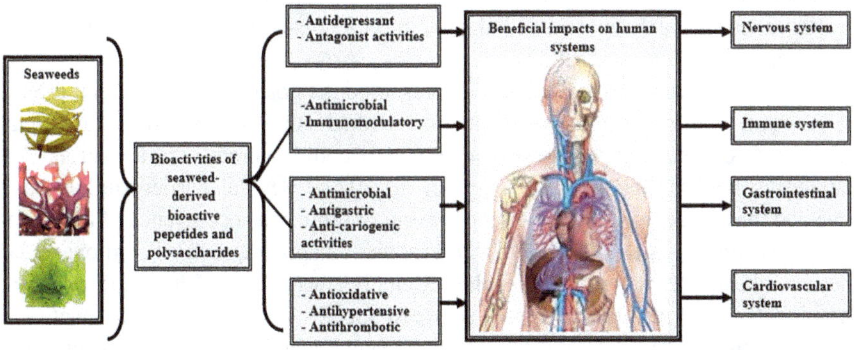

Fig. 5.2 Representation of various bioacivities of seaweed-derived peptides and polysaccharides and their beneficial impacts for human health

depending on their chemical structures (Rioux et al. 2007). Besides, these seaweeds-derived bio-compounds are recognized by FDA as safe (GRAS) compounds for oral administration. Therefore, these bioactive compounds can contribute to the gut's human health owing to their water-binding characteristics and thus their impacts in the reduction of the colonic transit times, which can be advantageous in the inhibition and the prevention of colonic cancer (Lafarga et al. 2020).

Furthermore, it has been revealed that the polysaccharides extracted from *Gracilaria rubra* have no impacts on simulated small intestine digestion (Di et al. 2018). Besides, previous works have proven than the red seaweed polysaccharides, isolated from *Porphyra haitanensis*, when applied with saliva can provoke a stimulation of small intestinal fluids, and gastric acid without the modification or the degradation in the molecular weight of these bioactive compounds (Xie and Cheong 2021). This study has revealed that these seaweed polysaccharides cannot be degenerated in the digestive tract. Hence, the incapability to digest some seaweed polysaccharides can due the lack of their appropriate enzymes in the upper gastrointestinal tract (Qiu et al. 2022).

In this context, earlier *in vivo* works have demonstrated the fermentation and digestion behaviors of seaweed polysaccharides and their excellent prebiotic activities. Consequently, several polysaccharides are considered as excellent food ingredients or biological and natural health products (Rioux et al. 2007). Furthermore, it has been revealed that the laminarins as seaweed polysaccharide contribute to the modulation of the intestinal metabolisms through their impacts on the microbiology and biochemistry of human gut microflora (Lafarga et al. 2020). Moreover, it has been demonstrated *in vivo* studies that the laminarins can induce a regulation of the pro and anti-inflammatory cytokines (Lafarga et al. 2020). Further, another *in vivo* works have revealed that the application of fucoidan and laminarins can reduce the concentration of *E. coli* during suckling and also can enhance the immunoglobulin amounts during the pregnancy and lactation periods (Lafarga et al. 2020).

Therefore, the seaweed polysaccharides have excellent effects on gut microbiota and the production of short-chain fatty acids, which consider the principal bacterial fermentation metabolites. In fact, they can regulate the composition of gut microbiota and thus prevent its dysbiosis or imbalance, which related to several human health concerns, such as cancer, diabetes, immune diseases, obesity, inflammatory bowel disorders (Qiu et al. 2022).

On the other hand, several works have reported the positive and beneficial effects of seaweed bioactive peptides for human health. In this context, these functional ingredients can present nutritional benefits that could decrease the appearance of dangerous health challenges (Bizzaro et al. 2022). Several earlier studies have revealed that the seaweed peptides can prevent or inhibit the occurrence of the metabolic syndromes which related to several medical disorders, such as cardiovascular disease, obesity, hypertension, diabetes … etc. (Lafarga et al. 2020). In this regard, the bioactive peptides isolated from seaweed species can be applied as functional substance in healthy diets and treat, and prevent several diverse chronic diseases (Brown et al. 2014; Cherry et al. 2019).

Furthermore, previous study performed for colorectal and stomach has demonstrated that bioactive peptides isolated from diverse groups of seaweeds have anti-tumorigenic effects (Olivares-Bãnuelos et al. 2019; Minami et al. 2020). Further, these bioactive compounds have revealed their impacts on the regulation of diabetes and obesity. Indeed, the use of *Saccharina japonica* and *Ascophyllum nodosum* in the obese individuals can decrease the glucose contents in blood after a four-week (Bermano et al. 2020; Kim et al. 2008). Besides, the seaweed-derived peptides have been characterized by their role in the prevention of cardiovascular disorders (Murai et al. 2019).

Thereby, seaweed-derived peptides and polysaccharides provide positive and excellent human health impacts, including improvement of gut health, immune regulation, reduction of the risk both cardiovascular and osteoporosis, protection against radiation, decrease of blood pressure, glucose, and fat, as well as prevention and inhibition of Alzheimer diseases.

5 Food Applications of Seaweed-Derived Bioactive Peptides and Polysaccharides

Marine macroalgae are considered as a considerable healthy food for humans due to their richness with many bioactive compounds, such as polysaccharides and peptides, which recognized as dietary fibers. Therefore, they have been increasingly studied and attracted a great interest in the last years. The seaweed-derived bioactive substances showed various biological properties, and hence, can be applied in different industrial areas, such as food industry. They characterized by many biophysical properties, including gelling, thickening that are indispensable for the production of functional foods (Raja et al. 2022). In this context, the nutritional and organoleptic attributes of the food products produced depending on the characteristics of seaweeds used and their bioactive compounds. Hence, diverse seaweeds can be used to manufacture different functional food products, including meat and poultry processing, canned fish, dairy products, desserts, beverages, frozen, ice-cream, and jelly (Demarco et al. 2022) (Table 5.3).

Meat and poultry and their products have a considerable role in the food diet of consumers, principally in advanced countries. However, the exaggerated consumption of meat and meat products is generally correlated to health risks, including colon cancer and cardiovascular diseases. Consequently, the production of meat-based functional foods can reduce the negative and dangerous health effects related to the consumption of meat and meat products and enhance their nutritive value (Raja et al. 2022). In this regard, seaweeds provide a considerable nutritional value due to its richness on dietary fibers, high proteins, peptides, carbohydrates, polysaccharides, and low lipid amount. So, the application and integration of the whole seaweeds or their extracts into meat and poultry products can improve the sensorial, nutritional, and physicochemical attributes of meat. Such use can also decrease both

Table 5.3 Relevant research on applications of seaweed and their bioactive compounds in different foods

Food products	Seaweeds	Mains findings	References
Meat products			
Pork products	*Himanthalia elongate* (Sea Spaghetti)	The incorporation of seaweed alginates as polysaccharides in food matrix can prevent the thermal deterioration of the protein fractions	Fernandez-Martin et al. (2009)
		Such incorporation can increase the elasticity and hardness and improve the water/oil retention ability	
		The addition of Sea Spaghetti can improve the antioxidant activity *in situ*	Cofrades et al. (2011) and López-López et al. (2009)
		Improve the n-3 PUFA amount	
		Increase the mineral levels (e.g. Ca, Mn, K, Mg…)	
		Provide bioactive compounds and dietary fibers in food matrix and thus can improve the health benefits	
	Undaria pinnatifida (Wakame)	The incorporation of Wakame can decrease the thrombogenic index	
		Enhance the n-3 PUFA levels, but decrease the n-6/n-3 PUFA ratio	
		Enhance the mineral levels (e.g. Ca, Mn, K, Mg…)	
	Porphyra umbilicalis (Nori)	The addition of Nori can reduce the thrombogenic index	
		Incorporation of Nori improved the amounts of amino acids	
		Enhance the mineral concentrations	
		Increase the n-3 PUFA	
	Laminaria digitate	The laminarin and fucoidan as seaweed polysaccharides present an excellent antioxidant activity of the functional cooked minced meat products	Moroney et al. (2013)
Beef products	*Himanthalia elongate* (Sea Spaghetti)	The incorporation of seaweeds can enhance the antioxidant activity	Cox and Abu-Ghannam (2013a, b)
		Inhibit the microbial growth and lipid oxidation	
		Enhance the sensory and physico-chemical attributes	
Chicken products	*Undaria pinnatifida* (Wakame)	The addition of seaweed can improve the sensory attributes of chicken meat and prevent the lipid peroxidation during storage	Sasaki et al. (2008)

(continued)

Table 5.3 (continued)

Food products	Seaweeds	Mains findings	References
Seafood products	*Eucheuma*	The seaweed addition into fish can enhance the sensory attributes	Senthil et al. (2005)
	Fucus vesiculosus	The application of seaweed into fish muscle can demonstrate an excellent antioxidant activity via DPPH and ORAC assays and contribute to the inhibition of lipid oxidation	Jónsdóttir et al. (2015)
Milk	*Ascophyllum nodosum*	The incorporation of seaweed enhances the shelf-life and quality of milk	O'Sullivan et al. (2011)
	Fucus vesiculosus		
Bread	*Kappaphycus alvarezii*	The incorporation of seaweed enhances the dough absorption and the organoleptic characteristics (e.g. firmness)	Mamat et al. (2014)

microbial growth and oxidation process (Roohinejad et al. 2017). Therefore, the incorporation of seaweed-derived bioactive compounds, such as functional bioactive peptides and polysaccharides, in meats and their products augment the shelf life and maintain their quality during their processing and storage (Gullón et al. 2020). In this regard, the application of these phytochemical compounds must be efficient at low amount, preserve the shelf life of food products during processing, do not be harmful to the consumer, do not influence their organoleptic properties, such as taste, flavor, and appearance (Lorenzo et al. 2018).

Thereby, the seaweed-derived peptides and polysaccharides can contribute to the enhanced of the fat and water binding properties that chewiness and the stability of meat structure. Besides, a previous research study has revealed that the incorporation of the macroalgae bioactive compounds in pork patties can extremely enhance the protein stability and inhibit the lipid oxidation compared to the control patties formulated with synthetic antioxidants, during storage for 180 days and at 4 °C (Raja et al. 2022). Moreover, the production of chicken nuggets by the application of the macroalgae extract as fat alternative can significantly enhance the water retention properties, fat, and cooking yield due to the presence of polysaccharides, like the carrageenan compared to the ordinary chicken nuggets (Raja et al. 2022). Besides, the application of fucoidan and laminarin isolated from seaweeds can enhance the antioxidative capacity of functional cooked meat products and thus improve the human antioxidant defense systems (Roohinejad et al. 2017). Previous works have revealed that the incorporation of seaweed extracts into seafood products can enhance their textural and sensorial properties (Roohinejad et al. 2017). Therefore, the application of seaweeds-derived bioactive compounds into meats and poultry can contribute to the inhibition of negative impacts of these food products.

On the other hand, the oxidative reactions can affect the appearance and safety of lipid foods and cause a reduction of their nutritional value. However, the use of macroalgae bioactive compounds, such as ulvan as polysaccharide can delay and inhibit the oxidation and peroxidation processes due to their excellent antioxidant activity and therefore it can be utilized in the food industry (Pangestuti and Kurnianto 2017). Further, the seaweed bioactive peptides have a role fundamental in food packaging and in the inhibition of oxidative stress, which considered as one of the responsible agents for food spoilage (Gómez-Estaca et al. 2014). Besides, these bioactive compounds have an excellent potential as food preservative owing to their antimicrobial ability against several microbial strains, and thus can lead to preserve the food matrix (Cunha and Pintado 2022).

The dairy products characterized by the presence of calcium in the form of casein, which cannot be reabsorbed by the people suffer from the lack of casein-degrading enzymes. Therefore, the incorporation of seaweed bioactive compounds in various dairy products, such as yoghurt, fermented milk products, and diverse forms of cheeses (cottage cheese, smoked cheese, Quarg Fresh cheese, processed cheese, and Appenzeller cheese) can enhance their sensory and nutritional characteristics (Raja et al. 2022). Furthermore, the application of seaweed powder into bakery products can contribute to the improve their nutritional attributes without influencing the organoleptic properties and overall acceptability. Therefore, the application of seaweed-based bioactive compounds can contribute to enhance the nutritional value of food products and improve their microbiological, antioxidant, and sensorial properties.

6 Future Trends and Challenges

In the last decades, the exploitation of seaweeds has been extremely increasing around the world. These marine microscopic organisms characterized by useful components, including macro- and micronutrients, which have an important potential as alternative ressources to those chemicals. The seaweed-derived compounds, including proteins, bioactive peptides, carbohydrates, polysaccharides, pigments, vitamins, and minerals have been exploited for their functional properties. Therefore, the bioactive peptides and polysaccharides have been proven to possess a wide range of biological activities, like antioxidant, antimicrobial, anti-hypertensive, anti-inflammatory, antitumoral, anticancer activities. Indeed, theses bioactive compounds have multiple advantages and several health-promoting impacts. So, they can be incorporated into food products, such as meat, dairy products, beverages, and bread. Indeed, they used as dietary fibers due to their functional and prebiotic properties.

On the other hand, the production and application of seaweed-derived bioactive substances surpass the challenges and limitations. However, further studies need to be performed to more understand the impacts the incorporation of each seaweed-derived bioactive compounds on some diseases, including inflammatory and cancer disorders. Although several studies have investigated the antimicrobial activity *in vitro* of these functional substances, there are few works were evaluated their activity in food products. Besides, there are some challenges in their commercialization due to the possibility sensory effects when they incorporated into foods. Further, the potential contamination of seaweed products with heavy metals and industrial waste still require to be investigated before these ressources can be applied in food and pharmacological industries.

7 Conclusions

In the last decades, the seaweeds have received great attention due to their riches on interesting bioactive compounds, such as bioactive peptides, pigments, polysaccharides, PUFAs, phenolic compounds, which characterized by their functional and nutritional values. Indeed, several seaweeds-derived bio-compounds have demonstrated a wide range of biological properties both *in vivo* or *in vitro*. Therefore, it has been reported that the bioactive peptides and polysaccharides isolated from macroalgae groups can exhibit many biological activities, including antioxidants, anti-hypertensive, antimicrobial, anti-diabetic, anti-inflammatory, anti-coagulant, anticancer, anti-tumor. Concequently, several research works have investigated the relationship between the chemical structures of these bioactive compounds and their biological properties in order to develop functional food ingredients or supplements. Thereby, the seaweed-derived peptides and polysaccharides are widely utilized in the food applications for diverse properties to enhance the nutritional and qualities of food products. Indeed, they can be used as emulsifier, stabilizer, thickening ingredients in the food industry. Besides, these bioactive compounds have demonstrated their excellent roles in the improvement of human health.

Acknowledgments This work was supported by the Scientific Research Project Office of Cukurova University under contract no: FBA-2022-15153.

References

Admassu H, Gasmalla MA, Yang R, Zhao W (2018) Identification of bioactive peptides with α-amylase inhibitory potential from enzymatic protein hydrolysates of red seaweed (*Porphyra* spp). J Agric Food Chem 66:4872–4882

Agyei D, Danquah MK (2012) Rethinking food-derived bioactive peptides for antimicrobial and immunomodulatory activities. Trends Food Sci Technol 23:62–69

Alves A, Duarte ARC, Mano JF, Sousa RA, Reis RL (2012a) PDLLA enriched with ulvan particles as a novel 3D porous scaffold targeted for bone engineering. J Supercrit Fluids 65:32–38

Alves A, Pinho ED, Neves NM, Sousa RA, Reis RL (2012b) Processing ulvan into 2D structures: cross-linked ulvan membranes as new biomaterials for drug delivery applications. Int J Pharma 426(1):76–81

Arterburn LM, Oken HA, Hoffman JP, Bailey-Hall E, Chung G, Rom D, Hamersley J, McCarthy D (2007) Bioequivalence of docosahexaenoic acid fromdifferent algal oils in capsules and in a DHA-fortified food. Lipids 42(11):1011

Banwo K, Olojede AO, Adesulu-Dahunsi AT, Verma DK, Thakur M, Tripathy S, Singh S, Patel AR, Gupta AK, Aguilar GN, Utama GL (2021) Functional importance of bioactive compounds of foods with Potential Health Benefits: a review on recent trends. Food Biosci 43:101320

Baweja P, Kumar S, Sahoo D, Levine I (2016) Biology of seaweeds. In: Fleurence J, Levine I (eds) Seaweed in health and disease prevention. Elsevier/Academic, London, pp 41–106

BeMiller J, Whistler R (2012) Industrial gums: polysaccharides and their derivatives. Elsevier, San Diego

Bermano G, Stoyanova T, Hennequart F, Wainwright CL (2020) Chapter 8: Seaweed-derived bioactives as potential energy regulators in obesity and type 2 diabetes. In: Du G (ed) Advances in pharmacology, vol 87. Academic, pp 205–256

Berri M, Olivier M, Holbert S, Dupont J, Demais H, Le Goff M et al (2017) Ulvan from *Ulva armoricana* (Chlorophyta) activates the PI3K/Akt signalling pathway via TLR4 to induce intestinal cytokine production. Algal Res 28:39–47

Biancarosa I, Espe M, Bruckner C, Heesch S, Liland N, Waagbø R, Torstensen B, Lock EJ (2017) Amino acid composition, protein content, and nitrogen-to-protein conversion factors of 21 seaweed species from Norwegian waters. J Appl Phycol 29(2):1001–1009

Bizzaro G, Vatland AK, Pampanin DM (2022) The One-Health approach in seaweed food production. Environ Int 158:106948

Bixler HJ, Porse H (2011) A decade of change in the seaweed hydrocolloids industry. J Appl Phycol 23:321–335. https://doi.org/10.1007/s10811-010-9529-3

Bleakley S, Hayes M (2017) Algal proteins: extraction, application, and challenges concerning production. Food Secur 6(5):33

Bonfanti C, Cardosob C, Afonso C, Matos J, Garcia T, Tanni S, Bandarra NM (2018) Potential of microalga *Isochrysis galbana*: bioactivity and bioaccessibility. Algal Res 29:242–248

Bouhlal R, Haslin C, Chermann JC, Colliec-Jouault S, Sinquin C, Simon G et al (2011) Antiviral activities of sulfated polysaccharides isolated from *Sphaerococcus coronopifolius* (Rhodophytha, Gigartinales) and *Boergeseniella thuyoides* (Rhodophyta, Ceramiales). Mar Drugs 9:1187–1209

Brown EM, Allsopp PJ, Magee PJ, Gill CI, Nitecki S, Strain CR et al (2014) Seaweed and human health. Nutr Rev 72(3):205–216

Brownlee IA, Allen A, Pearson JP, Dettmar PW, Havler ME, Atherton MR, Onsøyen E (2005) Alginate as a source of dietary fiber. Crit Rev Food Sci Nutr 45(6):497–510

Campo VL, Kawano DF, Silva DBD, Carvalho I (2009) Carrageenans: biological properties, chemical modifications and structural analysis: a review. Carbohydr Polym 77:167–180

Cao D, Lv X, Xu X, Yu H, Sun X, Xu N (2017) Purification and identification of a novel ACE inhibitory peptide from marine alga *Gracilariopsis lemaneiformis* protein hydrolysate. Eur Food Res Technol 243(10):1829–1837

Cassani L, Gomez-Zavaglia A, Jimenez-Lopez C, Lourenço-Lopes C, Prieto MA, Simal-Gandara J (2020) Seaweed-based natural ingredients: Stability of phlorotannins during extraction, storage, passage through the gastrointestinal tract and potential incorporation into functional foods. Food Res Int 137:109676

Cermeño M, Stack J, Tobin PR, O'Keeffe MB, Harnedy PA, Stengel DB et al (2019) Peptide identification from a: Porphyra dioica protein hydrolysate with antioxidant, angiotensin converting enzyme and dipeptidyl peptidase IV inhibitory activities. Food and Funct 10(6):3421–3429

Chen K, Roca M (2018) *In vitro* bioavailability of chlorophyll pigments from edible seaweeds. J Funct Foods 41:25–33

Cherry P, O'Hara C, Magee PJ, McSorley EM, Allsopp PJ (2019) Risks and benefits of consuming edible seaweeds. Nutr Rev 77(5):307–329

Choi DS, Athukorala Y, Jeon YJ, Senevirathne M, Cho KR, Kim SH (2007) Antioxidant activity of sulfated polysaccharides isolated from *Sargassum fulvellum*. Prev Nutr Food Sci 12:65–73

Cian RE, Garzón AG, Ancona DB, Guerrero LC, Drago SR (2015) Hydrolyzates from *Pyropia columbina* seaweed have antiplatelet aggregation, antioxidant and ACE I inhibitory peptides which maintain bioactivity after simulated gastrointestinal digestion. LWT-Food Sci Technol 64:881–888

Ciancia M, Fernandez PV, Leliaert F (2020) Diversity of sulfated polysaccharides from cell walls of coenocytic green algae and their structural relationships in view of green algal evolution. Front Plant Sci 11:Article 554585

Cofrades S, López-López I, Ruiz-Capillas C, Triki M, Jiménez-Colmenero F (2011) Quality characteristics of low-salt restructured poultry with microbial transglutaminase and seaweed. Meat Sci 87(4):373–380

Costa LS, Fidelis GP, Cordeiro SL, Oliveira RM, Sabry DA, Câmara RBG, Nobre LTDB, Costa MSSP, Almeida-Lima J, Farias EHC, EL Rocha HAOL (2010) Biological activities of sulfated polysaccharides from tropical seaweeds. Biomed Pharmacother 64(1):21–28

Cox S, Abu-Ghannam N (2013a) Incorporation of Himanthalia elongata seaweed to enhance the phytochemical content of breadsticks using Response SurfaceMethodology (RSM). Int Food Res J 20(4):1537–1545

Cox S, Abu-Ghannam N (2013b) Enhancement of the phytochemical and fibre content of beef patties with Himanthalia elongata seaweed. Int J Food Sci Technol 48(11):2239–2249

Cui K, Tai W, Shan X, Hao J, Li G, Yu G (2018) Structural characterization and anti-thrombotic properties of fucoidan from Nemacystus decipiens. Int J Biol Macromol 120:1817–1822

Cunha L, Grenha A (2016) Sulfated seaweed polysaccharides as multifunctional materials in drug delivery applications. Mar Drugs 14(3):42

Cunha SA, Pintado ME (2022) Bioactive peptides derived from marine sources: Biological and functional properties. Trends Food Sci Technol 119:348–370

Dash M, Samal SK, Bartoli C, Morelli A, Smet PF, Dubruel P, Chiellini F (2014) Biofunctionalization of ulvan scaffolds for bone tissue engineering. ACS Appl Mater Interfaces 6(5):3211–3218

Delaney A, Frangoudes K, Ii SA (2016) Society and Seaweed: understanding the past and present. In: Fleurence J, Levine I (eds) Seaweed in health and disease prevention. Elsevier/Academic, London

Demarco M, de Moraes JO, Matos AP, Derner RB, Neves FF, Tribuzi G (2022) Digestibility, bioaccessibility and bioactivity of compounds from algae. Trends Food Sci Technol 121:114–128

Deng Z, Liu Y, Wang J, Wu S, Geng L, Sui Z et al (2018) Antihypertensive effects of two novel angiotensin I-converting enzyme (ace) inhibitory peptides from gracilariopsis lemaneiformis (Rhodophyta) in spontaneously hypertensive rats (SHRs). Mar Drugs 16(9)

Deniaud-Bouët E, Hardouin K, Potin P, Kloareg B, Herve C (2017) A review about brown algal cell walls and fucose-containing sulfated polysaccharides: cell wall context, biomedical properties and key research challenges. Carbohydr Polym 175:395–408

Di T, Chen G, Sun Y, Ou S, Zeng X, Ye H (2017) Antioxidant and immunostimulating activities in vitro of sulfated polysaccharides isolated from *Gracilaria rubra*. J Funct Foods 28:64–75

Di T, Chen G, Sun Y, Ou S, Zeng X, Ye H (2018) *In vitro* digestion by saliva, simulated gastric and small intestinal juices and fermentation by human fecal microbiota of sulfated polysaccharides from *Gracilaria rubra*. J Funct Foods 40:18–27

Domínguez-González R, Romarís-Hortas V, García-Sartal C, Moreda-Pineiro A, Barciela-Alonso MC, Bermejo-Barrera P (2010) Evaluation of an *in vitro* method to estimate trace elements bioavailability inedible seaweeds. Talanta 82(5):1668–1673

Echave J, Otero P, Garcia-oliveira P, Munekata PES, Pateiro M, Lorenzo JM, Simal-Gandara J, Prieto MA (2022) Seaweed-derived proteins and peptides: promising marine bioactives. Antioxidant 11:176

Ermakova S, Men'shova R, Vishchuk O, Kim SM, Um BH, Isakov V et al (2013) Water-soluble polysaccharides from the brown alga *Eisenia bicyclis*: Structural characteristics and antitumor activity. Algal Res 2:51–58

Fernandez-Martin F, Lopez-Lopez I, Cofrades S, Colmenero FJ (2009) Influence of adding Sea Spaghetti seaweed and replacing the animal fat with olive oil or a konjac gel on pork meat batter gelation. Potential protein/alginate association. Meat Sci 83(2):209–217

Ferraces-Casais P, Lage-Yusty MA, de Quirós ARB, López-Hernández J (2012) Evaluation of bioactive compounds in fresh edible seaweeds. Food Anal Methods 5(4):828–834

Ferrara L (2020) Seaweeds: a food for our future. J Food Chem Nanotechnol 6:56–64

Fidelis GP, Camara RB, Queiroz M, Santos Pereira Costa MS, Santos PC, Rocha HA et al (2014) Proteolysis, NaOH and ultrasound-enhanced extraction of anticoagulant and antioxidant sulfated polysaccharides from the edible seaweed, *Gracilaria birdiae*. Molecules 19:18511–18526

Fitzgerald C, Gallagher E, O'Connor P, Prieto J, Mora-Soler L, Grealy M et al (2013) Development of a seaweed derived platelet activating factor acetylhydrolase (PAF-AH) inhibitory hydrolysate, synthesis of inhibitory peptides and assessment of their toxicity using the Zebrafish larvae assay. Peptides 50:119–124

Fitzgerald C, Aluko RE, Hossain M, Rai DK, Hayes M (2014) Potential of a renin inhibitory peptide from the red seaweed *Palmaria palmata* as a functional food ingredient following confirmation and characterization of a hypotensive effect in spontaneously hypertensive rats. J Agric Food Chem 62:8352–8356

Fleurence J (1999) Seaweed proteins: biochemical, nutritional aspects and potential uses. Trends Food Sci Technol 10:25–28

Fleurence J (2016) Seaweeds as food. In: Fleurence J, Levine I (eds) Seaweed in health and disease prevention. Elsevier/Academic, London

Francisco J, Horta A, Pedrosa R, Afonso C, Cardoso C, Bandarra NM, Gil MM (2020) Bioaccessibility of antioxidants and fatty acids from *Fucus spiralis*. Food Secur 9:440

Furuta T, Miyabe Y, Yasui H, Kinoshita Y, Kishimura H (2016) Angiotensin I converting enzyme inhibitory peptides derived from phycobiliproteins of dulse Palmaria palmate. Mar Drugs 14:32

Ganesan AR, Tiwari U, Rajauria G (2019) Seaweed nutraceuticals and their therapeutic role in diseaseprevention. Food Sci Human Wellness 8:252–263

Garcia-Vaquero M, Hayes M (2016) Red and green macroalgae for fish and animal feed and human functional food development. Food Rev Int 32(1):15–45

Garcia-Vaquero M, Rajauria G, O'Doherty J, Sweeney T (2017) Polysaccharides from macroalgae: recent advances, innovative technologies and challenges inextraction and purification. Food Res Int 99:1011–1020

Gómez-Estaca J, López-de-Dicastillo C, Hernandez-Munoz P, Catala R, Gavara R (2014) Advances in antioxidant active food packaging. Trends Food Sci Technol 35(1):42–51

Guerra A, Etienne-Mesmin L, Livrelli V, Denis S, Blanquet-Diot S, Alric M (2012) Relevance and challenges in modeling human gastric and small intestinal digestion. Trends Biotechnol 30(11):591–600

Gullón B, Gagaoua M, Barba FJ, Gullón P, Zhang W, Lorenzo JM (2020) Seaweeds as promising resource of bioactive compounds: Overview of novel extraction strategies and design of tailored meat products. Trends Food Sci Technol 100:1–18

Gupta S, Abu-Ghannam N (2011) Bioactive potential and possible health effects of edible brown seaweeds. Trends Food Sci Technol 22:315–326

Guzmán F, Wong G, Román T, Cardenas C, Alvarez C, Schmitt P et al (2019) Identification of antimicrobial peptides from the microalgae *tetraselmis suecica* (kylin) butcher and bactericidal activity improvement. Mar Drugs 17(8):453

Hadj AH, Lajili S, Ben SR, Le CD, Bouraoui A, Majdoub H (2015) Physicochemical characterization and pharmacological evaluation of sulfated polysaccharides from three species of Mediterranean brown algae of the genus *Cystoseira*. Daru J Pharm Sci 23:1–8

Haque E, Chand R (2008) Antihypertensive and antimicrobial bioactive peptides from milk proteins. Eur Food Res Technol 227(1):7–15

Harb TB, Chow F (2022) An overview of beach-cast seaweeds: Potential and opportunities for the valorization of underused waste biomass. Algal Res 62:102643

Harnedy PA, O'Keeffe MB, FitzGerald RJ (2017) Fractionation and identification of antioxidant peptides from an enzymatically hydrolysed Palmaria palmata protein isolate. Food Res Int 100:416–422

Hayashi K, Nakano T, Hashimoto M, Kanekiyo K, Hayashi T (2008) Defensive effects of a fucoidan from brown alga *Undaria pinnatifida* against herpes simplex virus infection. Int Immunopharmacol 8:109–116

Hentati F, Tounsi L, Djomdi D, Pierre G, Delattre C, Ursu AV, Fendri I, Abdelkafi S, Michaud P (2020) Bioactive Polysaccharides from Seaweeds. Molecules 25:3152

Heo SJ, Park EJ, Lee KW, Jeon YJ (2005) Antioxidant activities of enzymatic extracts from brown seaweeds. Bioresour Technol 96:1613–1623

Hmelkov AB, Zvyagintseva TN, Shevchenko NM, Rasin AB, Ermakova SP (2017) Ultrasound-assisted extraction of polysaccharides from brown alga *Fucus evanescens*. Structure and biological activity of the new fucoidan fractions. J Appl Phycol 30:2039–2046

Holdt SL, Kraan S (2011) Bioactive compounds in seaweed: functional food applications and legislation. J Appl Phycol 23:543–597

Huang X, Zhou H, Zhang H (2006) The effect of Sargassum fusiforme polysaccharide extracts on vibriosis resistance and immune activity of the shrimp, Fenneropenaeus chinensis. Fish Shellfish Immunol 20(5):750–757

Imbs TI, Skriptsova AV, Zvyagintseva TN (2015) Antioxidant activity of fucosecontaining sulfated polysaccharides obtained from Fucus evanescens by different extraction methods. J Appl Phycol 27:545–553

Inanli AG, Aksun Tümerkan ET, El Abed N, Regenstein JM, Özogul F (2020) The Impact of chitosan on seafood quality and human health: a review. Trends Food Sci Technol 97:404–416

Indumathi P, Mehta A (2016) A novel anticoagulant peptide from the Nori hydrolysate. J Funct Foods 20:606–617

Jeong HS, Venkatesan J, Kim SK (2013) Hydroxyapatite-fucoidan nanocomposites for bone tissue engineering. Int J Biol Macromol 57:138–141

Jiao K, Gao J, Zhou T, Yu J, Song H, Wei Y et al (2019) Isolation and purification of a novel antimicrobial peptide from *Porphyra yezoensis*. J Food Biochem 43(7):Article e12864

Jiménez-Escrig A, Gomez-Ordonez E, Ruperez P (2011) Seaweed as a source of novel nutraceuticals: sulfated polysaccharides and peptides. In: Kim SK (ed) Advances in food and nutrition research: marine medicinal foods, vol 64. Elsevier, London, pp 325–337

Jin G, Kim G (2011) Rapid-prototyped PCL/fucoidan composite scaffolds for bone tissue regeneration: design, fabrication, and physical/biological properties. J Mater Chem 21(44):17710–17718

Jónsdóttir R, Geirsdóttir M, Hamaguchi PY, Jamnik P, Kristinsson HG, Undeland I (2015) The ability of *in vitro* antioxidant assays to predict the efficiency of a cod protein hydrolysate and brown seaweed extract to prevent oxidation in marine food model systems. J Sci Food Agric 96(6):2125–2135

Kadam SU, Tiwari BK, O'Donnell CP (2013) Application of novel extraction technologies for extraction of bioactives from marine algae. J Agric Food Chem 61:4667–4675

Kadam SU, O'Donnell CP, Rai DK, Hossain MB, Burgess CM, Walsh D et al (2015) Laminarin from Irish Brown seaweeds Ascophyllum nodosum and Laminaria hyperborea: ultrasound assisted extraction, characterization and bioactivity. Mar Drugs 13:4270–4280

Kazir M, Abuhassira Y, Robin A, Nahor O, Luo J, Israel A, Golberg A, Livney YD (2019) Extraction of proteins from two marine macroalgae, *Ulva* sp. and *Gracilaria* sp., for food application, and evaluating digestibility, amino acid composition and antioxidant properties of the protein concentrates. Food Hydrocoll 87:194–203

Kendel M, Wielgosz-Collin G, Bertrand S, Roussakis C, Bourgougnon N, Bedoux G (2015) Lipid composition, fatty acids and sterols in the seaweeds *Ulva armoricana*, and *Solieria chordalis* from Brittany (France): an analysis from nutritional, chemotaxonomic, and antiproliferative activity perspectives. Mar Drugs 13:5606–5628

Kidgell JT, Magnusson M, de Nys R, Glasson CRK (2019) Ulvan: A systematic review of extraction, composition and function. Algal Res 39:101422

Kim MS, Kim JY, Choi WH, Lee SS (2008) Effects of seaweed supplementation on blood glucose concentration, lipid profile, and antioxidant enzyme activities in patients with type 2 diabetes mellitus. Nutr Res Pract 2(2):62–67

Knutsen S, Myslabodski D, Larsen B, Usov A (1994) A modified system of nomenclature for red algal galactans. Bot Mar 37:163–170

Kwak JY (2014) Fucoidan as a marine anticancer agent in preclinical development. Mar Drugs 12(2):851–870

Lafarga T, Acien-Fernandez FG, Garcia-Vaquero M (2020) Bioactive peptides and carbohydrates from seaweed for food applications: natural occurrence, isolation, purification, and identification. Algal Res 48:101909

Lahaye M, Robic A (2007) Structure and function properties of Ulvan, a polysaccharide from green seaweeds. Biomacromolecules 6:1765–1774

Lee KY, Mooney DJ (2012) Alginate: properties and biomedical applications. Prog Polym Sci 37(1):106–126

Lee JS, Jin GH, Yeo MG, Jang CH, Lee H, Kim GH (2012) Fabrication of electrospun biocomposites comprising polycaprolactone/fucoidan for tissue regeneration. Carbohyd Polym 90(1):181–188

Leiro JM, Castro R, Arranz JA, Lamas J (2007) Immunomodulating activities of acidic sulphated polysaccharides obtained from the seaweed *Ulva rigida C. Agardh*. Int Immunopharmacol 7(7):879–888

Li W, Jiang N, Li B, Wan M, Chang X, Liu H, Zhang L, Yin S, Qi H, Liu S (2018) Antioxidant activity of purified ulvan in hyperlipidemic mice. Int J Biol Macromol 113:971–975

López-López I, Bastida S, Ruiz-Capillas C, Bravo L, Larrea MT, Sánchez-Muniz F et al (2009) Composition and antioxidant capacity of low-salt meat emulsion model systems containing edible seaweeds. Meat Sci 83(3):492–498

Lorbeer AJ, Charoensiddhi S, Lahnstein J, Lars C, Franco CM, Bulone V, Zhang W (2017) Sequential extraction and characterization of fucoidans and alginates from *Ecklonia radiata, Macrocystis pyrifera, Durvillaea potatorum*, and *Seirococcus axillaris*. J Appl Phycol 29:1515–1526

Lorenzo JM, Munekata PES, Sant'Ana AS, Carvalho RB, Barba FJ, Toldrá F et al (2018) Main characteristics of peanut skin and its role for the preservation of meat products. Trends Food Sci Technol 77:1–10

Lourenço SO, Barbarino E, De-Paula JC, Pereira LOS, Marquez UML (2002) Amino acid composition, protein content and calculation of nitrogen-to-protein conversion factors for 19 tropical seaweeds. Phycol Res 50(3):233–241

Mabeau S, Fleurence J (1993) Seaweed in food products: biochemical and nutritional aspects. Trends Food Sci Technol 4:103–107

Makkar HP, Tran G, Heuzé V, Giger-Reverdin S, Lessire M, Lebas F, Ankers P (2016) Seaweeds for livestock diets: a review. Anim Feed Sci Technol 212:1–17

Malaguti M, Dinelli G, Leoncini E, Bregola V, Bosi S, Cicero AF, Hrelia S (2014) Bioactive peptides in cereals and legumes: agronomical, biochemical and clinical aspects. Int J Mol Sci 15(11):21120–21135

Mamat H, Matanjun P, Ibrahim SM, Amin S, Abdul Hamid M, Rameli A (2014) The effect of seaweed composite flour on the textural properties of dough and bread. J Appl Phycol 26(2):1057–1062

Mao X, Bai L, Fan X, Zhang X (2017) Anti-proliferation peptides from protein hydrolysates of Pyropia haitanensis. J Appl Phycol 29:1623–1633

Marsham S, Scott GW, Tobin ML (2007) Comparison of nutritive chemistry of a range of temperate seaweeds. Food Chem 100(4):1331–1336

Minami Y, Kanemura S, Oikawa T, Suzuki S, Hasegawa Y, Nishino Y et al (2020) Associations of Japanese food intake with survival of stomach and colorectal cancer: a prospective patient cohort study. Cancer Sci 111(7):2558–2569

Mišurcová L, Machů L, Orsavova J (2011) Seaweed minerals as nutraceuticals. Adv Food Nutr Res 64:371–390

Mohan K, Ganesan AR, Muralisankar T, Jayakumar R, Sathishkumar P, Uthayakumar V, Chandirasekar R, Revathi N (2020) Recent insights into the extraction, characterization, and bioactivities of chitin and chitosan from insects. Trends Food Sci Technol 105:17–42

Moroney N, O'Grady M, O'Doherty J, Kerry J (2013) Effect of a brown seaweed (*Laminaria digitata*) extract containing laminarin and fucoidan on the quality and shelf-life of fresh and cooked minced pork patties. Meat Sci 94(3):304–311

Murai U, Yamagishi K, Sata M, Kokubo Y, Saito I, Yatsuya H et al (2019) Seaweed intake and risk of cardiovascular disease: the Japan Public Health Center–based prospective (JPHC) study. Am J Clin Nutr 110(6):1449–1455

O'Sullivan A, O'Callaghan Y, O'Grady M, Queguineur B, Hanniffy D, Troy D et al (2011) *In vitro* and cellular antioxidant activities of seaweed extracts prepared from five brown seaweeds harvested in spring from the west coast of Ireland. Food Chem 126(3):1064–1070

Olivares-Bãnuelos T, Gutierrez-Rodríguez AG, Mendez-Bellido R, Tovar-Miranda R, Arroyo-Helguera O, Juarez-Portilla C et al (2019) Brown seaweed *Egregia menziesii*'s cytotoxic activity against brain cancer cell lines. Molecules 24(2):260

Øverland M, Mydland LT, Skrede A (2019) Marine macroalgae as sources of protein and bioactive compounds in feed for monogastric animals. J Sci Food Agric 99(1):13–24

Ozogul F, El Abed N, Ceylan Z, Ocak E, Ozogul Y (2021) Nano-technological approaches for plant and marine-based polysaccharides for nano-encapsulations and their applications in food industry. In: Toldrá F (ed) Advances in food and nutrition research, vol 97. Elsevier

Pampanin DM, Larssen E, Provan F, Sivertsvik M, Ruoff P, Sydnes MO (2012) Detection of small bioactive peptides from Atlantic herring (*Clupea harengus* L.). Peptides 34(2):423–426

Pangestuti R, Kurnianto D (2017) Green seaweeds-derived polysaccharides ulvan: Occurrence, medicinal value and potential applications. In: Venkatesan J, Anil S, Kim SK (eds) Seaweed polysaccharides. isolation, biological and biomedical applications. Elsevier, Oxford, pp 205–221

Patarra RF, Paiva L, Neto AI, Lima E, Baptista J (2011) Nutritional value of selected macroalgae. J Appl Phycol 23(2):205–208

Peñalver R, Lorenzo JM, Ros G, Amarowicz R, Pateiro M, Nieto G (2020) Seaweeds as a functional ingredient for a healthy diet. Mar Drugs 18:301

Peña-Rodríguez A, Mawhinney TP, Ricque-Marie D, Cruz-Suárez LE (2011) Chemical composition of cultivated seaweed *Ulva clathrata* (Roth) *C. Agardh*. Food Chem 129(2):491–498

Peng Y, Hu J, Yang B, Lin X-P, Zhou X-F, Yang XW, Yonghong Liu Y (2015) Chemical composition of seaweeds. In: Tiwari BK, Troy DJ (eds) Seaweed sustainability food and non-food applications. Elsevier

Pereira L, Critchley AT, Amado AM, Ribeiro-Claro PJ (2009) A comparative analysis of phycocolloids produced by underutilized versus industrially utilized carrageenophytes (Gigartinales, Rhodophyta). J Appl Phycol 21:599–605

Pliego-Cortés H, Wijesekara I, Lang M, Bourgougnon N, Bedoux G (2020) Current knowledge and challenges in extraction, characterization and bioactivity of seaweed protein and seaweed-derived proteins. In: Bourgougnon (ed) Advances in botanical research, vol 95. Elsevier, pp 289–326

Pradhan B, Bhuyan PP, Patra S, Nayak R, Behera PK, Behera C, Behera AK, Ki JS, Jena M (2022) Beneficial effects of seaweeds and seaweed-derived bioactive compounds: Current evidence and future prospective. Biocatal Agric Biotechnol 39:102242

Prajapati VD, Maheriya PM, Jani GK, Solanki HK (2014) Carrageenan: a natural seaweed polysaccharide and its applications. Carbohyd Polym 105:97–112

Puvaneswary S, Talebian S, Raghavendran HB, Murali MR, Mehrali M, Afifi AM et al (2015) Fabrication and *in vitro* biological activity of βTCP-chitosan-fucoidan composite for bone tissue engineering. Carbohyd Polym 134:799–807

Qi H, Huang L, Liu X, Liu D, Zhang Q, Liu S (2012) Antihyperlipidemic activity of high sulfate content derivative of polysaccharide extracted from *Ulva pertusa* (Chlorophyta). Carbohyd Polym 87(2):1637–1640

Qin Y (2018) Seaweed hydrocolloids as thickening, gelling, and emulsifying agents in functional food products. In: Qin Y (ed) Bioactive seaweeds for food applications. Elsevier, pp 135–152

Qiu SM, Aweya JJ, Liu X, Liu Y, Tang S, Zhang W, Cheong KL (2022) Bioactive polysaccharides from red seaweed as potent food supplements: a systematic review of their extraction, purification, and biological activities. Carbohydr Polym 275:118696

Raja K, Kadirvel V, Subramaniyan T (2022) Seaweeds, an aquatic plant-based protein for sustainable nutrition – a review. Future Foods 5:100142

Raposo MFDJ, De Morais AMB, De Morais RMSC (2015) Marine polysaccharides from algae with potential biomedical applications. Mar Drugs 13:2967–3028

Renn D (1997) Biotechnology and the red seaweed polysaccharide industry: status, needs and prospects. Trends Biotechnol 15(1):9–14

Rhein-Knudsen N, Meyer AS (2021) Chemistry, gelation, and enzymatic modification of seaweed food hydrocolloids. Trends Food Sci Technol 109:608–621

Righini H, Francioso O, Di Foggia M, Quintana AM, Roberti R (2020) Preliminary study on the activity of phycobiliproteins against *Botrytis cinerea*. Mar Drugs 18:600

Rinaudo M (2007) Seaweed polysaccharides. In: Kamerling JP (ed) Comprehensive glycoscience from chemistry to systems biology, vol 2. Elsevier, London, pp 691–735

Rioux L, Turgeon SL, Beaulieu M (2007) Characterization of polysaccharides extracted from brown seaweeds. Carbohydr Polym 3:530–537

Rizk AM (1997) Fatty acid composition of twelve algae forms the coastal zone of Qatar. Plant Foods Hum Nutr 51:27–34

Roohinejad S, Koubaa M, Barba FJ, Saljoughian S, Amid M, Greiner R (2017) Application of seaweeds to develop new food products with enhanced shelf-life, quality and health-related beneficial properties. Food Res Int 99:1066–1083

Saha D, Bhattacharya S (2010) Hydrocolloids as thickening and gelling agents in food: a critical review. J Food Sci Technol 47(6):587–597

Salehi B, Sharifi-Rad J, Seca AML, Pinto DC, Michalak I, Trincone A, Mishra AP, Nigam M, Zam W, Martins N (2019) Current trends on seaweeds: looking at chemical composition, phytopharmacology, and cosmetic applications. Molecules 24:4182

Sanjeewa KKA, Kang N, Ahn G, Jee Y, Kim YT, Jeon YJ (2018) Bioactive potentials of sulfated polysaccharides isolated from brown seaweed *Sargassum* spp in related to human health applications: a review. Food Hydrocoll 81:200–208

Santoso J, Gunji S, Yoshie-Stark Y, Suzuki T (2006) Mineral contents of Indonesian seaweeds and mineral solubility affected by basic cooking. Food Sci Technol Res 12(1):59–66

Sasaki K, Ishihara K, Oyamada C, Sato A, Fukushi A, Arakane T et al (2008) Effects of fucoxanthin addition to ground chicken breast meat on lipid and colour stability during chilled storage, before and after cooking. Asian-Aust J Anim Sci 21(7):1067–1072

Senthil MA, Mamatha B, Mahadevaswamy M (2005) Effect of using seaweed (Eucheuma) powder on the quality of fish cutlet. Int J Food Sci Nutr 56(5):327–335

Shofia SI, Jayakumar K, Mukherjee A, Chandrasekaran N (2018) Efficiency of brown seaweed (*Sargassum longifolium*) polysaccharides encapsulated in nanoemulsion and nanostructured lipid carrier against colon cancer cell lines HCT 116. RSC Adv 8:15973–15984

Šimat V, Elabed N, Kulawik P, Ceylan Z, Jamroz E, Yazgan H, Čagalj M, Regenstein JM, Özogul F (2020) Recent advances in marine-based nutraceuticals and their health benefits. Mar Drugs:18

Sivagnanavelmurugan M, Ramnath GK, Thaddaeus BJ, Palavesam A, Immanuel G (2015) Effect of *Sargassum wightii* fucoidan on growth and disease resistance to Vibrio parahaemolyticus in *Penaeus monodon* post-larvae. Aquac Nutr 21(6):960–969

Stiger-Pouvreau V, Bourgougnon N, Deslandes E (2016) Carbohydrates from seaweeds. In: Fleurence J, Levine I (eds) Seaweed in health and disease prevention. Elsevier/Academic, London, pp 223–247

Sudharsan S, Giji S, Seedevi P, Vairamani S, Shanmugam A (2018) Isolation, characterization and bioactive potential of sulfated galactans from *Spyridia hypnoides* (Bory) Papenfuss. Int J Biol Macromol 109:589–597

Suetsuna K, Nakano T (2000) Identification of an antihypertensive peptide from peptic digest of wakame (*Undaria pinnatifida*). J Nutr Biochem 11(9):450–454

Thanigaivel S, Vickram S, Saranya V, Huma A, Saud A, Modigunta JKR, Anbarasu Rajasekhar K et al (2022) Seaweed polysaccharide mediated synthesis of silver nanoparticles and its enhanced disease resistance in *Oreochromis mossambicus*. J King Saud Univ Sci 34:101771
Thomes P, Rajendran M, Pasanban B, Rengasamy R (2010) Cardioprotective activity of Cladosiphon okamuranus against isoproterenol induced myocardial infraction in rats. Phytomedicine 18:52–57
Upadhyay R, Ghosal D, Mehra A (2012) Characterization of bread dough: Rheological properties and microstructure. J. Food Eng 109(1):104–113
Usov AI, Zelinsky ND (2013) Chemical structures of algal polysaccharides. In: Domínguez H (ed) Functional ingredients from algae for foods and nutraceuticals. Woodhead Publishing Limited
Venkatesan J, Lowe B, Anil S, Manivasagan P, Kheraif AAA, Kang KH, Kim SK (2015) Seaweed polysaccharides and their potential biomedical applications. Starch-Stärke 67:381–390
Venkatesan JS, Anil S, Kim SK (2017) Introduction to seaweed polysaccharides. In: Venkatesan J, Anil S, Kim SK (eds) Seaweed polysaccharides. isolation, biological and biomedical applications. Elsevier, Oxford, pp 205–221
Vieira EF, Soares C, Machado S, Correia M, Ramalhosa MJ, Oliva-teles MT, Carvalho AP, Domingues VF, Antunes F, Oliveira TAC, Morais S, Delerue-Matos C (2018) Seaweeds from the Portuguese coast as a source of proteinaceous material: total and free amino acid composition profile. Food Chem 269:264–275
Vijayabaskar P, Vaseela N, Thirumaran G (2012) Potential antibacterial and antioxidant properties of a sulfated polysaccharide from the brown marine algae *Sargassum swartzii*. Chin J Nat Med 10(6):421–428
Vo TS, Ryu BM, Kim SK (2013) Purification of novel anti-inflammatory peptides from enzymatic hydrolysate of the edible microalgal *Spirulina maxima*. J Funct Foods 5(3):1336–1346
Wang S, Zhao S, Uzoejinwa BB, Zheng A, Wang Q, Huang J, Abomohra AEF (2020) A state-of-the-art review on dual purpose seaweeds utilization for wastewater treatment and crude bio-oil production. Energy Convers Manag 222:113253
Westermeier R, Murúa P, Patiño DJ, Muñoz L, Ruiz A, Müller DG (2012) Variations of chemical composition and energy content in natural and genetically defined cultivars of Macrocystis from Chile. J Appl Phycol 24:1191–1201
Wijesinghe WAJP, Jeon YJ (2012) Biological activities and potential industrial applications of fucose rich sulfated polysaccharides and fucoidans isolated from brown seaweeds: A review. Carbohydr Polym 88:13–20
Wijesinghe WAJP, Athukorala Y, Jeon YJ (2011) Effect of anticoagulative sulfated polysaccharide purified from the enzyme-assistant extract of a brown seaweed Ecklonia Cava investor rats. Carbohydr Polym 86:917–921
Xie XT, Cheong KL (2021) Recent advances in marine algae oligosaccharides: structure, analysis, and potential prebiotic activities. Crit Rev Food Sci Nutr:1–16
Yaich H, Ben Amira A, Abbes F, Bouaziz M, Besbes S, Richel A et al (2017) Effect of extraction procedures on structural, thermal and antioxidant properties of ulvan from *Ulva lactuca* collected in Monastir coast. Int J Biol Macromol 105:1430–1439
Zheng LX, Chen XQ, Cheong KL (2020) Current trends in marine algae polysaccharides: the digestive tract, microbial catabolism, and prebiotic potential. Int J Biol Macromol 151:344–354
Zimmermann MB (2008) Research on iodine deficiency and goiter in the 19th and early 20th centuries. J Nutr 138(11):2060–2063

Chapter 6
Seaweed as a Novel Feed Resource Including Nutritional Value and Implication Product Quality Animal Health

B. K. K. K. Jinadasa, Margareth Øverland, G. D. T. M. Jayasinghe, and Liv Torunn Mydland

1 Introduction

According to the United Nations Department of Economic and Social Affairs' world population prospects, 2019 revision, the current world population of 7.7 billion is expected to reach 8.5 billion in 2030 (10% increase), 9.7 billion in 2050 (26% increase) and 10.9 billion in 2100 (42% increase) (UNDESA 2019). This represents a challenge to providing food for a growing population, but the cultured meat industry and technology have not yet proven the capability to work out major upcoming problems with food security and the environment when meat demands (Post and Hocquette 2017; Henchion et al. 2021). Feeding a balanced diet with all nutrients is essential for high and sustainable livestock production. However, it seems challenging to supply high-quality feed (and fodder) and negatively impact food crops such as corn and soybean (Ayele et al. 2021). Thus, this will make competition between animal feed and human food plus biofuel production. Hence, macro algae,

B. K. K. K. Jinadasa (✉)
Analytical Chemistry Laboratory, National Aquatic Resources Research & Development Agency (NARA), Colombo, Sri Lanka

Department of Food Science and Technology (DFST), Faculty of Livestock, Fisheries & Nutrition (FLFN), Wayamba University of Sri Lanka, Makandura, Gonawila, Sri Lanka

M. Øverland · L. T. Mydland
Department of Animal and Aquacultural Sciences, Faculty of Biosciences, Norwegian University of Life Sciences, Ås, Norway

G. D. T. M. Jayasinghe
Analytical Chemistry Laboratory, National Aquatic Resources Research & Development Agency (NARA), Colombo, Sri Lanka

Universite de Pau et des Pays de l'Adour, E2S UPPA, CNRS, IPREM, Pau, France

F. Ozogul et al. (eds.), *Seaweeds and Seaweed-Derived Compounds*,
https://doi.org/10.1007/978-3-031-65529-6_6

commonly named seaweed or marine algae are one of the best alternatives for conventional animal feed (Makkar et al. 2016).

Seaweed are multicellular, macroscopic, eukaryotic, and autotrophic organisms and taxonomically categorized into three main groups based on the colour of thallus i.e., Chlorophyceae (green algae), Rhodophyceae (red algae), and Phaeophyceae (brown algae) (Leandro et al. 2020a). The most common genera are, *Ulva*, *Codium*, *Enteromorpha*, *Chaetomorpha* and *Cladophora* for green algae, *Ascophyllum*, *Laminaria*, *Saccharina*, *Macrocystis*, *Nereocystis*, and *Sargassum* for brown algae and *Pyropia*, *Porphyra*, *Chondrus* and *Palmaria* for red algae (Makkar et al. 2016). The responsible colour pigment for the brown colour is Phaeophyta, the red colour is phycobilins and chlorophyll a and b, carotenes, and xanthophylls for the green colour (Øverland et al. 2019).

The use of seaweed as feedstuff allows the provision of numerous proteins, lipids, vitamins, minerals, and several other important compounds such as pigments, antioxidants and antimicrobials (Morais et al. 2020; Jayasinghe et al. 2018). These nutrient levels vary with species, the season of harvest, geographic origin, and environmental conditions (Øverland et al. 2019). Despite the wide range of utilization of seaweed as feed ingredients, there are some disadvantages and challenges to the digestibility of seaweed by animals. However, literature data confirm that seaweeds can play an important role in animal feeding (Michalak and Mahrose 2020). In the present work chapter, we are discussing the nutritional value of seaweed and its implication on product quality of animal health as a novel feed resource.

2 Nutritional Properties of Seaweed as Feed

2.1 Proximate Composition

Seaweed are rich in different nutritional profiles which are presented in Table 6.1. As usual, moisture compromised a large share of the seaweed followed by ash, fiber and protein. The protein content in seaweeds varies from 5% to 47% of dry basic while this depends on the species, habitat, maturity and environmental conditions. Moreover, this amount incorporates amino acids, especially glycine, alanine, arginine, proline, glutamic, and aspartic acids (Černá 2011). Afonso et al. (2021) noted that the reported protein content varied with the analytical method. Most of the researchers used a conversion factor of 6.25 and hence the protein content was reported as overestimated value. Therefore, Lourenço et al. (2002) proposed a new N-protein conversion factor based on the ratio of amino acid residues to total nitrogen and the value range of the value 3.75 (red algae), 5.72 (brown algae), and 4.92 as an average factor for seaweed. In *Agarophyton vermiculophyllum*, the protein content (dw) showed a mean value of 17.0 ± 4.5%, ranging from 21.5 ± 2.1% in September to 6.2 ± 0.3% in August showing notorious variations along the season (Afonso et al. 2021). In another study in Sweeden, *Saccharina latissima* (brown

Table 6.1 Proximate composition of some selected seaweed

Species	Location	Moisture (%)	Protein (%)	Lipid (%)	Ash (%)	Fibre (%)	Ref.
Ulva lactuca	Sri Lanka	81.4	15.4	1.4	24.5	13.3	Jayasinghe et al. (2018)
S. wighitii	″	85.6	7.1	1.3	26.5	10.0	″
S. turbinaria	″	85.4	7.5	0.5	20.5	15.4	″
G. verrucosa	″	86.4	3.7	1.3	21.9	6.2	″
K. alvarezii	″	79.9	5.6	1.0	26.3	5.3	″
H. japonica	Hong Kong	–	19.0	1.4	22.1	53.2[a]	Wong and Cheung (2000)
H. charoides	″	–	18.4	1.5	22.8	50.3[a]	″
U. lactuca	″	–	7.1	1.6	21.3	55.4[a]	″
G. acerosa	India	12.2	0.6	0.03	0.1	13.5[a]	Syad et al. (2013)
S. wightii	″	22.4	1.5	0.03	0.3	17[a]	″
C. sinuosa	Iran	84.9	10.1	1.5	39.3	9.5	Tabarsa et al. (2012)
D. dichotoma	″	89.0	17.7	2.9	27.0	10.5	″
P. pavonica	″	87.0	11.8	1.8	33.1	11.6	″
Caulerpa racemosa	Malaysia	92.0	10.5	0.2	10.6	11.3	Ahmad et al. (2012)
Turbinaria conoides	″	83.8	7.4	0.6	21.4	29.6	″
Kappaphycus alvarezii	″	79.8	5.4	0.2	23.3	4.5	″
Eucheuma denticulatum	″	84.5	7.7	0.5	28.8	5.2	″
Gracilaria verrucosa	″	85.5	11.7	0.3	6.1	7.8	″
Gracilaria changii	″	5.32[b]	12.6	0.3	40.3	29.4	Chan and Matanjun (2017)
Ascophyllum nodosum	Spain	11.08[b]	8.7	3.6	30.9	–	Lorenzo et al. (2017)
Fucus vesiculosus	″	11.23[b]	13.0	3.8	20.7	–	″
Bifurcaria bifurcata	″	7.0[b]	8.9	6.5	31.7	–	″
Ulva lactuca	Egypt	10.7	23.1	0.3	29.9	12.3	Abudabos et al. (2013)
Ulva lactuca	″	8.9–14.8	16.8–20.1	3.1–4.1	17.6–23.2	–	Khairy and El-Shafay (2013)
Jania rubens	″	3.2–5.1	9.8–12.9	1.5–2.4	39.3–50.5	–	″

(continued)

Table 6.1 (continued)

Species	Location	Moisture (%)	Protein (%)	Lipid (%)	Ash (%)	Fibre (%)	Ref.
Pterocladia capillacea	"	9.2–10.2	17.4–23.8	1.8–2.7	13.0–23.7	_	"
Porphyra columbina	Argentina	12.79[b]	24.6	0.3	6.5	48.0[a]	Cian et al. (2014)

[a]Total dietary fibre
[b]Dry weight basis

algae) showed protein concentration peaked at August 110 ± 40 and lowest 30 ± 10 mg/g (dw) (Vilg et al. 2015).

Seaweed generally contains a low amount of lipids than fish, hence it is considered a low-fat food and a potential source of functional lipids due to its large stock (Afonso et al. 2021). Most seaweed contains up to 1–5% total lipids on dw basis (Miyashita et al. 2013). However, Gosch et al. (2012), reported that brown seaweed such as *Dictyota bartayresii*, *Spatoglossum macrodontum*, and *Dictyota dichotoma* have the highest total lipid contents (10–12% dw). The amount of lipids in seaweed also varies as a protein with several intrinsic and extrinsic factors. The seaweed from the cold-water area has significantly higher lipids content than the seaweed from tropical area (Susanto et al. 2019). As an example, Susanto et al. (2016) reported higher lipid content in 4 cold water brown seaweed species from Japan (37.42–78.69 mg/g, dw) than the 3 tropical brown seaweed from Indonesia (36.71–50.15 mg/g, dw).

Seaweed are rich in carbohydrates and its concentration is up to 76% of the algae dry weight basis. These carbohydrates (polysaccharides) are mostly found in the extracellular matrix (cell wall) and the main types are cellulose, hemicellulose, agar, carrageenan, sulfated fucans and alginates (Rioux et al. 2017). Dietary fiber is recognized as an important element for food and feed for healthy nutrition. According to the literature, seaweed are rich in dietary fiber and it accounted for up to 38% of dry weight basis (Fleurence 2016). Moreover, (Praveen et al. 2019), reported a very high dietary fiber level (60.64 ± 2.2%) in *Enteromorpha compressa* from India. There is no universal definition for dietary fiber, but it consists of a series of compounds i.e., oligosaccharides and polysaccharides, such as cellulose, hemicellulose, pectic substances, gums, resistant starch, and inulin (Peñalver et al. 2020).

Seaweed proteins are rich with amino acid such as glycine, alanine, arginine, proline, glutamic, and aspartic acids (Table 6.2). The amino acid profile of seaweed represent almost half of the essential amino acid and it is very close to amino acid profile of egg protein (Černá 2011). In general, seaweeds have total essential amino acids (EAA) (mean = 45.7% total amino acid, TAA) compared to fishmeal (43.4% TAA) and soybean meal (46.0% TAA). However, seaweed have substantially lower concentration of methionine and lysine than soyabean or fish meal. Comparatively, red seaweed have good quality protein and amino acid profile than the brown and green seaweed (Angell et al. 2016). Fatty acid profile of selected seaweed species is

Table 6.2 Amino acid profile of some selected seaweed species (all values are expressed as % from a total amino acid in dry weight basis)

Type of amino acid	A	B	C	D	E	F	G
Alanine	0.40 ± 0.03	0.66 ± 0.02	6.78 ± 0.06	6.09 ± 0.14	4.65 ± 0.11	8.4 ± 0.2	16.6 ± 0.1
Arginine	2.85 ± 0.23	2.28 ± 0.13	5.49 ± 0.07	3.31 ± 0.25	3.23 ± 0.16	6.4 ± 0.2	4.8 ± 0.0
Aspartic acid	3.66 ± 0.19	3.40 ± 0.23	9.69 ± 0.04	8.45 ± 0.10	5.95 ± 0.09	12.1 ± 0.3	11.9 ± 0.0
Cystine	–	–	2.90 ± 0.13	2.13 ± 0.39	0.96 ± 0.53	–	–
Glutamic acid	3.17 ± 0.14	3.52 ± 0.14	10.34 ± 0.04	15.38 ± 0.05	6.64 ± 0.08	13.5 ± 0.5	14.7 ± 0.1
Glycine	<LOD	<LOD	5.44 ± 0.07	3.77 ± 0.22	3.37 ± 0.15	6.4 ± 0.2	5.3 ± 0.0
Histidine	1.80 ± 0.02	6.20 ± 0.05	1.72 ± 0.22	1.23 ± 0.67	1.06 ± 0.48	1.8 ± 0.1	1.8 ± 0.0
Isoleucine	1.66 ± 0.03	1.98 ± 0.04	3.47 ± 0.11	2.51 ± 0.33	2.60 ± 0.20	4.2 ± 0.1	3.6 ± 0.0
Leucine	1.95 ± 0.07	2.08 ± 0.08	5.55 ± 0.07	4.41 ± 0.19	4.42 ± 0.12	8.0 ± 0.0	6.6 ± 0.0
Lysine	3.42 ± 0.04	2.71 ± 0.01	4.65 ± 0.08	3.20 ± 0.26	3.00 ± 0.17	5.5 ± 0.2	5.0 ± 0.0
Methionine	0.20 ± <0.01	0.29 ± 0.02	2.44 ± 0.15	1.98 ± 0.42	2.07 ± 0.25	2.2 ± 0.1	2.1 ± 0.0
Phenylalanine	1.03 ± 0.01	1.40 ± <0.01	3.70 ± 0.10	2.79 ± 0.30	2.88 ± 0.18	5.6 ± 0.4	4.6 ± 0.0
Proline	1.49 ± 0.02	1.09 ± <0.01	6.44 ± 0.06	5.05 ± 0.16	2.34 ± 0.22	4.7 ± 0.1	4.9 ± 0.0
Serine	4.67 ± 0.13	5.06 ± 0.36	3.78 ± 0.10	2.93 ± 0.28	2.45 ± 0.21	5.5 ± 0.1	4.7 ± 0.0
Taurine	0.31 ± 0.01	0.30 ± 0.01	–	–	–	–	–
Threonine	1.43 ± 0.03	1.11 ± 0.10	4.02 ± 0.09	3.53 ± 0.24	2.71 ± 0.19	5.5 ± 0.1	5.4 ± 0.0
Tryptophan	–	–	1.40 ± 0.11	1.65 ± 0.11	0.71 ± 0.02	–	–
Tyrosine	0.70 ± 0.03	1.05 ± 0.02	2.53 ± 0.15	1.43 ± 0.58	1.39 ± 0.37	3.5 ± 0.2	2.9 ± 0.0
Valine	2.08 ± 0.01	2.11 ± 0.01	5.17 ± 0.07	3.76 ± 0.22	3.31 ± 0.15	6.4 ± 0.1	5.1 ± 0.1
Protein (%)	19.5 ± 0.16	24.7 ± 0.24	36.4	25.1	–	22.6 ± 0.1	9.6 ± 0.1
Ref.	Vieira et al. (2018)	Vieira et al. (2018)	Mišurcová et al. (2014)	Mišurcová et al. (2014)	Mišurcová et al. (2014)	Biancarosa et al. (2017)	Biancarosa et al. (2017)

A: *Chondrus crispus*, B: *Gracilaria* sp., C: *Palmaria palmata* (dried seaweed), D: *Laminaria japonica* (dried seaweed), E: *Undaria pinnatifid* (dried seaweed), F: *Ulva lactuca*, G: *Laminaria digitata*

Table 6.3 Lipid content and fatty acid profile of some selected seaweed species (Individual fatty acid expressed as % of total fat amount)

Fatty acid	A	B	C	D	E	F	G
12:0	–	–	2.57 ± 0.20	–	–	–	–
14:0	4.42 ± 2.06	3.01 ± 0.23	0.95 ± 0.28	2.54 ± 0.70	1.53	1.34 ± 0.17	0.6 ± 0.2
14:1	–	–	–	0.38 ± 0.12	0.23	0.30 ± 0.03	0.4 ± 0.1
14:2	–	–	–	–	–	–	1.6 ± 0.1
15:0	–	–	–	0.34 ± 0.07	0.23	0.24 ± 0.01	0.4 ± 0.1
16:0	25.14 ± 2.90	31.13 ± 1.54	20.74 ± 0.81	7.17 ± 2.14	9.84	11.96 ± 0.54	42.0 ± 0.2
16:1	2.64 ± 0.38	2.63 ± 0.07	2.46 ± 0.12	2.21 ± 0.65	0.13	0.24 ± 0.02	ND
16:1(n-7)	–	–	–	0.62 ± 0.09	1.05	1.91 ± 0.37	1.9 ± 0.1
16:1(n-9)	–	–	–	0.28 ± 0.02	0.48	0.34 ± 0.04	2.7 ± 0.1
16:2(n-4)	–	–	–	0.05 ± 0.07	3.39	0.13 ± 0.05	–
16:2(n-6)	–	–	–	0.27 ± 0.06	0.39	0.82 ± 0.38	–
16:3(n-3)	–	–	–	0.16 ± 0.08	0.07	3.34 ± 1.71	–
16:3(n-6)	–	–	–	0.42 ± 0.10	0.22	0.16 ± 0.04	–
16:4(n-3)	–	–	–	0.31 ± 0.13	0.74	0.43 ± 0.03	6.4 ± 0.1
17:0	–	–	–	0.06 ± 0.09	–	0.21 ± 0.05	–
18:0	–	–	3.61 ± 0.28	0.78 ± 0.16	0.35	0.50 ± 0.09	1.0 ± 0.1
18:1	–	–	11.18 ± 0.76	–	–	–	1.9 ± 0.1
18:1(n-7)	–	–	–	–	–	17.3 ± 0.1	–
18:1(n-9)	11.17 ± 0.41	13.49 ± 0.23	–	4.65 ± 1.38	4.48	2.41 ± 0.32	ND
18:2(n-3)	–	–	–	–	–	–	8.4 ± 0.1
18:2(n-6)	6.71 ± 0.36	5.92 ± 0.26	2.69 ± 0.30	0.92 ± 0.33	3.03	2.75 ± 0.39	3.7 ± 0.1
18:3(n-6)	2.39 ± 0.21	1.13 ± 0.09	1.31 ± 0.14	0.49 ± 0.17	3.31	0.77 ± 0.19	–
18:3(n-3)	5.36 ± 0.33	9.23 ± 0.44	–	1.31 ± 0.57	0.52	6.14 ± 2.42	0.5 ± 0.1
18:4(n-3)	13.78 ± 0.61	8.53 ± 0.45	–	3.42 ± 1.21	0.45	0.64 ± 0.13	8.6 ± 0.1
18:4(n-6)	8.46 ± 0.19	12.70 ± 0.50	–	–	–	–	–

20:0	–	–	–	0.39 ± 0.13		0.10 ± 0.05	–
20:3(n-6)	2.42 ± 0.56	1.04 ± 0.04	–	–	0.15	0.28 ± 0.09	–
20:4(n-3)	–	–	–	–	–	0.17 ± 0.11	–
20:4(n-6) (ARA)	–	–	6.56 ± 0.48	–	–	1.17 ± 0.23	–
20:5(n-3) (EPA)	1.79 ± 0.57	4.37 ± 0.20	4.96 ± 1.19	–	0.6	1.51 ± 0.33	–
21:1	–	–	–	0.80 ± 0.29	–	–	–
22:0	–	–	–	–	0.16	0.52 ± 0.14	2.3 ± 0.1
22:4n-6	–	–	–	–	2.77	–	–
22:5n-3	–	–	0.65 ± 0.04	–	–	–	–
22:6(n-3) (DHA)	ND	ND	12.35 ± 0.83	–	0.18	–	–
24:0	–	–	–	0.01 ± 0.03	0.08	0.90 ± 0.17	0.2 ± 0.1
Total lipids (%, dw)	5.02 ± 0.12	3.67 ± 0.09	1.46 ± 0.38 to 2.94 ± 0.94	7.98 ± 1.4	5.75 ± 1.2	12.1 ± 0.34	2.6
Ref.	Susanto et al. (2016)	Susanto et al. (2016)	Tabarsa et al. (2012)	Gosch et al. (2012)	Gosch et al. (2012)	Gosch et al. (2012)	Kendel et al. (2015)

A: *Sargassum crassifolium*, B: *Padina australis*, C: *Colpomenia sinuosa*, D: *Dictyota bartayresii*, E: *Cladophora patentiramea*, F: *Derbesia tenuissima*, G: *Ulva armoricana* summarised in Table 6.3. In generally, seaweed contained all type of fatty acid, and its abundance are varied with the species, environmental condition, and life cycle stage. Especially temperature has significant influence on the fatty acid level, hence tropical seaweed contained lower fatty acid profile compared with the cold-water seaweed (Leandro et al. 2020b). In usually, predominant fatty acids were C16:0 (palmitoleic acid), C14:0 (myristic acid), 18:4(n-3) (stearidonic acid), and 18:1(n-9) (elaidic acid) (Table 6.3). A large of amount of lipid fraction of fatty acid of marine algae composed of poly unsaturated fatty acids (PUFAs) i.e., ω-3 and ω-6 (Leandro et al. 2020b; Denis et al. 2010).

Seaweeds are rich in both water (B1, B2, B12, C) and fat (β-carotene with vitamin A activity, vitamin E) soluble vitamins. The production of the vitamin will depend on several factors, especially on their exposure to sunlight, salinity and water temperature (Škrovánková 2011). Compared to the land base vegetable, seaweed such as *Eisenia arborea* (34.4 mg/100 g DW) is rich in vitamin C (Hernández-Carmona et al. 2009). The amount of vitamin A in *Codium fragile* and *Gracilaria chilensis* is relatively higher than the carrot (Viera et al. 2018). Moreover, seaweed considers a remarkable source for minerals (8–40% of the dry weight). Seaweed are rich in elements i.e., I, Fe, Mg, Ca, P, Na, K and some of these minerals are lacking in land vegetable (Leandro et al. 2020b). Nevertheless, seaweed accumulate a higher amount of non-essential trace element in contrast to land-based plant due to their polysaccharides having anionic carboxyl, sulfate and phosphate linkages that are found to be an appropriate site for heavy metals retention (Ganesan et al. 2019).

3 Application of Seaweed in Feed

Seaweed has significant potential as a feed and feeds ingredient due to its high growth rate and rich in useful metabolites (neutraceutical and pharmaceutical) and minerals (Morais et al. 2020). These feeds nourish the aquaculture finfish, shellfish, and terrestrial animals. Several research groups have been working on seaweed-based feed with different animals in various aspects i.e., health, meat quality, shelf life, and productivity. Apart from the nutritional advancement that seaweed can add to diets, supplemented feed also significantly changed the gastrointestinal flora, improved immunity, enhanced growth performance, and increased milk and meat quality and yield in livestock (Rajauria 2015). This may be the one factor in selecting seaweed as a feed ingredient. For the optimization of seaweed utilization as a commercialization feed ingredient, the species should fulfill several requirements such as a suitable nutritional profile, high digestibility, less influence of anti-nutritional factors, low production cost, availability through the year, and sufficient quantity availability for commercialization (Wan et al. 2019).

3.1 Aquaculture

3.1.1 Fish

With the increase in the global population and living standards, food systems are under growing pressure and developing the oceans have debatably become more important, especially since aquaculture become the top of the picture. Therefore, it has emerging demand for the feed industry *i.e.*, production volume, safety and quality and cost. Soy protein and fish meal is a key ingredients of the protein source. Due to several reasons and a deficit of the available stock, seaweed offers a novel and added-value dietary ingredient in formulated diets for fish (Smárason et al. 2019; Emblemsvåg et al. 2020). One of the earliest reported studies on the seaweed addition of finfish was by Nakagawa et al. (1984). In this study, fish meal was replaced by *Ulva pertusa* meal (dried and milled) and tested with black sea bream (*Acanthopagrus schlegeli*). They reported that the addition of 10% seaweed into formulated diet produced elevated protein efficiency without changing other growth parameters. The same research group studied the effect of *Ulva* meal on the lipid metabolomics of black sea bream (Nakagawa et al. 1987). They fed fish for 143 days in indoor tanks with a diet supplement with 10% *Ulva* meal. Authors found that *Ulva* meal resulted in an increase in total body lipids and influenced the composition of triglyceride.

There have been several seaweed feed supplement studies on finfish available and much of them focus on maintaining a balanced diet formulation replaced with a fish meal or soy meal by a considerable amount of seaweed, approximately 5–30% (Wan et al. 2019). Siddik et al. (2015), studied the potential use of seaweed (*Enteromorpha intestinalis*, gutweed) as a food ingredient substitution for fishmeal for Nile Tilapia (*Oreochromis niloticus*). They have tested a 0–50% replacement rate and observed the highest growth rate and cost reduction with a 20% replacement diet for Tilapia fry production. A number of studies have examined the use of *Ulva* species as a replacement seaweed source. The study of Shpigel et al. (2017) replaced 100% fish meal with *Ulva lactuca* and studied the growth performance and cost of gilthead seabream (*Sparus aurata*) juvenile in an integrated multi-trophic aquaculture system. There were no growth reduction and no difficulty in digestibility observed. Additionally, authors found no negative effect on fish performance up to 14.6% of *Ulva* addition. Several studies mentioned that good growth performance with 15% *Ulva rigida* (*Sparus aurata*) (Vizcaíno et al. 2016), 2.5% *U. ohnoi* (Atlantic Salmon, *Salmo salar*) (Norambuena et al. 2015) and >10% *U. onhoi* (sea bass, *Dicentrarchus labrax* and *S. aurata*) (Martínez-Antequera et al. 2021). However, Abdel-Warith et al. (2016) noted that African catfish *Clarias gariepinus* fed with 20–30% *U. lactuca* meal had poorer growth performance and feed utilization compared with the control diet (10% *U. lactuca*). From the literature, Ragaza et al. (2015) fed juvenile Japanese flounder (*Paralichthys olivaceus*) with red seaweed (*Eucheuma denticulatum*). According to blood chemical parameters and growth performance, the best result was achieved with 3% replacement. Another feeding trial was done by Sotoudeh and Jafari (2017) and fed rainbow trout, *Oncorhynchus mykiss*, with different percentages of red seaweed, *Gracilaria pygmaea*. They proposed that the 6% red seaweed was a useful amount when considering specific growth rate and feed conversion ratio. The study of Sotoudeh and Jafari (2017) reported that supplementation of experiment fed rainbow trout, *Oncorhynchus mykiss*, with red seaweed, *Gracilaria pygmaea*, and concluded that the supplementary feed was not affect to the whole-body composition and haematological parameters of juvenile. However, considering the weight gain, authors proposed dietary supplement with 6% of *G. pygmaea* useful for the rainbow trout juvenile. There was in contrast and interesting difference finding was reported by Shapawi and Zamry (2016) fed the Asian seabass juvenile with three seaweeds *Kappaphycus alvarezii* (KA), *Eucheuma denticulatum* (ED) and *Sargassum polycystum* (SP) meal as dietary ingredients. They reported that growth performance and feed conversion ratio was not significantly difference with seaweed inclusion in the fish diet and fish carcass composition was varied among the treatment.

Not only the growth performance, but also seaweed have been used in the fish culture industry as an ingredient to improve fish flesh characteristic, quality, and functions. Tissue pigmentation and immunological response were analyzed when *Gracilaria vermiculophylla* fed to rainbow trout (*Oncorhynchus mykiss*). Fish fed with 10% of the red seaweed diet showed higher carotenoid content in the skin (16.7 μg/g) while lower content in the flesh (0.23 μg/g). However, a 5% seaweed diet improved the innate immune response of fish i.e., the highest peroxide content, alternative complement (ACH50), and lysozyme activities. The latter authors suggested that fish fed up to 5% seaweed diet is best for higher inclusion levels impairs growth (Araújo et al. 2016). Effect of 3 seaweed (*Gracilariopsis persica*, *Hypnea flagelliformis*, *Sargassum boveanum*) on growth performance, fillet composition, digestive capacity, serum biochemical parameters, and expression of somatotropic axis genes of rainbow trout (*Oncorhynchus mykiss*) were tested by Marhamati et al. (2022). There was no significant effect on weight gain, feed conversion ratio, specific growth rate, and protein digestibility, but H. *flagelliformis*, *S. boveanum* fed fish showed improved colour in the fillet. Silva-Brito et al. (2021) also observed higher skin yellowness in gilthead seabream (*Sparus aurata*) fish fed with a 2.5% agar waste inclusion diet.

Gracilariopsis persica (GP), *Hypnea flagelliformis* (HF) and *Sargassum boveanum* (SB) were used as dietary inclusion of rainbow trout (*Oncorhynchus mykiss*) and investigated the effect of Immunomodulation, antioxidant enhancement and immune genes up-regulation. The results revealed that serum immune indices and head kidney antioxidant status and immune-related genes expression and the levels depend on the time and dose manner. There could be prebiotic potential use of seaweed supplements in the diet of aquaculture fish. Nazarudin et al. (2020) researched this with *Sargassum polycystum* as a supplement in diets for Asia sea bass (*Lates calcarife*) fingerlings. Except for better growth parameters, they observed pure colonies of descendant Gram-positive, none-spore forming, cocci and rod-shaped, catalase and oxidase negative bacteria, especially *Lactobacillus paracasei* subspecies paracasei in the intestine of the fish reared with the seaweed supplemented diet.

3.1.2 Shrimp

There are some nutritional studies with seaweed meals or seaweed extracts in dietary inclusion to ascertain their promising efficacy as functional, nutritional, and nutraceutical supplements in shrimp feed. Two red species, *Hypnea cervicornis* and *Cryptonemia crenulate* meal were tested as a protein source for white shrimp *Litopenaeus* vannamei, when there was an increase in feed conversion when the levels of algae were increased (da Silva and Barbosa 2009). Moreover, Cárdenas et al. (2015) conducted a feeding trial to evaluate the digestibility and growth rate of the same shrimp species with commercial seaweed diets. The results indicated that the inclusion of 4–8% of brown and green algae meal supported adequate growth and survival of juvenile shrimp. Omont et al. (2019) also studied the inclusion effect of three seaweed namely *Ulva lactuca*, *Eisenia* sp. and *Porphyra* sp., were tested with white shrimp. They proposed 5% inclusion of seaweed has resulted from an increase in enzyme activities and especially Ulva was suggested to promote shrimp growth productivity. The finding of Elizondo-González et al. (2018) also supported to this and they also confirmed that *U. lactuca* gave better growth performance and also better bioremediation properties in shrimp aquaculture.

The addition of the seaweed into the feed also increases the protein efficiency ratio in shrimp aquaculture and this ratio depends on the seaweed species. For instance, Cárdenas et al. (2015) examined that feeding white leg shrimp with commercial seaweed meals Nutrigreen improved the protein efficiency ratio (PER). Anh et al. (2018) found seaweed formulated feed with 10% and 20% of green seaweed *Cladophora* showed that significantly improved PER in post-larval tiger shrimp. The reason was that seaweed metabolites are helpful in the assimilation of proteins and can also modulate lipid metabolism (Banerjee et al. 2010). Some studies have shown that seaweed extraction could be beneficial in the prevention or treatment of diseases and improve health conditions. More recent work by Salehpour et al. (2021) showed that fucoidan extracted from brown seaweed, *Cystoseira trinodis* tested for the effect of white spot syndrome virus (WSSV) on white shrimp. They found that 0.4% fucoidan supplemented diet had enhanced the growth, immunological and biochemical parameters, and the expression of immune-related genes, as well as boosted the resistance against WSSV. Several studies have been published to examine the impact of several seaweed extractions on WSSV in different shrimp species (Immanuel et al. 2012; Wongprasert et al. 2014; Mariot et al. 2021; Schleder et al. 2020).

3.1.3 Mollusc

In general mollusc aquaculture has little or no external use of compound feed, however mollusc preferred seaweed as the main food in some stages of the life cycle (Rajauria 2015; Nazarudin et al. 2020). But, some land-based farming such as abalone (*Haliotis midae*) aquaculture in South Africa has used wild-harvested kelp (Ecklonia maxima) as a major feed source (Robertson-Andersson et al. 2009).

Likewise, O'Mahoney et al. (2014) practiced the mixed species of seaweed (*Laminaria digitata* meal, *Palmaria palmata* meal and *Ulva lactuca* meal) for Japanese abalone (Haliotis discus hannai), and proved that seaweed is good alternatives for fishmeal in abalone aquaculture. Similarly, in another study, an integrated system was practiced culture of fish, seaweed and Japanese abalone. The fish effluent drained into the microalgae culture (*Ulva lactuca* or *Gracilaria conferta*) and the algal production fed to the abalone. The total abalone yield was 9.4 kg/year and they concluded that the doubling abalone: fish yield ratio from 0.3 to 0.6 (Neori et al. 2000).

In the aquatic environment, the chemical cues act as intermediaries in different Intra and interspecific ecological relations between organisms especially species for the settlement of larvae, feeding and foraging behavior, prey selection, maturity and quality of meat (Nocchi et al. 2017). As an e.g. Nocchi et al. (2017) experimentally proved that sesquiterpene (+)-elatol from red seaweed (*Laurencia dendroidea*) attracted the sea hare (*Aplysia brasiliana*). Moreover, they found that the chemical cue of red seaweed allowed young individuals of sea hare to associate with *L. dendriidea*. Not only the growth improvement but also seaweed acts against cadmium (Cd) toxicity. A recent study reported by Luo et al. (2019) fed the abalone with seaweed (*Gracilaria lemaneiformis*) as the main dietary source and observed selenium (Se) enriched *Gracilaria* protects abalone against Cd toxicity.

3.1.4 Other Aquatic Species

The higher survival rate, and fast growth rate, enhance consumer preference (like meat colour) are the key elements of fish feed (Rajauria 2015). Several studies pointed out the use of seaweed as a feed ingredient for other aquatic species. Li et al. (2022) investigated the nutritional potential of different algae ingredients (*Sargassum polycystum*, *S. thunbergii*, *S. horneri*, *Enteromorpha prolifera* and *Macrocystis pyrifera*) powder for feed sea cucumber (*Apostichopus japonicas*). The seaweed powder was mixed with sea mud and fed to sea cucumber in 60-day trial and measured the growth, body composition, and digestive enzyme activities. The body weight increased significantly and the highest specific growth rate (1.31%/day) was recorded when fed with *S. horneri*. The highest ingestion rate was observed when fed with *M. pyrifera* residue (1.5 g/g, day). However, the constant amino acid composition was observed during the study period. In another study, Xia et al. (2012a) observed that the feeding preference indices changed with time. According to the observation, the preference for *Sargassum thunbergii* rose significantly changed from being rejected to somewhat preferred over 30 days. In conclusion, they recommended a diet with 50% *Laminaria japonica* was the best combination of sea cucumber culture. Except for the growth, ingestion rate and digestibility, seaweed feed cause to reduce ammonia-nitrogen production (Xia et al. 2012b). In this study lower ammonia nitrogen production was observed in *Apostichopus japonicus* fed the *Ulva lactuca* or *Laminaria japonica* diet compared to the *Sargassum thunbergii*, *Sargassum polycystum* and *Zostera marina*.

The gonad colour, size, texture and flavour of sea urchin is very important in market acceptability and seaweed can play important role in artificial feeding (Lourenço et al. 2022). Onomu et al. (2020), studied the gonads enhancement of *Tripneustes gratilla* 12 weeks fed fresh *Ulva rigida* (50:50 mixture of fresh *U. rigida* and *Gracilaria gracilis*, fresh *G. gracilis* and a formulated diet 20 U (containing 20% *U. rigida*), and in the final 6 weeks (phase 2) of the study, the diet was changed to a formulated feed (20 U diet). They found that the growing sea urchins initially on seaweed-only diets does not affect somatic growth, and a switch to a diet of high-protein formulated feed containing seaweed (*U. rigida*) is gifted of speedy growth of high quality sea urchin gonads. In another study, Dworjanyn et al. (2007), examined the effects of algae addition feeding stimulants to artificial diets for sea urchins (*Tripneustes gratilla*). They prepared three artificial feeds containing 5% of three types of seaweed namely, *Ecklonia radiata*, *Sargassum linearifolium* and *Ulva lactuca*. *T. gratilla* consumed more *Ecklonia* and *Sargassum* diet than control diet while juvenile sea urchins grew significantly faster on a wet weight basis on *Sargassum* diet.

3.2 Poultry (Egg & Broilers)

Poultry production is an important area of the agricultural economy and healthy nations. Feed production with easily available biomass such as seaweed plays a significant role to reduce the cost of the feed and enhance health. In addition to that seaweed influence the positively immune status and nutrient quality, especially in reducing lipid and cholesterol level even though there is some disadvantage (Morais et al. 2020; Michalak and Mahrose 2020). Abudabos et al. (2013), found that supplements of 1% and 3% *Ulva lactuca* showed the lowest serum total lipid, cholesterol and uric acid concentrations. Moreover, 3% *U. lactuca* treated birds received higher dressing percentage and breast muscle yield. Another study conducted by Petrolli et al. (2019), and observed the broiler chicken fed by *Schizotrichium* ssp. showed supplements changed the fatty acid profile, enriching the omega 3 fatty acid series, primarily the docosahexaenoic acid (DHA). Seaweed affects the growth performance and blood profiles including serum immunoglobulin in broilers. Choi et al. (2014), fed the broiler chick during the 5 week time with brown seaweed (*Undaria pinnatifida*) by-product and seaweed fusiforme (*Hizikia fusiformis*) and found positive effects on growth performance and immune response. Studies with red seaweed, *Palmaria palmata*, with added to the broiler diet and found greater body weight increase, linearity related to the level of albumin, creatinine, uric acid, and white blood cells. In addition to that, the seaweed had a beneficial effect on the fecal microbes as *Lactobacillus* sp. counts increased and *Escherichia coli* and *Salmonella* sp. count decreased. There is more study showing that seaweeds as sources of nutrients and favorable bioactive compounds can promote sustainable production of functional poultry products (Matshogo et al. 2020, 2021; Mohammadigheisar et al. 2020; Stokvis et al. 2022).

There were several studies in the last few decades with the objectives of enhancing egg characterization and egg quality. Kulshreshtha et al. (2014), studied the egg quality of layer hens fed with red seaweed (*Chondrus crispus* and *Sarcodiotheca gaudichaudii*). The results were shown that short-chain fatty acids, including acetic acid, propionic acid, n- butyric acid, and i-butyric acid were significantly higher in seaweed supplemented-fed group. In addition, seaweed acted potential prebiotic to improve performance, egg quality, and overall gut health in layer hens. Another study was conducted by Guo et al. (2020) to evaluate the effect of marine delivery polysaccharide (MDP) from seaweed Enteromorpha on productive performance, egg quality, antioxidant capacity, and jejunal morphology in late-phase laying hens. The dietary MDP improved liver catalase activity, villus height, villus height/crypt depth ratio, egg quality, antioxidant capacity, and jejunal morphology. Abo El-Maaty et al. (2021), studied the effect of calcium nanoparticles with biocompatible *Sargassum latifolium* algae extract supplementation on egg quality and found that supplementary powder up to 1.5 g/kg to the diet of laying hens improved the eggshell thickness, shell weight percentage shell weight per unit surface.

3.3 Ruminants

The use of seaweeds as constituents of ruminant diets can be an alternative to conformist feedstuffs. Recent in-vitro and in-vivo studies proposed that seaweed can alleviate enteric methane (CH_4) emissions from ruminants when added to diets. Especially red seaweed are rich in protein and halogenated compounds while brown seaweed is rich in Phlorotannin and those compounds decreased methane production (Min et al. 2021). Three red seaweed, 3 brown seaweed and 1 green seaweed were tested for rumen nutrition and methane production by Molina-Alcaide et al. (2017). Among the studied species, the autumn collection showed lower methane concentration than the spring collection. Another study conducted by Glasson et al. (2022), and highlighted that seaweed species (*Asparagopsis taxiformis* and *Asparagopsis armata*) cattle and sheep inhibit methanogenesis by up to 98% and this was resulted to increase the global demand. It has been experimentally demonstrated that red seaweed (*Asparagopsis taxiformis*) decreased methane up to 40% and 98% without changing meat eating quality (Kinley et al. 2020).

Nutrition profile is very important in dairy cattle, especially early lactation period. Rey-Crespo et al. (2014), used seaweed from the Galician coast as a source of micromineral supplements, especially iodine. They observed the algae supplement significantly improved animal minerals status mainly iodine and selenium. Moreover, the researcher proposed the requirement of novel research on antioxidant, anti-microbial, and immunomodulatory activities, particularly on milk production, reproductive function and resistance to diseases. As similarly, iodine content was observed in cow's milk samples fed with seaweed (*Ascophyllum nodosum* and *Laminaria digitata*), but same time high arsenic (As) content was also observed in the cows fed with high seaweed amount. The feed supplement containing brown seaweed (*Ascophyllum nodosum*) extract increased the antioxidant activities of goats (Kannan et al. 2007). Seaweed provides functional polysaccharides, and they are a good source of vitamins, minerals, and phenolic compounds. Samarasinghe et al. (2021), studied the effect of feeding milk supplemented with *Ulva lactuca*, *Ascophyllum nodosum*, or *Saccharina latissima* on the systemic immune status of preweaning dairy calves. They observed that feeding milk accompanied with dried seaweed improved plasma concentrations of variables related to the innate immune response in preweaning dairy calves. It should be noted here that according to Maheswari et al. (2021), 2.5% seaweed (*Cappaphycus alvarezii*, *Gracilaria salicornia* and *Turbinaria conoides*) based feed supplement of lactating Murrah buffaloes showed the improvement of antioxidant status, immunity and milk yield. de Lima et al. (2019), succeeded in demonstrating that seaweed (*Gracilaria birdiae*) may alleviate to reduce the environmental stress faced by dairy goats without significant effect on milk production.

Even though there was some positive impact, many seaweed may negatively impact the ruminant if exceeded the supplement level. Machado et al. (2016), reported that seaweed had some interference with ruminant digestion, especially decreasing total gas production and volatile fatty acid amount. Moreover, seaweeds caused the reduction of organic matter digestibility (Rjiba-Ktita et al. 2019), and fiber digestibility and caused to increase in heavy metal i.e. arsenic (As) concentration (Cabrita et al. 2016). Even some researchers reported that seaweed supplements had no influence on the milk yield (Antaya et al. 2019; Karatzia et al. 2012), and there was further evidence that extreme seaweed supplements may cause to decrease in the milk yield (Roque et al. 2019).

3.4 Rabbits

Seaweed is one of the alternative solutions to feed the rabbit meat industry after the ban on the prophylactic use of antibiotics in animal feed (Al-Soufi et al. 2022). Several investigations have been carried out on the application of seaweed as additives for rabbits' feeds. Rossi et al. (2020a), investigated the growth performance and meat quality of growing rabbits with brown seaweeds (*Laminaria* spp.) and plant extracts mixed feed. The cholesterol level decreased (41% than control), enhanced the α-tocopherol and retinol content and improved the sensory attribute of the rabbits meet fed 0.6% of the natural feed additive. The development of gut health has been shown to have the significant potential to decrease the negative economic impact of the rabbit farming. The growth performances, carcass features and meat quality parameters from growing rabbit were investigated after fed with two levels of dietary brown seaweed (*Laminaria* spp) and plant polyphenols by Rossi et al. (2020b). New Zealand White rabbits (144 individuals) were allotted for this study and researchers observed the vitamin A, E and oxidative stability enhancement in treated groups, while the cholesterol content tended to be lower than control. Moreover, the polysaccharides from seaweed extract help to enhanced the rabbit's plasma testosterone concentration and improved various sperm motility parameters (Okab et al. 2013; Vizzari et al. 2021).

4 Limitation & Future Perspectives

There is increasing interest in seaweed as an alternative sustainable feed resource. Like most other alternatives, seaweeds do not come without any problems. The main challenge with seaweed is seasonal variation in chemical composition and thus a variable nutritional profile. The difference is mainly related to the mineral content, as well as the content of polysaccharides and crude protein (Polat and Ozogul 2013; Marinho et al. 2019). In addition, the bioactive compounds and antioxidant capacity of the seaweed also changed during the season (Mansur et al. 2020; Roleda et al. 2019). Hence, these compositional changes can be a challenge when using seaweed as a feed resource.

Another major constraint is that seaweed contains heavy metals such as mercury, lead, cadmium and arsenic. The level of trace metal has shown temporal and spatial variations with the species (Banach et al. 2020). Seaweed can also contain hazard compounds such as pesticide residues, dioxins and polychlorinated biphenyls (PCBs), brominated flame retardants, polycyclic aromatic hydrocarbons (PAHs), pharmaceuticals, marine biotoxins, allergens, micro-plastics, other pathogenic bacteria, norovirus, and hepatitis E virus and the contamination levels depend on where seaweed are cultivation (Banach et al. 2020). Therefore, more risk assessment studies are needed to understand the bioavailability of such contaminants via the feeding pathway.

Use of seaweed in modern feed formulation will require targeted processing to upgrade the nutritional value, especially when used in diets for monogastric animals. Seaweed contains many bioactive components with potential health benefits in both for food and feed applications. Hence, functional foods or feeds containing seaweed extracts have great potential. However, some natural metabolites can adversely affect growth performance and health of animals such as use of red seaweed *Asparagopsis* spp. in diets for ruminants to reduce methane emission. It has been reported that this seaweed can reduce the rumen methane production by 98% where the active agent responsible for this is bromoform, but which is also considered a carcinogenic (Kinley et al. 2020).

In this chapter, we reviewed the latest application, development, and drawbacks in the research on seaweed as a valuable component to feed for livestock and aquaculture animals. Even though there are some constraints and limitations, there are many advantages of using seaweed in animal nutrition as pointed out i.e., improved growth performance, egg and meat quality, and health, as well as environmental sustainability (Michalak and Mahrose 2020).

References

Abdel-Warith A-WA, Younis E-SMI, Al-Asgah NA (2016) Potential use of green macroalgae *Ulva lactuca* as a feed supplement in diets on growth performance, feed utilization and body composition of the African catfish, *Clarias gariepinus*. Saudi J Biol Sci 23:404–409. https://doi.org/10.1016/j.sjbs.2015.11.010

Abo El-Maaty HA, El-Khateeb AY, Al-Khalaifah H, Hamed ESAE, Hamed S, El-Said EA, Mahrose KM, Metwally K, Mansour AM (2021) Effects of ecofriendly synthesized calcium nanoparticles with biocompatible *Sargassum latifolium* algae extract supplementation on egg quality and scanning electron microscopy images of the eggshell of aged laying hens. Poult Sci 100:675–684. https://doi.org/10.1016/j.psj.2020.10.043

Abudabos AM, Okab AB, Aljumaah RS, Samara EM, Abdoun KA, Al-Haidary AA (2013) Nutritional value of green seaweed (*Ulva lactuca*) for broiler chickens. Ital J Anim Sci 12:e28. https://doi.org/10.4081/ijas.2013.e28

Afonso C, Correia AP, Freitas MV, Baptista T, Neves M, Mouga T (2021) Seasonal changes in the nutritional composition of *Agarophyton vermiculophyllum* (Rhodophyta, Gracilariales) from the center of Portugal. Food Secur 10:1145

Ahmad F, Sulaiman MR, Saimon W, Yee CF, Matanjun P (2012) Proximate compositions and total phenolic contents of selected edible seaweed from Semporna, Sabah, Malaysia. Borneo Sci 31:74–83

Al-Soufi S, García J, Muíños A, López-Alonso M (2022) Marine macroalgae in rabbit nutrition—a valuable feed in sustainable farming. Animals 12:2346

Angell AR, Angell SF, de Nys R, Paul NA (2016) Seaweed as a protein source for mono-gastric livestock. Trends Food Sci Technol 54:74–84. https://doi.org/10.1016/j.tifs.2016.05.014

Anh NTN, Hai TN, Hien TTT (2018) Effects of partial replacement of fishmeal protein with green seaweed (*Cladophora* spp.) protein in practical diets for the black tiger shrimp (*Penaeus monodon*) postlarvae. J Appl Phycol 30:2649–2658. https://doi.org/10.1007/s10811-018-1457-7

Antaya NT, Ghelichkhan M, Pereira ABD, Soder KJ, Brito AF (2019) Production, milk iodine, and nutrient utilization in Jersey cows supplemented with the brown seaweed *Ascophyllum nodosum* (kelp meal) during the grazing season. J Dairy Sci 102:8040–8058. https://doi.org/10.3168/jds.2019-16478

Araújo M, Rema P, Sousa-Pinto I, Cunha LM, Peixoto MJ, Pires MA, Seixas F, Brotas V, Beltrán C, Valente LMP (2016) Dietary inclusion of IMTA-cultivated *Gracilaria vermiculophylla* in rainbow trout (*Oncorhynchus mykiss*) diets: effects on growth, intestinal morphology, tissue pigmentation, and immunological response. J Appl Phycol 28:679–689. https://doi.org/10.1007/s10811-015-0591-8

Ayele J, Tolemariam T, Beyene A, Tadese DA, Tamiru M (2021) Assessment of livestock feed supply and demand concerning livestock productivity in Lalo Kile district of Kellem Wollega Zone, Western Ethiopia. Heliyon 7:e08177. https://doi.org/10.1016/j.heliyon.2021.e08177

Banach JL, Hoek-van den Hil EF, van der Fels-Klerx HJ (2020) Food safety hazards in the European seaweed chain. Compr Rev Food Sci Food Saf 19:332–364. https://doi.org/10.1111/1541-4337.12523

Banerjee K, Mitra A, Mondal K (2010) Cost-effective and eco-friendly shrimp feed from red seaweed *Catenella repens* (Gigartinales: Rhodophyta). Curr Biotica 8:23–43

Biancarosa I, Espe M, Bruckner CG, Heesch S, Liland N, Waagbø R, Torstensen B, Lock EJ (2017) Amino acid composition, protein content, and nitrogen-to-protein conversion factors of 21 seaweed species from Norwegian waters. J Appl Phycol 29:1001–1009. https://doi.org/10.1007/s10811-016-0984-3

Cabrita ARJ, Maia MRG, Oliveira HM, Sousa-Pinto I, Almeida AA, Pinto E, Fonseca AJM (2016) Tracing seaweeds as mineral sources for farm-animals. J Appl Phycol 28:3135–3150. https://doi.org/10.1007/s10811-016-0839-y

Cárdenas JV, Gálvez AO, Brito LO, Galarza EV, Pitta DC, Rubin VV (2015) Assessment of different levels of green and brown seaweed meal in experimental diets for whiteleg shrimp

(*Litopenaeus vannamei*, Boone) in recirculating aquaculture system. Aquac Int 23:1491–1504. https://doi.org/10.1007/s10499-015-9899-2

Černá M (2011) Chapter 24 – Seaweed proteins and amino acids as nutraceuticals. In: Kim S-K (ed) Advances in food and nutrition research. Academic, pp 297–312. https://doi.org/10.1016/B978-0-12-387669-0.00024-7

Chan PT, Matanjun P (2017) Chemical composition and physicochemical properties of tropical red seaweed, *Gracilaria changii*. Food Chem 221:302–310. https://doi.org/10.1016/j.foodchem.2016.10.066

Choi YJ, Lee SR, Oh JW (2014) Effects of dietary fermented seaweed and seaweed fusiforme on growth performance, carcass parameters and immunoglobulin concentration in broiler chicks. Asian Australas J Anim Sci 27:862–870. https://doi.org/10.5713/ajas.2014.14015

Cian RE, Fajardo MA, Alaiz M, Vioque J, González RJ, Drago SR (2014) Chemical composition, nutritional and antioxidant properties of the red edible seaweed *Porphyra columbina*. Int J Food Sci Nutr 65:299–305. https://doi.org/10.3109/09637486.2013.854746

da Silva RL, Barbosa JM (2009) Seaweed meal as a protein source for the white shrimp Litopenaeus vannamei. J Appl Phycol 21:193–197. https://doi.org/10.1007/s10811-008-9350-4

de Lima RN, de Souza JBF, Batista NV, de Andrade AKS, Soares ECA, dos Santos Filho CA, da Silva LA, Coelho WAC, de Macedo Costa LL, de Oliveira Lima P (2019) Mitigating heat stress in dairy goats with inclusion of seaweed *Gracilaria birdiae* in diet. Small Rumin Res 171:87–91. https://doi.org/10.1016/j.smallrumres.2018.11.008

Denis C, Morançais M, Li M, Deniaud E, Gaudin P, Wielgosz-Collin G, Barnathan G, Jaouen P, Fleurence J (2010) Study of the chemical composition of edible red macroalgae *Grateloupia turuturu* from Brittany (France). Food Chem 119:913–917. https://doi.org/10.1016/j.foodchem.2009.07.047

Dworjanyn SA, Pirozzi I, Liu W (2007) The effect of the addition of algae feeding stimulants to artificial diets for the sea urchin *Tripneustes gratilla*. Aquaculture 273:624–633. https://doi.org/10.1016/j.aquaculture.2007.08.023

Elizondo-González R, Quiroz-Guzmán E, Escobedo-Fregoso C, Magallón-Servín P, Peña-Rodríguez A (2018) Use of seaweed *Ulva lactuca* for water bioremediation and as feed additive for white shrimp *Litopenaeus vannamei*. PeerJ 6:e4459. https://doi.org/10.7717/peerj.4459

Emblemsvåg J, Kvadsheim NP, Halfdanarson J, Koesling M, Nystrand BT, Sunde J, Rebours C (2020) Strategic considerations for establishing a large-scale seaweed industry based on fish feed application: a Norwegian case study. J Appl Phycol 32:4159–4169. https://doi.org/10.1007/s10811-020-02234-w

Fleurence J (2016) Chapter 5 – Seaweeds as food. In: Fleurence J, Levine I (eds) Seaweed in health and disease prevention. Academic, San Diego, pp 149–167. https://doi.org/10.1016/B978-0-12-802772-1.00005-1

Ganesan AR, Tiwari U, Rajauria G (2019) Seaweed nutraceuticals and their therapeutic role in disease prevention. Food Sci Human Wellness 8:252–263. https://doi.org/10.1016/j.fshw.2019.08.001

Glasson CRK, Kinley RD, de Nys R, King N, Adams SL, Packer MA, Svenson J, Eason CT, Magnusson M (2022) Benefits and risks of including the bromoform containing seaweed *Asparagopsis* in feed for the reduction of methane production from ruminants. Algal Res 64:102673. https://doi.org/10.1016/j.algal.2022.102673

Gosch BJ, Magnusson M, Paul NA, de Nys R (2012) Total lipid and fatty acid composition of seaweeds for the selection of species for oil-based biofuel and bioproducts. GCB Bioenergy 4:919–930. https://doi.org/10.1111/j.1757-1707.2012.01175.x

Guo Y, Zhao Z-H, Pan Z-Y, An L-L, Balasubramanian B, Liu W-C (2020) New insights into the role of dietary marine-derived polysaccharides on productive performance, egg quality, antioxidant capacity, and jejunal morphology in late-phase laying hens. Poult Sci 99:2100–2107. https://doi.org/10.1016/j.psj.2019.12.032

Henchion M, Moloney AP, Hyland J, Zimmermann J, McCarthy S (2021) Review: trends for meat, milk and egg consumption for the next decades and the role played by livestock systems in the global production of proteins. Animal 15:100287. https://doi.org/10.1016/j.animal.2021.100287

Hernández-Carmona G, Carrillo-Domínguez S, Arvizu-Higuera DL, Rodríguez-Montesinos YE, Murillo-Álvarez JI, Muñoz-Ochoa M, Castillo-Domínguez RM (2009) Monthly variation in the chemical composition of *Eisenia arborea* J.E. Areschoug. J Appl Phycol 21:607–616. https://doi.org/10.1007/s10811-009-9454-5

Immanuel G, Sivagnanavelmurugan M, Marudhupandi T, Radhakrishnan S, Palavesam A (2012) The effect of fucoidan from brown seaweed *Sargassum wightii* on WSSV resistance and immune activity in shrimp *Penaeus monodon* (Fab). Fish Shellfish Immunol 32:551–564. https://doi.org/10.1016/j.fsi.2012.01.003

Jayasinghe GDTM, Jinadasa BKKK, Chinthaka SDM (2018) Nutritional composition and heavy metal content of five tropical seaweeds. Open Sci J Anal Chem 3:17–22

Kannan G, Saker KE, Terrill TH, Kouakou B, Galipalli S, Gelaye S (2007) Effect of seaweed extract supplementation in goats exposed to simulated preslaughter stress. Small Rumin Res 73:221–227. https://doi.org/10.1016/j.smallrumres.2007.02.006

Karatzia M, Christaki E, Bonos E, Karatzias C, Florou-Paneri P (2012) The influence of dietary *Ascophyllum nodosum* on haematologic parameters of dairy cows. Ital J Anim Sci 11:e31. https://doi.org/10.4081/ijas.2012.e31

Kendel M, Wielgosz-Collin G, Bertrand S, Roussakis C, Bourgougnon N, Bedoux G (2015) Lipid composition, fatty acids and sterols in the seaweeds *Ulva armoricana*, and *Solieria chordalis* from Brittany (France): an analysis from nutritional, chemotaxonomic, and antiproliferative activity perspectives. Mar Drugs 13:5606–5628

Khairy HM, El-Shafay SM (2013) Seasonal variations in the biochemical composition of some common seaweed species from the coast of Abu Qir Bay, Alexandria, Egypt. Oceanologia 55:435–452. https://doi.org/10.5697/oc.55-2.435

Kinley RD, Martinez-Fernandez G, Matthews MK, de Nys R, Magnusson M, Tomkins NW (2020) Mitigating the carbon footprint and improving productivity of ruminant livestock agriculture using a red seaweed. J Clean Prod 259:120836. https://doi.org/10.1016/j.jclepro.2020.120836

Kulshreshtha G, Rathgeber B, Stratton G, Thomas N, Evans F, Critchley A, Hafting J, Prithiviraj B (2014) Feed supplementation with red seaweeds, *Chondrus crispus* and *Sarcodiotheca gaudichaudii*, affects performance, egg quality, and gut microbiota of layer hens. Poult Sci 93:2991–3001. https://doi.org/10.3382/ps.2014-04200

Laramore SE, Wills PS, Hanisak MD (2022) Seasonal variation in the nutritional profile of *Ulva lactuca* produced in a land-based IMTA system. Aquac Int 30:3067–3079. https://doi.org/10.1007/s10499-022-00950-3

Leandro A, Pereira L, Gonçalves AMM (2020a) Diverse applications of marine macroalgae. Mar Drugs 18:17

Leandro A, Pacheco D, Cotas J, Marques JC, Pereira L, Gonçalves AMM (2020b) Seaweed's bioactive candidate compounds to food industry and global food security. Life 10:140

Li X, Wang Y, Jiang X, Li H, Liu T, Ji L, Sun Y (2022) Utilization of different seaweeds *(Sargassum polycystum*, *Sargassum thunbergii*, *Sargassum horneri*, *Enteromorpha prolifera*, *Macrocystis pyrifera*, and the residue of *M. pyrifera*) in the diets of sea cucumber *Apostichopus japonicus* (Selenka, 1867). Algal Res 61:102591. https://doi.org/10.1016/j.algal.2021.102591

Lorenzo JM, Agregán R, Munekata PES, Franco D, Carballo J, Şahin S, Lacomba R, Barba FJ (2017) Proximate composition and nutritional value of three macroalgae: *Ascophyllum nodosum*, *Fucus vesiculosus* and *Bifurcaria bifurcata*. Mar Drugs 15:360

Lourenço SO, Barbarino E, De-Paula JC, Pereira LODS, Marquez UML (2002) Amino acid composition, protein content and calculation of nitrogen-to-protein conversion factors for 19 tropical seaweeds. Psychol Res 50:233–241. https://doi.org/10.1046/j.1440-1835.2002.00278.x

Lourenço S, Raposo A, Cunha B, Pinheiro J, Santos PM, Gomes AS, Ferreira S, Gil MM, Costa JL, Pombo A (2022) Temporal changes in sex-specific color attributes and carotenoid concentration in the gonads (roe) of the purple sea urchin (*Paracentrotus lividus*) provided dry feeds supplemented with β-carotene. Aquaculture 560:738608. https://doi.org/10.1016/j.aquaculture.2022.738608

Luo H, Wang Q, He Z, Wu Y, Long A, Yang Y (2019) Protection of dietary selenium-enriched seaweed *Gracilaria lemaneiformis* against cadmium toxicity to abalone *Haliotis discus hannai*. Ecotoxicol Environ Saf 171:398–405. https://doi.org/10.1016/j.ecoenv.2018.12.105

Machado L, Magnusson M, Paul NA, Kinley R, de Nys R, Tomkins N (2016) Dose-response effects of *Asparagopsis taxiformis* and *Oedogonium* sp. on in vitro fermentation and methane production. J Appl Phycol 28:1443–1452. https://doi.org/10.1007/s10811-015-0639-9

Maheswari M, Das A, Datta M, Tyagi AK (2021) Supplementation of tropical seaweed-based formulations improves antioxidant status, immunity and milk production in lactating Murrah buffaloes. J Appl Phycol 33:2629–2643. https://doi.org/10.1007/s10811-021-02473-5

Makkar HPS, Tran G, Heuzé V, Giger-Reverdin S, Lessire M, Lebas F, Ankers P (2016) Seaweeds for livestock diets: a review. Anim Feed Sci Technol 212:1–17. https://doi.org/10.1016/j.anifeedsci.2015.09.018

Mansur AA, Brown MT, Billington RA (2020) The cytotoxic activity of extracts of the brown alga *Cystoseira tamariscifolia* (Hudson) Papenfuss, against cancer cell lines changes seasonally. J Appl Phycol 32:2419–2429. https://doi.org/10.1007/s10811-019-02016-z

Marhamati A, Vazirzadeh A, Chisti Y (2022) Seaweed-based diets lead to normal growth, improved fillet color but a down-regulated expression of somatotropic axis genes in rainbow trout (*Oncorhynchus mykiss*). Aquaculture 554:738183. https://doi.org/10.1016/j.aquaculture.2022.738183

Marinho GS, Sørensen A-DM, Safafar H, Pedersen AH, Holdt SL (2019) Antioxidant content and activity of the seaweed *Saccharina latissima*: a seasonal perspective. J Appl Phycol 31:1343–1354. https://doi.org/10.1007/s10811-018-1650-8

Mariot LV, Bolívar N, Coelho JDR, Goncalves P, Colombo SM, do Nascimento FV, Schleder DD, Hayashi L (2021) Diets supplemented with carrageenan increase the resistance of the Pacific white shrimp to WSSV without changing its growth performance parameters. Aquaculture 545:737172. https://doi.org/10.1016/j.aquaculture.2021.737172

Martínez-Antequera FP, Martos-Sitcha JA, Reyna JM, Moyano FJ (2021) Evaluation of the inclusion of the green seaweed *Ulva ohnoi* as an ingredient in feeds for gilthead sea bream (*Sparus aurata*) and European sea bass (*Dicentrarchus labrax*). Animals 11:1684

Matshogo TB, Mnisi CM, Mlambo V (2020) Dietary green seaweed compromises overall feed conversion efficiency but not blood parameters and meat quality and stability in broiler chickens. Agriculture 10:547

Matshogo TB, Mlambo V, Mnisi CM, Manyeula F (2021) Effect of pre-treating dietary green seaweed with fibrolytic enzymes on growth performance, blood indices, and meat quality parameters of cobb 500 broiler chickens. Livest Sci 251:104652. https://doi.org/10.1016/j.livsci.2021.104652

Michalak I, Mahrose K (2020) Seaweeds, intact and processed, as a valuable component of poultry feeds. J Mar Sci Eng 8:620

Min BR, Parker D, Brauer D, Waldrip H, Lockard C, Hales K, Akbay A, Augyte S (2021) The role of seaweed as a potential dietary supplementation for enteric methane mitigation in ruminants: challenges and opportunities. Anim Nutr 7:1371–1387. https://doi.org/10.1016/j.aninu.2021.10.003

Mišurcová L, Buňka F, Vávra Ambrožová J, Machů L, Samek D, Kráčmar S (2014) Amino acid composition of algal products and its contribution to RDI. Food Chem 151:120–125. https://doi.org/10.1016/j.foodchem.2013.11.040

Miyashita K, Mikami N, Hosokawa M (2013) Chemical and nutritional characteristics of brown seaweed lipids: a review. J Funct Foods 5:1507–1517. https://doi.org/10.1016/j.jff.2013.09.019

Mohammadigheisar M, Shouldice VL, Sands JS, Lepp D, Diarra MS, Kiarie EG (2020) Growth performance, breast yield, gastrointestinal ecology and plasma biochemical profile in broiler chickens fed multiple doses of a blend of red, brown and green seaweeds. Br Poult Sci 61:590–598. https://doi.org/10.1080/00071668.2020.1774512

Molina-Alcaide E, Carro MD, Roleda MY, Weisbjerg MR, Lind V, Novoa-Garrido M (2017) In vitro ruminal fermentation and methane production of different seaweed species. Anim Feed Sci Technol 228:1–12. https://doi.org/10.1016/j.anifeedsci.2017.03.012

Morais T, Inácio A, Coutinho T, Ministro M, Cotas J, Pereira L, Bahcevandziev K (2020) Seaweed potential in the animal feed: a review. J Mar Sci Eng 8:559

Nakagawa H, Kasahara S, Sugiyama T, Wada I (1984) Usefulness of ulva-meal [*Ulva pertusa*] as feed supplementary in cultured black sea bream [*Acanthopagrus shlegeli*]. Aquiculture 32:20–27

Nakagawa H, Kasahara S, Sugiyama T (1987) Effect of Ulva meal supplementation on lipid metabolism of black sea bream, *Acanthopagrus schlegeli* (Bleeker). Aquaculture 62:109–121

Nazarudin MF, Yusoff F, Idrus ES, Aliyu-Paiko M (2020) Brown seaweed *Sargassum polycystum* as dietary supplement exhibits prebiotic potentials in Asian sea bass *Lates calcarifer* fingerlings. Aquac Rep 18:100488. https://doi.org/10.1016/j.aqrep.2020.100488

Neori A, Shpigel M, Ben-Ezra D (2000) A sustainable integrated system for culture of fish, seaweed and abalone. Aquaculture 186:279–291. https://doi.org/10.1016/S0044-8486(99)00378-6

Nocchi N, Soares A, Souto M, Fernández J, Martin M, Pereira R (2017) Detection of a chemical cue from the host seaweed Laurencia dendroidea by the associated mollusc Aplysia brasiliana. PLoS One 12:e0187126

Norambuena F, Hermon K, Skrzypczyk V, Emery JA, Sharon Y, Beard A, Turchini GM (2015) Algae in fish feed: performances and fatty acid metabolism in juvenile Atlantic salmon. PLoS One 10:e0124042

O'Mahoney M, Rice O, Mouzakitis G, Burnell G (2014) Towards sustainable feeds for abalone culture: evaluating the use of mixed species seaweed meal in formulated feeds for the Japanese abalone, Haliotis discus hannai. Aquaculture 430:9–16. https://doi.org/10.1016/j.aquaculture.2014.02.036

Okab AB, Samara EM, Abdoun KA, Rafay J, Ondruska L, Parkanyi V, Pivko J, Ayoub MA, Al-Haidary AA, Aljumaah RS, Peter M, Lukac N (2013) Effects of dietary seaweed *(Ulva lactuca)* supplementation on the reproductive performance of buck and doe rabbits. J Appl Anim Res 41:347–355. https://doi.org/10.1080/09712119.2013.783479

Omont A, Quiroz-Guzman E, Tovar-Ramirez D, Peña-Rodríguez A (2019) Effect of diets supplemented with different seaweed extracts on growth performance and digestive enzyme activities of juvenile white shrimp *Litopenaeus vannamei*. J Appl Phycol 31:1433–1442. https://doi.org/10.1007/s10811-018-1628-6

Onomu AJ, Vine NG, Cyrus MD, Macey BM, Bolton JJ (2020) The effect of fresh seaweed and a formulated diet supplemented with seaweed on the growth and gonad quality of the collector sea urchin, *Tripneustes gratilla*, under farm conditions. Aquac Res 51:4087–4102. https://doi.org/10.1111/are.14752

Øverland M, Mydland LT, Skrede A (2019) Marine macroalgae as sources of protein and bioactive compounds in feed for monogastric animals. J Sci Food Agric 99:13–24. https://doi.org/10.1002/jsfa.9143

Peñalver R, Lorenzo JM, Ros G, Amarowicz R, Pateiro M, Nieto G (2020) Seaweeds as a functional ingredient for a healthy diet. Mar Drugs 18:301. https://doi.org/10.3390/md18060301

Petrolli TG, Petrolli OJ, Pereira ASC, Zotti CA, Romani J, Villani R, Leite F, Zanandréa FM (2019) Effects of the dietary supplementation with a microalga extract on broiler performance and fatty-acid meat profile. Braz J Poult Sci 21:eRBCA-2018

Polat S, Ozogul Y (2013) Seasonal proximate and fatty acid variations of some seaweeds from the northeastern Mediterranean coast. Oceanologia 55:375–391. https://doi.org/10.5697/oc.55-2.375

Post MJ, Hocquette JF (2017) Chapter 16 – New sources of animal proteins: cultured meat. In: Purslow PP (ed) New aspects of meat quality. Woodhead Publishing, pp 425–441. https://doi.org/10.1016/B978-0-08-100593-4.00017-5

Praveen MA, Parvathy KK, Jayabalan R, Balasubramanian P (2019) Dietary fiber from Indian edible seaweeds and its in-vitro prebiotic effect on the gut microbiota. Food Hydrocoll 96:343–353. https://doi.org/10.1016/j.foodhyd.2019.05.031

Ragaza JA, Koshio S, Mamauag RE, Ishikawa M, Yokoyama S, Villamor SS (2015) Dietary supplemental effects of red seaweed *Eucheuma denticulatum* on growth performance, carcass composition and blood chemistry of juvenile Japanese flounder, *Paralichthys olivaceus*. Aquac Res 46:647–657. https://doi.org/10.1111/are.12211

Rajauria G (2015) Chapter 15 – Seaweeds: a sustainable feed source for livestock and aquaculture. In: Tiwari BK, Troy DJ (eds) Seaweed sustainability. Academic, San Diego, pp 389–420. https://doi.org/10.1016/B978-0-12-418697-2.00015-5

Rey-Crespo F, López-Alonso M, Miranda M (2014) The use of seaweed from the Galician coast as a mineral supplement in organic dairy cattle. Animal 8:580–586. https://doi.org/10.1017/S1751731113002474

Rioux L-E, Beaulieu L, Turgeon SL (2017) Seaweeds: a traditional ingredients for new gastronomic sensation. Food Hydrocoll 68:255–265. https://doi.org/10.1016/j.foodhyd.2017.02.005

Rjiba-Ktita S, Chermiti A, Valdés C, López S (2019) Digestibility, nitrogen balance and weight gain in sheep fed with diets supplemented with different seaweeds. J Appl Phycol 31:3255–3263. https://doi.org/10.1007/s10811-019-01789-7

Robertson-Andersson DV, Potgieter M, Hansen J, Bolton JJ, Troell M, Anderson RJ, Halling C, Probyn T (2009) Integrated seaweed cultivation on an abalone farm in South Africa. In: Borowitzka MA, Critchley AT, Kraan S, Peters A, Sjøtun K, Notoya M (eds) Nineteenth international seaweed symposium: proceedings of the 19th international seaweed symposium, held in Kobe, Japan, 26–31 March, 2007. Springer, Dordrecht, pp 129–145. https://doi.org/10.1007/978-1-4020-9619-8_18

Roleda MY, Marfaing H, Desnica N, Jónsdóttir R, Skjermo J, Rebours C, Nitschke U (2019) Variations in polyphenol and heavy metal contents of wild-harvested and cultivated seaweed bulk biomass: health risk assessment and implication for food applications. Food Control 95:121–134. https://doi.org/10.1016/j.foodcont.2018.07.031

Roque BM, Salwen JK, Kinley R, Kebreab E (2019) Inclusion of *Asparagopsis armata* in lactating dairy cows' diet reduces enteric methane emission by over 50 percent. J Clean Prod 234:132–138. https://doi.org/10.1016/j.jclepro.2019.06.193

Rossi R, Vizzarri F, Ratti S, Palazzo M, Casamassima D, Corino C (2020a) Effects of long-term supplementation with brown seaweeds and polyphenols in rabbit on meat quality parameters. Animals 10:2443

Rossi R, Vizzarri F, Chiapparini S, Ratti S, Casamassima D, Palazzo M, Corino C (2020b) Effects of dietary levels of brown seaweeds and plant polyphenols on growth and meat quality parameters in growing rabbit. Meat Sci 161:107987. https://doi.org/10.1016/j.meatsci.2019.107987

Salehpour R, Amrollahi Biuki N, Mohammadi M, Dashtiannasab A, Ebrahimnejad P (2021) The dietary effect of fucoidan extracted from brown seaweed, *Cystoseira trinodis* (C. Agardh) on growth and disease resistance to WSSV in shrimp *Litopenaeus vannamei*. Fish Shellfish Immunol 119:84–95. https://doi.org/10.1016/j.fsi.2021.09.005

Samarasinghe MB, Sehested J, Weisbjerg MR, Vestergaard M, Hernández-Castellano LE (2021) Milk supplemented with dried seaweed affects the systemic innate immune response in preweaning dairy calves. J Dairy Sci 104:3575–3584. https://doi.org/10.3168/jds.2020-19528

Schleder DD, Blank M, Peruch LGB, Poli MA, Gonçalves P, Rosa KV, Fracalossi DM, do Nascimento Vieira F, Andreatta ER, Hayashi L (2020) Impact of combinations of brown seaweeds on shrimp gut microbiota and response to thermal shock and white spot disease. Aquaculture 519:734779. https://doi.org/10.1016/j.aquaculture.2019.734779

Shapawi R, Zamry AA (2016) Response of Asian seabass, *Lates calcarifer* juvenile fed with different seaweed-based diets. J Appl Anim Res 44:121–125. https://doi.org/10.1080/09712119.2015.1021805

Shpigel M, Guttman L, Shauli L, Odintsov V, Ben-Ezra D, Harpaz S (2017) *Ulva lactuca* from an integrated multi-trophic aquaculture (IMTA) biofilter system as a protein supplement in gilthead seabream (*Sparus aurata*) diet. Aquaculture 481:112–118. https://doi.org/10.1016/j.aquaculture.2017.08.006

Siddik MAB, Rahman MM, Anh NTN, Nevejan N, Bossier P (2015) Seaweed, *Enteromorpha intestinalis*, as a diet for Nile tilapia *Oreochromis niloticus* fry. J Appl Aquac 27:113–123. https://doi.org/10.1080/10454438.2015.1008286

Silva-Brito F, Alexandrino DAM, Jia Z, Mo Y, Kijjoa A, Abreu H, Carvalho MF, Ozório R, Magnoni L (2021) Fish performance, intestinal bacterial community, digestive function and skin and fillet attributes during cold storage of gilthead seabream (*Sparus aurata*) fed diets supplemented with *Gracilaria* by-products. Aquaculture 541:736808. https://doi.org/10.1016/j.aquaculture.2021.736808

Škrovánková S (2011) Chapter 28 – Seaweed vitamins as nutraceuticals. In: Kim S-K (ed) Advances in food and nutrition research. Academic, pp 357–369. https://doi.org/10.1016/B978-0-12-387669-0.00028-4

Smárason BÖ, Alriksson B, Jóhannsson R (2019) Safe and sustainable protein sources from the forest industry—the case of fish feed. Trends Food Sci Technol 84:12–14. https://doi.org/10.1016/j.tifs.2018.03.005

Sotoudeh E, Jafari M (2017) Effects of dietary supplementation with red seaweed, *Gracilaria pygmaea*, on growth, carcass composition and hematology of juvenile rainbow trout, *Oncorhynchus mykiss*. Aquac Int 25:1857–1867. https://doi.org/10.1007/s10499-017-0158-6

Stokvis L, Rayner C, van Krimpen MM, Kals J, Hendriks WH, Kwakkel RP (2022) A proteolytic enzyme treatment to improve *Ulva laetevirens* and *Solieria chordalis* seaweed co-product digestibility, performance, and health in broilers. Poult Sci 101:101777. https://doi.org/10.1016/j.psj.2022.101777

Susanto E, Fahmi AS, Abe M, Hosokawa M, Miyashita K (2016) Lipids, fatty acids, and fucoxanthin content from temperate and tropical brown seaweeds. Aquat Procedia 7:66–75. https://doi.org/10.1016/j.aqpro.2016.07.009

Susanto E, Fahmi AS, Hosokawa M, Miyashita K (2019) Variation in lipid components from 15 species of tropical and temperate seaweeds. Mar Drugs 17:630. https://doi.org/10.3390/md17110630

Syad AN, Shunmugiah KP, Kasi PD (2013) Seaweeds as nutritional supplements: analysis of nutritional profile, physicochemical properties and proximate composition of *G. acerosa* and *S. wightii*. Biomed Prev Nutr 3:139–144. https://doi.org/10.1016/j.bionut.2012.12.002

Tabarsa M, Rezaei M, Ramezanpour Z, Robert Waaland J, Rabiei R (2012) Fatty acids, amino acids, mineral contents, and proximate composition of some brown seaweeds. J Phycol 48:285–292. https://doi.org/10.1111/j.1529-8817.2012.01122.x

UNDESA (2019) World population prospects 2019: highlights. Department of Economic and Social Affairs, Population Division

Vieira EF, Soares C, Machado S, Correia M, Ramalhosa MJ, Oliva-teles MT, Paula Carvalho A, Domingues VF, Antunes F, Oliveira TAC, Morais S, Delerue-Matos C (2018) Seaweeds from the Portuguese coast as a source of proteinaceous material: total and free amino acid composition profile. Food Chem 269:264–275. https://doi.org/10.1016/j.foodchem.2018.06.145

Viera I, Pérez-Gálvez A, Roca M (2018) Bioaccessibility of marine carotenoids. Mar Drugs 16:397

Vilg JV, Nylund GM, Werner T, Qvirist L, Mayers JJ, Pavia H, Undeland I, Albers E (2015) Seasonal and spatial variation in biochemical composition of *Saccharina latissima* during a potential harvesting season for Western Sweden. Bot Mar 58:435–447. https://doi.org/10.1515/bot-2015-0034

Vizcaíno AJ, Mendes SI, Varela JL, Ruiz-Jarabo I, Rico R, Figueroa FL, Abdala R, Moriñigo MÁ, Mancera JM, Alarcón FJ (2016) Growth, tissue metabolites and digestive functionality in *Sparus aurata* juveniles fed different levels of macroalgae, *Gracilaria cornea* and *Ulva rigida*. Aquac. Res 47:3224–3238

Vizzari F, Massányi M, Knížatová N, Corino C, Rossi R, Ondruška Ľ, Tirpák F, Halo M, Massányi P (2021) Effects of dietary plant polyphenols and seaweed extract mixture on male-rabbit semen: quality traits and antioxidant markers. Saudi J Biol Sci 28:1017–1025. https://doi.org/10.1016/j.sjbs.2020.11.043

Wan AHL, Davies SJ, Soler-Vila A, Fitzgerald R, Johnson MP (2019) Macroalgae as a sustainable aquafeed ingredient. Rev Aquac 11:458–492. https://doi.org/10.1111/raq.12241

Wong KH, Cheung PCK (2000) Nutritional evaluation of some subtropical red and green seaweeds: part I — proximate composition, amino acid profiles and some physico-chemical properties. Food Chem 71:475–482. https://doi.org/10.1016/S0308-8146(00)00175-8

Wongprasert K, Rudtanatip T, Praiboon J (2014) Immunostimulatory activity of sulfated galactans isolated from the red seaweed *Gracilaria fisheri* and development of resistance against white spot syndrome virus (WSSV) in shrimp. Fish Shellfish Immunol 36:52–60. https://doi.org/10.1016/j.fsi.2013.10.010

Xia S, Zhao P, Chen K, Li Y, Liu S, Zhang L, Yang H (2012a) Feeding preferences of the sea cucumber *Apostichopus japonicus* (Selenka) on various seaweed diets. Aquaculture 344–349:205–209. https://doi.org/10.1016/j.aquaculture.2012.03.022

Xia S, Yang H, Li Y, Liu S, Zhou Y, Zhang L (2012b) Effects of different seaweed diets on growth, digestibility, and ammonia-nitrogen production of the sea cucumber *Apostichopus japonicus* (Selenka). Aquaculture 338–341:304–308. https://doi.org/10.1016/j.aquaculture.2012.01.010

Chapter 7
Synthesis of Nanoparticles from Seaweeds and Their Biopotency

Johnson Marimuthu Alias Antonysamy, Shivananthini Balasundaram, Silvia Juliet Iruthayamani, and Vidyarani George

1 Introduction

Based on the pigments, nutrients and chemical composition, Seaweeds (marine macroalgae) can be broadly classified into Rhodophyta (red algae), Phaeophyta (brown algae) and Chlorophyta (green algae). Globally algal diversity harbors 72,500 taxa of which 44,000 have previously described. Seaweeds benefit human health because it contains a lot of inorganic and organic components. Since the time immemorial, seaweeds are employed as food, medicine, feed and fertilizer. Seaweeds possess more than 60 trace elements with a varied concentration much higher than the land plants. In addition, seaweeds also possess protein, iodine, bromine, vitamins and substances with medicinal properties. The secondary metabolites of seaweeds are widely employed in food, confectionary, textile, pharmaceutical, dairy and paper industries as gelling, stabilizing and thickening agents. The available literature suggest that, nearly 150 are consumed as food and 250 seaweeds are commercially utilized at the global.

In the modern science and technology, nanobiotechnology is growing as one of the most promising area. The notable and achievable development in Nanotechnology is the synthesis and application of nanoparticles in biology and medicine. Because of potential applications in drug delivery, antibacterial, antifungal, antiviral activity, bio sensing, biological labeling, catalysis, detection of genetic disorders, gene therapy and DNA sequencing and their unique physical and chemical properties nanoparticles have been extensively studied (Thirumurugan et al. 2010). The top-down and bottom-up methods are commonly employed for the synthesis of nanoparticles (Balantrapu and Goia 2009). The large materials are slowly broken down into small (nano) sized materials in top-down method, while in bottom-up method,

J. M. Alias Antonysamy (✉) · S. Balasundaram · S. J. Iruthayamani · V. George
Centre for Plant Biotechnology, Department of Botany, St. Xavier's College (Autonomous), Tirunelveli, Tamil Nadu, India

F. Ozogul et al. (eds.), *Seaweeds and Seaweed-Derived Compounds*,
https://doi.org/10.1007/978-3-031-65529-6_7

atoms or molecules are bring together to form the molecular structures (nanometer). For chemical and biological synthesis of nanoparticles, the bottom-up approach is employed. Various methods are employed for the metal nanoparticles synthesis and stabilization process, generally chemical and mechanical methods are most accepted one by the researchers (Tripathi et al. 2010). Next to that electrochemical techniques (Patakfalvi and Dekany 2010), photochemical reactions in reverse micelles and currently via green chemistry methods are employed for the synthesis of nanoparticles (Taleb et al. 1998). Due to clean, biocompatible, non-toxic and eco-friendly nature, the nanoparticle synthesized via biological methods are often recognized as one of most accepted method for phytosynthesis of nanoparticles. In plant kingdom, due to its availability and medicinal properties, seaweed associated biosynthesis of nanoparticles are rapid, cost effective, ecofriendly and showed advantages over other microscopic organisms.

To date, metallic and metallic oxide based nanoparticles are synthesized and prepared from noble metals, *i.e.* silver (Ag), platinum (Pt), gold (Au) and palladium (Pd) (Norziah and Ching 2002). Using the seaweeds as reducing agents silver, gold, copper, zinc, iron, palladium, platinum nanoparticles are prepared (Fig. 7.1). The size and shape of nanoparticle is governed by four factors viz., metal salt concentration, reducing agent concentration, temperature and pH (Fig. 7.2).

So far, there are several algal mediated metal nanoparticles have been synthesized. The present review aims to summarize the synthesized metal and metallic oxide based nanoparticles, synthesis methods and characterization of the synthesized nanoparticles from seaweeds and their potential biological applications at a glance.

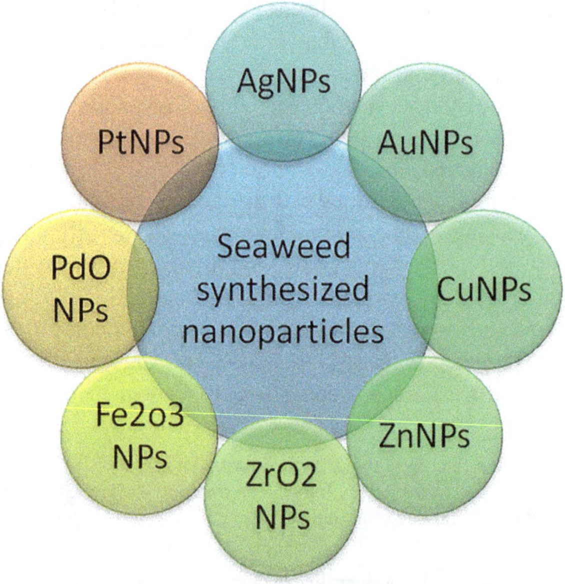

Fig. 7.1 Various types of NPs synthesized from seaweeds

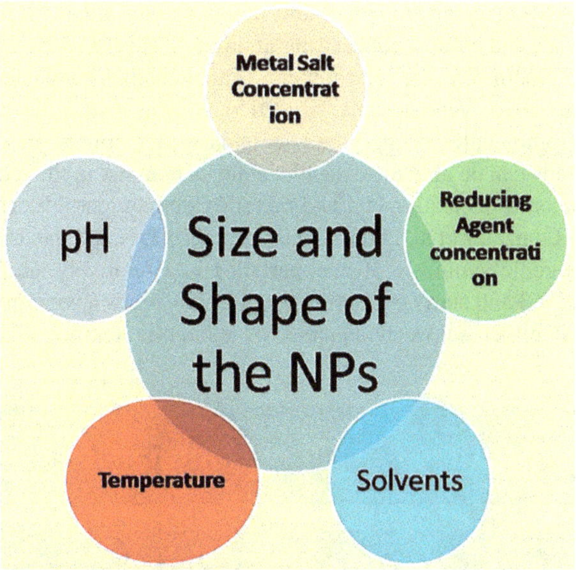

Fig. 7.2 Various factors influence on size and shape of NPs

2 Silver Nanoparticles Synthesis

Silver (Ag) is given top priority of choice over the other metals and explains its significance in the research field of biological systems, living organisms and medicine (Parashar et al. 2009). The most important advantages of silver nanoparticles are the surface plasmon resonance energy of silver because of that silver nanoparticleis above with other metal nanoparticles (*e.g.* gold and copper). In the recent days, a number of reports are available on marine algal mediated silver nanoparticles synthesis using green/biological methods *viz.*, *Sargassum wightii* (Govindaraju et al. 2009), *Enteromorpha compressa* (Dhanalakshmi et al. 2012), *Ulva lactuca* (Bharathi Raja et al. 2012), *Sargassum plagiophyllum, Turbinaria conoides* and *Ulva reticulata* (Rajeshkumar et al. 2012a), *Padina tetrastromatica* (Rajeshkumar et al. 2012b), *Padina pavonica* (Sahayaraj et al. 2012), *Urospora* sp. (Suriya et al. 2012), *Sargassum longifolium* (Saraniya Devi et al. 2013), *Cystophora moniliformis* (Prasad et al. 2013), *Sargassum longifolium* (Rajeshkumar et al. 2014), *Hypnea muciformis* (Saraniya Devi and Valentin Bhimba 2014) and *Sargassum wightii* (Sunitha et al. 2015), Gracillaria biridae (Aragao et al. 2016), *Ulva fasciata (*Fatah Hamouda et al. 2018), *Portieria hornemannii* (Ramamoorthy et al. 2019), *Halymenia porphyroides* (Manam and Subbaiah 2020).

The biogenic synthesis of nanoparticles from seaweeds is believed to be a clean, nontoxic, cost effective and ecofriendly green chemistry approach with the potential to deliver a wide variety of particle sizes, morphologies, compositions, and physicochemical properties. The seaweeds possess various types of polysaccharides viz., agar, alginate, carrageenan, fucoidan, and laminarin and other biomolecules viz., proteins, enzymes and other secondary metabolites. The existence of these bioactive materials has fascinated the researchers (Panchanathan and Se-Kwon 2015; Ermakova et al. 2015; Venkatesan et al. 2015). Recent studies have revealed that many of the biomolecules primary and secondary metabolites present in the cell walls of seaweeds can be able to act as biocatalysts to assist in the reduction of precursor metal salts to nucleate metal and metal oxide nanoparticles (Mahdavi et al. 2013; Romero-González et al. 2003; Kumar et al. 2013; Nagarajan and Kuppusamy 2013). The biogenic synthesis of nanoparticles is a relatively simple process as depicted in Fig. 7.3 and starts by mixing a metal salt solution with an aqueous solution containing either seaweed aqueous or ethanolic extract. Reduction starts

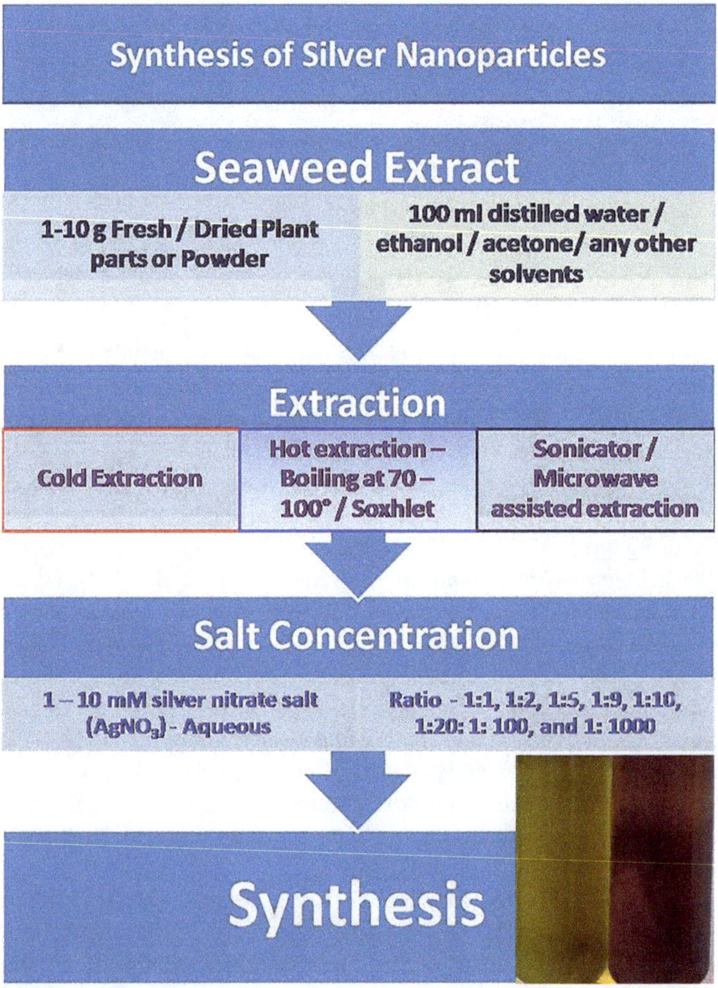

Fig. 7.3 Synthesis of silver nanoparticles from seaweeds

immediately and a color change (Paddy straw yellow colour/Dark yellowish brown/ Brown) in the reaction mixture indicate the formation of nanoparticles. Table 7.1 summarizes the silver nanoparticle synthesis from seaweeds.

The authors adopted aqueous (water), methanol, chloroform: methanol, ethanol as solvents for the extraction of reducing agents from the thallus of studied seaweeds. Among these various solvents, water is used by more than 80% of the authors, distilled water as source of extraction. More than 80% of the authors adopted the hot extraction method for the extraction of reducing agents using the distilled water and other solvents. The authors are used 1 mM $AgNO_3$ salt concentration for the silver nanoparticles synthesis and achieved their goal. Except the nanoparticle of *Gracilaria birdiae* and *Sargassum tenerrimum* (90 °C), *Urospora* sp., *Pterocladia capillacae* (Pc), *Jania rubins* (Jr), *Ulva faciata* (Uf), and *Colpmenia sinusa* (Cs) (70 °C), *Sargassum plagiophyllum, Ulva reticulata* and *Enteromorpha compressa, Gracilaria corticata* and *Padina* sp. (60 °C) all other synthesized nanoparticles from seaweeds require room temperature for the synthesis of nanoparticles.

3 Gold Nanoparticle Synthesis

Next to silver, Gold nanoparticles (AuNPs) are highlighted due to their low toxicity and physico-chemical properties viz., compatability, adaptability and high surface area to volume ratio (Ginzburg et al. 2018; Mieszawska et al. 2013). Biosynthesis of AuNPs using seaweeds is considered a simple, low-cost, and eco-friendly approach. Figure 7.4 explains the synthesis of gold nanoparticles from seaweeds. Besides, nanoparticles of different metals from seaweeds also reported such as Gold nanoparticles from *Sargassum wightii* (Singaravelu et al. 2007), *Turbinaria conoides* (Vijayaraghavan et al. 2011), *Laminaria japonica* (Ghodake and Lee 2011), *Stoechospermum marginatum* (Arockia Aarthi Rajathi et al. 2012), *Egregia sp.* (Colin et al. 2018) palladium nanoparticle synthesis from *Sargassum bovinum*, Copper oxide nanoparticle synthesis using *Bifurcaria bifurcate* (Abboud et al. 2014) and *Cystoseira trinodis* (Gu et al. 2018), *Sargassum polycystum* (Ramasamy et al. 2016), Synthesis of Fe_3O_4-NPs from *Sargassum acinarium* (El-Kassas et al. 2016), Platinum nanoparticles from *Padina gymnospora* (Ramkumar et al. 2017b) and Zinc oxide nanoparticles from *Gracillaria edulis* (Priyadharshini et al. 2014). Table 7.2 explains the different methods of gold nanoparticle synthesis from seaweeds.

GNPs have a wide variety of applications and possess reconfigurable Surface Plasmon resonance vibrations. The unique surface plasmon renounce (SPR) of GNPs, facilitating and en routing for bio-labeling. GNPs possess distinctive electrical, magnetic, catalytic and optical properties (Abdelghany et al. 2019; Augustine et al. 2016; Bar et al. 2009), these properties are responsible for the compatability and adaptability of GNPs.

Table 7.1 Summary of silver nanoparticles synthesized using seaweeds

Botanical name of the seaweed	Solvent	Extraction method	Salt conc. & ratio	Time	Temperature	References
Ulva rigida (green alga), Cystoseira myrica (brown alga), Gracilaria foliifera (red alga)	Aqueous	Hot plate	10^{-3} M	15 min	70 °C	Algotiml et al. (2022)
Colpomenia sinuosa C. mediterranea	n-hexane	Fractionation	1:10 w:v	24 h	25 ± 2 °C	Ghareeb et al. (2022)
Halymenia porphyriformis and Solieria robusta	Aqueous	Hot plate	8:1	24 h	60–65 °C	Khan et al. (2022)
C. racemosa	Aqueous	Cold	1:10	30 min	50 °C	Thanigaivel et al. (2022)
Enteromorpha compressa	Aqueous	Hot	1 mM AgNO3	1 h	Room temperature	Ramkumar et al. (2017a)
Laurencia catarinensis	Aqueous, Ethanol, Chloroform	Hot	10^{-3} M aqueous AgNO3 solution	3 h	Room temperature	Abdel-Raouf et al. (2018)
Codium capitatum	Aqueous	Hot	1 mM AgNO3	48 h	Room temperature	Kannan et al. (2013)
Palmaria decipiens	Aqueous	Hot	0.005 M	8 h	Room temperature	González-Ballesteros et al. (2018)
Desmarestiam enziesii	Aqueous	Hot	0.005 M	8 h	Room temperature	González-Ballesteros et al. (2018)
Gracilaria birdiae	Polysaccharide extraction	Cold	1 mM	30 min	90 °C	De Aragão et al. (2019)
Sargassum polycystum	Methanol	Soxhlet	1 M	24 h	Room temperature	Thangaraju et al. (2012)
Padina tetrastromatica	Aqueous	Hot	0.25 mM	48 h	25 °C	Kayalvizhi et al. (2014)
Turbinaria ornata	Aqueous	Hot	0.25 mM	48 h	25 °C	Kayalvizhi et al. (2014)
Chaetomorpha aerea	Aqueous	Hot	1 mM	48 h		Sivakumar et al. (2018)
Portieriahor nemannii	Aqueous	Hot	1 mM	1 h	Room temperature	Ramamoorthy et al. (2019)

Gracilaria fergusonii	Aqueous	Hot	3 mM	8 h	Room temperature	Udhaya et al. (2015)
Ulva fasciata	Aqueous	Hot	0.1 mM	1 h	40 °C	Hamouda et al. (2018)
Turbinaria conoides	Aqueous	Hot	1 mM	1 h	Room temperature	Rajeshkumar et al. (2012a)
Padina tetrastromatica	Aqueous	Hot	1 mM	72 h	Room temperature	Jegadeeswaran et al. (2012)
Padina gymnospora	Aqueous	Hot	0.004 mM	–	30 °C	Shiny et al. (2013)
Halymenia porphyroides	Aqueous	Hot	10^{-3} M aqueous AgNO3 solution	24 h	Room temperature	Manam and Subbaiah (2020)
Amphiroa anceps	Aqueous	Hot	1 mM	–	Room temperature	Roy and Anantharaman (2018)
Urospora sp.	Aqueous	Magnetic stirrer condition	1 mM	15 min	70 °C	Suriya et al. (2012)
Spatoglossum asperum	Chloroform: Methanol (1:1)	Hot	1×10^{-1} M	–	Room temperature	Jothirethinam et al. (2015)
Hedophyllum sessile	Chloroform: Methanol (1:1)	Hot	1×10^{-1} M	–	37 °C	Jothirethinam et al. (2015)
Ulothrix flacca, *Ulva fasciata* and *Caulerpa taxifolia*	Methanol	Soxhlet	1 mM	10 min	12 °C	Rao and Boominathan (2015)
Padina sp.	Aqueous	Intracellular extraction method	0.01 mM	30 min	60 °C	Bhuyar et al. (2020)
Hypnea cervicornis	Aqueous	Hot	1 mM	30 min	Room temperature	Leela and Anchana Devi (2017)
Pterocladia capillacae (Pc), *Jania rubins* (Jr), *Ulva faciata* (Uf), and *Colpmenia sinusa* (Cs)	Aqueous	Hot	0.1 mM	1 h	70 °C	El-Rafiea et al. (2013)
Sargassum polycystum	Aqueous	Hot	1 mM	–	20 °C	Subramanian et al. (2016)

(continued)

Table 7.1 (continued)

Botanical name of the seaweed	Solvent	Extraction method	Salt conc. & ratio	Time	Temperature	References
Spyridia filamentosa	Aqueous	Hot	1 mM	30 min	37 °C	Valarmathi et al. (2019)
Gelidiella acerosa	Aqueous	Hot	1 mM	24 h	Room temperature	Thiruchelvi et al. (2020)
Chaetomorpha linum	Aqueous	Hot	1 mM	30 min	37 °C	Kannan et al. (2013)
Gracilaria corticata	Aqueous	Hot	1 mM	20 min	60 °C	Kumar et al. (2013)
Sargassum tenerrimum	Aqueous	Hot	1 mM	20 min	90 °C	Kumar et al. (2012)
Sargassum plagiophyllum, *Ulva reticulata* and *Enteromorpha compressa.*	Aqueous, Methanol	Cold	1 mM	15 min	60 °C	Dhanalakshmi et al. (2012)
Padina tetrastromatica	Aqueous	Hot	1 mM	24 h	Room temperature	Rajeshkumar et al. (2012b)
Laurencia aldingensis and *Laurenciella* sp.	Aqueous	Exhaustive	0.500 g 0.85 m	20 min	Room temperature	Vieira et al. (2015)
Cladophora fascicularis	Aqueous	Hot	1 mM	12	Room temperature	Subramanian et al. (2016)
Caulerpa racemosa	Aqueous	Hot	10^{-3} M aqueous AgNO3	24 h	Room temperature	Valarmathi et al. (2019)
Sargassum wightii	Aqueous	Hot	–	24 h	28 °C	Thiruchelvi et al. (2020)
Sargassum angustifolium	Aqueous	Hot	1 mM	2 h	Room temperature	Kannan et al. (2013)
Ulva flexuosa	Aqueous	Ultrasonic irradiation method	0.1 M	24	25 °C	Kumar et al. (2013)

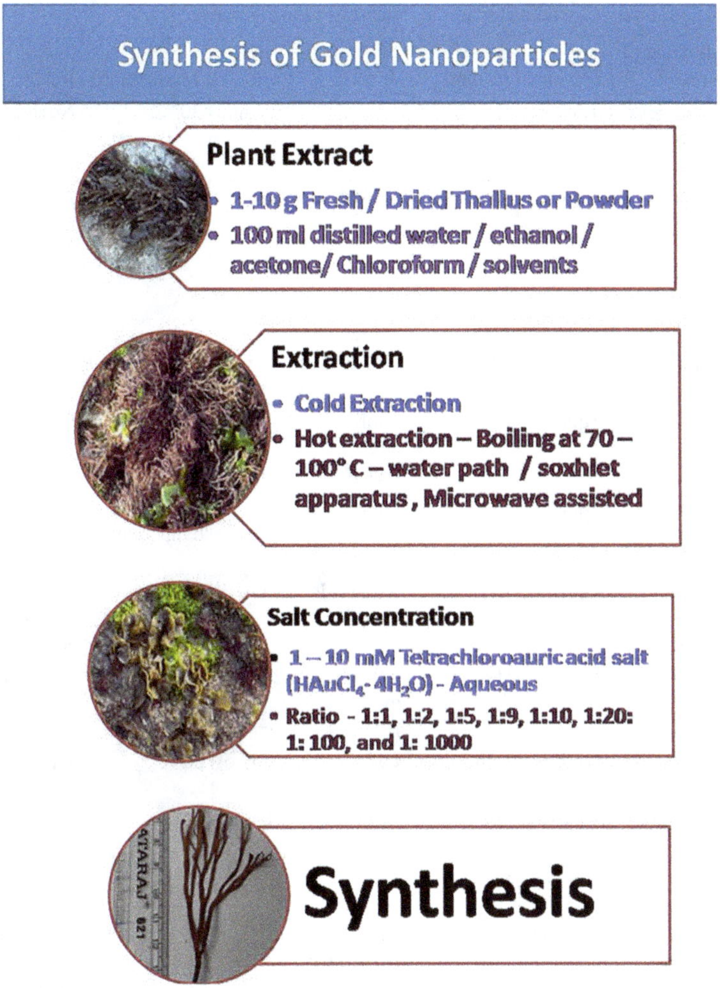
Synthesis of Gold Nanoparticles
Plant Extract
1-10 g Fresh / Dried Thallus or Powder
100 ml distilled water / ethanol / acetone/ Chloroform / solvents
Extraction
Cold Extraction
Hot extraction – Boiling at 70 – 100° C – water path / soxhlet apparatus, Microwave assisted
Salt Concentration
1 – 10 mM Tetrachloroauric acid salt (HAuCl4·4H2O) - Aqueous
Ratio - 1:1, 1:2, 1:5, 1:9, 1:10, 1:20: 1: 100, and 1: 1000
Synthesis

Fig. 7.4 Synthesis of gold nanoparticles from seaweeds

The formation of gold nanoparticles by the reduction of aqueous Au3+ ions occurred using ethanol/Aqueous $AuCl_4$ solution (1 mM $AuCl_4$). Formation of gold nanoparticles through reduction of Chloroauric acid into Au3+ ion by the reducing agent is known to be associated with colour change (Ruby red) of the reaction mixture at stirring condition or might be in normal condition in the different temperature.

Most of the authors adopted aqueous (distilled water) as solvents for the extraction of reducing agents from the thallus of studied seaweeds except *Galaxaura elongata*. The hot extraction method was accepted by all the authors for the extraction of reducing agents using the distilled water and ethanol. The authors used 1–10 mM $HAuCl_4$ salt concentration for the gold nanoparticles synthesis and achieved their

Table 7.2 Summary of gold nanoparticles synthesized using seaweeds

Botanical name of the seaweed	Solvent	Extraction method	Salt conc. & ratio	Time	Temperature	References
Chondrus crispus Gelidium corneum Porphyra linearis	Aqueous	Hot	–	24 and 48 h	30 °C	Noelia et al. (2022)
Ulva rigida, Cystoseira myrica, Gracilaria foliifera	Aqueous	Hot	10^{-3} M	24	45 °C	Algotiml et al. (2022)
Padina tetrastromatica	Aqueous	Magnetic stirrer method	1 mM	72	Room temperature	Rajeshkumar et al. (2020)
T. conoides	Aqueous	Hot	1 mM aqueous AuCl4	–	Room temperature	Ramakrishna et al. (2016)
S. tenerrimum	Aqueous	Hot	1 mM aqueous AuCl4	–	Room temperature	Ramakrishna et al. (2016)
Cystoseira baccata	Aqueous	Hot	0.01 mM HAuCl4	24 h	Room temperature	González-Ballesteros et al. (2017)
Gracilaria verrucosa	Aqueous	Hot	0.0199 $molL^{-1}$ HAuCl4	20 min	60 °C	Chellamuthu et al. (2019)
Halymenia dilatata	Aqueous	Hot	1 M HAuCl4	12 h	60 °C	Vinosha et al. (2019)
Stoechospermum marginatum	Aqueous	Hot	1 mM HAuCl4	10 min	–	Rajathi et al. (2012)
Sargassum myriocystum	Aqueous	Hot	1 mM HAuCl4	15 min	76 °C	Stalin Dhas et al. (2012)
Turbinaria ornata	Aqueous	Hot	1 mM HAuCl4	4 h	32 ± 1 °C	Ashokkumar and Vijayaraghavan (2016)
Gracilaria edulis	Aqueous	Hot	1 mM	–	Room temperature	Abideen and Sankar (2015)
Syringodium isoetifolium	Aqueous	Hot	1 mM	–	Room temperature	Abideen and Sankar (2015)
Galaxaura elongata	Ethanol	Hot	10^3 M HAuCl4 aqueous solution	3 h	Room temperature	Abdel-Raouf et al. (2017)
Palmaria decipiens	Aqueous	Hot	0.01 M HAuCl4	24 h	Room temperature	González-Ballesteros et al. (2018)

(continued)

Table 7.2 (continued)

Botanical name of the seaweed	Solvent	Extraction method	Salt conc. & ratio	Time	Temperature	References
Desmarestiam enziessi	Aqueous	Hot	0.01 M HAuCl4	72 h	Room temperature	González-Ballesteros et al. (2018)
Sargassum plagiophyllum	Aqueous	Hot	1×10^{-3} M HAuCl4	5 min	60 °C	Stalin Dhas et al. (2020)

goal. Except the nanoparticle of *Sargassum myriocystum* (Cs) (76 °C), *Sargassum plagiophyllum*, *Gracilaria verrucosa* and *Halymenia dilatata* (60 °C), all other synthesized nanoparticles from seaweeds require room temperature for the synthesis of gold nanoparticles from seaweeds.

In addition to silver and gold, the copper, zinc, iron, zirconia, palladium, platinum and oxide nanoparticles are synthesized using seaweeds (Table 7.3). PtNPs is having an excellent oxidation and reduction properties (Hikosaka et al. 2008). Due to the application as catalysis, photocatalysis, optical and biological nanosensors and electrode nanomaterials, the polymer doped PtNPs and other metal NPs are gained interest (Kajita et al. 2007; Tian et al. 2007; Formo et al. 2008; Kang et al. 2008; Chu et al. 2010; Liu et al. 2015). The formation of platinum nanoparticles is performed by mixing the aqueous extracts of seaweeds and 0.001 M platinum chloride (H_2PtCl_6) solution in stirring condition for 10 min. Formation of platinum nanoparticles through reduction of platinum chloride (H_2PtCl_6) into Pt^+ ion by the reducing agent (aqueous extracts of seaweed thallus) is known to be associated with colour change (Yellow to dark brown colour) of the reaction mixture. Ramkumar et al. (2017b) suggested a method for the synthesis of platinum nanoparticles using *Padina gymnospora*.

In addition to biological application, the nanoparticles showed their application environmental science also. Zinc oxide nanocrystalline materials are environmentally flexible and favorable material with potential applications in pollutants removal appliances i.e., removal of dyes, toxic organic components to reduce the sulphur and arsenic from water (Karunakaran and Dhanalakshmi 2008; Tiwari et al. 2008; Zhang et al. 2010) irradiated under the UV light, solar cells, gas sensors and electric materials (Belaidi et al. 2009; Zong et al. 2010; Straumal et al. 2011). Semi-conductive zinc oxide nanomaterials hold the attractive characteristic properties of non toxic, thermal stability, high chemical, large binding energy, wide band gap, and high piezoelectric property (Fortunato et al. 2005). Furthermore, ZnO NPs has shown prospective application in biological sciences as nano-medicine, bio-sensor, biological labeling (Yoon and Kim 2006), gene and drug delivery (Xiong 2013; Vijayakumar et al. 2016; Malaikozhundan et al. 2017), antifungal (Applerot et al. 2009), antibacterial agents (Sharma et al. 2010), anti biofilm (Vijayakumar et al. 2017; Thaya et al. 2016), larvicidal, acaricidal, pediculocidal properties (Kirthi et al. 2011; Benelli 2016a, b; Ashokan et al. 2017) and anti-diabetic activities (Alkaladi et al. 2014).

Table 7.3 Summary of copper, zinc, iron, palladium, platinum and zirconia nanoparticles synthesized using seaweeds

Botanical name of the seaweed	Type of NPs	Solvents	Extraction methods	Salt concentration	Time	Temperature	References
Sargassum polycystum	Copper oxide	Aqueous	Hot	1 mM CuSo4	–	RT	Sri Vishnu Priya et al. (2016)
Macrocystis pyrifera	Copper oxide	Aqueous	–	100 mM CuSo4	48 h	45 °C	Araya-Castro et al. (2021)
Ulva lactuca	Zinc oxide	Aqueous	Hot	1 mM zinc acetate	3 h	70 °C	Ishwarya et al. (2017)
Padina gymnospora	SCZ	Aqueous	Ultrasonication	0.25 M ZnCl2 solution and 0.1 M CdCl2	24 h	RT	Rajaboopathi and Thambidurai (2017)
Gracilaria edulis	Iron oxide	Aqueous	Microwave oven irradiation	FeCl3 (0.1 mol/L)	90 min	RT	Subhashini et al. (2018)
Kappaphycus alvarezii	Iron oxide	Aqueous	Hot extract	Fe3+ and Fe2+ with a 2:1 M ratio	1 h	RT	Yew et al. (2016)
Padina pavonica	Iron oxide	Aqueous	Hot extract	FeCl3 (0.1 mol/L)	90 min	RT	El-Kassas et al. (2016)
Sargassum acinarium	Iron oxide	Aqueous	Hot extract	FeCl3 (0.1 mol/L)	90 min	RT	El-Kassas et al. (2016)
Sargassum myriocystum	Zinc oxide	Aqueous	Hot extract	1 mM zinc nitrate	10 min	80 °C	Nagarajan and Kuppusamy (2013)
Hypnea musciformis	Zinc oxide	Aqueous	Hot extract	0.2 M zinc nitrate	2	60 °C	Yousefzadi et al. (2017)
Sargassum muticum	Zinc oxide	Aqueous	Hot extract	(Zn(Ac)2 2H2O) (2 mM)	3–4 h	70 °C	Azizi et al. (2014)
Sargassum wightii	Zirconia	Aqueous	Combustion route	5 mM of ZrO(NO3)2.xH2O	20 min	400 ± 10 °C	Kumaresan et al. (2018)
Padina boryana	Palladium	Aqueous	Ultrasonication	10 mM disodium tetrachloropalladate (II) (Na2PdCl4)	2 h	60 °C	Sonbol et al. (2021)
Dictyota indica	Palladium oxide	Aqueous	Hot	1 mM PdCl2 solution	2 h	60 °C	Shargh et al. (2018)
Padina gymnospora	Platinum	Aqueous	Hot	0.001 M platinum chloride (H2PtCl6).	10 min	–	Ramkumar et al. (2017b)

The formation of zinc oxide nanoparticles by the reduction of aqueous zinc nitrate and zinc acetate dehydrate occurred using aqueous extract and using hot extraction method in the concentration of 0.2 M, 1–2 mM zinc nitrate and zinc acetate dehydrate. Formation of zinc oxide nanoparticles through reduction of zinc nitrate and zinc acetate dehydrate into Zn^{2+}ion and zinc oxide by the reducing agent is known to be associated with colour change (dark brown to a pale white color) of the reaction mixture at stirring condition or might be in normal condition in the different temperature. The available literature confirms the synthesis of Zinc oxide Nanoparticles of *Sargassum muticum*, *Hypnea musciformis*, *Sargassum myriocystum* and *Ulva lactuca* (Fig. 7.5).

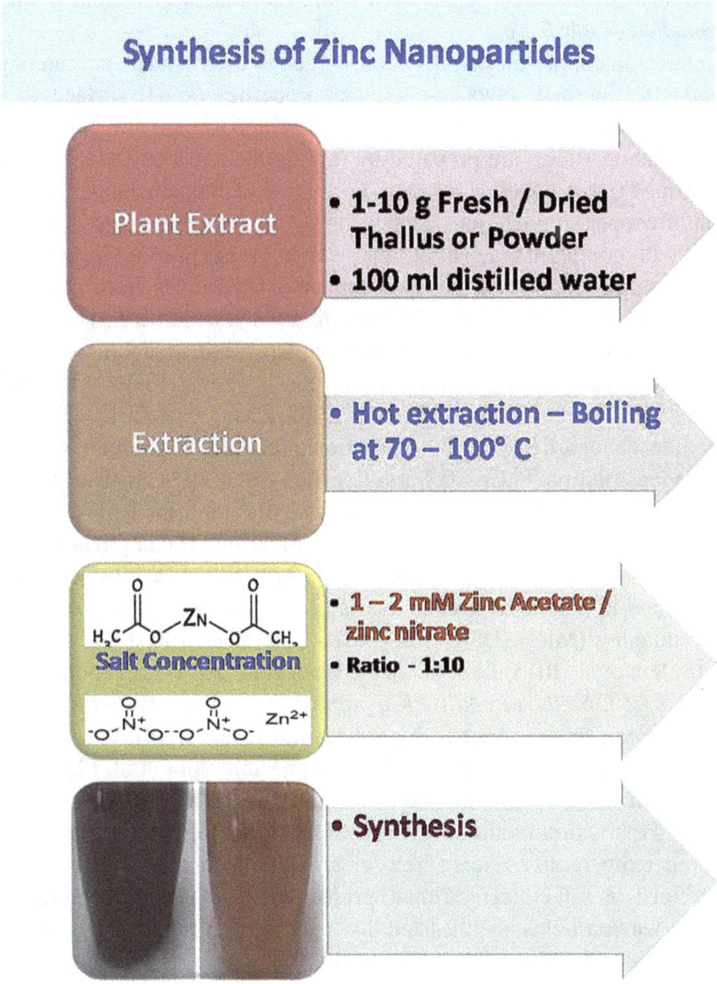

Fig. 7.5 Synthesis of zinc nanaoparticles from *Sargassum muticum*, *S. myriocystum*, *Hypnea musciformis* and *Ulva lactuca.* (Source: Nagarajan and Kuppusamy 2013; Azizi et al. 2014; Yousefzadi et al. 2017)

Mostly the Palladium (Pd) is used in the fields of environmental and material science as a transition metal. They act as catalytic converters, dental crowns, fuel cells, coating materials etc. (Higashi et al. 2017). Palladium oxide nanoparticles are act as a catalyst for many organic and inorganic reactions at ambient temperature and involved in the oxidation and regeneration process. They are used to eliminate environmental contaminants, due to the specific surface area high reactivity and adsorption. Formation of palladium and palladium oxide nanoparticles through reduction of palladium chloride, disodium tetrachloropalladate and dehydrate into palladium ions by the reducing agent (Thallus of Seaweeds) is known to be associated with colour change (yellow to dark brown) of the reaction mixture at stirring condition or might be in normal condition in the different temperature. So far two Palladium nanoparticles are synthesized using the thallus of *Padina boryana* and *Dictyota indica* (Table 7.3).

The interest in copper-based NPs is raised due to their optical, conducting, magnetic, catalytic, thermal, physico-chemical properties (small surface to volume ratio) and antibiotic activity (Huang et al. 1997; Khodashenas and Ghorbani 2014). The copper nanoparticles are prepared by mixing the aqueous extract of seaweeds and 1–100 mM $CuSo_4$ aqueous solution. Formation of copper nanoparticles through reduction of copper sulphate into copper ion by the reducing agent is known to be associated with colour change (light blue to light green) at room temperature. The available literature confirmed that two copper nanoparticles from seaweeds viz., *Sargassum polycystum* and *Macrocystis pyrifera* CuNPs (Table 7.3).

Due to the superparamagnetic, biocompatible, biodegradable properties, the researchers focused their attention on the synthesis of Fe_3O_4-NPs (Mahdavian and Mirrahimi 2010; Hu et al. 2006; Zhao et al. 2009; Zhang et al. 2013). The synthesis of iron oxide nanoparticles by the reduction process using aqueous extract and metal aqueous solution $FeCl_3$ (0.1 mol/L), Fe3+ and Fe2+ with a 2:1 M ratio. Synthesis of iron oxide nanoparticles was visually confirmed by colour change (brick red) of the reaction mixture at stirring condition. The application of Fe_3O_4-NPs as catalysis (Gawande et al. 2013; Sharad et al. 2014), biosensors (Kavitha et al. 2013) and in magnetic storage media (Terris and Thomson 2005), magnetic resonance imaging (MRI) (Haw et al. 2010; Qiao et al. 2009) and targeted drug delivery (Salem et al. 2015; Li et al. 2012; Wani et al. 2014) are reported. The iron nanoparticles of *Gracilaria edulis*, *Kappaphycus alvarezii*, *Padina pavonica* and *Sargassum acinarium* are reported by Subhashini et al. (2018), Yew et al. (2016), and El-Kassas et al. (2016) respectively. Except Subhashini et al. (2018) all other authors adopted the hot extraction method. Subhashini et al. (2018) used the microwave assisted extraction method for extraction. All authors performed the reaction in the room temperature. Except Yew et al. (2016) all other authors employed 0.1 mol/L $FeCl_3$ as salt concentration to prepare the Iron oxide nanoparticles.

Zirconia nanoparticles synthesized by means of different physico-chemical methods such as sol-gel synthesis (Xu et al. 2003), aqueous precipitation method (Southon et al. 2002) and hydrothermal methods (Noh et al. 2003) requires high temperatures and pressures. Zirconia is widely used as catalyst in numerous applications (Yamaguchi 1994). Zirconia nanoparticles are also able to possess remarkable antimicrobial property (Gowri et al. 2014). Zirconia nanoparticles can be applied as piezoelectric, electro-optic and dielectric materials because of their good optical and electronic properties (Somiya et al. 1988). The synthesis of zirconia nanoparticles is followed the combustion route (Kumaresan et al. 2018). 5 mM of $ZrO(NO_3)_2.xH_2O$ salt is mixed with 1 g aqueous extract of seaweeds/plant samples and milled well with mortar and pestle for 20 min. After well grained, reaction mixture was kept inside the box furnace at 400 ± 10 °C. The prepared sample was subjected to further calcination in muffle furnace at 400 °C for 5 h (Kumaresan et al. 2018).

4 Characterization of Phytosynthesized Nanoparticles

Due to improvement in the sophisticated instrumentation facilities, various analytical techniques are adopted to characterize the synthesized nanoparticles of seaweeds. Various types of spectroscopic and microscopic analysis are employed to characterize the phytosynthesized nanoparticles. The literatures suggest that the appearance of paddy straw yellow to slight brownish-yellow color is an indicator of AgNPs synthesis. The literatures recommended SPR peak for AgNPs is ranged from 380 to 450 nm (Johnson et al. 2017). The phytosynthesized silver nanoparticles of seaweeds showed the SPR Peak value around 380–500 nm (Table 7.4). In addition, the UV-Vis spectral analyses are adopted to confirm the size stability of synthesized AgNPs (Srikar et al. 2016). The literatures recommended SPR peak for AuNPs is ranged from 380 to 450 nm (Poinern 2014). The phytosynthesized nanoparticles of studied seaweeds absorption spectra of AuNPs are within the range of 500–550 nm (Table 7.4). The literatures recommended SPR peak for ZnONPs is ranged from 370 to 400 nm (Vennila and Jesurani 2017). The phytosynthesized nanoparticles of studied seaweeds absorption spectra of ZnONPs are within the range of 325–372 nm (Table 7.4). The literatures suggested SPR peak for Fe_3O_4NPs is ranged from 300 to 520 nm (Abdallah and Al-Haddad 2021). The phytosynthesized nanoparticles of studied seaweeds absorption spectra of Fe_3O_4NPs are between 402 and 415 nm (Table 7.4). The literatures recommended SPR peak for PdNPs is ranged from 280 to 400 nm (Siddiqi and Husen 2016). The phytosynthesized nanoparticles of studied seaweeds absorption spectra of PdNPs are within the range of 260–420 nm (Table 7.4). The purity and crystalline nature of the phytosynthesised nanoparticles are confirmed through X-ray diffraction (XRD), which explains a rough idea of the particle size, determined by the Debye-Scherer equation (Ullah et al. 2017). The Joint Committee on Powder Diffraction Standards (JCPDS) is employed as standard pattern to confirm the gold (JCPDS 04-0784, with lattice constant$^{\alpha}$ = 4.078 Å), silver (JCPDS 04-0783 with lattice constant$^{\alpha}$ = 4.086 Å), iron (JCPDS-05-0661) Palladium (JCPDS no 46-1043) and Zinc (JCPDS, card No. 89-7102). The phytosynthesized nanoparticles are confirmed their crystalline nature with the comparison of Joint Committee on Powder Diffraction Standards data. The authors confirmed the silver, gold, palladium, platinum, zinc and iron and oxide

Table 7.4 Characterization profile of seaweed nanoparticles

Seaweed name	Type of nanoparticle	UV–Vis (nm)	FTIR/ functional group	XRD value	EDX	SEM/TEM	Size in nm	SHAPE	Reference
Enteromorpha compressa	Ag NPs	421	O=H and C=O, C–OH	111, 200, 211	Silver	TEM	4–24	○	Ramkumar et al. (2017a)
Laurencia catarinensis	Ag NPs	400	N–H, C–H, C–C, C≡C, –NH, –CH, C=O, C–N, C–OH, C=CH2, C=C	–	–	TEM	77.7	▭ □	Abdel-Raouf et al. (2018)
Codium capitatum	Ag NPs	422 and 425	(NH) C=O, –C–O–C, C=C, C=N, C–N	–	3 keV 63.4 ± 1.6%	–		–	Kannan et al. (2013)
Palmaria decipiens	Ag NPs	425	N–H, O–H (CH) (CO) (CO) (C–OH)	–	–	TEM	7.0 ± 1.2	○	González-Ballesteros et al. (2018)
Desmarestia menziesii	Ag NPs	405	N–H, O–H (CH) (CO) (CO) (SO3) (C–OH)	–	–	TEM	17.8 ± 2.6		González-Ballesteros et al. (2018)
Gracilaria birdiae	Ag NPs	410	OH and CH	–	–	–			De Aragão et al. (2019)
Sargassum polycystum	Ag NPs	420		111, 200, 211		–	5–7	○	Thangaraju et al. (2012)

Padina tetrastromatica	Ag NPs		C=O and –C–OH, –OH	–	–		20–90	□	Kayalvizhi et al. (2014)
Turbinari aornata	Ag NPs		C=O and –C–OH, –OH	–	–			□	Kayalvizhi et al. (2014)
Chaetomorpha aerea	Ag NPs	380	R–N=C=S_	220	Ag			○	Sivakumar et al. (2018)
Portieria hornemannii	Ag NPs	430	–	111, 200, 220, 311	–		34.1–39.8		Ramamoorthy et al. (2019)
Gracilaria fergusonii	Ag NPs	442	O–H, –NH, C–N	111, 200, 220, 311	Ag		18–45	○	Udhaya et al. (2015)
Ulva fasciata	Ag NPs	446, 428 & 426	Amide band in proteins	–	Ag—46%	TEM	9–20	○	Hamouda et al. (2018)
Turbinaria conoides	Ag NPs	420	N–H, C–O, C–N, C–I, C–Cl, C–Br	111, 200 and 220	–		96	○	Rajeshkumar et al. (2012a)
Padina tetrastromatica	Ag NPs	426	NH, C=C, CH3, CH	220	Ag, Cu, Cl, O		20	○	Jegadeeswaran et al. (2012)
Padina gymnospora	Ag NPs	410	–	–	–	–	25–40	○	Shiny et al. (2013)
Halymenia porphyroides	Ag NPs	430.5	O–H, =C–H, C–H, C=N,	100, 110, 111, 200 and 220	–		32	○	Manam and Subbaiah (2020)
Amphiroa anceps	Ag NPs	420	N–H, C–H, C=O, O–H, N–O, C–C	111, 200 and 220	–		10–80	○	Roy and Anantharaman (2018)
Urospora sp.	Ag NPs	430	(–OH) (C=O) (C–O)	111, 200 and 220	–	–	20–30	○	Suriya et al. (2012)

(continued)

Table 7.4 (continued)

Seaweed name	Type of nanoparticle	UV–Vis (nm)	FTIR/ functional group	XRD value	EDX	SEM/TEM	Size in nm	SHAPE	Reference
Spatoglossum asperum	Ag NPs	400	N–H, C–N, C–X	111, 200	–	–		–	Annakodi et al. (2015)
Hedophyllum sessile	Ag NPs	400	C–O, C–N, C–X	111, 200 and 220	–	–		–	Annakodi et al. (2015)
Padina sp.	Ag NPs	445	N–H, N–H, O–H, C≡C–, C–H, N–H N–O, NO2, or C–H, 0 C–N C–H (–CH2X), C–N	–	48.34% Ag.		33.75	–	Bhuyar et al. (2020)
Hypnea cervicornis	Ag NPs	430	O–H group C–H, C=O C=C, C–N, C–N	–	–	Spherical		○	Leela and Anchana Devi (2017)
Pterocladia capillacae (Pc), *Jania rubins* (Jr), *Ulva faciata* (Uf), and *Colpomenia sinuosa* (Cs)	Ag NPs	407–424	O–H, C–H, –C–O–O–, S=O, C–N, –C–O–, –C–O–SO4	–	–	–	7–20		El-Rafiea et al. (2013)
Sargassum polycystum	Ag NPs	405	C=O, C–O	111, 200, 220, 311	2.5–3.5 keV		~28	○	Subramanian et al. (2016)
Spyridia filamentosa	Ag NPs	420	CO, CC, –OH, –N–H, C–N	111, 142, 200, 220, 311	Ag, Cl	–	20–30	○	Valarmathi et al. (2019)

Gelidiella acerosa	Ag NPs	404.1	CO, CC, –OH, –N–H, C–N	–	–	–		–	Thiruchelviet al. (2020)
Chaetomorpha linum	Ag NPs	422	–C=C –C–O, –C–O–C and C=C	–	–		3–44	–	Ragupathi Raja Kannan et al. (2013)
Gracilaria corticata	Ag NPs	420	O–H	–	–	–	18–46	○	Kumar et al. (2013)
Sargassum tenerrimum	Ag NPs	420	N–H, CO	–	–	–	20	○	Kumar et al. (2012)
Sargassum plagiophyllum, *Ulva reticulata* and *Enteromorpha compressa*	Ag NPs	420	N–H, O–H, C≡N, C–H		–	Sp—20 Ur and Ec—40–50		○	Dhanalakshmi et al. (2012)
Padina tetrastromatica	Ag NPs	460	N–H, O–H, C–N	111, 200, 220, 311	–	–	14		Rajeshkumar et al. (2012b)
Laurencia aldingensis and *Laurenciella* sp.	Ag NPs	440	–	111, 200, 220, 222, 311	Ag	–	5 to 10	○ ⬡ △	Vieira et al. (2015)
Cladophora fascicularis	Ag NPs	427	N–H, O–H, C=O, –C=O	111, 200, 220, 311	Ag, Cl, C, O	–	45	○	Subramanian et al. (2016)
Caulerpa racemosa	Ag NPs	413	(NH)=O, OH, C–N	111, 200, 220, 311			10	○ △	Valarmathiet al. (2019)
Sargassum wightii	Ag NPs	439	(NH)=O, O–H, C–N, C–H	111, 200, 220, 311	–	–	15–20	○	Thiruchelvi et al. (2020)
Sargassum angustifolium	Ag NPs	406	–	–	–	–	32.54 nm	○	Kannan et al. (2013)

(continued)

Table 7.4 (continued)

Seaweed name	Type of nanoparticle	UV–Vis (nm)	FTIR/ functional group	XRD value	EDX	SEM/TEM	Size in nm	SHAPE	Reference
Ulva flexuosa	Ag NPs	480	O–H, N–H, –PH, CH, COO	–	Ag, Cl, C, O	–	4.93–6.70		Kumar et al. (2013)
T. conoides	AuNPs	540	$–NH_2$, –OH, O–H, N–H	–	–		27.5	○ △	Ramakrishna et al. (2016)
Halymenia dilatata	Hd-AuNPs	529	OH, CN	111, 200, 220, 311	Au, Cu		16	○ △	Vinosha et al. (2019)
Stoechospermum marginatum	AuNPs	550	$–NH_2$, CH–O–H C–OH, C–O–H	111, 200, 220, 311	Au, Br, Cu, Fe, and Sr		40 and 85	○△ ⬡	Rajathi et al. (2012)
S. tenerrimum	AuNPs	547	COO–, –OH, O–H, N–H, C–N	–	–		–		Ramakrishna et al. (2016)
Cystoseira baccata	AuNPs	532	C–OH, NH, OH, CH, –SO3	–	Ag, Ca, K, C, Na, Cu		8.4 ± 2.2	○	González-Ballesteros et al. (2017)
Gracilaria verrucosa	AuNPs	520	N–H, C=O, OH	111, 200, 220, 311				○ ⬡ ⬠ △	Chellamuthu et al. (2019)
Sargassum myriocystum	AuNPs	533	(O–H), (C=C), (C–H), (C–N), –OH	111, 200, 220, 311			10–23	○ △	Stalin Dhas et al. (2012)
Turbinaria ornata	AuNPs	525	O–H, C–C, C–H	111, 200, 220, 311, 322	Au, O, C		7–11	○	Ashokkumar and Vijayaraghavan (2016)

Gracilaria edulis	AuNPs		C–H "oop", C–N, C–C, N–O, N–H, C–H, O–H	–	–		71	○	Abideen and Sankar (2015)
Galaxaura elongata	AuNPs	~535	–N–H	–	–		3.85–77.13 nm	○ △	Abdel-Raouf et al. (2017)
Palmaria decipiens	AuNPs	548	N–H, O–H (CH) (CO) (SO_3) (C–OH)	–	–		36.8 ± 5.3	○	González-Ballesteros et al. (2018)
Desmarestiam enziessi	AuNPs	527	N–H, O–H (CH) (CO) (SO_3) (C–OH)	–	–		11.5 ± 3.3	○	González-Ballesteros et al. (2018)
Sargassum plagiophyllum	AuNPs	532	(O–H) (C–H) –C≡C–s (C=O) (C–N) (C–N) or (C–O)	111, 200, 220, 311	–		65.87		Stalin Dhas et al. (2020)
Ulva lactuca	ZnONPs.	325	C–C, O–H	100, 002, 101, 102, 110, 103, 200	Zinc (78%)		15	△ ⬡ ▭	Ishwarya et al. (2017)

(continued)

Table 7.4 (continued)

Seaweed name	Type of nanoparticle	UV–Vis (nm)	FTIR/ functional group	XRD value	EDX	SEM/TEM	Size in nm	SHAPE	Reference
Sargassum myriocystum	ZnO NP	372	O–H and C=O, C–H	100, 002, 101, 102, 110, 103, 200, 112 and 004	Zinc (52%) and oxygen (48%)		96	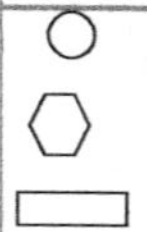	Nagarajan and Kuppusamy (2013)
Hypnea musciformis	Hy-ZnONps	370	C=N, O–H, C–H, Fe–O	100, 002, 101, 102, 110, 103, 112 and 201	Zn, O		26–35		Yousefzadi et al. (2017)
Sargassum muticum	ZnO NPs	334	C–O–SO3, C–O, C=O, (NH) C=O	101	–		30–57		Azizi et al. (2014)
Padina gymnospora	SCZ nanoparticles	670	(COO–), O–H, C–C, C–H	100, 002, 101, 102, 110 and 103 CdO 111, 200, 220, 311 and 222	Carbon (C), oxygen (O), zinc (Zn) and cadmium (Cd)		20–50	.	Rajaboopathi and Thambidurai (2017)
Gracilaria edulis	Fe_3O_4 NPs	410	N–H, C=O, N=O	44.80°	Iron (72.11%)		20–100		Subhashini et al. (2018)

Kappaphycus alvarezii	Fe_3O_4	400	–CH2, O–H, C–H, C–O–SO_3, Fe–O	200, 311, 400, 422, 511, 440, and 533	–		14.7		Yew et al. (2016)
Padina pavonica	Fe_3O_4	402	OH, C–H, C=O, C–I, N–H	–	Na, Cl, C,O, Fe, Cu, Zn, P, S		10–19.5		El-Kassas et al. (2016)
Sargassum acinarium	Fe_3O_4	415	OH, C–H, C=O, N–H, C–C	–	Na, Cl, Fe, P, S, Si, Ca, Mg, K		21.6–27.4		El-Kassas et al. (2016)
Sargassum wightii	ZrO2 NPs	277	OH, C–O, Zr–O, C=O	111, 200, 202, 220, 311	–		5		Kumaresan et al. (2018)
Padina boryana	Pd-NPs	293	C–H, OH,	111, 200, 220, 311	C, O, and Cl Pd was 28.31%		5–20	○	Hana Sonbol et al. (2021)
Dictyota indica	PdNP	420	C=C, OH, C–H, C–C and C–O, C–Br	200, 220, 311, 322, 400	–		–	○	Shargh et al. (2018)
Padina gymnospora	PdNP	260, 370	CH, C=C, C–H, C–C(O)–C, C–N, C–OH, C–Cl	111, 200, 220	–		–	–	Ramkumar et al. (2017b)

nanoparticles formation from the seaweeds using the lattice planes proposed by Joint Committee on Powder Diffraction Standards (Table 7.4).

The chemical and metal composition of AgNPs, AuNPs, ZnONPs, Fe_3O_4NPs, PdNPs is verified by energy-dispersive X-ray spectroscopy (EDX) (Shah et al. 2015). The available literature on phytosynthesized seaweeds mediated nanoparticles also confirmed the metal composition by EDX. The percentage of silver composition is ranged from 2.36% to 87.86% (3KeV), gold is ranged from 41.85% to 48.23% (2KeV), iron is 64.34%, for copper—0.9 KeV, ruthenium—2–3 KeV (Table 7.4).

Fourier transform infrared spectroscopy (FT-IR) employed to determine the functional atoms or groups bound to the surface of nanoparticles (Dahoumane et al. 2017). The following functional groups viz., O–H, N–H, C–H, C=0, C–N, C–X, P–O, S=O, C=C, –C–O–C, C≡N, $-NH_2$, –C–O–O, C–I, C–Br presence/absence/position are explained the silver, gold, zinc, palladium, iron nanoparticle synthesis (Table 7.4). The functional constituent's presence in the nanoparticle confirms the role of reducing agents and capping process. To study the size and shape of the nanoparticles the Scanning electron microscopy (SEM), transmission electron microscopy (TEM), high-resolution transmission electron microscopy (HRTEM) and atomic force microscopy (AFM) are employed (Marquis et al. 2009; Quester et al. 2013; Azharuddin et al. 2019; Khanna et al. 2019). Most of the researchers used the SEM and TEM for the measurement of nanoparticle size and shape (Table 7.4). In addition, the SEM and TEM employed to obtain 2D and 3D images of the phytosynthesized nanoparticles (Quester et al. 2013). The available literature on the seaweeds nanoparticles authenticated the application of different electron microscopy on the characterization of silver, gold, zinc, palladium, platinum, SCZ and iron nanoparticles of seaweeds (Table 7.4).

5 Biological Application of Nanoparticles of Seaweeds

A nutchell of the reported biological application of phytosynthesized nanoparticles of seaweeds are summarized in Figs. 7.6, 7.7, 7.8, 7.9, 7.10, and 7.11. The complete report is presented in Table 7.5.

6 Conclusion

The biologist, pharmacist, biotechnologist and biochemist are attracted and generated interest towards the biogenic synthesis of metal and metal oxide nanoparticles recently. The wide range of pharmaceutical and biomedical applications of seaweeds and nanoparticles are attracted the biologist and the unique physico-chemical properties of nanoparticle make them highly desirable. The available literatures confirm that biogenic synthesis of nanoparticles from marine algae has the potential

Antibacterial potential of Green Seaweeds
• Enteromorpha compressa (Ramkumar et al. 2017a, b)
• Urospora sps.(Suriya et al. 2012)
• Caulerpa racemosa (Valarmathi et al. 2019)
• Ulothrix flacca, Ulva fasciata and Caulerpa taxifolia (Rao et al. 2015)
• Ulva faciata (Hamouda et al. 2018)
• Cladophora fascicularis (Subramanian et al. 2016)
• Ulva flexuosa (Kumar et al. 2013)
• Ulva lactuca (Ishwarya et al. 2017)
• Ulva faciata (El-Rafie et al. 2013)

Fig. 7.6 Antibacterial potential of phytosytnhtsized nanoparticles of Green seaweeds

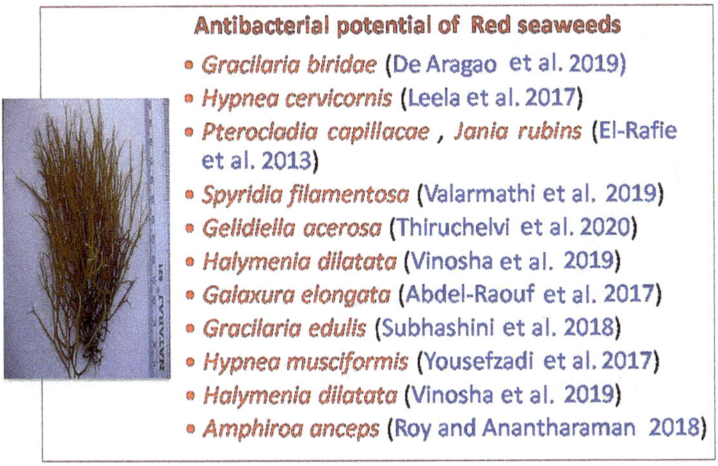

Antibacterial potential of Red seaweeds
• Gracilaria biridae (De Aragao et al. 2019)
• Hypnea cervicornis (Leela et al. 2017)
• Pterocladia capillacae , Jania rubins (El-Rafie et al. 2013)
• Spyridia filamentosa (Valarmathi et al. 2019)
• Gelidiella acerosa (Thiruchelvi et al. 2020)
• Halymenia dilatata (Vinosha et al. 2019)
• Galaxura elongata (Abdel-Raouf et al. 2017)
• Gracilaria edulis (Subhashini et al. 2018)
• Hypnea musciformis (Yousefzadi et al. 2017)
• Halymenia dilatata (Vinosha et al. 2019)
• Amphiroa anceps (Roy and Anantharaman 2018)

Fig. 7.7 Antibacterial potential of phytosytnhtsized nanoparticles of Red seaweeds

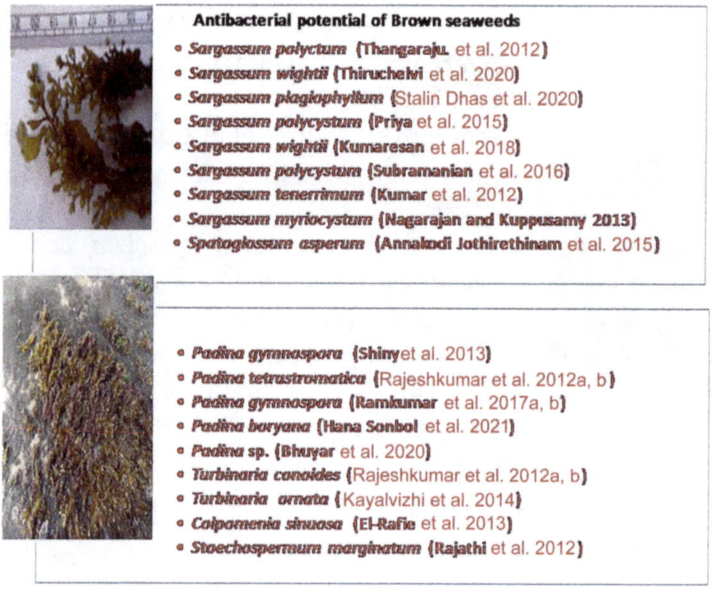

Fig. 7.8 Antibacterial potential of phytosytnhtsized nanoparticles of Brown seaweeds

Fig. 7.9 Bio potential of phytosytnhtsized nanoparticles of seaweeds

Fig. 7.10 Bio potential of phytosytnhtsized nanoparticles of seaweeds

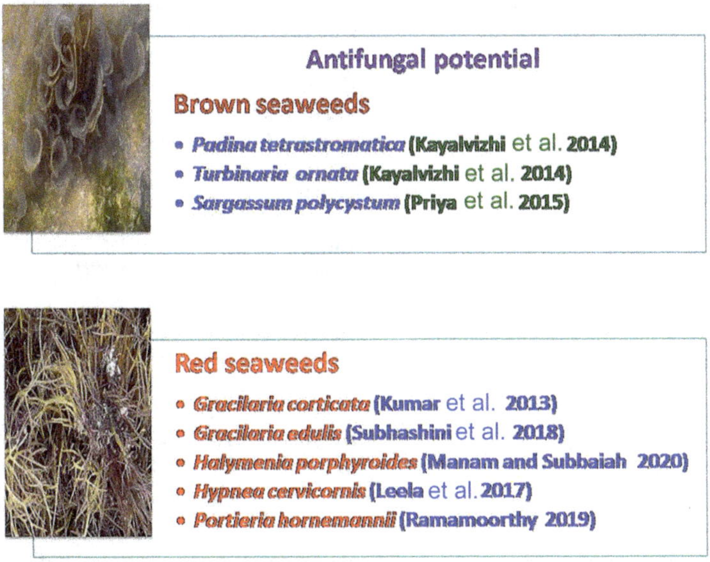

Antifungal potential
Brown seaweeds
Padina tetrastromatica (Kayalvizhi et al. 2014)
Turbinaria ornata (Kayalvizhi et al. 2014)
Sargassum polycystum (Priya et al. 2015)
Red seaweeds
Gracilaria corticata (Kumar et al. 2013)
Gracilaria edulis (Subhashini et al. 2018)
Halymenia porphyroides (Manam and Subbaiah 2020)
Hypnea cervicornis (Leela et al. 2017)
Portieria hornemannii (Ramamoorthy 2019)

Fig. 7.11 Anti-fungal activity of phytosytnhtsized nanoparticles of seaweeds

Table 7.5 Bio-potentials of synthesized nanoparticles of seaweeds

Seaweed	NPs	Biopotency	Reference
Enteromorpha compressa	AgNPs, spherical, 4–24 nm	**Antibacterial—***Escherichia coli, Klebsiella pneumoniae, Pseudomonas sp., Staphylococcus aureus* and *Salmonella paratyphi* (10.5–12.0 mm) Antifungal—*Aspergillus flavus, A. niger, A. ochraceus, A. terreus and Fusarium moniliforme* (9.2–10.2 mm) **Cytotoxicity potential** IC_{50}–95.35 mg mL 1.	Ramkumar et al. (2017a)
Palmaria decipiens	Ag@PD 7.0 ± 1.2 nm	**Antioxidant activity** **ROS activity** (mg ascorbic acid/g seaweed) 43.66 ± 0.57 **DPPH scavenging activity** (EC_{50} mg/mL) 127.21	González-Ballesteros et al. (2018)
Desmarestiam enziesii	Spherical, Ag@DM, 17.8 ± 2.6	**Antioxidant activity** **ROS activity** (mg ascorbic acid/g seaweed) 126.35 ± 0.90 **DPPH scavenging activity** (EC50 mg/mL) 68.70	González-Ballesteros et al. (2018)
Gracilaria birdiae	AgNPs, spherical, 20.2 and 94.9 nm	**Antibacterial potential** *S. aureus* ATCC 29213 and E. coli ATCC 25922 MICs—*S. aureus*—162.5; *E. coli*—81.2	De Aragão et al. (2019)
Oscillatoria sp.	AgNPs, spherical, 10 nm in size	**Antibacterial potential** *S. aureus* ATCC29213, *E. coli* ATCC 11775, *E. coli* ATCC 35218, *P. aeruginosa* ATCC27853, *Citrobacter* sp., *S. typhi* ATCC 14028 and *Bacillus cereus* Inhibition zone—13.0–24.0 mm. **Antibiofilm potential** *Pseudomonas aeruginosa* ATCC 27853 *Citrobacter* sp.	Adebayo-Tayo et al. (2019)
Sargassum polycystum	AgNPs, spherical, 5–7 nm	**Antibacterial potential** *Pseudomonas aeruginosa, Klebsiellapneumoniae, Escherichia coli* and *Staphylococcus aureus.* Maximum zone of inhibition—*S. aureus* (38 mm). IC_{50} seaweed extract—300 mg mL^{-1} IC_{50} silver nanoparticles—135 mg mL^{-1}	Thangaraju et al. (2012)

(continued)

Table 7.5 (continued)

Seaweed	NPs	Biopotency	Reference
Padina tetrastromatica	AgNPs, cubical, 18–90 nm	**Antimicrobial potential** *Escherichia coli*, *Staphylococcus aureus*, *Salmonella typhi* and *Pseudomonas aeruginosa* Fungi—*Candida* albicans, *Alternaria alternata*, *Penicillium italicum* and *Fusarium equiseti.* Maximum inhibition zone—*S. aureus* (20 ± 2.4) and *C. albicans*—18 ± 2.1)	Kayalvizhi et al. (2014)
Turbinaria ornata	AgNPs, cubical, 20–90 nm	**Antimicrobial potential** *Escherichia coli*, *Staphylococcus aureus*, *Salmonella typhi* and *Pseudomonas aeruginosa* Fungi—*Candida* albicans, *Alternaria alternata*, *Penicillium italicum* and *Fusarium equiseti.* Maximum inhibition zone—*S. aureus* (20 ± 2.5) and *P. italicum* (14 ± 1.6)	Kayalvizhi et al. (2014)
Chaetomorpha aerea	AgNPs, cubical	**Seed germination—Phytotoxicity**	Sivakumar et al. (2018)
Portieria hornemannii	AgNPs, spherical, 16 nm	**Antifungal activity** *Beauveria bassiana* and *Metahizium anisopliae* pathogens—100 μl AgNPs—maximum zone of inhibition against *B. bassiana* (22.6 mm) and *M. anisopliae* (21.0 mm)	Ramamoorthy et al. (2019)
Ulvafasciata	AgNPs, spherical, 11–37 nm	**Antibacterial potential** *Escherichia coli* ATTCC 8739 and *Proteus mirabilis* ATTCC 9240, *Micrococcus leutus* and *Kocuria varians*. Highest inhibition—*Micrococcus leutus*, followed by *Escherichia coli* ATTCC 8739	Hamouda et al. (2018)
Turbinaria conoides	AgNPs, spherical, average 96 nm	**Antibacterial potential** *Bacillus subtilis* (MTCC3053) and *Klebsiella planticola* (MTCC2277) *Klebsiella planticola*—20.67 ± 0.668 (300 μl)	Rajeshkumar et al. (2012a)
Padina gymnospora	AgNPs, spherical, 25–40 nm	**Antibacterial potential** *Bacillus cereus* and *Escherichia coli*. *B. cereus* (13.06 ± 0.40)	Shiny et al. (2013)

(continued)

Table 7.5 (continued)

Seaweed	NPs	Biopotency	Reference
Halymenia porphyroides	AgNPs, Spherical,34.3–80 nm	**Antifungal activity** *M. nanum* (2 ± 0.002 mm) and *T. mentagrophytes* (2 ± 0.001 mm) *R. microspores* (4 ± 0.002 mm)	Manam and Subbaiah (2020)
Amphiro anceps	AgNPs, spherical, 10–80 nm	**Antibacterial potential** *Escherichia coli* (0.95 ± 0.057 cm), *Klebsiella pneumonia* (0.63 ± 0.05 cm), **Phytotoxicity—seed germination** Seed germination—75% *Abelmoschus esculentus* and 80% *Raphanus sativus* var. *longipinnatus*	Roy and Anantharaman (2018)
***Urospora* sp.**	AgNPs, spherical, 10–20 nm	**Antibacterial potential** *Staphylococcus aureus*, *Escherichia coli*, *Pseudomonas aeruginosa*, *Klebsiella pneumoniae* and *Bacillus subtilis*. 100 µl silver nanoparticles—*S. aureus* (23 mm), *Bacillus subtilis* (20 mm), and *E. coli* (18 mm)	Suriya et al. (2012)
Spatoglossum asperum	AgNPs, crystalline structure	**Antimicrobial potential** Xanthomonas axonopodispv. citri and X. oryzaepv. oryzae) and a fungus Ustilaginoid eavirens	Jothirethinam et al. (2015)
Hedophyllum sessile	AgNPs, crystalline structure	**Antimicrobial potential** Xanthomonas axonopodispv. citri and X. oryzaepv. oryzae) and a fungus Ustilaginoid eavirens	Jothirethinam et al. (2015)
Ulothrix flacca, *Ulva fasciata* and *Caulerpa taxifolia*	AgNPs	**Antimicrobial potential** Four gram positive and four gram negative pathogens *Ulva fasciata* silver nanoparticles (57 µg/mL)—26.3 ± 0.6 mm against *Listeria monocytogenes*. IC_{50}—*Ulva fasciata*—15 µg/mL, *Caulerpa taxifolia*—27 µg/mL and *Ulothrix flacca* 23 µg/mL	Rao and Boominathan (2015)

(continued)

Table 7.5 (continued)

Seaweed	NPs	Biopotency	Reference
Padina sp.	AgNPs, 25–60 nm	**Antimicrobial potential** *Staphylococcus aureus*, *Bacillus subtilis*, *Pseudomonas aeruginosa*, *Salmonella typhi*, and *Escherichia coli* *Staphylococcus aureus*—15.17 ± 0.58 mm and *Pseudomonas aeruginosa* 13.33 ± 0.76 mm	Bhuyar et al. (2020)
Hypnea cervicornis	AgNPs, spherical	**Antimicrobial potential** *Staphylococcus aureus*, *Streptococcus spp*, *Escherichia coli*, *Klebsiella pneumoniae*, *Salmonella spp* and *Pseudomonas aeruginosa.* *Staphylococcus aureus*—higher **Antifungal activity** *Aspergillus niger* and *Candida albicans* maximum inhibition—*Aspergillus niger* (13 nm) **Antioxidant activity** **Antidiabetic activity**	Leela and Anchana Devi (2017)
Pterocladia capillacae (Pc), *Jania rubins* (Jr), *Ulva faciata* (Uf), and *Colpmenia sinusa* (Cs)	AgNPs, spherical, with size of 7, 7, 12, and 20 nm	**Antimicrobial activity** *E. coli*, *S. aureus*	El-Rafiea et al. (2013)
Sargassum polycystum	AgNPs, spherical, ~28 nm	**Antimicrobial activity** The AgNPs—higher inhibition (78.2% at a concentration of 500 µg/mL); aqueous extract (69.8% at a concentration of 500 µg/mL) **Antioxidant assay** AgNPs—DPPH radical scavenging (78.2% of inhibition), reducing power (0.18 ± 0.006) and total antioxidant activities (59.2 ± 0.54%) **In vitro cytotoxicity assay** IC_{50}—20 µg/mL against human colon cancer (HT-29) cells	Subramanian et al. (2016)

(continued)

Table 7.5 (continued)

Seaweed	NPs	Biopotency	Reference
Spyridia filamentosa	AgNPs, spherical, 20–30 nm	**Antimicrobial activity** *Staphylococcus* sp. (KC688883) and *Klebsiella* sp. (KC899845). AgNPs—*Klebsiella* sp. (63.4%) **Cytotoxicity assay** AgNPs showed strong toxic activity to MCF-7 cell lines	Valarmathi et al. (2019)
Gelidiella acerosa	AgNPs	**Antimicrobial activity** *Pseudomonas aeruginosa* and *Escherichia coli*. The AgNPs—*Pseudomonas aeruginosa*—inhibition	Thiruchelvi et al. (2020)
Gracilaria corticata	AgNPs, spherical, 18–46 nm	**Antifungal activity** C. albicans (12 mm) and C. glabrata (11 mm) at 30 μl concentration.	Ragupathi Raja Kannan et al. (2013)
Sargassum tenerrimum	AgNPs, spherical, 20 nm	**Antimicrobial activity** *Vibrio cholerae* (18 mm)	Kumar et al. (2012)
Padina tetrastromatica	AgNPs, 14 nm	**Antimicrobial activity** *Pseudomonas sps, E.coli, Bacillus subtilis and Klebsiella planticola.*	Rajeshkumar et al. (2012b)
L. aldingensis and Laurenciella sp.	AgNPs, spherical, 10 nm	**Cytotoxicity assay** IC_{50} for L. aldingensis—1.07 ± 1.36 0.97 ± 1.62 for AgNPs synthesized Laurenciella sp. extracts	Vieira et al. (2015)
Cladophora fascicularis	AgNPs, spherical, 45 nm	**Antimicrobial activity** A. hydrophila—19 mm at 150 μg/mL, **Cytotoxic study** CfAgNPs—not acutely toxic to Artemia nauplii (150 μg/mL),	Subramanian et al. (2016)
Caulerpa racemosa	AgNPs, spherical, 10 nm	**Antimicrobial activity** *P. mirabilis*—14 mm and *S. aureus*—7 mm	Valarmathi et al. (2019)
Sargassum wightii	AgNPs, spherical, 15–20 nm	**Antibacterial activity** Maximum ZI *S. aureus* (12 mm)	Thiruchelvi et al. (2020)
Sargassum angustifolium	AgNPs, spherical, 32.54 nm	**Acute toxicity testing** The maximum acceptable toxicant concentration (MATC) of silver nanoparticles 7.95 (24 h), 5.22 (48 h), 3.06 (72 h) and 1.13 (96 h) mg/L	Kannan et al. (2013)
Ulva flexuosa	AgNPs, 4.93–6.70 nm	**Antibacterial activity**	Kumar et al. (2013)

(continued)

Table 7.5 (continued)

Seaweed	NPs	Biopotency	Reference
Cystoseira baccata	AuNPs, spherical, 8.4 ± 2.2 nm	**In vitro anticancer activity** Caco-2 and HT-29, IC_{50}—79.03 M for Caco-2 and 49.61 M for HT-29.	González-Ballesteros et al. (2017)
Gracilaria verrucosa	AuNPs, spherical,oval, octahedral, pentagonal, rhombus and triangular, <20 nm	**Cell culture and biocompatibility studies** HEK 293—80% of cell viability	Chellamuthu et al. (2019)
Halymenia dilatata	AuNPs were triangular, spherical, 16 nm	**In vitro antioxidant activities** DPPH radical scavenging (13.68 ± 0.40 to 58.77 ± 0.85%), reducing power (16.34 ± 0.63 to 62.45 ± 0.60%) and total antioxidant (12.45 ± 0.50 to 56.67 ± 0.74%) **Cytotoxicity assay** IC_{50} HdAuNPs—22.62 µg/mL against HT-29 cancer cells. **Antibacterial assay** *Aeromonas hydrophila*—21 mm	Vinosha et al. (2019)
Stoechospermum marginatum	AuNPs, spherical,18.5–93.7 nm	**Antibacterial assay** *E. faecalis* (11 mm), *K. pneumoniae* (6 mm).	Rajathi et al. (2012)
Gracilaria edulis	AuNPs, spherical, 71 and 110 nm	**Anti-diabetic activity** Glucose inhibition assay—98.75% at 400 mg/ml; inhibition of α amylase activity—78.75% at 400 µg/mL	Abideen and Sankar (2015)
Syringodium isoetifolium	AuNPs, spherical, 71 and 110 nm	**Antibacterial activity** Maximum ZI—10 mm *E. coli.* Glucose inhibition assay—45.25% at 400 µg/mL; inhibition of α amylase activity—77.25% at 400 µg/mL	Abideen and Sankar (2015)
Galaxaura elongata	AuNPs, Spherical, 3.85–77.13 nm	**Antibacterial activity** Maximum zones of inhibition (17–16 mm)—*E. coli*, *K. pneumoniae*; *S. aureus* and *P. aeruginosa* (13 mm).	Abdel-Raouf et al. (2017)
Sargassum plagiophyllum	AuNPs, spherical, 65.87 nm	**Antibacterial activity** *Escherichia coli* (ATCC 25922) and *Salmonella typhi* (ATCC 6539)	Stalin Dhas et al. (2020)

(continued)

Table 7.5 (continued)

Seaweed	NPs	Biopotency	Reference
Sargassum polycystum	CuONPs	**Antibacterial activity** *Pseudomonas aeruginosa* (15 ± 0.5 mm), *Shigella dysenteriae* (6 ± 0.5 mm) **Antifungal activity** *Aspergillus niger* (20 ± 0.5 mm) **Cytotoxicity assay** MCF-7 breast cancer cell lines IC_{50}—61.25 μg mL^{-1}.	*Ramaswamy et al. (*2016*)*
Ulva lactuca	ZnONPs, sponge like nanoparticles, 15 nm	**Larvicidal activity** 100% mortality of *Aedes aegypti* fourth instar larvae—50 μg/mL within 24 h. **Biofilm reduction activity** *B. licheniformis* (90%), *B. pumilis* (89%), *E. coli* (90%), *P. vulgaris* (91%) during light exposure. In the dark, *B. licheniformis* (82%), *B. pumilis* (84%), *E. coli* (80%), *P. vulgaris* (79%)	Ishwarya et al. (2017)
Padina gymnospora	SCZ, 20–50 nm	**Photocatalytic activity** SCZ nanoparticles—99.57% degradation within 30 min the cycle stability of SCZ nanoparticles—94.1% after three cycles.	Rajaboopathi and Thambidurai (2017)
Gracilaria edulis	Cube shaped,Fe3O4 NPs, 20–100 nm	**Antibacterial activity** *Pseudomonas aeruginosa*—15 ± 0.5 mm, *K. pneumonia*—3 ± 0.5 mm. **Antifungal activity** *Aspergillus nidulans*—16 ± 0.5 mm and *Aspergillus oryzae*—12 ± 0.5 mm.	Subhashini et al. (2018)
Padina pavonica	Fe3O4 NPs	**Pb removal efficiency** 91% of Pb removal 75 min	El-Kassas et al. (2016)
Sargassum acinarium	Fe3O4 NPs	**Pb removal efficiency** 78% of Pb removal—75 min	El-Kassas et al. (2016)
Sargassumm yriocystum	Zinc oxide NPs, rectangle, spherical, triangle, radial and spheres, 96–110 nm	***Antibacterial activity*** *C. albicans*	Nagarajan and Kuppusamy (2013)
Hypnea musciformis	Hy-ZnONps, 26–35 nm,	**Antibacterial activity** *S. aureus* and *E. coli*	Yousefzadi et al. (2017)
Sargassum wightii	Zirconia NPs, 4.8 nm	**Antibacterial activity**	Kumaresan et al. (2018)

(continued)

Table 7.5 (continued)

Seaweed	NPs	Biopotency	Reference
Padina boryana	Pd-NPs, 5–20 nm	**Biofilm inhibition** *P. mirabilis* (36.9%) >*S. aureus* (27.2%) >*A. pitti* (21.36%), *P. aeruginosa* (15.63%) >*E. fergusonii* (12.36%) >*A. enteropelogenes* (8.9%) at 31.25 μg/mL concentration. **Antibacterial activity**	Hana Sonbol et al. (2021)
Padina gymnospora	P-NPs	**Antibacterial activity** 2 ± 0.05 and 15.6 ± 0.17 mm zone of inhibition **Cytotoxicity and Anticrustacean activity** LC_{50}-200 ± 6.3 mg mL^{-1}. Comparatively, PVP/PtNPsnanocomposite **Anticrustacean activity**—A. salina—LC_{50} 100 ± 4 mg mL^{-1}. **Haemolysis assay** 56.9% (600 mg $mL^{-1)}$, 70.7% (900 mg mL^{-1}) and 93% (1200 mg mL^{-1}) **Antioxidant activity** 31.17 ± 0.85% DPPH free radical scavenging activity	Ramkumar et al. (2017b)

to deliver a facile, green, ecofriendly and cost effective approach. The metabolites and active principles present in the marine alga may be responsible for the reduction and capping. By comparing the diversity wealth and biological application, only small number of marine algae has been studied to date. For the green chemist, phycologist and biotechnologist huge opportunities are available for biogenic synthesis of metal and metal oxide nanoparticles and to determine their biological application.

References

Abboud Y, Saffaj T, El Bouari A, Brouzi K, Tanane O, Ihssane B (2014) Biosynthesis, characterization and antimicrobial activity of copper oxide nanoparticles (CONPs) produced using brown alga extract (*Bifurcaria bifurcata*). Appl Nanosci 4:571–576. https://doi.org/10.1007/s13204-013-0233-x

Abdallah AM, Al-Haddad RMS (2021) Optical and morphology properties of the Magnetite (Fe3O4) nanoparticles prepared by green method. J Phys Conf Ser 1829:012022. https://doi.org/10.1088/1742-6596/1829/1/012022

Abdelghany AM, Oraby AH, Farea MO (2019) Influence of green synthesized gold nanoparticles on the structural, optical, electrical and dielectric properties of (PVP/SA) blend. Phys B Condens Matter 560:162–173

Abdel-Raouf NR, Alharbi M, Al-Enazi NM, Alkhulaififi MM, Ibraheem IBM (2018) Rapid biosynthesis of silver nanoparticles using the marine red alga *Laurencia catarinensis*and their characterization. Beni-Suef Univ J Basic Appl Sci 7:150–157

Abdel-Raouf NR, Al-Enazi NM, Ibraheem IBM (2017) Green biosynthesis of gold nanoparticles using *Galaxaura elongate* and characterization of their antibacterial activity. Arab J Chem 10:S3029–S3039

Abideen S, Sankar MV (2015) In-vitro screening of antidiabetic and antimicrobial activity against green synthesized AgNO3 using seaweeds. J Nanomed Nanotechnol S6:001. https://doi.org/10.4172/2157-7439.S6-001

Algotiml R, Gab-Alla A, Seoudi R, Abulreesh H, El-Readi M, Elbanna K (2022) Anticancer and antimicrobial activity of biosynthesized Red Sea marine algal silver nanoparticles, vol 12. Natureportfolio, pp 2421–2439

Alkaladi A, Abdelazim AM, Afifi M (2014) Antidiabetic activity of zinc oxide and silver nanoparticles on streptozotocin, induced diabetic rats. Int J Mol Sci 15:2015–2023

Annakodi J, Prathiba S, Shanthi N, Arunkumar K (2015) Green synthesized silver nanoparticles prepared from the antimicrobial crude extracts of two brown seaweeds against plant pathogens. Am J Nanotechnol 6(2):31–39

Applerot G, Lipovsky A, Dror R, Perkas N (2009) Enhanced antibacterial activity of nanocrystalline ZnO due to increased ROS-mediated cell injury. Adv Funct Mater 19:842–852

Aragao AP, Oliveira TM, Quelemes PV, Perfeito MLG, Araujo MC, Santiago JAS, Cardoso VS, Quaresma P, Leite JRSA, Silva DA (2016) Green synthesis of silver nanoparticles using the seaweed Gracilaria birdiae and their antibacterial activity. Arab J Chem 12(8). https://doi.org/10.1016/j.arabjc.2016.04.014

Araya-Castro K, Chao TC, Durán-Vinet B, Cisternas C, Ciudad G, Rubilar O (2021) Green synthesis of copper oxide nanoparticles using protein fractions from an aqueous extract of brown algae *Macrocystis pyrifera*. PRO 9:78. https://doi.org/10.3390/pr9010078

Ashokan AP, Paulpandi M, Dinesh D, Murugan K, Vadivalagan C, Benelli G (2017) Toxicity on dengue mosquito vectors through *Myristicafragrans*-synthesized zinc oxide Nanorods, and their cytotoxic effects on liver cancer cells (HepG2). J Clust Sci 28:205–226

Ashokkumar T, Vijayaraghavan K (2016) Brown seaweed-mediated biosynthesis of gold nanoparticles. J Environ Biotechnol Res 2(1):45–50

Augustine R, Augustine A, Kalarikkal N, Thomas S (2016) Fabrication and characterization of biosilver nanoparticles loaded calcium pectinatenano-micro dual-porous antibacterial wound dressings. Prog Biomater 5:223–235

Azharuddin M, Zhu GH, Das D, Ozgur E, Uzun L, Turner APF, Patra HK (2019) A Repertoire of biomedical applications of noble metal nanoparticles. Chem Comm 49. https://doi.org/10.1039/C9CC01741K

Azizi S, Ahmad MB, Namvar F, Mohamad R (2014) Green biosynthesis and characterization of zinc oxide nanoparticles using brown marine macroalga Sargassum muticum aqueous extract. Mater Lett 116:275–277

Balantrapu K, Goia D (2009) Silver nanoparticles for printable electronics and biological applications. J Mater Res 24(9):2828–2836

Bar H, Bhui DK, Sahoo GP, Sarkar P, Pyne S, Misra A, De SP (2009) Green synthesis of silver nanoparticles using latex of *Jatrophacurcas*. Colloids Surf A Physicochem Eng Asp 339:134–139

Belaidi A, Dittrich T, Kieven D, Tornow J, Schwarzburg K, Kunst M, Allsop N, Lux-Steiner MC, Gavrilov S (2009) ZnO-nanorods arrays for solar cells with extremely thin sulfidic absorber. Sol Energy Mater Sol Cells 93:1033–1036

Benelli G (2016a) Green synthesized nanoparticles in the fight against mosquito-borne diseases and cancer a brief review. Enzyme Microb Technol 95:58–68

Benelli G (2016b) Plant-mediated biosynthesis of nanoparticles as an emerging tool against mosquitoes of medical and veterinary importance: a review. Parasitol Res 115:23–34

Bharathi Raja S, Suriya J, Sekar V, Rajasekaran R (2012) Biomimetic of silver nanoparticles by *Ulvalactuca*seaweed and evaluation of its antibacterial activity. Int J Pharm Pharm Sci 4(3):139–143

Bhuyar P, Rahim MHA, Sundararaju S, Ramaraj R, Maniam GP, Govindan N (2020) Synthesis of silver nanoparticles using marine macroalgae Padina sp. and its antibacterial activity towards pathogenic bacteria. Benisuef Uni J Basic Appl Sci 9:3. https://doi.org/10.1186/s43088-019-0031-y

Chellamuthu C, Balakrishnan R, Patel P, Shanmuganathan R, Pugazhendhi A, Ponnuchamy K (2019) Gold nanoparticles using red seaweed *Gracilariaverrucosa*: green synthesis, characterization and biocompatibility studies. Process Biochem 02:009

Chu X, Wu B, Xiao C, Zhang X, Chen J (2010) A new amperometric glucose biosensor based on platinum nanoparticles/polymerized ionic liquid-carbon nanotubes nanocomposites. Electrochim Acta 55:2848–2852

Colin JA, Pech-Pech IE, Oviedo M et al (2018) Gold nanoparticles synthesis assisted by marine algae extract: biomolecules shells from a green chemistry approach. Chem Phys Lett 708:210–215. https://doi.org/10.1016/j.cplett.2018.08.022

Dahoumane SA, Jeffryes C, Mechouet M, Agathos SN (2017) Biosynthesis of inorganic nanoparticles: A fresh look at the control of shape, size and composition. Bioengineering (Basel) 4(1):14. https://doi.org/10.3390/bioengineering4010014

De Aragão AP, De Oliveira TM, Quelemes PV, Perfeito MLG, Arau'jo MC, Santiago JDAS, Cardoso VS, Quaresma P, Leite JRDSDA, Da Silva DA (2019) Green synthesis of silver nanoparticles using the seaweed Gracilariabirdiae and their antibacterial activity. Arab J Chem 12:4182–4188

Dhanalakshmi PK, Azeez R, Rekha R, Poonkodi S, Thangaraju N (2012) Synthesis of silver nanoparticles using green and brown seaweeds. Phykos 42(2):39–45

El-Kassas HY, Mohamed AAM, Samiha MG (2016) Green synthesis of iron oxide (Fe3O4) nanoparticles using two selected brown seaweeds: characterization and application for lead bioremediation. Acta Oceanol Sin 35:89–98. https://doi.org/10.1007/s13131-016-0880-3

El-Rafiea HM, El-Rafie MH, Zahran MK (2013) Green synthesis of silver nanoparticles using polysaccharides extracted from marine macro algae. Carbohydr Polym 96:403–410

Ermakova S, Kusaykin M, Trincone A, Tatiana Z (2015) Are multifunctional marine polysaccharides a myth or reality? Front Chem 3:39. https://doi.org/10.3389/fchem.2015.00039, 2-s2.0-84969508719

Formo E, Lee E, Campbell D, Xia Y (2008) Functionalization of electrospun TiO2 nanofibers with Pt nanoparticles and nanowires for catalytic applications. Nano Lett 8:668–672

Fortunato E, Barquinha P, Pimentel A, Gonçalves A, Marques A, Pereira L, Martins R (2005) Recent advances in ZnO transparent thin film transistors. Thin Solid Films 487:205–211

Gawande MB, Branco PS, Varma RS (2013) Nano-magnetite (Fe3O4) as a support for recyclable catalysts in the development of sustainable methodologies. Chem Soc Rev 42(8):3371–3393

Ghareeb R, El-Din N, Maghraby D, Ibrahim D, Abdel-Megeed A, Abdelsalam N (2022) Nematicidal activity of seaweed synthesized silver nanoparticles and extracts against Meloidogyne incognita on tomato plants. Natureportfolia 12:3841–3857

Ghodake G, Lee DS (2011) Biological synthesis of gold nanoparticles using the aqueous extract of the Brown algae *Laminaria japonica*. J Nanoelectron Optoelectron 6:1–4

Ginzburg AL, Truong L, Tanguay RL, Hutchison JE (2018) Synergistic toxicity produced by mixtures of biocompatible gold nanoparticles and widely used surfactants. ACS Nano 12:5312–5322

González-Ballesteros N, González-Rodríguez JB, Rodríguez-Argüelles MC, Lastra M (2018) New application of two Antarctic macroalgae *Palmariadecipiens and Desmarestiamenziesii* in the synthesis of gold and silver nanoparticles. Pol Sci 15:49–54

González-Ballesteros N, Prado-López S, Rodríguez-González JB, Lastra M, Rodríguez-Argüelles MC (2017) Green synthesis of gold nanoparticles using brown algae *Cystoseirabaccata:* its activity in colon cancer cells. Colloids Surf B: Biointerfaces 153:190–198

Govindaraju K, Kiruthiga V, Ganesh Kumar V, Singaravelu G (2009) Extracellular synthesis of silver nanoparticles by a marine alga, *Sargassum wightii* Grevilli and their antibacterial effects. J Nanosci Nanotechnol 9:5497–5501

Gowri S, Rajiv Gandhi R, Sundrarajan M (2014) Structural, optical, antibacterial and antifungal properties of zirconia nanoparticles by biobased protocol. J Mater Sci Technol 30:782–790

Gu H, Chen X, Chen F, Zhou X, Parsaee Z (2018) Ultrasound-assisted biosynthesis of CuO-NPs using brown alga *Cystoseiratrinodis*: characterization, photocatalytic AOP, DPPH scavenging and antibacterial investigations. Ultrason Sonochem 41:109–119

Hamouda RAE, El-Mongy MA, Eid KF (2018) Antibacterial activity of silver nanoparticles using *Ulva fasciata* extracts as reducing agent and sodium dodecyl sulfate as stabilizer. Int J Pharmacol 14(3):359–368

Haw CY, Mohamed F, Chia CH, Radiman S, Zakaria S, Huang NM, Lim HN (2010) Hydrothermal synthesis of magnetite nanoparticles as MRI contrast agents. Ceram Int 36(4):1417–1422

Higashi A, Kishikawa N, Ohyama K, Kuroda N (2017) A simple and highly selective fluorescent sensor for palladium based on benzofuran-2-boronic acid. Tetrahedron Lett 58(28):2774–2778

Hikosaka K, Kim J, Kajita M, Kanayama A, Miyamoto Y (2008) Platinum nanoparticles have an activity similar to mitochondrial NADH:ubiquinone oxidoreductase. Colloids Surf B Biointerfaces 66:195–200

Hu FQ, Wei L, Zhou Z, Ran YL, Li Z, Gao MY (2006) Preparation of biocompatible magnetite nanocrystals for in vivo magnetic resonance detection of cancer. Adv Mater 18(19):2553–2556

Huang HH, Yan FQ, Kek YM, Chew CH, Xu GQ, Ji W, Oh PS, Tang SH (1997) Synthesis, characterization, and nonlinear optical properties of copper nanoparticles. Langmuir 13:172–175

Ishwarya R, Vaseeharan B, Kalyani S, Banumathi B, Govindarajan M, Alharbi NS, Kadaikunnan S, Al-anbr MN, Khaled JM, Facile GB (2017) Facile green synthesis of zinc oxide nanoparticles using *Ulva lactuca* seaweed extract and its evaluation of photocatalytic, antibiofilm and larvicidal activity: impact on mosquito morphology and biofilm architecture. J Photochem Photobiol B Biol 178:249–258

Jegadeeswaran P, Rajeshwari S, Venckatesh R (2012) Green synthesis of silver nanoparticles from extract of *Padina tetrastromatica* leaf. Dig J Nanomater Biostruct 7(3):991–998

Johnson M, Amutha S, Shibila T, Narayanan J (2017) Green synthesis of silver nanoparticles using Cyathea nilgirensis Holttum and their cytotoxic and phytotoxic potentials. Particul Sci Technol 36(3)

Jothirethinam A, Prathiba S, Shanthi N, Arunkumar K (2015) Green synthesized silver nanoparticles prepared from the antimicrobial crude extracts of two brown seaweeds against plant pathogens. Curr Res Nanotechnol 6(2):31–39

Kajita M, Hikosaka K, Iitsuka M, Kanayama A, Toshima N, Miyamoto Y (2007) Platinum nanoparticle is a useful scavenger of superoxide anion andhydrogen peroxide. Free Radic Res 41:615–626

Kang X, Mai Z, Zou X, Cai P, Mo J (2008) Glucose biosensors based on platinum nanoparticles deposited carbon nanotubes in sol-gel chitosan/silica hybrid. Talanta 74:879–886

Kannan RRR, Stirk WA, Van Staden J (2013) Synthesis of silver nanoparticles using the seaweed *Codiumcapitatum P.C. Silva (Chlorophyceae)*. S Afr J Bot 86:1–4

Karunakaran C, Dhanalakshmi R (2008) Photocatalytic performance of particulate semiconductors under natural sunshine-oxidation of carboxylic acids. Sol Energy Mater Sol Cells 92:588–593

Kavitha AL, Prabu HG, Babu SA, Suja SK (2013) Magnetite nanoparticleschitosan composite containing carbon paste electrode for glucose biosensor application. J Nanosci Nanotechnol 13(1):98–104

Kayalvizhi K, Asmathunisha N, Vasuki S, Kathiresan K (2014) Purification of silver and gold nanoparticles from two species of brown seaweeds (*Padina tetrastromatica* and *Turbinariaornata*). J Med Plants Stud 2(4):32–37

Khan K, Hanif U, Liaqat I, Shaheen S, Awan U, Ishtiaq S, Pereira L, Bahadur S, Khan M (2022) Application of green silver nanoparticles synthesized from the red seaweeds Halymenia porphyriformis and Solieria robusta against oral pathogenic bacteria by using microscopic technique. Front Biosci (Elite Ed) 14(2):13–22

Khanna P, Kaur A, Goyal D (2019) Algae-based metallic nanoparticles: Synthesis, characterization and applications. J Microbiol Methods 163:105656

Khodashenas B, Ghorbani HR (2014) Synthesis of copper nanoparticles: an overview of the various methods. Korean J Chem Eng 31:1105–1109

Kirthi AV, Rahuman AA, Rajakumar G, Marimuthu S, Santhoshkumar T, Jayaseelan C, Velayutham K (2011) Acaricidal, pediculocidal and larvicidal activity of synthesized ZnO nanoparticles using wet chemical route against blood feeding parasites. Parasitol Res 109:461–472

Kumar P, Senthamil Selvi S, Govindaraju M (2013) Seaweed-mediated biosynthesis of silver nanoparticles using *Gracilariacorticata* for its antifungal activity against *Candida spp*. Appl Nanosci 3:495–500

Kumar P, Senthamil Selvi S, Lakshmi Prabha A, Prem Kumar K, Ganeshkumar RS, Govindaraju M (2012) Synthesis of silver nanoparticles from *Sargassum tenerrimum* and screening phytochemicals for its antibacterial activity. Nano Biomed Eng 4(1):12–16

Kumaresan M, Vijai Anand K, Govindaraju K, Tamilselvan S, Ganesh Kumar V (2018) Seaweed *Sargassumwightii* mediated preparation of zirconia (ZrO2) nanoparticles and their antibacterial activity against gram positive and gram negative bacteria. Microb Pathog 124:311–315

Leela K, Anchana Devi C (2017) A study on the applications of silver nanoparticles synthesized usingaqueous extract and purified secondary metabolites of seaweed *Hypneacervicornis*. IOSR J Pharm 7(10):46–61

Li XL, Li H, Liu GQ, Deng ZW, Wu SL, Li PH, Xu ZS, Xu HB, Chu PK (2012) Magnetite-loaded fluorine-containing polymeric micelles for magnetic resonance imaging and drug delivery. Biomaterials 33(10):3013–3025

Liu M, Liu L, Gao W, Su M, Ge Y, Shi L, Zhang H, Dong B, Li CY (2015) Nanoparticle mediated micromotor motion. Nanoscale 7:4949–4955

Mahdavi M, Namvar F, Ahmad MB, Mohamad R (2013) Green biosynthesis and characterization of Magnetic Iron Oxide (Fe3O4) nanoparticles using seaweed (Sargassum muticum) aqueous extract. Molecules 18:5954–5964

Mahdavian AR, Mirrahimi MAS (2010) Efficient separation of heavy metal cations by anchoring-polyacrylic acid on superparamagnetic magnetite nanoparticles through surface modification. Chem Eng J 159(1):264–271

Malaikozhundan B, Vaseeharan B, Vijayakumar S, Pandiselvi K, Kalanjiam MR, Murugan K, Benelli G (2017) Biological therapeutics of *Pongamiapinnata*coated zinc oxide nanoparticles against clinically important pathogenic bacteria, fungi and MCF-7 breast cancer cells. Microb Pathog 104:268–277

Manam VK, Subbaiah M (2020) Biosynthesis and characterization of silver nanoparticles from marine macroscopic red seaweed *Halymenia porphyroides Boergesen (Crypton)* and its antifungal efficacy against dermatophytic and non-dermatophytic fungi. Asian J Pharm Clin Res 13(8):174–181

Marquis BJ, Love SA, Braun KL, Haynes CL (2009) Analytical methods to assess nanoparticles toxicity. Analyst 134:425–439

Mieszawska AJ, Mulder WJM, Fayad ZA, Cormode DP (2013) Multifunctional gold nanoparticles for diagnosis and therapy of disease. Mol Pharm 10:831–847

Nagarajan S, Kuppusamy KA (2013) Extracellular synthesis of zinc oxide nanoparticle using seaweeds of gulf of Mannar, India. J Nanobiotechnology 11(39):1–11

Noelia G, Lara D, Mariano L, Maria G, Antonella C, Franca B, Carmen M, Rosana S (2022) Immunomodulatory and Antitumoral activity of gold nanoparticles synthesized by red algae aqueous extracts. Mar Drugs 20:182

Noh H, Seo D, Kim H, Lee J (2003) Synthesis and crystallization of anisotropic shaped ZrO2 nanocrystalline powders by hydrothermal process. Mater Lett 57:2425–2431

Norziah MH, Ching YC (2002) Nutritional composition of edible seaweeds *Gracilaria changgi*. Food Chem 68:69–76

Panchanathan M, Se-Kwon K (2015) Introduction to marine bioenergy. Marine Bioenergy. CRC Press, Boca Raton, FL, USA, pp 3–12

Parashar V, Prashar R, Sharma B, Pandey AC (2009) *Parthenium* leaf extract mediated synthesis of silver nanoparticles: a novel approach towards weed utilization. Digest J Nanomater Biostruct 4:45–50

Patakfalvi R, Dekany I (2010) Preparation of silver nanoparticles in liquid crystalline systems. Colloid Polym Sci 280(5):461–470

Poinern E (2014) A laboratory course in nanoscience and nanotechnology. In: Taylor and Francis. https://doi.org/10.1201/b17753

Prasad TN, Rao VSK, Ravi N (2013) Phyconanotechnology: synthesis of silver nanoparticles using brown marine algae *Cystophoramoniliformis* and their characterisation. J Appl Phycol 25(1):177–182

Priyadharshini RI, Prasannaraj G, Geetha N, Venkatachalam P (2014) Microwave mediated extracellular synthesis of metallic silver and zinc oxide nanoparticles using macro-algae (*Gracilariaedulis*) extracts and its anticancer activity against human PC3 cell lines. Appl Biochem Biotechnol 174(8):2777–2790

Qiao RR, Yang CH, Gao MY (2009) Superparamagnetic iron oxide nanoparticles: from preparations to in vivo MRI applications. J Mater Chem 19(35):6274–6293

Quester K, Avalos-Borja M, Vilchis-Nestor AR, Camacho-Lopez MA, Castro-Longoria E (2013) SERS properties of different sized and shaped gold nanoparticles biosynthesized under different environmental conditions by Neurospora crassa extract. PLoS ONE 8(10):e77486. https://doi.org/10.1371/journal.pone.0077486

Rajaboopathi S, Thambidurai S (2017) Green synthesis of seaweed surfactant based CdO-ZnO nanoparticles for better thermal and photocatalytic activity. Curr Appl Phys 17(12):1622–1638. https://doi.org/10.1016/j.cap.2017.09.006

Rajathi AAF, Parthiban C, Ganesh Kumar V, Anantharaman P (2012) Biosynthesis of antibacterial gold nanoparticles using brown alga, *Stoechospermummarginatum* (kützing). Spectrochim Acta A Mol Biomol Spectrosc 99:166–173

Rajeshkumar S, Kannan C, Annadurai G (2012a) Green synthesis of silver nanoparticles using marine brown algae *Turbinaria Conoides* and its antibacterial activity. Int J Pharm Bio Sci 3(4):502–510

Rajeshkumar S, Kannan C, Annadurai G (2012b) Synthesis and characterization of antimicrobial silver nanoparticles using marine brown seaweed *Padina tetrastromatica*. Drug Invent Today 4(10):511–513

Rajeshkumar S, Malarkodi C, Paulkumar K, Vanaja M, Gnanajobitha G, Annadurai G (2014) Algae mediated green fabrication of silver nanoparticles and examination of its antifungal activity against clinical pathogens. Int J Met 2014:692643. https://doi.org/10.1155/2014/692643

Rajeshkumar S, Sherif MH, Chelladurai M, Ponnanikajamideen M, Arasu MV, Al-Dhabi NA, Roopan SM (2020) Cytotoxicity behaviour of response surface model optimized gold nanoparticles by utilizing fucoidan extracted from Padina tetrastromatica. J Mol Struct 1228(8):129440

Ramakrishna M, Babu DR, Gengan RM, Chandra S, Nageswara Rao G (2016) Green synthesis of gold nanoparticles using marine algae and evaluation of their catalytic activity. J Nanostructure Chem 6:1–13

Ramamoorthy R, Vanitha S, Krishnadev P (2019) Green synthesis of silver nanoparticles using red seaweed *Portieria hornemannii* (Lyngbye) P.C. silva and its antifungal activity against silkworm (*Bombyx mori* L.) Muscardine pathogens. J Pharmacogn Phytochem 8(3):3394–3398

Ramaswamy SVP, Narendhran S, Sivaraj R (2016) Potentiating effect of ecofriendly synthesis of copper oxide nanoparticles using brown alga: antimicrobial and anticancer activities. Bull Mater Sci 39:361–364

Ramkumar VS, Pugazhendhi A, Gopalakrishnan K, Sivagurunathan P, Saratale GD, Bao Dung TN, Kannapiran E (2017a) Biofabrication and characterization of silver nanoparticles using aqueous extract of seaweed *Enteromorpha compressa* and its biomedical properties. Biotechnol Rep 14:1–7

Ramkumar VS, Pugazhendhi A, Prakash S, Ahila NK, Vinoj G, Selvam S, Kumar G, Kannapiran E, Babu Rajendran R (2017b) Synthesis of platinum nanoparticles using seaweed *Padinagymnospora* and their catalytic activity as PVP/PtNPsnanocomposite towards biological applications. Biomed Pharmacother 92:479–490

Rao BV, Boominathan M (2015) Antibacterial activity of silver nanoparticles of seaweeds. Am J Adv Drug Deliv 3(06):296–307

Romero-González ME, Williams CJ, Gardiner PHE, Gurman SJ, Habesh S (2003) Spectroscopic studies of the Biosorption of Gold (III) by dealginated seaweed waste. Environ Sci Technol 37(18):4163–4169

Roy S, Anantharaman P (2018) Biosynthesis of silver nanoparticle by *Amphiroaanceps (Lamarck)* decaisne and its biomedical and ecological implications. J Nanomed Nanotechnol 9:2. https://doi.org/10.4172/2157-7439.1000492

Sahayaraj K, Rajesh S, Rathi JM (2012) Silver nanoparticles biosynthesis using marine alga *Padina pavonica* (L.) and its microbicidal activity. Dig J Nanomater Biostruct 7(4):1557–1567

Salem M, Xia Y, Allan A, Rohani S, Gillies ER (2015) Curcumin-loaded, folic acid-functionalized magnetite particles for targeted drug delivery. RSC Adv 5(47):37521–37366

Saraniya Devi J, Valentin Bhimba B (2014) Antibacterial and antifungal activity of silver nanoparticles synthesized using *Hypnea muciformis*. Biosci Biotechnol Res Asia 11(1):235–238

Saraniya Devi J, Valentin Bhimba B, Magesh Peter D (2013) Production of biogenic silver nanoparticles using *Sargassum longifolium* and its applications. Indian J Geomarine Sci 42(1):125–130

Shah M, Fawcett D, Sharma S, Tripathy SK, Poinern GEJ (2015) Green synthesis of metallic nanoparticles via biological entities. Materials (Basel) 8(11):7278–7308

Sharad NS, Swapnil RB, Ganesh RM, Samadhan SK, Dinesh KM, Shashikant BB, Anuj KR, Nenad B, Orlando MNDT, Radek Z, Rajender SV, Manoj BG (2014) Iron oxide-supported copper oxide nanoparticles (nanocat-Fe-CuO): magnetically recyclable catalysts for the synthesis of pyrazole derivatives, 4-methoxyaniline, and ullmann-type condensation reactions. ACS Sustain Chem Eng 2(7):1699–1706

Shargh AY, Sayadi MH, Heidari A (2018) Green biosynthesis of palladium oxide nanoparticles using *Dictyota indica* seaweed and its application for adsorption. J Water Environ Nanotechnol 3(4):337–347

Sharma D, Rajput J, Kaith BS, Kaur M, Sharma S (2010) Synthesis of ZnO nanoparticles and study of their antibacterial and antifungal properties. Thin Solid Films 519:1224–1229

Shiny PJ, Amitava M, Chandrasekaran N (2013) Marine algae mediated synthesis of the silver nanoparticles and its antibacterial efficiency. Int J Pharm Pharm Sci 5(2):239–241

Siddiqi KS, Husen A (2016) Green synthesis, characterization and uses of Palladium/Platinum nanoparticles. Nanoscale Res Lett 11(1):482. https://doi.org/10.1186/s11671-016-1695-z

Singaravelu G, Arockiamary JS, Ganesh Kumar V, Govindaraju K (2007) A novel extracellular synthesis of monodisperse gold nanoparticles using marine alga, *Sargassum wightii Greville*. Colloids Surf B Biointerfaces 57:97–101

Sivakumar SR, Krishnamoorthi R, Malathi K (2018) Silver nanoparticle synthesized from *Chaetomorpha aerea* Kutz, enhancement for seed germination in *Solanum lycopersicum L.* crop. Pharma Innov J 7(11):184–191

Somiya S, Yamamoto N, Yanagina H (1988) Science and technology of zirconia (III), vol 24A, 24B. American Ceramic Society, Westerville

Sonbol H, Ameen F, Yahya SA, Almansob A, Alwakeel S (2021) *Padina boryana* mediated green synthesis of crystalline palladium nanoparticles as potential nanodrug against multidrug resistant bacteria and cancer cells. Sci Rep 11:5444. https://doi.org/10.1038/s41598-021-84794-6

Southon PD, Baotlett JR, Woolfrey JL, Ben-Nissan B (2002) Formation and characterization of an aqueous zirconium hydroxide colloid. Chem Mater 14:4313–4319

Sri Vishnu Priya R, Narendhran S, Rajeshwari S (2016) Potentiating effect of ecofriendly synthesis of copper oxide nanoparticlesusing brown alga: antimicrobial and anticancer activities. Bull Mater Sci 39(2):361–364

Srikar SK, Giri DD, Pal DB, Mishra PK, Upadhyay SN (2016) Green synthesis of silver nanoparticles: A review. Green Sustain Chem 6(1):34–56

Stalin Dhas T, Ganesh Kumar V, Stanley Abraham L, Karthick V, Govindaraju K (2012) *Sargassum myriocystum* mediated biosynthesis of gold nanoparticles. Spectrochim Acta A Mol Biomol Spectrosc 99:97–101

Stalin Dhas T, Sowmiya P, Ganesh Kumara V, Ravia M, Suthindhiranb K, Francis Borgioc J, Narendrakumard V, Ramesh Kumard V, Karthicka V, Vineeth Kumar CM (2020) Antimicrobial effect of *Sargassum plagiophyllum* mediated gold nanoparticles on *Escherichia coli and Salmonella typhi*. Biocatal Agric Biotechnol 26:101627. https://doi.org/10.1016/j.bcab.2020.101627

Straumal BB, Mazilkin AA, Protasova SG, Myatiev AA, Straumal PB, Goering E (2011) Amorphous grain boundary layers in the ferromagnetic nano grained ZnO films. Thin Solid Films 520:1192–1194

Subhashini G, Ruban P, Daniel T (2018) Biosynthesis and characterization of magnetic (Fe3o4) iron oxide nanoparticles from a red seaweed *Gracilaria edulis* and its antimicrobial activity. Int J Adv Sci Res 3(10):184–189

Subramanian P, Periyannan R, Gandhi V, Ganesan R, Ramar M, Prabhu NM (2016) A green route to synthesis silver nanoparticles using *Sargassum polycystum* and its antioxidant and cytotoxic effects: an in vitro analysis. Mater Lett 189:196–200

Sunitha S, Nageswara Rao A, Stanley Abraham L, Dhayalan E, Thirugnanasambandam R, Ganesh Kumar V (2015) Enhanced bactericidal effect of silver nanoparticles synthesized using marine brown macro algae. J Chem Pharm Res 27(3):191–195

Suriya J, Bharathi Raja S, Sekar V, Rajasekaran R (2012) Biosynthesis of silver nanoparticles and its antibacterial activity using seaweed *Urospora* sp. Afr J Biotechnol 11(58):12192–12198

Taleb A, Petit C, Pileni MP (1998) Optical properties of self assembled 2D and 3D superlattices of silver nanoparticles. J Phys Chem B 102(12):2214–2220

Terris BD, Thomson T (2005) Nanofabricated and self-assembled magnetic structures as data storage media. J Phys D Appl Phys 38(12):R199–R222

Thanigaivel S, Vickram S, Saranya V, Ali H, Alarifi S, Modigunta J, Anbarasu K, Lakshmipathy R, Rohini K (2022) Seaweed polysaccharide mediated synthesis of silver nanoparticles and its enhanced disease resistance in Oreochromis mossambicus. J King Saud Univ Sci 34:101771

Thangaraju N, Venkatalakshmi RP, Chinnasamy A, Kannaiyan P (2012) Synthesis of silver nanoparticles and the antibacterial and anticancer activities of the crude extract of*sargassum-polycystum C. Agardh*. Nano Biomed Eng 4(2):89–94

Thaya R, Malaikozhundan B, Vijayakumar S, Sivakamavalli J, Jeyasekar R, Shanthi S, Vaseeharan B, Palaniappan R (2016) Chitosan coated Ag/ZnOnanocomposite and their antibiofilm, antifungal and cytotoxic effects on marine macrophages. Microb Pathog 100:124–132. https://doi.org/10.1016/j.micpath.2016.09.010

Thiruchelvi R, Jayashree P, Mirunaalini K (2020) Synthesis of silver nanoparticle using marine red seaweed *Gelidiella acerosa*. A complete study on its biological activity and its characterization. Mater Today Proc 37(2):1693–1698

Thirumurugan A, Jiflin GJ, Rajagomathi G, Neethu Anns T, Ramachandran S, Jaiganesh R (2010) Synthesis of gold nanoparticles of *Azadirachta indica* leaf extract. Int J Biol Technol 1(1):75–77

Tian N, Zhou Z-Y, Sun S-G, Ding Y, Wang ZL (2007) Synthesis of tetrahexahedral platinum nanocrystalswith high-index facets and high electro-oxidation activity. Science 316:732–735

Tiwari D, Behari J, Sen P (2008) Application of nanoparticles in waste water treatment. World Appl Sci J 3:417–433

Tripathi RM, Saxena A, Gupta N, Kapoor H, Singh RP (2010) High antibacterial activity of silver nanoballs against *E. coli* MTCC 1302, *S. typhimurium* MTCC 1254, *B. subtilis* MTCC 1133 and *P. aeruginosa* MTCC 2295. Dig J Nanomater Biostruct 5(2):323–330

Udhaya CI, Paul JJ, Lawrence RA (2015) Seaweed mediated biosynthesis of silver nanoparticles using *Gracilariafergusonii J. Ag*. collected from Hare Island, Thoothukudi in the South East Coast of Tamil Nadu, India. IJPB 17:24

Ullah AKMA, Kibria AKMF, Akter M, Khan MNI, Maksud MA, Jahan RA, Firoz SH (2017) Synthesis of Mn3O4 nanoparticles via a facile gel formation route and study of their phase and structural transformation with distinct surface morphology upon heat treatment. J Saudi Chem Soc 21:830–836

Valarmathi N, Ameen F, Almansob A, Kumar P, Arunprakash S, Govarthanan M (2019) Utilization of marine seaweed *Spyridia filamentosa* for silver nanoparticles synthesis and its clinical applications. Mater Lett 263:127244. https://doi.org/10.1016/j.matlet.2019.127244

Venkatesan J, Lowe B, Anil S, Manivasagan P, Kheraif AAA, Kang K, Kim S (2015) Seaweed polysaccharides and their potential biomedical applications. Starch, Biosynthesis Nutrition Biochemical 67(5–6):381–390

Vennila S, Jesurani S (2017) Eco-friendly green synthesis and characterization of stable ZnO nanoparticles using small gooseberry fruits extracts. Int J ChemTech Res 10:271–275

Vieira AP, Stein EM, Andreguetti DX, Colepicolo P, Ferreira AMDC (2015) Preparation of silver nanoparticles using aqueous extracts of the red algae Laurencia aldingensis and Laurenciella sp. and their cytotoxic activities. J Appl Phycol 28:2615–2622. https://doi.org/10.1007/s10811-015-0757-4

Vijayakumar S, Malaikozhundan B, Shanthi S, Vaseeharan B, Thajuddin N (2017) Control of biofilm forming clinically important bacteria by green synthesized ZnO nanoparticles and its ecotoxicity on *Ceriodaphniacornuta*. Microb Pathog 107:88–97

Vijayakumar S, Vaseeharan B, Malaikozhundan B, Shobiya M (2016) *Laurus nobilis* leaf extract mediated green synthesis of ZnO nanoparticles: characterization and biomedical applications. Biomed Pharmacother 84:1213–1222

Vijayaraghavan K, Mahadevan A, Sathishkumar M, Pavagadhi S, Balasubramanian R (2011) Biosynthesis of Au (0) from Au (III) via bisorption and bioreduction using brown marine algae *Turbinaria conoides*. Chem Eng J 167:223–227

Vinosha M, Palanisamy S, Muthukrishnan R, Selvam S, Kannapiran E, You S, Prabhu NM (2019) Biogenic synthesis of gold nanoparticles from *Halymenia dilatata* for pharmaceutical applications: antioxidant, anti-cancer and antibacterial activities. Process Biochem 85:212–219. https://doi.org/10.1016/j.procbio.2019.07.013

Wani KD, Kadu BS, Mansara P, Gupta P, Deore AV, Chikate RC, Poddar P, Dhole SD, Kaul-Ghanekar R (2014) Synthesis, characterization and in vitro study of biocompatible cinnamaldehyde functionalized magnetite nanoparticles (CPGF Nps) for hyperthermia and drug delivery applications in breast cancer. PLoS One 9(9):e107315. https://doi.org/10.1371/journal.pone.0107315

Xiong HM (2013) ZnO nanoparticles applied to bioimaging and drug delivery. Adv Mater 25:5329–5335

Xu H, Qin DH, Yang Z, Li HL (2003) Fabrication and characterization of highly ordered zirconia nanowire arrays by sol-gel template method. Mater Chem Phys 80:524–528

Yamaguchi T (1994) Application of ZrO2 as a catalyst and a catalyst support. Catal Today 20:199–217

Yew YP, Shameli K, Miyake M, Kuwano N, Ahmad Khairudin NBB, Mohamad SEB, Lee KX (2016) Green synthesis of magnetite (Fe3O4) nanoparticles using seaweed (*Kappaphycus alvarezii*) extract. Nanoscale Res Lett 11(1):276. https://doi.org/10.1186/s11671-016-1498-2

Yoon SH, Kim DJ (2006) Fabrication and characterization of ZnO films for biological sensor application of FPW device. In: 15th IEEE international symposium on the applications of ferroelectrics, Sunset Beach. IEEE, pp 322–325

Yousefzadi M, Kokabi M, Ebrahimi SN, Zarei M (2017) Green synthesis of zinc oxide nanoparticles using seaweed aqueous extract and evaluation of antibacterial and ecotoxicological activity. J Persian Gulf 8(27):61–72

Zhang J, Xiao X, Nan J (2010) Hydrothermal-hydrolysis synthesis and photocatalytic properties of nano-TiO2 with an adjustable crystalline structure. J Hazard Mater 176:617–622

Zhang L, Dong WF, Sun HB (2013) Multifunctional superparamagnetic iron oxide nanoparticles: design, synthesis and biomedical photonic applications. Nanoscale 5(17):7664–7684

Zhao H, Saatchi K, Häfeli UO (2009) Preparation of biodegradable magnetic microspheres with poly(lactic acid)-coated magnetite. J Magn Magn Mater 321(10):1356–1363

Zong Y, Cao Y, Jia D, Bao S, Lu Y (2010) Facile synthesis of Ag/ZnOnanorodsusingAg/C cables as templates and their gas-sensing properties. Mater Lett 64:243–245

Chapter 8
Pharmacological Activities of Seaweeds

Johnson Marimuthu alias Antonysamy, Shivananthini Balasundaram, Vidyarani George, and Silvia Juliet Iruthayamani

1 Introduction

Seaweeds or marine algae are primitive plants without any real root, stem and leaves hence they belong to the division of Thallophyta in the plant kingdom. They are macroscopic, attached or freely floating plants and form one of the essential marine living, renewable resources. Seaweeds have an amazing variety of beautiful shapes, colors and sizes and found in all of the world's oceans. Seaweeds are classified into red algae (Rhodophyceae), green algae (Chlorophyceae) and Brown algae (Phaeophyceae) based on their pigment contents, anatomical and morphological characters. Globally algal diversity harbors 72,500 taxa of which 44,000 have previously described. Marine algal community possess a huge source of compounds e.g., Polysaccharides, polyunsaturated fatty acids (PUFA), plant growth harmones, pigments, proteins, lipids, peptides, minerals and vitamins (Husni et al. 2016) with potent biological activities. In addition to that, they have fast and efficient growth particularly in comparison to terrestrial plants and in other cases they may be readily available as byproducts of aquaculture industry (Kindleysides et al. 2012). In ancient times (300 BC), seaweeds are used to treat goiter and other glandular problems and also it was used by romans in the treatment of wounds, burns and rashes. They were also reportedly used in the treatment of dysmenorrhea (Hocman 1989). Recent findings reported the presence of anti-hypertensive, anti-hyperlipidemic, anti-coagulant, apoptotic activities (Lee and Han 2012), antioxidant, anti-inflammatory and anticancer (Khalid et al. 2018) activities in seaweeds. Seaweeds controlled diabetes by regulating the glucose-instigated oxidative stress and by suppressing starch hydrolyzing enzymes (Unnikrishnan et al. 2015). Seaweeds have been used in the food manufacturing industry because of its nutritional benefits and

J. M. alias Antonysamy (✉) · S. Balasundaram · V. George · S. J. Iruthayamani
Centre for Plant Biotechnology, Department of Botany, St. Xavier's College (Autonomous), Palayamkottai, Tamil Nadu, India

F. Ozogul et al. (eds.), *Seaweeds and Seaweed-Derived Compounds*,
https://doi.org/10.1007/978-3-031-65529-6_8

also used as medicine for various purposes. Currently, in a antiviral effectiveness test an edible seaweeds extract significantly surpassed remdesivir, the present standard antiviral agent used to combat COVID-19 disease and comparably Heparin also inhibit the SARS-CoV-2 infection in mammalian cells (Seema 2012; Kwon et al. 2020). Seaweeds are the most important source of non-animal polysaccharides and are the only resource for industrially important polymers. Agar and carrageenan from red seaweeds, alginate, fucoidan and laminarin from brown seaweeds are some important industrial application polymers. These polysaccharides widely used in food industries as gelling agents, thickening and stable excipients for control release of drugs (Seema 2012). Besides these, seaweeds polysaccharides possess excellent biological activities hence they are being widely used in the pharmaceutical and biomedical sectors (Ngo and Se-Kwon 2013; Raja et al. 2016; Sanjeewa et al. 2018). Therefore this review attempts to explore the various therapeutic activities of seaweeds from the published literatures of different pharmacological activities of seaweeds.

2 Antibacterial Activity

The incidence of microbial infections has increased fiercely during the last few decades. The emergence of new multi-drug resistance bacterial strains and high mortality rate becomes one of the menacing health problems worldwide (Visvesvara et al. 2007; Rose et al. 2001). Different drug molecules are being used to fight against those microorganisms. Antibiotics, metal ions, and various quaternary ammonium compounds are antimicrobial compounds being used worldwide however these compounds are claimed to be associated with antibiotic resistance, complex chemical synthesis, environmental pollution, and high cost (Levy and Marshall 2002; Buffet-Bataillon et al. 2012). Different seaweeds have been used against predator defense in preceding times and could be a promising antibacterial agent (Nygaard et al. 1992; Vairappan et al. 2004).

3 Antibacterial Activity of Green Seaweeds

The available literature indicated the antibacterial potential of green seaweeds viz., *Enteromorpha intestinalis* (Imran et al. 2021), *Caulerpa racemosa* (Belkacemi et al. 2020), *Ulva fasciata* (Fayzi et al. 2020), *Ulva lactuca* (Vikneshan et al. 2020), *Ulva fasciata, Ulva lactuca, Chladophora vagabunda, Caulepa taxifolia, Chaetomorpha antennina and Chaetomorpha linum* (Agbaje-Daniels et al. 2020), *Caulerpa peltata, Caulerpa scalpelliformis, Ulva lactuca* (Bharath et al. 2020), *Caulerpa racemosa* (Louiza Belkacemi et al. 2020), *Enteromorpha intestinalis* (Hasan et al. 2019), *Ulva lactuca* (O'Keeffe et al. 2019), *Codium elongatum, Enteromorpha ramulosa, Ulva rigida, Ulva sp.* (Bouhraoua et al. 2018), *Ulva conglobate, Ulva fasciata, Chaetomorpha antennina, Codium cylindricum* (Peihang et al. 2018), *Ulva lactuca* (Mishra 2018), *Ulva pertus, Ulva prolifera* (Li et al. 2018), *Ulva lactuca* (Ghalem and Zouaoui 2018), *Acrosiphonia sonderi, Ulva lactuca* (Chingizova et al. 2017), *Caulerpa serrulata, Ulva fasciata* (Kumari et al. 2017), *Cladophora socialis* (Moubayed et al. 2017), *Ulva lactuca* (Pragnesh et al. 2017), *Halimeda gracilis, Caulerpa serrulata, Ulva lactuca, Ulva fasciata, Caulerpa scalpelliformis, Caulerpa taxifolia, Chaetomorpha crassa, Enteromorpha flexuosa* (Sasikala and Ramani 2017), Ulva lactuca (Kavitha et al. 2017; El-Shouny et al. 2017), *Halimeda macroloba, Halimeda tuna, Enteromorpha sp.* and *Acetabularia acetabulum* (Sivaramakrishnan et al. 2017), *Ulva lactuca* (Deveau et al. 2016), *Valoniopsis pachynema* (Dhinakaran et al. 2016), *Dictyosphaeria cavernosa, Acetabularia calyculus* (Karthick et al. 2015), *Ulva lactuca, Codium fragile* (El Wahidi et al. 2015), *Ulva lactuca, Ulva fasciata, Ulva intestinalis* (Abdel-Khaliq et al. 2014), *Uva lactuca, Enteromorpha intestinals* (Alshalmani et al. 2014), *Enteromorpha compressa, Enteromorpha intestinalis, Ulva fasciata, Ulva lactuca, Chaetomorpha antennina* (Saranya et al. 2014), *Ulva lactuca, Enteromorpha compressa* (Elnabris et al. 2013), *Ulva faciata* (Radhika et al. 2013a), *Ulva lactuca* (Saidani et al. 2012), *Ulva lactuca, Codium fragile* (Águila-Ramírez et al. 2012), *Halimeda opuntia* (Selim 2012), *Codium adhaerens* (Seenivasan et al. 2012), *Enteromorpha compressa* and *Chaetomorpha antennina* (Latha and Latha 2011), *Halimeda macroloba* (Boonchum et al. 2011), *Ulva fasciata* (Priyadharshini et al. 2011), *Enteromorpha spirulina* (Cox et al. 2010), *Cladophora glomerata, Ulva lactuca, Ulva reticulata* (Mansuya et al. 2010), *Ulva lactuca* (Vallinayagam et al. 2009), *Calorpha peltada* (Kolanjinathan et al. 2009), *Bryopsis muscosa, Chaetomorpha linum, Cladophora rupestris, Codium bursa, Codium coralloides, Codium vermilara, Flabellia petiolata, Halimeda tuna, Palmophyllum crissum, Ulva rigida, Valonia macrophysa* (Salvador et al. 2007), *Ulva faciata* (Selvin and Lipton 2004). The reported antibacterial activities of the green seaweeds are due the presence of secondary metabolites viz., phenol, tannin, flavonoid and steroids etc.

4 Antibacterial Activity of Brown Seaweeds

The brown seaweeds also equally possessed the antibacterial properties and are reported by phycologists viz., *Undaria pinnatifida, Laminaria japonica, Sargassum fusiforme, Eisenia bicyclis* (Čmiková et al. 2022), *Padina tetrastromatica, Sargassum ilicifolium* (AftabUddin et al. 2020), *Padina australis* (Klimjit et al. 2021), *Bifurcaria bifurcata* (Fayzi et al. 2020), *Sargassum wightii* (Anitha et al. 2020), *Turbinaria ornata, Sargassum polycystum, Padina tetrastromatica* (Bharath et al. 2020), *Dictyota barteyresiana* (Durairaj and Andiyappan 2020), *Fucus serratus, Ascophyllum nodosum* (O'Keeffe et al. 2019), *Sargassum wightii* (Shibu 2019), *Sargassum cinereum, Turbinaria turbinate, Cystoseira myrica, Hormophysa cuneiformis* (Osman et al. 2019), *Bifurcaria bifurcata, Fucus spiralis, Laminaria digitata, Cystoseira humilis, Sargassum muticum, Sargassum vulgaris* (Bouhraoua et al. 2018), *Endarachne binghamiae, Grateloupia turuturu, Padina arborescens, Colpomenia sinuosa, Sargassum hemiphyllum, Sargassum vachellianum, Pachydictyon coriaceum* (Peihang et al. 2018), *Colpomenia sinuosa, Padina pavonia, Cystoseira barbata* and *Sargassum vulgare* (Mohy El-Din and Mohyeldin 2018), *Sargassum wightii* (Mishra 2018), *Padina, Turbinaria sps.* (Sayegh 2018), *Ishige okamurae, Sargassum fusiforme* (Li et al. 2018), *Padina pavonica, Sargassum vulgare, Dictyota dichotoma* (Saleh and Al-mariri 2018), *Dictyota dichotoma* (Ghalem and Zouaoui 2018), *Eualaria fistulosa, Undaria pinnatifida, Costaria costata, Saccharina japonica, Saccharina cichorioides, Cystoseira crassipes, Sargassum pallidum, Coccophora langsdorfii, Fucus evanescens, Desmarestia viridis, Scytosiphon lomentaria* (Chingizova et al. 2017), *Laminaria digitata and Undaria pinnatifida* (Corato et al. 2017), *Hydroclathrus clathratus* (Vimala and Poonghuzhali 2017), *Sargassum latifolium B* and *Sargassum platycarpum* (Moubayed et al. 2017), *Sargassum tenerrimum, Sargassum cinctum, Sargassum myriocystum, Padina boergesenii* (Pragnesh et al. 2017), *Sargassum swartzii, Sargassum wightii, Stoechospermum marginatum, Turbinaria ornate* (Sasikala and Ramani 2017), *Petalonia fascia* (El-Shouny et al. 2017), *Sargassum wightii, S.oligocystum, S.vulgare, Turbinaria ornata, Padina tetrastromatica, Stoechospermum marginatum* (Rani et al. 2016), *Sargassum swartzii* (Dhinakaran et al. 2016), *Sargassum vulgare, Sargassum fusiforme and Padina pavonia* (El Shafay et al. 2015), *Caulerpa corynephora, Caulerpa scalpelliformis, Chaetomorpha antennia, Enteromorpha compressa, Halimeda macroloba, Ulva fasciata and Ulva lactuca* (Babu et al. 2015), *Galaxura marginata* (Karthick et al. 2015), *Spathoglossum asperum, Turbinaria conoides* (Valentina et al. 2015), *Bifurcaria bifurcata, Fucus vesiculosus, Laminaria ochroleuca, Cystoseira compressa, Cystoseira brachycarpa* (El Wahidi et al. 2015), *Cystoseira compressa* (Alshalmani et al. 2014) *Turbinaria conoides, Sargassum sp.* (Anjum et al. 2014), *Sargassum wightti* (Venugopal et al. 2014; Radhika et al. 2013b), *Padina gymnospora* (Chander et al. 2014; Saranya et al. 2014), *Cystoseira tamaricifolia* and *Padina pavonica* (Saidani et al. 2012), *Ascophyllum nodosum, Sargassum muticum, Pelvetia canaliculata, Fucus spiralis, Laminaria hyperborean, Sargassum filipendula, Sargassum stenophyllum* (Peres et al. 2012), *Dictyota flabellate, Padina concrescens* (Águila-Ramírez et al. 2012), *Cystoseira humilis, C. compressa, Cladostephus spongiosus* (Zbakh et al. 2012), *Sargassum wightii* (Seenivasan et al. 2012; Mansuya et al. 2010), *Turbinaria conoides, Padina gymnospora and Sargassum tenerrimum* (Manivannan et al. 2011), *Turbinaria conoides, Sargassum binderi* (Boonchum et al. 2011), *Ascophyllum nodosum, Laminaria digitata* (Qiao 2010), *Alaria esculenta, Fucus vesiculosus, Fucus sp., Ecklonia maxima* (Andreea et al. 2010), *Laminaria digitata, Laminaria saccharina, Himanthalia elongata* (Cox et al. 2010). *Colpomenia sinuosa, Hydroclathrus clathratus, Dictyota dichotoma, Padina australis, P. minor, Sargassum polycystum and Turbinaria conoides* (Kantachumpoo and Chirapart 2010), *Sargassum ilicifolium, Padina tetrastromatica* (Subbarangaiah et al. 2010), *Sargassum weightii* and *Padina gymnospora* (Vallinayagam et al. 2009), *Hydroclothres sp* (Kolanjinathan et al. 2009), *Sargassum sp.* (Jayanta Kumar et al. 2008), *Colpomenia sinuosa, Cystoseira barbata, Cystoseira brachycarpa,*

Cystoseira compressa, Cystoseira mediterranea, Cystoseira tamariscifolia, Dictyopteris polypodioides, Dictyota dichotoma, Dictyota spiralis, Hapalospongidion macrocarpum, Padina pavonica, Scytosiphon lomentaria, Stypocaulon scoparium, Taonia atomaria, Zanardinia typus (Salvador et al. 2007). The presence of tannin, phenolics, steroids and flavonoids confirmed the antibacterial potential of the brown seaweeds.

5 Antibacterial Activity of Red Seaweeds

The antibacterial potenatil of *Palmaria palmata* (Čmiková et al. 2022), *Hypnea musciformis* (Imran et al. 2021), *Kappaphycus alvarezii* (Bhuyar et al. 2020), *Corallina officinalis, Corallina elongata* (Fayzi et al. 2020), *Gracilaria corticata, Grateloupia lithophila, Acanthophora spicifera* (Bharath et al. 2020), *Hypnea musciformis* (Hasan et al. 2019), *Polysiphonia lanosa* (O'Keeffe et al. 2019), *Laurencia papillosa, Actinotrichia fragilis* (Osman et al. 2019), *Osmundea pinnatifida, Gelidium Sp1, Hypnea musciformis, Plocamium cartilagineum, Gelidium pulchellum, Gracilaria multipartita, Ellisolandia elongata, Coralina officinalis, Bornetia secundiflora, Gelidium Sp2, Gracilaria cervicornis, Halopitys incurvus, Gymnocongrus norvrgicus* (Bouhraoua et al. 2018), *Caloglossa sp., Caula anthus okamurai, Gelidium planiusculum, Gelidium kintaroi, Chondracanthus intermedius, Ahnfeltiopsis hainanensis, Ahnfeltiopsis guangdongensis, Ahnfeltiopsis masudai, Polysiphonia hainanensis, Hypnea chordacea, Hypnea japonica, Hypnea boergesenii, Laurencia okamurai, Laurencia chinensis* (Peihang et al. 2018), *Gracillaria edulis, Gracillaria corticata* (Mishra 2018), *Hypnea sps.* (Sayegh 2018), *Osmundea pinnatifida* (Silva et al. 2018), *Ceramium kondoi, Delesseria serrulata, Neorhodomela larix subsp. aculeata, Ptilota filicina, Chondrus pinnulatus, Mazzaella laminarioides, Tichocarpus crinitus, Palmaria stenogona* (Chingizova et al. 2017), *Porphyra umbilicalis, Eucheuma denticulatum and Gelidium pusillum* (Corato et al. 2017), *Gracilaria edulis* (Kumari et al. 2017; Pragnesh et al. 2017), *Jania rubens, Gracilaria corticata* (Sasikala and Ramani 2017), *Gelidium spinosum* (El-Shouny et al. 2017), *Ceramium rubrum* (El Shafay et al. 2015), *Portieria hornemanni, Corallina sp.* (Karthick et al. 2015), *Gelidium sesquipedale, Gelidium attenuatum, Chondrus crispus* (El Wahidi et al. 2015), *Kappaphycus alvarezii* (Madhavarani and Ramanibai 2014), *Corallina sp* (Alshalmani et al. 2014), *Gracilaria sp.* (Anjum et al. 2014), *Grateloupia lithophila, and Hypnea valentiae* (Saranya et al. 2014), *Kappaphycus alvarezii, Kappaphycus striatum* (Prasad et al. 2013), *Jania rubens* (Elnabris et al. 2013), *Gellidella acerosa, Gracillaria verrucosa* and *Hypnea musciformis* (Krishnapriya et al. 2013), *Gracilaria corticata* (Radhika et al. 2013b), *Rhodomella confervoides* (Saidani et al. 2012), *Gracilaria edulis, Laurencia dendroidea* (Peres et al. 2012), *Laurencia johnstonii, Gymnogongrus martinensis* (Águila-Ramírez et al. 2012), *Gymnogongrus patens, Plocamium coccineum, Asparagopsis armata, Centroceras clavulatum, Gracilaria confervoïdes, G. bursa- pastoris, Hypnea musciformis,*

Alsidium corallinum (Zbakh et al. 2012), *Sarconema filiforme* (Selim 2012), *Acanthophora spicifera* (Seenivasan et al. 2012; Pandian et al. 2011), *Amphiroa sp.* (Boonchum et al. 2011), *Palmaria palmata, Chondrus crispus* (Cox et al. 2010), *Alsidium corallinum, Centroceras clavulatum, Callithamnion granullatum, Ceramium rubrum, Halopitys incurvus, Osmundea pinnatifida, Pterosiphonia complanata, Boergeseniella thuyoides, Gelidium attenatum, Gelidium latifolium, Gelidium pusillum, Gelidium pulchellum, Gelidium sesquipedale, Gelidium spinulosum, Pterocladea capillacea, Chondrocanthus acicularis, Caulacanthus ustulatus, Gracilaria confervoides, Gracilaria multipartite, Gymnogongrus patens, Hypnea musciformis, Plocamium cartilagineum, Plocamium coccineum, Sphaerococcus coronopifolius, Asparagopsis armata, Rhodymenea pseudopalmata* (Bouhlal et al. 2010), *Gracilaria corticata, Kappaphycus alvarezii* (Mansuya et al. 2010), *Gracilaria corticata* (Subbarangaiah et al. 2010), *Gracilaria edulis* (Vallinayagam et al. 2009), *Gracilaria edulis* (Kolanjinathan et al. 2009), *Hypnea musciformis* (Selvin and Lipton 2004) are reported. The secondary metabolites existence validated the antibacterial potential of red seaweeds.

6 Gram Positive Bacteria

Among the various gram positive organism, the following organism *Bacillus subtilis, Bacillus cereus, Bacillus megaterium, Bacillus pumilus, Clavibacter species, Coryne bacterium sps., Corynebacterium diphtheria, Corynebacterium michiganese, Enterococcus sps., Enterococcus faecalis, Listeria monocytogenes, Micrococcus luteus, Mycobacterium, Mycobacterium aurum, Propionibacterium acnes, Staphylococcus aeruginosa, Staphylococcus aureus, Staphylococcus epidermidis, Staphylococcus saprophyticus, Staphylococcus xylosus, Streptococcus anginosus, Streptococcus faecalis Streptococcus mutans, Streptococcus pneumonia, Streptococcus pyogenes, Streptococcus sp., Streptococcusfaecalis, Streptococcuspneumonia* are employed to screen the antibacterial potential of seaweeds.

7 Gram Negative Bacteria

The gram negative bacteria viz., *Acinetobacter calcoaceticus, Aeromonas hydrophila, Aeromonas liquefaciens, Enterobacter aerogenes, Erwinia amylovora, Erwinia cartovora, Eschericia coli, Haemophilus influenza, Klebsiella pneumonia, Klebsiella sps., Lactobacillus, Lactobacillus acidophilus, Neisseria meningitides, Pseudomonas aeruginosa, Proteous vulgaris, Proteus mirabilis, Proteus sp., Pseudomonas aureus, Pseudomonas fluorescens, Pseudomonas putida, Pseudomonas sp., Ralstonia sps., Salmonella abony, Salmonella paratyphi, Salmonella sp., Salmonella typhii, Seratia marcescens, Serratia sps., Shewanella sps., Shigella bodii, Shigella dysentriae, Shigella flexneri, Shigella sp., V. alginolyticus, V. vulnificus, V. alginolyticus, V. cholera, V. harveyi, V. mimicus, V. parahaemolyticus, V. splendidus, V. vulnificus, Vibrio fischeri, Vibrio fluvialis, Vibrio harveyi, Xanthomonas species, Yersinia enterocolitica* are employed to screen the antibacterial potential of seaweeds.

8 Fungi

Similarly, the seaweeds displayed the antifungal potential against *Aspergillus flavus, Amanita longipes, Alternaria dauci, Absidia corymbifer, Alternaria alternate, Alternaria infectoria, Aspergillus clavatus, Aspergillus flavus, Aspergillus fumigatus, Aspergillus niger, Aspergillus sps., Aspergillus tetreus, Botrytis cinerea, Candida albicans, Candida glabrata, Candida parapsilosis, Candida sps., Candida utili, Colletotrichum lagenarium, Cryptococcus neoformans, Fusarium culmorum, Fusarium oxysporum, Fusarium roseum, Fusarium solani, Fusarium sps., Geotrichum candidum, Microsorum gypseum, Monilinia laxa, Mucor circinelloides, Mucor racemosus, Mucor ramaniannus, Nomuraea rileyi, Penicillin notatum, Penicillium digitatum, Penicillium expansum, Penicillium marneffei, Penicillium parasiticus, Penicillium sp., Rhizoctonia solanii, Rhizopus oryzae, Rhizopus sp., Rhizopus stolonifer, Saccharomyces cerevisiae, Stachybotrys chartarum, Syncephalastrum racemosum, Trichoderma viride, Trichophyton sp.* Numbers of microorganisms used are represented in Fig. 8.1.

In this review, a total of 44 species of Green seaweeds, 99 species of brown seaweeds and 96 species of red seaweeds are compiled (Fig. 8.2). Disc diffusion method is most common method adopted by the authors. In addition various

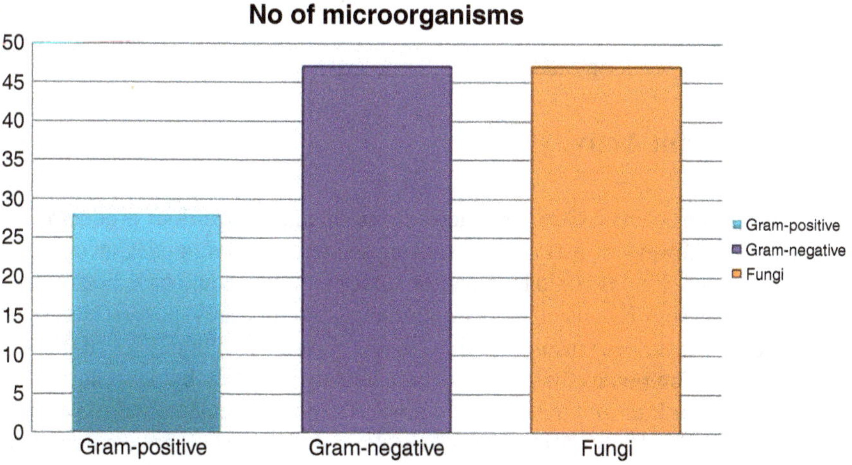

Fig. 8.1 Number of microorganisms used for antibacterial activity

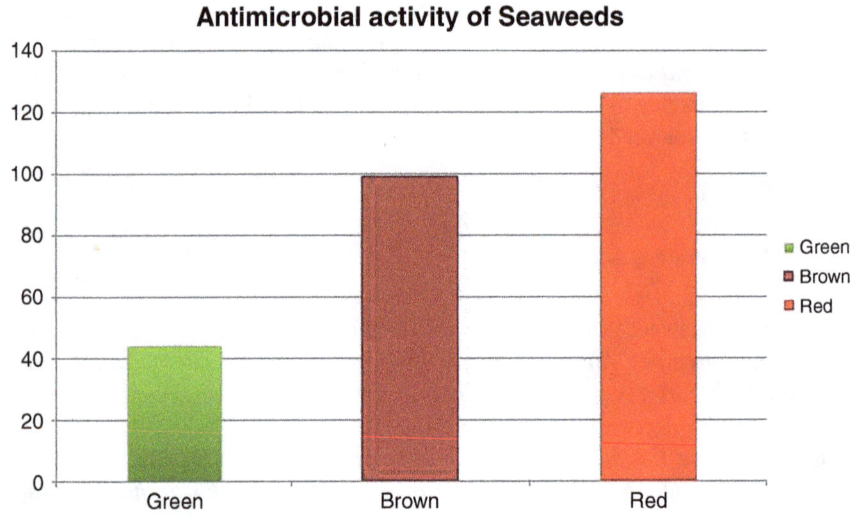

Fig. 8.2 Antimicrobial activity of seaweeds

methods, viz., Cylinder plate method, 96-well method, Minimum Inhibitory concentration, Maximum Inhibitory Concentration, Cup-plate method, Radial growth assay, Direct Bioautography assay are also employed by the authors. Various types of solvents viz., methanol, ethanol, ethyl acetate, chloroform, aqueous, butanol, n-hexane, n-butanol, dichloromethane, benzene, acetone, hexane, dimethyl ether, petroleum ether are employed to extract the metabolites from seaweeds. Among these, methanol is predominantly used for extraction. Among the organism, *Staphylococcus aureus, Eschericia coli* and *Pseudomonas aeruginosa* are the mostly used against all seaweed extracts. Among the seaweeds red seaweeds showed more resistance against all the microorganisms used followed by brown seaweeds and green seaweeds.

9 Antioxidant Activity

Reactive oxygen species (ROS) produced as intermediate hazardous product in the biochemical pathways of metabolism causing oxidative stress impact such as protein denaturation, lipid peroxidation and/or DNA conjugation and have been related with many diseases like cancer, diabetic, cardiovascular diseases, post-ischemic and neural degradation, Parkinson's and Alzheimer disease (Pinteus et al. 2017). Antioxidant is compound that prevents or alleviates diseases by scavenging free radicals to stop or interrupt the oxidation process in cells. Butylatedhydroxytoluene (BHT), Butyatedhydroxyanisole (BHA), and Propyl gallate (PG) are commercially available synthetic antioxidants which are reported to have carcinogenic effect (Agregan et al. 2017). Seaweeds have been used as natural source of antioxidants because of the existence of compounds such as polyphenols, terpenoids, alkaloids, saponins, tannins and steroids (Kumar et al. 2019; Ragunathan et al. 2019). The antioxidant activity of Green, Brown, Red seaweeds are tabulated in Tables 8.1, 8.2, and 8.3.

22 green seaweeds, 42 brown seaweeds and 50 redseaweeds are accounted for antioxidant activity. DPPH, Reducing Power assay, ABTS+, ORAC, Phosphomolybdenum assay, total antioxidant, total phenolic content, DHA power reducing, Nitric oxide radical scavenging, lipid peroxidation, xanthine oxidase inhibitory, AchE inhibitory, Hydrogen peroxide, β-carotene bleaching assay and glutathione S-transferase activity are different assays used for evaluate the antioxidant activity of seaweeds. DPPH assay is mostly adopted by phycologists followed by Reducing power assay, Total antioxidant.

10 Anticancer and Cytotoxicity Activity

Cancer is a deadly disease that is considered to be a major public health issue around the world (El-Beltagi et al. 2019; Hamed et al. 2019). The formation of a tumor mass is a result of uncontrolled cell development spreads into the surrounding tissues (Hussain et al. 2016). Recent research focused on the anticancer potential of natural compounds derived from seaweeds, as well as the signaling pathways involved in anticancer activity Because of the existence secondary metabolites having no hazardous effects, they have seen a lot of progress in the treatment of numerous diseases, including cancer (Gutierrez-Rodriguez et al. 2018). Anticancer and Cytotoxic activity of Green, Brown, Red seaweeds are tabulated in Tables 8.4, 8.5, and 8.6.

Anticancer and cytotoxic activities of 25 green seaweeds, 45 red seaweeds and 47 Brown seaweeds are summarized. Brine Shrimp Lethality Bioassay, 3-(4,5-dimethylthiazol-2-yl-)-2,5-diphenyl-2H-tetrazolium bromide (MTT assay), Sulforhodomine B assay, Neutral red dye assay, Resazurin Reduction test (RRT), DNA Fragmentation, Tryohan Blue methods are adopted. Among this, MTT assay is chiefly used one. For extract, aqueous, methanol, chloroform, hexane, ethyl acetate, ethanol, acetone, dichloromethane. Of which, methanol is highly used.

11 Anti-inflammatory Activity

Gastrointestinal ulcer is a common major digestive system disorder affecting millions of peoples. Anti-inflammatory activity involves the inhibition of the activity of inflammatory cells or inhibition of the production of inflammatory mediators. Steroidal anti-inflammatory drugs (SAID) and non-steroidal anti-inflammatory drugs (NSAID) are commonly used as therapeutic drugs for these diseases and cause sideeffects such as Gastric discomfort, nausea and vomiting etc. In addition

Table 8.1 Antioxidant activity of Green seaweeds

Name of the seaweed	Solvent/extract	Method/assay	IC_{50}/%	Reference
Caulerpa racemosa, *Ulva (Enteromorpha) intestinalis*	Methanol	DPPH assay	54.16 ± 0.14% and 56.83 ± 0.81% inhibition at 400µg/ml for *C. racemosa* and *U. intestinalis* respectively.	Aftabuddin et al. (2020)
Halimeda opuntia	Ethanol, methanol	DPPH assay	IC_{10} (mg/mL) 48.12 ± 2.9	Toufiqul et al. (2020)
Caulerpa racemosa var. cylindracea	Methanol, chloroform, hexane	DPPH assay	Chloroform extract IC_{50} [(1.98 ± 0.08) mg/mL]	Belkacemi et al. (2020)
		ABTS assay	Chloroform extract [(1.66 ± 0.05) mg/mL]	
		β-carotene bleaching (BCB) assay	Hexane, chloroform extract IC_{50} = (0.39 ± 0.02) mg/mL and (0.43 ± 0.00) mg/mL	
Ulva lactuca	Acetone, ethanol, methanol, distilled water	DPPH, reducing power and total antioxidant capacity assay	0.24 ± 0.006 to 1.18 ± 0.004 (mg AAE/g DW)	Ismail et al. (2020)
Chaetomorpha antennina	Methanol, dichloromethane	TPC, Total antioxidant activity and DPPH assay	0.872 ± 0.040 GAE/gm	Maithili and Ramanathan (2020)
Caulerpa lentillifera	Ethanol	DPPH assay, metal chelating activity	DPPH EC_{50} (mg/mL) 0.55 ± 0.02–22.45 ± 0.90	Pechroj et al. (2020)
Ulva rigida	Aqueous	DPPH, hydroxyl radical scavenging, Fe^{2+} chelation assay	DPPH IC_{50} 278.90 ± 4.51 Fe^{2+} chelation 320.92 ± 5.67	Olasehinde et al. (2019)

Codium sp.	Crude polysaccharide	Total antioxidant activity	(85.53 ± 0.25%)	Kalliswari et al. (2016)
		Reducing power	(0.248 ± 0.45%)–(1.579 ± 0.32%)	
		H_2O_2 assay	79.34 ± 0.18%	
		DPPH	71.18 ± 0.54%	
		ABTS	69.74 ± 0.49%	
		Hydroxyl radical scavenging activity	70.44 ± 0.33%	
		Superoxide anion scavenging activity	66.43 ± 0.27%	
		Nitric oxide scavenging assay	65.74 ± 0.18%	
Ulva lactuca	Polysacharide extraction	ABTS, hydroxyl radical scavenging, nitrite scavenging, reducing power assay	ABTS 3.59 ± 0.11 Hydroxyl radical 80.07% ± 2.17% mg/mL at 4 mg/ml	He et al. (2016)
Enteromorpha compressa, Ulva fasciata	Petroleum ether, ethyl acetate, methanol	TPC, total reducing power, nitric oxide and H_2O_2 scavenging assay	127.65µg/ml	Mohapatra et al. (2016)
Codium adhaerens, Codium tomentosum, Codium vermilara and *Ulva compressa*	Methanol, dichloromethane	TPC, DPPH, ORAC assay	DPPH 49.2 – 1000 (IC_{50} µg/ml) TPC (mg GAE/g extract) 0.26 ± 0.001–397.23 ± 0.02 ORAC (µmol TE/g extract) 33.24 ± 20.63–4184.16 ± 130.60	Pinteus et al. (2016)

(continued)

Table 8.1 (continued)

Name of the seaweed	Solvent/extract	Method/assay	IC_{50}/%	Reference
Caulerpa racemosa	Hexane, dichloromethane, ethyl acetate, acetone, methanol	DPPH, nitric oxide and hydroxyl radical scavenging assays	Ranging from 10.96 ± 0.81% to 75.17 ± 2.04%	Chia et al. (2015)
Enteromorpha compressa, Enteromorpha intestinalis, Ulva fasciata, Ulva lactuca, Chaetomorpha antennina	Acetone	Total antioxidant activity, DPPH, hydroxyl radicals scavenging, reducing power assay	Reducing power E. compressa minimum (0.69 ± 0.02) *Hydroxyl C. antennina* (28%) *E. compressa* (31%) *E. intestinalis* (31%)	Saranya et al. (2014)
Ulva clathrata, Ulva prolifera	80% methanol	DPPH assay	*U. clathrata* IC_{50} – 0.715 ± 0.078 mg mL^{-1} *U. prolifera* 3.101 ± 0.107	Farasat et al. (2013)
Ulva fasciata	Ethanol	DPPH assay	39.79%	Radhika et al. (2013a)
Caulerpa racemosa, Caulerpa peltata, Caulerpa taxifolia, Chlorodesmis fastigiata, Codium elongatum	Methanol	DPPH, reducing power and metal ion chelating assays	DPPH IC_{50} 1.024 ± 0.012 mg mL to 3.105 ± 0.048 mg mL^{-1} Ferrous chelating assay IC_{50} 0.042 ± 0.0008–9.425 ± 1.034 mg mL^{-1}	Vinayak et al. (2011)
Enteromorpha spirulina	Ethanol, methanol, acetone	DPPH assay	EC_{50} 50.00 ± 0.72	Cox et al. (2010)

Table 8.2 Antioxidant activity of Brown seaweeds

Name of the seaweed	Solvent/extract	Method/assay	$IC_{50\%}$/%	Reference
Laminaria japonica, Undaria pinnatifida, Eisenia bicyclis, Sargassum fusiforme	Ethanol	DPPH assay	0.00–2641.34 TEAC	Čmiková et al. (2022)
Padina pavonica	Polysaccharide	DPPH assay, reducing power	63%	Arunkumar et al. (2021)
Dictyota barteyresiana	Methanol, ethanol, benzene, acetone	DPPH assay	51.78 ± 0.01–91.66 ± 0.01	Durairaj and Andiyappan (2020)
		Phospho molybdate assay	17.29 ± 0.02–82.45 ± 0.02	
Turbinaria decurrens, Padina pavonica, Sargassum muticum and *Sargassum acinarium*	Acetone, ethanol, methanol, distilled water	DPPH assay, reducing power and total antioxidant capacity assay	*T. decurrens* IC_{50} 40.66 mg/mL	Ismail et al. (2020)
Dictyota cervicornis, Sargassum polycystum, Padina australis	Ethanol	DPPH assay, metal chelating activity	DPPH EC_{50} (mg/mL) 0.55 ± 0.02–22.45 ± 0.90	Pechroj et al. (2020)
Ecklonia maxima	Aqueous	DPPH assay, hydroxyl radical scavenging assay, Fe^{2+} chelation assay	DPPH – IC_{50} 181.66 ± 0.36	Olasehinde et al. (2019)
Cystoseira barbata	Acetone (70%), methanol, water	DPPH, ABTS assays and reducing power assay	ABTS EC_{50} 13.9 ± 0.2–22.1 ± 0.3µg/mL	Trifan et al. (2019)
Turbinaria conoides	n-hexane, MeOH/EtOAc solvent system 2H-pyranoids (1–3)	DPPH and ABTS+ radical scavenging assays	IC_{50} 0.54–0.69 mg mL^{-1}	Chakraborty and Dhara (2019)

(continued)

Table 8.2 (continued)

Name of the seaweed	Solvent/extract	Method/assay	$IC_{50\%}$/%	Reference
Endarachne binghamiae, Grateloupia turuturu, Padina arborescens, Colpomenia sinuosa, Sargassum hemiphyllum, Sargassum vachellianum, Pachydictyon coriaceum	*Ethanol*	*TPC, FRAP assay*	3.3 ± 0.1–102.6 ± 3.6 (mg GAE/g dry extract)	Peihang et al. (2018)
Sargassum vulgare	Ethyl acetate	DPPH and ABTS+ assay	93.33% and 90.67%	Shreadah et al. (2018)
Sargassum latifolium B, Sargassum platycarpum	Methanol	DPPH assay	66%	Moubayed et al. (2017)
Padina tetrastromatica	Powder	Reduced glutathione (GSH), vitamin C, vitamin E	GSH (U/mg protein) 35.21 ± 2.46	Ponnanikajamideen et al. (2017)
Ascophyllum nodosum (ANE), Bifurcaria Bifurcata (BBE), Fucus vesiculosus (FVE)	Aqueous	ABTS, DPPH, FRAP and ORAC assay	DPPH result FVE – $EC_{38.9}$ 4.2 mg/mL followed by BBE – $EC_{23.6}$ 7.5 mg/mL and ANE – $EC_{4.3}$ 11.5 mg/mL	Agregán et al. (2017)
Durvillaea Antarctica	Polysaccharide extraction	ABTS, hydroxyl, nitrite scavenging activity, reducing power assay	1.0–4.0 mg/ml	He et al. (2016)
Turbinaria conoides	Petroleum ether, ethyl acetate, methanol	TPC, total reducing power, nitric oxide and H_2O_2 scavenging assay	IC_{50} (μg/ml) 141.35	Mohapatra et al. (2016)

Fucus spiralis, Halopteris filicina, Sacchorhiza polyschides, Bifurcaria bifurcata, Padina pavonica, Colpomenia peregrina, Dictyota dichotoma, Sargasssum muticum, Cystoseira tamariscifolia, Cladostephus spongiosus, Cystoseira usneoides, Taonia atomaria, Sargassum vulgare and Stypocaulon scoparium	Methanol, dichloromethane	TPC	0.26 ± 0.001–397.23 ± 0.02 (mg GAE/g extract)	Pinteus et al. (2016)
		DPPH	49.2 – 1000 (IC_{50} μg/ml)	
		ORAC assay	33.24 ± 20.63–4184.16 ± 130.60 (μmol TE/g extract)	
Padina tetrastromatica, Turbinaria ornate	Hexane, dichloromethane, ethyl acetate, acetone, methanol	DPPH, superoxide, nitric oxide, hydroxyl radical scavenging assays	Ranging from 10.96 ± 0.81% to 75.17 ± 2.04%	Chia et al. (2015)
Padina gymnospora	Acetone	Total antioxidant activity, DPPH, hydroxyl radicals scavenging, reducing power assay	Reducing power (2.678 ± 0.03) (91%) for hydroxyl Marks highest	Saranya et al. (2014)
Cystoseira crinita, Cystoseira sedoides, Cystoseira compressa	Aqueous	DPPH assay	IC_{50} (μg/mL) 12.0 ± 0.7–75.0 ± 0.8	Mhadhebi et al. (2014)
Padina boergesenii	Methanol	Phosphomolybdenum, DPPH, Nitric oxide scavenging assay	58.5 ± 0.48%	Jeevitha et al. (2014)
Cystoseira crinita	Chloroform, methanol, ethyl acetate	DPPH assay	MeOH IC_{50} 107 ± 0.001	Mhadhebi et al. (2011a)
		FRAP assay	Ethyl acetate 4.62 ± 0.004	

(continued)

Table 8.2 (continued)

Name of the seaweed	Solvent/extract	Method/assay	$IC_{50\%}$/%	Reference
Cystoseira sedoides	Chloroform, ethyl acetate, methanol	TPC, reducing power, DPPH assay	121µg/mL	Mhadhebi et al. (2011b)
Laminaria digitata, Laminaria saccharina, Himanthalia elongate	Ethanol, methanol, acetone	DPPH radical scavenging assay	50% inhibition	Cox et al. (2010)
Sargassum sp.	Methanol	DPPH	54.93	Patra et al. (2008)
		Hydroxyl radicals scavenging assay	32.66	
		Glutathione-S-transferase activity	72.54	

Table 8.3 Antioxidant activity of Red seaweeds

Name of the seaweed	Solvent/extract	Method/assay	IC_{50}/%	Reference
Palmaria palmate	Ethanol	DPPH assay	0.00–2641.34 TEAC	Čmiková et al. (2022)
Portieria hornemannii, Spyridia hypnoides, Asparagopsis taxiformis, Centroceras clavulatum	Polysaccharide	DPPH, reducing power assay	*Spyridia hypnoides* and *A. taxiformis* (31%) *C. clavulatum* 28%	Arunkumar et al. (2021)
Pterocladia capillacea	Acetone, ethanol, methanol, distilled water	DPPH, reducing power and total antioxidant capacity assay	0.16 ± 0.009–1.01 ± 0.001 (mg AAE/g DW)	Ismail et al. (2020)
Kappaphycus alvarezii	Ethanol	DHA power reducing assay	85%	Manimegalai and Bakiya Lakshmi (2020)
Kappaphycus alvarezii	Ethanol	TPC, ferric reducing power	20.25 ± 0.03 mg GAE/g	Bhuyar et al. (2020)
Eucheuma cottonii	Ethanol, distilled water, hot distilled water	DPPH, and α-amylase activity assay	The highest percent inhibition on DPPH free radical scavenging and α-amylase used ethanol solvent (34.27%) and (59.33%),	Prasasty et al. (2019)
Gracilaria beckeri, Gelidium pristoides	Aqueous	DPPH, hydroxyl radical, Fe^{2+} chelation assay	IC_{50} 186.71 ± 6.53 DPPH Fe 334.45 ± 6.53, 303.28 ± 2.39	Olasehinde et al. (2019)
Laurencia dendroidea	80% methanol, n-hexane, chloroform, ethyl acetate, butanol and water	DPPH, nitric oxide radical scavenging, metal chelating assays	IC_{50} (μg/ml) 75.48 ± 2.07–700.79 ± 4.07	Nguyen et al. (2019)
Jania rubens, Kappaphycus alvarezii	Ethanol, dichloromethane, n-hexane	ABTS+, lipid peroxidation inhibitory, Fe^{+2} chelating asaays	IC_{50} 0.22 mg/mL	Kajal and Vamshi (2018)

(continued)

Table 8.3 (continued)

Name of the seaweed	Solvent/extract	Method/assay	IC_{50}/%	Reference
Caloglossa sp., *Caula anthus, okamurai Gelidium planiusculum, Gelidium* sp, *Gelidium kintaroi, Chondracanthus intermedius, Ahnfeltiopsis hainanensis, Ahnfeltiopsis guangdongensis, Ahnfeltiopsis masudai, Polysiphonia hainanensis, Hypnea chordacea, Hypnea japonica, Hypnea boergesenii, Laurencia okamurai, Laurencia chinensis*	Ethanol	TPC, FRAP assay	3.3 ± 0.1–102.6 ± 3.6 (mg GAE/g dry extract)	Peihang et al. (2018)
Gracilaria changii, G. manilaensis and *Gracilaria* sp.	Methanol	DPPH, xanthine oxidase inhibitory, AChE inhibitory activity, TLC bioautographic method)	DPPH 8.4 ± 0.001 to 93.6 ± 0.3 (quercitin) Xanthine oxidase inhibition 8.2 ± 0.1%–72.4 ± 1.6 AChE inhibitory activity 0.6–10.0	Andriani et al. (2016)
Gracilaria salicornia (GS) *Gracilaria corticata* (GC)	Methanol	DPPH assay	IC_{50} (mg/ml) GS – 0.73 GC – 0.54	Ghannadi et al. (2016)

Gracilaria lemaneiformis (GLP), *Sarcodia ceylonensis* (SCP)	Polysacharide extraction	ABTS, hydroxyl radical scavenging, nitrite scavenging, reducing power assays	Hydroxyl radical 83.33 ± 2.31% at 4 mg/ml Nitrite scavenging activity- could not determined	He et al. (2016)
Gracilaria verrucosa	Petroleum ether, ethyl acetate, methanol	TPC, total reducing power, nitric oxide and H_2O_2 scavenging assays	IC_{50} (μg/ml) 168.96	Mohapatra et al. (2016)
Asparagopsis armata, *Ceramium ciliatum*, *Plocamium cartilagineum*, *Corallina elongata*, *Porphyra linearis*, *Gelidium pulchellum*, *Jania rubens*, *Nitophyllum punctatum*, *Sphaerococcus coronopifolius*	Methanol, dichloromethane	TPC	0.26 ± 0.001 – 397.23 ± 0.02 (mg GAE/g extract)	Pinteus et al. (2016)
		DPPH	49.2 – >1000 (IC_{50} μg/ml)	
		ORAC assay	33.24 ± 20.63–4184.16 ± 130.60 (μmol TE/g extract)	
Amansia multifida, *Meristiella echinocarpa*	70% ethanol	DPPH, FRAP, FIC, BCB assay	DPPH radical scavenging and ferrous ion chelating showed values of 60% and 17%, respectively	Alencar et al. (2014)
Hypnea valentiae, *Grateloupia lithophila*	Acetone	Total antioxidant activity, DPPH, hydroxyl radicals scavenging assay, reducing power assay	54% in ydroxyl radical scavenging assay	Saranya et al. (2014)
Gracilaria corticata	Ethanol	DPPH, reducing power assay	IC_{50} 1.93 mg/ml	Sreejamole and Greeshma (2013)
Amphiora corallina, *Acanthophora spicifera*, *Hypnea valentiae*	Ethanol	DPPH assay	86.13%	Radhika et al. (2013a)
Palmaria palmata, *Chondrus crispus*	Ethanol, methanol, acetone	DPPH assay	EC_{50} 25.00 ± 0.35, 10.00 ± 0.88	Cox et al. (2010)

Table 8.4 Anticancer and cytotoxicity activity of Green seaweeds

Name of the seaweed	Solvent	Method	Cell line	Result	Reference
Halimeda opuntia	Methanol	MTT assay	MCF-7, MDA-MBA-231, HT-29, HepG2, 3T3	IC_{50} of 25.14 ± 1.02 g/mL	Nazarudin et al. (2022)
Valoniopsis pachynema, Codium iyengarii, Enteromorpha intestinales, and Caulerpa scalpelliformis	70% ethanol	MTT assay	NCI-H460 cells (human carcinoma lung cell)	IC_{50} value for *E. intestinalis* 188.600 ± 0.039µg/mL, *C. scalpelliformis* 149.283 ± 3.326µg/mL	Rasheed et al. (2021)
Caulerpa racemosa, Ulva (Enteromorpha) intestinalis	Methanol	BSLB	–	LC_{50} 1012.86 and 695.41µg/ml	AftabUddin et al. (2020)
Chaetomorpha antennina	Methanol, dichloromethane	BSLB	–	LC_{50} −47 ± 11.12	Maithili and Ramanathan (2020)
Halimeda opuntia	50% ethanol, 70% methanol	BSLB	–	LC_{50} 94.24µg/ml	Toufiqul et al. (2020)
Ulva Lactuca, Caulerpa racemosa, Halimeda opuntina, Caulerpa sertularioides, Chaetomorpha antennina, Chaetomorpha crassa	Aqueous	BSLB, MTT assay	Mouse fibroblasts (L929) cell line	*C. antennina* 34.04µg/µl *U. lactuca*; 19.54µg/µl for cell line	Premarathna et al. (2020)
Ulva fasciata	Methanolic extract	MTT assay	Vero cells and Hep2 cell lines	402.16µg/ml	Alphonse et al. (2019)
Caulerpa lentillifera	Distilled water, dyed silk fabric	MTS assay	Mouse embryonic cells, SH-SY5Y cell line	101.92%	Kadir et al. (2016)
Codium adhaerens, Codium tomentosum, Codium vermilara and Ulva compressa	Methanol, dichloromethane	MTT assay	MCF-7 cells	*U. compressa* 91.42 ± 2.38% viable cells	Pinteus et al. (2016)

Cladophoropsis sp.	Ethanol	MTT assay	Breast cancer cell line MDA-MB-231 (ER−), MCF-7 (ER+), and T-47D (ER+)	IC_{50} values 66.48 ± 4.96, 150.86 ± 51.56, and >400μg/ml, respectively.	Erfani et al. (2015)
Halimeda opuntia	Methanol	BSLB	–	LC_{50} = 192.3μg	Selim (2012)
Ulva fasciata	Ethanol	BSLB	–	LC_{50} = 724μg	Ayesha et al. (2010)
Avrainvillea cf. digitata sp., Halimeda incrassata Halimeda tuna, Pencillus dumetosus, Pencillus lamourouxii, Rhipocephalus phoenix brevifolius, Udotea conglutinata, Udotea flabellum	Dichloromethane:methanol (7:3)	(MTT) and sulforhodamine B (SRB) assays	Normal canine kidney (MDCK) cells, human laryngeal carcinoma (Hep-2) cells, human cervical adenocarcinoma (HeLa) cells, and human nasopharyngeal carcinoma (KB) cells	($CC_{50} \leq 30\mu g\ mL^{-1}$)	Moo-Puc et al. (2009)

Table 8.5 Cytotoxic activity of Brown seaweeds

Name of the seaweed	Solvent	Method	Cell line	Result	Reference
Valoniopsis pachynema, Codium iyengarii, Enteromorpha intestinales, and Caulerpa scalpelliformis	70% ethanol	MTT assay	NCI-H460 cells (human carcinoma lung cell)	*S. variable* IC_{50} value 187.182 ± 2.517	Rasheed et al. (2021)
Padina antillarum, Sargassum illicifolium, Sargassum Polycystem, Turbinaria ornata, Sargassum illicifolium, Stoechospermum polypodioides	Aqueous	Brine shrimp lethality assay, MTT assay	Mouse fibroblasts (L929) cell line	*S. polycystum* (12.80μg/μl) *P. antillarum*; 18.24μg/μl	Premarathna et al. (2020)
Padinaaustralis	Hexane, dichloromethane, butanol and water	GFPMA and MTT methods	Cervical cancer cell line, Hela, and human umbilical vein endothelial cells, HUVEC	IC_{50} value of hexane, dichloromethane, butanol and water partitions were 2.0, 20, 19.7, and 182.7μg/ml	Vaseghi et al. (2019)
Sargassum polycystum	n-hexane, ethyl acetate, chloroform, ethanol	MTT cell proliferation assay	HCT-116 and lung-A549 cancer cells	21.3–33.4μg/mL	Arsianti et al. (2019)

Cystoseira barbata	70% acetone, methanol, water	MTT assay	Pulmonary adenocarcinoma A549, colorectal adenocarcinoma HT-29, mammary adenocarcinoma (MCF-7 cells) and one non-tumor mammary epithelial cell line (MCF-10A)	$IC_{50} = 72.12 \pm 1.53$µg/mL	Trifan et al. (2019)
Turbinaria ornata and Padina australis	Ethanol	Resazurin reduction test (RRT)	–	IC_{50} of 530.53 and 528.78 ppm	Canoy and Bitacura (2018)
Sargassum glaucescens	Methanol, chloroform, ethyl acetate and hexane	MTT and trypan blue methods, DNA fragmentation (electrophoresis method)	MCF-7 and HT-29 cancer cells	630.8 ± 16.37 and 774.01 ± 28.07µg/ml	Taheri et al. (2018)
Sargassum wightii, Sargassum duplicatum and Sargassum tenerrimum	Methanol	MTT assay	HCC cells (Hep 3B) and normal fibroblast cells (Vero)	*S. weightii* CTC_{50} 10.20 ± 1.06 and 180.65 ± 2.87µg/ml against vero cells and Hep 3B respectively	Gunasekaran et al. (2017)
Sargassum fluitans	Polysaccharide (acetone, methanol)	Neutral red dye assay	Vero cell lines	*S. fluitans* ($EC_{50} = 42.8$µg/ml)	Bedoux et al. (2017)
Sargassum sp.	Distilled water, dyed silk fabric	MTS assay	Mouse embryonic cells, SH-SY5Y cell line	Extract-522.38% dyed silk fabric-192.67%	Kadir et al. (2016)

(continued)

Table 8.5 (continued)

Name of the seaweed	Solvent	Method	Cell line	Result	Reference
Fucus spiralis, Halopteris filicina, Sacchorhiza polyschides, Bifurcaria bifurcata, Padina pavonica, Colpomenia peregrina, Dictyota dichotoma, Sargasssum muticum, Cystoseira tamariscifolia, Cladostephus spongiosus, Cystoseira usneoides, Taonia atomaria, Sargassum vulgare and Stypocaulon scoparium	Methanol, dichloromethane	MTT assay	MCF-7 cells	*Sargassum muticum* (87.87 ± 1.88% viable cells)	Pinteus et al. (2016)
Cystoseira myrica, Iyengaria stellate, Colpomenia sinousa	Ethanol	MTT assay	Breast cancer cell line MDA-MB-231 (ER−), MCF-7 (ER+), and T-47D (ER+)	IC_{50} > 400 (μg/ml)	Erfani et al. (2015)
Padina tetrastromatica, Turbinaria ornatata	Hexane, dichloromethane (DCM), ethyl acetate (EA), acetone and methanol	MTT assay	Human breast cell line, 184B5, and the human breast adenocarcinoma cell line, MCF-7	IC_{50} (μg/mL) *P. tetrastromatica* 130.0 ± 1.72 *T. ornatata* 240.0 ± 1.89	Chia et al. (2015)
Padina boergesenii	Methanol	MTT assay	Liver cancer cell line Hep G2	IC_{50} 1.67 mg/ml	Jeevitha et al. (2014)
Sargassum sp	Ethanol	MTT assay	MCF-7 (breast cancer) and Hep-2 (liver cancer) cell lines	IC_{50} 200–250μg/ml	Mary et al. (2012)

Dictyota dichotoma var. velutricata, D. hauckiana, D. indica, Iyengaria stellata, Jolyna laminarioides, Sargassum ilicifolium, S. lanceolatum	Ethanol	Brine shrimp lethality assay	–	*D. indica* LC_{50} = 143μg	Ayesha et al. (2010)
Dictyota caribaea, Lobophora variegata, Padina perindusiata, Sargassum fluitans, Turbinaria turbinata	Dichloromethane:methanol (7:3)	(MTT) and sulforhodamine B (SRB) assays	Normal canine kidney (MDCK) cells, human laryngeal carcinoma (Hep-2) cells, human cervical adenocarcinoma (HeLa) cells, and human nasopharyngeal carcinoma (KB) cells	*T. turbinata* exhibited high cytotoxic and antiproliferative activity against KB cells (CC_{50} = 23.94μg mL^{-1}, IG_{50} = 29.84μg mL^{-1}) respectively	Moo-Puc et al. (2009)
Sargassum thunbergii, Dictyopteris divaricata	Ethyl acetate	MTT assay	HL-60 human promyelocytic leukaemia cell line, HT-29 (human colon carcinoma cell line), B16F10 (murine melanoma cell line), A549 (human lung cancer cell line)	IC_{50} (μg/mL) 23.46 and 15.15	Kim et al. (2009)

Table 8.6 Cytotoxic activity of Red seaweeds

Name of the seaweed	Solvent/extract	Method/assay	Cell line	Result	Reference
Gracilaria corticata, Acanthophora spicifera, Gelidiopsis variabilis, Jania adhaereus	Aqueous	BSLB, MTT assay	Mouse fibroblasts (L929) cell line	*A. spicifera* (LC_{50} = 0.072μg/μl). *J. adhaereus* 50.70–7.304% cell viability	Premarathna et al. (2020)
Acanthophora Spicifera	Methanol	MTT assay, comet assay, cytotoxicity assay	Human Colon cancer cell adenocarcinoma (HT-29 cell line)	83.9 ± 7.1%	Kumar and Prakash (2019)
Gracilaria verrucosa	Hexane, ethyl acetate, chloroform and ethanol	MTT assay	Cervical HeLa cells	IC_{50} values for hexane – 14.94μg/mL, chloroform – 15.74μg/mL, ethyl acetate – 16.18μg/mL, and ethanol IC_{50} 19.43μg/mL.	Dewi et al. (2018)
Gracilaria corticata	Aqueous extract	MTT assay	Human gastric adenocarcinoma (AGS), human colon cancer (RKO) and human hepatocellular carcinoma (HepG2) cell lines	74%, 69%, and 68%	Mazaheri et al. (2017)
Rhodymenia pseudopalmata, Solieria filiformis, Hydropuntia cornea	Polysaccharide (acetone, methanol)	Neutral red dye assay	Vero cell lines	*Solieria filiformis* (EC_{50} = 136.0μg/ml)	Bedoux et al. (2017)

Asparagopsis armata, Ceramium ciliatum, Plocamium cartilagineum, Corallina elongata, Porphyra linearis, Gelidium pulchellum, Jania rubens, Nitophyllum punctatum and Sphaerococcus coronopifolius	Methanol, dichloromethane	MTT assay	MCF-7 cells	85%	Pinteus et al. (2016)
Gracilaria foliifera, Gracilaria salicornia, Gracilariopsis longssima, Hypnea flagelliformis, Laurencia papillosa, Botryocladia leptopoda	Ethanol	MTT assay	Breast cancer cell line MDA-MB-231 (ER−), MCF-7 (ER+), and T-47D (ER+)	IC_{50} values of *G. foliifera* on MDA-MB-231, MCF-7, and T-47D cell lines were 74.89 ± 21.71, 207.81 ± 12.07 and 203.25 ± 30.98µg/ml, respectively.	Erfani et al. (2015)
Eucheuma cottonii	Methanol	MTT assay	HeLa (human cervix adeno carcinoma), human lung carcinoma cell line (SKLU-1), human colon carcinoma cell line (HCT-116), and fibroblast	0.005 ± 0.003–0.743 ± 0.093	Lee et al. (2015)
Amansia multifida and Meristiella echinocarpa	70% ethanol	BSLB	–	LC_{50} 484.2 and 281.9µg mL^{-1}	Alencar et al. (2014)
Gracilaria corticata	Ethanol	BSLB	–	LC_{50} 1.081 mg/ml	Sreejamole and Greeshma (2013)

(continued)

Table 8.6 (continued)

Name of the seaweed	Solvent/extract	Method/assay	Cell line	Result	Reference
Sarconema filiforme	Methanol	BSLB	–	LC_{50} = 328.9μg	Selim (2012)
Melanothamnus afaqhusainii	Ethanol	BSLB	–	LC_{50} = 190μg	Ayesha et al. (2010)
Agardhiella sp., Bryothamnion triquetrum, Ceranium nitens, Champia salicornioides, Eucheuma isiforme, Gracilaria caudata, Gracilaria cervicornis, Gracilaria damaecornis, Gracilaria sp., Halymenia floresii, Heterosiphonia gibbesii, Hydropuntia cornea, Jania capillacea, Laurencia microcladia Melanothamnus somalensis, Coelarthrum muelleri, Laurencia obtusa, Gracilaria corticata, and Gelidium pusillum	Dichloromethane:methanol (7:3)	(MTT) and sulforhodamine B (SRB) assays	Normal canine kidney (MDCK) cells, human laryngeal carcinoma (Hep-2) cells, human cervical adenocarcinoma (HeLa) cells, and human nasopharyngeal carcinoma (KB) cells	*B. triquetrum* extract showed the highest cytotoxic and selective activity against Hep-2 cells (CC_{50} = 8.29μg mL^{-1}, SI = 12.04).	Moo-Puc et al. (2009)

Table 8.7 Anti-inflammatory activity of Green seaweeds

Name of the seaweed	Solvent/ extract	Method/assay	Result	Reference
Codium flabellatum	70% methanol	Carrageenan induced rat paw oedema method	1.99 ± 0.04–3.61 ± 0.14	Yasmeen et al. (2021)
Caulerpa lentillifera	95% ethanol	Nitric acid suppressing assay	IC_{50} value of 117.54 ± 7.94µg/mL	Pechroj et al. (2020)
Ulva fasciata	Methanol	Albumin denaturation, antiproteinase, hypotonicity-induced haemolysis and anti-lipoxygenase assays	Hemolysis inhibition 72%	Alphonse et al. (2019)
Chaetomorpha linum, Rhizoclonium riparium, Ulva intestinalis, U. lactuca, U. prolifera	Aqueous extract	Cyclooxygenase (COX-2) inhibition assay	31% and 45% of COX-2 inhibition.	Ripol et al. (2018)
Ulva reticulate	Methanol	Acute inflammatory test, carrageenan-induced hind paw edema in rats, chronic inflammation	21.61% at 350 mg/kg body weight dose	Hong et al. (2011)
Ulva linza	Ethanol	Mouse ear edema and erythema induced by phorbol myristate acetate.	IC_{50} values 20, 26, and 31 mg ml^{-1}	Khan et al. (2008)

they are mainly used to reduce the symptoms of the disease without treating or preventing the inflammatory and ulcerogenic processes (Bhatia et al. 2015). Due to this limitation, researchers focused on the discovery and development of new bioactive natural products with anti-inflammatory and antinociceptive properties. The anti-inflammatory activities of green, brown and red seaweeds are tabulated in Tables 8.7, 8.8, and 8.9.

Anti-inflammatory activities of 10 green seaweeds, 18 brown seaweeds and 10 red seaweeds are studied. Carrageenan induced rat paw oedema method is dominantly used for assay. Other assays are COX, LOX, Myleperoxidase, nitrate levels inhibition, writhing induced by acetic acid, Mouse ear edema test, Phospholipase A2 activity, Nitric acid suppressing, albumin denaturation, antiproteinase, hypotonicity induced haemolysis, Acute inflammatory test, erythema induced by phorbol-myristate acetate, Human red blood cell membrane stabilization assay, protein denaturation, formalin-induced paw oedema assay, Lactate dehydrogenase cytotoxicity assay, Chronic inflammation, ethyl phenylpropiolate (EPP)-ear edema etc. Methanol solvent is used in high level. Ethanol, aqueous, benzene, acetone, ethyl acetate, chloroform, dichloromethane are also used to evaluate anti-inflammatory activity.

Table 8.8 Anti-inflammatory activity of Brown Seaweeds

Name of the seaweed	Solvent/extract	Method/assay	Result	Reference
Calpomenia sinuosa	Ethanol	Anti-lipooxygenase activity	42.5%	Shobier et al. (2022)
Sargassum wightii, S. swartzii and Cystoseira indica	70% methanol	Carrageenan induced rat paw oedema method	1.99 ± 0.04–3.61 ± 0.14	Yasmeen et al. (2021)
Turbinaria ornata, Sargassum polycystum, Padina tetrastromatica	Ethanol	Human red blood cell (HRBC) membrane stabilization method.	IC_{50} values of *T. oranta* extract 377.28µg/mL	Bharath et al. (2020)
Dictyota barteyresiana	Methanol, ethanol, benzene, acetone	Protein denaturation assay	71.50 ± 0.01	Durairaj and Andiyappan (2020)
Dictyota cervicornis, Sargassum polycystum, Padina australis	95% ethanol	Nitric acid suppressing assay	IC_{50} value of 19.97 ± 1.48µg/mL	Pechroj et al. (2020)
Sargassum wightii	Methanol	Formalin induced paw oedema assay	60%	Rajakumari and Kaleeswari (2019)
Sargassum weightii	Ethyl acetate:methanol and chloroform	Cyclooxygenase (COX-1 and COX-2) inhibition assays by 2,7-dichlorofluorescein method	Angiotensin converting enzyme-I inhibitory activity (IC_{50} 0.084 mg/mL), anti-COX-1, 2, and 5-LOX (IC_{50} 0.03–0.05 mg/mL) and DPP-4 inhibitory (IC_{50} ~0.013 mg/mL)	Maneesh et al. (2017)
Sargassum weightii	Ethanol	Albumin denaturation assay.	IC_{50} 2.5 mg/ml	Begum and Hemalatha (2017)
Cystoseira crinita, Cystoseira sedoides and *Cystoseira compressa*	Aqueous extract	Carrageenan induced rat paw oedema method	70.9–82.1%	Mhadhebi et al. (2014)

Sargassum patens	Ethyl acetate	Lactate dehydrogenase (LDH) cytotoxicity assay, determination of NO, prostaglandin E2 (PGE2), Interleukin-6 (IL-6), tumour necrosis factor-α (TNF-α) production	50μg ml^{-1} ethanol extract suppressed 25% LPS induced (TNF-α) production	Kim et al. (2013)
Sargassum swartzii	Methanol	Acute inflammatory test, carrageenan-induced hind paw edema in rats, chronic inflammation	52.12% at 175 mg/kg body weight dose	Hong et al. (2011)
Turbinaria conoides	Aqueous	Ethyl phenylpropiolate (EPP)-induced ear edema, carrageenan-induced hind paw edema, irritation test	12.73–80.43%	Boonchum et al. (2011)
Cystoseira crinita	Chloroform, ethyl acetate and methanol	Carrageenan induced rat paw oedema	84.96%	Mhadhebi et al. (2011a)
Undaria pinnatifida	Ethanol	Mouse ear edema and erythema induced by phorbol myristate acetate.	IC_{50} values 10, 15, and 18 mg ml^{-1}	Khan et al. (2008)
Sargassum fulvellum and *Sargassum thunbergii*	Dichloromethane, ethanol, boiling water	Yeast-induced pyrexia, tail-flick test, And phorbol myristate acetate-induced inflammation (edema, erythema, and blood flow)	The dichloromethane extract (0.4 mg/ear) of *Sargassum Fulvellum* inhibited an inflammatory symptom of mouse ear edema by 79.1%. The ethanol extract (0.4 mg/ear) of *Sargassum thunbergii* also Inhibited edema by 72.1%.	Kang et al. (2008)

Table 8.9 Anti-inflammatory activity of Red seaweeds

Name of the seaweed	Solvent/extract	Method/assay	Result	Reference
Coelarthrum muelleri, Melanothamnus afaqhusainii and Solieria robusta	70% methanol	Carrageenan induced rat paw oedema method	1.99 ± 0.04–3.61 ± 0.14	Yasmeen et al. (2021)
Gracillaria corticata	Ethanol	Inhibition of COX, LOX, and Myleoperoxidase and nitrate levels	53%	Abraham et al. (2018)
Gracilaria salicornia, Gracilaria edulis, Gracilaria corticata, Gracilaria fergusonii, Gracilaria verrucosa, Gracilaria corticata var. *cylindrical*	Distilled water	Heat induced haemolysis	43.81–95.55%	Chalini et al. (2017)
Porphyra vietnamensis	Aqueous, alcoholic fractions	Paw edema induced by carrageenan, writhing induced by acetic acid assay	75.49%	Bhatia et al. (2015)
Gracilaria edulis	Aqueous, methanol	Carrageenan induced rat paw oedema assay	58.20%	Vijayalakshmi (2015)
Dichotomaria obtusata	Methanol	Mouse ear edema test, phospholipase A_2 activity assay, acetic acid induced writhing assay	ED_{50} 4.87μg/ear.	Delgado et al. (2013)

12 Anti-diabetic Activity

Diabetes mellitus is one of chronic diseases caused by disorder of the carbohydrate metabolism. It is predicted as the ninth major cause of death worldwide considering for 1.5 million deaths (WHO 2019). Diabetes in regularly associated with blindness, heart attack, kidney failure, stroke and leg amputation. Diabetic mellitus categorized into two major types (Type 1 and Type 2). Diabetes was managed by diminishing postprandial hyperglycaemia by suspending the absorption of glucose with the help of carbohydrate hydrolyzing enzymes, α-amylase and α-glucosidase in digestive tract. The release of glucose from dietary multifarious carbohydrate can be delayed by α-glucosidase and obstruct the glucose absorption resulting in a diminished postprandial plasma glucose level and decreased postprandial hyperglycaemia (Kumar and Sudha 2012). The anti-diabetic activities of seaweeds are tabulated in Table 8.10.

Table 8.10 Antidiabetic activity of seaweeds

Name of the seaweed	Solvent	Method	Result	Reference
Antidiabetic activity of Green seaweeds				
Enteromorpha compressa, Ulva fasciata	Petroleum ether (pet. ether), ethyl acetate and methanol	α-glucosidase inhibitory assay	Ulva faciata IC_{50} 69.122 (μg/ml)	Mohapatra et al. (2016)
Halimeda macroloba	Distilled water	α-glucosidase inhibitory assay	70.58% inhibition at 40.22 mg mL^{-1}	Chin et al. (2014)
Anti-diabetic activity of Brown seaweeds				
Himanthalia elongata, Saccharina latissima Laminaria saccharina, Laminaria digitata, Fucus vesiculosus, Ascophyllum nodosum and Alaria esculenta	Water and ethanol	α-glucosidase and dipeptidyl peptidase 4 (DPP-4) enzymatic activity, block sodium glucose transporter-2 (SGLT-2) activity and stimulate glucagon-like peptide-1 (GLP-1) secretion and synthesis	13.9 ± 0.2% to 89.5 ± 0.4% *Alaria esculenta* and water extracts of *Laminaria digitata* strongly inhibited DPP-4 activity by 91.3 ± 0.1% and 90.0 ± 0.2%	Calderwood et al. (2021)
Turbinaria decurrens, Padina pavonica, Sargassum muticum and Sargassum acinarium	80% acetone, ethanol, methanol, distilled water	In vitro α-glucosidase inhibition, in vitro α-amylase inhibition	IC_{50} value of 4.37 mg/ml	Ismail et al. (2020)
Cystoseira myrica, Hormophysa cuneiformis, Sargassum cinereum, Turbinaria turbinate	80% methanol	α-glucosidase inhibitory assay	676.9 ± 2.5	Osman et al. (2019)
Undaria pinnatifida	Acetone	α-glucosidase inhibition assay	IC_{50} 0.08 ± 0.002 mg/mL	Zaharudin et al. (2018)

(continued)

Table 8.10 (continued)

Name of the seaweed	Solvent	Method	Result	Reference
Padina tetrastromatica	Aqueous extract	STZ influenced plasma glucose level test	Glucose 114.79 ± 3.03 (mg/dL) Insulin 24.21 ± 1.46 (IU/ml) STZ & *P. tetrastromatica* (500 mg/kg) Glucose (mg/dL) 156.84 ± 4.89 Insulin (IU/ml) 19.37 ± 2.13	Ponnanikajamideen et al. (2017)
Sargassum wightii	Ethanol	*α*-glucosidase inhibition, α-amylase inhibition assay, antiglycation activity	*α-amylase* (IC_{50}) value of 8 mg/ml, *α-glucosidase* (IC_{50}) 6 mg/ml.	Begum and Hemalatha (2017)
Laminaria digitata and Undaria pinnatifida	Methanol, acetone, water	α-amylase inhibition assay	α-amylase *L. digitata* IC_{50} 0.74 ± 0.02 mg/ml and *U. pinnatifida* 0.81 ± 0.03 mg/ml	Zaharudin et al. (2017)
Sargassum polycystum and Sargassum wightii	Petroleum ether, benzene, ethyl acetate, acetone and methanol	*α*-glucosidase inhibition, α-amylase inhibition assay, dipeptidyl peptidase-IV inhibition assay	*s. weightii α*-amylase (IC_{50} 378.3μg/ml) and *α*-glucosidase (IC_{50} 314.8μg/ml) DPP-IV- (IC_{50} 38.27μg/ml) *S. polycystum* *α*-amylase (IC_{50} 438.5μg/ml) and *α*-glucosidase (IC_{50} 289.7μg/ml) and DPP-IV (36.94μg/ml).	Unnikrishnan et al. (2015)
Padina sulcata, Sargassum binderi and Turbinaria conoides	Distilled water	α-glucosidase inhibitory assay	67.38% for *T. conoides*, 41.10% for *P. sulcata* and 24.18% for *S. binderi*	Chin et al. (2014)
Anti-diabetic activity of Red seaweeds				
Actinotrichia fragilis, Laurencia papillosa	80% methanol	α-glucosidase inhibitory assay	IC_{50} 920 ± 1.3	Osman et al. (2019)

(continued)

Table 8.10 (continued)

Name of the seaweed	Solvent	Method	Result	Reference
Ceramium rubrum	85% methanol	Glucose uptake assay, α-amylase inhibitory assay	IC_{50} 750µg/mL 43.9% at 400µg/mL concentrations	Sneha et al. (2019)
Eucheuma cottonii	Ethanol, distilled water, hot distilled water	α-amylase activity assay	Ethanol shows significant antioxidant activity	Prasasty et al. (2019)
Laurencia dendroidea	80% methanol, n-hexane, chloroform, ethyl acetate, butanol and water	α-glucosidase inhibitory assay	IC_{50} value of 8.14µg/mL	Nguyen et al. (2019)
Halymenia durvilae	Methanol extract, n-hexane, chloroform, and water fraction	α-glucosidase inhibitory assay	IC_{50} 4.34 ± 0.32 mg mL^{-1}	Sanger et al. (2019)
Spyridia filamentosa, Grateloupia lithophila and Hypnea musciformis	Methanol	α-amylase inhibitory assay, α-glucosidase inhibitory assay	*S. filamentosa* (IC_{50} = 58.02 and 66.06µg/ml), *G. lithophila* (IC_{50} = 53.01 and 58.02µg/ml) and *H. musciformis* (IC_{50} = 48.01 and 51.02µg/ml)	Brabakaran and Thangaraju (2018)
Acanthophora spicifera	Ethanol	Glucose 6 phosphate dependent spectrophotometric method	148.32 ± 3.18 mg%	Radhika and Priya (2015)

3 green seaweeds, 22 brown seaweeds, 10 redseaweeds are studied for antidiabetic activity. α-glucosidase inhibitory assay, glucose uptake assay, glucose 6-phosphate dependent spectrophotometric method are used. Among this, α-glucosidase inhibitory assay is widely used.

13 Hepatoprotectivity

The liver regulates many important metabolic functions viz., detoxification of the exogenous xenobiotics, viral infections, chronic alcoholism and drugs and endogenous bile pigments. Liver damage causes warping of these metabolic functions (Wolf 1999). Hepatic injury mostly involves oxidative stress and gives rise to diseases from steatosis to chronic hepatitis, fibrosis, cirrhosis, and hepatocellular carcinoma (Anup and Shivanandappa 2010). Hepatoprotective drugs are quite limited despite advancements in modern medicine. The development of new molecules for hepatic injury remains a challenge in the field of drug development (Ahmad et al. 2016). Many species exhibiting hepatoprotective and antioxidant activities have been isolated for plant based drugs. Silymarin is a well-known potential hepatoprotective drug isolated from Silybum marianum. Hepatoprotectivity activities of seaweeds are tabulated in Table 8.11.

Hepatoprotectivity activities of 1 green seaweeds, 15 brown seaweeds, 6 red seaweeds are summarized. CCl-4 induced, Chromium induced, Rifampicin induced, acetaminophen induced, streptozotocin induced, doxorubicin induced methods are utilized for assessing the activity. Ethanol, methanol, aqueous, acetone, alcohol, diethyl ether solvents are used.

14 Conclusion

Seaweeds are used in medicines, cosmetics, energy, fertilizers and industrial agar and alginate biosynthesis and also used as fodder. Seaweeds hire a source of highly bioactive secondary metabolites that could act as key medicinal components. Due to the presence of minerals, vitamins, phenols, polysaccharides, and sterols, as well as several other bioactive compounds, it consider to have antioxidant, anti-inflammatory, anti-cancer, antimicrobial and anti-diabetic activities (Fig. 8.3). The available literatures confirmed the presence of various pharmacological activities of seaweeds however further research required to isolate the specific bioactive compounds from the crude extracts of seaweeds. The isolated active principles may act as alternative drug for various biological ailments.

Table 8.11 Hepatoprotectivity activity of seaweeds

Name of the seaweed	Solvent	Method	Result	Reference
Caulerpa lentillifera	Methanol	Acetaminophen-induced liver toxicity in *Danio rerio, histology analysis*	At 10–30μg/L concentration, it prevented the progression of hepatic damage caused by 10μM APAP	Kimberly et al. (2020)
Hepatoprotectivity activity of Brown seaweeds				
Sargassum muticum	Methanol	Streptozotocin-(STZ-) induced hepatic injury	Treatment at 200 and 500 mg minimized the glucose level and protect liver.	Safhi et al. (2019)
Sargassum fluitans	Ethanol	Acetaminophen (APAP) in Balb/c mice to induce acute damage; carbon tetrachloride (CCl4) in Wistar rats to induce chronic damage.	50 mg/kg seaweed extract reduced this APAP- and CCl4-induced elevation to normal levels.	Carlos et al. (2018)
Sargassum variegatum, *Sargassum tenerrimum* and *Sargassum binderi*	Ethanol	CCl4-induced toxicity, acetaminophen-induced hepatotoxicity	*S. variegatum* showed highest activity by reducing the elevated level of hepatic enzymes, bilirubin, serum glucose, triglyceride with restoration of cholesterol and urea and creatinine	Hira et al. (2016)
Ecklonia stolonifera, *Ecklonia cava*, *Eisenia bicyclis* and *Pelvetia siliquosa*	95% ethanol	Doxorubicin-induced hepatotoxicity	(EC_{50}) values of 2.0, 2.5, 3.0, and 15.0μg/ml	Jung et al. (2014)
Sargassum dentifolium	Ethanol	(CCl4)-induced hepatitis in rats.	Albumin levels are decreased while, significant increase ($p < 0.05$) in serum marker enzymes AST, ALT, ALP and bilirubin levels in group II (CCl4 intoxicated rats)	Madkour et al. (2012)
Padina boergesenii	Diethyl ether	CCl4-induced oxidative damage and liver fibrosis in rats	*P. boergesenii* extract (150 mg/kg)-decreased in the liver weight (3.57 ± 0.33) g	Karthikeyan et al. (2010)

(continued)

Table 8.11 (continued)

Name of the seaweed	Solvent	Method	Result	Reference
Sargassum polyctum	Alcoholic extract	Acetaminophen induced oxidative stress	Oral pretreatment with plant alcoholic extract (200 mg/kgy wt/day for a period of 15 days) showed hepatoprotectivity	Raghavendran et al. (2004)
Myagropsis myagroides, Sargassum henslowianum and S. siliquastrum	Aqueous	(CCl4)-induced liver injury in the rat	1.25 ml/kg of CCl4 was able to produce increased levels of serum glutamic pyruvic transaminase (GPT) and glutamic oxaloacetic transminase (GOT)	Wong et al. (2000)
Hepatoprotectivity activity of Red seaweeds				
Gracilaria corticata	Ethanol	CCl4 treated change liver cell lines.	48%	Abraham et al. (2018)
Bryothamnion triquetrum	Aqueous extract	CCl4-induced oxidative damage in Wistar rats	The pre-treatment with *B. triquetrum* (200 mg extract/kg), led to 1.93 times reduction in TBARS liver levels with respect to control approximately while in animals treated with ferulic acid was 1.55 times (0.45 ± 0.10 nmol/ mg protein)	Novoa et al. (2019)
Hypnea muciformis	Ethanol	CCl4 induced hepatotoxicity in male albino rats	87.5% hepatoprotective activity	Bupesh et al. (2012)
Portieria hornemannii	Methanol	Chromium-induced oxidative damage in Wistar rats	The animals treated with methanol extracts (200 mg/kg) prevent the leakage of the Enzymes into the blood.	Subbiah et al. (2016)
Gracillaria crassa and *Laurencia papillosa*	Acetone	Rifampicin induced oxidative damage in albino rats	*L. pappilosa* (81%) at 200 mg/kg	Senthil and Murugan (2013)

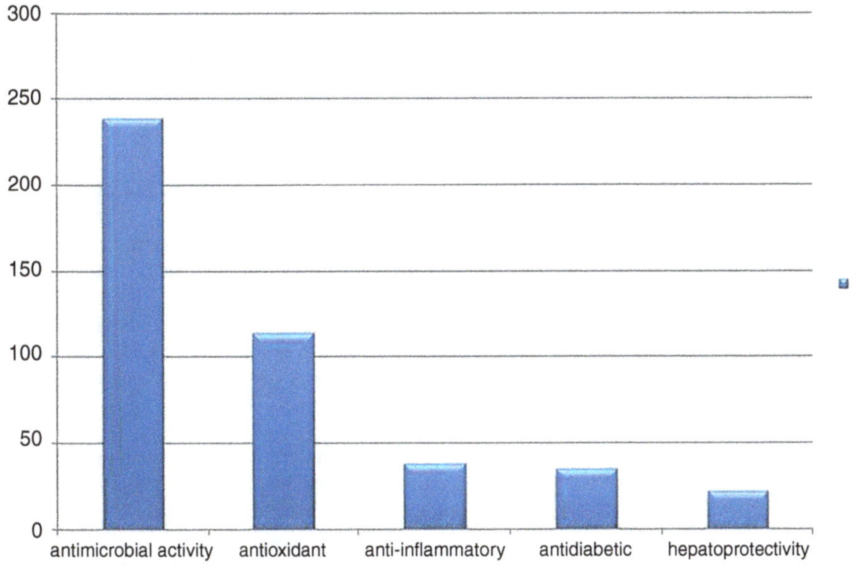

Fig. 8.3 Pharmacological activity of seaweeds

References

Abdel-Khaliq A, Hassan HM, Rateb ME et al (2014) Antimicrobial activity of three *Ulva* species collected from some Egyptian Mediterranean seashores. Int J Eng Res Gen Sci 2(5):648–669

Abraham J, Sheeba DG, Gomathy S (2018) Hepatoprotective and anti-inflammatory activity of marine red algae *Gracillaria corticata*. Biosci Discov 9(2):232–236

AftabUddin S, Akter S, Hossen S et al (2020) Antioxidant, antibacterial and cytotoxic activity of *Caulerpa racemosa* (Forsskål) J. Agardh and *Ulva (Enteromorpha) intestinalis* L. Bangladesh. J Sci Ind Res 55(4):237–244

Agbaje-Daniels F, Adeleye A, Nwankwo D et al (2020) Antibacterial activities of selected green seaweeds from West African coast. EC Pharmacol Toxicol 8(4):84–92

Agregán R, Munekata PE, Domınguez R et al (2017) Proximate composition, phenolic content and *in vitro* antioxidant activity of aqueous extracts of the seaweeds *Ascophyllum nodosum*, *Bifurcaria bifurcata* and *Fucus vesiculosus*. Effect of addition of the extracts on the oxidative stability of canola oil under accelerated storage conditions. Food Res Int 99(3):986–994. https://doi.org/10.1016/j.foodres.2016.11.009

Águila-Ramírez RN, González AA, Hernández-Guerrero CH, Bárbara GA, Borges-Souza JM, Véron B, Josephine P, Claire H (2012) Antimicrobial and antifouling activities achieved by extracts of seaweeds from Gulf of California, Mexico. Hidrobiológica 22(1):8–15

Ahmad I, Aqil F, Owais M (2016) Modern phytomedicine: turning medicinal plants into drugs. Wiley, New York

Alencar DBD, Da Silva SR, Pires-Cavalcante MS et al (2014) Antioxidant potential and cytotoxic activity of two red seaweed species, *Amansia multifida* and *Meristiella echinocarpa*, from the coast of Northeastern Brazil. Ann Braz Acad Sci 86(1):251–263

Alphonse JKM, Asha KRT, Immanuel G (2019) Evaluation of anticancer and anti-inflammatory potential of methanolic extract of green seaweed *Ulva fasciata*. Int J Pharm Biol Sci-IJPBSTM 9(2):379–387

Alshalmani SK, Zobi NH, Bozakouk IH (2014) Antibacterial activity of Libyan seaweed extracts. Int J Pharm Sci Res-IJPSR 5(12):5425–5429

Andreea C, Oana A, Beatrice L (2010) Antifungal activity of macroalgae extracts. Scientific Papers, UASVM Bucharest. Series A LIII:442–447

Andriani Y, Syamsumira DF, Yee TC et al (2016) Biological activities of isolated compounds from three edible Malaysian red seaweeds, *Gracilaria changii*, *G. Manilaensis* and *Gracilaria sp.* Nat Prod Commun 11(8):1117–1120

Anitha C, Akila, Muninathan N et al (2020) Studies on phytochemical analysis and antimicrobial activity of *Sargassum Wightii*. Medico-Legal Update 20(4):2340–2343

Anjum A, Aruna G, Noorjahan CM (2014) Phytochemical analysis and antibacterial activity of selected seaweeds from sea coast of Mandapam, Tamilnadu. Indian J Appl Microbiol 17(1):50–58

Anup S, Shivanandappa T (2010) Hepatoprotective effect of the root extract of Decalepis hamiltonii against carbon tetrachloride-induced oxidative stress in rats. Food Chem 118:411–417

Arsianti A, Fadilah F, Bahtiar A et al (2019) Phytochemistry profile and *in vitro* cytotoxicity of seaweed macroalgae *Sargassum polycystum* against colon HCT-116 and lung A-549 cancer cells. Int J Green Pharm 13(2):141–146

Arunkumar K, Raja R, Sameer Kumar VB et al (2021) Antioxidant and cytotoxic activities of sulfated polysaccharides from five different edible seaweeds. J Food Meas Charact 15:567–576

Ayesha H, Sultana V et al (2010) *In vitro* cytotoxicity of seaweeds from Karachi coast on brine shrimp. Pak J Bot 42(5):3555–3560

Babu A, Johnson M, Raja PD (2015) Bio-efficacy of green seaweeds from south east coast of Tamil Nadu, India. J Microbiol Exp 2(2):80–85. https://doi.org/10.15406/jmen.2015.02.00045

Bedoux G, Caamal-Fuentes E, Boulho R et al (2017) Antiviral and cytotoxic activities of polysaccharides extracted from four tropical seaweed species. Nat Prod Commun 12(6):807–811

Begum SMFM, Hemalatha S (2017) Characterization, *in silico* and *in vitro* determination of antidiabetic and anti-inflammatory potential of ethanolic extract of *sargassum wightii*. Asian J Pharm Clin Res 10(4):297–301

Belkacemi L, Belalia M, Djendara AC et al (2020) Antioxidant and antibacterial activities and identification of bioactive compounds of various extracts of *Caulerpa racemosa* from Algerian coast. Asian Pac J Trop Biomed 10(2):87–94

Bharath B, Pavithra AN, Divya A et al (2020) Chemical composition of ethanolic extracts from some seaweed species of the South Indian coastal zone, their antibacterial and membrane-stabilizing activity. Russ J Mar Biol 46(5):370–378

Bhatia S, Sharma K, Sharma A et al (2015) Anti-inflammatory, analgesic and antiulcer properties of *Porphyra vietnamensis*. Avicenna J Phytomed 5(1):69–77

Bhuyar P, Rahim MH, Sundararaju S et al (2020) Antioxidant and antibacterial activity of red seaweed; *Kappaphycus alvarezii* against pathogenic bacteria. Global J Environ Sci Manage 6(1):47–58

Boonchum W, Peerapornpisal Y, Kanjanapothi D et al (2011) Antimicrobial and anti-inflammatory properties of various seaweeds from the gulf of Thailand. Int J Agric Biol 13:100–104

Bouhlal R, Riadi H, Martínez J et al (2010) The antibacterial potential of the seaweeds (Rhodophyceae) of the Strait of Gibraltar and the Mediterranean Coast of Morocco. Afr J Biotechnol 9(38):6365–6372

Bouhraoua J, Lakhdar F, Mabrouki S et al (2018) Antibacterial activity of 23 seaweeds from the Coast of El Jadida Morocco against *Fusarium culmorum* and *Alternria alternata*. Int J Pharm Sci Rev Res 51(2):53–60

Brabakaran A, Thangaraju N (2018) In vitro evaluation of methanolic extract of red seaweeds against α-amylase and α-glucosidase enzyme inhibitory activity Asian. J Pharm Pharmacol 4(3):339–342

Buffet-Bataillon S, Tattevin P, Bonnaure-Mallet M, Jolivet-Gougeon (2012) An emergence of resistance to antibacterial agents: the role of quaternary ammonium compounds-a critical review. Int J Antimicrob Agents 5:381–389

Bupesh G, Amutha C, Vasanth S et al (2012) Hepatoprotective efficacy of *Hypnea muciformis* ethanolic extract on CCl4 induced toxicity in rats. Braz Arch Biol Technol 55(6):857–863

Calderwood D, Rafferty E, Fitzgerald C et al (2021) Profiling the activity of edible European macroalgae towards pharmacological targets for type 2 diabetes mellitus. Applied Phycology 2(1):10–21. https://doi.org/10.1080/26388081.2020.1852519

Canoy JL, Bitacura JG (2018) Cytotoxicity and antiangiogenic activity of *Turbinaria ornata* Agardh and *Padina australis* Hauck ethanolic extracts, Hindawi Anal Cell Pathol. 2018:3709491, 8 pages. https://doi.org/10.1155/2018/3709491

Carlos QN, Rangel-Méndez J, Ortiz-Tello A et al (2018) A *Sargassum fluitans* Borgesen ethanol extract exhibits a hepatoprotective effect *in vivo* in acute and chronic liver damage models. BioMed Res Int 2018:6921845, 9 pages. https://doi.org/10.1155/2018/6921845

Chakraborty K, Dhara S (2019) First report of substituted 2*H*-pyranoids from brown seaweed *Turbinaria conoides* with antioxidant and anti-inflammatory activities. Nat Prod Res:2–11. https://doi.org/10.1080/14786419.2019.1578761

Chalini K, Johnson M, Adaikalaraj G et al (2017) Anti-inflammatory activity of aqueous extracts of *Gracilaria*. Int J Curr Pharm Res 9(5):17–19

Chander MP, Veeraragavam S, Vijayachari P (2014) Antimicrobial and hemolytic activity of seaweed *Padina gymnospora* from South Andaman, Andaman and Nicobar Islands of India. Int J Curr Microbiol App Sci 3(6):364–369

Chia YY, Kanthimathi MS, Khoo KS et al (2015) Antioxidant and cytotoxic activities of three species of tropical seaweeds. BMC Complement Altern Med 15:339. https://doi.org/10.1186/s12906-015-0867-1

Chin YX, Lim PE, Phang SM et al (2014) Anti-diabetic potential of selected Malaysian seaweeds 5th congress of the International Society for Applied Phycology. J Appl Phycol. https://doi.org/10.1007/s10811-014-0462-8

Chingizova EA, Skriptsova AV, Anisimov MM et al (2017) Antimicrobial activity of marine algal extracts. Int J Phytomed 9(1):113–122

Čmiková N, Galovičová L, Miškeje M, Borotová P, Kluz M, Kačániová M (2022) Determination of antioxidant, antimicrobial activity, heavy metals and elements content of seaweed extracts. Plan Theory 11:1493. https://doi.org/10.3390/plants11111493

Corato UD, Salimbeni R, Pretis AD et al (2017) Antifungal activity of crude extracts from brown and red seaweeds by a supercritical carbon dioxide technique against fruit postharvest fungal diseases. Postharvest Biol Technol 131:16–30

Cox S, Abu-Ghannam N, Gupta S (2010) An assessment of the antioxidant and antimicrobial activity of six species of edible Irish seaweeds. Int Food Res J 17:205–220

Delgado NG, Vázquez AIF, Sánchez HC et al (2013) Anti-inflammatory and antinociceptive activities of methanolic extract from red seaweed *Dichotomaria obtusata*. Braz J Pharm Sci 49(1):65–73

Deveau AM, Miller-Hope Z, Lloyd E et al (2016) Antimicrobial activity of extracts from macroalgae *Ulva lactuca* against clinically important *Staphylococci* is impacted by lunar phase of macroalgae harvest. Lett Appl Microbiol 62(5):363–371

Dewi MK, Arsianti A, Zagloel CRZ et al (2018) *In vitro* evaluation of seaweed *Gracilaria verrucosa* for cytotoxic activity against cervical HeLa cells. Pharmacogn J 10(5):1007–1011

Dhinakaran DI, Rajalakshmi R, Sivakumar T et al (2016) Antimicrobial activities and bioactive metabolites from marine algae *Valoniopsis pachynema* and *Sargassum swartzii*. JPRPC 4(1):19–26

Durairaj SB, Andiyappan BR (2020) Screening of phytochemicals, antibacterial, antioxidant and anti-inflammatory activity of *Dictyota barteyresiana* seaweed extracts. Asian J Biol Life Sci 9(1):20–26

El Shafay SM, Ali SS, El Sheekh M (2015) Antimicrobial activity of some seaweeds species from Red Sea, against multidrug resistant bacteria. Egypt J Aquat Res 42:65–74. https://doi.org/10.1016/j.ejar.2015.11.006

El Wahidi M, El Amraoui B, El Amraoui M et al (2015) Screening of antimicrobial activity of macroalgae extracts from the Moroccan Atlantic coast. Ann Pharm Fr. https://doi.org/10.1016/j.pharma.2014.12.005

El-Beltagi HS, Mohamed HI, Abdelazeem AS, Youssef R, Safwat G (2019) GC-MS analysis, antioxidant, antimicrobial andanticancer activities of extracts from *Ficus sycomorus* fruits and leaves. Not Bot Horti Agrobot Cluj-Napoca 47:493–505

Elnabris KJ, Elmanama AA, Chihadeh WN (2013) Antibacterial activity of four marine seaweeds collected from the coast of Gaza Strip, Palestine?. Mesopot J Mar Sci 28(1):81–92

El-Shouny WA, Gaafar RM, Ismail GA et al (2017) Antibacterial activity of some seaweed extracts against multidrug resistant urinary tract bacteria and analysis of their virulence genes. Int J Curr Microbiol App Sci 6(11):2569–2586

Erfani N, Nazemosadat Z, Moein M (2015) Cytotoxic activity of ten algae from the Persian Gulf and Oman sea on human breast cancer cell lines MDA-MB-231, MCF-7 and T-47D. Pharm Res 7(2):133–137

Farasat M, Khavari-Nejad R-A, Nabavi SMB et al (2013) Antioxidant properties of two edible green seaweeds from northern coasts of the Persian Gulf. Jundishapur J Nat Pharm Prod 8(1):47–52

Fayzi L, Askarne L, Cherifi O et al (2020) Comparative antibacterial activity of some selected seaweed extracts from Agadir coastal regions in Morocco. Int J Curr Microbiol App Sci 9:390–399

Ghalem BR, Zouaoui B (2018) Antibacterial activity of diethyl ether and chloroform extracts of seaweeds against Escherichia coli and Staphylococcus aureus. Int J Avian Wild 3(4):310–313

Ghannadi A, Shabani L, Yegdaneh A (2016) Cytotoxic, antioxidant and phytochemical analysis of *Gracilaria* species from Persian Gulf. Adv Biomed Res 5:139

Gunasekaran S, Vinoth Kumar T, LakshmanaSenthil S, Suganya P, Rincy Y, Amrutha C (2017) Screening of in vitro cytotoxic activity of brown seaweeds againsthepatocellular carcinoma. J Appl Pharm Sci 7(5):51–55. https://doi.org/10.7324/JAPS.2017.70509

Gutierrez-Rodriguez AG, Juarez-Portilla C, Olivares-Banuelos T et al (2018) Anticancer activity of seaweeds. Drug Discov Today 23:434–447

Hamed MM, Abd El-Mobdy MA, Kamel MT et al (2019) Phytochemical and biological activities of two asteraceae plants *Senecio vulgaris* and *Pluchea dioscoridis* L. Pharmacol Online 2:101–121

Hasan M, Imran Md AS, Bhuiyan FR (2019) Phytochemical constituency profiling and antimicrobial activity screening of seaweeds extracts collected from the Bay of Bengal Sea coasts. Conference: National Biotechnology Fair

He J, Xu Y, Chen H et al (2016) Extraction, structural characterization and potential antioxidant activities of the polysaccharides from four seaweeds. Int J Mol Sci 17:1988

Hira K, Sultana V, Ara J et al (2016) Hepatoprotective potential of three *sargassum* species from Karachi coast against carbon tetrachloride and acetaminophen intoxication. J Coast Life Med 4(1):10–13

Hocman G (1989) Prevention of cancer: vegetables and plants. Comp Biochem Physiol 93(2):201–212

Hong DD, Hien HM, Hoang TLA (2011) Studies on the analgesic and anti-inflammatory activities of *Sargassum swartzii* (Turner) C. Agardh (Phaeophyta) and *Ulva reticulata* Forsskal (Chlorophyta) in experiment animal models. Afr J Biotechnol 10(12):2308–2314

Husni A, Anggara FP, Isnansetyo A et al (2016) Blood glucose level and lipid profile of streptozotozin-induced diabetic rats treated with Sargassum polystum extract. J Biol Sci 16(3):58–64

Hussain E, Wang LJ, Jiang B et al (2016) A review of the components of brown seaweeds as potential candidates in cancer therapy. RSC Adv 6:12592–12610

Imran MAS, Bhuiyan FR, Ahmed SR et al (2021) Phytochemical constituency profiling and antimicrobial activity screening of seaweeds extracts collected from the Bay of Bengal sea coasts. J Adv Biotechnol Exp Ther 4(1):25–34

Ismail GA, Gheda SF, Abo-Shady AM, Abdel-Karim OH (2020) *In vitro* potential activity of some seaweeds as antioxidants and inhibitors of diabetic enzymes. Food Sci Technol Campinas 40(3):681–691

Jayanta Kumar P, Rath SK, Jena K (2008) Evaluation of antioxidant and antimicrobial activity of seaweed (Sargassum sp.) Extract: A Study on Inhibition of Glutathione-S-Transferase Activity. Turk J Biol 32:119–125

Jeevitha K, Damahe J, Das S et al (2014) *In vitro* antioxidant and cytotoxic activity of brown alga *Padina Boergesenii*. Int J Drug Dev Res 6(2):110–119

Jung HA, Kimb J-I, Choungc SY et al (2014) Protective effect of the edible brown alga *Ecklonia stolonifera* on doxorubicin-induced hepatotoxicity in primary rat hepatocytes. J Pharm Pharmacol 66:1180–1188

Kadir MIA, Ahmad MR, Ismail A et al (2016) Investigations on the cytotoxicity, neurotoxicity and dyeing performances of natural dye extracted from *Caulerpa lentillifera* and *Sargassum* sp. seaweeds. Adv Appl ScI 1(3):46–52. https://doi.org/10.11648/j.aas.20160103.11

Kajal C, Vamshi KR (2018) *Invitro* bioactive analysis and antioxidant activity of two species of seaweeds from the Gulf of Mannar. Nat Prod Res 32(22):2729–2734. https://doi.org/10.1080/14786419.2017.1375923

Kalliswari G, Mahendran S, Subalakshmi P et al (2016) Purification, characterization and antioxidant activity of green seaweed Codium sp. Adv Pharmacol Pharm 4(2):16–21

Kang JY, Khan MNA, Park MNH et al (2008) Antipyretic, analgesic, and anti-inflammatory activities of the seaweed *Sargassum fulvellum* and *Sargassum thunbergii* in mice. J Ethnopharmacol 116(2008):187–190

Kantachumpoo A, Chirapart A (2010) Components and antimicrobial activity of polysaccharides extracted from Thai brown seaweeds. Kasetsart J (Nat Sci) 44:220–233

Karthick P, Mohanraju R, Kada NM et al (2015) Antibacterial activity of seaweeds collected from South Andaman, India. J Algal Biomass Utln 6(1):33–36

Karthikeyan R, Somasundaram ST, Manivasagam T et al (2010) Hepatoprotective activity of brown alga Padina boergesenii against CCl4 induced oxidative damage in Wistar rats. Asian Pac J Trop Med 2010:696–701

Kavitha K, Rynghang JS, Peter JD (2017) Antimicrobial activity of seaweed – Ulva lactuca common bacterial pathogens, *Staphyococcus aureus* and *Eschericia coli*. Indian J Appl Microbiol 20(1):42–46

Khalid S, Abbas M, Saeed F et al (2018) Therapeutic potential of seaweed bioactive compounds. Seaweed Biomater 2. https://doi.org/10.5772/intechopen.74060

Khan MNA, Choi JS, Lee MC et al (2008) Anti-inflammatory activities of methanol extracts from various seaweed species. J Environ Biol 29(4):465–469

Kim KN, Ham YM, Moon JY et al (2009) In vitro cytotoxicity activity of *Sargassum thunbergii* and *Dictyopteris divaricata* (Jeju seaweeds) on the HL-60 tumour cell line. Int J Pharmacol 5(5):298–306

Kim KN, Kim J, Yoon WJ, Yang H-M et al (2013) Inhibitory effect of *Sargassum patens* on inflammation and melanogenesis. Indian J Pharmacol 9(8):524–532

Kimberly DC, Marquina REM, Nolasco ADG et al (2020) Evaluation of the hepatoprotective effect of methanolic extract of *Caulerpa lentillifera* against acetaminophen-induced liver toxicity in juvenile zebrafish (*Danio rerio*). Jurnal Illmiah Farmasi 16(1):1–89

Kindleysides S, Quek SY, Miller MR (2012) Inhibition of fish oil oxidation and the radical scavenging activity of New Zealand seaweed extracts. Food Chem 133:1624–1631

Klimjit A, Praiboon J, Tiengrim S et al (2021) Phytochemical composition and antibacterial activity of brown seaweed, *Padina australis* against human pathogenic bacteria. J Fish Environ 45(1):9–22

Kolanjinathan K, Ganesh P, Govindarajan M (2009) Antibacterial activity of ethanol extracts of seaweeds against fish bacterial pathogens. Eur Rev Med Pharmacol Sci 13:173–177

Krishnapriya MV, Milton MCJ, Arulvasu C (2013) Evaluation of antibacterial properties of selected red seaweeds from Rameshwaram, Tamil Nadu, India. J Acad Ind Res 1(11):667–670

Kumar RR, Prakash KJ (2019) Cell viability and cytotoxic effect of *acanthophora spicifera* (red seaweed) on ht29 colon cancer cell line. IJPSR 10(7):3318–3324
Kumar S, Sudha S (2012) Evaluation of alpha-amylase and alpha-glucosidase inhibitory properties of selected seaweeds from Gulf of Mannar. Int Res J Pharm 3:128–130
Kumar JGS, Umamaheswari S, Kavimani S, Ilavarasan R (2019) Pharmacological potential of green algae Caulerpa: a review. Int J Pharm Sci Res 10:1014–1024
Kumari SS, Kumar VD, Priyanka B (2017) Antifungal efficacy of seaweed extracts against fungal pathogen of silkworm, *Bombyx mori* L. Int J Agric Res 12:123–129
Kwon PS, Oh H, Kwon SJ et al (2020) Sulfated polysaccharides effectively inhibit SARS-CoV-2 *in vitro*. Cell Discov 6:50
Latha D, Latha KPJH (2011) Antimicrobial activity of the chloroform extracts of the chlorophycean seaweeds *Enteromorpha compressa* and *Chaetomorpha antennina*. Int Res J Microbiol (IRJM) 2(8):249–252
Lee CW, Han JS (2012) Hypoglycemic effect of Sargassum ringgoldianum extract in STZ-induced diabetic mice. Prev Nutr Food Sci 17(1):8–13
Lee JW, Wang JH, Ng KM et al (2015) *In-vitro* anticancer activity of *eucheuma cottonii* extracts against HeLa cell line, humn lung carcinoma cell line (SK-LU-1), human colon carcinoma cell line (HCT-116), and fibroblast. IJCMS 1(2):69–73
Levy SB, Marshall B (2002) Antibacterial resistance worldwide: causes, challenges and responses. Nat Med 12:122–129
Li Y, Sun S, Pu X et al (2018) Evaluation of antimicrobial activities of seaweed resources from Zhejiang Coast, China. Sustainabilty 10:2158. https://doi.org/10.3390/su10072158
Madhavarani A, Ramanibai R (2014) In-vitro antibacterial activity of *Kappaphycus alvarezii* extracts collected from Mandapam Coast, Rameswaram, Tamil Nadu. Int J Innov Res Sci Eng Technol 3(1):8436–8440
Madkour FF, Khalil WF, Dessouki AA (2012) Protective effect of ethanol extract of *Sargassum dentifolium* (Phaeophyceae) in carbon tetrachloride induced hepatitis in rats. Int J Pharm Pharm Sci 4(3):637–641
Maithili SS, Ramanathan G (2020) Evaluation of antioxidant and their cytotoxic properties on secondary metabolites from green seaweed. Inf Res 9(1):754–770
Maneesh A, Chakraborty K, Makkar F (2017) Pharmacological activities of brown seaweed *Sargassum wightii* (Family Sargassaceae) using different *invitro* models. Int J Food Prop 20(4):931–945. https://doi.org/10.1080/10942912.2016.1189434
Manimegalai V, Bakiya Lakshmi SV (2020) A study on phytochemical and pharmacological activity of marine algae (*Kapphaphycus alverizii)*. Int J Anal Exp Modal Anal 12(8):2089–2099
Manivannan K, Karthikai Devi G, Anantharaman P et al (2011) Antimicrobial potential of selected brown seaweeds from Vedalai coastal waters, Gulf of Mannar. Asian Pac J Trop Biomed 1:114–120
Mansuya P, Aruna P, Sridhar S et al (2010) Antibacterial activity and qualitative phytochemical analysis of selected seaweeds from Gulf of Mannar region. J Exp Sci 1(8):23–26
Mary JS, Vinotha P, Pradeep AM (2012) Screening for *in vitro* cytotoxic activity of seaweed, *Sargassum sp.* against Hep-2 and MCF-7 cancer cell lines. Asian Pac J Cancer Prev 13:6073–6076
Mazaheri G, Fazilati M, Nazem H (2017) Survey of antioxidant and cytotoxic activities of *Gracilaria corticata* (a red seaweed), against RKO, AGS and HepG2 human cancer cell lines. J Biol Today's World 6(12):269–273
Mhadhebi L, Laroche-Clary A, Robert J et al (2011a) Anti-inflammatory, anti-proliferative and anti-oxidant activities of organic extracts from the Mediterranean seaweed, *Cystoseira crinita*. Afr J Biotechnol 10(73):16682–16690
Mhadhebi L, Laroche-Clary A, Robert J et al (2011b) Antioxidant, anti-inflammatory, and antiproliferative activities of organic fractions from the Mediterranean brown seaweed *Cystoseira sedoides*. Can J Physiol Pharmacol 89:911–921

Mhadhebi L, Laroche-Clary A, Robert J et al (2014) Antioxidant, anti-inflammatory, and antiproliferative effects of aqueous extracts of three Mediterranean brown seaweeds of the genus *Cystoseira*. Iran J Pharm Res 13(1):207–220

Mishra AK (2018) Sargassum, Gracilaria and Ulva exhibit positive antimicrobial activity against human pathogens. Open Access Libr J 5:e4258. https://doi.org/10.4236/oalib.1104258

Mohapatra L, Bhattamisra SK, Chandra Panigrahy R et al (2016) Evaluation of the antioxidant, hypoglycaemic and antidiabetic activities of some seaweed collected from the east coast of India. Biomed Pharmacol J 9(1):365–375

Moo-Puc R, Robledo D, Freile-Pelegrín Y (2009) *In vitro* cytotoxic and antiproliferative activities of marine macroalgae from Yucatán, Mexico. Cienc Mar 35(4):345–358

Moubayed NMS, Al Houri HJ, Al Khulaifi MM et al (2017) Antimicrobial, antioxidant properties and chemical composition of seaweeds collected from Saudi Arabia (Red Sea and Arabian Gulf). Saudi J Biol Sci 24:162–169

Mohy El-Din SM, Mohyeldin MM (2018) Component analysis and antifungal activity of the compounds extracted from four brown seaweeds with different solvents at different seasons. J Ocean Univ China 17:1178–1188. https://doi.org/10.1007/s11802-018-3538-2

Nazarudin MF, Yasin ISM, Mazli NAIN et al (2022) Preliminary screening of antioxidant and cytotoxic potential of green seaweed, Halimeda opuntia (Linnaeus) Lamouroux. Saudi J Biol Sci 29(2022):2698–2705

Ngo D-H, Se-Kwon K (2013) Sulfated polysaccharide as bioactive agents from marine algae. Int J Biol Macromol 62:70–75

Nguyen TH, Nguyen TH, Nguyen VM et al (2019) Antidiabetic and antioxidant activities of red seaweed *Laurencia dendroidea*. Asian Pac J Trop Biomed 9(12):501–509

Novoa AJV, Silva AMO, Mancini DAP et al (2019) Hepatoprotective properties from the seaweed *Bryothamnion triquetrum* (S.G.Gmelin) M.A.Howe againstCCl4-induced oxidative damage in rats. J Pharm Pharmacogn Res 7(1):31–46

Nygaard K, Lunestad BT, Hektoen H, Berge JA, Hormazabal V (1992) Resistance to oxytetracycline, oxolinic acid and furazolidone in bacteria from marine sediments. Aquaculture 104(1–2):31–36

O'Keeffe E, Hughes H, McLoughlin P (2019) Antibacterial activity of seaweed extracts against plant pathogenic bacteria. J Bacteriol Mycol 6(3):1105

Olasehinde TA, Olaniran AO, Okoh AI (2019) Phenolic composition, antioxidant activity, anticholinestrase potential and modulatory effects of aqueous extracts of some seaweeds on β-amyloid aggregation and disaggregation. Pharm Biol 57(1):460–469

Osman NAHK, Siam AA, El-Manawy IM et al (2019) Anti-microbial and anti-diabetic activity of six seaweeds collected from the Red Sea, Egypt. Catrina 19(1):55–60

Pandian P, Selvamuthukumar S, Manavalan R et al (2011) Screening of antibacterial and antifungal activities of red marine algae *Acanthaphora spicifera* (Rhodophyceae). Biomed Sci Res 3(3):444–448

Patra JK, Rath SK, Jena K (2008) Evaluation of antioxidant and antimicrobial activity of seaweed (Sargassum sp.) extract: a study on inhibition of glutathione-S-transferase activity. Turk J Biol 32:119–125

Pechroj S, Potiparsat K, Nangam O et al (2020) Comparative evaluation of antioxidant and anti-inflammatory activities of four seaweed species from the east coast of the Gulf of Thailand. Naresuan Phayao J 13(3):11–21

Peihang XU, Huaqiang TAN, Weiguang JIN et al (2018) Antioxidative and antimicrobial activities of intertidal seaweeds and possible effects of abiotic factors on these bioactivities. J Oceanol Limnol 36(6):2243–2256. https://doi.org/10.1007/s00343-019-7046-z

Peres JCF, de Carvalho LR, Goncalez et al (2012) Evaluation of antifungal activity of seaweed extracts. Cienc Agrotec Lavras 36(3):294–299

Pinteus S, Silva J, Alves C et al (2016) Cytoprotective effect of seaweeds with high antioxidant activity from the Peniche coast (Portugal). Food Chem 218:591–599. https://doi.org/10.1016/j.foodchem.2016.09.067

Pinteus S, Silva J, Alves C (2017) Cryoprotective effect of seaweeds with high antioxidant activity from the Peniche Coast (Portugal). Food Chem 218:591–599

Ponnanikajamideen M, Suneetha V, Rajeshkumar S (2017) Antidiabetic, antihyperlipedimic and antioxidant activity of marine brown seaweed *Padina tetrastromatica*. J Chem Pharm Sci: JCPS 10(1):379–384

Pragnesh ND, Parul NM, Vijay RR et al (2017) Antibacterial and antioxidant activity of Sulphated polysaccharides from selected seaweeds of Vadinar coast. Acta Chim Pharm Indica 7(3):111

Prasad MP, Shekhar S, Babhulkar AP (2013) Antibacterial activity of seaweed (*Kappaphycus*) extracts against infectious pathogens. Afr J Biotechnol 12(20):2968–2971

Prasasty VD, Haryani B, Hutagalung RA et al (2019) Evaluation of antioxidant and anti diabetic activities from red seaweed (*Eucheuma cottonii*) extract. Sys Rev Pharm 10(1):276–288

Premarathna AD, Ranahewa TH, Wijesekera SK et al (2020) Preliminary screening of the aqueous extracts of twenty-three different seaweed species in Sri Lanka with in-vitro and in-vivo assays. Heliyon 6:e03918

Priyadharshini S, Bragadeeswaran S, Prabhu K et al (2011) Antimicrobial and hemolytic activity of seaweed extracts Ulva fasciata (Delile 1813) from Mandapam, Southeast coast of India. Asian Pac J Trop Biomed 2011:S37–S39

Qiao J (2010) Antibacterial effect of extracts from two Icelandic algae (Ascophyllum nodosum and Laminaria digitata) Final Project Report 2010. https://www.grocentre.is/static/gro/publication/226/document/jin-proofreadcaitlin-and-ski-mfd.pdf

Radhika D, Priya R (2015) Assessment of anti-diabetic activity of some selected seaweeds. Eur J Biomed Pharm Sci 2(6):151–154

Radhika D, Veerabahu C, Priya R (2013a) Invitro studies on antioxidant and haemagglutination activity of some selected seaweeds. Int J Pharm Pharm Sci 5(1):152–155

Radhika D, Veerabahu C, Vijayalakshmi M et al (2013b) Antibacterial effect of some species of seaweeds from different stations off the Gulf of Mannar coast-India. Int J Biol Pharm Res 4(11):783–787

Raghavendran HRB, Sathivel A, Devaki T (2004) Hepatoprotective nature of seaweed alcoholic extract on acetaminophen induced hepatic oxidative stress. J Health Sci 5(1):42–46

Ragunathan V, Pandurangan J, Ramakrishnan T (2019) Gas chromatography mass spectrometry analysis of methanol extracts from marine red seaweed *Gracilaria corticata*. Pharmacogn J 11:547–554

Raja R, Hemaiswarya S, Arunkumar K et al (2016) Antioxidant activity and lipid profile of three seaweeds of Faro, Portugal. Braz J Bot 39(1):9–17

Rajakumari M, Kaleeswari B (2019) Anti-inflammatory, antipyretic, wound healing properties of marine brown algae *Sargassum wightii*. Int J Sci Res 8(1):50–53

Rani V, Jawahar P, Shakila RJ et al (2016) Antibacterial activity of some brown seaweeds of Gulf of Mannar, south east coast of India. J Pharm BioSci 4(3):14–21

Rasheed I, Tabassum A, Aliya R (2021) Cytotoxicity of seaweeds counter to human carcinoma lung cells. Int J Biol Biotechnol 18(2):263–265

Ripol A, Cardoso C, Afonso C et al (2018) Composition, anti-inflammatory activity and bioaccessibility of green seaweeds from fish pond aquaculture. Nat Prod Commun 13(5):603–608

Rose JB, Epstein PR, Lipp EK, Sherman BH, Bernard SM, Patz JA (2001) Climate variability and change in the United States: potential impacts on water-and foodborne diseases caused by microbiologic agents. Environ Health Perspect 109(2):211–221

Safhi MM, Alam MF, Sivakumar SM et al (2019) Hepatoprotective potential of Sargassum muticum against STZ-induced diabetic liver damage in Wistar rats by inhibiting cytokines and the apoptosis pathway. Anal Cell Pathol 2019:7958701, 8 pages. https://doi.org/10.1155/2019/7958701

Saidani K, Bedjou F, Benabdesselam F et al (2012) Antifungal activity of methanolic extracts of four Algerian marine algae species. Afr J Biotechnol 11(39):9496–9500

Saleh B, Al-mariri A (2018) Antifungal activity of crude seaweed extracts collected from Latakla Coast, Syria. J Fish Aquat Sci 13:49–55

Salvador N, Garreta AG, Lavelli L et al (2007) Antimicrobial activity of Iberian macroalgae. Sci Mar 71(1):101–113
Sanger G, Rarung LK, Damongilala LJ et al (2019) Phytochemical constituents and antidiabetic activity of edible marine red seaweed (*Halymenia durvilae*). IOP Conf Ser Earth Environ Sci 278(2019):012069
Sanjeewa KKA, Kang N, Ahn G et al (2018) Bioactive potentials of sulfated polysaccharides isolated from brown seaweed *Sargassum* spp in related to human health applications: a review. Food Hydrocoll 81:200–208
Saranya C, Parthiban C, Anantharaman P (2014) Evaluation of antibacterial and antioxidant activities of seaweeds from Pondicherry coast. Adv Appl Sci Res 5(4):82–90
Sasikala C, Ramani GD (2017) Comparative study on antimicrobial activity of seaweeds. Asian J Pharm Clin Res 10(12):384–386
Sayegh F (2018) Antimicrobial activity of some seaweed collected from South East Coast of Jeddah, Saudi Arabia. Int J Pharm Res Allied Sci 7(2):153–159
Seema P (2012) Therapeutic importance of sulfated polysaccharides from seaweeds: updating the recent findings. Biotech 2(3):171–185
Seenivasan R, Rekha M, Indu H et al (2012) Antibacterial activity and phytochemical analysis of selected seaweeds from Mandapam coast. India J Appl Pharm Sci 2(10):159–169
Selim SA (2012) Antimicrobial, antiplasmid and cytotoxicity potentials of marine algae *Halimeda opuntia* and *Sarconema filiforme* collected from Red Sea Coast World Academy of Science. Eng Technol 61:1154–1159
Selvin J, Lipton AP (2004) Biopotentials of Ulvafasciataand Hypneamusciformis collected from the peninsular coast of India. J Mar Sci Technol 12:1–6
Senthil KA, Murugan A (2013) Antiulcer, wound healing and hepatoprotective activities of the seaweeds Gracilaria crassa, Turbinaria ornata and Laurencia papillosa from the southeast coast of India. Braz J Pharm Sci 49(4):670–677
Shibu A (2019) Antifungal activity of marine brown algae *Sargassum wightii* collected from Gulf of Mannar biosphere reserve. Int J Recent Sci Res 10(12E):36608–36610
Shobier AH, Ismail MM, Hassan SWM (2022) Variation in anti-inflammatory, anti-arthritic, and antimicrobial activities of different extracts of common Egyptian seaweeds with an emphasis on their phytochemical and heavy metal contents. Biol Trace Elem Res 201:2071–2087. https://doi.org/10.1007/s12011-022-03297-1
Shreadah MA, Abd El Moneam NM, Al-Assar SA et al (2018) Phytochemical and pharmacological screening of *Sargassium vulgare* from Suez Canal. Egypt Food Sci Biotechnol 27(4):963–979. https://doi.org/10.1007/s10068-018-0323-3
Silva P, Fernandes C, Barros L et al (2018) The antifungal activity of extracts of *Osmundea pinnatifida*, an edible seaweed, indicates its usage as a safe environmental fungicide or as a food additive preventing post-harvest fungal food contamination. Food Funct 9:6187–6195
Sivaramakrishnan T, Biswas L, Shalini B et al (2017) Analysis of proximate composition and *in-vitro* antibacterial activity of selected green seaweeds from South Andaman Coast of India. Int J Curr Microbiol App Sci 6(12):1739–1749
Sneha K, Vishnu Priya V, Gayathri R (2019) An *in vitro* study on the antidiabetic effect on red sea weed (*Ceramium rubrum*). Drug Invent Today 12(6):1244–1246
Sreejamole KL, Greeshma PM (2013) Antioxidant and Brineshrimp cytotoxicity activities of ethanolic extract of red algae *Gracilaria corticata* (J Agardh) J Agardh. Indian J Nat Prod Resour 4(2):233–237
Subbarangaiah G, Lakshmi P, Manjula E (2010) Antimicrobial activity of seaweeds Gracillaria, Padina and Sargassum sps. on clinical and phytopathogens. Int J Chem Anal Sci 1(6):114–117
Subbiah M, Bhuvaneswari S, Kalandar A et al (2016) Hepatoprotective efficiency of methanol extract of red algae against chromium-induced oxidative damage in Wistar rats. J Coast Life Med 4(7):541–546

Taheri A, Ghaffari M, Bavi Z et al (2018) Cytotoxic effect of seaweed (*Sargassum glaucescens*) extract against breast (MCF-7) and colorectal (HT-29) cancer cell lines. Feyz Med Sci 22(3):292–301

Toufiqul I, Tamim A, Md Abdul A, Md Farzanoor R, Mohammad NH, Alam MM (2020) Bioactive compounds screening and in vitro appraisal of potential antioxidant and cytotoxicity of Cladophoropsis sp. isolated from the Bay of Bengal. EC Pharmacology and Toxicology 8(10):19–31

Trifan A, Vasincu A, Luca SV et al (2019) Unravelling the potential of seaweeds from the Black Sea coast of Romania as bioactive compounds sources. Part I: Cystoseira barbata (Stackhouse) C. Agardh. Food Chem Toxicol 134:110820

Unnikrishnan PS, Suthindhiran K, Jayasri MA (2015) Antidiabetic potential of marine algae by inhibiting key metabolic enzymes. Front Life Sci 8:148–159

Vairappan CS, Kawamoto T, Miwa H, Suzuki M (2004) Potent antibacterial activity of halogenated compounds against antibiotic-resistant bacteria. Planta Med 70(11):1087–1090

Valentina J, Poonguzhali TV, Nisha LLJL (2015) Antibacterial and antioxidant activity, total phenolics and flavonoid contents of brown seaweeds from southeast coast of India. Int J Adv Res 3(5):1131–1135

Vallinayagam K, Arumugam R, Ragupathi RR et al (2009) Antibacterial activity of some selected seaweeds from Pudumadam coastal regions. Global J Pharmacol 3(1):50–52

Vaseghi G, Zakeri N, Mazloumfard F et al (2019) Evaluation of the antituberculosis and cytotoxic potential of the seaweed Padina Australis. Iran J Pharm Sci 15(1):29–38

Venugopal AKB, Thirumalairaj VK, Durairaj G et al (2014) Microbicidal activity of crude extracts from *Sargassum wightii* against *Bacillus cereus*. Int Curr Pharm J 3(10):326–327

Vijayalakshmi S (2015) Screening and anti-inflammatory ativity of methanolic and aqueous extracts of seaweed *Gracilaria edulis*. Int J Mod Chem Appl Sci 2(4):248–250

Vikneshan M, Kumar RS, Mangaiyarkarasi R et al (2020) Antimicrobial activity of *Ulva lactuca*, green algae, against common oral pathogens. SBV J Basic Clin Appl Health Sci 3(4):168–170

Vimala T, Poonghuzhali TV (2017) *In vitro* antimicrobial activity of solvent extracts of marine brown alga, *Hydroclathrus clathratus* (C. Agardh) M. Howe from Gulf of Mannar. J Appl Pharm Sci 7(04):157–162

Vinayak RC, Sudha SA, Chatterji A (2011) Bio-screening of a few green seaweeds from India for their cytotoxic and antioxidant potential. J Sci Food Agric 91:2471–2476

Visvesvara GS, Moura H, Schuster FL (2007) Pathogenic and opportunistic free-living amoebae: *Acanthamoeba spp., Balamuthia mandrillaris, Naegleriafowleri,* and *Sappiniadiploidea.* FEMS Immunol Med Microbiol 50:1–26

WHO (2019) https://openknowledge.fao.org/server/api/core/bitstreams/97409d09-2f8e-4712-b11e-60105d89959b/content

Wolf PL (1999) Biochemical diagnosis of liver disease. Indian J Clin Biochem 14:59–90

Wong CK, Ooi VEC, Ang PO (2000) Protective effects of seaweeds against liver injury caused by carbon tetrachloride in rats. Chemosphere 41(2000):173–176

Yasmeen A, Ibrahim M, Hasan MM et al (2021) Phycochemical analyses and pharmacological activities of seven macroalgae of Arabian Sea (Northern coast line). Pak J Pharm Sci 34(3):963–969

Zaharudin N, Salmeán AA, Dragsted LO (2017) Inhibitory effects of edible seaweeds, polyphenolics and alginates on the activities of porcine pancreatic α-amylase. Food Chem 245:1196–1203. https://doi.org/10.1016/j.foodchem.2017.11.027

Zaharudin N, Staerk D, Dragsted LO (2018) Inhibition of α-glucosidase activity by selected edible seaweeds and fucoxanthin. Food Chem 270:481–486. https://doi.org/10.1016/j.foodchem.2018.07.142

Zbakh H, Chiheb H, Bouziane H, Sánchez VM, Riadi H (2012) Antibacterial activity of benthic marine algae extracts from the mediterranean coast of morocco. J Microbiol Biotechnol Food Sci 2:219–222

Chapter 9
Macroalgal Nutraceuticals and Phycotherapeutants

Tejal K. Gajaria, Darshee Baxi, Elizabeth Robin, Parth Pandya, and A. V. Ramachandran

Abbreviations

pg	Peta grams
CO_2	Cabon dioxide
COVID	Corona virus disease
DW	Dry weight
FW	Fresh weight
UV	Ultraviolet radiation
NPN	Non-protein nitrogen
DM	Dry matter
EFSA	European Food Safety Authority
ASEAN	Association of Southeast Asian Nations
As	Arsenic
B	Boron
Cd	Cadmium
Ca	Calcium
Co	Cobalt
Cr	Cromium
I	Iodine
Fe	Iron
Pb	Lead
Mo	Molybdenum
Mg	Magnesium
Mn	Manganese
Ni	Nickel
K	Pottasium

T. K. Gajaria · D. Baxi · E. Robin · P. Pandya · A. V. Ramachandran (✉)
Division of Biomedical and Life Sciences, School of Science, Navrachana University, Vadodara, Gujarat, India
e-mail: tejal.gajaria@nuv.ac.in; Darsheeb@nuv.ac.in; elizabethr@nuv.ac.in; parthp@nuv.ac.in; avramachandran@nuv.ac.in

F. Ozogul et al. (eds.), *Seaweeds and Seaweed-Derived Compounds*,
https://doi.org/10.1007/978-3-031-65529-6_9

Se Selenium
Na Sodium
Zn Zinc

1 Introduction

The marine autotrophs contribute to ~40% of total primary production on earth owing to the fixation of 30–60 pg (1 pg = 10^{15} g) of organic carbon annually and play a key role in structuring habitat for nearshore benthic communities (Al Duarte and Cebrih 1996). The seaweed biome performs numerous ecosystem services, such as providing shelter to diverse epiphytes, vertebrate and invertebrate organisms, dampening waves, feeding and nursery grounds, removing excess nutrients and serving as a huge carbon sink for the absorption of anthropogenic CO_2 emission (Sondak et al. 2017). Apart from other marine autotrophs, seaweeds have been utilized extensively as food by various coastal communities around the globe. Taxonomically, seaweeds are classified into three groups based on their pigmentation; Rhodophyceae, Chlorophyceae and Phaeophyceae representing red, green and brown colours. Ancient civilizations such as those from South Asian countries, especially Indonesia, Philippines, Korea, Japan and China, have a great history of seaweed utilization in foods.

In contrast, western countries are primarily involved in phycocolloids production, i.e., carrageenan, agar, alginate, porphyrin and ulvan (Mohamed et al. 2012). Marine macroalgae are considered healthy, nourishing and low-calorie food. Consumable seaweeds are known to be rich in proteins (Gajaria and Mantri 2022), minerals (Circuncisão et al. 2018), vitamins, bioactive antioxidants, soluble dietary fibres, phytochemicals and polyunsaturated fatty acids (Biris-Dorhoi et al. 2020). However, previously seaweeds were reported for use only as emulsifying agents such as gelling and thickening chemicals in the food or pharmaceutical industries. Recent reports have established their potential as complementary and alternative medicine. Seaweeds have been reported with therapeutic properties for promoting health and disease management, including antioxidant, antimicrobial, antihyperlipidemic, immunomodulatory, antiobesity, antiestrogenic, antihypertensive, neuroprotective, antidiabetic, anti-inflammatory, anticancer, anticoagulant, thyroid-stimulating and tissue healing properties *in vivo* (Mohamed et al. 2012). The active compounds providing a spectrum of activities include sulphated polysaccharides, polyphenols, flavonoids, pigments, active peptides and fatty acids with proven benefits against degenerative metabolic diseases (Mohamed et al. 2012). However, the biggest challenge in developing marine-origin organic pharmaceuticals/nutraceuticals is that the biochemical composition of seaweeds varies with not only species but with geographic area, season, sampling methods, storage condition and storage condition environmental conditions as well (Hafting et al. 2015). Considering the ever-expanding world population and un avoidable climatic catastrophe, seaweeds are considered among the top sustainable solutions contributing to global food security (Cai et al. 2021).

2 Therapeutics: Potential of Seaweeds in Complementary and Alternative Medicine

Modern medicine poses major therapeutic challenges in the current times. Major concerns that need immediate intervention are increasing drug resistance and toxicity besides, suppression of immune reactions and secondary effects on non-target organs (Upadhyay et al. 2021; Mukherjee et al. 2021). There is a need to look for more sustainable health therapeutics that would help target the diseases at multiple sites for sustainable outcomes. Diseases like cancer and diabetes are heterogeneous in nature with multiple etiologies; hence, in the last few years, critical concerns have been raised about the existing drug regimen. Also, the fact that individuals present multiple pathophysiologies simultaneously further complicates the scenario (Upadhyay et al. 2021; Mukherjee et al. 2021). This was evident in the recent COVID pandemic, wherein patients with co-morbidities had greater challenges and required an individual intervention plan to manage the treatment (Sanyaolu et al. 2020). The pandemic also witnessed the application of complementary and alternative medicine and the repurposing of the available drug library (Badria 2022; Khan and Al-Balushi 2021; Alam et al. 2021). Such needs indicate the changing scenario in health management worldwide and emphasize the need for newer approaches to alternative medicines.

Several herbal medications have been formulated over the last few years to meet this ever-increasing need for holistic disease management. One of the promising candidates in this regard for the current decade is the exploration of marine bioactives (Biris-Dorhoi et al. 2020). Marine plants have been less known; thus, evaluating their potential will be very important when searching for alternatives to synthetic medication (Patel 2012). Marine algae are one of those groups that possess promising attributes worth exploring for such treatments; however, in-depth scientific understanding is required to validate such hypotheses further. Both fresh and marine forms of brown, red and green algae have showcased a prominent potential for detailed evaluation (Biris-Dorhoi et al. 2020). One of the major attributes of formulating these algal forms is that most of these are being thoroughly evaluated for pilot scale cultivation to meet the needs of a constant supply of therapeutic formulations (Rorrer et al. 1995; Hafting et al. 2015).

2.1 Role of Seaweeds in the Management of Metabolic Disorders

Metabolic disorders have multiple etiologies; hence, single target-based drug regimens have often failed to prove effective. Diet, genetic variations, lifestyle, lack of exercise and sedentary life, socioeconomic status and irregular work shifts contribute to metabolic dyshomeostasis (Rochlani et al. 2017; Saklayen 2018; Patti et al. 2018). Thus, counter mechanisms for several aspects are required to meet the holistic management of these disorders like cardiovascular, diabetes, obesity etc. Out of all the above factors, dietary interventions have proven the best alternatives for managing these disorders (Aude et al. 2004; Minehira and Tappy 2002). Diet management and the introduction of functional foods have shown promising outcomes for these disorders. In this regard, the marine sulfated polysaccharide has been very important in treating these metabolic imbalances. These sulfated polysaccharides are found more in marine-origin plant species, and they have served as good options for treating metabolic disorders (Pradhan et al. 2020). The human body cannot utilize these sulfated polysaccharides directly, but they get absorbed through the action of the gut microbiome. The gut microbiome has been recently explored for its relatedness to the onset of metabolic disorders (Dabke et al. 2019). Thus, directly and indirectly, such supplements help maintain the human body's metabolic homeostasis and maintain a healthy microbiome. Several sulfated polysaccharides from different algal sources have been effective in diseases like obesity, diabetes, and other disorders (Ben Abdallah Kolsi et al. 2015; Gomez-Zavaglia et al. 2019).

Marine Sulfated polysaccharides from *Cymodocea nodosa* and *Ascophyllum nodosum* are known to reduce weight in obese animal models, regulate gut microbiota, improve nutrient absorption and energy metabolism and regulate serum lipid and inflammatory cytokines (Ben Abdallah Kolsi et al. 2015; Chen et al. 2018). Studies have also shown the effect of fucoidan as an alpha-glucosidase inhibitor, which is important in reducing hyperglycemia. High molecular weight fucoidan isolated from *F. vesiculosus* has also been reported to have a metabolic regulatory activity to reduce hyperglycemia (Mabate et al. 2021). There are also reports of metabolic syndrome-related signalling pathways and gene expression changes upon adding marine sulfated polysaccharides to the diet. Seaweeds are also reported to have protective effects against breast cancer progression via their action on estrogen receptors (Pradhan et al. 2020; Pangestuti and Kurnianto 2017; Mišurcová et al. 2012).

2.2 Role of Seaweeds in the Management of Infectious Diseases

Infectious diseases are on the rise in their extent of spread and their lethality. Several synthetic antibiotic classes have been evaluated and developed to combat infectious diseases. One of the major study areas is influenza, and there is a pressing need to look for suitable alternative medicines to combat such viral infectious agents. Several algal polysaccharides have been studied and reported for their promising results as antivirals (Kwon et al. 2020; Takebe et al. 2013; O'Keefe et al. 2010; Shi et al. 2017). One of them is carrageenans which have been used even for intranasal use and have no reported side effects. Studies have shown that carrageenans created a physical barrier in the nasal cavity against respiratory viruses, including several influenza virus strains. They have been reported to prevent adsorption, penetration and replication of the viruses when administered in the initial phase of infection (Shi et al. 2017; Higgins et al. 2020). Fucoidans have also been noted to affect viral attachment and replication directly (Pradhan et al. 2020).

They also promote antiviral immunity, antioxidant protection and reduce inflammation. Lecithins and algal polyphenols have also been associated with similar actions on viral infectious diseases (Biris-Dorhoi et al. 2020). Apart from these effects, seaweed extracts have also been tested for their antimicrobial action on several pathogenic bacterial forms, including *S. aureus*, *K. pneumoniae* and other disease-causing microbes (Pérez et al. 2016).

2.3 Seaweeds, Immune Modulation and Gut Microbiome

Immuno modulation through the gut microbiome is an unexplored area. Recent studies indicate that marine algal polysaccharides are important supplements that can be evaluated for their role in the healthy gut microbiome and their relation to the improved immune potential (Reilly et al. 2008; Shannon et al. 2021; Praveen et al. 2019). The prebiotic effects of the dietary fibres help enhance gut microbial activity by fermentation and inhibition potential. These seaweed derivatives act as prebiotics for formulating food ingredients for immunomodulatory applications (Praveen et al. 2019). Some specific studies for the brown alga *H. fusiforme* have proved that the polysaccharides from this alga activate murine macrophages and splenocytes (Jeong et al. 2015). However, such studies and other similar studies have not yet been able to identify the precise mechanism of action that could provide scope for future studies. The maintenance of the gut microbiome has taken the central seat in evaluating its impact on the immune system. Algal supplements are new promising candidates that might bridge the gap created due to unhealthy lifestyles and dietary habits in the current times, and thus might be a key supplement for maintaining gut health to promote a good immune system in the body (Fig. 9.1).

3 Classes of Phycotherapeutants

Phycotherapeutants mean healing or curative agents or medicines developed from seaweeds (algae). The different classes of phycotherapeutants are pigments, sulphated carbohydrates, proteins, polyphenols and flavonoids, which will be explained in detail in this chapter.

Macroalgae or seaweeds are relatively unexplored promising sources of novel molecules for the food industry, including peptides and carbohydrates for their use as functional foods and nutraceuticals. The discovery of marine algal metabolites showing biological activity has increased significantly in the last decades. Seaweeds have proved to be a valuable source of structurally diverse bioactive compounds (Lafarga et al. 2020).

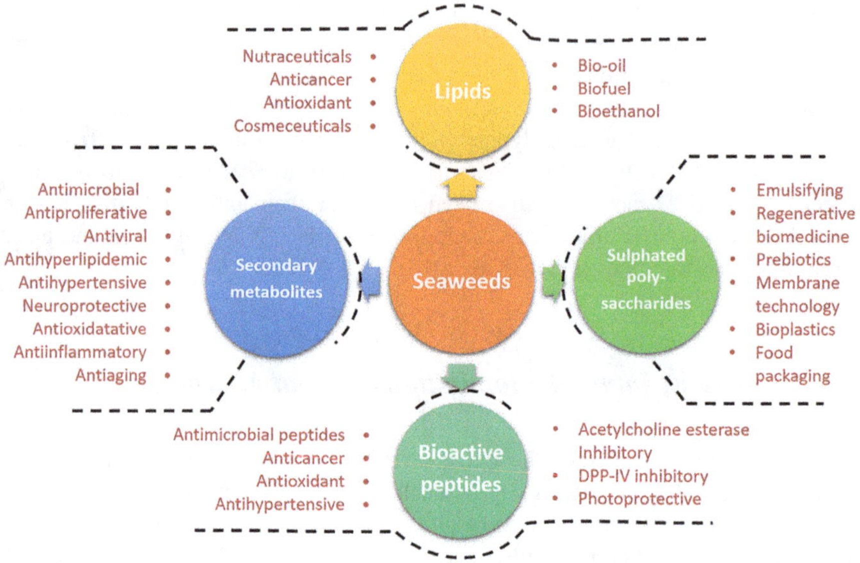

Fig. 9.1 Macroalgae based bioactive compounds and their respective applications discovered so far

3.1 Pigments

Marine macroalgae are a diverse group of multicellular, plant-like protists that can be classified into brown (Phaeophyta), green (Chlorophyta) and red (Rhodophyta) algae. The pigment responsible for the brown colour of Phaeophyta is fucoxanthin, the red colour of Rhodophyta comes from phycobilins, and several pigments like chlorophyll a and b, carotenes, and xanthophylls are responsible for the green colour of Chlorophyta (Manivasagan et al. 2018; Rodriguez-Amaya 2016). The chemical composition of macroalgae varies considerably between species and with the season of harvest, growth habitat, and environmental conditions. Even within a small geographic area, growth rate and chemical composition may vary depending on abiotic and biotic factors, e.g., harvest season, sunlight, salinity, depth in the sea, local water currents, or closeness to aquacultural plants. Reported ranges in proximate composition of brown, green, and red macroalgae are shown in the table below (Øverland et al. 2019) (Table 9.1).

Carotenoids are orange/red pigments that absorb light energy and then pass it on to chlorophyll, therefore playing a secondary role in photosynthesis. Carotenoids help the light-gathering potential of the algae. There is some variation in seaweeds' chlorophyll and carotenoid contents depending on the ultraviolet radiation (UV) levels throughout the year. Both chlorophyll and carotenoid pigments possess antioxidant and chemopreventive properties. The main algal pigments in commercial use comprise beta-carotene, astaxanthin, lutein, phycocyanin, chlorophyll, and fucoxanthin (Biris-Dorhoi et al. 2020; Manivasagan et al. 2018).

Table 9.1 Ranges of proximate composition of marine macroalgae[a]

Chemical constituent	Brown macroalgae[b]	Green macroalgae[c]	Red macroalgae[d]
Water, g kg^{-1} of wet biomass	610–940	780–920	720–910
Crude protein[e], g kg^{-1}	24–168	32–352	64–376
Crude lipids, g kg^{-1}	3–96	3–28	2–129
Polysaccharides, g kg^{-1}	380–610	150–650	360–660
Ash, g kg^{-1}	150–450	110–550	120–422

[a]Values are in g kg^{-1} of DM unless otherwise specified
[b]Values are for typical brown macroalgal species: e.g. *Laminaria*, *Saccharina*, *Fucus*, *Ascophyllum*, *Alaria*, *Pelvetia* and *Undaria* spp.
[c]Values are for typical green macroalgal species: e.g., *Ulva*, *Cladophora*, and *Enteromorpha* spp.
[d]Values are for typical red macroalgal species: e.g., *Palmaria*, *Chondrus*, *Porphyra*, *Vertebrata*, and *Gracilaria* spp.
[e]All values for CP have been recalculated using the recommended nitrogen-to-protein factor of five

3.2 *Sulphated Carbohydrates*

The marine macroalgae contain many different complex carbohydrates and polysaccharides. Brown macroalgae mainly contain alginates, sulphated fucoidans, and laminarin; green macroalgae contain xylans and sulphated galactan (ulvan), and red macroalgae contain agars, carrageenans, xylans, sulphated galactan, and porphyrans (Lakshmi et al. 2020). Algal polysaccharides such as alginates, carrageenan, fucoidan, and laminarin, are noted to show a wide spectrum of biological activities like antioxidant, antithrombotic, anti-inflammatory, and neuroprotective (Biris-Dorhoi et al. 2020). The cell walls of marine macroalgae lack lignin, but they have 'lignin-like' compounds, and true lignin has been reported in some species. Compared to terrestrial plants, where lignin is important for rigidity, the cell walls of macroalgae are more flexible. The main structural components are alginate and fucoidan in brown, xylan and ulvan in green, and carrageenans in red macroalgae. The main storage components are laminarin in brown algae and floridean starch (amylopectin) in green and red species. Another main difference from cell walls in terrestrial plants is the presence of many unique polysaccharides that can be sulphated, methylated, acetylated, or pyruvylated. Finally, macroalgae can also contain sugar alcohols such as mannitol. Especially in some brown species, the mannitol content can be up to 25% of the dry weight.

Sulfated polysaccharides are common in the cell walls of seaweeds. Many marine algal cell walls are rich in sulphated polysaccharides; for example, carrageenan (red algae), ulvan (green algae) and fucoidan (brown algae). Carrageenan is the most important compared to fucoidan and ulvan, with wide applications as an emulsifier, stabilizer or thickener. Carrageenan, a natural phycocolloid, is a primary additive used in many dairy products like yoghurts, flavoured milkshakes, flans, jellies, ice creams, beers and meat products (e.g. hams), as a thickening, emulsifier or stabilizing agent (Leandro et al. 2020a). Fucoidan is available commercially from various cheap sources and has been studied recently for producing drugs, medicines and functional foods. Ulvan, the less known of the group, displays several physicochemical and biological features of potential interest for food, pharmaceutical, agricultural and chemical applications. Studies have shown that marine algal sulphated polysaccharides have potential application in pharmaceutics owing to their biological activities (Lakshmi et al. 2020).

A sulfated galactan was isolated from red seaweed, *Grateloupia indica*, with strong blood anticoagulant activity (Sen et al. 1994). Because of their biological and physicochemical properties, sulfated polysaccharides like carrageenan, fucoidan and ulvan are helpful in designing drug delivery systems (Cunha and Grenha 2016). These polysaccharides possess antibacterial and antiviral activity along with antitumoral and immunomodulatory potential. Alginate, another polysaccharide finds application in pharmaceutical formulations as excipient. Polymers of alginate has great potential in drug formulation because of their non-toxicity. To meet the demand as applicant in pharmaceutical and biomedical areas, they hold great potential as they can be tailor-made (Leandro et al. 2020b). A galactan sulfate (GS) was also isolated from an aqueous extract of the red seaweed Aghardhiella tenera, showing antiviral properties (Witvrouw et al. 2016).

3.3 Proteins

Comparing the protein content of macroalgae reported in different studies seems difficult because of methodological differences (Gajaria and Mantri 2022). Nitrogen is found in proteins, nucleic acids, and other organic compounds such as chlorophyll. Macroalgae also contain significant amounts of inorganic non-protein nitrogen (NPN; e.g., ammonia, nitrate, and nitrite). Spectroscopic methods can aid in protein determinations. Many proteins from macroalgae are difficult to extract and contain several coloured substances that may influence the measurements. One gram of meal from algae with the highest protein levels (e.g., *Enteromorpha intestinalis*, *Palmaria palmata*, and *Vertebrata lanosa*) was equal to or higher than the amounts of all the essential amino acids from rice, corn, and wheat. The lysine content is reportedly three to nine times higher than in the latter (Biris-Dorhoi et al. 2020).

The protein content of brown macroalgae was generally low (usually below 150 g kg^{-1} of dry matter (DM)). In contrast, green and red macroalgae have a higher protein content on a DM basis. Some red macroalgae, such as *Porphyra* spp., have protein levels comparable to soybean. Compared with fishmeal, the lysine proportion is lower in macroalgae but is usually higher in red than in brown and green species. Many macroalgal species are low in histidine, but the methionine content can be relatively high in many species. Macroalgae usually contain high levels of glutamic acid, which is present in free and protein-bound forms and contributes to macroalgae's typical taste (umami). Macroalgae also contain many bioactive amino acids and peptides [e.g., taurine, carnosine, and glutathione (Øverland et al. 2019). The seaweed varieties with a high protein content can serve as ingredients in the manufacturing process of different foods. Porphyra species find use in the famous sushi preparations. For example, species such as *Ulva pertusa*, *Enteromorpha* sp., and *Monostroma* sp. (protein levels of 26, 19, and 20% dw, respectively) are mixed to create a food product called "aonori", a protein-rich product consumed by people in Japan. In Europe and Canada, *Palmaria palmata* finds use as a food ingredient. It can be prepared as dry flakes and used to make different functional products due to its high protein content (up to 35% dw) (Biris-Dorhoi et al. 2020).

3.4 Polyphenols and Flavonoids

Phenolic compounds in macroalgae vary from simple molecules, such as phenolic and cinnamic acids or flavonoids, to the more complex phlorotannin polymeric structures. Phenolic compounds from macroalgae possess anti-Alzheimer, anti-inflammatory, anti-allergic, anti-proliferative, antioxidant, antiobesity, and bactericidal activities. Hydroxybenzoic acid derivatives, such as gallic acid, are commonly reported as constituents of different green, red, and brown macroalgae species (Santos et al. 2019).

The phenolic compounds in marine algae can protect against stress conditions by acting as antioxidant compounds. Polyphenolic compounds such as catechins and flavonols are the most dominant polyphenols in all classes of macroalgae: Chlorophyta (green algae), Rhodophyta (red algae) and Phaeophyta (brown algae) (Cikoš et al. 2019). The most researched seaweed polyphenol class is the phlorotannins, specifically synthesized by brown seaweeds; there are other polyphenolic compounds, such as bromophenols, flavonoids, phenolic terpenoids, and mycosporine-like amino acids (Cotas et al. 2020). Brown macroalgae contain relatively large amounts of α-, β-, γ-, and δ-tocopherols. In contrast, red and green algae contain detectable levels of α-tocopherol, with only traces of the other tocopherols (Øverland et al. 2019).

Ulva species possess antioxidant potential, which has future applications in medicine, dietary supplement, cosmetic or food industries. Algae generally have higher antioxidant activity due to a higher content of nonenzymatic antioxidant components, such as ascorbic acid, reduced glutathione, phenols and flavonoids. As a result, many marine bio-sources have attracted attention in searching for natural bioactive compounds to develop new drugs and healthy foods in the last decades (Farasat et al. 2013). *Gracilaria bursa-pastoris* can be a source of natural antioxidant compounds due to its phenolic content. Hence, *G. bursa-pastoris* can be valuable for developing therapeutic products, food supplements, and pharmaceutical applications (Ramdani et al. 2017). The antioxidant effect is mostly due to polyphenols, particularly phlorotannins, oligomers or polymers of phloroglucinol. Some polyphenolic compounds also show antioxidant effects, such as catechins, flavonols and flavonol glycosides, which have been identified in methanolic extracts of red and brown algae (Mahendran et al. 2021).

4 Challenges in the Use of Macroalgae-Based Nutraceuticals

4.1 Processed vs. Raw Biomass

Algae are a rich source of diverse compounds in biofuels, health supplements, pharmaceuticals, and cosmetics besides applications in wastewater treatment and atmospheric CO_2 mitigation. Macroalgae and microalgae produce various bio-products, including polysaccharides, proteins, lipids, pigments, proteins, vitamins, bioactive compounds, and antioxidants (Brennan and Owende 2010). Due to the diverse applicability of algal biomass, research mainly targets upstream, midstream, and downstream processes. Upstream processing on microalgal cultivation focuses on the maximization of biomass production. Midstream processing intends to harvest microalgae from cultivation media, dry the collected biomass, and rupture the cell walls before extraction. The last part deals with downstream processing, the most crucial step, and aims to extract and purify the bioproduct(s) from algal biomass (Manirafasha et al. 2016).

Additionally, the indigenous use of algae as food sources is a prehistoric practice. Many green algae species have been used as food from olden times (Krause-Jensen et al. 2007). A Systematic cultivation practice started only a few decades ago when it became clear that food would be scarce due to the fast-growing world population (Borowitzka 1998). Thus, due to the chemical composition, algae serve as an excellent food source. Their other important bioproducts can be exploited greatly as natural supplements and potent antimicrobials (Becker 2007). The first commercial large-scale microalgae cultivation was of *Chlorella* sp. in Japan in the 1960s (Iwamoto 2007), whereas the first large-scale seaweed cultivation was of *Eucuma cottonii* in 1967 in the Phillipines (Porse and Rudolph 2017). The total seaweed production was 34.7 million tonnes compared to only 56,546 tonnes of total microalgae production in 2019 (FAO 2021). In addition to seaweeds, *Spirulina* and *Chlorella*-based products are very well known to be a rich source of proteins, fatty acids, pigments, minerals (e.g., calcium, magnesium, phosphorus, zinc, copper, and iron), bioavailable forms of B_{12}, C, and E (Wells et al. 2016; Buono et al. 2014). As indicated by clinical trials, both the supplements reveal promising bioactive properties encompassing immunomodulatory, antihypertensive, antilipidemic, and hypoglycemic effects (Kim et al. 2016; Juszkiewicz et al. 2018; Nielsen et al. 2010) but no such report is available in case of seaweed derived suppliments.

4.2 Potential Toxicity

Earlier studies have reported that plant-based food supplements may often contain increased levels of toxic heavy metals like cadmium (Cd) or lead (Pb). Aplausible reason could be contamination of the ambient environment from which the biomass was originally derived, and the bioaccumulation tendency of plant species. Among the members of the entire algal domain, few macroalgae, as well as microalgae, are frequently reported for their uptake and accumulation of toxic metals, hence providing a strong possibility of the presence of toxic metals in algae-based food supplements owing to the chemical composition of culture medium comprising of various minerals and metal ions. Marine macroalgae from the domain of Phaeophyceae are thoroughly studied for their bio assimilation of heavy metals, including a few biologically valuables such as iodine. The foremost study reported the presence of nickel, zinc, copper, cadmium, lead, chromium and manganese from *Fucus vesiculosus* and *Ascophyllum nodosum* collected from the Menai Straits, North Wales, in 1976 (Foster 1976). These were the decades from when seaweed research and applications gained tremendous momentum, leading to global concerns about their heavy metal contents. Soon after the preliminary studies from Foster in 1976, Sirota and Uthe in 1978 reported the contents of arsenic, mercury, cadmium, lead, copper, zinc and selenium from Dulse, which is regarded as one of the most protein-rich red seaweed to date (Sirota and Uthe 1979). Followed by the global alert call, a comprehensive study was conducted that included 17 seaweed species from Goa, the Western coast of India, comprising six Chlorophyceae, seven Phaeophyceae and four Rhodophyceae specimens for cobalt, copper, iron, manganese, nickel, led and zinc contents (Agadi et al. 1978). The observations from these studies led to the inspiration for utilizing algae as an indicator species for heavy metal pollution *in-situ* and the explorations of chemical compounds responsible for chelation and accumulation of heavy metals in algal tissues. The understanding from these studies helped develop heavy metal scavenging composites for bioremediation in the early 1980s that remains one of the thurst areas of environmental engineering and management (Bryan and Hummerstone 1973; Luoma et al. 1982). These physiological functions also greatly affect the labelling of seaweed-based packaged products. As studied in the consumer markets of selected European countries by Besada et al. (2009) and Filippini et al. (2021), iodine content was found to be much higher than the permitted values, followed by inorganic arsenic and cadmium contents (Besada et al. 2009; Filippini et al. 2021).

In addition to the presence of heavy metals, the form of heavy metal is also crucial in considering their toxicity, as in the case of some algal supplements which contained pentavalent arsenic. Although this form of arsenic has been found bound with organic compounds, showing the least toxicity to humans; a consideration of the safety limits as per the EFSA (0.24 to 0.38 μg kg^{-1} bw per day for adults), with a daily intake of even three gms of such contaminated food supplement may result in detrimental health effects (Rzymski et al. 2019; EFSA 2016). Therefore, the potential monitoring enforcement should consider not only algae-based formulations but all food supplements to ensure consumer safety.

4.3 Environmental Factors Influencing Biochemical Composition

Different methods are employed to culture macro and microalgae from different environmental conditions. Most importantly, they require light as an energy source to convert the absorbed water and CO_2 into biomass through photosynthesis (Badger 1994). Products from photosynthesis accumulate in various forms, such as cell components or storage materials, and vary from 20 to 50% of their total biomass (Chisti 2007). Apart from this, algae also need nitrogen and phosphorus as major nutrients, which account for 10–20% of algal biomass (Khan et al. 2018). Other requirements for growth include the necessary amount of Na, Mg, Ca, and K; micronutrients, such as Mo, Mn, B, Co, Fe, and Zn; and other trace elements. The algal cells undergo different phases in the growth process (e.g., lag, exponential, stationary, death). Different species may vary in biochemical composition due to changes in the available biotic and abiotic factors. However, all species require essential nutrients, carbon, nitrogen, phosphorus and iron source for growth and reproduction. With varying geo-graphical conditions, the tissue composition and nutritional qualities also largely differ. In addition to the natural environmental influences, anthropogenic activities have also contributed largely in the modelling of coastal marine autotrophs in recent times. The excess of phosphates and nitrates flowing through the farmlands and fishery industries has led to the repetitive occurences of massive golden and green tides globally. Such biomass has proved detrimental to the coastal communities as well as for tourist visitations, thereby significantly impacting the respective economy. However, these free floating seaweed mats actually absorb excess nutrients at the same time from marine water. A systematic study reporting their possible influence on coastal biodiversity (flora and fauna) is yet to be undertaken.

4.4 Sustainable Value-Chain

Currently, there is a global impetus in the production of seaweeds. However, unprecedent climate change is posing newer challenges for both the producers and the environment. Seaweed aquaculture plays a vital role in the economy of the Asian region (Cai et al. 2021; Mantri et al. 2020; Sahu et al. 2020). Asia's contribution to global seaweed production in 2019 was estimated at 97% of the total 35.8 million tonnes produced (FAO, IFAD, UNICEF, WFP, WHO 2021). For instance, the cultivation of the *eucheumatoid* seaweeds *Kappaphycus* and *Eucheuma* is a major contributor to the economy, food supply, and rural livelihoods in the ASEAN region (Cai et al. 2021; Mantri et al. 2016). However, compared to the commercially valuable phycocolloidal species, the cultivation of therapeutically valued seaweed/algal species is scarce. Therefore, resource limitation seems to be a crucial bottleneck for developing algae-based therapeutics.

5 Future Prospects

Marine macroalgal cultivation has proven advantages compared to freshwater species in terms of land and freshwater utilization. In addition, it does not require additional nutritive inputs during the culturing period, leaving only the handling, transportation and capital costs to be managed for year-round production. However, changing environments and increasing climatic catastrophes are a vital threat to the on-shore cultivation activities. In addition, sustainable off-shore cultivation practices are still in infancy, which is one of the potential reasons for an underdeveloped seaweed-based food and nutraceutical market. Furthermore, several seaweed species rich in unique metabolites are highly sensitive to their growth environments, i.e., light intensity, salinity, wave intensity and nutrient enrichment, due to which replication of their entire life-cycle is highly limited. In addition, few classes of rich bioactives belonging to Rhodophyceae and Phaeophyceae have considerably lower growth rates than the conventional seaweeds that make such species non-viable at large scale for the recovery of therapeutically active ingredients. The future of marine macroalgae-derived active pharmaceutical ingredients largely relies on advancements in upstream processing compared to downstream recovery.

References

Agadi VV, Bhosle NB, Untawale AG (1978) Metal concentration in some seaweeds of Goa (India). Bot Mar 21:247–250. https://doi.org/10.1515/BOTM.1978.21.4.247/MACHINEREADABLECITATION/RIS

Al Duarte C, Cebrih J (1996) The fate of marine autotrophic production. Limnol Ocean 41:1758–1766

Alam S, Sarker MMR, Afrin S, Richi FT, Zhao C, Zhou JR, Mohamed IN (2021) Traditional herbal medicines, bioactive metabolites, and plant products against COVID-19: update on clinical trials and mechanism of actions. Front Pharmacol 12:1248. https://doi.org/10.3389/FPHAR.2021.671498/BIBTEX

Aude YW, Mego P, Mehta JL (2004) Metabolic syndrome: dietary interventions. Curr Opin Cardiol 19:473–479. https://doi.org/10.1097/01.HCO.0000134610.68815.05

Badger MR (1994) The role of carbonic anhydrase in photosynthesis. Annu Rev Plant Physiol Plant Mol Biol 45:369–392. https://doi.org/10.1146/annurev.arplant.45.1.369

Badria FA (2022) Perspective chapter: repurposing natural products to target COVID-19: molecular targets and new avenues for drug discovery. Antivir Drugs [Working Title]. https://doi.org/10.5772/INTECHOPEN.103153

Becker W (2007) Microalgae in human and animal nutrition. In: Handbook of microalgal culture. Wiley, pp 312–351. https://doi.org/10.1002/9780470995280.ch18

Ben Abdallah Kolsi R, Ben Gara A, Chaaben R, El Feki A, Paolo Patti F, El Feki L, Belghith K (2015) Anti-obesity and lipid lowering effects of Cymodocea nodosa sulphated polysaccharide on high cholesterol-fed-rats. Arch Physiol Biochem 121:210–217. https://doi.org/10.3109/13813455.2015.1105266

Besada V, Andrade JM, Schultze F, González JJ (2009) Heavy metals in edible seaweeds commercialised for human consumption. J Mar Syst 75:305–313. https://doi.org/10.1016/J.JMARSYS.2008.10.010

Biris-Dorhoi ES, Michiu D, Pop CR, Rotar AM, Tofana M, Pop OL, Socaci SA, Farcas AC (2020) Macroalgae – a sustainable source of chemical compounds with biological activities. Nutrients 12(10):3085. https://doi.org/10.3390/NU12103085

Borowitzka MA (1998) Algae as food. Microb Ferment Foods:585–602. https://doi.org/10.1007/978-1-4613-0309-1_18

Brennan L, Owende P (2010) Biofuels from microalgae – a review of technologies for production, processing, and extractions of biofuels and co-products. Renew Sust Energ Rev 14:557–577. https://doi.org/10.1016/J.RSER.2009.10.009

Bryan GW, Hummerstone LG (1973) Brown seaweed as an indicator of heavy metals in estuaries in South-West England. J Mar Biol Assoc UK 53:705–720. https://doi.org/10.1017/S0025315400058902

Buono S, Langellotti AL, Martello A, Rinna F, Fogliano V (2014) Functional ingredients from microalgae. Food Funct 5:1669–1685. https://doi.org/10.1039/C4FO00125G

Cai J, Lovatelli A, Aguilar-Manjarrez J, Cornish L, Dabbadie L, Desrochers A, Diffey S, Garrido Gamarro E, Geehan J, Hurtado A, Lucente D, Mair G, Miao W, Potin P, Przybyla C, Reantaso M, Roubach R, Tauati M, Yuan X (2021) Seaweeds and microalgae: an overview for unlocking their potential in global aquaculture development. FAO Fish Aquac Circ. https://doi.org/10.4060/CB5670EN

Chen L, Xu W, Chen D, Chen G, Liu J, Zeng X, Shao R, Zhu H (2018) Digestibility of sulfated polysaccharide from the brown seaweed Ascophyllum nodosum and its effect on the human gut microbiota in vitro. Int J Biol Macromol 112:1055–1061. https://doi.org/10.1016/J.IJBIOMAC.2018.01.183

Chisti Y (2007) Biodiesel from microalgae. Biotechnol Adv 25:294–306. https://doi.org/10.1016/J.BIOTECHADV.2007.02.001

Cikoš AM, Jurin M, Čož-Rakovac R, Jokić S, Jerković I (2019) Update on monoterpenes from red macroalgae: isolation, analysis, and bioactivity. Mar Drugs 17. https://doi.org/10.3390/MD17090537

Circuncisão AR, Catarino MD, Cardoso SM, Silva AMS (2018) Minerals from macroalgae origin: health benefits and risks for consumers. Mar Drugs 16. https://doi.org/10.3390/MD16110400

Cotas J, Leandro A, Pacheco D, Gonçalves AMM, Pereira L (2020) A comprehensive review of the nutraceutical and therapeutic applications of red seaweeds (Rhodophyta). Life 10. https://doi.org/10.3390/LIFE10030019

Cunha L, Grenha A (2016) Sulfated seaweed polysaccharides as multifunctional materials in drug delivery applications. Mar Drugs 14. https://doi.org/10.3390/MD14030042

Dabke K, Hendrick G, Devkota S (2019) The gut microbiome and metabolic syndrome. J Clin Invest 129:4050–4057. https://doi.org/10.1172/JCI129194

EFSA (2016) The 2014 European Union report on pesticide residues in food. EFSA J 14(2016):e04611. https://doi.org/10.2903/J.EFSA.2016.4611

FAO (2021) Seaweeds and microalgae: an overview for unlocking their potential in global aquaculture development, FAO Fisheries and Aquaculture Circular NFIA/C1229 (En). Food and Agriculture Organization. https://doi.org/10.4060/cb5670en. Accessed 2 July 2022

FAO, IFAD, UNICEF, WFP, WHO (2021) The state of food security and nutrition in the world 2021. FAO, IFAD, UNICEF, WFP and WHO, Rome. https://doi.org/10.4060/cb4474en

Farasat M, Khavari-Nejad RA, Nabavi SMB, Namjooyan F (2013) Antioxidant properties of two edible green seaweeds from northern coasts of the persian gulf, Jundishapur. J Nat Pharm Prod 8. https://doi.org/10.5812/jjnpp.7736

Filippini M, Baldisserotto A, Menotta S, Fedrizzi G, Rubini S, Gigliotti D, Valpiani G, Buzzi R, Manfredini S, Vertuani S (2021) Heavy metals and potential risks in edible seaweed on the market in Italy. Chemosphere 263:127983. https://doi.org/10.1016/J.CHEMOSPHERE.2020.127983

Foster P (1976) Concentrations and concentration factors of heavy metals in brown algae. Environ Pollut 10:45–53. https://doi.org/10.1016/0013-9327(76)90094-X

Gajaria TK, Mantri VA (2022) Emerging trends on the integrated extraction of seaweed proteins: challenges and opportunities. Sustain Glob Resour Seaweeds 2(2):219–234. https://doi.org/10.1007/978-3-030-92174-3_11

Gomez-Zavaglia A, Prieto Lage MA, Jimenez-Lopez C, Mejuto JC, Simal-Gandara J (2019) The potential of seaweeds as a source of functional ingredients of prebiotic and antioxidant value. Antioxidants 8:406. https://doi.org/10.3390/antiox8090406

Hafting JT, Craigie JS, Stengel DB, Loureiro RR, Buschmann AH, Yarish C, Edwards MD, Critchley AT (2015) Prospects and challenges for industrial production of seaweed bioactives. J Phycol 51:821–837. https://doi.org/10.1111/JPY.12326

Higgins TS, Wu AW, Illing EA, Sokoloski KJ, Weaver BA, Anthony BP, Hughes N, Ting JY (2020) Intranasal antiviral drug delivery and Coronavirus disease 2019 (COVID-19): a state of the art review. Otolaryngol Neck Surg:019459982093317. https://doi.org/10.1177/0194599820933170

Iwamoto H (2007) Industrial production of microalgal cell-mass and secondary products – major industrial species: chlorella. Handb Microalgal Cult:253–263. https://doi.org/10.1002/9780470995280.CH11

Jeong SC, Jeong YT, Lee SM, Kim JH (2015) Immune-modulating activities of polysaccharides extracted from brown algae Hizikia fusiforme. Biosci Biotechnol Biochem 79:1362–1365. https://doi.org/10.1080/09168451.2015.1018121

Juszkiewicz A, Basta P, Petriczko E, Machaliński B, Trzeciak J, Łuczkowska K, Skarpańska-Stejnborn A (2018) An attempt to induce an immunomodulatory effect in rowers with spirulina extract. J Int Soc Sports Nutr 15:9. https://doi.org/10.1186/S12970-018-0213-3

Khan SA, Al-Balushi K (2021) Combating COVID-19: the role of drug repurposing and medicinal plants. J Infect Public Health 14:495–503. https://doi.org/10.1016/J.JIPH.2020.10.012

Khan AH, Levac E, van Guelphen L, Pohle G, Chmura GL (2018) The effect of global climate change on the future distribution of economically important macroalgae (seaweeds) in the Northwest Atlantic. Facets 3:275–286. https://doi.org/10.1139/FACETS-2017-0091/ASSET/IMAGES/MEDIUM/FACETS-2017-0091F4.GIF

Kim S, Kim J, Lim Y, Kim YJ, Kim JY, Kwon O (2016) A dietary cholesterol challenge study to assess Chlorella supplementation in maintaining healthy lipid levels in adults: a double-blinded, randomized, placebo-controlled study. Nutr J 15. https://doi.org/10.1186/S12937-016-0174-9

Krause-Jensen D, Middelboe AL, Carstensen J, Dahl K (2007) Spatial patterns of macroalgal abundance in relation to eutrophication. Mar Biol 152:25–36. https://doi.org/10.1007/S00227-007-0676-2/FIGURES/8

Kwon PS, Oh H, Kwon SJ, Jin W, Zhang F, Fraser K, Hong JJ, Linhardt RJ, Dordick JS (2020) Sulfated polysaccharides effectively inhibit SARS-CoV-2 in vitro. Cell Discov 6:50. https://doi.org/10.1038/s41421-020-00192-8

Lafarga T, Acién-Fernández FG, García-Vaquero M (2020) Bioactive peptides and carbohydrates from seaweed for food applications: natural occurrence, isolation, purification, and identification. Algal Res 48. https://doi.org/10.1016/J.ALGAL.2020.101909

Lakshmi DS, Sankaranarayanan S, Gajaria TK, Li G, Kujawski W, Kujawa J, Navia R (2020) A short review on the valorization of green seaweeds and ulvan: feedstock for chemicals and biomaterials. Biomol Ther 10:1–20. https://doi.org/10.3390/biom10070991

Leandro A, Pacheco D, Cotas J, Marques JC, Pereira L, Gonçalves AMM (2020a) Seaweed's bioactive candidate compounds to food industry and global food security. Life 10:140. https://doi.org/10.3390/LIFE10080140

Leandro A, Pereira L, Gonçalves AMM (2020b) Diverse applications of marine macroalgae. Mar Drugs 18:17. https://doi.org/10.3390/MD18010017

Luoma SN, Bryan GW, Langston WJ (1982) Scavenging of heavy metals from particulates by Brown seaweed. Mar Pollut Bull 13:394–396

Mabate B, Daub CD, Malgas S, Edkins AL, Pletschke BI (2021) Fucoidan structure and its impact on glucose metabolism: implications for diabetes and cancer therapy. Mar Drugs 19. https://doi.org/10.3390/MD19010030

Mahendran S, Maheswari P, Sasikala V, Rubika JJ, Pandiarajan J (2021) In vitro antioxidant study of polyphenol from red seaweeds dichotomously branched gracilaria Gracilaria edulis and robust sea moss Hypnea valentiae. Toxicol Rep 8:1404–1411. https://doi.org/10.1016/J.TOXREP.2021.07.006

Manirafasha E, Ndikubwimana T, Zeng X, Lu Y, Jing K (2016) Phycobiliprotein: potential microalgae derived pharmaceutical and biological reagent. Biochem Eng J 109:282–296. https://doi.org/10.1016/J.BEJ.2016.01.025

Manivasagan P, Bharathiraja S, Santha Moorthy M, Mondal S, Seo H, Dae Lee K, Oh J (2018) Marine natural pigments as potential sources for therapeutic applications. Crit Rev Biotechnol 38:745–761. https://doi.org/10.1080/07388551.2017.1398713

Mantri VA, Eswaran K, Shanmugam M, Ganesan M, Veeragurunathan V, Thiruppathi S, Reddy CRK, Seth A (2016) An appraisal on commercial farming of Kappaphycus alvarezii in India: success in diversification of livelihood and prospects. J Appl Phycol 291(29):335–357. https://doi.org/10.1007/S10811-016-0948-7

Mantri VA, Kavale MG, Kazi MA (2020) Seaweed biodiversity of India: reviewing current knowledge to identify gaps, challenges, and opportunities. Divers 12:13. https://doi.org/10.3390/D12010013

Minehira K, Tappy L (2002) Dietary and lifestyle interventions in the management of the metabolic syndrome: present status and future perspective. Eur J Clin Nutr 5612(56):1264–1269. https://doi.org/10.1038/sj.ejcn.1601645

Mišurcová L, Škrovánková S, Samek D, Ambrožová J, Machů L (2012) Health benefits of algal polysaccharides in human nutrition. In: Advances in food and nutrition research. Academic, pp 75–145. https://doi.org/10.1016/B978-0-12-394597-6.00003-3

Mohamed S, Hashim SN, Rahman HA (2012) Seaweeds: a sustainable functional food for complementary and alternative therapy. Trends Food Sci Technol 23:83–96. https://doi.org/10.1016/J.TIFS.2011.09.001

Mukherjee R, Pandya P, Baxi D, Ramachandran AV (2021) Endocrine disruptors–'food' for thought. Proc Zool Soc 74:432–442. https://doi.org/10.1007/S12595-021-00414-1/TABLES/2

Nielsen CH, Balachandran P, Christensen O, Pugh ND, Tamta H, Sufka KJ, Wu X, Walsted A, Schjørring-Thyssen M, Enevold C, Pasco DS (2010) Enhancement of natural killer cell activity in healthy subjects by Immulina®, a Spirulina extract enriched for Braun-type lipoproteins. Planta Med 76:1802–1808. https://doi.org/10.1055/S-0030-1250043

O'Keefe BR, Giomarelli B, Barnard DL, Shenoy SR, Chan PKS, McMahon JB, Palmer KE, Barnett BW, Meyerholz DK, Wohlford-Lenane CL, McCray PB (2010) Broad-spectrum in vitro activity and in vivo efficacy of the antiviral protein Griffithsin against emerging viruses of the family Coronaviridae. J Virol 84:2511–2521. https://doi.org/10.1128/jvi.02322-09

Øverland M, Mydland LT, Skrede A (2019) Marine macroalgae as sources of protein and bioactive compounds in feed for monogastric animals. J Sci Food Agric 99:13–24. https://doi.org/10.1002/JSFA.9143

Pangestuti R, Kurnianto D (2017) Green seaweeds-derived polysaccharides Ulvan: occurrence, medicinal value and potential applications. In: Seaweed polysaccharides: isolation, biological and biomedical applications. Elsevier, pp 205–221. https://doi.org/10.1016/B978-0-12-809816-5.00011-6

Patel S (2012) Therapeutic importance of sulfated polysaccharides from seaweeds: updating the recent findings. 3 Biotech 2:171–185. https://doi.org/10.1007/s13205-012-0061-9

Patti AM, Al-Rasadi K, Giglio RV, Nikolic D, Mannina C, Castellino G, Chianetta R, Banach M, Cicero AFG, Lippi G, Montalto G, Rizzo M, Toth PP (2018) Natural approaches in metabolic syndrome management. Arch Med Sci 14:422. https://doi.org/10.5114/AOMS.2017.68717

Pérez MJ, Falqué E, Domínguez H (2016) Antimicrobial action of compounds from marine seaweed. Mar Drugs 14. https://doi.org/10.3390/md14030052

Porse H, Rudolph B (2017) The seaweed hydrocolloid industry: 2016 updates, requirements, and outlook. J Appl Phycol 295(29):2187–2200. https://doi.org/10.1007/S10811-017-1144-0

Pradhan B, Patra S, Nayak R, Behera C, Dash SR, Nayak S, Sahu BB, Bhutia SK, Jena M (2020) Multifunctional role of fucoidan, sulfated polysaccharides in human health and disease: a journey under the sea in pursuit of potent therapeutic agents. Int J Biol Macromol 164:4263–4278. https://doi.org/10.1016/j.ijbiomac.2020.09.019

Praveen MA, Parvathy KRK, Balasubramanian P, Jayabalan R (2019) An overview of extraction and purification techniques of seaweed dietary fibers for immunomodulation on gut microbiota. Trends Food Sci Technol 92:46–64. https://doi.org/10.1016/J.TIFS.2019.08.011

Ramdani M, Elasri O, Saidi N, Elkhiati N, Taybi FA, Mostareh M, Zaraali O, Haloui B, Ramdani M (2017) Evaluation of antioxidant activity and total phenol content of Gracilaria bursa-pastoris harvested in Nador lagoon for an enhanced economic valorization. Chem Biol Technol Agric 4. https://doi.org/10.1186/S40538-017-0110-Z

Reilly P, O'Doherty JV, Pierce KM, Callan JJ, O'Sullivan JT, Sweeney T (2008) The effects of seaweed extract inclusion on gut morphology, selected intestinal microbiota, nutrient digestibility, volatile fatty acid concentrations and the immune status of the weaned pig. Animal 2:1465–1473. https://doi.org/10.1017/S1751731108002711

Rochlani Y, Pothineni NV, Kovelamudi S, Mehta JL (2017) Metabolic syndrome: pathophysiology, management, and modulation by natural compounds. Ther Adv Cardiovasc Dis 11:215–225. https://doi.org/10.1177/1753944717711379

Rodriguez-Amaya DB (2016) Natural food pigments and colorants. Curr Opin Food Sci 7:20–26. https://doi.org/10.1016/J.COFS.2015.08.004

Rorrer GL, Modrell J, Zhi C, Yoo HD, Nagle DN, Gerwick WH (1995) Bioreactor seaweed cell culture for production of bioactive oxylipins. J Appl Phycol 72(7):187–198. https://doi.org/10.1007/BF00693067

Rzymski P, Budzulak J, Niedzielski P, Klimaszyk P, Proch J, Kozak L, Poniedziałek B (2019) Essential and toxic elements in commercial microalgal food supplements. J Appl Phycol 31:3567–3579. https://doi.org/10.1007/s10811-018-1681-1

Sahu SK, Mantri VA, Zheng P, Yao N (2020) Algae biotechnology. Encycl Mar Biotechnol:1–31. https://doi.org/10.1002/9781119143802.CH1

Saklayen MG (2018) The global epidemic of the metabolic syndrome. Curr Hypertens Rep 202(20):1–8. https://doi.org/10.1007/S11906-018-0812-Z

Santos JP, Torres PB, dos Santos DYAC, Motta LB, Chow F (2019) Seasonal effects on antioxidant and anti-HIV activities of Brazilian seaweeds. J Appl Phycol 31:1333–1341. https://doi.org/10.1007/s10811-018-1615-y

Sanyaolu A, Okorie C, Marinkovic A, Patidar R, Younis K, Desai P, Hosein Z, Padda I, Mangat J, Altaf M (2020) Comorbidity and its impact on patients with COVID-19. Sn Compr Clin Med 2:1069–1076. https://doi.org/10.1007/S42399-020-00363-4

Sen AK, Das AK, Banerji N, Siddhanta AK, Mody KH, Ramavat BK, Chauhan VD, Vedasiromoni JR, Ganguly DK (1994) A new sulfated polysaccharide with potent blood anti-coagulant activity from the red seaweed Grateloupia indica. Int J Biol Macromol 16:279–280. https://doi.org/10.1016/0141-8130(94)90034-5

Shannon E, Conlon M, Hayes M (2021) Seaweed components as potential modulators of the gut microbiota. Mar Drugs 19:358. https://doi.org/10.3390/MD19070358

Shi Q, Wang A, Lu Z, Qin C, Hu J, Yin J (2017) Overview on the antiviral activities and mechanisms of marine polysaccharides from seaweeds. Carbohydr Res 453–454:1–9. https://doi.org/10.1016/j.carres.2017.10.020

Sirota GR, Uthe JF (1979) Heavy metal residues in dulse, an edible seaweed. Aquaculture 18:41–44. https://doi.org/10.1016/0044-8486(79)90099-1

Sondak CFA, Ang PO Jr, Beardall J, Bellgrove A, Min Boo S, Gerung GS, Hepburn CD, Diem Hong D, Hu Z, Kawai H, Largo D, Ae Lee J, Lim P-E, Mayakun J, Nelson WA, Hyun Oak J, Phang S-M (2017) Carbon dioxide mitigation potential of seaweed aquaculture beds (SABs). J Appl Phycol 29:2363–2373. https://doi.org/10.1007/s10811-016-1022-1

Takebe Y, Saucedo CJ, Lund G, Uenishi R, Hase S, Tsuchiura T, Kneteman N, Ramessar K, Tyrrell DLJ, Shirakura M, Wakita T, McMahon JB, O'Keefe BR (2013) Antiviral lectins from red and blue-green algae show potent in vitro and in vivo activity against hepatitis C virus. PLoS One 8:e64449. https://doi.org/10.1371/journal.pone.0064449

Upadhyay K, Patel F, Ramachandran AV, Robin E, Baxi D (2021) Breaching the barriers of chemotherapeutics for breast cancer with alternative medicine. J Endocrinol Reprod 25:23–35. https://doi.org/10.18311/JER/2021/27792

Wells ML, Potin P, Craigie JS, Raven JA, Merchant SS, Helliwell KE, Smith AG, Camire ME, Brawley SH (2016) Algae as nutritional and functional food sources: revisiting our understanding. J Appl Phycol 292(29):949–982. https://doi.org/10.1007/S10811-016-0974-5

Witvrouw M, Este JA, Mateu MQ, Reymen D, Andrei G, Snoeck R, Ikeda S, Pauwels R, Bianchini NV, Desmyter J, De Clercq E (2016) Activity of a sulfated polysaccharide extracted from the red seaweed Aghardhiella Tenera against human immunodeficiency virus and other enveloped viruses. 5:297–303. https://doi.org/10.1177/095632029400500503

Chapter 10
Seaweeds Polyphenolic Compounds: A Marine Potential for Human Skin Health

Ratih Pangestuti, Puji Rahmadi, Evi Amelia Siahaan, Yanuariska Putra, and Se-Kwon Kim

1 Introduction

Seaweeds, commonly known as marine algae, are a beneficial resource of bioactive chemicals with a range of biological functions, benefits for health and medicinal value. These bioactive compounds include sulfated polysaccharides, natural pigments (chlorophylls, carotenoids, phycobilliproteins), bioactive peptides and polyphenolic compounds (Wijesekara et al. 2010; Pangestuti and Kim 2011a). Polyphenolic compounds are secondary metabolites. It can be divided into a number of groups according to the phenol rings number and the structural components connecting these rings. Various form of polyphenolic compounds have been extracted and reports from brown seaweeds, including phlorotannins, mycosporine-like amino acids (MMAs), phenolic terpenoids, flavonoids, and bromophenols. Phlorotannins are the main polyphenolics found in brown seaweed among other polyphenolic substances. Phlorotannins were yielded from a variety brown seaweed species, including *Ecklonia cava*, *E. stolonifera*, *Eisenia byciclis*, *Sargassum*

R. Pangestuti (✉) · E. A. Siahaan
Research Center for Food Processing and Technology, National Research and Innovation Agency (BRIN), Yogyakarta, Indonesia
e-mail: rati008@brin.go.id

P. Rahmadi
Research Center for Oceanography, National Research and Innovation Agency (BRIN), Jakarta, Indonesia

Y. Putra
Research Center for Conservation of Marine and Inland Water Resources, National Research and Innovation Agency (BRIN), Cibinong, Indonesia

S.-K. Kim
Department of Marine Sciences and Convergence Engineering, College of Science and Technology, Hanyang University, Seoul, Republic of Korea

F. Ozogul et al. (eds.), *Seaweeds and Seaweed-Derived Compounds*,
https://doi.org/10.1007/978-3-031-65529-6_10

thunbergia, *S. piluliferum*, *Hizikia fusiforme*, *Endarachne binghamiae*, and *Laminaria sp*. In many studies, it was found that *E. cava* contained more total phenol contents compared to other species. Meanwhile, flavonoids, MAAs, phenolic terpenoids, and bromophenols make up the majority component of the green and red seaweeds (Cotas et al. 2020; Pangestuti et al. 2018, 2019). The MAAs are water-soluble, low molecular weight (LMW), and strongly UVA and UVB absorbent chemicals. MAA typically have molecular weights of 400 Da or less (Miyamoto et al. 2014). MAAs were were utilized as passive shielding materials, dispersing absorbed radiation energy as harmless heat rather than initiating photochemical processes (Karsten et al. 2005). Their maximum absorbance ranged between 310 and 360 nm reliant on the chemical structure (Pangestuti et al. 2018; Bhatia et al. 2011). According to the structural perspective, MAAs are characterized by a cyclohexenone ring that link to amino alcohol, amino acid, or amino group substituents (Singh et al. 2020).

Most polyphenolic compounds isolated from seaweeds possess a vast variation of biological activities such as antioxidant, anti-viral, anti-inflammatory, anticancer, neuroprotective, anti-microbial, and hepatoprotective capabilities (Pangestuti et al. 2018, 2019; Pangestuti and Kim 2011b, 2013; Kim and Pangestuti 2011a; Pangestuti and Wibowo 2013). Not only restricted to those activities but the health benefit effects of seaweed polyphenolic compounds for human skins are getting more recognition. Currently, seaweed polyphenolic compounds with health benefit claims have been applied in skincare as well as cosmetic products available in the market. The chemical variety and distinctive capabilities of seaweed polyphenolic compounds have been a fascinating subjects in recent years in the skincare and cosmetic industry. This contribution objective is to present an extensive understanding of seaweed phenolic compounds, especially phlorotannins and MAAs, providing critical information about these compounds current and potential status, headlining the potential activities for human skins, commercial utilization in industries, and the potential for novel product development.

2 Potential Skin Health Benefit Effects

2.1 Antioxidant Activity

Polyphenolic compounds have been known as the most potent antioxidant compounds in seaweeds. Large numbers of studies have demonstrated antioxidant activity of seaweed polyphenolic compounds, including DPPH assays, ABTS, FRAP. In addition, many studies have used antioxidant assays in the assay-guided purification process (Wijesekara et al. 2010; Pangestuti and Kim 2011a; Vinayak et al. 2011).

We have summarized brown seaweeds extract rich in polyphenolic compounds that showed potential antioxidant activity in UV irradiated skin cells (i.e., HaCaT and HDF). Many studies have correlated polyphenolic compounds in seaweeds and their antioxidant activity. Our previous study analyzed the correlation of total sugar, protein, TFC, TPC, and antioxidant activity of seaweeds (Pangestuti et al. 2019). Our results showed that antioxidant activity negatively correlated with total sugar. However, there was found to be a favorable relationship between antioxidant activity and phenolic contents. The Pearson correlation value between total phenolic content and DPPH scavenging assay was 0.983 ($p < 0.01$), indicating an excellent correlation. In addition, many studies have reported a strong connection among the phenolic content of seaweed and their antioxidant activity (Table 10.1).

Polyphenols are bioactive compounds that contain multiple phenolic group (hydroxyl group linked to benzene ring). We believe that their antioxidant and radical-scavenging capacities in skin cells are strongly related. A free radical or other reactive species can receive electrons from the OH group of an aromatic ring that acts as an electron supporter. This explains why signal transduction pathways, including the phosphoinositide 3-*kinase/protein kinase* B (PI3K/PKB) and mitogen-activates protein kinases (MAPKs) signaling pathway were not stimulated, which is caused by suppressing of reactive oxygen species (ROS) and ROS-mediated degradation to the macromolecules.

In addition to phlorotannins, mycosporine like amino acids (MAAs) also showed skin protection by ROS scavenging activity. MAA from seaweeds such as mycosporine-glycine, porphyra-334, shinorine, asterina-330, and palythine has been examined in numerous assays for its ROS scavenging activity. These include β-carotene/linoleate bleaching technique, the ABTS+ radical scavenging method, the oxygen radical absorbance capacity (ORAC-Fluorescein) method, the superoxide radical scavenging capacity assay, and ROS scavenging (De la Coba et al. 2009; Ryu et al. 2014; Athukorala et al. 2016). Generally, MAAs have shown potent antioxidant properties. Meanwhile, the precise mechanism remains uncertain and further studies are needed on the antioxidant mechanisms of MAAs. Additionally, seaweed extract rich in MAAs have been demonstrated as potential antioxidants (Athukorala et al. 2016; Gacesa et al. 2018). Furthermore, porphyra-334 and shinorine have been reported as potential catalysts of the cytoprotective Keap1-Nrf2 pathway. Porphyra-334 and shinorine showed enhanced transcriptional regulation in human dermal fibroblast (HDF) cells during UVR-induced oxidative stress. Additionally, both porphyra-334 and shinorine exhibited potential *in-vitro* antioxidant activities.

Seaweed-derived polyphenolic compounds showed biological properties by various actions which include high UV absorption, protection of macromolecules from damage, stimulation of endogenous antioxidant enzymes, and radical scavenging activity.

Table 10.1 Antioxidant activity of seaweed extracts and phlorotannins in UV irradiated skin cells

Seaweeds species	Origin	Type of extract/ polyphenolics	Model	Mechanisms	Ref
Sargassum muticum	Korea	Ethanol (80%); Ethyl Acetat fraction	UVB irradiation HaCaT	(↓) intracellular ROS; (↑) antioxidant enzyme	Piao et al. (2011)
Sargassum glaucescens	Taiwan	Aqueous extrac	UVA irradiation HaCaT	(↓) intracellular ROS; (↑) antioxidant enzyme	Li et al. (2019)
Fucus spiralis	Portugal	Aqueous extract; Ethyl Acetat; Ethanol; Cyclohexane; Diethylether	UVB irradiation HaCaT	(↑) antioxidant enzyme; (↓) intracellular ROS	Freitas et al. (2020)
Iridaea laminarioides	Chile	Acetone	UVB irradiation *Danio rerio* embryo	Existence of healthy embryo (91.7%)	Guinea et al. (2012)
Undaria crenata	Korea	Ethanol (80%)	UVB irradiation HaCaT	(↓) hydroxyl radical; (↓) superoxide radical; (↓) intracellular ROS	Hyun et al. (2013)
Carpomitra costata	Korea	Ethanol (80%)	UVB irradiation HaCaT	(↑) antioxidant enzyme; (↓) hydroxyl radical; (↓) superoxide radical; (↓) intracellular ROS	Zheng et al. (2016)
Ecklonia stolonifera	Korea	Ethanol (80%)	UVA irradiation HDF	(↓) intracellular ROS	Jun et al. (2020)
Ecklonia cava	Korea	Methanol extract; Phloroglucinol; Triphlorethol-A	UVB irradiation HaCaT	(↓) hydroxyl & superoxide radical, intracellular ROS; (↑) SOD, GSH	Ko et al. (2011)
Ecklonia cava	Korea	Fucodiphlorethol G	UVB irradiation HaCaT	(↓) DPPH, intracellular ROS;	Kim et al. (2014) and Ham et al. (2007)
Porphyra yezoensis	NA	Shinorine	UVA irradiation HDF	Activated (Keap1) pathway; (↑) SOD, GSH-Px,CAT) (↓) malondialdehyde	Gacesa et al. (2018) and Ying et al. (2019)
Gloiopeltis furcate	NA	Porphyra-334	UVA irradiation HDF	Activated (Keap1) pathway; (↑) SOD, GSH-Px,CAT) (↓) malondialdehyde	Ryu et al. (2014), Gacesa et al. (2018), and Ying et al. (2019)

2.2 Anti-Inflammatory Activity

The host's reaction to a number of stimuli, including physical injury, UV irradiation, microbial invasion, and immunological responses is greatly influenced by the presence of inflammation. In human skins, UV irradiation may cause both local and systemic inflammation. Macrophages provide the first line of protection during inflammatory responses in the existence of a stimulus like ultraviolet radiation (UVR), which generates multiple proinflammatory mediators, such like reactive oxygen and nitrogen species like nitrogen monoxide radical ($NO^{\bullet}$), prostaglandin E_2 (PGE_2), cytokines, and others. In the ordinary circumstances, the delivery of these

Table 10.2 Polyphenolic compounds obtained from brown and red seaweeds with anti-inflammatory properties

Polyphenolic compounds	Seaweeds	Anti-photoaging	Ref
Eckol	*Ecklonia cava; E. stolonifera*	Suppress NF-κB stimulation, (↓) Pro-inflammatory mediators	Ko et al. (2011), Joe et al. (2006), and Lee et al. (2018)
Dieckol	*E. stolonifera*	Suppress NF-κB stimulation, (↓) Pro-inflammatory mediators	Joe et al. (2006), Lee et al. (2018), Ha et al. (2019), and Wang et al. (2020a)
Phloroglucinol	*E. cava*	(↓) Pro-inflammatory mediators	Ko et al. (2011), Kim et al. (2012), Piao et al. (2014), Im et al. (2016), Piao et al. (2015a), and Park et al. (2019)
Diphlorethohydroxycarmalol	*Ishige okamurae*	Suppress MAPKs stimulation; (↓) Pro-inflammatory mediators	Piao et al. (2015b) and Wang et al. (2020b, c)
Fucofuroeckol-A	*Eisenia bicyclis*	(↓)UVB-induced mast cell Degranulation's cytokine generation	Vo et al. (2018)
Shinorine	*Porphyra* sp	Suppress NF-κB, MAPKs stimulation (↓) Pro-inflammatory mediators	Becker et al. (2016)
Pophyra-334	*Porphyra* sp	Suppress NF-κB, MAPKs stimulation (↓) Pro-inflammatory mediators	Becker et al. (2016)

mediators is critical, appearing quickly and greatly severe on injury, but lasting just a brief of period before the detrimental stimuli are resolved.

Anti-inflammatory activities of polyphenolic compounds in seaweeds, including phlorotannins and MAAs have also been investigated (Table 10.2). Treatments with Porphyra-334 treatments inhibited the cyclooxygenase-2 (COX-2) generation, which is principal cytotoxic mediators involved in the mammalian immune response (Pangestuti et al. 2011). Furthermore, treatment with Porphyra-334 and Shinorine in lipopolysaccharides-induced macrophage cells demonstrated prospective anti-inflammation capabilities (Torsdottir et al. 1991). On the other hand, the generation of pro-inflammatory mediators triggered by NF-κB signaling pathway activation could be suppressed by the utilization of MAAs (Becker et al. 2016). These findings are supported by the fact that Poprhyra-334 therapy prevents UV-irradiated mice's NF-κB and MAPKs signal transduction pathways from being stimulated (Terazawa et al. 2020). Furthermore, several intracellular signaling pathways contribute to inflammatory responses. In contrast, two of the essential signaling molecules associated with inflammatory reactions are NF-κB and MAPKs (Rui et al. 2019).

In addition, anti-inflammatory potential of phlorotannins has attracted great attention. The effect of phlorotannins and phlorotannins rich extracts from brown seaweeds have been reports by many studies. The majority of the phlorotannins have been shown to suppress cellular adhesion molecule expression like IL-6, IL-1β, TNF alpha, NO and PGE_2. In numerous skin cell lines stimulated by UV-irradiation, the suppressive effects of phlorotannins such as eckol, dieckol, phloroglucinol, diphlorethohydroxycarmalol, fucofuroeckol-A over MAPKs and NF-κB have been observed. Taken together, these studies demonstrate polyphenolic compounds as prospective anti-inflammatory agents in human skin activated by UV-irradiation.

2.3 Photoprotective Activity

Photoprotective activity of seaweeds polyphenols especially phlorotannins and MAAs have been reported by many studies. Compared to other brown seaweed species, phlorotannins from *Ecklonia cava* is one of the most explored. In Japan and the Korean peninsula, Ecklonia cava is a popularly consumed seaweed that is widely distributed. Phlorotannins from *Ecklonia cava* have been investigated for their preventive effects against UVB-induced oxidative stress. It was found that major phlorotannins in Ecklonia cava are eckstolonol, triphlorethol A, phloroglucinol, and dieckol (Ko et al. 2011). Phlorotannin's capability to scavenge free radicals has been demonstrated ini numerous studies to be significantly correlated with their ability to prevent photoaging. An electron is provided to free radicals or ROS by the hydroxyl (–OH) group attached to the aromatic ring. As a result, signal transduction pathways including MAPKs and NF-κB are not activated, which helps to minimize damage to macromolecules caused by ROS.

Polyphenolic compounds are highly effective as active anti-photoaging agents as a result of their numerous effects as antioxidants, MMP inhibitors, anti-inflammatory agents, and down-regulators of pro-apoptotic mediators. They have no hazardous effect up to a specific dosage, indicating that they can be utilized in skincare and cosmetic commodity as safe anti-photoaging agents. Other biological activity of polyphenolic compounds, such as anti-microbial activity, demonstrate their utility as natural preserving agent in skincare and cosmetic commodity. Thus, in addition to their anti-photoaging properties, they demonstrate significant potential for usage as skincare and cosmetic agents including additional prospective skin benefits.

The MAAs have been reported as one among the most powerful naturally occurring UVA-absorbers (Pangestuti et al. 2018). Presently, MAAs have been discovered in more than 500 different seaweed species (Kim et al. 2021; Sun et al. 2020). MAAs anti-photoaging capabilities are mediated by their photoprotecting action through ultraviolet radiation absorption, as well as their antioxidant activity, anti-inflammatory, radical scavenging, macromolecular damage prevention, MMP inhibitor and another prospective antiphotoaging activities. In HaCaT cells, seaweed-derived MAAs demonstrated photoprotective activity against DNA damage induced by UVB exposure (Suh et al. 2017). The summary of NF-κB and MAPKs signaling pathways suppression by phlorotannins and MAAs is shown in Fig. 10.1.

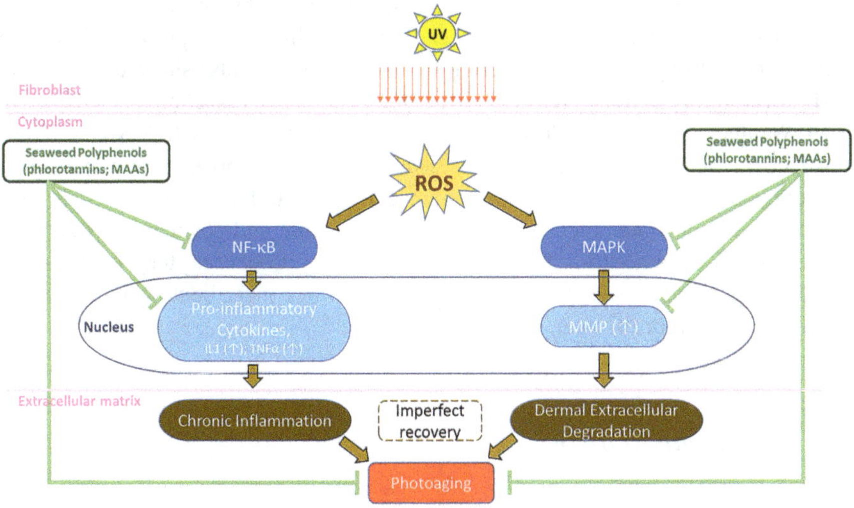

Fig. 10.1 NF-κB and MAPKs signaling pathways inhibition by seaweed polyphenols. Abbreviations: mycosporine-like amino acids (MMAs), reactive oxygen species (ROS), mitogen activated protein kinase (MAPK), nuclear factor κB (NF-κB), interleukin-1 (IL1), tumor necrosis factor α (TNFα), matrix metalloproteinase (MMP), increased (↑)

2.4 Matrix Metalloproteinase (MMP) Inhibitory Activity

The UV radiation (UVR) stimulated the response of complex signaling cascade in human skin, either directly and or indirectly (Xu and Fisher 2005). This process starts with the generation of chemical energy through the absorption of electromagnetic energy via cellular chromophores. Furthermore, these stimulated the activation of NF-κB and activator protein-1 (AP-1) by reactive oxygen species (ROS) generated by chromophores (Ichihashi et al. 2009; Panich et al. 2016). The presence of AP-1 induces matrix metalloproteinases (MMPs) development such as stromelysin-1, gelatinase A, as well as collagenase-1 that particularly breakdown connective tissues like elastin and collagen, and the skin collagen may circumstantially suppress (Pandel et al. 2013). Collagen is a primary fibrillar protein in the extracellular matrix (ECM) which supply the cell with a supporting framework and in charge for the skin's fitness, flexibility, and moisture (Reilly and Lozano 2021). As a result, collagen and ECM contribute a critical part in the condition, appearance and aging of the skin. Anti-photoaging effects may enhance through the inhibition of collagenase-1 and stromelysin-1 synthesis by Porphyra-334. Porphyra-334 treatment causes dermal fibroblast cells to produces more elastin, type I collagen, and procollagen, which are all components of ECM (Ryu et al. 2014; Orfanoudaki et al. 2020). Advanced glycation end products (AGEs) were similarly inhibited by Porphyra-334 (Orfanoudaki et al. 2020). The results revealed that Porphyra-334 treatment preserved the structural integrity of collagen fibers by UVR absorption.

Phlorotannins including eckol, dieckol, phlrooglucinol, triphlorethol-A, and diphlorethohydroxycarmalol isolated from brown seaweeds also showed substantial MMP-1 inhibitory activity (Ko et al. 2011; Joe et al. 2006; Lee et al. 2018; Ha et al. 2019; Wang et al. 2020a). Phlorotannins could be employed as safe MMP-1 inhibitors in cosmetic and skin care products because they exhibit no adverse effects up to a specific concentration. Another biological activity of phlorotannins, such as antimicrobial, demonstrates their utility as natural skincare and cosmetic product preservatives. Thus, in addition to their anti-photoaging properties, they demonstrate significant potential for usage as skincare and cosmetic ingredients with additional potential benefits for the skin.

2.5 Anti-Acne Activity

Acne is a frequent skin problem. It is a complex and prolonged inflammatory condition that affects the skin's pilosebaceous unit, which involves androgen-induced sebum hyperplasia, alteration of keratinization pattern, hormonal changes, and *Propionibacterium acnes* occupation (Sinha et al. 2014). Acne treatments consist of several methods, i.e. minimising sebum production, discharging dead skin cells, and antibiotics (Kim and Pangestuti 2011b). Compared to other therapies, antibiotics are generally administered to treat acne, however antibiotic resistance is becoming

Table 10.3 Anti-acne activity of phlorotannins

Phlorotannins	Seaweeds	Bacteria		Ref
Dieckol	*Ecklonia cava*	*Propionibacterium acnes*	MIC: 39 μg/mL	Choi et al. (2014)
Phlorofucofuroeckol A	*Ecklonia cava*	*Propionibacterium acnes*	MIC: 39 μg/mL	Choi et al. (2014)
Fucofuroeckol-A	*Eisenia bicyclis*	*Propionibacterium acnes*	MIC: 64 μg/mL	Lee et al. (2014)
Eckol	*Eisenia bicyclis*	Inhibit *Propionibacterium acnes*-induced HaCaT cells	–	Eom et al. (2017)

another problems in dermatologic illnesses. Therefore, novel remedies are in great demand, and an ethnopharmacological approach to finding novel plant sources of anti-acne medications could help bridge the gap in currently available treatments. Seaweeds phlorotanins isolated from *Ecklonia cava* and *Eisenia bicyclis* (Table 10.3) have been demonstrated to inhibit *Propionibacterium acnes* growth with minimal prohibitory concentrations (MIC) from 39 to 64 μg/mL. It has been report elsewhere that the correlations between bacterial proteins and phlorotannins were thought to be key in phlorotannins antiacne effects (Eom et al. 2012). It was demonstrated that phlorotannins isolated from seaweeds carry anti-bacterial properties; however, less number of studies is focusing on anti-acne properties of these compounds. Therefore, future studies is required, particularly emphasizing the anti-acne capabilities of phlorotannins and the underlying mechanisms.

2.6 *Skin Brightening Activity*

Around 15% of the global population is interested in skin lightening products, with Asia dominating the market. Melanin is particularly in charge for the coloration of human eyes, skin, and hair, and it is generated in a 1:36 ratio by epidermal melanocytes and basal keratinocytes. Melanocytes manufacture melanin by responding to ultraviolet B (UVB) exposure. This melanin-producing stage is called melanogenesis. This complex process is introduced by the crucial enzyme tyrosinase (TYR, monophenol or *o*-diphenol, dihydroxyl-Lphenylalanine, oxidoreductase, EC 1.14.18.1, syn. polyphenol oxidase), which catalyzed the oxidation of L-tyrosine to dopaquinone (DQ).

Since last decade, a substantial number of tyrosinase inhibitor either synthetic or natural origin has been reported. Tyrosinase inhibitors including arbutin, azelaic acid, hydroquinones and kojic acid have been tested in pharmaceuticals and skin care for their capability to prevent malanogenesis. However, the usages of some of tyrosinase inhibitor are limited due to low stability, poor skin penetration, insufficient activity, and side effect such as cytotoxicity potential. As a result, novel tyrosinase inhibitors from natural sources with minimum side effects, safe, high efficiency and penetration are required.

Phlorotannins derived from brown seaweeds have been demonstrated to carry tyrosinase inhibitory activity (Table 10.4). Based on the IC_{50} value, 7-phloroeckol showed substantial tyrosinase inhibitory activity compared to other phlorotannins, arbutin and kojic acid (Yoon et al. 2009). Suppression of L-tyrosine mediated by mushroom tyrosinase tests has been used to identify 7-phloroeckol's tyrosinase inhibitory activity. Mushroom tyrosinase activity was unaffected by preincubation with 7-phloroeckol in the absence of L-tyrosine, which seems to indicate that 7-phloroeckol are inhibitor rather than an inactivators of the mushroom tyrosinase.

Table 10.4 Potential tyrosinase inhibitory activity of brown seaweeds phlorotannins

Phlorotannins	Brown seaweeds	Type of inhibition	IC_{50}	Ref
Phloroglucinol	*Ecklonia stolonifera*	Competitive	NA	Yoon et al. (2009), Kang et al. (2004), and Heo et al. (2009)
Eckstolonol	*Ecklonia stolonifera*	Competitive	NA	Kang et al. (2004)
Eckol	*Ecklonia stolonifera*	Non-competitive	NA	Kang et al. (2004), Heo et al. (2009), and Kim et al. (2019)
Phlorofucofuroeckol	*Ecklonia stolonifera*	Non-competitive	NA	Kang et al. (2004)
Dieckol	*Ecklonia stolonifera*	Non-competitive	NA	Kang et al. (2004) and Heo et al. (2009)
974-A	*Ecklonia stolonifera*	Competitive	1.57 ± 0.08 μM; 3.56 ± 0.22 μM	Manandhar et al. (2019)
Dioxinodehydroeckol	*Ecklonia cava*, *Ecklonia stolonifera*	NA	NA	Yoon et al. (2009) and Lee et al. (2012)
7-Phloroeckol	*Ecklonia cava*	Non-competitive	0.85 μM	Yoon et al. (2009)
4-Hydroxyphenethyl alcohol	*Hizikia fusiformis*	NA	NA	Jang et al. (2014)
Octaphlorethol A	*Ishige foliacea*	NA	NA	Kim et al. (2015)
Triphlorethol A	*Ecklonia cava*	NA	NA	Kim et al. (2019)
2-phloroeckol	*Ecklonia cava*	Competitive	7.0 ± 0.2 μM	Kim et al. (2019)
Phlorofucofuroeckol	*Ecklonia cava*	NA	7.0 ± 0.2 μM	Kim et al. (2019)
2-TP-6,6′-bieckol	*Ecklonia cava*	Competitive	8.8 ± 0.1 μM	Kim et al. (2019)
6,8′-bieckol	*Ecklonia cava*	NA	NA	Kim et al. (2019)
8,8′-bieckol	*Ecklonia cava*	NA	NA	Kim et al. (2019)

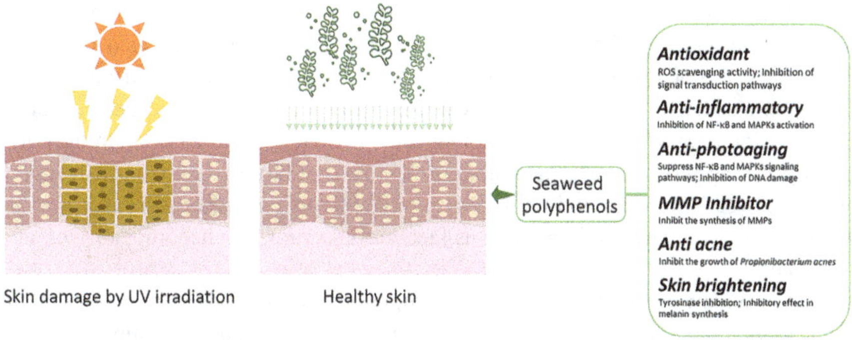

Fig. 10.2 Seaweed polyphenols potential benefits for skin health. Abbreviations: Reactive Oxygen Species (ROS), Nuclear Factor κB (NF-κB), Mitogen Activated Protein Kinases (MAPK), Matrix Metalloproteinases (MMP)

Overall, phlorotannins inhibited tyrosinase in two ways: competitive and noncompetitive. Competitive inhibition is developed when phlorotannins binds to a free enzyme and stops it from attaching to its substrate. Due to the competition for the same site, phlorotannins as inhibitors and L-tyrosine as substrate cannot coexist in this circumstance. Meanwhile, noncompetitive inhibition occurs when phlorotannins binds to enzyme at a site other than the active site. In addition, some phlorotannins also showed inhibitory effect in melanin synthesis. Collectively, it can be suggested that phlorotannins may provide a structural templates for designing and developing novel tyrosinase inhibitors as effective anti-dark spots, melisma, freckles, and skin lightening agents in skin care as well as cosmetics applications. Overall, seaweed polyphenols potential benefits for skin health are illustrated in Fig. 10.2.

3 Current Development, Potency and Future Perspectives of Seaweeds Polyphenolic Compounds

The nutrition, skincare and cosmetic industries have risen to grow as one of the economy's most rapidly expanding and profitable sectors (Hu et al. 2020). Demands for new and innovative skincare products with skin health benefits are continuously growing. Therefore, since last decade's scientists around the worlds have tried to find natural origin, novel bioactive substances with health benefit for the skin. Currently, seaweeds based products with skin health benefit claims have been extended and become commercially-available products in the market.

In 2020, a group of scientists from Marinova Ltd. from Australia reported that the development of sirtuin 1 (SIRT1) could be elevated by the application of *Fucus vesiculosus* extracts, which contain 30% polyphenol and 60% fucoidan. SIRT1 is a protein that has been found to have anti-aging and longevity-promoting properties.

The effectiveness of the extracts in variety of tested application was also determined by scientific analysis. The reduction of age spots is greatly influenced by *Fucus vesiculosus* extract, which also improves protection, brightness, and comforting (Fitton et al. 2015). These brown seaweed extracts are marketed by the Australian Biotech Company (Marinova's). Another anti-photoaging extract containing polyphenolic compounds are also available in the market (Table 10.5). Currently, many brown seaweed extracts rich in phlorotannins (*Ecklonia cava*, *Saccharina japonica*) were used in skin care products such as face mask, serum, essence, moisturizer, and sunscreen.

Due to wide variety of polyphenolic compounds in seaweeds, this wide range of skin health benefits, the innovations of seaweeds polyphenolic compounds will

Table 10.5 Commercial skin care products with seaweeds polyphenolic compounds as their active substances

Commercial name	Seaweeds species	Active ingredients	Health benefits claim	Ref
Helionori®-(Gelyma, French)	*Poprphyra umbilicalis*	Palythine, Porphyra 334 and Shinorine	Photoprotective DNA protection Prevent sunburn	Gelyma Helionori (n.d.)
Helioguard365 (Mibelle Biochemistry, Switzerland)	*Poprphyra umbilicalis*	Porphyra-334 & Shinorine	Photoprotective	Mibellebiochemistry Helioguard™ 365 (n.d.)
Algae gorria; Alga marris (Laboratoires de biarritz, French)	*Poprphyra umbilicalis*	NA	Photoprotective	Laboratoires-biarritz Suncare organic (n.d.)
Maritech bright - (Marinova, Australia)	*Fucus vesiculosus*	Fucoidan-polyphenol complex	Brightening, anti-oxidant, anti-ageing, balancing	Marinova Marinova Product Portofolio (n.d.)
Maritech synergy (Marinova, Australia)	*Fucus vesiculosus*	Fucoidan & polyphenol complex	Antioxidant, anti-inflammation, rejuvenating	Marinova Marinova Product Portofolio (n.d.)
Vinanza®; Wakame bioactive gel (New Zealand Extracts Limited)	*Undaria pinnatifidia*	Fucoidan, Fucoxanthin & Polyphenols	Anti-ageing, antioxidant, skin energising	Vinanza Wakame Bioactive Gel (n.d.)
FucoSkin® (Hi-Q, Taiwan)	*Laminaria japonica*	Oligo fucoidan	Shooting, anti-ageing, anti-inflammation	Hi-Q Ocean Re-New Anti-Aging Series (n.d.)
Hydra Blue® (Repechage, United Kingdom)	*Laminaria digitata Ascophyllum nosodum*	Fucoidan	Anti-oxidant, moisturizing, nourishing	Science of Seaweed (n.d.)
Grown Alchemist® (United States)	*Fucus* sp.	Fucoidan	Anti-oxidant, hydrating, skin energizing	Grown Alchemist (n.d.)

continuously grow and attracting more researchers, industries as well as consumers consideration. Seaweeds polyphenolic compounds are suggested as active ingredients in supplements, sun creams, photoaging control creams, moisturizers in skincare and cosmetics. Furthermore, eco-friendly technology to isolate seaweed polyphenolic compounds is required to increase market availability of eco-friendly products.

4 Conclusions

Seaweeds polyphenolic compounds comprise diverse structures with bioactivities and potential benefits for human's skin. Those remarkable bioactive properties show that polyphenolic compounds are valuable natural resources to be applied in skin care, cosmetics and supplements. There are still significant prospects for developing seaweed polyphenolic compounds in the skincare and cosmetics industry. As a result, development seaweeds polyphenolic compounds for human skin is critical and presents a challenges for scientists, seaweed producers, and skincare or cosmetic manufacturers.

References

Athukorala Y, Trang S, Kwok C, Yuan YV (2016) Antiproliferative and antioxidant activities and mycosporine-like amino acid profiles of wild-harvested and cultivated edible Canadian marine red macroalgae. Molecules 21(1):119

Becker K, Hartmann A, Ganzera M, Fuchs D, Gostner JM (2016) Immunomodulatory effects of the mycosporine-like amino acids shinorine and porphyra-334. Mar Drugs 14(6):119

Bhatia S, Garg A, Sharma K, Kumar S, Sharma A, Purohit AP (2011) Mycosporine and mycosporine-like amino acids: a paramount tool against ultra violet irradiation. Pharmacogn Rev 5(10):138–146

Choi J-S, Lee K, Lee B-B, Kim Y-C, Kim YD, Hong Y-K, Cho KK, Choi IS (2014) Antibacterial activity of the phlorotanninsdieckol and phlorofucofuroeckol-A from Ecklonia cava against Propionibacteriumacnes. Bot Sci 92(3):425–431

Cotas J, Leandro A, Monteiro P, Pacheco D, Figueirinha A, Gonçalves AM, da Silva GJ, Pereira L (2020) Seaweed phenolics: from extraction to applications. Mar Drugs 18(8):384

De la Coba F, Aguilera J, Figueroa FL, De Gálvez M, Herrera E (2009) Antioxidant activity of mycosporine-like amino acids isolated from three red macroalgae and one marine lichen. J Appl Phycol 21(2):161–169

Eom S-H, Kim Y-M, Kim S-K (2012) Antimicrobial effect of phlorotannins from marine brown algae. Food Chem Toxicol 50(9):3251–3255

Eom SH, Lee EH, Park K, Kwon JY, Kim PH, Jung WK, Kim YM (2017) Eckol from Eisenia bicyclis inhibits inflammation through the Akt/NF-κB signaling in Propionibacterium acnes-induced human keratinocyte Hacat cells. J Food Biochem 41(2):e12312

Fitton JH, Dell'Acqua G, Gardiner V-A, Karpiniec SS, Stringer DN, Davis E (2015) Topical benefits of two fucoidan-rich extracts from marine macroalgae. Cosmetics 2(2):66–81

Freitas R, Martins A, Silva J, Alves C, Pinteus S, Alves J, Teodoro F, Ribeiro HM, Gonçalves L, Petrovski Ž (2020) Highlighting the biological potential of the brown seaweed Fucus spiralis for skin applications. Antioxidants 9(7):611

Gacesa R, Lawrence KP, Georgakopoulos ND, Yabe K, Dunlap WC, Barlow DJ, Wells G, Young AR, Long PF (2018) The mycosporine-like amino acids porphyra-334 and shinorine are antioxidants and direct antagonists of Keap1-Nrf2 binding. Biochimie 154:35–44

Gelyma Helionori (n.d.) Natural sun protection thanks to Marine UVA Filters. February 7. http://www.gelyma.com/helionori.html

Grown Alchemist. https://grownalchemist.com/products/hydra-restore-cream-cleanser-olive-leaf-plantago-extract

Guinea M, Franco V, Araujo-Bazán L, Rodríguez-Martín I, González S (2012) In vivo UVB-photoprotective activity of extracts from commercial marine macroalgae. Food Chem Toxicol 50(3–4):1109–1117

Ha JW, Song H, Hong SS, Boo YC (2019) Marine alga ecklonia cava extract and dieckol attenuate prostaglandin E2 production in HaCaT keratinocytes exposed to airborne particulate matter. Antioxidants 8(6):190

Ham Y-M, Baik J-S, Hyun J-W, Lee N-H (2007) Isolation of a new phlorotannin, fucodiphlorethol G, from a brown alga Ecklonia cava. Bull Korean Chem Soc 28(9):1595–1597

Heo S-J, Ko S-C, Cha S-H, Kang D-H, Park H-S, Choi Y-U, Kim D, Jung W-K, Jeon Y-J (2009) Effect of phlorotannins isolated from Ecklonia cava on melanogenesis and their protective effect against photo-oxidative stress induced by UV-B radiation. Toxicol In Vitro 23(6):1123–1130

Hi-Q Ocean Re-New Anti-Aging Series (n.d.) Skincare products. http://www.fucoidanhiq.com/mhe32

Hu Y, Zeng H, Huang J, Jiang L, Chen J, Zeng Q (2020) Traditional Asian herbs in skin whitening: the current development and limitations. Front Pharmacol 11:982

Hyun YJ, Piao MJ, Ko MH, Lee NH, Kang HK, Yoo ES, Koh YS, Hyun JW (2013) Photoprotective effect of Undaria crenata against ultraviolet B-induced damage to keratinocytes. J Biosci Bioeng 116(2):256–264

Ichihashi M, Ando H, Yoshida M, Niki Y, Matsui M (2009) Photoaging of the skin. Anti Aging Med 6(6):46–59

Im AR, Nam KW, Hyun JW, Chae S (2016) Phloroglucinol reduces photodamage in hairless mice via matrix metalloproteinase activity through MAPK pathway. Photochem Photobiol 92(1):173–179

Jang M-S, Park H-Y, Nam K-H (2014) Whitening effects of 4-hydroxyphenethyl alcohol isolated from water boiled with Hizikia fusiformis. Food Sci Biotechnol 23(2):555–560

Joe M-J, Kim S-N, Choi H-Y, Shin W-S, Park G-M, Kang D-W, Kim YK (2006) The inhibitory effects of eckol and dieckol from Ecklonia stolonifera on the expression of matrix metalloproteinase-1 in human dermal fibroblasts. Biol Pharm Bull 29(8):1735–1739

Jun E-S, Kim YJ, Kim H-H, Park SY (2020) Gold nanoparticles using Ecklonia stolonifera protect human dermal fibroblasts from UVA-induced senescence through inhibiting MMP-1 and MMP-3. Mar Drugs 18(9):433

Kang HS, Kim HR, Byun DS, Son BW, Nam TJ, Choi JS (2004) Tyrosinase inhibitors isolated from the edible brown alga Ecklonia stolonifera. Arch Pharm Res 27(12):1226–1232

Karsten U, Friedl T, Schumann R, Hoyer K, Lembcke S (2005) Mycosporine-like amino acids and phylogenies in green algae: prasiola and its relatives from the trebouxiophyceae (chlorophyta) 1. J Phycol 41(3):557–566

Kim S-K, Pangestuti R (2011a) Potential role of marine algae on female health, beauty, and longevity. Adv Food Nutr Res 64:41–55

Kim S-K, Pangestuti R (2011b) Biological properties of cosmeceuticals derived from marine algae. CRC Press, Boca Raton, pp 191–200

Kim KC, Piao MJ, Cho SJ, Lee NH, Hyun JW (2012) Phloroglucinol protects human keratinocytes from ultraviolet B radiation by attenuating oxidative stress. Photodermatol Photoimmunol Photomed 28(6):322–331

Kim KC, Piao MJ, Zheng J, Yao CW, Cha JW, Kumara MHSR, Han X, Kang HK, Lee NH, Hyun JW (2014) Fucodiphlorethol G purified from Ecklonia cava suppresses ultraviolet B radiation-induced oxidative stress and cellular damage. Biomol Ther 22(4):301

Kim K-N, Yang H-M, Kang S-M, Ahn G, Roh SW, Lee W, Kim D, Jeon Y-J (2015) Whitening effect of octaphlorethol A isolated from Ishige foliacea in an in vivo zebrafish model. J Microbiol Biotechnol 25(4):448–451

Kim JH, Lee S, Park S, Park JS, Kim YH, Yang SY (2019) Slow-binding inhibition of tyrosinase by Ecklonia cava phlorotannins. Mar Drugs 17(6):359

Kim SY, Cho WK, Kim H-I, Paek SH, Jang SJ, Jo Y, Choi H, Lee JH, Moh SH (2021) Transcriptome profiling of human follicle dermal papilla cells in response to Porphyra-334 treatment by RNA-Seq. Evid Based Complement Alternat Med 2021:6637513

Ko S-C, Cha S-H, Heo S-J, Lee S-H, Kang S-M, Jeon Y-J (2011) Protective effect of Ecklonia cava on UVB-induced oxidative stress: in vitro and in vivo zebrafish model. J Appl Phycol 23(4):697–708

Laboratoires-biarritz Suncare organic (n.d.) Alga Maris®. February 7. https://www.laboratoires--biarritz.com/fr/

Lee MS, Yoon HD, Kim JI, Choi JS, Byun DS, Kim HR (2012) Dioxinodehydroeckol inhibits melanin synthesis through PI3K/Akt signalling pathway in α-melanocyte-stimulating hormone-treated B16F10 cells. Exp Dermatol 21(6):471–473

Lee J-H, Eom S-H, Lee E-H, Jung Y-J, Kim H-J, Jo M-R, Son K-T, Lee H-J, Kim JH, Lee M-S (2014) In vitro antibacterial and synergistic effect of phlorotannins isolated from edible brown seaweed Eisenia bicyclis against acne-related bacteria. Algae 29(1):47–55

Lee J-W, Seok JK, Boo YC (2018) Ecklonia cava extract and dieckol attenuate cellular lipid peroxidation in keratinocytes exposed to PM10. Evid Based Complement Alternat Med 2018:8248323

Manandhar B, Wagle A, Seong SH, Paudel P, Kim H-R, Jung HA, Choi JS (2019) Phlorotannins with potential anti-tyrosinase and antioxidant activity isolated from the marine seaweed Ecklonia stolonifera. Antioxidants 8(8):240

Marinova Marinova Product Portofolio. Februa https://www.marinova.com.au/product-portfolio/

Mibellebiochemistry Helioguard™ 365 (n.d.) A natural UV-screening active to protect against photo-aging. February 7. https://mibellebiochemistry.com/helioguardtm-365

Miyamoto KT, Komatsu M, Ikeda H (2014) Discovery of gene cluster for mycosporine-like amino acid biosynthesis from Actinomycetales microorganisms and production of a novel mycosporine-like amino acid by heterologous expression. Appl Environ Microbiol 80(16):5028–5036

Orfanoudaki M, Hartmann A, Alilou M, Gelbrich T, Planchenault P, Derbré S, Schinkovitz A, Richomme P, Hensel A, Ganzera M (2020) Absolute configuration of mycosporine-like amino acids, their wound healing properties and in vitro anti-aging effects. Mar Drugs 18(1):35

Pandel R, Poljšak B, Godic A, Dahmane R (2013) Skin photoaging and the role of antioxidants in its prevention. Int Sch Res Notices 2013:930164

Pangestuti R, Kim S-K (2011a) Biological activities and health benefit effects of natural pigments derived from marine algae. J Funct Foods 3(4):255–266

Pangestuti R, Kim S-K (2011b) Neuroprotective effects of marine algae. Mar Drugs 9(5):803–818

Pangestuti R, Kim S-K (2013) Marine-derived bioactive materials for neuroprotection. Food Sci Biotechnol 22(5):1–12

Pangestuti R, Wibowo S (2013) Prospects and health promoting effects of brown algal-derived natural pigments. Squalen Bull Mar Fish Postharvest Biotechnol 8(1):37–46

Pangestuti R, Bak SS, Kim SK (2011) Attenuation of pro-inflammatory mediators in LPS-stimulated BV2 microglia by chitooligosaccharides via the MAPK signaling pathway. Int J Biol Macromol 49(4):599–606

Pangestuti R, Siahaan E, Kim S-K (2018) Photoprotective substances derived from marine algae. Mar Drugs 16(11):399

Pangestuti R, Getachew AT, Siahaan EA, Chun B-S (2019) Characterization of functional materials derived from tropical red seaweed Hypnea musciformis produced by subcritical water extraction systems. J Appl Phycol 31(4):2517–2528
Panich U, Sittithumcharee G, Rathviboon N, Jirawatnotai S (2016) Ultraviolet radiation-induced skin aging: the role of DNA damage and oxidative stress in epidermal stem cell damage mediated skin aging. Stem Cells Int 2016:7370642
Park C, Cha H-J, Hong SH, Kim G-Y, Kim S, Kim H-S, Kim BW, Jeon Y-J, Choi YH (2019) Protective effect of phloroglucinol on oxidative stress-induced DNA damage and apoptosis through activation of the Nrf2/HO-1 signaling pathway in HaCaT human keratinocytes. Mar Drugs 17(4):225
Piao MJ, Yoon WJ, Kang HK, Yoo ES, Koh YS, Kim DS, Lee NH, Hyun JW (2011) Protective effect of the ethyl acetate fraction of Sargassum muticum against ultraviolet B–irradiated damage in human keratinocytes. Int J Mol Sci 12(11):8146–8160
Piao MJ, Ahn MJ, Kang KA, Kim KC, Zheng J, Yao CW, Cha JW, Hyun CL, Kang HK, Lee NH (2014) Phloroglucinol inhibits ultraviolet B radiation-induced oxidative stress in the mouse skin. Int J Radiat Biol 90(10):928–935
Piao MJ, Ahn MJ, Kang KA, Kim KC, Cha JW, Lee NH, Hyun JW (2015a) Phloroglucinol enhances the repair of UVB radiation-induced DNA damage via promotion of the nucleotide excision repair system in vitro and in vivo. DNA Repair 28:131–138
Piao MJ, Kumara MHSR, Kim KC, Kang KA, Kang HK, Lee NH, Hyun JW (2015b) Diphlorethohydroxycarmalol suppresses ultraviolet B-induced matrix metalloproteinases via inhibition of JNK and ERK signaling in human keratinocytes. Biomol Ther 23(6):557
Reilly DM, Lozano J (2021) Skin collagen through the lifestages: importance for skin health and beauty. Plast Aesthet Res 8:2
Rui Y, Zhaohui Z, Wenshan S, Bafang L, Hu H (2019) Protective effect of MAAs extracted from Porphyra tenera against UV irradiation-induced photoaging in mouse skin. J Photochem Photobiol B Biol 192:26–33
Ryu J, Park S-J, Kim I-H, Choi YH, Nam T-J (2014) Protective effect of porphyra-334 on UVA-induced photoaging in human skin fibroblasts. Int J Mol Med 34(3):796–803
Science of Seaweed. https://www.beautyfromthesea.co.uk/pages/science-of-seaweed-1
Singh DK, Pathak J, Pandey A, Singh V, Ahmed H, Rajneesh, Kumar D, Sinha RP (2020) Chapter 15 – Ultraviolet-screening compound mycosporine-like amino acids in cyanobacteria: biosynthesis, functions, and applications. In: Singh PK, Kumar A, Singh VK, Shrivastava AK (eds) Advances in cyanobacterial biology. Academic, pp 219–233
Sinha P, Srivastava S, Mishra N, Yadav NP (2014) New perspectives on antiacne plant drugs: contribution to modern therapeutics. Biomed Res Int 2014:301304
Suh S-S, Oh SK, Lee SG, Kim I-C, Kim S (2017) Porphyra-334, a mycosporine-like amino acid, attenuates UV-induced apoptosis in HaCaT cells. Acta Pharma 67(2):257–264
Sun Y, Zhang N, Zhou J, Dong S, Zhang X, Guo L, Guo G (2020) Distribution, contents, and types of mycosporine-like amino acids (MAAs) in marine macroalgae and a database for MAAs based on these characteristics. Mar Drugs 18(1):43
Terazawa S, Nakano M, Yamamoto A, Imokawa G (2020) Mycosporine-like amino acids stimulate hyaluronan secretion by up-regulating hyaluronan synthase 2 via activation of the p38/MSK1/CREB/c-Fos/AP-1 axis: MAAs stimulate the secretion of HA via HAS2. J Biol Chem 295(21):7274–7288
Torsdottir I, Alpsten M, Holm G, Sandberg A-S, Tölli J (1991) A small dose of soluble alginate-fiber affects postprandial glycemia and gastric emptying in humans with diabetes. J Nutr 121(6):795–799
Vinanza Wakame Bioactive Gel. https://www.nzextracts.com/uploads/images/brochures/NZExtracts-Vinanza-Wakame-BioactiveGel.pdf
Vinayak RC, Sabu A, Chatterji A (2011) Bio-prospecting of a few brown seaweeds for their cytotoxic and antioxidant activities. Evid Based Complement Alternat Med 2011:673083

Vo TS, Kim S-K, Ryu B, Ngo DH, Yoon N-Y, Bach LG, Hang NTN, Ngo DN (2018) The suppressive activity of fucofuroeckol-a derived from brown algal Ecklonia stolonifera Okamura on UVB-induced mast cell degranulation. Mar Drugs 16(1):1

Wang L, Lee W, Jayawardena TU, Cha S-H, Jeon Y-J (2020a) Dieckol, an algae-derived phenolic compound, suppresses airborne particulate matter-induced skin aging by inhibiting the expressions of pro-inflammatory cytokines and matrix metalloproteinases through regulating NF-κB, AP-1, and MAPKs signaling pathways. Food Chem Toxicol 146:111823

Wang L, Kim HS, Je J-G, Oh JY, Kim Y-S, Cha S-H, Jeon Y-J (2020b) Protective effect of diphlorethohydroxycarmalol isolated from Ishige okamurae against particulate matter-induced skin damage by regulation of NF-κB, AP-1, and MAPKs signaling pathways in vitro in human dermal fibroblasts. Molecules 25(5):1055

Wang L, Kim HS, Oh JY, Je JG, Jeon Y-J, Ryu B (2020c) Protective effect of diphlorethohydroxycarmalol isolated from Ishige okamurae against UVB-induced damage in vitro in human dermal fibroblasts and in vivo in zebrafish. Food Chem Toxicol 136:110963

Wijesekara I, Pangestuti R, Kim SK (2010) Biological activities and potential health benefits of sulfated polysaccharides derived from marine algae. Carbohydr Polym 84(1):14–21

Xu Y, Fisher GJ (2005) Ultraviolet (UV) light irradiation induced signal transduction in skin photoaging. J Dermatol Sci Suppl 1(2):S1–S8

Ying R, Zhang Z, Zhu H, Li B, Hou H (2019) The protective effect of mycosporine-like amino acids (MAAs) from Porphyra yezoensis in a mouse model of UV irradiation-induced photoaging. Mar Drugs 17(8):470

Yoon NY, Eom T-K, Kim M-M, Kim S-K (2009) Inhibitory effect of phlorotannins isolated from Ecklonia cava on mushroom tyrosinase activity and melanin formation in mouse B16F10 melanoma cells. J Agric Food Chem 57(10):4124–4129

Li Z-Y, Yu C-H, Lin Y-T, Su H-L, Kan K-W, Liu F-C, Chen C-T, Lin Y-T, Hsu H-F, Lin Y-H (2019) The potential application of spring Sargassum glaucescens extracts in the moisture-retention of keratinocytes and dermal fibroblast regeneration after UVA-irradiation. Cosmetics 6(1):17

Zheng J, Hewage SM, Piao MJ, Kang KA, Han X, Kang H, Yoo E, Koh Y, Lee N, Ko C (2016) Photoprotective effect of carpomitra costata extract against ultraviolet B-induced oxidative damage in human keratinocytes. J Environ Pathol Toxicol Oncol 35:1

Chapter 11
Cosmetic and Dermatological Application of Seaweed: Skincare Therapy-Cosmeceuticals

Cengiz Gokbulut

Abbreviations

AP-1	Activator protein-1
BEPFs	Biological Effective Protection Factors
CAGR	compound annual growth rate
COX	Cyclooxygenase
DNA	Deoxyribonucleotide
EC	European Commission
EGFR	Epidermal growth factor receptor
ERK	Extracellular signal regulated-kinase
ES-GNPs	*Ecklonia stolonifera* gold nanoparticles
EtOH	Ethanol
HDF	Human dermal fibroblasts
HDFN	Human dermal fibroblast, neonatal
HSPB1	Heat-shock protein B1
IL-1R	Interleukin-1R
iNOS	Inducible nitric oxide synthase
LMW	Low molecule weight
LPS	Lipopolysaccharide
MAAs	Mycosporine-Like Amino Acids
MAPK	Mitogen-activated protein kinase
MDA	Malondialdehyde
MeOH	Methanol
MMP	Matrix metalloproteinase
MPO	Myeloperoxidase

C. Gokbulut (✉)
Department of Medical Pharmacology, Faculty of Medicine, Balikesir University, Turkey, Balikesir

F. Ozogul et al. (eds.), *Seaweeds and Seaweed-Derived Compounds*,
https://doi.org/10.1007/978-3-031-65529-6_11

MSH	Melanocyte-stimulating hormones
NADPH	Nicotinamide adenine dinucleotide phosphate
NF-κB	Nuclear factor-κB
Nrf2	Nuclear factor erythroid 2-related factor 2
PGE_2	Prostaglandin E_2
RHS	Reconstructed human skin
ROS	Reactive oxygen species
TBARS	Thiobarbituric acid reactive substances
TGFβRII	Transforming growth factor beta receptor II
THP-1	Human leukaemia monocytic cell line
TNFα-R	Tumor necrosis factor alpha-R
UV	Ultra violet
UVA	Ultra violet-A
UVB	Ultra violet-B
UVR	Ultra violet radiation

1 Introduction

Etymologically "cosmetics" has been derived from the Greek word "komaō/κομάω", a word whose meanings range from cosmic order to personal adornment (Resinski 1998). Cosmetic products have attracted the attention of people in every period from prehistoric times to the present and have important roles in human history. Archaeological research has revealed that the first findings of cosmetic products belong to the Egyptian period. Remains of containers of face paints and ointments from this period provide evidence of the use of cosmetic products since 4000 BC. In addition, the importance that the Egyptians gave to their hair and facial appearance is evident in the Ebers Papyrus, especially because they were concerned about their physical appearance (Blanco-Dávila 2000).

The first purpose of the cosmetic product was to provide hygiene and health advantages (Chaudhri and Jain 2014; Amberg and Fogarassy 2019). However, their use also has benefits for healthcare or the prevention or delaying of the aging of the skin. In addition, the prevention of skin aging is a relatively new approach to cosmetics. The history of cosmetic products has been paralleled by many significant technological developments in chemistry, pharmaceuticals, materials and packaging innovations (Chaudhri and Jain 2014).

Looking and feeling good is a basic need in human life and a huge deal in today's modern society. In today's world, cosmetic products have become a very basic and daily need for people. With the use of cosmetic products, everyone desires to look clean, younger, well-groomed and beautiful, and to be admired and respected in social life (Sherrow 2001). In this respect, many kinds of cosmetic products such as perfumes, lotions and creams have been developed and used to be admired by others, to protect from natural conditions, to cover the appearance of wrinkles, and scars on their skin, and to mask bad odors.

In the historical period of cosmetic product development, although it was only for moisturizing, softening the skin or for make-up purposes, today's changing cultural and social understandings, and the dizzying progress of science and technology have led researchers to new searches. They offer cosmetics obtained by using new raw materials and applying new technologies to the service of consumers. The adaptation of modern carrier systems to cosmetics, the discovery of new cosmetic formulations and new active and natural ingredients have opened a new era in the cosmetic industry (Millikan 2001). Cosmetic products containing active ingredients are usually formulated as creams, ointments, emulsions, solutions or powders. Nowadays, consumers have increasingly demanded cosmetic products containing natural ingredients with active functions on their skin and would prefer to pay more for a product that promises more skin benefits (Draelos 2019). Therefore, the world cosmetics market tends to rely more on natural products or ingredients due to the belief that chemical substances may be toxic and harmful to the skin following prolonged exposure.

The products applied externally to the skin can be divided into drugs and cosmetics. Medicines are used in the prevention and treatment of diseases. A cosmetic product is defined as "any substance or mixture intended to be placed in contact with the external parts of the human body (epidermis, hair system, nails, lips and external genital organs) or with the teeth and the mucous membranes of the oral cavity with a view exclusively or mainly to cleaning them, perfuming them, changing their appearance, protecting them, keeping them in good condition or correcting body odors" according to Regulation European Commission (EC) 1223/2009 (EC 2009). However, there is another products type related with cosmetics called as "cosmeceuticals". The 'cosmeceuticals' word was first used by Albert Kligman in 1984, to have an expert definition of products offering both cosmetic and therapy value (Draelos 2005). Cosmeceuticals can be defined as relatively a novel class of products, which is a combination of cosmetics and pharmaceuticals, including natural products such as extracts made from algae, seaweeds and sea minerals that often possess some antioxidants and UV protection properties (Ojha and Tiwari 2016). Cosmeceuticals are preparations that have a cosmetic effect by positively changing the structure and functions of the skin and its appendages through physiological effects. In addition, like a barrier that protects the organism from external influences, the skin is an aesthetically important organ that plays a role in temperature regulation and perception of the senses. Cosmeceuticals can be listed as products containing natural moisturizing factors, protein-containing substances, ceramides, alpha-hydroxy acids, vitamins, and retinoic acid (Kligman 2000). Wide cooperation is required in fields such as pharmaceutical technology, pharmacology, biotechnology, chemistry, toxicology and food technology for the optimum development and safe use of cosmetic or cosmeceutical products (Faria-Silva et al. 2020).

Since cosmetic products are the most demanded and popular products, the cosmetic industry and market have been growing intensively in recent years. The global cosmetics market was valued at US$532 billion in 2017 and is estimated to reach US$806 billion by 2023, with a compound annual growth rate (CAGR) of 7.14% from 2018 to 2023 (Bilal and Iqbal 2020). In this context, Europe is the largest cosmetic market in the world with €72 billion, followed by the United States (€38 billion) and Japan (€29 billion) (Couteau and Coiffard 2016).

For protecting the skin against the deleterious effects of ultraviolet radiation (UVR) and to increase the integrity of the skin barrier, many studies have shown that safe and effective new bioactive compounds can be developed from natural sources instead of synthetic compounds (Cavinato et al. 2017).

Seaweed or macroalgae is the common name for a variety of different species of marine plants that can be found in oceans and other water environments like rivers and lakes around the world. Seaweeds are considered the most abundant source of natural polysaccharides, carotenoids, amino acids, flavonoids and minerals that promote skin health (Kim et al. 2018). Therefore, various bioactive compounds including metabolites and pigments obtained from different seaweed species (Table 11.1) have attracted great interest and are known for promoting distinct skincare properties of interest such as anti-aging, anti-inflammatory, antioxidant, and anti-melanogenic effects for the development of a variety of cosmetic products as active ingredients.

In this chapter, the possible uses of the important components obtained from various seaweeds (excluding phenolic compounds that are included in the previous chapter) as cosmetic or cosmeceutical ingredients have been explored and discussed regarding their various skin-beneficial properties and potential roles in skincare and cosmetic products.

2 UV-Induced Skin Damage and Photoaging

The skin protects the body against constant attacks from external factors, stress and pathogens (Slominski et al. 2008). While the skin has only a few millimeters in thickness, it is the largest and one of the most complex organs representing approximately 10–15% of the total body weight and has a surface area of approximately 1.6–2.0 m^2 in the human body (Hombach-Klonisch et al. 2019). The skin acts as a physical and biological barrier throughout life to protect the body against water loss as well as environmental threats such as pathogens, harmful chemicals and physical effects (eg extreme of temperature and UVR).

Aging is an inevitable and natural process that increases over time. As people get older; the biological activity of the cells in the body decreases, their regeneration features slow down, the detoxification mechanism decreases, and the immune system becomes more ineffective (Surowiak et al. 2014). Skin aging can be divided into two types, age-related aging and photoaging (Habib et al. 2014). The formation of premature aging changes in our skin by UVR is called "photoaging" (Surowiak et al. 2014). Photoaging describes the process of complex skin changes caused by prolonged exposure to UVR.

Although small amounts of UVR can have many beneficial effects, single prolonged exposure to the sun causes erythema and sunburn; on the other hand, repetitive exposure to the sun could be an aggressive factor for premature skin aging

Table 11.1 Some seaweed-derived bioactive compounds that are used or have the potential to be used as cosmetics/cosmeceuticals and their biological activities

Bioactive compounds	Some Seaweed species produced	Biological activities	References
Carotenoids			
Astaxanthin	*Chlorella vulgaris, Haematococcus pluvialis Chlorella zofingiensis*	Antioxidants, UV-photoprotective, anti-inflammatory, eye/skin health, anti-aging property, nutraceutical, cosmeceutical	Catanzaro et al. (2020b), Zhang et al. (2020), Afzal et al. (2019), Kuedo et al. (2016), Ito et al. (2018), Tominaga et al. (2017), Chalyk et al. (2017), Hama et al. (2012)
Fucoxanthin	*Odontella aurita, Phaeodactylum tricornutum, Isochrysis aff. galbana*	Antioxidant, anti-inflammatory, anti-pigmentary, skin protective, UV-photoprotective, anti-viral, anti-cancer, anti-allergic	Peng et al. (2011), Tavares et al. (2020a), Kou et al. (2020), Rodriguez-Luna et al. (2018), Shimoda et al. (2010), Urikura et al. (2011), Rokkaku et al. (2013)
Zeaxanthin and Lutein	*Chlorella pyrenoidosa Botryococcus braunii Botryococcus brauni*	Eye and skin health	Roberts et al. (2009), Palombo et al. (2007), Juturu et al. (2016), Obana et al. (2020), Roberts (2013)
Siphonaxanthin	*Codium fragile*	Antioxidant, anti-cancer, anti-inflammatory	Ganesan et al. (2011), Manabe et al. (2020), Zheng et al. (2020)
Neoxanthin	*Codium tomentosum Codium intricatum*	Anti-cancer, antioxidant	Kotake-Nara et al. (2005), Dall'Osto et al. (2007)
Violaxanthin	*Dunaliella tertiolecta Chlorella ellipsoide*	Anti-cancer, anti-inflammatory	Pasquet et al. (2011), Soontornchaiboon et al. (2012)
Canthaxanthin	*Coelastrella striolata Var. multistriata Chlorella vulgaris Chlorella zofingiensis*	Anti-cancer, anti-pigmentation, UV-photoprotective	Gensler et al. (1990), Kim and Lee (2012), Tronnier (1984), Chan et al. (2009), Silke and Wilhelm (2007), Katsumura et al. (1996)

(continued)

Table 11.1 (continued)

Bioactive compounds	Some Seaweed species produced	Biological activities	References
β-carotene	*Dunaliella salina* *Chlorella zofingiensis* *Spirulina platensis* Gracillaria sp.	Vitamin A precursor, antioxidant, anti-cancer, prevents macular degeneration, skin burn from UV rays, psoriasis, pharmaceutical, cosmeceutical	Kavalappa et al. (2019), Sangeetha et al. (2009), Briand et al. (1985), Mathewsroth (1982)
Polysaccharides:			
Fucoidan	*Sargasum siliquastrum* *Undaria pinnatifida* *Chaetoceros sp.* *Odontella aurita*	Antioxidant, anti-inflammatory, anti-aging, UV-photoprotective, Skin tissue regeneration/wound healing, anti-cancer, immunomodulator	Ahmad et al. (2021), Mhadhebi et al. (2012), Shiao et al. (2020), Chen et al. (2021a), Ku et al. (2008), Moon et al. (2008), Jing et al. (2021), Fernando et al. (2021), Song et al. (2014), Karami et al. (2021), Chen et al. (2019a)
Alginate (Alginic acid)	*Sargassum wightii* *Sargassum horneri* *Echinacea purpurea* *Padina boryana*	Skin tissue regeneration/wound healing, anti-inflammatory, antioxidant, skincare, hair follicle regeneration, skin bioengineering, cosmeceutical	Bialik-Was et al. (2021), Nozari et al. (2021), Sarithakumari et al. (2013), Jayawardena et al. (2020), Dong et al. (2021), Solovieva et al. (2021), Fernando et al. (2018), Bin Bae et al. (2019)
Porphyran	*Porphyra yezoensis*	Antioxidant, anti-inflammatory, anti-cancer	Isaka et al. (2015), Liu et al. (2019)
Ulvan	*Ulva rigida* *Ulva pertusa*	Skin bioengineering, cosmeceutical, antioxidant, anti-cancer	Morelli et al. (2019), Leiro et al. (2007), Madub et al. (2021), Chen et al. (2021b), Kidgell et al. (2019)
Carrageenan	*Chondrus cripus* *Gigartina stellata* *Eucheuma cottoni*	Cosmeceutical, antioxidant, UV-photoprotective	Infante and Campos (2021), Rukmanikrishnan et al. (2020), Thevanayagam et al. (2014)

(continued)

Table 11.1 (continued)

Bioactive compounds	Some Seaweed species produced	Biological activities	References
Mycosporine-like amino acids (MAAs)	*Porphyra umbilicalis*	UV-photoprotective, antioxidant, DNA protective	Schmid et al. (2006), Daniel (2004)
	Porphyra yezoensis *Porphyra tenera*	UV-photoprotective	Ying et al. (2019), Rui et al. (2019)
	Chlamydomonas hedleyi	UV-photoprotective, anti-inflammatory, anti-aging	Suh et al. (2014)
Lipids			
Eicosapentaenoic (EPA), Docosahexaenoic (DHA)	*Pyropia yezoensis* *Liagora boergesenii* *Tetraselmis sp.* *Nannochloropsis sp.* *Porphyridium sp.*	Antioxidant, anti-inflammatory, anti-aging, UV-photoprotective, skin healing	Kim et al. (2006), Danno et al. (1993), Pilkington et al. (2016), Balvers et al. (2010), Rahman et al. (2011), Arantes et al. (2016), Candreva et al. (2019), Cezar et al. (2019)
Eicosatetraenoic acid (ETA) polyunsaturated ω-3 fatty acids		Anti-inflammatory	Goto et al. (2015)
Phlorotannins			
Phlorofucofuroeckol	*Eisenia arborea* *Ecklonia kurome* *Ecklonia cava*	Anti-allergic, anti-inflammatory, antioxidant, anti-cancer	Sugiura et al. (2006), Kim et al. (2011, 2016), Lee et al. (2018a, 2020)
Eckol	*Ecklonia cava* *Eisenia bicyclis* *Ecklonia stolonifera* *Ecklonia kurome*	Antioxidant, anti-inflammatory, anti-aging, anti-cancer	Oh et al. (2018), Zhang et al. (2019), Joe et al. (2006), Heo et al. (2009)
Bieckol	*Ecklonia kurome*	Antioxidant, anti-inflammatory, anti-cancer	Oh et al. (2018), Park et al. (2015), Yang et al. (2014)
Dieckol	*Ecklonia stolonifera* *Ecklonia kurome*	Antioxidant, anti-inflammatory, UV-photoprotective, anti-aging, anti-cancer	Pyeon et al. (2021), Li et al. (2021a, b), Yang et al. (2016, 2020), Wang et al. (2020a, 2021), Jang et al. (2015), Xiao et al. (2021), Xu et al. (2021), Lee et al. (2018b)

(continued)

Table 11.1 (continued)

Bioactive compounds	Some Seaweed species produced	Biological activities	References
Triphlorethol-A	*Ecklonia cava*	Antioxidant, UV-photoprotective,	Kang et al. (2008), Piao et al. (2012a), Ko et al. (2011)
Fucodiphlorethol G	*Ecklonia cava*	Antioxidant, UV-photoprotective	Kim et al. (2014a)
Fucofuroeckol-A	*Ecklonia stolonifera Okamura*	Antioxidant	Vo et al. (2018)
Eckstolonol	*Ecklonia cava*	UV-photoprotective,	Jang et al. (2012)
Diphlorethohydroxycarmalol	*Ishige okamurae*	Antioxidant, UV-photoprotective, anti-inflammatory	Heo et al. (2008, 2010), Wang et al. (2020b), Han et al. (2014b)
Phloroglucinol	*Ecklonia cava Ecklonia maxima Carpophyllum flexuosum*	Antioxidant, anti-inflammatory, anti-allergic	Kim and Kim (2010), Daikonya et al. (2002)
Phlorotannin mixture	*Cystoseira compressa Hizikia fusiforme, Halidrys siliquosa, Himanthalia elongata, Fucus serratus,*	Antioxidant, anti-inflammatory, anti-bacterial, UV-photoprotective	Manandhar et al. (2019), Barbosa et al. (2017), Ferreres et al. (2012), Tang et al. (2020), Baek et al. (2021), Yoon et al. (2009)

(Marrot and Meunier 2008) and skin cancer (Solano 2020; Sambandan and Ratner 2011). Photoaging is a very complex process. Although both UVA and UVB rays play a role in the photoaging process; it plays a fundamental role as UVA can reach deeper into the dermis and its rays reach the earth more than 10 times compared to UVB. UVB is mainly absorbed by cellular DNA in the epidermis and causes skin damage through the formation of cyclobutane pyrimidine dimers. UVB is responsible for sunburns, photo-carcinogenesis and immunosuppression (Han et al. 2014a).

Many chemical reactions are responsible for the formation of photoaging. These include intra- or intercellular mediators (EGFR, IL-1R and TNFα-R), activation of the intracellular signal propagation cascade (eg AP-1, NF-kB), collagen damage via matrix metalloproteinase-1 (MMP1) and damage through inflammatory infiltrates (Surowiak et al. 2014). UV radiation reduces collagen synthesis and increases collagenolytic MMP synthesis in skin fibroblasts, which causes deep corrugation in photoaged skin. However, there are still unknown points in the detailed mechanism of photoaging (Baron and Suggs 2014). UVA radiation causes the formation of reactive oxygen radicals in keratinocytes, which are the basic cells of the epidermis. UVR activates the release of cell membrane fatty acids and regulates their metabolism to turn into eicosanoids via cyclooxygenase and lipoxygenase enzyme pathways (Yin et al. 2015).

When skin is exposed to UV irradiation, acute inflammatory changes such as erythema, edema, followed by pigmentation or tanning, and chronic changes such as premature aging, immunosuppression, or photo-carcinogenesis may occur depending on the duration of exposure (Balupillai et al. 2015; Nishisgori 2015). Long-term exposure to UVR from sunlight is one of the major causes of premature skin aging (photoaging) and its symptoms include wrinkling, loss of skin tone, resilience, and moisture, mottled pigmentation (hypo- or hyper-pigmentation) and melanoma (Dahmane et al. 2015). Therefore, it is important to provide effective photoprotection for the reduction of ultraviolet (UV) radiation-induced photoaging and other skin damage.

Sunscreen products have been developed to prevent these harmful effects caused by UVR. These products exert their effects by directly blocking or absorbing UVR. In addition to delaying skin aging, efficient sun protection can also make a great contribution to the prevent UVR—induced other deleterious effects such as DNA damage, photo-carcinogenesis, and suppression of the immune system. Research on the development of topical and systemically acting photoprotective agents increases, due to the severity and number of disorders related to UVR exposure. Since sunscreen cosmetic products include compounds that can absorb UVR, they provide effective protection against its harmful effects (Saewan and Jimtaisong 2015). UVB light filtering agents such as para-amino benzoic acid derivatives, cinnamates, salicylates, octocrylene and ensulizol and UVA filters such as benzophenones, butyl methoxybenzoyl and meradimate are used in products used against skin aging (Singh et al. 2013). In addition, there are many bioactive substances obtained from various natural sources to reduce oxidative damage and accelerate skin aging caused by UV radiation.

3 Seaweeds as a Potential Source of Skincare Compounds

Nature is a source of a large group of structurally unique natural compounds that are mainly found in various plants and organisms including animals and microorganisms in terrestrial and marine environments. Since prehistoric times, humans have used various natural products obtained from plants, animals and marine organisms as medicines to alleviate and treat diseases. The marine environment is one of the most diverse and richest ecosystems on earth and harbours an enormous number of different living creatures with large genetic diversity and multiple bioactive compounds. Although many marine bioactive substances have the potential to be used as remedies in the treatment of various diseases and as cosmetics or cosmeceuticals to prevent the skin from external harmful effects, they can also be used as experimental tools or food supplements (Kijjoa and Sawangwong 2004; Lichota and Gwozdzinski 2018; Newman and Cragg 2016).

Seaweeds or marine macroalgae are aquatic simple plant-like and photosynthetic organisms also known as consumable sea vegetables. They are found worldwide and grow in marine environments along with coastal ecosystems and can also be found in fresh waters like lakes and rivers. Seaweeds are taxonomically classified into three major groups such as *Rhodophyceae* (red algae), *Chlorophyceae* (green algae) and *Phaeophyceae* (brown algae), based on their pigment types (Wang et al. 2015).

The global production of aquatic plants, comprising seaweeds has increased dramatically in the last 50 years. Globally, 2.2 million tons of seaweed were produced in 1969, including wild collection and cultivation. After half a century, wild production remained at 1.1 million tons, while cultivation production, which constitutes 97 per cent of the world's seaweed production, increased to 35.8 million tons in 2019 (FAO 2021). Asia is the largest cultivator of seaweeds in the world with more than 90% being contributed by China, Indonesia, Korea and the Philippines (FAO 2021). There are thousands of seaweed species. The most cultivated species are the brown algae *Saccharina japonica* (Japanese kelp or kombu) and *Undaria pinnatifida* (Japanese wakame), the green algae *Monostroma* spp. and *Enteromorpha spp.* and the red algae *Eucheuma* sp., *Gracilaria* spp., *Kappaphycus* spp., and *Porphyra* spp. (Japanese nori) totalling ~90% of the cultivated fresh weight (Luning and Pang 2003).

Over recent decades, seaweeds or macroalgae and their bioactive compounds have received particular attention from researchers and got importance as experimental tools because of seaweeds are major resources for bioactive compounds with a wide variety of applications in many fields including medicine (Catanzaro et al. 2020a, b), cosmetics (Jesumani et al. 2019), food (Qiu et al. 2022), aquaculture (Moreira et al. 2021), environmental sciences (Michalak 2020) and agriculture (Daniel and Fabio 2020). For these reasons, the commercial production of seaweeds has been dramatically increasing.

Seaweeds have a wide range of applications due to their valuable bioactive compounds and potent bioactivity. Many studies have revealed that the bioactive compounds derived from seaweeds including phenolic compounds, polysaccharides, pigments, sterols, proteins, peptides, and amino acids have antioxidant, antitumor, anti-inflammatory, antiaging, anti-pigmentation, antilipidemic, antibacterial and antiallergic properties (Raposo et al. 2015). Various seaweeds have been used in a wide range of foods, nutritional supplements, feed and feed additives, pharmaceuticals, cosmeceuticals, and cosmetics, and are often claimed to have beneficial effects on human health.

Moreover, many *in vitro* and *in vivo* studies have indicated that various seaweeds and related compounds can provide a sort of skin beneficial activities including antioxidant, anti-photoaging, anti-inflammatory, anti-lipidemic, antitumor, and anti-allergic effects. For example, extracts obtained from various seaweed species are shown to have strong potential antioxidant and anti-photoaging activity (Table 11.3). Moreover, seaweed polysaccharides including polysaccharide-rich extract, laminarin, fucoidan, and carrageenan exerted potential anti-photoaging activity mediated by intracellular ROS scavenging activity in UV-exposed human cell culture and experimental animal models (Kim et al. 2018; Slominski et al. 2008; Hombach-Klonisch et al. 2019; Surowiak et al. 2014). Furthermore, other seaweed-derived compounds such as carotenoids including fucoxanthin and astaxanthin, and mycosporine-like amino acids (MAAs) are also well-known natural UV

Table 11.2 Examples of some commercial seaweed-derived cosmetical ingredients available in the market and their supplier and biological activities

Seaweed species produced	Trade names	Supplier	Biological activities	References
Codium tomentosum	Codiavelane™ BG PF	Seppic	Anti-aging, moisturizing	SEPPIC (2022a)
Laminaria digitate Pelvetia canaliculata	Bioenergizer™ P BA PF	Seppic	Hair care, skincare, UV-photoprotective	SEPPIC (2022b)
Alaria esculenta	Alariane™	Seppic	Hair care	https://cosmetics.specialchem.com/product/i-seppic-alariane
Asparagopsis armata	Aspar'age™	Seppic	Anti-aging; skincare	SEPPIC (2022c)
Ulva lactuca	Akomarine™ Sea Lettuce	Akott	Moisturizing, anti-cellulitis, slimming	https://cosmetics.specialchem.com/product/i-akott-akomarine-sea-lettuce
Fucus vesiculosus	Extrapone™ Seaweed	Symrise	Anti-aging, anti-cellulite, Moisturizing	https://cosmetics.specialchem.com/product/i-symrise-extrapone-seaweed
Undaria pinnatifida	Pheofiltrat™ Undaria HG	Codif	Moisturizing, smoothing and slimming agent	https://cosmetics.specialchem.com/product/i-codif-pheofiltrat-undaria-hg
Laminaria saccharina	Actipone™	Symrise	Anti-aging, anti-wrinkle, anti-stress and anti-inflammatory	https://cosmetics.specialchem.com/product/i-symrise-actipone-laminaria-saccharina
Enteromorpha compressa	Enteline™	Sumitomo corporation	Anti-acne and soothing agent	https://cosmetics.specialchem.com/product/i-presperse-sumitomo-corporation-enteline-2
Undaria Pinnatifida	Ephemer™	Seppic	Anti-aging, antioxidants	https://cosmetics.specialchem.com/product/i-seppic-ephemer
Not specified	Fucosorb™	Sensient Cosmetic Tech.	Anti-inflammatory and moisturizing agent, skincare	https://cosmetics.specialchem.com/product/i-sensient-cosmetic-technologies-fucosorb-wp

(continued)

Table 11.2 (continued)

Seaweed species produced	Trade names	Supplier	Biological activities	References
Chondrus crispus	Genugel™ X-902-02	CP-Kelco	Skincare hair care	https://cosmetics.specialchem.com/product/i-cp-kelco-genugel-x-902-02
Brown seaweeds	Gioactive™ F-1000	Greaf	Moisturizing, skin recovering, skin conditioner, anti-aging and anti-wrinkle	https://cosmetics.specialchem.com/product/i-greaf-gioactive-f-1000
Fucus vesiculosus	Gmoist™ Sea-Gel-H	Greaf	Moisturizing, skincare	https://cosmetics.specialchem.com/product/i-greaf-gmoist-sea-gel-h
Enteromorpha compressa	Homeostatine™ M.S.	Provital	Anti-aging, anti-wrinkle, moisturizing agents	https://cosmetics.specialchem.com/product/i-provital-homeostatine-m-s
Hypnea musciformis	Hypneane™ BG	Sumitomo corporation	Moisturizing, skincare, UV-photoprotective	https://cosmetics.specialchem.com/product/i-presperse-sumitomo-corporation-hypneane-bg
Fucus Vesiculosus	Maritech™ Bright	Marinova	Anti-aging, antioxidants Lightening/Whitening Agents smoothness	https://cosmetics.specialchem.com/product/i-marinova-maritech-bright
Not specified	Oceagen™ LS 8424	BASF	Anti-wrinkle, skincare, moisturizing	https://cosmetics.specialchem.com/product/i-basf-oceagen-ls-8424
Not specified	OceanDerMX™ Lift & Firm	Organic Bioactives	Anti-free radical, skincare, anti-wrinkle	https://cosmetics.specialchem.com/product/i-organic-bioactives-oceandermx-lift-firm
Ascophyllum nodosum	Pheofiltrat™ Ascophyllum HG/A	Codif	Regenerating/revitalizing, anti-acne	https://cosmetics.specialchem.com/product/i-codif-pheofiltrat-ascophyllum-hg-a
Laminaria digitata	Phycosaccharide™ AG	Codif	Anti-aging, skincare, UV-photoprotective	https://cosmetics.specialchem.com/product/i-codif-phycosaccharide-ag
Chondrus crispus	Rhodofiltrat™ Chondrus HG	Codif	Skincare, anti-inflammatory and decongestant	https://cosmetics.specialchem.com/product/i-codif-rhodofiltrat-chondrus-hg
Cystoseira amentacea	Sea™ HEATHER	Gelyma	Anti-inflammatory, skincare, anti-aging, UV-photoprotective	https://cosmetics.specialchem.com/product/i-gelyma-sea-heather

Chondrus Crispus	Seamollient™ W	BASF	Softening, skincare, moisturizing	https://cosmetics.specialchem.com/product/i-basf-seamollient-w
Laminaria Digitata	Seanergilium™	BASF	Skincare, facial cleansing, whitening	https://cosmetics.specialchem.com/product/i-basf-seanergilium-bg
Pelvetia Canaliculata	Xylishine™	Seppic	Moisturizing, hair repairing and shining	https://cosmetics.specialchem.com/product/i-seppic-xylishine
Laminaria Digitata	Vitaplex™ LS 9799	BASF	Anti-stress/Relaxing, hair growth promoters/anti-hair loss	https://cosmetics.specialchem.com/product/i-basf-vitaplex-ls-9799
Not specified	TroyCare™ FP99	Troy Corporation	Hair care, Perfumes & fragrances Skincare (Facial care, Facial cleansing, sun care	https://cosmetics.specialchem.com/product/i-troy-corporation-troycare-fp99
Fucus Vesiculosus	Sublimalg™ LS 9700	BASF	Peeling, brightness, smoothness	https://cosmetics.specialchem.com/product/i-basf-sublimalg-ls-9700
Not specified	SPD Superphycol™ D	Sumitomo corporation	Antioxidants Anti-aging Agents, skincare, UV-photoprotective	https://cosmetics.specialchem.com/product/i-presperse-sumitomo-corporation-spd-superphycol-d

Table 11.3 Summary of the studies on the photoprotective or skincare activities of extracts obtained from various seaweed species

Source algae species	Extraction method	Experimental design or assay	Health-promoting benefits	Findings	References
Curdiea racovitzae *Iridaea cordata*	MeOH:H2O (1:4; v/v)	UVA irradiated HaCaT and fibroblast cell cultures	UV-photoprotective Antioxidant	The extracts showed antioxidant activity against superoxide radicals, decreased ROS generation and protected the fibroblasts by absorbing UVB and UVA rays.	Rangel et al. (2020)
Fucus spiralis	EtOH:H_2O (70/30, v/v) Ethyl acetate and Aqueous fraction of the extract	UVB irradiated HaCaT and RAW264.7 cell cultures	UV-photoprotective Antioxidant	The fraction of the extract reduced ROS production and showed great antioxidant activity.	Freitas et al. (2020)
Sargassum vachellianum	EtOH (90%) extract	Spectroscopy	UV-photoprotective Antioxidant	The extracts showed strong UVA and UVB absorption and had antioxidant, whitening and moisture-retaining activities.	Jesumani et al. (2020)
Ecklonia stolonifera	EtOH (80%) extract gold nanoparticles	UVA-irradiated human dermal fibroblasts	UV-photoprotective Antioxidant	The extract significantly inhibited UVA-induced ROS levels and G1 arrest. Besides, ES-GNPs significantly downregulated the transcription and translation of MMP 1/-3.	Jun et al. (2020)
Sargassum glaucescens	Aqueous extracts	UVA- irradiated human keratinocytes and CCD-966SK fibroblasts	UV-photoprotective Antioxidant Anti-aging	The extract protected dermal fibroblasts against oxidative stress and damage from UVA irradiation. The extract repaired the scratch wound of dermal fibroblasts by promoting cell proliferation and stimulating cell viability and cytoskeletal scaffolding-related gene expressions in keratocytes.	Li et al. (2019)
Sargassum cristafolium	EtOH extract	UVA irradiated HeLa cell line and BALBL/c mice	UV-photoprotective Antioxidant	The extract exerts cytoprotective effects by the presence of MAA-palythene, which probably contributes to the inhibition of DNA damage by UVA irradiation.	Prasedya et al. (2019)

Seaweed fulvescens	EtOH (70%) extract	A mouse model of *Dermatophagoides farinae* body-induced atopic dermatitis and HaCaT cell cultures	Anti-Atopic Dermatitis	The extract suppressed the production of proinflammatory cytokines and reduced the phosphorylation of signal transducer and activator of transcription 1.	Gil et al. (2019)
Caulerpa sp.	EtOH extract	UVB-irradiated male Wistar mice	Antioxidant	The extract has an antioxidant activity for the prevention of photo-aging through its inhibitory activity of MMP-1 and prevention of oxidative cellular DNA damage.	Wiraguna et al. (2018)
Porphyra umbilicalis	Cosmetic formula (5% extract) with Ginkgo biloba, vitamins	UVA-and UVB-irradiated mice	Cell renewal Anti-apoptosis Anti-aging	The combination prevents UVA- and UVB-induced DNA damage and inflammation.	Mercurio et al. (2015)
Carpomitra costata	EtOH (80%) extract	UVB irradiated HaCaT cell culture	UV-photoprotective Antioxidant	The extract reduced superoxide anion, hydroxyl radical, and UVB-stimulated intracellular reactive oxygen species (ROS) levels.	Zheng et al. (2016)
Sargassum muticum	Ethyl acetate fraction of EtOH (80%) extract	UVB irradiated hairless mouse skin and HaCaT cell culture	UV-photoprotective Antioxidant	Oral treatment of the extract improved the wrinkles and prevented the increased epidermal thickness and collagen bundle formation in the UVB-exposed hairless mice. The pre-treatment also inhibited the UVB-induced upregulation in the expression and activity of MMP-1 in human keratinocytes.	Song et al. (2016)

(continued)

Table 11.3 (continued)

Source algae species	Extraction method	Experimental design or assay	Health-promoting benefits	Findings	References
Porphyra yezoensis	EtOH extract (80%)/Chl/MeOH/ H_2O (2/1/0.9)	UVB irradiated HaCaT cell culture	UV-photoprotective Anti-aging	The extract protects UVB-exposed skin cells by increasing cell proliferation and by enhancing apoptosis of damaged cells, via activating JNK and ERK signalling pathways.	Kim et al. (2014c)
Gelidium amansii *Cirsium Japonicum*	MeOH extract the following fermentation	UVB-irradiated HS 68 DF& SKH-1 hairless mice	Skin protective Anti-wrinkle Anti-photoaging	The extract mixture exerts potent anti-photoaging activities by improving wrinkle formation and dryness.	Kim et al. (2014b)
Polyopes affinis	EtOH (80%) extract	UVB irradiated HaCaT cell culture	Antioxidant Anti-apoptosis UV-photoprotective	The extract significantly decreased cellular damage, UVB-induced apoptosis, intracellular ROS and superoxide radicals.	Hyun et al. (2014)
Undaria crenata	EtOH (80%) extract	UVB irradiated HaCaT cell culture	Antioxidant UV-photoprotective	The extract decreased oxidative stress and UVB-stimulated apoptosis by reducing in apoptotic bodies and nuclear and DNA fragmentation, resulting in the recovery of cell viability.	Hyun et al. (2013)
Polysiphonia morrowii	EtOH (80%) extract	UVB irradiated HaCaT cell culture	Antioxidant Anti-apoptosis	The extract reduced UVB-induced apoptosis and protects the skin cells against UVB-induced oxidative stress.	Piao et al. (2012b)
Chondracanthus tenellus	EtOH (80%) extract	UVB irradiated HaCaT cell culture	Antioxidant Anti-apoptosis UV-photoprotective	The extract protected the skin cells by absorbing UVB rays, decreasing the degree of injury resulting from UVB-induced oxidative stress and reducing UVB-induced apoptosis.	Piao et al. (2012c)

Bonnemaisonia hamifera	EtOH (80%) extract	UVB irradiated HaCaT cell culture	Antioxidant Anti-apoptosis UV-photoprotective	The extract exerted scavenging activity against intracellular reactive oxygen species (ROS), reduced UVB-induced apoptosis, decreased DNA damage and elevated levels of 8-isoprostane and protein carbonyls resulting from UVB-induced oxidative stress.	Piao et al. (2012d)
Alaria esculenta		Normal human dermic fibroblasts	Anti-aging	The extract was able to counterbalance the disequilibrium in progerin production observed during ageing.	Verdy et al. (2011)

photo-protectant, antioxidant and anti-inflammatory agents. In addition, various seaweed-derived compounds have been also available in the market for using an active ingredient in sunscreen, anti-photoaging, moisturizer, anti-wrinkle, skincare and other cosmetic products (Table 11.2).

3.1 Seaweeds Extracts as Potential Skincare or Anti-Photoaging Substances

Various seaweeds have been used in a wide range of foods, nutritional supplements, feed and feed additives, pharmaceuticals, cosmeceuticals, and cosmetics and are often claimed to have beneficial effects on human health. Seaweeds have been used as an alternative remedy for skin-related diseases since ancient times. They have several bioactive compounds with important qualities such as low cytotoxicity and low allergen properties for potential use in cosmetic and cosmeceutical products (Morais et al. 2021). Recently, the use of seaweed-derived ingredients in the cosmetics and cosmeceutical industry has attracted special interest as a result of the many scientific investigations that revealed the potential skincare or skin protective activities of their bioactive compounds (Jesumani et al. 2019; Pangestuti et al. 2021). It has been already demonstrated that several valuable bioactive compounds — including proteins, vitamins, lipids, carotenoids (such as astaxanthin and fucoxanthin), mycosporine-like amino acids (MAAs), polysaccharides (such as fucoidan, laminarin and carrageenan) and phenolic molecules— derived from various seaweeds possess high potential bioactivities and can be used as active ingredients in the cosmetic industry (Table 11.1) (Couteau and Coiffard 2016; Pangestuti et al. 2021; Lopez-Hortas et al. 2021). In this case, these active ingredients are added to the skincare products, because they have a variety of health benefit activities such as anti-photoaging, anti-wrinkling, antioxidant anti-inflammatory, moisturizing, anti-allergic, anti-acne, antimicrobial and whitening or melanin-inhibitory. Seaweeds also provide an important source of thickening ingredients and water-binding agents for cosmetic or cosmeceutical products. Therefore, seaweeds or their bioactive compounds are used in a wide range of products in cosmetics for skincare, hair care, and oral care as well as deodorants and make-up products (Pangestuti et al. 2021; Aryee et al. 2018).

Ultraviolet radiation (UVR) is one of the major stress factors for most phototrophic organisms found in land and aqua ecosystems. Seaweeds are found in intertidal zone to a depth of 150 m and are highly exposed and susceptible to solar UVR. Hence, to prevent or decrease the harmful effects of UVR, they synthesize several photoprotective compounds. These compounds can be used to protect the skin from photodamage induced by ultraviolet B (UVB) and ultraviolet A (UVA) rays.

The UV filters could be used in cosmetic or skincare formulas to protect the skin from UVR-induced photodamage. Many commercial UV filter products available on the market not only contain synthetic UV filters but also contain bioactive compounds or extracts from natural sources. In addition, cosmetic skincare products containing natural anti-aging ingredients are more effective in overcoming the undesirable effects of UVR and have become more demanded by consumers in recent years. Current isolation or extraction techniques of photoprotective compounds from seaweed have been in rapid progress. The extraction or isolation technique and seaweed species dramatically affect the content of the bioactive compounds and their photoprotective activities (Pangestuti et al. 2021).

Many studies with *in vitro* and *in vivo* experimental models have indicated that extracts or purified compounds obtained from various seaweed species have photoprotective or skincare activities and could be used as an active agent incorporated in cosmetic formulas (Table 11.3). It is accepted that these activities are mainly mediated by antioxidant properties, radical scavenging activity and UV absorption capacity of bioactive compounds of the extracts. The methanolic extracts of *Gelidium amansii* and *Cirsium Japonicum* showed skin protective or anti-photoaging activity in UVB-irradiated hairless mice by improving wrinkle formation and dryness of the skin (Kim et al. 2014b). It has been indicated that a cosmetic formula including a 5% extract of *Porphyra umbilicalis* in combination with *Ginkgo biloba* and vitamins exerts anti-apoptosis and anti-photoaging activity by preventing UVA- and UVB-induced DNA damage and inflammation in UVA- and UVB-irradiated mice (Mercurio et al. 2015). Moreover, the ethanolic extract of *Caulerpa sp.* had a potential antioxidant activity for the protection of UVR-induced photo-aging through its inhibitory activity of MMP-1 and prevention of oxidative DNA damage (Wiraguna et al. 2018). Similarly, in UVA irradiated HeLa cell culture and BALBL/c mice models, it has been indicated that an extract obtained from *Sargassum cristafolium* also showed UV-photoprotective and cytoprotective activities by the presence of MAA-palythene, which probably contributes the inhibition of DNA damage by UVA irradiation (Prasedya et al. 2019). Besides, apart from the UV-photoprotective activity, the ethanolic extract of Seaweed fulvescens showed anti-atopic dermatitis activity by suppressing the production of proinflammatory cytokines and reduced the phosphorylation of signal transducer and activator of transcription 1 in a mouse model of *Dermatophagoides farina*-induced atopic dermatitis and HaCaT cell culture model (Gil et al. 2019). Guinea and co-workers investigated the protective potential of extracts against UV irritation obtained from 21 commercial macroalgae species from different aquatic ecosystems of the world using a zebrafish embryo model (Guinea et al. 2012). The results indicated that the phenolic extracts obtained from *Porphyra columbina*, *Macrocystis pyrifera*, *Sarcothalia radula* and *Gigartina skottsbergii* exhibited the highest photoprotective effect compared to the other seaweed species tested. Eventually, the authors emphasized that this photoprotective activity was attributable to the total phenolic and MAA contents of the extracts.

Several studies also have been conducted with i*n vitro* human fibroblast and keratinocyte cell culture models to reveal the UV-photoprotective potential or skincare activity of various seaweed extracts (Table 11.3). Recently, the potential photoprotective activity and toxicity of methanolic extracts obtained from two red seaweed species *Curdiea racovitzae* and *Iridaea cordata* have been assessed by UVA irradiated HaCaT and fibroblast and cell culture models (Rangel et al. 2020).

Both extracts have been considered to have no cytotoxic or irritant effects. Furthermore, the extract from *C. racovitzae* exerts higher photoprotective activity compared with an extract from *Iridaea cordata*, and this difference can be attributed to the total amount of the MAAs content since *C. racovitzae* extract (150.17 µg/mg) had much higher total MAAs content than *Iridaea cordata* extract had (60.78 µg/mg). Therefore, the extracts from these seaweed species could be used in skincare and cosmetic products as an antiphotoaging ingredient to protect the skin from the harmful effects of UVR (Rangel et al. 2020). Freitas and co-workers reported that different fractions extracted from the brown seaweed *Fucus spiralis* reduced ROS production and showed great antioxidant activity in UVA-irradiated HaCaT and human fibroblast cell culture models. These bioactivities are attributable to the high phlorotannin content of the fractions that could be used as an ingredient in cosmetic products (Freitas et al. 2020).

The photoprotective or anti-photoaging activities of the ethanol extract of seaweeds, *Sargassum glaucescens* (Li et al. 2019), *Carpomitra costata* (Zheng et al. 2016), *Porphyra yezoensis* (Kim et al. 2014c), *Polyopes affinis* (Hyun et al. 2014), *Undaria crenata* (Hyun et al. 2014) have been indicated in various *in vitro* cell culture models using ultraviolet UVA or UVB-irradiated cultured human keratinocytes (HaCaT) (Table 11.3). Studies by Piao and co-workers have also demonstrated in human HaCaT keratinocytes that ethanolic extracts (80%) from three different seaweed species, *Polysiphonia morrowii, Chondracanthus tenellus, Bonnemaisonia hamifera* decreased the degree of UVB-induced apoptosis and oxidative stress by scavenging ROS and absorbing UVB photons, thereby reducing dermal cell injury (Piao et al. 2012b, c, d).

Heo and co-workers investigated the potential inhibitory activity of ethanol extracts (80%) obtained from 21 different seaweeds on melanogenesis by tyrosinase inhibitory effect in a human fibroblast cell model (Heo et al. 2010). The authors emphasized that the extract from *Ishige okamurae* showed significant inhibitory effects against tyrosinase and melanin synthesis and has potential whitening and photoprotective activities on UVB-induced cell damages which might be used in the cosmetic industry.

Therefore, the studies in the literature have shown that the extracts obtained from various seaweed are still good candidates for future skincare products in the cosmetic industry, as they have great UV-light protective and antioxidant activities.

3.2 *Bioactive Compounds Derived from Seaweed as Skincare Agents*

3.2.1 Polysaccharides

Seaweeds have attracted a special interest with their rich bioactive sulphated polysaccharide content, which has the potential to be used in pharmaceutic, cosmetic, microbiology and biotechnology fields, in addition to being a good food source

(Renn 1997). The chemical structures and amounts of polysaccharides vary according to the source of seaweed species. Brown algae mainly contain laminarins (up to 32–35%) and fucoidans; red species are mainly rich in carrageenans and porphyrans; and green species typically contain ulvans (Ngo et al. 2011; Perez et al. 2016). The major polysaccharides found in seaweeds have been broadly used in pharmaceutical, biomedical, cosmetic, and cosmeceutical production, due to their unique physicochemical and bioactive properties, such as wound healing, anti-cancer, antioxidant, antibacterial, anti-inflammatory or immunostimulatory activities (Lee et al. 2017). Furthermore, polysaccharides in cosmetic products possess important skin protective properties such as hair conditioners, moisturizers, emulsifiers, and wound-healing and thickening agents (Jesumani et al. 2019).

Polysaccharides Rich Extracts

Polysaccharide compounds obtained from seaweeds have a great water-holding capacity which can be used as a moisturizer and humectant in cosmetic products. It was indicated that the polysaccharides extracted from *Laminaria japonica* have higher hydrating and moisturizing activities than hyaluronic acid and the formulation prepared in combination with the extract provided an improvement in skin moisture (Pimentel et al. 2018; Kalasariya et al. 2021). Besides, the moisture absorption and retention properties of polysaccharides extracted from five algae including three green algae *Codium fragile*, *Enteromorpha linza* and *Bryopsis plumos*, one red alga *Porphyra haitanensis* and one brown alga *Saccharina japonica* were investigated by Wang and co-workers (2013). The results of the study indicated that the polysaccharides with lower molecular weight extracted from brown seaweed exhibited the greatest moisture-absorption and moisture-retention activities in the polysaccharides studied, suggesting the sulphate content and molecular weight plays an important role in the moisture-retention capacity of skin. In another study, Shao and co-workers reported that sulphated polysaccharide extracted from the green algae *Ulva fasciata* Delile showed a greater capability both in the moisture-absorption and moisture-retention properties for 4 days compared with glycerol application (Shao et al. 2015). Similarly, the polysaccharides extracted from *Sargassum vachellianum* had great skin protective activity mediated by their antioxidant, whitening, moisture-retaining and UV- photoprotective abilities (Jesumani et al. 2020). Polysaccharides obtained from *Sargassum fusiforme* were also shown to alleviate UVB-induced oxidative stress in hairless mice by reducing the levels of reactive oxygen species (ROS) and malondialdehyde (MDA) (Ye et al. 2018). Furthermore, the treatment also suppressed the levels of matrix metalloproteinase (MMP)-1 and 9. In addition, Wang and co-workers investigated the potential skin protective activity of sulphated polysaccharides obtained from *Hizikia fusiforme and Ecklonia maxima* using *in vitro* in human dermal fibroblasts or keratinocytes and *in vivo* in zebrafish models (Wang et al. 2018, 2019, 2020c). The findings of the studies suggested that sulphated polysaccharides-rich extracts had great *in vitro* and *in vivo* UV-protective effects and most of the polysaccharides rich extracts were able to prevent ROS production and downregulated matrix metalloproteinases (MMPs) expression (Ye et al. 2018; Wang et al. 2020c).

In addition, polysaccharides-rich extracts enhanced skin beauty and maintained the skin in good condition especially due to their extreme moisturizing properties when added to skincare or cosmetic product formulations as an active ingredient (Pangestuti et al. 2021). Furthermore, it is accepted that some polysaccharides might also improve the stability and sensorial properties of skincare and cosmetic products (Pangestuti et al. 2021).

Laminarin

Laminarin (also known as laminaran or leucosin) is a non-toxic polysaccharide derived bioactive molecule from various brown algae species, such as *Laminaria digitate*, *Ascophyllum nodosum Laminarina hyperborean* and *Laminaria japonica*, and exhibits antioxidant and antimicrobial activities (Kadam et al. 2015). Laminarin is a class of low-molecular-weight storage β-glucans and contains a β-(1-3)-linked glucan backbone with β-(1-6)-linked side chains of various lengths and distributions (Rioux et al. 2007). The bioactivities of laminarin can be enhanced or modified using various chemical techniques including irradiation, oxidation, reduction and sulphation. It was indicated that the molecular weight of laminarin was significantly reduced with the formation of carbonyl groups during irradiation. The carbonyl groups of laminarin are mainly related to its increased antioxidant activity and lead to enhanced antioxidant activity (Choi et al. 2011).

Various studies with experimental animal and human keratinocyte cell models have indicated that laminarin treatment showed photo-protective activity against the harmful effects of UVB exposure (Table 11.4). The effect of laminarin on the activity of metalloproteinase (MMP-1) has been investigated following abdominal administration on the photoaging skin of UVA and UVB-irradiated mice (Li et al. 2013). The laminarin administrations considerably enhanced the thickness of the dermis, tissue inhibitor MMP-1 level and reduced the expression and release of MMP-1. Besides, it can regulate the metabolism of collagen photoaging skins by adjusting the activity of matrix metalloproteinase. In a recent study, it has been indicated that topical laminarin pre-treatment significantly decreased skin damage with enhanced epithermal thickness and the destruction of dermal collagen fibers by decreasing oxidative stress and enhancing antioxidant enzymes in UVB-irradiated mouse skin (Ahn et al. 2020). Based on the findings of the study, the authors suggest that laminarin could be a useful compound for the development of sunscreen cosmetic products to protect the skin from UVB-induced photodamage.

Furthermore, in an in vitro study, the antioxidant and mitochondrial effects of laminarin obtained from *Laminaria digitata* were investigated in human dermal fibroblasts and epidermal keratinocytes (Ozanne et al. 2020). The finding of the study suggested that a low concentration (10 μg/ml) of laminarin showed antioxidant protection and a positive effect of β-(1,3)-glucan on cutaneous cells under inflammatory conditions with environmental factors. However, laminarin

Table 11.4 The studies investigating the photoprotective or skincare activities of laminarin

Source	Experimental design or assay	Health-promoting benefits	Usage	References
Laminarin (from Sigma)	Male ICR mice exposed to UVB irradiation	UV-photoprotective	Topical laminarin decreased ROS production and increased the expression of antioxidant enzymes.	Ahn et al. (2020)
Laminarin obtained from *Laminaria digitata*	Human dermal fibroblasts adult (HDFa) and epidermal keratinocytes (NHEK)	Skin protective	Laminarin showed antioxidant protection and a positive effect of β-(1,3)-glucan on cutaneous cells under inflammatory conditions with environmental factors.	Ozanne et al. (2020)
Laminarin (abdominal injection)	Kunming SPF mice exposed to UVB and UVA irradiations	UV-photoaging	Laminarin significantly increased the thickness of the dermis, tissue inhibitor MMP-1 level and reduced the expression and release of MMP.	Li et al. (2013)
Laminarin (from *Cystoseira barbata)* containing cream	Rats with full-thickness wounds	Antioxidant, Antibacterial Wound healing	Laminarin-containing cream showed wound-healing effects and noticeable antioxidant and antibacterial properties.	Sellimi et al. (2018)

treatments at higher concentrations affect negatively the metabolic activity in skin keratinocytes and fibroblast cells (Ozanne et al. 2020).

Different from its antiaging activity, the wound-healing effect of laminarin-based creams was investigated following topical treatment in full-thickness wounds induced on rats (Sellimi et al. 2018). Topical laminarin treatment induced the wound healing process in rats by improving the wound contraction, accelerating re-epithelization and protecting the cutaneous cells against free radical-mediated oxidative stress, as well as protecting the cells against bacterial infections. Various studies have reported that laminarin has health-beneficial effects such as anti-aging, wound healing, anti-inflammatory and antioxidant activity. Therefore, to evaluate the potential use of laminarin as a cosmetic product, further experimental studies should be carried out on its solubility, efficacy, pharmacokinetics, skin penetration ability and possible adverse effects.

Fucoidan

Fucoidans, sulphated polysaccharides, are found in the cell wall matrix of several brown seaweed species including Cladosiphon sp. (mozuku), *Undaria pinnatifida* (wakame), *Fucus vesiculosus* (bladderwrack), and *Laminaria japonica* (kombu) (Senthilkumar et al. 2013).

Various experimental studies indicated that fucoidan has numerous health benefit properties (Luthuli et al. 2019; Rani et al. 2021) including, anti-tumor (Ale et al. 2011a; Chiang et al. 2021), anti-viral (Wang et al. 2017), anti-coagulant (Zhu et al. 2010), immunoregulation (Yoo et al. 2019), inhibition of enzymes (Kim et al. 2014d), and anti-inflammatory effects (Lean et al. 2015). These distinctive biological activities make fucoidan a potential pharmaceutical, cosmeceutical or nutraceutical candidate in the drug, cosmeceutical and food industries (Vo and Kim 2013; Ale et al. 2011b). However, many factors including species of seaweed, purity, molecular weight, sugar composition, sulphation degree, glycosidic linkage, co-extracted impurities, and branching site dramatically affect the biological activities of this compound (Zayed et al. 2020).

Also, there has been growing interest in fucoidan because it has the potential to protect against skin damage and the harmful effects of UVB exposure. The photoprotective, anti-inflammatory and anti-melanogenesis activities of fucoidan have been investigated in UVB-irradiated human keratinocyte (HaCaT) and fibroblast (HS68 and CCD-25Sk) cell cultures and *in vivo* models (Table 11.5). More recently, Obluchinskay and co-workers have investigated the anti-inflammatory potential of fucoidan following topical treatment of a fucoidan-based cream in rats with carrageenan-induced edema (Obluchinskaya et al. 2021). Topical treatment of the fucoidan-based cream inhibited carrageenan-induced cutaneous edema and ameliorated mechanical allodynia in rats suggesting that fucoidan could be considered a promising anti-inflammatory formulation.

An *in vivo* study showed that dietary fucoidan supplements decreased metalloproteinase 1 levels and had an effective suppressor of inflammation in UVB-irradiated mouse skin (Maruyama et al. 2015). Fucoidan was shown to block UVB-introduced expression of matrix metalloproteinase (MMP)-1 and thus enhanced the synthesis of type I procollagen (Moon et al. 2008, 2009). Besides, it was indicated that fucoidan extracted from *Undaria pinnatifida* had strong anti-inflammatory activity in human HaCaT keratinocytes exposed to UVB and UVA radiation and can improve UV-induced skin photoaging through inhibition of ROS production by the alleviation of mitochondrial dysfunction by regulating the SIRT-1/PGC-1α signalling pathway (Jing et al. 2021). Besides, Fucoidan obtained from *Sargassum confusum* increased cytoprotective effects by Nrf2 activation and inhibition of apoptosis and decreased oxidative stress in UVB-irradiated human keratinocytes (Fernando et al. 2020a). Furthermore, skin protective effects of a fucoidan from *Sargassum horneri* were shown in fine-dust-induced HaCaT keratinocytes by lowering inflammatory responses and defective skin barrier functions. The findings of this study also suggested that NF-κB and MAPK pathways might involve in fucoidan-mediated anti-inflammatory effects. These studies suggest that fucoidan

Table 11.5 The studies investigating the skin beneficial activities of fucoidan

Source or pharmaceutic forms	Experimental design or assay	Health-promoting benefits	Usage	References
Fucoidan (from *Fucus vesiculosus*) -Based Cream	Wistar albino rats	Anti-Inflammatory	Topical cream containing fucoidan inhibited carrageenan-induced edema and ameliorated mechanical allodynia.	Obluchinskaya et al. (2021)
Fucoidan from *Undaria pinnatifida*	HaCaT cells and HFF-1 cells exposed to UVB and UVA radiation	UV-photoprotective	UV-induced skin photoaging can be improved by Fucoidan through inhibition of ROS production via the alleviation of mitochondrial dysfunction.	Jing et al. (2021)
Silibinin-loaded chitosan–fucoidan hydrogel	Ex-vivo with UVB-irradiated hairless mice skin	UV-photoprotective	The formulation reduced the inflammation mediators and prevented acanthosis, hyperkeratosis, and infiltration of neutrophils into the dermis by UVB radiation.	Ali Karami et al. (2021)
Fucoidan fraction (SHC4-6) from *Sargassum horneri*	Fine-dust-induced HaCaT keratinocytes	Skin protective	Fucoidan fractions displayed protective effects by reducing inflammation and deterioration of the skin barrier.	Fernando et al. (2021)
Fucoidan (SCFC4) from *Sargassum confusum*	UVB-irradiated human HaCaT keratinocytes	UVB-photoprotective	Fucoidan decreased oxidative stress in UVB-irradiated human keratinocytes increasing cytoprotective effects.	Fernando et al. (2020a)
An LMW fucoidan fraction (SHC4) from *Sargassum horneri*	UVB-irradiated human HaCaT keratinocytes	UVB-photoprotective	Fucoidan prevented UVB-stimulated apoptotic cell formation, sub-G1 accumulation and DNA damage.	Fernando et al. (2020b)

(continued)

Table 11.5 (continued)

Source or pharmaceutic forms	Experimental design or assay	Health-promoting benefits	Usage	References
Fucoidan from *Hizikia fusiforme*	UVB-irradiated human keratinocytes (HaCaT)	Anti-Photoaging Anti-Melanogenesis	Fucoidan showed strong anti-photoaging and anti-melanogenesis activities.	Wang et al. (2020d)
Fucoidans (from *Sargassum hemiphyllum*) *with different MW*	UVB-irradiated human foreskin fibroblast (Hs68)	UVB-photoprotective	LMW fucoidan displayed protective effects against UVB irritation by inhibiting UVB-induced transcription factor activator protein-1 (AP-1)-stimulated transcription of MMP genes.	Hwang et al. (2017)
LMW Fucoidans from Undaria pinnatifida	B16BL6 melanoma cells	Anti-Melanogenesis	It showed radical scavenging properties that induce higher anti-melanogenesis activity.	Park and Choi (2017)
Fucoidan from *Fucus vesiculosus*	CCD-25Sk human fibroblasts HaCaT human keratinocytes	Reconstruction of skin	Fucoidan significantly increased the expression of α6-integrin and has positive effects on epidermal reconstruction.	Song et al. (2014)
Fucoidan from *Undaria pinnatifida*	NC/Nga mice Human keratinocyte	Anti-atopic dermatitis	Fucoidan was shown to improve atopic dermatitis-like conditions as effective as dexamethasone.	Yang (2012)
Dietary Mekabu Fucoidan	UVB-irradiated mice skin	UV-photoprotective	Fucoidan treatment decreased metalloproteinase 1 levels by suppression of inflammation in mice.	Maruyama et al. (2015)
Fucoidan (from Sigma)	UVB-Induced MMP-1 Expression human skin fibroblast (HS68)	UV-photoprotective	Fucoidan prevents UVB radiation-induced MMP-1 expression by inhibiting the ERK pathways.	Moon et al. (2008)

(continued)

Table 11.5 (continued)

Source or pharmaceutic forms	Experimental design or assay	Health-promoting benefits	Usage	References
Fucoidan (from Sigma)	UVB-induced human skin fibroblasts (HS68)	UV-photoprotective	Fucoidan prevent UVB-induced MMP-1 expression and inhibits the down-regulation of type I procollagen synthesis	Moon et al. (2009)
Fucoidan (from Sigma)	UVB-induced human skin fibroblasts (HS68)	UV-photoprotective	Fucoidan decreased UVB-induced ROS generation and could display skin photoprotection activity against UVB-stimulated oxidative stress.	Ku et al. (2010)
Fucoidan (from Sigma)	UVB-induced human skin fibroblasts (HS68)	UV-photoprotective	Fucoidan pre-treatment significantly improved the cell viability and inhibited the production of MMP-1 in UVB-induced skin fibroblasts.	Lee et al. (2007)
Fucoidan fraction 7" (Calbiochem) and a cruder preparation of fucoidan (Fluka)	Human skin fibroblasts	Wound healing	Fucoidan has beneficial properties in the treatment of wound healing.	O'Leary et al. (2004)

may have potential therapeutic activity against skin damage and the harmful effects of UVB exposure and could be used as an ingredient in the pharmaceutical and cosmeceutical industries to prevent or treat skin photoaging.

Previous studies have indicated that the molecular weight of fucoidan markedly affects its biological properties and plays an important role in its photoprotective activity. It has been shown that low molecular weight fucoidans have stronger activities compared with those of high molecular weight fucoidans (Park et al. 2017; Fernando et al. 2020b). Supporting these results, Hwang and co-workers showed that low molecular fucoidan extracted from *S. hemyphyllum* exerts more effective UVB-photoprotective activity in UVB-induced HS 68 cells by preventing UVB damage through effects on AP-1, TGFβRII, and MMPs compared with high molecular weight fucoidan (Hwang et al. 2017).

The biological activity of fucoidan on the proliferation of fibroblasts and the reconstruction of a skin equivalent was investigated in human keratinocyte (HaCaT) and fibroblast (CCD-25Sk) cultures (Song et al. 2014). The finding of the study indicated that fucoidan significantly increased the expression of α6-integrin, but β1-integrin, type 1 collagen, elastin and fibronectin were not significantly affected. This study suggested that fucoidan has beneficial activities on epidermal reconstruction and therefore fucoidan could be useful in reconstructing the skin equivalent (Song et al. 2014).

Some of the health-beneficial effects of fucoidan are thought to be due to its interactions with growth factors such as a basic fibroblast growth factor (bFGF) (Matou et al. 2002) and transforming growth factor-b (TGF-b) (McCaffrey et al. 1994). Furthermore, it was shown that fucoidan increased fibroblast proliferation in the presence of the transforming growth factor-β1 (TGF-β1), indicating that fucoidan may modulate growth factor-dependent pathways in the cell biology of tissue repair and promote wound healing (O'Leary et al. 2004). These findings all have indicated that fucoidan has UV-photoprotective and skincare activities and highlighted the potential use of this compound in the pharmaceutical and cosmeceutical fields.

Carrageenans

Carrageenans are a type of natural polysaccharide obtained from edible red seaweeds. They are the generic name for the family of water-soluble, gel-forming, viscosifying sulphated galactans with an ester sulphate content of 15–40% (w/w) and found in the cell wall and intercellular tissue matrix in various red seaweed species such as Eucheuma spp, *Chondrus crispus* (Irish moss), and *Gigartina stellate* (Goncalves et al. 2002). Carrageenans can be classified into three types such as iota (ι), lambda (λ) and kappa (κ) carrageenans based on their gelling properties and solubility in potassium chloride (Michel et al. 2006). Polysaccharide carrageenans obtained from seaweed are widely used not only in the food industry and biotechnological research but also in the pharmaceutical and cosmetics industry due to their biocompatibility, high viscosity, gelling capacity, high molecular weight and UVR absorbing ability.

Carrageenans can soften, soothe and moisturize the skin in addition to their anti-inflammatory, demulcent, emollient, expectorant, laxative and nutritive properties (Matsui et al. 2003; Pangestuti and Kim 2014; Necas and Bartosikova 2013). In the cosmetic industry, carrageenans are used as an excipient to improve the lubricity and softness of cosmetic products. These compounds are also incorporated into the formulations of sunscreens, anti-aging and facial creams, and soap due to their thickening with water-binding properties (Necas and Bartosikova 2013). Although extensively used as an excipient in skincare products, health-beneficial activities of carrageenan such as a photoprotective agent for skin have not been extensively explored. A few studies have been carried out to assess the cytotoxic, photoprotective and anti-oxidative effects of carrageenan in UVB-induced human keratinocytes (HaCaT cells). Thevanayagam and co-workers reported that λ, κ, and ι- carrageenans showed significant photoprotective activity against deleterious effects of UVB-induced apoptosis and scavenging free radicals in human keratocytes. The ability to protect against UVB suggests that carrageenans have potential application as a photoprotective agent in addition to just being used as an excipient (Thevanayagam et al. 2014). Photoprotective effects effect of κ-carrageenan oligosaccharide (κ-ca3000) and collagen peptide complex were also investigated in UV-irradiated human keratinocytes (HaCaT) and mouse embryonic fibroblasts (Ren et al. 2010). The complex exerted antioxidant properties that efficaciously alleviated UV-induced cell damage and skin photoaging by suppressing cell apoptosis and expression of MMP-1 through the MAPKs signalling pathways. Therefore, the κ-carrageenan and collagen peptide complex has the potential to be used in the pharmaceutical and cosmetic industries as it has potential anti-aging and protective effects on UV-induced skin cell damage (Ren et al. 2010).

In addition, previous studies have shown that the anti-photoaging activity of these polysaccharides is also associated with modulations of inflammatory responses. Carrageenans can induce the activation of proinflammatory mediators including tumor necrosis factor (TNF)-, interleukin (IL)-6, cyclooxygenase-2 (COX-2) and inducible nitric oxide synthase (iNOS) (Nantel et al. 1999). Furthermore, it was demonstrated that COX-2 expression is an important factor for the survival and proliferation of keratinocytes following UV irradiation and the inhibition of COX-2 expression reduced epidermal keratinocytes proliferation (Tripp et al. 2003). it is believed that the modulation of inflammatory responses and antioxidant properties of carrageenans could play an important role in their anti-photoaging activity (Pangestuti et al. 2021).

Recently, apart from their photoprotective effect, Iwasaki and Watarai have investigated the effects of oral λ-carrageenan treatment in a hapten-induced atopic dermatitis-like model in mice (Iwasaki and Watarai 2020). The findings suggested that oral λ-carrageenan alleviates skin symptoms by suppressing the immunological response to the allergen and might be useful for the prevention or treatment of atopic dermatitis.

3.2.2 Pigments (Carotenoids)

Over the last decades, seaweeds have been the focus of a growing interest among researchers, manufacturers, and entrepreneurs in the fields of food and nutrition because of the capacity of certain seaweed species to produce high-value products including carotenoids. Among the various natural pigments, carotenoids comprise an important group of more than 600 structurally different compounds (Krinsky et al. 2004). Carotenoids are compounds that impart their bright colors to a variety of fruits and vegetables, flowers, as well as some insects, birds, and marine animals (Gouveia and Empis 2003).

Carotenoids are lipophilic essential natural pigments produced by various photosynthetic organisms such as plants, bacteria, fungi and algae. These compounds can be chemically classified as carotenes, such as α-carotene, β-carotene, lycopene, and xanthophylls (Jackson et al. 2008). According to the type of pigments, seaweeds are divided into three groups: brown (carotenoids), green (chlorophylls a, b, and c) and red (phycobilins as phycoerythrin). Carotenoids are used as natural color enhancers in the food, pharmaceutical and cosmetic industries. In addition, these compounds also have important properties that make them suitable for skincare, such as free radical scavenging, inhibiting melanogenesis, and premature aging or photoprotection (Christaki et al. 2013).

Carotenoids derived from seaweeds, such as astaxanthin and fucoxanthin are characterized by excellent antioxidant and photoprotective activities and could be used in the cosmetic industry as a leading natural resource and an innovative potential functional ingredient.

Fucoxanthin

Fucoxanthin is one of the major xanthophyll carotenoids contained in brown algae such as Wakame (*Undaria pinnatifida*) and Kombu (*Laminaria japonica*) (Haugan et al. 1995). It has a high commercial value in the global market, and is used in biological activities such as anti-aging (Kang et al. 2020), anti-obesity (Gammone and D'Orazio 2015), antioxidant (Foo et al. 2017), anti-angiogenic (Kang et al. 2020), anti-inflammatory (Heo et al. 2012), anti-cancer (Chen et al. 2019b), photoprotective (Pangestuti et al. 2021; Wijesekara et al. 2011), and neuro-protective activity (Silva et al. 2018). Unlike astaxanthin, fucoxanthin exerts more effective photoprotective activity after topical application since it hardly reaches an effective concentration in the skin following oral administration (Hashimoto et al. 2009).

Many studies have indicated that fucoxanthin has greater antioxidant activity and photoprotective properties compared to most terrestrial-derived antioxidants and carotenoids (Rodriguez-Luna et al. 2018). Fucoxanthin has standard polyene chain links to functional groups, such as an allenic bond, and hydroxyl, epoxy, carbonyl, and carboxyl moieties in the terminal rings. The complex and unusual allenic bond and 5,6-monoepoxide in its molecule play an important role in ROS scavenging activity and explain the ability to quench singlet oxygen and scavenge free radicals (Zhang et al. 2015; Sachindra et al. 2007). In addition, the electron-rich status of fucoxanthin also makes this molecule a strong radical scavenger compound (Miyashita et al. 2020).

UV-photoprotective properties of fucoxanthin have been investigated by various in vitro and in vivo studies (Table 11.6). In a recent trial, *Phaeodactylum tricornutum* concentrate containing ≥50% fucoxanthin (PT-FX50) has been used as the raw material to prepare a wrinkle care cream in 21 women subjects (aged 35–50 years) for 8 weeks (Kang et al. 2020). The results of the study have indicated that the product displayed considerable efficacy and excellent antiwrinkle and skin moisturizing effects after 4 weeks of treatment without any adverse effects.

Furthermore, considerable ROS scavenging activity of fucoxanthin has been shown in studies with HaCaT cell-culture models. Zheng and co-workers demonstrated that the protective activity of fucoxanthin against hydrogen peroxide-induced cell death HaCaT cell-culture models was mediated by the down-regulation of apoptosis-promoting mediators such as B-cell lymphoma-2-associated x protein,

Table 11.6 The studies investigating the photoprotective or skincare activities of fucoxanthin

Pharmaceutical forms	Experimental design or assay	Health-promoting benefits	Findings	References
Phaeodactylum tricornutum concentrate containing ≥50% fucoxanthin (PT-FX50) used as the raw material to prepare a wrinkle care cream	In vitro: Human dermal fibroblast, neonatal (HDFN) cells. In vivo: Women subjects (aged 35–50 years, n = 21)	Anti-aging	The cream showed considerable efficacy as it significantly increased skin moisture and elasticity after 4 weeks of treatment. The product displayed excellent antiwrinkle and moisturizing effects.	Kang et al. (2020)
Trans-fucoxanthin (≥95% püre from Sigma-Aldrich	Reconstructed human skin changes in inflammation (IL-1α, IL-6, IL-8), homeostasis (EGFR, HSPB1) and metabolism (NAT1)	Skin protective	Fucoxanthin exerts skin protective effects following topical application.	Tavares et al. (2020a)
(≥95%) from Sigma	3T3 mouse fibroblast culture vs. full-thickness reconstructed human skin (RHS)	UV-photoprotective	Fucoxanthin displayed UV-photoprotective activity *in vivo.*	Tavares et al. (2020b)
(≥95%) from Sigma	Skin biopsy samples from volunteers. COMET and tetrazolium-based colorimetric (MMT) tests	Skin protective	Fucoxanthin presented a therapeutic potential for dermatological and cosmetic applications.	Denis et al. (2019)
Fucoxanthin (from Sigma) loaded cream	In vitro (THP-1 macrophages and HaCaT cells) and in vivo (UVB-induced skin erythema in mice)	UV-photoprotective Antioxidant activity	Fucoxanthin provides antioxidant and anti-inflammatory activities by downregulating COX-2 expression following UVB radiation. Fucoxanthin prevents epidermal hyperplasia and UVB-induced skin erythema.	Rodriguez-Luna et al. (2018)

(continued)

Table 11.6 (continued)

Pharmaceutical forms	Experimental design or assay	Health-promoting benefits	Findings	References
Fucoxanthin (from Sigma) in combination with rosmarinic acid	UVB-irradiated HaCaT keratinocyte	UV-photoprotective Anti-inflammatory	The combination showed photo-protective effects by enhancing the Nrf2 signalling pathway and down-regulating the inflammation component (NRLP3).	Rodriguez-Luna et al. (2019)
Fucoxanthin (from Santa Cruz Biotechnolgy)	In vitro human HaCaT keratinocytes	Antioxidant activity	Fucoxanthin protects human keratinocytes against oxidative damage by inhibiting apoptosis and scavenging ROS.	Zheng et al. (2013)
Ethanol and silica gel column chromatography extract of *Laminaria japonica* (Kombu)	Guinea-pigs, male hairless mice and mice with B16 melanoma (JCRB0202)	Anti-pigmentary	Topical or oral fucoxanthin displayed anti-pigmentary effect in UVB-induced melanogenesis. Topical fucoxanthin reduced mRNA expression of COX-2, endothelin receptor A, neurotrophin receptor (NTR), melanocortin 1 receptor and PGE receptor 1.	Shimoda et al. (2010)
Undaria pinnatifida (brown alga)	HOS-HR-1 hairless mice. Fucoxanthin solution (0.001%) 2 h each time before UVB irradiation (5 times a week) for 10 weeks.	Anti-aging	Topical fucoxanthin treatment prevented the skin photoaging by reduced UVB-induced epidermal hypertrophy, VEGF, and MMP-13 expression in the epidermis and thiobarbituric acid reactive substances (TBARS) in the skin.	Urikura et al. (2011)
Ethanol and silica gel column chromatography extract from *Sargassum siliquastrum*	UVB induced cell injury in human fibroblast. MTT and COMET assays	UV-photoprotective	Fucoxanthin increased the cell survival rate and reduced oxidative stress via ROS scavenging activity.	Heo and Jeon (2009)

(continued)

Table 11.6 (continued)

Pharmaceutical forms	Experimental design or assay	Health-promoting benefits	Findings	References
Methanol and silica gel column chromatography *extract of Undaria pinnatifida* (brown alga)	Human dermal fibroblasts (HDF) and A431 cells Mild exposure to ultraviolet (UV) radiation	UV-photoprotective	Fucoxanthin displayed protective effects against UV-induced damage by the promotion of skin barrier formation through induction of filaggrin.	Matsui et al. (2016)
Methanol extract of *Ishige okamurae* Yendo,	Reduction in levels of proinflammatory mediator activation and the suppression of MAPK phosphorylation in RAW 264.7 cells	Anti-inflammatory activity	Fucoxanthin exerted anti-inflammatory properties by reducing the amounts of pro-inflammatory mediators.	Kim et al. (2010)

caspase-9, and caspase-3 and the up-regulation of apoptosis inhibitor, B-cell lymphoma-2 (Zheng et al. 2013). The photoprotective activity of fucoxanthin extracted from brown seaweeds *Sargassum siliquastrum* has been shown in UVB-irradiated human fibroblast (Heo and Jeon 2009). The pre-treatment of fucoxanthin provided increasing the cell survival rate and reduced oxidative stress via ROS scavenging activity in a dose-dependent manner. After exposure to UVB irradiation, skin cells face an intense oxidative reaction that gives rise to photodamage and photoaging. The UV-photoprotective activity of fucoxanthin is thought to be mainly due to its strong antioxidant activity due to its singlet oxygen quenching and ROS scavenging effects (Heo and Jeon 2009; Pangestuti and Kim 2011). However, a later study indicated that fucoxanthin might exert its photoprotective effects by promotion of skin barrier formation through the induction of filaggrin, unrelated to the quenching of ROS. Filaggrin is a protein fiber that acts as an essential mechanical support for the assembly of keratin filaments and regulates epidermal homeostasis and is associated with the degree of skin damage. The levels of structural proteins for the epidermal permeability barrier, including filaggrin (filament aggregating protein) markedly decrease in aged skin. The protective effects of fucoxanthin against UV-induced sunburn in human dermal fibroblasts (HDF) have been demonstrated by the promotion of skin barrier formation through the induction of filaggrin. The investigators suggested that alternative protective mechanisms of fucoxanthin might be exerted by the promotion of skin barrier formation through the induction of UV-sensitive gene expression (Matsui et al. 2016).

Chronic UV irradiation induces skin angiogenesis and causes DNA damage, increases oxidative stress, and reactive oxygen species (ROS) in the epidermal cells, and leads to premature aging and wrinkle formation (Rastogi et al. 2010; Chung and

Eun 2007). Continuous exposure to UV-irradiation caused increasing inflammation, cytokines, chemokines, and skin aging by an increase in nicotinamide adenine dinucleotide phosphate (NADPH) oxidase and generating ROS (Ansary et al. 2021). It is well-described that the acute dermal response to UV irradiation is inflammation, such as erythema and edema, and DNA and mitochondrial damage caused by ROS (Mittal et al. 2014). Inflammatory stimuli could trigger matrix metalloproteinase (MMP) which leads to photoaging, and when skin exposures the UVB irritation, keratinocytes which represent the first target act as sentinels, and initiate the signalling cascade. These events address the stress and stimulate the production of pro-inflammatory mediators such as PGE_2, TNF-α, IL-1, IL-6, and NO (Pangestuti et al. 2021). The anti-inflammatory and antioxidant activities of fucoxanthin were investigated both in vivo (with TNF-α-stimulated HaCaT keratinocytes, LPS-stimulated THP-1 macrophages and UVB-irradiated HaCaT cells) and in vitro (with TPA-induced epidermal hyperplasia in mice) skin models by Rodriguez-Luna and co-workers (Rodriguez-Luna et al. 2018). In addition, the synergistic activity of fucoxanthin and rosmarinic acid from *Rosmarinus officinalis* has been also demonstrated in UVB-irradiated HaCaT (Rodriguez-Luna et al. 2019). The findings of the studies indicated that fucoxanthin improved TPA-induced hyperplasia, by reducing skin edema, epidermal thickness, MPO activity and COX-2 expression. In addition, UVB-induced erythema in mice was reduced with fucoxanthin treatment by down-regulation of inflammatory cytokines (TNF-α and IL-6) as well as up-regulation of HO-1 protein via Nrf-2 pathway.

Despite the high antioxidant or skin protective properties, oral fucoxanthin treatment does not result in efficient dermal concentrations. Moreover, since fucoxanthin has high lipophilicity and great molecular weight, there are some difficulties in its topical administration. As known, a photoprotective compound must diffuse across the stratum corneum and tight junctions to achieve effective permeability. Thus, the composition of the formulation will affect the permeation of the active compound. The permeation profiles showed that the cream formulation was the most favorable vehicle for topical administration of fucoxanthin as a penetration enhancer after testing several vehicles such as hydrogel, cream, and ointment commonly used for enhancing dermal penetration (Rodriguez-Luna et al. 2018). Fucoxanthin-loaded cream was administered topically in mice with UVB-induced skin erythema and provided UVB- photoprotective activity by downregulating COX-2 expression and the up-regulation of HO-1 protein via Nrf-2 pathway after UVB irradiation (Rodriguez-Luna et al. 2018).

In addition, it has been indicated that topical treatment of fucoxanthin (0.001%) before UVB radiation decreased epidermal hypertrophy and MMP-13 expression in the epidermis and exerted considerable ROS scavenging and potential anti-angiogenic activities in hairless mice with UVB-induced epidermal hypertrophy (Urikura et al. 2011).

The toxicity and protective effects of topical fucoxanthin treatment were investigated upon UVB irradiation reconstructed human skin (RHS) (Tavares et al. 2020a, b). The results of the studies indicated that fucoxanthin ameliorated proinflammatory cytokines (IL-6 and IL-8) and has protective potential for the skin without any adverse effects. In addition, Denis and co-workers also investigated the effects of fucoxanthin on human skin fibroblasts by the tetrazolium-based colorimetric (MTT test) and COMET assays. According to the observation of the study, the authors suggested that fucoxanthin has therapeutic potential for cosmetic and dermatological applications (Denis et al. 2019).

Many endogenous substances such as melanocyte-stimulating hormones (MSH), prostaglandins, and common cytokines and exogenous factors such as UVB exposure, toxic compounds and environmental pollutants can trigger melanocyte receptors and start melanogenesis. Chronic exposure to UVB radiation causes dermal inflammation and activates fibroblasts. UVB exposure can stimulate keratinocytes to increase melanocyte and therefore, melanin production by secreting various growth factors, cytokines, and hormones including inducible nitric oxide synthase (iNOS) (Imokawa et al. 1998). Shimoda and co-workers investigated the anti-pigmentary activity of fucoxanthin in Guinea pigs and hairless mice following oral and topical treatments (Shimoda et al. 2010). The findings of the study indicated that fucoxanthin inhibited tyrosinase activity and exerted anti-pigmentary activity in UVB-induced melanogenesis after both treatment routes in UV-irradiated guinea pigs. The anti-melanogenic activity of topical fucoxanthin (1%) treatment may be due to the suppression of prostaglandin (PGE_2) synthesis and expression of melanogenic receptors in melanocytes (Shimoda et al. 2010).

Therefore, fucoxanthin might be a useful pharmaceutical or cosmetic agent that reduces the health benefits effects of oxidative stress on the skin. Although experimental studies have shown that fucoxanthin has great potential to be used as a UV-photoprotective agent, it has not yet given conclusive results due to the lack of appropriate clinical trials. Further research is needed to demonstrate the cause-effect relationship between fucoxanthin application and the prevention or protection of UV irradiation-mediated skin damage.

Astaxanthin

Astaxanthin is a xanthophyll carotenoid naturally biosynthesized by some yeasts, bacteria and various algae and consumed by many marine animals, such as salmon, trout, crab, krill, lobster and shrimp (Higuera-Ciapara et al. 2006). In nature, the highest levels of astaxanthin are produced by the single-celled alga, *Haematococcus pluvialis* (Huangfu et al. 2013).

Astaxanthin has a strong anti-oxidative activity by scavenging free radicals. Astaxanthin has been reported to be a ten-fold more potent antioxidant than other carotenoids such as lutein, zeaxanthin, canthaxanthin, and β-carotene (Naguib 2000). Astaxanthin shares a similar chemical structure with β-carotene but has 40 times higher antioxidant capacity, as its polar ionone rings on both ends can quench free radicals and other reactive oxygen species (ROS), and the thirteen conjugated double, polyunsaturated bonds can remove high-energy electrons (Davinelli et al. 2018). While many experimental and clinical trials have shown its bioactivities, particularly due to its excellent antioxidant and anti-inflammatory properties, astaxanthin has acquired a great economical potential for human use, such as a potent food ingredient for nutrition or a supplement for health as well as in pharmaceutic and cosmetic industries (Jannel et al. 2020).

Astaxanthin derived from microalgae *H. pluvialis* has been approved as a dietary supplement for human consumption in Europe, Japan, and the USA (Davinelli et al. 2018). The European Food Safety Authority (EFSA) advised 0.034 mg/kg BW of astaxanthin (2.38 mg/d in a 70-kg person) for an acceptable daily intake (ADI) (EFSA. Efsa Panel on Dietetic Products, Nutrition Allergies 2014). Numerous human clinical trials have been done on the safety of Astaxanthin. These studies indicated that astaxanthin did not produce any clinical toxicity signs and significant changes in biochemical and hematological parameters in healthy adults following *per os* administration at a dose of 20 mg/d for 4 weeks (Satoh et al. 2009) or 6 mg/d for 8 weeks (Spiller and Dewell 2003). No adverse effects or toxicity were observed in healthy men following daily oral administration of a food supplement containing 8 mg of astaxanthin for 3 months (Karppi et al. 2007). Besides, it has been reported that astaxanthin (40 mg) was well-tolerated in 32 healthy subjects with three mild adverse effects reported 48 h post-treatment (Odeberg et al. 2003). However, the dose of astaxanthin from *H. pluvialis* for humans has been approved by the Food and Drug Administration (FDA) up to 12 mg/d and up to 24 mg/d for no more than 30 days (Visioli and Artaria 2017). Besides, recently, the EFSA recommended that a consumption of 8 mg astaxanthin per day as food supplements is safe for adults even in combination with the high exposure estimate to astaxanthin from the diet (Turck et al. 2020).

In several clinical trials and experimental animal studies, various dermatological health-promoting effects of astaxanthin have been reported including anti-aging or anti-photoaging, UV-photoprotective and anti-inflammatory actions, suppression of hyper-pigmentation and inhibitions of melanin synthesis (Table 11.7). Several *in vivo* and *in vitro* experimental studies have indicated that astaxanthin treatment provides beneficial dermatologic effects such as antioxidant, anti-inflammatory, UV-photoprotective and anti-photocarcinogenetic activities (Kuedo et al. 2016; Yoshihisa et al. 2016; Park et al. 2018; Komatsu et al. 2017; Li et al. 2018; Park and Song 2021). Rao and co-workers investigated the potential beneficial activity of astaxanthin and astaxanthin esters extracted from the Green Alga *Haematococcus pluvialison* in albino rats with UV-DMBA-induced skin carcinogenesis (Rao et al. 2013). The findings indicated that astaxanthin had antioxidative activity and showed effective inhibition of skin cancer and tyrosinase. Furthermore, it was indicated that pre-treatment of human keratinocytes (HaCaT) with astaxanthin (5 μM) before UVB exposure or topical application of ASX gel (0.02%) after chronic UVB irradiation in mice inhibited oxidative DNA damage and apoptosis (Wiraguna et al. 2018; Yoshihisa et al. 2014). Besides, the protective activities of the astaxanthin-rich extract against UVA-induced DNA alterations in UVA-induced DNA damage were also shown in human melanocytes (HEMAc, skin fibroblasts (1BR-3),) and intestinal CaCo-2 cells (Lyons and O'Brien 2002). Hama and co-workers investigated the potential protective effects of topical liposomal formulation of astaxanthin on UVB-induced skin damage in male hairless mice. The results of the study

Table 11.7 Clinical studies investigating the skincare activities of astaxanthin

Pharmaceutical forms	Dosage regime	Duration	Study design	Subjects	Health-promoting benefits	Findings	References
Capsules	4 mg/d	10 weeks	A randomized, double-blind, placebo-controlled, parallel-group comparison trial	22 volunteers	UV-photoprotective	The astaxanthin showed UV-photoprotective effects compared with the placebo.	Ito et al. (2018)
Capsules	4 mg/d	4 weeks	Monitoring of oxidative stress and skin aging parameters of the subjects	17 men and 14 women age range (40–80)	Anti-aging Facial skin rejuvenation	Daily Astaxanthin treatment displayed a strong antioxidant effect resulting in facial skin rejuvenation.	Chalyk et al. (2017)
Capsules	6 or 12 mg/d	16 weeks	Randomized, double-blind, parallel-group, placebo-controlled	65 healthy female voluntaries	Anti-aging	Wrinkle depths, skin elasticity, and moisture were stable for both astaxanthin dose groups, while significantly worsened wrinkle parameters and skin moisture content were observed in the placebo group. Astaxanthin supplementation could prevent age-related skin deterioration and maintain skin conditions by its anti-inflammatory effect.	Tominaga et al. (2017)
Astaxanthin containing drink	3 mg/d	8 weeks	Randomized, double-blind, placebo-controlled	20 females (30–50-year-old)	Skincare	Astaxanthin had protection activity on the skin barrier and is considered to increase skin moisture content and provided skin elasticity and skin texture.	Tsukahara et al. (2016)

(continued)

Table 11.7 (continued)

Pharmaceutical forms	Dosage regime	Duration	Study design	Subjects	Health-promoting benefits	Findings	References
Capsules	2 mg/d and collagen tablets	12 weeks	Randomized, double-blind, placebo-controlled	44 healthy females	Anti-aging	Dietary astaxanthin supplementation combined with collagen provide improvements on elasticity and barrier integrity of photo-aged facial skin.	Yoon et al. (2014)
Capsules and solution	6 mg/d capsule and 2 ml (78.9 μM solution) per day topical application	8 weeks	Open-label non-controlled study	30 healthy females and 36 healthy males	Anti-aging	Daily oral astaxanthin and its topical treatment provided improvements of skin condition in all layers in not only women but also men.	Tominaga et al. (2012)
Capsules	2 or 8 mg/d	8 weeks	Randomized double-blind, controlled study	14 healthy females (averaged 21.5-year-old)	Antioxidant Anti-inflammatory	Dietary astaxanthin improves immune response and reduces a DNA damage biomarker and acute-phase protein and in females.	Park et al. (2010)
Capsules	4 mg/d	6 weeks	A single-blind placebo-controlled clinic test. Assessment of skin condition by subjects, dermatologist and instrumental analysis	49 healthy females	Anti-aging	A noticeable improvement in skin condition with oral astaxanthin supplementation.	Yamashita (2005)

suggested that liposomal astaxanthin treatment showed great protective activity against UVB-induced skin damage and inhibited collagen degradation and melanin production through its efficient antioxidant activity (Hama et al. 2012). In another investigation, oral supplementation of astaxanthin extracted from *Haematococcus pluvialis* decreased the visible signs of UV-induced photoaging by reducing the wrinkle formation and skin thickening, and increasing collagen density in UV-irradiated skin of HR-1 hairless mice (Li et al. 2020). It was also observed that the treatment also prevented the reduction of capillaries diameter indicating an association between photoaging and capillary regression in UV-irradiated skin.

However, the beneficial effects obtained in experimental animal studies cannot be directly translated into the same effects in humans, due to the different bioavailability and pharmacokinetic dispositions of astaxanthin between species. For this reason, several clinical trials have been performed to investigate the potential skincare and anti-photoaging, anti-inflammatory and anti-pigmentation effects of astaxanthin supplementation (Table 11.7). In a clinical trial, the effects of oral administration of a capsule containing 4 mg astaxanthin have been evaluated on UV-induced skin deterioration for 10 weeks in a double-blind placebo-controlled study with 23 healthy Japanese participants (Ito et al. 2018). The astaxanthin group showed increased minimal erythema dose and had a reduced loss of skin moisture in the UVB irradiated area compared with the placebo. Besides, astaxanthin also reduced UV-induced moisture loss in healthy subjects. These findings indicated that astaxanthin has protective effects against UV-induced skin deterioration and could be used to maintain skin health and as a sunburn protector.

To assess the effect of daily supplementation of astaxanthin on the skin parameters and aging-related changes on the surface of the stratum corneum in humans, Chalyk and co-workers (2017) performed a study in which 31 participants (17 men and 14 women) consumed 4 mg of astaxanthin per day for 4 weeks. The findings of this study suggested that continuous consumption of astaxanthin for only 4 weeks resulted in changes in the residual skin surface component consistent with the reversal of aging process in participants. Also, continuous astaxanthin supplementation produces a potent antioxidant activity, which results in facial skin rejuvenation particularly evident in obese individuals (Chalyk et al. 2017).

In a placebo-controlled, double-blind, randomized study, involving 59 healthy female participants, two groups of 22 and 19 individuals received 6 or 12 mg of astaxanthin oral supplementation per day for 16 weeks, respectively and 18 subjects served as a placebo group (Tominaga et al. 2017). Skin elasticity, wrinkle depths, and moisture were stable for both astaxanthin dose groups, while significantly worsened wrinkle parameters and skin moisture content were observed in the placebo group after 16 weeks. In another clinical trial, both oral supplementation (6 mg) and topical (2 ml, 78.9 μM solution) administrations of daily astaxanthin in 30 healthy female and 36 healthy male participants for 8 weeks improve skin condition in all layers and can improve the skin condition both female and men (Tominaga et al. 2012). Therefore, these studies have suggested that astaxanthin supplementation could prevent age-related skin deterioration and protect the skin from environmental-induced damage through its anti-inflammatory effect (Tominaga et al. 2017).

Besides, a randomized, double-blind, placebo-controlled study using dietary astaxanthin (2 mg/d) supplementation combined with collagen hydrolysate was performed on 44 healthy females for 12 weeks. The findings have indicated that astaxanthin supplementation can improve elasticity and barrier integrity in photoaged facial skin without any adverse signs (Yoon et al. 2014).

A recent review has aimed to conduct a systemic review and meta-analysis to evaluate the dermatological health benefits of astaxanthin in humans, based on currently published clinical studies (Zhou et al. 2021). In this evaluation, the researchers concluded that oral and/or topical applications of astaxanthin can improve the signs of skin aging by increasing skin moisture content and elasticity, reducing facial wrinkles and sebum oil, due to its antioxidant, anti-inflammatory and beneficial effects on skin barrier integrity. Moreover, it was emphasized that *per os* administration could be more sustained and pronounced compared with the topical route.

3.2.3 Mycosporine-Like Amino Acids (MAAs)

Mycosporine-like amino acids (MAAs) are colorless, low-molecular-weight (<400 ~Da) and water-soluble molecules that strongly absorb UVA and UVB radiations in the wavelength range 310–365 nm (Miyamoto et al. 2014). Considering their chemical structure, these compounds consist of a cyclohexenime ring conjugated with two amino acids, amino alcohols, or amino group substitutions. They are widely distributed and accumulated in different organisms such as eukaryotic (microalgae, yeasts, and fungi), various marine macroalgae species, dinoflagellates, cyanobacteria, corals and many other marine invertebrates (Sinha et al. 2007). These MAAs have great ability as sunscreen compounds to protect these organisms against the harmful effects of environmental UV radiation.

Mycosporine-like amino acid compounds have been reported to have multiple types of activity, such as UV-radiation absorbing, anti-photoaging and antioxidant activities, as a result of their phenolic hydroxyl structure, which can effectively dispel ROS (Rui et al. 2019). Hence, MAAs have attracted great attention in the cosmetic industry due to their effective UVAbsorbing capacity and ability to prevent UV-induced skin damage (De la Coba et al. 2009).

Mycosporine-like amino acids occur naturally in seaweeds. Also, red and brown algae have higher MAA content compared to other seaweed classes (Frohnmeyer and Staiger 2003). Seaweeds-derived MAAs have great UV-radiation absorption capacity and anti-photoaging activities (Singh et al. 2008), which provide efficient protection against UV irradiation-induced damage by inhibiting the expression of the MMP gene by suppressing the MAPK signal pathway (Chen et al. 2016).

The efficacy of a cream formulation containing MAAs extracted from red algae *Porphyra umbilicalis* in the prevention of photoaging induced by UVA irradiation was investigated in 20 women as well as *in vitro* human keratinocytes (HaCaT) and fibroblast cultures (Schmid et al. 2006). It was reported that MAAs-rich extract had abilities to reduce the harmful effects on the UVA-irradiated skin by antioxidant and DNA protective activities of the extract. Therefore, the author suggested that the daily use of the cream containing MAAs seems to be a highly effective way to sustain good skin health as well as prevent premature skin aging (Schmid et al. 2006). In addition, Daniel ad co-workers reported that sunscreen cream containing 0.005% MAA extracted from *Porphyra umbilicalis* also prevented UV-induced photodamage as efficiently as a cream containing 1% reference synthetic UVA and 4% UVB filters (Daniel 2004). In another study, it was found that a sunscreen product in combination with MAAs indicated the same Sun Protecting Factor (SPF) and UVB Biological Effective Protection Factors (BEPFs) as reference sunscreens but slightly lower UVA-BEPFs (De la Coba et al. 2019).

The photoprotective effects of MAAs extracted from *Porphyra yezoensis* and *Porphyra tenera* were investigated in mice with UV irradiation-induced photoaging (Ying et al. 2019; Rui et al. 2019). The findings indicated that MAAs rich extracts prevented UV-induced photoaging by inhibiting the reduction of endogenous antioxidant enzymes, inflammatory cytokines and the expression of interstitial collagenase (MMP-1). In another study, MAAs extracted from green algae *Chlamydomonas hedleyi* were also shown to have a protective effect against UV-irradiation on human keratinocytes (HaCaT) by modulating the expression of genes associated with oxidative stress, inflammation and skin aging (Suh et al. 2014).

Besides, the anti-inflammatory activity of seaweed-derived MAAs has been also shown in UV-irradiated HaCaT cell culture (Cho et al. 2014). In addition, the treatment of seaweed-derived MAAs such as porphyra-334 and shinorine in LPS-stimulated macrophages showed potential anti-inflammatory activity. While MAAs treatment significantly inhibited the release of pro-inflammatory mediators which were mediated through NF-κB signalling pathway (Becker et al. 2016). Furthermore, it was indicated that porphyra-334 treatment prevented COX-2 expression and the main cytotoxic mediators participating in the innate response in mammalian species (Pangestuti et al. 2011).

The investigations in the literature have indicated that the anti-photoaging activities of seaweed-derived MAAs are not only mediated by their photoprotective activity by absorbing UVR, but also by a potent antioxidant, anti-inflammatory, radical scavenging, macromolecule damage-protection, and MMP inhibitor activities (Pangestuti et al. 2021). Collectively, the treatment with MAAs has been able to provide healthier skin by preventing skin wrinkle depth, anti-photoaging, elasticity and roughness. These skin benefit effects of MAAs indicated that these compounds can be used in cosmetic products as skincare ingredients.

4 Conclusion

Recent investigations in literature have provided important evidence that seaweeds and seaweed-derived compounds play an important role in skin health. These compounds have attracted more attention and commercial importance due to their many skin-beneficial properties such as photoprotection, skin whitening, antiwrinkle, moisturizing, antioxidant, anti-inflammatory, antimicrobial and anticancer activities. Seaweeds and seaweed-derived compounds include a wide range of bioactive substances and are a great natural source of raw materials as active ingredients for the preparation of cosmeceutical and pharmaceutical products due to their potential skin benefits In addition, they are commercially available in the market, and many cosmeceutical companies have been already using these compounds in many cosmetic products. However, more detailed studies are required to fully understand the mechanism of action of these bioactive molecules since some seaweed-derived substances have not been fully investigated. In addition, there are not enough clinical studies yet for cosmetic products containing these compounds. Hence, further investigations, clinical studies and evaluations are also required to improve the quality of these cosmetic formulations, which will be beneficial to provide consumer safety.

References

Afzal S, Garg S, Ishida Y, Terao K, Kaul SC, Wadhwa R (2019) Rat Glioma cell-based functional characterization of anti-stress and protein deaggregation activities in the marine carotenoids, Astaxanthin and Fucoxanthin. Mar Drugs 17(3):189

Ahmad T, Eapen MS, Ishaq M, Park AY, Karpiniec SS, Stringer DN et al (2021) Anti-inflammatory activity of Fucoidan extracts in vitro. Mar Drugs 19(12):702

Ahn JH, Kim DW, Park CW, Kim B, Sim H, Kim HS et al (2020) Laminarin attenuates ultraviolet-induced skin damage by reducing superoxide anion levels and increasing endogenous antioxidants in the dorsal skin of mice. Mar Drugs 18(7):345

Ale MT, Maruyama H, Tamauchi H, Mikkelsen JD, Meyer AS (2011a) Fucoidan from Sargassum sp and Fucus vesiculosus reduces cell viability of lung carcinoma and melanoma cells in vitro and activates natural killer cells in mice in vivo. Int J Biol Macromol 49(3):331–336

Ale MT, Mikkelsen JD, Meyer AS (2011b) Important determinants for Fucoidan bioactivity: a critical review of structure-function relations and extraction methods for Fucose-containing sulfated polysaccharides from Brown Seaweeds. Mar Drugs 9(10):2106–2130

Ali Karami M, Sharif Makhmalzadeh B, Pooranian M, Rezai A (2021) Preparation and optimization of silibinin-loaded chitosan–fucoidan hydrogel: An in vivo evaluation of skin protection against UVB. Pharm Dev Technol 26(2):209–219

Amberg N, Fogarassy C (2019) Green consumer behavior in the cosmetics market. Resources 8(3):137

Ansary TM, Hossain M, Kamiya K, Komine M, Ohtsuki M (2021) Inflammatory molecules associated with ultraviolet radiation-mediated skin aging. Int J Mol Sci 22(8):3974

Arantes EL, Dragano N, Ramalho A, Vitorino D, de Souza GF, Lima MHM et al (2016) Topical docosahexaenoic acid (DHA) accelerates skin wound healing in rats and activates GPR120. Biol Res Nurs 18(4):411–419

Aryee ANA, Agyei D, Akanbi TO (2018) Recovery and utilization of seaweed pigments in food processing. Curr Opin Food Sci 19:113–119

Baek S, Cao L, Lee H, Lee Y, Lee S (2021) Effects of UV and heating on the stability of Fucoxanthin, total Phlorotannin and total antioxidant capacities in Saccharina japonica ethanol extract and solvent fractions. Appl Sci Basel 11(17):7831

Balupillai A, Prasad RN, Ramasamy K, Muthusamy G, Shanmugham M, Govindasamy K et al (2015) Caffeic acid inhibits UVB-induced inflammation and photocarcinogenesis through activation of peroxisome proliferator-activated receptor-γ in mouse skin. Photochem Photobiol 91(6):1458–1468

Balvers MGJ, Verhoeckx KCM, Plastina P, Wortelboer HM, Meijerink J, Witkamp RF (2010) Docosahexaenoic acid and eicosapentaenoic acid are converted by 3T3-L1 adipocytes to N-acyl ethanolamines with anti-inflammatory properties. Biochim Biophys Acta Mol Cell Biol Lipids 1801(10):1107–1114

Barbosa M, Lopes G, Ferreres F, Andrade PB, Pereira DM, Gil-Izquierdo A et al (2017) Phlorotannin extracts from Fucales: marine polyphenols as bioregulators engaged in inflammation-related mediators and enzymes. Algal Res Biomass Biofuels Bioproducts 28:1–8

Baron ED, Suggs AK (2014) Introduction to photobiology. Dermatol Clin 32(3):255–266

Becker K, Hartmann A, Ganzera M, Fuchs D, Gostner JM (2016) Immunomodulatory effects of the mycosporine-like amino acids shinorine and porphyra-334. Mar Drugs 14(6):119

Bialik-Was K, Pluta K, Malina D, Majka TM (2021) Alginate/PVA-based hydrogel matrices with Echinacea purpurea extract as a new approach to dermal wound healing. Int J Polym Mater Polym Biomater 70(3):195–206

Bilal M, Iqbal H (2020) New insights on unique features and role of nanostructured materials in cosmetics. Cosmetics 7(2):24

Bin Bae S, Nam HC, Park WH (2019) Electrospraying of environmentally sustainable alginate microbeads for cosmetic additives. Int J Biol Macromol 133:278–283

Blanco-Dávila F (2000) Beauty and the body: the origins of cosmetics. Plast Reconstr Surg 105(3):1196–1204

Briand G, Foultier MT, Amory MC, Barriere H, Litoux P, Bousquet B et al (1985) Comparative-study of 2 natural carotenoids (Beta-Carotene and Fucoxanthine) on the evolution of Papillomas and Photoinduced cutaneous cancers in the hairless mouse. Annales De Biologie Clinique 43(4):693

Candreva T, Kühl CMC, Burger B, Dos Anjos MBP, Torsoni MA, Consonni SR et al (2019) Docosahexaenoic acid slows inflammation resolution and impairs the quality of healed skin tissue. Clin Sci 133(22):2345–2360

Catanzaro E, Calcabrini C, Bishayee A, Fimognari C (2020a) Antitumor potential of marine and freshwater lectins. Mar Drugs 18(1):11

Catanzaro E, Bishayee A, Fimognari C (2020b) On a beam of light: photoprotective activities of the marine carotenoids Astaxanthin and Fucoxanthin in suppression of inflammation and cancer. Mar Drugs 18(11):544

Cavinato M, Waltenberger B, Baraldo G, Grade CVC, Stuppner H, Jansen-Dürr P (2017) Plant extracts and natural compounds used against UVB-induced photoaging. Biogerontology 18(4):499–516

Cezar TLC, Martinez RM, Rocha C, Melo CPB, Vale DL, Borghi SM et al (2019) Treatment with maresin 1, a docosahexaenoic acid-derived pro-resolution lipid, protects skin from inflammation and oxidative stress caused by UVB irradiation. Sci Rep 9(1):1–14

Chalyk NE, Klochkov VA, Bandaletova TY, Kyle NH, Petyaev IM (2017) Continuous astaxanthin intake reduces oxidative stress and reverses age-related morphological changes of residual skin surface components in middle-aged volunteers. Nutr Res 48:40–48

Chan KC, Mong MC, Yin MC (2009) Antioxidative and anti-inflammatory neuroprotective effects of Astaxanthin and Canthaxanthin in nerve growth factor differentiated PC12 cells. J Food Sci 74(7):H225–HH31

Chaudhri SK, Jain NK (2014) History of cosmetics. Asian J Pharm 3(3):164

Chen T, Hou H, Fan Y, Wang S, Chen Q, Si L et al (2016) Protective effect of gelatin peptides from pacific cod skin against photoaging by inhibiting the expression of MMPs via MAPK signalling pathway. J Photochem Photobiol B Biol 165:34–41

Chen Y, Li XN, Gan XS, Qi JM, Che B, Tai ML et al (2019a) Fucoidan from Undaria pinnatifida Ameliorates epidermal barrier disruption via Keratinocyte differentiation and CaSR level regulation. Mar Drugs 17(12):660

Chen WW, Zhang HJ, Liu Y (2019b) Anti-inflammatory and apoptotic signalling effect of Fucoxanthin on Benzo(A)Pyrene-induced lung cancer in mice. J Environ Pathol Toxicol Oncol 38(3):239–251

Chen BR, Hsu KT, Hsu WH, Lee BH, Li TL, Chan YL et al (2021a) Immunomodulation and mechanisms of fucoidan from Cladosiphon okamuranus ameliorates atopic dermatitis symptoms. Int J Biol Macromol 189:537–543

Chen J, Zeng W, Gan J, Li Y, Pan Y, Li J et al (2021b) Physicochemical properties and antioxidation activities of ulvan from Ulva pertusa Kjellm. Algal Res 55:102269

Chiang CS, Huang BJ, Chen JY, Chieng WW, Lim SH, Lee W et al (2021) Fucoidan-based nanoparticles with inherently therapeutic efficacy for cancer treatment. Pharmaceutics 13(12):1986

Cho MJ, Jung HS, Song MY, Seo HH, Kulkarni A, Suh SS et al (2014) Effect of sun screen utilizing Porphyra-334 derived from ocean algae for skin protection. J Korea Academ Indus Cooper Soc 15(7):4272–4278

Choi JI, Kim HJ, Lee JW (2011) Structural feature and antioxidant activity of low molecular weight laminarin degraded by gamma irradiation. Food Chem 129(2):520–523

Christaki E, Bonos E, Giannenas I, Florou-Paneri P (2013) Functional properties of carotenoids originating from algae. J Sci Food Agric 93(1):5–11

Chung JH, Eun HC (2007) Angiogenesis in skin aging and photoaging. J Dermatol 34(9):593–600

Couteau C, Coiffard L (2016) Seaweed application in cosmetics. In: Seaweed in health and disease prevention. Elsevier, pp 423–441

Dahmane R, Pandel R, Trebse P, Poljsak B (2015) The role of sun exposure in skin aging. In: Sun exposure: risk factors, protection practices and health effects, vol 2015. Nova Science Publishers, Inc, Hauppauge, pp 1–40

Daikonya A, Katsuki S, Wu J-B, Kitanaka S (2002) Anti-allergic agents from natural sources (4): anti-allergic activity of new phloroglucinol derivatives from Mallotus philippensis (Euphorbiaceae). Chem Pharm Bull 50(12):1566–1569

Dall'Osto L, Cazzaniga S, North H, Marion-Poll A, Bassi R (2007) The Arabidopsis aba4-1 mutant reveals a specific function for neoxanthin in protection against photooxidative stress. Plant Cell 19(3):1048–1064

Daniel S (2004) UVA sunscreen from red algae for protection against premature skin aging. Cosmetic and Toiletries Manufacture worldwide. Food Chem 3:139

Daniel E, Fabio G (2020) An assessment of seaweed extracts: innovation for sustainable agriculture. Agronomy-Basel 10(9):1433

Danno K, Ikai K, Imamura S (1993) Anti-inflammatory effects of eicosapentaenoic acid on experimental skin inflammation models. Arch Dermatol Res 285(7):432–435

Davinelli S, Nielsen ME, Scapagnini G (2018) Astaxanthin in skin health, repair, and disease: a comprehensive review. Nutrients 10(4):522

De la Coba F, Aguilera J, De Galvez MV, Alvarez M, Gallego E, Figueroa FL et al (2009) Prevention of the ultraviolet effects on clinical and histopathological changes, as well as the heat shock protein-70 expression in mouse skin by topical application of algal UVAbsorbing compounds. J Dermatol Sci 55(3):161–169

De la Coba F, Aguilera J, Korbee N, de Gálvez MV, Herrera-Ceballos E, Álvarez-Gómez F et al (2019) UVA and UVB photoprotective capabilities of topical formulations containing mycosporine-like amino acids (MAAs) through different biological effective protection factors (BEPFs). Mar Drugs 17(1):55

Denis E, Papurina T, Koliada A, Vaiserman A (2019) Evaluation of the stimulating and protective effects of Fucoxanthin against human skin fibroblasts: an in vitro study. Clin Dermatol (Wilmington) 2(1):1–8

Dong K, Wang XY, Shen Y, Wang YY, Li BB, Cai CL et al (2021) Maintaining inducibility of dermal follicle cells on silk fibroin/sodium alginate scaffold for enhanced hair follicle regeneration. Biology-Basel 10(4):269

Draelos ZD (2005) The future of cosmeceuticals: an interview with Albert Kligman, MD, PhD. Dermatologic Surg 31(7):890

Draelos ZD (2019) Cosmeceuticals: What's real, what's not. Dermatol Clin 37(1):107–115

EC (2009) Cosmetic products: regulation (EC) No 1223/2009 of the European Parliement and of the Council of 30 November 2009. L-342/59–L-/209. Official Journal of the European Union

EFSA. Efsa Panel on Dietetic Products, Nutrition Allergies (2014) Scientific opinion on the safety of astaxanthin-rich ingredients (AstaREAL A1010 and AstaREAL L10) as novel food ingredients. EFSA J 12(7):3757

FAO (2021, March) FAO global fishery and aquaculture production statistics. FishStatJ; www.fao.org/fishery/statistics/software/fishstatj/en

Faria-Silva C, Ascenso A, Costa AM, Marto J, Carvalheiro M, Ribeiro HM et al (2020) Feeding the skin: a new trend in food and cosmetics convergence. Trends Food Sci Technol 95:21–32

Fernando IPS, Jayawardena TU, Sanjeewa KKA, Wang L, Jeon YJ, Lee WW (2018) Anti-inflammatory potential of alginic acid from Sargassum horneri against urban aerosol-induced inflammatory responses in keratinocytes and macrophages. Ecotoxicol Environ Saf 160:24–31

Fernando IPS, Dias MKHM, Madusanka DMD, Han EJ, Kim MJ, Jeon YJ et al (2020a) Fucoidan refined by Sargassum confusum indicate protective effects suppressing photo-oxidative stress and skin barrier perturbation in UVB-induced human keratinocytes. Int J Biol Macromol 164:149–161

Fernando IPS, Dias MKHM, Madusanka DMD, Han EJ, Kim MJ, Jeon YJ et al (2020b) Human Keratinocyte UVB-protective effects of a low molecular weight Fucoidan from Sargassum horneri purified by step gradient ethanol precipitation. Antioxidants 9(4):340

Fernando IPS, Dias MKHM, Madusanka DMD, Han EJ, Kim MJ, Heo SJ et al (2021) Low molecular weight fucoidan fraction ameliorates inflammation and deterioration of skin barrier in fine-dust stimulated keratinocytes. Int J Biol Macromol 168:620–630

Ferreres F, Lopes G, Gil-Izquierdo A, Andrade PB, Sousa C, Mouga T et al (2012) Phlorotannin extracts from Fucales characterized by HPLC-DAD-ESI-MSn: approaches to hyaluronidase inhibitory capacity and antioxidant properties. Mar Drugs 10(12):2766–2781

Foo SC, Yusoff FM, Ismail M, Basri M, Yau SK, Khong NMH et al (2017) Antioxidant capacities of fucoxanthin-producing algae as influenced by their carotenoid and phenolic contents. J Biotechnol 241:175–183

Freitas R, Martins A, Silva J, Alves C, Pinteus S, Alves J et al (2020) Highlighting the biological potential of the Brown Seaweed Fucus spiralis for skin applications. Antioxidants 9(7):611

Frohnmeyer H, Staiger D (2003) Ultraviolet-B radiation-mediated responses in plants. Balanc Damage Protect Plant Physiol 133(4):1420–1428

Gammone MA, D'Orazio N (2015) Anti-obesity activity of the marine carotenoid Fucoxanthin. Mar Drugs 13(4):2196–2214

Ganesan P, Noda K, Manabe Y, Ohkubo T, Tanaka Y, Maoka T et al (2011) Siphonaxanthin, a marine carotenoid from green algae, effectively induces apoptosis in human leukemia (HL-60) cells. BBA-Gen Subjects 1810(5):497–503

Gensler HL, Aickin M, Peng YM (1990) Cumulative reduction of primary skin tumor growth in UV-irradiated mice by the combination of retinyl palmitate and canthaxanthin. Cancer Lett 53(1):27–31

Gil TY, Kang YM, Eom YJ, Hong CH, An HJ (2019) Anti-atopic dermatitis effect of seaweed Fulvescens extract via inhibiting the STAT1 pathway. Mediat Inflamm 2019:3760934

Goncalves AG, Ducatti DRB, Duarte MER, Noseda MD (2002) Sulfated and pyruvylated disaccharide alditols obtained from a red seaweed galactan: ESIMS and NMR approaches. Carbohydr Res 337(24):2443–2453

Goto T, Urabe D, Isobe Y, Arita M, Inoue M (2015) Total synthesis of four stereoisomers of (5Z, 8Z, 10E, 14Z)-12-hydroxy-17, 18-epoxy-5, 8, 10, 14-eicosatetraenoic acid and their anti-inflammatory activities. Tetrahedron 71(43):8320–8332

Gouveia L, Empis J (2003) Relative stabilities of microalgal carotenoids in microalgal extracts, biomass and fish feed: effect of storage conditions. Innovative Food Sci Emerg Technol 4(2):227–233

Guinea M, Franco V, Araujo-Bazan L, Rodriguez-Martin I, Gonzalez S (2012) In vivo UVB-photoprotective activity of extracts from commercial marine macroalgae. Food Chem Toxicol 50(3–4):1109–1117

Habib MA, Salem SAM, Hakim SA, Shalan YAM (2014) Comparative immunohistochemical assessment of cutaneous cyclooxygenase-2 enzyme expression in chronological aging and photoaging. Photodermatol Photoimmunol Photomed 30(1):43–51

Hama S, Takahashi K, Inai Y, Shiota K, Sakamoto R, Yamada A et al (2012) Protective effects of topical application of a poorly soluble antioxidant astaxanthin liposomal formulation on ultraviolet-induced skin damage. J Pharm Sci 101(8):2909–2916

Han A, Chien AL, Kang S (2014a) Photoaging. Dermatol Clin 32(3):291–299

Han S-C, Kang N-J, Kang G-J, Koh Y-S, Hyun J-W, Lee N-H et al (2014b) 71: anti-inflammatory effect of diphlorethohydroxycarmalol (DPHC) isolated from lshige okamuarae in vitro and in vivo. Cytokine 70(1):44–45

Hashimoto T, Ozaki Y, Taminato M, Das SK, Mizuno M, Yoshimura K et al (2009) The distribution and accumulation of fucoxanthin and its metabolites after oral administration in mice. Br J Nutr 102(2):242–248

Haugan JA, Aakemann T, Liaaen-Jensen S, Britton G, Pfander H (1995) Example 2: macroalgae and microalgae. Carotenoid 1:215–226

Heo SJ, Jeon YJ (2009) Protective effect of fucoxanthin isolated from Sargassum siliquastrum on UVB induced cell damage. J Photochem Photobiol B-Biol 95(2):101–107

Heo S-J, Kim J-P, Jung W-K, Lee N-H, Kang H-S, Jun E-M et al (2008) Identification of chemical structure and free radical scavenging activity of diphlorethohydroxycarmalol isolated from a brown alga, Ishige okamurae. J Microbiol Biotechnol 18(4):676–681

Heo S-J, Ko S-C, Cha S-H, Kang D-H, Park H-S, Choi Y-U et al (2009) Effect of phlorotannins isolated from Ecklonia cava on melanogenesis and their protective effect against photo-oxidative stress induced by UVB radiation. Toxicol In Vitro 23(6):1123–1130

Heo S-J, Ko S-C, Kang S-M, Cha S-H, Lee S-H, Kang D-H et al (2010) Inhibitory effect of diphlorethohydroxycarmalol on melanogenesis and its protective effect against UVB radiation-induced cell damage. Food Chem Toxicol 48(5):1355–1361

Heo SJ, Yoon WJ, Kim KN, Oh C, Choi YU, Yoon KT et al (2012) Anti-inflammatory effect of fucoxanthin derivatives isolated from Sargassum siliquastrum in lipopolysaccharide-stimulated RAW 264.7 macrophage. Food Chem Toxicol 50(9):3336–3342

Higuera-Ciapara I, Felix-Valenzuela L, Goycoolea FM (2006) Astaxanthin: a review of its chemistry and applications. Crit Rev Food Sci Nutr 46(2):185–196

Hombach-Klonisch S, Klonisch T, Peeler J (2019) Sobotta clinical atlas of human anatomy, one volume. English Urban & Fischer

Huangfu J, Liu J, Sun Z, Wang M, Jiang Y, Chen Z-Y et al (2013) Antiaging effects of astaxanthin-rich alga Haematococcus pluvialis on fruit flies under oxidative stress. J Agric Food Chem 61(32):7800–7804

Hwang P-A, Yan M-D, Kuo K-L, Phan NN, Lin Y-C (2017) A mechanism of low molecular weight fucoidans degraded by enzymatic and acidic hydrolysis for the prevention of UVB damage. J Appl Phycol 29(1):521–529

Hyun YJ, Piao MJ, Ko MH, Lee NH, Kang HK, Yoo ES et al (2013) Photoprotective effect of Undaria crenata against ultraviolet B-induced damage to keratinocytes. J Biosci Bioeng 116(2):256–264

Hyun YJ, Piao MJ, Kim KC, Zheng J, Yao CW, Cha JW et al (2014) Photoprotective effect of a Polyopes affinis (Harvey) Kawaguchi and Wang (Halymeniaceae)-derived ethanol extract on human Keratinocytes. Trop J Pharm Res 13(6):863–871

Imokawa G, Yada Y, Morisaki N, Kimura M (1998) Biological characterization of human fibroblast-derived mitogenic factors for human melanocytes. Biochem J 330(3):1235–1239

Infante VHP, Campos PMBGM (2021) Application of factorial design in the development of cosmetic formulations with carrageenan and argan oil. Int J Phytocosmet Nat Ingred 8(1):4

Isaka S, Cho K, Nakazono S, Abu R, Ueno M, Kim D et al (2015) Antioxidant and anti-inflammatory activities of porphyran isolated from discolored nori (Porphyra yezoensis). Int J Biol Macromol 74:68–75

Ito N, Seki S, Ueda F (2018) The protective role of astaxanthin for UV-induced skin deterioration in healthy people—a randomized, double-blind, placebo-controlled trial. Nutrients 10(7):817
Iwasaki T, Watarai S (2020) Oral lambda-carrageenan intake alleviates skin symptoms in a hapten induced atopic dermatitis-like model. J Vet Med Sci 82(11):1639–1642
Jackson H, Braun CL, Ernst H (2008) The chemistry of novel xanthophyll carotenoids. Am J Cardiol 101(10):S50–SS7
Jang J, Ye B-R, Heo S-J, Oh C, Kang D-H, Kim JH et al (2012) Photo-oxidative stress by ultraviolet-B radiation and antioxidative defense of eckstolonol in human keratinocytes. Environ Toxicol Pharmacol 34(3):926–934
Jang SK, Lee DI, Kim ST, Kim GH, Park DW, Park JY et al (2015) The anti-aging properties of a human placental hydrolysate combined with dieckol isolated from Ecklonia cava. BMC Complement Altern Med 15:345
Jannel S, Caro Y, Bermudes M, Petit T (2020) Novel insights into the biotechnological production of Haematococcus pluvialis-derived astaxanthin: advances and key challenges to allow its industrial use as novel food ingredient. J Marine Sci Eng 8(10):789
Jayawardena TU, Sanjeewa KKA, Wang L, Kim WS, Lee TK, Kim YT et al (2020) Alginic acid from Padina boryana Abate particulate matter-induced inflammatory responses in Keratinocytes and dermal fibroblasts. Molecules 25(23):5746
Jesumani V, Du H, Aslam M, Pei PB, Huang N (2019) Potential use of seaweed bioactive compounds in skincare-a review. Mar Drugs 17(12):688
Jesumani V, Du H, Pei P, Aslam M, Huang N (2020) Comparative study on skin protection activity of polyphenol-rich extract and polysaccharide-rich extract from Sargassum vachellianum. PLoS One 15(1):e0227308
Jing RR, Guo KK, Zhong YL, Wang LS, Zhao JG, Gao BY et al (2021) Protective effects of fucoidan purified from Undaria pinnatifida against UV-irradiated skin photoaging. Ann Transl Med 9(14):1185
Joe MJ, Kim SN, Choi HY, Shin WS, Park GM, Kang DW et al (2006) The inhibitory effects of eckol and dieckol from Ecklonia stolonifera on the expression of matrix metalloproteinase-1 in human dermal fibroblasts. Biol Pharm Bull 29(8):1735–1739
Jun ES, Kim YJ, Kim HH, Park SY (2020) Gold nanoparticles using Ecklonia stolonifera protect human dermal fibroblasts from UVA-induced senescence through inhibiting MMP-1 and MMP-3. Mar Drugs 18(9):433
Juturu V, Bowman JP, Deshpande J (2016) Overall skin tone and skin-lightening-improving effects with oral supplementation of lutein and zeaxanthin isomers: a double-blind, placebo-controlled clinical trial. Clin Cosmet Investig Dermatol 9:325
Kadam SU, O'Donnell CP, Rai DK, Hossain MB, Burgess CM, Walsh D et al (2015) Laminarin from Irish Brown Seaweeds Ascophyllum nodosum and Laminaria hyperborea: ultrasound assisted extraction. Charact Bioact Marine Drugs 13(7):4270–4280
Kalasariya HS, Yadav VK, Yadav KK, Tirth V, Algahtani A, Islam S et al (2021) Seaweed-based molecules and their potential biological activities: An eco-sustainable cosmetics. Molecules 26(17):5313
Kang KA, Zhang R, Piao MJ, Ko DO, Wang ZH, Lee K et al (2008) Inhibitory effects of triphlorethol-A on MMP-1 induced by oxidative stress in human keratinocytes via ERK and AP-1 inhibition. J Toxic Environ Health A 71(15):992–999
Kang SY, Kang H, Lee JE, Jo CS, Moon CB, Ha J et al (2020) Antiaging potential of Fucoxanthin concentrate derived from Phaeodactylum tricornutum. J Cosmet Sci 71(2):53–64
Karami MA, Makhmalzadeh BS, Pooranian M, Rezai A (2021) Preparation and optimization of silibinin-loaded chitosan-fucoidan hydrogel: an in vivo evaluation of skin protection against UVB. Pharm Dev Technol 26(2):209–219
Karppi J, Rissanen TH, Nyyssönen K, Kaikkonen J, Olsson AG, Voutilainen S et al (2007) Effects of astaxanthin supplementation on lipid peroxidation. Int J Vitam Nutr Res 77(1):3–11
Katsumura N, Okuno M, Onogi N, Moriwaki H, Muto Y, Kojima S (1996) Suppression of mouse skin papilloma by canthaxanthin and beta-carotene in vivo. Possibility of the regres-

sion of tumorigenesis by carotenoids without conversion to retinoic acid. Nutr Cancer Int J 26(2):203–208

Kavalappa YP, Rudresh DU, Gopal SS, Shivarudrappa AH, Stephen NM, Rangiah K et al (2019) beta-carotene isolated from the marine red alga, Gracillaria sp. potently attenuates the growth of human hepatocellular carcinoma (HepG2) cells by modulating multiple molecular pathways. J Funct Foods 52:165–176

Kidgell JT, Magnusson M, de Nys R, Glasson CRK (2019) Ulvan: a systematic review of extraction, composition and function. Algal Res 39:101422

Kijjoa A, Sawangwong P (2004) Drugs and cosmetics from the sea. Mar Drugs 2(2):73–82

Kim M-M, Kim S-K (2010) Effect of phloroglucinol on oxidative stress and inflammation. Food Chem Toxicol 48(10):2925–2933

Kim Y-O, Lee S-M (2012) Effects of dietary inclusion of Spirulina, astaxanthin, canthaxanthin or paprika on the skin pigmentation of red-and white-colored fancy carp Cyprinus carpio var. koi. Korean J Fish Aquat Sci 45(1):43–49

Kim HH, Cho S, Lee S, Kim KH, Cho KH, Eun HC et al (2006) Photoprotective and anti-skin-aging effects of eicosapentaenoic acid in human skin in vivo. J Lipid Res 47(5):921–930

Kim KN, Heo SJ, Yoon WJ, Kang SM, Ahn G, Yi TH et al (2010) Fucoxanthin inhibits the inflammatory response by suppressing the activation of NF-kappa B and MAPKs in lipopolysaccharide-induced RAW 264 7 macrophages. Eur J Pharmacol 649(1–3):369–375

Kim AR, Lee M-S, Shin T-S, Hua H, Jang B-C, Choi J-S et al (2011) Phlorofucofuroeckol A inhibits the LPS-stimulated iNOS and COX-2 expressions in macrophages via inhibition of NF-κB, Akt, and p38 MAPK. Toxicol In Vitro 25(8):1789–1795

Kim KC, Piao MJ, Zheng J, Yao CW, Cha JW, Kumara MHSR et al (2014a) Fucodiphlorethol G purified from Ecklonia cava suppresses ultraviolet B radiation-induced oxidative stress and cellular damage. Biomol Ther 22(4):301

Kim HM, Lee DE, Park SD, Kim YT, Kim YJ, Jeong JW et al (2014b) Preventive effect of fermented Gelidium amansii and Cirsium japonicum extract mixture against UVB-induced skin photoaging in hairless mice. Food Sci Biotechnol 23(2):623–631

Kim S, You DH, Han T, Choi E-M (2014c) Modulation of viability and apoptosis of UVB-exposed human keratinocyte HaCaT cells by aqueous methanol extract of laver (Porphyra yezoensis). J Photochem Photobiol B Biol 141:301–307

Kim KT, Rioux LE, Turgeon SL (2014d) Alpha-amylase and alpha-glucosidase inhibition is differentially modulated by fucoidan obtained from Fucus vesiculosus and Ascophyllum nodosum. Phytochemistry 98:27–33

Kim J-J, Kang Y-J, Shin S-A, Bak D-H, Lee JW, Lee KB et al (2016) Phlorofucofuroeckol improves glutamate-induced neurotoxicity through modulation of oxidative stress-mediated mitochondrial dysfunction in PC12 cells. PLoS One 11(9):e0163433

Kim JH, Lee JE, Kim KH, Kang NJ (2018) Beneficial effects of marine algae-derived carbohydrates for skin health. Mar Drugs 16(11):459

Kligman D (2000) Cosmeceuticals. Dermatol Clin 18(4):609–615

Ko S-C, Cha S-H, Heo S-J, Lee S-H, Kang S-M, Jeon Y-J (2011) Protective effect of Ecklonia cava on UVB-induced oxidative stress: in vitro and in vivo zebrafish model. J Appl Phycol 23(4):697–708

Komatsu T, Sasaki S, Manabe Y, Hirata T, Sugawara T (2017) Preventive effect of dietary astaxanthin on UVA-induced skin photoaging in hairless mice. PLoS One 12(2):e0171178

Kotake-Nara E, Asai A, Nagao A (2005) Neoxanthin and fucoxanthin induce apoptosis in PC-3 human prostate cancer cells. Cancer Lett 220(1):75–84

Kou PT, Marraiki N, Elgorban AM, Du YW (2020) Fucoxanthin modulates the development of 7, 12-dimethyl benz (a) anthracene-induced skin carcinogenesis in Swiss Albino mice in vivo. Pharmacogn Mag 16(71):681–688

Krinsky NI, Mayne ST, Sies H (2004) Carotenoids in health and disease. CRC Press

Ku MJ, Lee MS, Moon HJ, Lee YH (2008) Antioxidation effects of polysaccharide fucoidan extracted from seaweeds in skin photoaging. FASEB J 22:647

Ku M-J, Lee M-S, Moon H-J, Lee Y-H (2010) Protective effects of fucoidan against UVB-induced oxidative stress in human skin fibroblasts. J Life Sci 20(1):27–32
Kuedo Z, Sangsuriyawong A, Klaypradit W, Tipmanee V, Chonpathompikunlert P (2016) Effects of Astaxanthin from Litopenaeus Vannamei on Carrageenan-induced Edema and pain behavior in mice. Molecules 21(3):382
Lean QY, Eri RD, Fitton JH, Patel RP, Gueven N (2015) Fucoidan extracts Ameliorate Acute Colitis. PLoS One 10(6):e0128453
Lee YH, Moon HJ, Lee SR, Jeong SH, Chang HK, Stonik VA et al (2007) Protective effects of fucoidan against UVB-induced aging in human skin fibroblasts. Wiley Online Library
Lee Y-E, Kim H, Seo C, Park T, Lee KB, Yoo S-Y et al (2017) Marine polysaccharides: therapeutic efficacy and biomedical applications. Arch Pharm Res 40(9):1006–1020
Lee S-S, Bang M-H, Jeon H-J, Hwang T, Yang S-A (2018a) Anti-inflammatory and anti-allergic effects of Phlorofucofuroeckol A and Dieckol isolated from Ecklonia cava. J Life Sci 28(10):1170–1178
Lee JW, Seok JK, Boo YC (2018b) Ecklonia cava extract and Dieckol attenuate cellular lipid peroxidation in Keratinocytes exposed to PM10. Evid Based Complement Alternat Med 2018:27
Lee YJ, Park JH, Park SA, Joo NR, Lee BH, Lee KB et al (2020) Dieckol or phlorofucofuroeckol extracted from Ecklonia cava suppresses lipopolysaccharide-mediated human breast cancer cell migration and invasion. J Appl Phycol 32(1):631–640
Leiro JM, Castro R, Arranz JA, Lamas J (2007) Immunomodulating activities of acidic sulphated polysaccharides obtained from the seaweed Ulva rigida C. Agardh Int Immunopharmacol 7(7):879–888
Li J, Xie L, Qin Y, Liang W-H, Mo M-Q, Liu S-L et al (2013) Effect of laminarin polysaccharide on activity of matrix metalloproteinase in photoaging skin. Zhongguo Zhong yao za zhi= Zhongguo Zhongyao Zazhi= China J Chinese Materia Medica 38(14):2370–2373
Li F-M, Liu Y, Liao J-F, Duan X-L (2018) The preliminary study on anti-photodamaged effect of astaxanthin liposomes in mice skin. Sichuan da xue xue bao Yi xue ban= J Sichuan Univ Med Sci Edit 49(5):712–715
Li Z-y, Yu C-H, Lin Y-T, Su H-L, Kan K-W, Liu F-C et al (2019) The potential application of spring Sargassum glaucescens extracts in the moisture-retention of keratinocytes and dermal fibroblast regeneration after UVA-irradiation. Cosmetics 6(1):17
Li X, Matsumoto T, Takuwa M, Saeed Ebrahim Shaiku Ali M, Hirabashi T, Kondo H et al (2020) Protective effects of astaxanthin supplementation against ultraviolet-induced photoaging in hairless mice. Biomedicines. 8(2):18
Li ZY, Wang YQ, Zhao J, Zhang H (2021a) Dieckol attenuates the nociception and inflammatory responses in different nociceptive and inflammatory induced mice model. Saudi J Biol Sci 28(9):4891–4899
Li A, Zhang L, Veeraraghavan V, Mohan S, Wang JD (2021b) Dieckol attenuates cell proliferation in Molt-4 leukemia cells via modulation of JAK/STAT3 signalling pathway. Pharmacogn Mag 17(73):45–50
Lichota A, Gwozdzinski K (2018) Anticancer activity of natural compounds from plant and marine environment. Int J Mol Sci 19(11):3533
Liu ZW, Gao TH, Yang Y, Meng FX, Zhan FP, Jiang QC et al (2019) Anti-cancer activity of Porphyran and carrageenan from red seaweeds. Molecules 24(23):4286
Lopez-Hortas L, Florez-Fernandez N, Torres MD, Ferreira-Anta T, Casas MP, Balboa EM et al (2021) Applying seaweed compounds in cosmetics, Cosmeceuticals and Nutricosmetics. Mar Drugs 19(10):552
Luning K, Pang SJ (2003) Mass cultivation of seaweeds: current aspects and approaches. J Appl Phycol 15(2–3):115–119
Luthuli S, Wu SY, Cheng Y, Zheng XL, Wu MJ, Tong HB (2019) Therapeutic effects of Fucoidan: a review on recent studies. Mar Drugs 17(9):487
Lyons NM, O'Brien NM (2002) Modulatory effects of an algal extract containing astaxanthin on UVA-irradiated cells in culture. J Dermatol Sci 30(1):73–84

Madub K, Goonoo N, Gimie F, Ait Arsa I, Schonherr H, Bhaw-Luximon A (2021) Green seaweeds ulvan-cellulose scaffolds enhance in vitro cell growth and in vivo angiogenesis for skin tissue engineering. Carbohydr Polym 251:117025

Manabe Y, Takii Y, Sugawara T (2020) Siphonaxanthin, a carotenoid from green algae, suppresses advanced glycation end product-induced inflammatory responses. J Nat Med 74(1):127–134

Manandhar B, Wagle A, Seong SH, Paudel P, Kim HR, Jung HA et al (2019) Phlorotannins with potential anti-Tyrosinase and antioxidant activity isolated from the marine seaweed Ecklonia stolonifera. Antioxidants 8(8):240

Marrot L, Meunier J-R (2008) Skin DNA photodamage and its biological consequences. J Am Acad Dermatol 58(5):S139–SS48

Maruyama H, Tamauchi H, Kawakami F, Yoshinaga K, Nakano T (2015) Suppressive effect of dietary Fucoidan on Proinflammatory immune response and MMP-1 expression in UVB-irradiated mouse skin. Planta Med 81(15):1370–1374

Mathewsroth MM (1982) Anti-tumor activity of Beta-Carotene, Canthaxanthin Phytoene. Oncology 39(1):33–37

Matou S, Helley D, Chabut D, Bros A, Fischer AM (2002) Effect of fucoidan on fibroblast growth factor-2-induced angiogenesis in vitro. Thromb Res 106(4–5):213–221

Matsui MS, Muizzuddin N, Arad S, Marenus K (2003) Sulfated polysaccharides from red microalgae have antiinflammatory properties in vitro and in vivo. Appl Biochem Biotechnol 104(1):13–22

Matsui M, Tanaka K, Higashiguchi N, Okawa H, Yamada Y, Tanaka K et al (2016) Protective and therapeutic effects of fucoxanthin against sunburn caused by UV irradiation. J Pharmacol Sci 132(1):55–64

McCaffrey TA, Falcone DJ, Vicente D, Du B, Consigli S, Borth W (1994) Protection of transforming growth factor β activity by heparin and fucoidan. J Cell Physiol 159(1):51–59

Mercurio DG, Wagemaker TAL, Alves VM, Benevenuto CG, Gaspar LR, Campos PM (2015) In vivo photoprotective effects of cosmetic formulations containing UV filters, vitamins, Ginkgo biloba and red algae extracts. J Photochem Photobiol B Biol 153:121–126

Mhadhebi L, Dellai A, Clary-Laroche A, Ben Said R, Robert J, Bouraoui A (2012) Anti-inflammatory and antiproliferative activities of organic fractions from the Mediterranean Brown Seaweed, Cystoseira Compressa. Drug Dev Res 73(2):82–89

Michalak I (2020) The application of seaweeds in environmental biotechnology. In: Seaweeds around the world: State of Art and perspectives, vol 95, pp 85–111

Michel G, Nyval-Collen P, Barbeyron T, Czjzek M, Helbert W (2006) Bioconversion of red seaweed galactans: a focus on bacterial agarases and carrageenases. Appl Microbiol Biotechnol 71(1):23–33

Millikan LE (2001) Cosmetology, cosmetics, cosmeceuticals: definitions and regulations. Clin Dermatol 19(4):371–374

Mittal M, Siddiqui MR, Tran K, Reddy SP, Malik AB (2014) Reactive oxygen species in inflammation and tissue injury. Antioxid Redox Signal 20(7):1126–1167

Miyamoto KT, Komatsu M, Ikeda H (2014) Discovery of gene cluster for mycosporine-like amino acid biosynthesis from Actinomycetales microorganisms and production of a novel mycosporine-like amino acid by heterologous expression. Appl Environ Microbiol 80(16):5028–5036

Miyashita K, Beppu F, Hosokawa M, Liu XY, Wang SZ (2020) Nutraceutical characteristics of the brown seaweed carotenoid fucoxanthin. Arch Biochem Biophys 686:108364

Moon HJ, Lee SR, Shim SN, Jeong SH, Stonik VA, Rasskazov VA et al (2008) Fucoidan inhibits UVB-induced MMP-1 expression in human skin fibroblasts. Biol Pharm Bull 31(2):284–289

Moon HJ, Lee SH, Ku MJ, Yu BC, Jeon MJ, Jeong SH et al (2009) Fucoidan inhibits UVB-induced MMP-1 promoter expression and down regulation of type I procollagen synthesis in human skin fibroblasts. Eur J Dermatol 19(2):129–134

Morais T, Cotas J, Pacheco D, Pereira L (2021) Seaweeds compounds: An ecosustainable source of cosmetic ingredients? Cosmetics 8(1):8

Moreira A, Cruz S, Marques R, Cartaxana P (2021) The underexplored potential of green macroalgae in aquaculture. Rev Aquac 14(1):5–26
Morelli A, Massironi A, Puppi D, Creti D, Domingo Martinez E, Bonistalli C et al (2019) Development of ulvan-based emulsions containing flavour and fragrances for food and cosmetic applications. Flavour Fragr J 34(6):411–425
Naguib YMA (2000) Antioxidant activities of astaxanthin and related carotenoids. J Agric Food Chem 48(4):1150–1154
Nantel F, Denis D, Gordon R, Northey A, Cirino M, Metters KM et al (1999) Distribution and regulation of cyclooxygenase-2 in carrageenan-induced inflammation. Br J Pharmacol 128(4):853–859
Necas J, Bartosikova L (2013) Carrageenan: a review. Veterinarni Medicina 58(4):187–205
Newman DJ, Cragg GM (2016) Drugs and drug candidates from marine sources: An assessment of the current "state of play". Planta Med 82(09/10):775–789
Ngo D-H, Wijesekara I, Vo T-S, Van Ta Q, Kim S-K (2011) Marine food-derived functional ingredients as potential antioxidants in the food industry: An overview. Food Res Int 44(2):523–529
Nishisgori C (2015) Current concept of photocarcinogenesis. Photochem Photobiol Sci 14(9):1713–1721
Nozari M, Gholizadeh M, Oghani FZ, Tahvildari K (2021) Studies on novel chitosan/alginate and chitosan/bentonite flexible films incorporated with ZnO nano particles for accelerating dermal burn healing: in vivo and in vitro evaluation. Int J Biol Macromol 184:235–249
Obana A, Gohto Y, Nakazawa R, Moriyama T, Gellermann W, Bernstein PS (2020) Effect of an antioxidant supplement containing high dose lutein and zeaxanthin on macular pigment and skin carotenoid levels. Sci Rep 10(1):1–12
Obluchinskaya ED, Pozharitskaya ON, Flisyuk EV, Shikov AN (2021) Formulation, optimization and in vivo evaluation of Fucoidan-based cream with anti-inflammatory properties. Mar Drugs 19(11):643
Odeberg JM, Lignell Å, Pettersson A, Höglund P (2003) Oral bioavailability of the antioxidant astaxanthin in humans is enhanced by incorporation of lipid based formulations. Eur J Pharm Sci 19(4):299–304
Oh S, Son M, Lee HS, Kim HS, Jeon YJ, Byun K (2018) Protective effect of Pyrogallol-Phloroglucinol-6,6-Bieckol from Ecklonia cava on monocyte-associated vascular dysfunction. Mar Drugs 16(11):441
Ojha KS, Tiwari BK (2016) Novel fermented marine-based products. In: Novel Food Fermentation Technologies. Springer, pp 235–622016
O'Leary R, Rerek M, Wood EJ (2004) Fucoidan modulates the effect of transforming growth factor (TGF)-beta(1) on fibroblast proliferation and wound repopulation in in vitro models of dermal wound repair. Biol Pharm Bull 27(2):266–270
Ozanne H, Toumi H, Roubinet B, Landemarre L, Lespessailles E, Daniellou R et al (2020) Laminarin Effects, a β-(1,3)-Glucan, on skin cell inflammation and oxidation. Cosmetics 7(3):66
Palombo P, Fabrizi G, Ruocco V, Ruocco E, Fluhr J, Roberts R et al (2007) Beneficial long-term effects of combined oral/topical antioxidant treatment with the carotenoids lutein and zeaxanthin on human skin: a double-blind, placebo-controlled study. Skin Pharmacol Physiol 20(4):199–210
Pangestuti R, Kim SK (2011) Biological activities and health benefit effects of natural pigments derived from marine algae. J Funct Foods 3(4):255–266
Pangestuti R, Kim S-K (2014) Biological activities of carrageenan. Adv Food Nutr Res 72:113–124
Pangestuti R, Bak S-S, Kim S-K (2011) Attenuation of pro-inflammatory mediators in LPS-stimulated BV2 microglia by chitooligosaccharides via the MAPK signalling pathway. Int J Biol Macromol 49(4):599–606
Pangestuti R, Shin KH, Kim SK (2021) Anti-Photoaging and potential skin health benefits of seaweeds. Mar Drugs 19(3):172
Park EJ, Choi JI (2017) Melanogenesis inhibitory effect of low molecular weight fucoidan from Undaria pinnatifida. J Appl Phycol 29(5):2213–2217

Park JW, Song H-S (2021) Effect of Astaxanthin on anti-inflammatory and anti-oxidative effects of Astaxanthin treatment for atopic dermatitis-induced mice. J Acupunct Res 38(4):293–299
Park JS, Chyun JH, Kim YK, Line LL, Chew BP (2010) Astaxanthin decreased oxidative stress and inflammation and enhanced immune response in humans. Nutr Metabol 7(1):1–10
Park MH, Heo SJ, Kim KN, Ahn G, Park PJ, Moon SH et al (2015) 6,6 '-Bieckol protects insulinoma cells against high glucose-induced glucotoxicity by reducing oxidative stress and apoptosis. Fitoterapia 106:135–140
Park JH, Choi SH, Park SJ, Lee YJ, Park JH, Song PH et al (2017) Promoting wound healing using low molecular weight Fucoidan in a full-thickness dermal excision Rat Model. Mar Drugs 15(4):112
Park JH, Yeo IJ, Han JH, Suh JW, Lee HP, Hong JT (2018) Anti-inflammatory effect of astaxanthin in phthalic anhydride-induced atopic dermatitis animal model. Exp Dermatol 27(4):378–385
Pasquet V, Morisset P, Ihammouine S, Chepied A, Aumailley L, Berard J-B et al (2011) Antiproliferative activity of violaxanthin isolated from bioguided fractionation of Dunaliella tertiolecta extracts. Mar Drugs 9(5):819–831
Peng J, Yuan JP, Wu CF, Wang JH (2011) Fucoxanthin, a marine carotenoid present in Brown seaweeds and diatoms: metabolism and bioactivities relevant to human health. Mar Drugs 9(10):1806–1828
Perez MJ, Falque E, Dominguez H (2016) Antimicrobial action of compounds from marine seaweed. Mar Drugs 14(3):52
Piao MJ, Zhang R, Lee NH, Hyun JW (2012a) Protective effect of triphlorethol-A against ultraviolet B-mediated damage of human keratinocytes. J Photochem Photobiol B Biol 106:74–80
Piao MJ, Kang HK, Yoo ES, Koh YS, Kim DS, Lee NH et al (2012b) Photo-protective effect of Polysiphonia morrowii Harvey against ultraviolet B radiation-induced Keratinocyte damage. J Korean Soc Appl Biol Chem 55(2):149–158
Piao MJ, Hyun YJ, Oh TH, Kang HK, Yoo ES, Koh YS et al (2012c) Chondracanthus tenellus (Harvey) hommersand extract protects the human keratinocyte cell line by blocking free radicals and UVB radiation-induced cell damage. In Vitro Cell Dev Biol Anim 48(10):666–674
Piao MJ, Hyun YJ, Cho SJ, Kang HK, Yoo ES, Koh YS et al (2012d) An ethanol extract derived from Bonnemaisonia hamifera scavenges ultraviolet B (UVB) radiation-induced reactive oxygen species and attenuates UVB-induced cell damage in human Keratinocytes. Mar Drugs 10(12):2826–2845
Pilkington SM, Gibbs NK, Costello P, Bennett SP, Massey KA, Friedmann PS et al (2016) Effect of oral eicosapentaenoic acid on epidermal Langerhans cell numbers and PGD 2 production in UVR-exposed human skin: a randomised controlled study. Exp Dermatol 25(12):962–968
Pimentel FB, Alves RC, Rodrigues F, Oliveira PP, MB. (2018) Macroalgae-derived ingredients for cosmetic industry—An update. Cosmetics 5(1):2
Prasedya ES, Syafitri SM, Geraldine BAFD, Hamdin CD, Frediansyah A, Miyake M et al (2019) UVA Photoprotective activity of Brown macroalgae Sargassum cristafolium. Biomedicines 7(4):77
Pyeon DB, Lee SE, Yoon JW, Park HJ, Park CO, Kim SH et al (2021) The antioxidant dieckol reduces damage of oxidative stress-exposed porcine oocytes and enhances subsequent parthenotes embryo development. Mol Reprod Dev 88(5):349–361
Qiu SM, Aweya JJ, Liu XJ, Liu Y, Tang SJ, Zhang WC et al (2022) Bioactive polysaccharides from red seaweed as potent food supplements: a systematic review of their extraction, purification, and biological activities. Carbohydr Polym 275:118696
Rahman MM, Kundu JK, Shin J-W, Na H-K, Surh Y-J (2011) Docosahexaenoic acid inhibits UVB-induced activation of NF-κB and expression of COX-2 and NOX-4 in HR-1 hairless mouse skin by blocking MSK1 signalling. PLoS One 6(11):e28065
Rangel KC, Villela LZ, Pereira KD, Colepicolo P, Debonsi HM, Gaspar LR (2020) Assessment of the photoprotective potential and toxicity of Antarctic red macroalgae extracts from Curdiea racovitzae and Iridaea cordata for cosmetic use. Algal Res Biomass Biofuels Bioproducts 50:101984

Rani V, Prabhu A, Venkatesan J, Kim S-K (2021) Seaweed Polysaccharides: promising molecules for biotechnological applications. In: Reference module in chemistry, molecular sciences and chemical engineering
Rao AR, Sindhuja HN, Dharmesh SM, Sankar KU, Sarada R, Ravishankar GA (2013) Effective inhibition of skin cancer, tyrosinase, and antioxidative properties by astaxanthin and astaxanthin esters from the green alga Haematococcus pluvialis. J Agric Food Chem 61(16):3842–3851
Raposo MFD, de Morais AMB, de Morais RMSC (2015) Marine polysaccharides from algae with potential biomedical applications. Mar Drugs 13(5):2967–3028
Rastogi RP, Kumar A, Tyagi MB, Sinha RP (2010) Molecular mechanisms of ultraviolet radiation-induced DNA damage and repair. J Nucleic Acids 2010:592980
Ren S-W, Li J, Wang W, Guan H-S (2010) Protective effects of κ-ca3000+ CP against ultraviolet-induced damage in HaCaT and MEF cells. J Photochem Photobiol B Biol 101(1):22–30
Renn D (1997) Biotechnology and the red seaweed polysaccharide industry: status, needs and prospects. Trends Biotechnol 15(1):9–14
Resinski R (1998) Cosmos and cosmetics: constituting an adorned female body in ancient Greek literature. University of California, Los Angeles
Rioux LE, Turgeon SL, Beaulieu M (2007) Characterization of polysaccharides extracted from brown seaweeds. Carbohydr Polym 69(3):530–537
Roberts RL (2013) Lutein, zeaxanthin, and skin health. Am J Lifestyle Med 7(3):182–185
Roberts RL, Green J, Lewis B (2009) Lutein and zeaxanthin in eye and skin health. Clin Dermatol 27(2):195–201
Rodriguez-Luna A, Avila-Roman J, Gonzalez-Rodriguez ML, Cozar MJ, Rabasco AM, Motilva V et al (2018) Fucoxanthin-containing cream prevents epidermal hyperplasia and UVB-induced skin erythema in mice. Mar Drugs 16(10):378
Rodriguez-Luna A, Avila-Roman J, Oliveira H, Motilva V, Talero E (2019) Fucoxanthin and Rosmarinic acid combination has anti-inflammatory effects through regulation of NLRP3 Inflammasome in UVB-exposed HaCaT Keratinocytes. Mar Drugs 17(8):451
Rokkaku T, Kimura R, Ishikawa C, Yasumoto T, Senba M, Kanaya F et al (2013) Anticancer effects of marine carotenoids, fucoxanthin and its deacetylated product, fucoxanthinol, on osteosarcoma. Int J Oncol 43(4):1176–1186
Rui Y, Zhaohui Z, Wenshan S, Bafang L, Hu H (2019) Protective effect of MAAs extracted from Porphyra tenera against UV irradiation-induced photoaging in mouse skin. J Photochem Photobiol B Biol 192:26–33
Rukmanikrishnan B, Rajasekharan SK, Lee J, Ramalingam S, Lee J (2020) K-Carrageenan/lignin composite films: biofilm inhibition, antioxidant activity, cytocompatibility, UV and water barrier properties. Mater Today Commun 24:101346
Sachindra NM, Sato E, Maeda H, Hosokawa M, Niwano Y, Kohno M et al (2007) Radical scavenging and singlet oxygen quenching activity of marine carotenoid fucoxanthin and its metabolites. J Agric Food Chem 55(21):8516–8522
Saewan N, Jimtaisong A (2015) Natural products as photoprotection. J Cosmet Dermatol 14(1):47–63
Sambandan DR, Ratner D (2011) Sunscreens: an overview and update. J Am Acad Dermatol 64(4):748–758
Sangeetha RK, Bhaskar N, Baskaran V (2009) Comparative effects of beta-carotene and fucoxanthin on retinol deficiency induced oxidative stress in rats. Mol Cell Biochem 331(1–2):59–67
Sarithakumari CH, Renju GL, Kurup GM (2013) Anti-inflammatory and antioxidant potential of alginic acid isolated from the marine algae, Sargassum wightii on adjuvant-induced arthritic rats. Inflammopharmacology 21(3):261–268
Satoh A, Tsuji S, Okada Y, Murakami N, Urami M, Nakagawa K et al (2009) Preliminary clinical evaluation of toxicity and efficacy of a new astaxanthin-rich Haematococcus pluvialis extract. J Clin Biochem Nutr 44(3):280–284
Schmid D, Schürch C, Zülli F (2006) Mycosporine-like amino acids from red algae protect against premature skin-aging. Euro Cosmet 9:1–4

Sellimi S, Maalej H, Rekik DM, Benslima A, Ksouda G, Hamdi M et al (2018) Antioxidant, antibacterial and in vivo wound healing properties of laminaran purified from Cystoseira barbata seaweed. Int J Biol Macromol 119:633–644

Senthilkumar K, Manivasagan P, Venkatesan J, Kim SK (2013) Brown seaweed fucoidan: biological activity and apoptosis, growth signalling mechanism in cancer. Int J Biol Macromol 60:366–374

SEPPIC Ingredients and formulas. Available online: https://www.seppic.com/en/wesource/codiavelane-bg-pf. Accessed date 9 Feb 2022a. https://www.seppic.com/en/wesource/codiavelane-bg-pf

SEPPIC Ingredients and formulas. Available online: https://www.seppic.com/en/wesource/bioenergizertm-p-bg-pf. Accessed date 9 Feb 2022b.

SEPPIC Ingredients and formulas. Available online: https://www.seppic.com/en/wesource/asparage. Accessed date 9 Feb 2022c.

Shao P, Shao J, Han L, Lv R, Sun P (2015) Separation, preliminary characterization, and moisture-preserving activity of polysaccharides from Ulva fasciata. Int J Biol Macromol 72:924–930

Sherrow V (2001) For appearance'sake: the historical encyclopedia of good looks, beauty, and grooming. Greenwood Publishing Group

Shiao WC, Kuo CH, Tsai YH, Hsieh SL, Kuan AW, Hong YH et al (2020) In vitro evaluation of anti-colon cancer potential of crude extracts of Fucoidan obtained from Sargassum Glaucescens pretreated by compressional-puffing. Appl Sci Basel 10(9):3058

Shimoda H, Tanaka J, Shan SJ, Maoka T (2010) Anti-pigmentary activity of fucoxanthin and its influence on skin mRNA expression of melanogenic molecules. J Pharm Pharmacol 62(9):1137–1145

Silke DS, Wilhelm S (2007) Astaxanthin and canthaxanthin in UV-protection: effects on gap junctional communication. Ann Nutr Metab 51:224

Silva J, Alves C, Pinteus S, Mendes S, Pedrosa R (2018) Neuroprotective effects of seaweeds against 6-hydroxidopamine-induced cell death on an in vitro human neuroblastoma model. BMC Complement Altern Med 18:58

Singh SP, Kumari S, Rastogi RP, Singh KL, Sinha RP (2008) Mycosporine-like amino acids (MAAs): chemical structure, biosynthesis and significance as UVAbsorbing/screening compounds. Indian J Exp Biol 46(1):7–17

Singh TK, Tiwari P, Singh CS, Prasad RK (2013) Cosmeceuticals: enhance the health and beauty of the skin. World J Pharm Res 2:1475–1485

Sinha RP, Singh SP, Häder D-P (2007) Database on mycosporines and mycosporine-like amino acids (MAAs) in fungi, cyanobacteria, macroalgae, phytoplankton and animals. J Photochem Photobiol B Biol 89(1):29–35

Slominski A, Wortsman J, Paus R, Elias PM, Tobin DJ, Feingold KR (2008) Skin as an endocrine organ: implications for its function. Drug Discov Today Dis Mech 5(2):e137–ee44

Solano F (2020) Photoprotection and skin pigmentation: melanin-related molecules and some other new agents obtained from natural sources. Molecules 25(7):1537

Solovieva EV, Teterina AY, Klein OI, Komlev VS, Alekseev AA, Panteleyev AA (2021) Sodium alginate-based composites as a collagen substitute for skin bioengineering. Biomed Mater 16(1):015002

Song YS, Li HL, Balcos MC, Yun HY, Baek KJ, Kwon NS et al (2014) Fucoidan promotes the reconstruction of skin equivalents. Korean J Physiol Pharmacol 18(4):327–331

Song JH, Piao MJ, Han X, Kang KA, Kang HK, Yoon WJ et al (2016) Anti-wrinkle effects of Sargassum muticum ethyl acetate fraction on ultraviolet B-irradiated hairless mouse skin and mechanistic evaluation in the human HaCaT keratinocyte cell line. Mol Med Rep 14(4):2937–2944

Soontornchaiboon W, Joo SS, Kim SM (2012) Anti-inflammatory effects of violaxanthin isolated from microalga Chlorella ellipsoidea in RAW 264.7 macrophages. Biol Pharm Bull 35(7):1137–1144

Spiller GA, Dewell A (2003) Safety of an astaxanthin-rich Haematococcus pluvialis algal extract: a randomized clinical trial. J Med Food 6(1):51–56

Sugiura Y, Matsuda K, Yamada Y, Nishikawa M, Shioya K, Katsuzaki H et al (2006) Isolation of a new anti-allergic phlorotannin, phlorofucofuroeckol-B, from an edible brown alga, Eisenia arborea. Biosci Biotechnol Biochem 70(11):2807–2811

Suh S-S, Hwang J, Park M, Seo HH, Kim H-S, Lee JH et al (2014) Anti-inflammation activities of mycosporine-like amino acids (MAAs) in response to UV radiation suggest potential anti-skin aging activity. Mar Drugs 12(10):5174–5187

Surowiak P, Gansukh T, Donizy P, Halon A, Rybak Z (2014) Increase in cyclooxygenase-2 (COX-2) expression in keratinocytes and dermal fibroblasts in photoaged skin. J Cosmet Dermatol 13(3):195–201

Tang JL, Wang WQ, Chu WH (2020) Antimicrobial and anti-quorum sensing activities of Phlorotannins from seaweed (Hizikia fusiforme). Front Cell Infect Microbiol 10:586750

Tavares RSN, Maria-Engler SS, Colepicolo P, Debonsi HM, Schafer-Korting M, Marx U et al (2020a) Skin irritation testing beyond tissue viability: Fucoxanthin effects on inflammation, homeostasis, and metabolism. Pharmaceutics 12(2):136

Tavares RSN, Kawakami CM, Pereira KD, do Amaral GT, Benevenuto CG, Maria-Engler SS et al (2020b) Fucoxanthin for topical administration, a phototoxic vs. photoprotective potential in a tiered strategy assessed by in vitro methods. Antioxidants 9(4):328

Thevanayagam H, Mohamed SM, Chu W-L (2014) Assessment of UVB-photoprotective and anti-oxidative activities of carrageenan in keratinocytes. J Appl Phycol 26(4):1813–1821

Tominaga K, Hongo N, Karato M, Yamashita E (2012) Cosmetic benefits of astaxanthin on humans subjects. Acta Biochim Pol 59(1):43–47

Tominaga K, Hongo N, Fujishita M, Takahashi Y, Adachi Y (2017) Protective effects of astaxanthin on skin deterioration. J Clin Biochem Nutr 61(1):33–39

Tripp CS, Blomme EAG, Chinn KS, Hardy MM, LaCelle P, Pentland AP (2003) Epidermal COX-2 induction following ultraviolet irradiation: suggested mechanism for the role of COX-2 inhibition in photoprotection. J Invest Dermatol 121(4):853–861

Tronnier H (1984) Protective effect of beta-carotene and canthaxanthin against UV reactions of the skin. Zeitschrift fur Hautkrankheiten 59(13):859–870

Tsukahara H, Matsuyama A, Abe T, Kyo H, Ohta T, Suzuki N (2016) Effects of intake of astaxanthin contained drink on skin condition. Jpn J Complement Altern Med 13(2):57–62

Turck D, Castenmiller J, de Henauw S, Hirsch-Ernst KI, Kearney J, Maciuk A et al (2020) Safety of astaxanthin for its use as a novel food in food supplements. EFSA J 18(2):e05993

Urikura I, Sugawara T, Hirata T (2011) Protective effect of Fucoxanthin against UVB-induced skin photoaging in hairless mice. Biosci Biotech Bioch 75(4):757–760

Verdy C, Branka JE, Mekideche N (2011) Quantitative assessment of lactate and progerin production in normal human cutaneous cells during normal ageing: effect of an Alaria esculenta extract. Int J Cosmet Sci 33(5):462–466

Visioli F, Artaria C (2017) Astaxanthin in cardiovascular health and disease: mechanisms of action, therapeutic merits, and knowledge gaps. Food Funct 8(1):39–63

Vo T-S, Kim S-K (2013) Fucoidans as a natural bioactive ingredient for functional foods. J Funct Foods 5(1):16–27

Vo TS, Kim SK, Ryu B, Ngo DH, Yoon NY, Bach LG et al (2018) The suppressive activity of Fucofuroeckol-A derived from Brown algal Ecklonia stolonifera Okamura on UVB-induced mast cell degranulation. Mar Drugs 16(1):1

Wang J, Jin W, Hou Y, Niu X, Zhang H, Zhang Q (2013) Chemical composition and moisture-absorption/retention ability of polysaccharides extracted from five algae. Int J Biol Macromol 57:26–29

Wang H-MD, Chen C-C, Huynh P, Chang J-S (2015) Exploring the potential of using algae in cosmetics. Bioresour Technol 184:355–362

Wang W, Wu JD, Zhang XS, Hao C, Zhao XL, Jiao GL et al (2017) Inhibition of Influenza A virus infection by Fucoidan targeting viral neuraminidase and cellular EGFR pathway. Sci Rep 7:40760

Wang L, Lee W, Oh JY, Cui YR, Ryu B, Jeon Y-J (2018) Protective effect of sulfated polysaccharides from celluclast-assisted extract of Hizikia fusiforme against ultraviolet B-induced skin damage by regulating NF-κB, AP-1, and MAPKs signalling pathways in vitro in human dermal fibroblasts. Mar Drugs 16(7):239

Wang L, Oh JY, Yang H-W, Kim HS, Jeon Y-J (2019) Protective effect of sulfated polysaccharides from a Celluclast-assisted extract of Hizikia fusiforme against ultraviolet B-induced photoaging in vitro in human keratinocytes and in vivo in zebrafish. Marine Life Sci Technol 1(1):104–111

Wang L, Lee W, Jayawardena TU, Cha SH, Jeon YJ (2020a) Dieckol, an algae-derived phenolic compound, suppresses airborne particulate matter-induced skin aging by inhibiting the expressions of pro-inflammatory cytokines and matrix metalloproteinases through regulating NF-kappa B, AP-1, and MAPKs signalling pathways. Food Chem Toxicol 146:111823

Wang L, Kim HS, Oh JY, Je JG, Jeon Y-J, Ryu B (2020b) Inhibitory effect of diphlorethohydroxycarmalol on melanogenesis and its protective effect against UVB radiation-induced cell damage. Food Chem Toxicol 136:110963

Wang L, Jayawardena TU, Yang HW, Lee HG, Jeon YJ (2020c) The potential of sulfated polysaccharides isolated from the Brown SeaweedEcklonia maximain cosmetics: antioxidant, anti-melanogenesis, and Photoprotective activities. Antioxidants. 9(8):724

Wang L, Oh JY, Kim YS, Lee HG, Lee JS, Jeon YJ (2020d) Anti-Photoaging and anti-Melanogenesis effects of Fucoidan isolated fromHizikia fusiformeand its underlying mechanisms. Mar Drugs 18(8):427

Wang L, Je JG, Yang HW, Jeon YJ, Lee S (2021) Dieckol, an algae-derived phenolic compound, suppresses UVB-induced skin damage in human dermal fibroblasts and its underlying mechanisms. Antioxidants 10(3):352

Wijesekara I, Pangestuti R, Kim SK (2011) Biological activities and potential health benefits of sulfated polysaccharides derived from marine algae. Carbohydr Polym 84(1):14–21

Wiraguna AAGP, Pangkahila W, Astawa INM (2018) Antioxidant properties of topical Caulerpa sp. extract on UVB-induced photoaging in mice. Dermatol Reports 10(2):7597

Xiao WM, Liu HY, Lei Y, Gao HW, Alahmadi TA, Peng HT et al (2021) Chemopreventive effect of dieckol against 7,12-dimethylbenz(a)anthracene induced skin carcinogenesis model by modulatory influence on biochemical and antioxidant biomarkers. Environ Toxicol 36(5):800–810

Xu JW, Yan Y, Wang L, Wu D, Ye NK, Chen SH et al (2021) Marine bioactive compound dieckol induces apoptosis and inhibits the growth of human pancreatic cancer cells PANC-1. J Biochem Mol Toxicol 35(2):e22648

Yamashita E (2005) The effects of a dietary supplement containing astaxanthin on skin condition. Food Style 9(9):72

Yang JH (2012) Topical application of fucoidan improves atopic dermatitis symptoms in NC/Nga mice. Phytother Res 26(12):1898–1903

Yang YI, Jung SH, Lee KT, Choi JH (2014) 8,8 '-Bieckol, isolated from edible brown algae, exerts its anti-inflammatory effects through inhibition of NF-kappa B signalling and ROS production in LPS-stimulated macrophages. Int Immunopharmacol 23(2):460–468

Yang G, Oh JW, Lee HE, Lee BH, Lim KM, Lee JY (2016) Topical application of Dieckol Ameliorates Atopic Dermatitis in NC/Nga mice by suppressing Thymic stromal Lymphopoietin production. J Invest Dermatol 136(5):1062–1066

Yang BX, Li Y, Yang ZB, Xue LG, Zhang MQ, Chen GL et al (2020) Anti-inflammatory and anti-cell proliferative effects of dieckol in the prevention and treatment of colon cancer induced by 1,2-dimethyl hydrazine in experimental animals. Pharmacogn Mag 16(72):851–858

Ye Y, Ji D, You L, Zhou L, Zhao Z, Brennan C (2018) Structural properties and protective effect of Sargassum fusiforme polysaccharides against ultraviolet B radiation in hairless Kun Ming mice. J Funct Foods 43:8–16

Yin H, Niki E, Uchida K (2015) Special issue on "Recent progress in lipid peroxidation based on novel approaches". Free Radic Res 49(7):813–815

Ying R, Zhang Z, Zhu H, Li B, Hou H (2019) The protective effect of mycosporine-like amino acids (MAAs) from Porphyra yezoensis in a mouse model of UV irradiation-induced photoaging. Mar Drugs 17(8):470

Yoo HJ, You DJ, Lee KW (2019) Characterization and immunomodulatory effects of high molecular weight Fucoidan fraction from the Sporophyll of Undaria pinnatifida in cyclophosphamide-induced immunosuppressed mice. Mar Drugs 17(8):447

Yoon NY, Eom T-K, Kim M-M, Kim S-K (2009) Inhibitory effect of phlorotannins isolated from Ecklonia cava on mushroom tyrosinase activity and melanin formation in mouse B16F10 melanoma cells. J Agric Food Chem 57(10):4124–4129

Yoon H-S, Cho HH, Cho S, Lee S-R, Shin M-H, Chung JH (2014) Supplementing with dietary astaxanthin combined with collagen hydrolysate improves facial elasticity and decreases matrix metalloproteinase-1 and-12 expression: a comparative study with placebo. J Med Food 17(7):810–816

Yoshihisa Y, Rehman MU, Shimizu T (2014) Astaxanthin, a xanthophyll carotenoid, inhibits ultraviolet-induced apoptosis in keratinocytes. Exp Dermatol 23(3):178–183

Yoshihisa Y, Andoh T, Matsunaga K, Rehman MU, Maoka T, Shimizu T (2016) Efficacy of astaxanthin for the treatment of atopic dermatitis in a murine model. PLoS One 11(3):e0152288

Zayed A, El-Aasr M, Ibrahim ARS, Ulber R (2020) Fucoidan characterization: determination of purity and physicochemical and chemical properties. Mar Drugs 18(11):571

Zhang H, Tang YB, Zhang Y, Zhang SF, Qu J, Wang X et al (2015) Fucoxanthin: a promising medicinal and nutritional ingredient. Evid Based Complement Alternat Med 2015:723515

Zhang MY, Guo J, Hu XM, Zhao SQ, Li SL, Wang J (2019) An in vivo anti-tumor effect of eckol from marine brown algae by improving the immune response. Food Funct 10(7):4361–4371

Zhang XJ, Li WJ, Dou X, Nan D, He GH (2020) Astaxanthin encapsulated in biodegradable calcium alginate microspheres for the treatment of hepatocellular carcinoma in vitro. Appl Biochem Biotechnol 191(2):511–527

Zheng J, Plao MJ, Keum YS, Kim HS, Hyun JW (2013) Fucoxanthin protects cultured human Keratinocytes against oxidative stress by blocking free radicals and inhibiting apoptosis. Biomol Ther 21(4):270–276

Zheng J, Hewage SRKM, Piao MJ, Kang KA, Han X, Kang HK et al (2016) Photoprotective effect of Carpomitra costata extract against ultraviolet B-induced oxidative damage in human Keratinocytes. J Environ Pathol Toxicol Oncol 35(1):11–28

Zheng JW, Manabe Y, Sugawara T (2020) Siphonaxanthin, a carotenoid from green algae Codium cylindricum, protects Ob/Ob mice fed on a high-fat diet against lipotoxicity by ameliorating somatic stresses and restoring anti-oxidative capacity. Nutr Res 77:29–42

Zhou X, Cao Q, Orfila C, Zhao J, Zhang L (2021) Systematic review and meta-analysis on the effects of astaxanthin on human skin ageing. Nutrients 13(9):2917

Zhu ZX, Zhang QB, Chen LH, Ren S, Xu PX, Tang Y et al (2010) Higher specificity of the activity of low molecular weight fucoidan for thrombin-induced platelet aggregation. Thromb Res 125(5):419–426

Chapter 12
Sustainable and Cost-Effective Management of Residual Aquatic Seaweed Biomass. Business Opportunity for Seaweeds Biorifineries

Monica Trif, Alexandru Vasile Rusu, Touria Ould Bellahcen, Ouafa Cherifi, and Maryam El Bakali

1 Introduction

Managing residual aquatic seaweed biomass in a sustainable and cost-effective manner requires a comprehensive approach that considers both environmental considerations and economic feasibility (Farghali et al. 2023). Utilization of residual aquatic seaweed biomass is an active area of research, and the specific applications and technologies may continue to evolve as further studies are conducted. Residual aquatic seaweed biomass refers to the remaining organic matter of seaweed after various processes such as harvesting, processing, or utilization. It represents the portion of seaweed biomass that is not used for immediate applications or products (Cotas et al. 2023). The residual aquatic seaweed biomass can be further

M. Trif
Food Research Department, Centre for Innovative Process Engineering GmbH, Stuhr, Germany

A. V. Rusu
Strategic Research Department, Biozoon Food Innovations GmbH, Bremerhaven, Germany

T. O. Bellahcen
Health and Environment Laboratory, Faculty of Sciences Ain Chock, Hassan II University of Casablanca, Casablanca, Morocco

O. Cherifi
Laboratory of Water Biodiversity and Climate Changes, Cadi Ayyad University, Marrakech, Morocco

National Center for Studies and Research on Water and Energy, Cadi Ayyad University, Marrakech, Morocco

M. El Bakali (✉)
Research Team in BV2MAP, FPL, Adelmalek Essaadi University, Tetouan, Morocco
e-mail: m.elbakali@uae.ac.ma

F. Ozogul et al. (eds.), *Seaweeds and Seaweed-Derived Compounds*,
https://doi.org/10.1007/978-3-031-65529-6_12

categorized into different fractions based on their composition and characteristics. These fractions may include the following: (a) *residue after processing*: After seaweed is harvested and processed to extract specific compounds or compounds of interest, there may be residual biomass left. For example, if seaweed is used to extract alginate, carrageenan, or other functional compounds, the remaining biomass after extraction is considered residual biomass. (b) *non-commercial species or parts*: Some seaweed species may not have significant commercial value due to their lower nutritional or functional properties. In such cases, these species or specific parts of the seaweed, such as the holdfast or stipe, may be considered residual biomass. (c) *byproducts or waste streams*: During the processing of seaweed, certain byproducts or waste streams may be generated. These could include parts of the seaweed that are not suitable for direct use, such as small or damaged fragments, as well as process residues like wash water or spent processing solutions.

Common strategies and practices that can be employed for the management of residual seaweed biomass could be listed as following:

I. Collection and Harvesting:

- Selective harvesting: Implement selective harvesting techniques to ensure only mature seaweed is collected, allowing the younger seaweed to continue growing and maintain the ecosystem's balance.
- Manual or mechanical harvesting: Utilize manual or mechanical methods to collect seaweed biomass efficiently and reduce labor costs.
- Spatial planning: Designate specific areas for seaweed cultivation and harvesting to minimize negative impacts on other marine organisms and ecosystems.

II. Resource Utilization:

- Food and feed applications: Seaweed biomass can be used as a food source for human consumption or processed into animal feed, maximizing its value and reducing waste.
- Bioenergy production: Seaweed biomass can be converted into biofuels, such as biogas or bioethanol, through anaerobic digestion or fermentation processes.
- Bioplastics and bio-based materials: Seaweed contains natural polymers that can be used in the production of biodegradable plastics and other bio-based materials, reducing reliance on fossil fuel-derived products.

III. Waste Management:

- Composting and fertilizer production: Seaweed biomass can be composted or processed into organic fertilizers, providing valuable nutrients for agricultural purposes.
- Soil erosion control: Use seaweed biomass as a natural mulch or ground cover to prevent soil erosion, particularly in coastal areas.

- Phytoremediation: Seaweed has the ability to absorb and remove pollutants from water, making it useful for phytoremediation projects to improve water quality (Rahhou et al. 2023).

IV. Research and Innovation:

- Collaboration and partnerships: Encourage collaboration between researchers, industry stakeholders, and policymakers to develop innovative solutions and exchange knowledge on sustainable seaweed biomass management.
- Technological advancements: Invest in research and development to improve cultivation techniques, optimize processing methods, and explore new applications for seaweed biomass.

V. Regulatory Support:

- Government incentives: Provide financial incentives, grants, or tax breaks to support the development of sustainable seaweed biomass management practices and encourage private sector engagement.
- Policy frameworks: Establish clear regulations and guidelines that promote sustainable seaweed cultivation, harvesting, and processing while safeguarding marine ecosystems and coastal communities.

By adopting such strategies, stakeholders can achieve a balance between sustainable management of residual seaweed biomass and cost-effective utilization, ensuring the long-term viability of seaweed as a valuable resource.

The inherent advantage of feedstock biopower and biomaterials production in the marine environment is that it does not compete with food production for land or fresh-water. From this marine ecosystem, macroalgae have been widely used already for centuries because they have high growth rate and have the potential to partly replace terrestrial biomass. Additionally, they are used as feedstock for different kind of materials used in different fields (Davis et al. 2003; Rioux et al. 2007; Abenavoli et al. 2016; Álvarez-Viñas et al. 2019; Mæhre et al. 2014). The global seaweed market has been experiencing significant growth in recent years due to the increasing demand for seaweed-based products in various industries. Currently, the global market for algae is approximately U.S. $782.9 million, which is expected to reach nearly U.S. $1.2 billion by 2027 (Gomez-Zavaglia et al. 2019). Algal cultivation is encouraged worldwide to meet the requirement of renewable biomass source for industrial implementation and for the production of sustainable food, feed, pharmaceutical products and energy. These valuable products can be a way to promising low-carbon economy. Moreover, there are numerous benefits associated with algae such as no additional freshwater or fertilizer is required for their cultivation, and even their production method does not occupy any extra agricultural space or terrestrial land (Duan et al. 2002). However, it generates leftovers and waste products. In order to maximize the biomass conversion efficiency and to reduce the amount of raw material needed within a nearly-zero waste flow is created. Biorefinery is a way to decrease negative impact on environment and create products with higher added value to get more economical and environmental benefits (Kostas et al. 2016a; Balina et al. 2017; Maneein et al. 2018).

With respect to algal bioactive compound, they are sensitive to extraction techniques based on the use of heat or solvent. There is a need to identify and develop new efficient extraction processes to better utilize the bioactive compounds present in seaweed. To this end, recent research has focused on the development of novel extraction techniques that are more efficient in terms of yield, time, and cost as well as being environmentally friendly and better able to preserve the activity of target compounds. Extraction technologies that use ultrasound, microwaves, enzymes, supercritical fluid, and pressurized liquid have been researched for extraction of bioactive compounds for Food and pharmaceutical applications (Kadam et al. 2013). On this basis, we embarked on a review of the literature entailing seaweeds in biorefinery and the eco-friendly ways to utilize all components of this biomass.

2 Chemical Composition of Seaweeds and the Most Useful Techniques for Their Extraction

Seaweeds, also called marine macroalgae, considered to be an excellent natural source of primary and secondary metabolites that could lead to the development of innovative food and novel compounds with a diverse array of biological activities (Álvarez-Viñas et al. 2019). It has been classified into three taxonomical groups according to their pigments: green, brown and red algae corresponding to Chlorophyta, Phaeophyta and Rhodophyta, respectively. According to some authors, Kostas et al. (2016b), it is essential to characterize the entire composition of the starting material in order to develop an effective biorefinery process, for any feedstock. In this way, for example, Maneein et al. (2018) highlight the importance of the structural and chemical composition of seaweed in order to enhance biofuel production. However, it is well known that the components in amounts vary greatly with species and environmental conditions (Fleurence 1999; Phillips and Williams 2009; Terasaki et al. 2009; Nomura et al. 2013; Álvarez-Viñas et al. 2019; Baghel et al. 2020; Torres et al. 2019). In addition, Thallus segments of macroalgae are also involved in these amounts variability (Tabassum et al. 2018; Øverland et al. 2019).

2.1 *Moisture and Salt*

Seaweeds have a higher water and salt contents than most terrestrial plants (Milledge and Harvey 2016a, b). However, the same authors highlight the negative impact of the high percentage of water content of macroalgae on the energy balance of applications that depend on dry biomass. This makes seaweed suitable for processes that can produce net energy gain based on the use of wet biomass but undesirable for direct combustion, pyrolysis, and gasification as mentioned by Horn et al. (2000) and Murphy et al. (2013).

2.2 *Polysaccharides*

Polysaccharides are found in abundance in red, green, and brown algae (Pérez et al. 2016). Cell walls are composed of various types. Agar, alginates, galactans, carrageenans, laminarans, fucoidan, and ulvans are the most common polysaccharides found in algae. Algal macromolecules are made up of different monosaccharides connected by glucosidic linkages, and some feature linear backbones with repeating disaccharide units (Vera et al. 2011).

A diverse array of different carbohydrate polymers, the most dominant component in seaweeds, exists between the three groups. For instance, green seaweeds mainly contain those found in terrestrial plants like cellulose, which is a crystalline polymer that comes in two forms: a-cellulose, which has a one chain triclinic structure, and b-cellulose, which has a two-chain monoclinic structure (Doh et al. 2020). They contain also monosulfated polysaccharides such as ulvan, sulfatedrhamnan and galactan most dominant in Ulva, Monostroma and Codium genera, respectively (Cho and You 2015). For the other ones, they contain polysaccharides which are unique. For example, the most predominant polysaccharide found in Rhodophyta is galactan which forms a network with cellulose that constitutes the cell wall matrix according to Hoek et al. (1995). Carrageenan describes a group of sulphated galactans that are linear in form and water soluble and have a basic structure. Additionally, laminarin and alginate are polysaccharides that are only present in brown seaweeds. Polysaccharides such as agar, carrageenan and alginates are of interest to many industrial sectors due to their physico-chemical, gelling or stabilizing properties. Different industries such as agri-food, cosmetics, pharmaceutical and textile have used them in a number of ways (Alba and Kontogiorgos 2018; Azizi et al. 2014; González-López et al. 2012; Namvar et al. 2014).

In contrast to terrestrial plants, seaweeds contain very little or no lignin in their composition, which makes lignocellulyticbiomass, based processing methods that have already been developed for terrestrial plant based feedstocks not suitable for seaweeds (Jung et al. 2013a).

Two methods are used for carrageenan extraction. They are resulting in isolation of refined or semi-refined powders (Rudolph 2000). For food-grade agar extraction, the strategy includes a step of washing followed by pre-treatments and hot water extraction. The syneresis method results in production of gels that contain nearly twice of agar yield than the freeze-thaw method (Phillips and Williams 2009). With respect to alginates, their extraction is based on conversion of all insoluble salts of alginic acid that are present within the cell wall of brown seaweed to the soluble (Na^+) salt of alginic that is then recuperated as alginic acid or calcium alginate. Details about these carbohydrate extractions are reported in Alba and Kontogiorgos (2018).

It is essential to pay attention to the carbohydrate content, especially, in the case of bioethanol production. Indeed, carbohydrate content facilitates the overall process efficiency calculations and their quantification is proportional to ethanol yields in biochemical conversion processes (Aden et al. 2002). More details about the quantification of these polymers can be found in Kostas et al. (2016a) research.

2.3 Pigment

The majority of algal pigments such as chlorophylls, carotenoids and phycobilins have been commercialized for many years for coloring purposes. For instance, all the phycobiliproteins have been used for decades as natural colorants in foods, cosmetic and in pharmaceutical products (Jespersen et al. 2005; Sekar and Chandramohan 2008). However, many researchers have reported their applications in human health (Sachindra et al. 2007; Sekar and Chandramohan 2008; Chang et al. 2011; Fung et al. 2013) where fucoxanthin remains the main macroalgal pigment under the spotlight of several industries (Kim et al. 2011). Thus, Phycoiliproteins and Fucoxanthin have shown several biological activities as antioxidative, anti-inflammatory, antiviral, and neuroprotective, antioxidant, anticancer, antiobesity, antidiabetic, and anti-photoagingactivity (Heo et al. 2005; Ragan and Glombitza 1986).

Among the seaweed pigments, there is fucoxanthin, which is an accessory pigment, belonging to the carotenoids named xanthophylls. It represents 10% of their total natural production (Kim et al. 2011; Kadam et al. 2013. Fucoxanthin, as other algal xanthophylls, is commonly extracted with hexane and other non-polar solvents, by liquid solvent extraction (Humphrey 2004; Hosikian et al. 2010). In order to minimize its degradation, alternative methods, such as the enzyme-assisted and microwave-assisted extractions and pressurized liquid extraction techniques, have been used (Pasquet et al. 2011; Billakanti et al. 2013). Red algae contain water soluble pigments called phycobiliproteins found in the cytoplasm or in the stroma of the chloroplasts, and are responsible of their color. R-phycoerythrin is the most common phycobiliprotein in many Rhodobionta, with levels, varying between 0.2 and 12% (Kim et al. 2013; Niu et al. 2006).

The extraction of phycobiliproteins is based on chemical and physical techniques for a primary isolation from the algae. The addition of other processes (freezing, sonication and homogenization, use of enzymes) is used in order to improve the extraction yield. Phycobiliproteins are then purified, usually by chromatographic methods, or by the use of novel techniques such as immuno-absorption and genetic recombination. All these techniques are well described by Kim et al. (2013) and Jespersen et al. (2005).

2.4 Proteins and Lipids

Most of seaweeds contain all the essential amino acids that have a promising potential for exploitation. An extensive review has found that these proteins have shown positive bioactive effects in the treatment of a broad range of human diseases (Holdt and Kraan 2011). Green and red macroalgae have relatively high protein contents, ranging from 9–26% or reaching up to 47% (w/w) dry weight, respectively (Fleurence 1999). With respect to the lipid fraction, it represents less than 5%. In the last years, fatty acid profile of seaweeds has attracted much attention due to their high amounts of polyunsaturated fatty acids (PUFA), such as α-linolenic (ALA, 18:3 n-3), octadecatetraenoic (18:4 n-3), arachidonic (AA, 20:4 n-6), eicosapentaenoic (EPA, 20:5 n-3) and docosahexaenoic (DHA, 22:6 n-3) acids. It is well known that this type of acids has important nutritional properties as well as beneficiary effects on human health. For example, PUFA possess anti-tumoural, antiviral and antiobesity properties, and they are further related with the prevention of cardiovascular diseases (Liang et al. 2014).

In terms of bioenergy generation, lipids produce more biogas than protein and carbohydrates. (Heaven et al. 2011; Weiland 2010; Zamalloa et al. 2011).

2.5 Phlorotannins

In marine brown algae, there is a specific class of polyphenols called phlorotannins, which have a less complex structure than terrestrial tannins and are represented by polymers of phloroglucinol. Their levels depend on the species and on the environmental factors (Pavia and Toth 2000). However, they can constitute up to 15% of the brown algae dry weight (Waterman and Mole 1994; Targett and Arnold 1998). The phlorotannins from marine algae have been less studied, especially in the field of animal nutrition; nevertheless, many authors elucidated multiple functions of this class such as antimicrobial and antioxidant activities (Horikawa et al. 1999; Hwang et al. 2006; Nakamura et al. 1996).

Traditionally, phlorotannins have been extracted using different solvents such as ethanol, methanol or aqueous acetone (Kadam et al. 2013; Liu and Gu 2012). The purification of these organic compounds is achieved by chromatographic techniques (Heo et al. 2009; Isaza Martínez and Torres Castañeda 2013). Due to the low stability of phlorotannins, the extracts are frequently obtained with nitrogen or by adding potassium metabisulfiteor ascorbic acid in order to prevent oxidation (Grosse-Damhues and Glombitza 1984; Lim et al. 2002).

2.6 Minerals

Macroalgae are rich in various minerals, higher mineral content than edible terrestrial plants and animals. They are known to contain a wide variety and high levels of certain minerals (this may vary form 8 to 40% of algal dry weight (DW) and have therefore been employed as mineral additives to feed and food supplements (Rupérez 2002). They have high content in many essential minerals, such as Na, Mg, P, K, I, Fe, and Zn (Bocanegra et al. 2009) Macroalgae also contain great amounts of trace elements (Rupérez 2002) and thus, algal-based supplement can supply to humans the daily demand of these elements (Indergaard 1991). The mineral content of seaweeds can vary depending on the species, growing conditions, and processing methods.

3 Biorefinery

3.1 Definition

A biorefinery is a facility that integrates various biomass conversion processes to produce a wide range of value-added products, including biofuels, biochemicals, and other bioproducts. It serves as a sustainable alternative to traditional petroleum-based refineries by utilizing renewable biomass feedstocks such as agricultural residues, energy crops, and algae. The primary objective of a biorefinery is to maximize the efficient use of biomass resources and minimize waste generation. It employs a combination of physical, chemical, and biological processes to convert biomass into multiple streams of products, which can be used in various industries such as transportation, chemicals, plastics, agriculture and Integrated Agriculture- Aquaculture (El Bakali and Aba 2023).

Kamm et al. (1998) first defined biorefinery as a complex system of sustainable, environment and resources-friendly technologies for the comprehensive utilization and the exploitation of biological raw materials. Different authors (Gravitis et al. 2008; Kamm and Kamm 2004) have also introduced other definitions of a biorefinery. As shown in the literature (Bioenergy 2009), researchers and technologists have also defined biorefineries based on the generation of feedstock.

3.2 Biopower and Biomaterial Production

Biorefineries can provide a significant contribution to sustainable development, generating added value to sustainable biomass use and producing a range of biopower (power, heat, fuels) and biomaterials (feed, food, chemicals, materials) at the same time. This requires optimal biomass conversion efficiency, thus minimizing feedstock requirements while strengthening, at the same time, economic viability of market sectors (De Jong and Jungmeier 2015).

3.2.1 Biopower

Worldwide, renewable energy is a part of energy programs. This tendency will increase because of the depletion of the fossil energy reserves coupled to the continuous increase of the oil product prices and with problems linked to the greenhouse effect. Indeed, biofuels are among the promising forms of renewable energy as they have good potential in reducing greenhouse gas (GHG) emissions by around 52% compared to conventional fossil fuels (Ng et al. 2017), as well as they can be produced from a wide variety of feedstock biomasses. Thus, they can improve energy security, trade balance and job opportunities. Thus, during these last decades new forms of renewable energy especially in liquid fuels transportation have been developed. For instance, successes were realized in the biofuel development such as bioethanol, biodiesel, hydrogen, biobutanol and the biogas leading in some cases to industrial level production. Biomass is currently the only abundant renewable energy source for the direct production of fuel. Nevertheless, there are still important gaps in research programs to enhance the process yields and to find new forms of non-food related resources (Kadam et al. 2013; Chen et al. 2015; Chisti 2013; Kerrison et al. 2015). Algae, therefore are considered as among the most potentially significant future sources of sustainable biofuels (Menetrez 2012), and have been described as potential sunlight-driven cell factories for the conversion of carbon dioxide to biofuels and chemical feedstocks (Bhateria and Dhaka 2014; Chisti 2007). Nevertheless, despite their obvious potential, there are no economically viable commercial-scale quantities of fuel from either micro- or macroalgae (Aresta et al. 2005; Milledge and Heaven 2014).

3.2.2 Biomaterials

Besides the well-established biopower production from microalgae and seaweeds, the broad range of their bioactive compounds, such as polysaccharides, proteins, lipids, pigments, polyphenols, minerals and vitamins find value-added applications in the human and animal food, pharmaceutical, cosmetics and agricultural industries (Heo et al. 2005; Vera et al. 2011; Liang et al. 2014; Pérez et al. 2016). The utilization of seaweeds for biomaterials is an active area of research and development, driven by their abundant availability, sustainability, and diverse properties. These biomaterials offer promising alternatives to synthetic materials in various industries, contributing to a more sustainable and eco-friendly approach. In addition, algal feedstock can be processed into many biomaterials used in dyes, paints, bioplastics, biopolymers, and nanoparticles, or as hydrochar and biochar in solid fuel cells and in agricultural field (Orejuela-Escobar et al. 2021).

4 Background Information Related to Biorefinery Concepts

The concept of a biorefinery aligns with the principles of the biobased economy, which seeks to replace fossil-based products with sustainable alternatives. It promotes resource efficiency, reduces greenhouse gas emissions, and contributes to the development of a more sustainable and circular economy. In order to produce energy and marketable products from biomasses, biorefineries generally integrate various biomass conversion technologies. These refineries have evolved over the last two decades mainly in three phases as described by Pande and Bhaskarwar (2012):

Phase I biorefineries, a single raw material is converted to a single product.

Phase II where multiple processing tools are used to convert a single raw material to obtain a broad range of products.

Phase III biorefineries, qualified as integrated biorefineries use a wide range of raw materials and technologies simultaneously or sequentially to produce a wide range of valuable products.

Some integrated biorefineries use various feedstock and technologies to produce biofuels as main products along with co-products such as platform chemicals, heat and power. De Jong and Jungmeier (2015) give a smartly summarized classification of biorefinery system related to different feedstocks used for the production of biopower and biomaterial products according to Bioenergy (2009).

5 Seaweed in Biorefinery

Seaweed biorefineries offer several promising business opportunities due to the numerous applications and benefits associated with seaweed-based products, but in the same time offer several environmental benefits, including carbon dioxide absorption, as seaweed grows rapidly and absorbs significant amounts of CO_2 from the atmosphere. Moreover, seaweed cultivation does not require freshwater or arable land, reducing the pressure on terrestrial resources.

Potential avenues to explore in the seaweed biorefinery industry:

I. Seaweed-based biofuels: Seaweed can be used to produce biofuels such as bioethanol, biogas, and biodiesel. These biofuels can be utilized as renewable alternatives to fossil fuels, offering lower carbon emissions and greater sustainability. Establishing a seaweed biorefinery focused on biofuel production can be a lucrative opportunity.

II. Nutraceuticals and functional foods: Seaweed is rich in various bioactive compounds, vitamins, and minerals that offer health benefits. Developing seaweed-based nutraceuticals, dietary supplements, and functional food products can tap into the growing market for natural and healthy products.

III. Cosmetics and personal care products: Seaweed extracts are known for their beneficial properties for the skin and hair. Utilizing seaweed in the production of cosmetics, skincare products, shampoos, and conditioners can attract consumers seeking natural and sustainable beauty solutions.

IV. Animal feed and aquaculture: Seaweed has great potential as a nutritious ingredient in animal feed formulations, especially for aquaculture species such as fish and shellfish. Seaweed-based animal feed can improve the sustainability and nutritional value of livestock and aquaculture production.

V. Bioplastics and biomaterials: Seaweed contains polysaccharides that can be used to produce biodegradable bioplastics and biomaterials. With the growing demand for eco-friendly packaging and sustainable materials, establishing a seaweed biorefinery focused on producing bioplastics can be a profitable venture.

VI. Fertilizers and soil conditioners: Seaweed extracts are known for their high nutrient content and ability to enhance soil fertility. Developing seaweed-based fertilizers and soil conditioners can cater to the increasing demand for organic and sustainable agriculture practices.

VII. Pharmaceutical and medical applications: Seaweed-based compounds have demonstrated potential therapeutic properties and can be used in the development of pharmaceutical drugs. Researching and producing seaweed-derived drugs and medical products can lead to valuable contributions to the healthcare industry.

When considering any business opportunity, it is essential to conduct thorough market research, assess the feasibility and scalability of the chosen venture, and develop a comprehensive business plan. Additionally, collaborations with researchers, industry experts, and potential customers can help drive innovation and market acceptance of seaweed-based products.

The development of seaweed biorefineries is gaining momentum due to the potential for sustainable production and the versatility of seaweed as a renewable resource. Ongoing research and technological advancements are focused on optimizing the efficiency and scalability of these biorefineries to further enhance the economic viability and environmental benefits of seaweed-based industries. The process of seaweed biorefining typically involves several steps:

- Harvesting: Seaweed is cultivated or harvested from natural sources such as coastal areas or dedicated seaweed farms. Sustainable harvesting practices are employed to ensure the long-term health of the seaweed population and marine ecosystems.
- Preprocessing: The harvested seaweed undergoes preprocessing steps, which may include washing, removing impurities, and drying. This step prepares the seaweed for further processing.
- Fractionation: Seaweed contains various components, including proteins, carbohydrates (such as alginate and agar), lipids, and minerals. Fractionation involves separating these different components using physical or chemical methods.
- Extraction: Specific compounds of interest, such as bioactive compounds or polysaccharides, are extracted from the fractionated seaweed. Extraction

techniques can involve solvent-based processes, enzymatic treatments, or physical separation methods.
- Conversion and refining: The extracted compounds can be further processed to produce different end products. For example, seaweed sugars can be fermented to produce bioethanol or other biofuels, while proteins can be used as food additives or in the development of biodegradable plastics.
- Waste utilization: Seaweed biorefineries aim to minimize waste and maximize resource utilization. Byproducts and waste streams generated during the biorefining process can be used for purposes such as fertilizer production or as animal feed supplements.

5.1 Bioenergy and Biomaterial Production

According to Menetrez (2012), seaweeds, as well as microalgae, are considered as among the most potentially significant future sources of sustainable biofuels. While microalgae has been explored mainly as substrate for biodiesel and bio-oils, macroalgae has been used predominantly in biogas and bioethanol production (Chisti 2007). With respect to macroalgae, there has been just few research directed towards developing feedstocks for fuels and/or producing fuels (Aresta et al. 2005). However, according to Lundquist et al. (2010), the use of seaweed feedstocks for non-fuel uses is 100 times bigger in wet tonnage terms than that for microalgae. In 2004, the overall utilisation of biomaterial products obtained from macroalgae was a multibillion-dollar industry (Smit 2004). Current uses of seaweeds include human foods, fertilisers, cosmetics ingredients and phycocolloids (Balina et al. 2017; Kraan 2013). The latters, mainly agars, carrageenans and alginates are widely used in food industry with a combined annual production of 86,000 tonnes (Holdt and Kraan 2011; Milledge 2012; Milledge et al. 2019.

5.2 Macroalgae Biorefinery Concept

Because seaweeds have the potential to substitute terrestrial crop feedstocks competing with food as a biofuel feedstock in the biogas and ethanol production infrastructures (Wei et al. 2013), many authors tried to develop different techniques in order to increase biopower yield and to use the entire seaweed biomass. Baghel et al. (2020), Prabhu et al. (2020) and Zollmann et al. (2019) gave examples of integrated biorefinery processes for the green, brown and red seaweeds, respectively. Figure 12.1 shows a seaweed based biorefinery model according to Ramirez (2017).

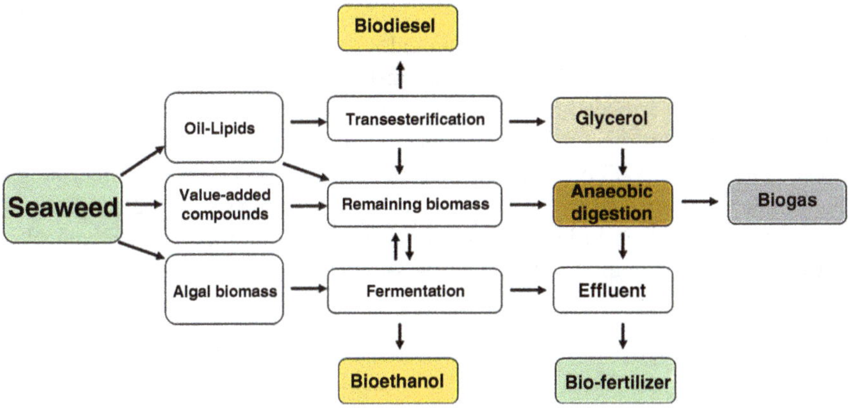

Fig. 12.1 Seaweed based biorefinery model (Ramírez 2015)

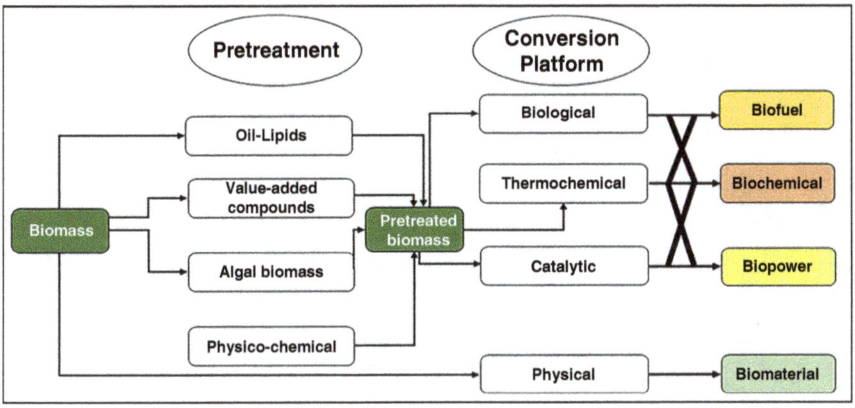

Fig. 12.2 Pretreatment and biomass conversion platform (Ng et al. 2017)

5.3 *Biorefinery Optimization Techniques*

5.3.1 Pre-Treatments of the Biomass (PTB)

According to Ng et al. (2017), prior to conversion technologies, all feedstocks (seaweeds and lignocellulosic biomasses) are required to be pre-treated because of the inconsistent characteristic of biomass. As shown in Fig. 12.2, each of the PTB and conversion platforms can be divided into 4 categories.

The authors Gao et al. (2015) in their study reported that the pre-treatments of biomass that modify the bioavailability of polysaccharides for their hydrolysis could have a positive impact on both rate and yields of biogas or ethanol enabling higher biofuel production. Continuing efforts have been made by many authors during the last decade, aiming to optimize pre-treatments to achieve better yields and lower costs (Jung et al. 2013a; Rodriguez-Jasso et al. 2011; Wei et al. 2013; Michalak 2018). For instance, Maneein et al. (2018) gave a review about seaweed pretreatment methods for enhanced biofuel production where he focuses on anaerobic digestion and fermentation. He reported that processes that use the entire biomass rather than just the fermentable sugars have more favorable energy return on investments. Figure 12.3 provides mechanisms involved in producing either biogas (Fig. 12.3a) or bioethanol (Fig. 12.3b) where the hydrolysis of the polymers to sugars is the first stage in both cases. More details about different mechanical of pretreatments techniques are givenin this review.

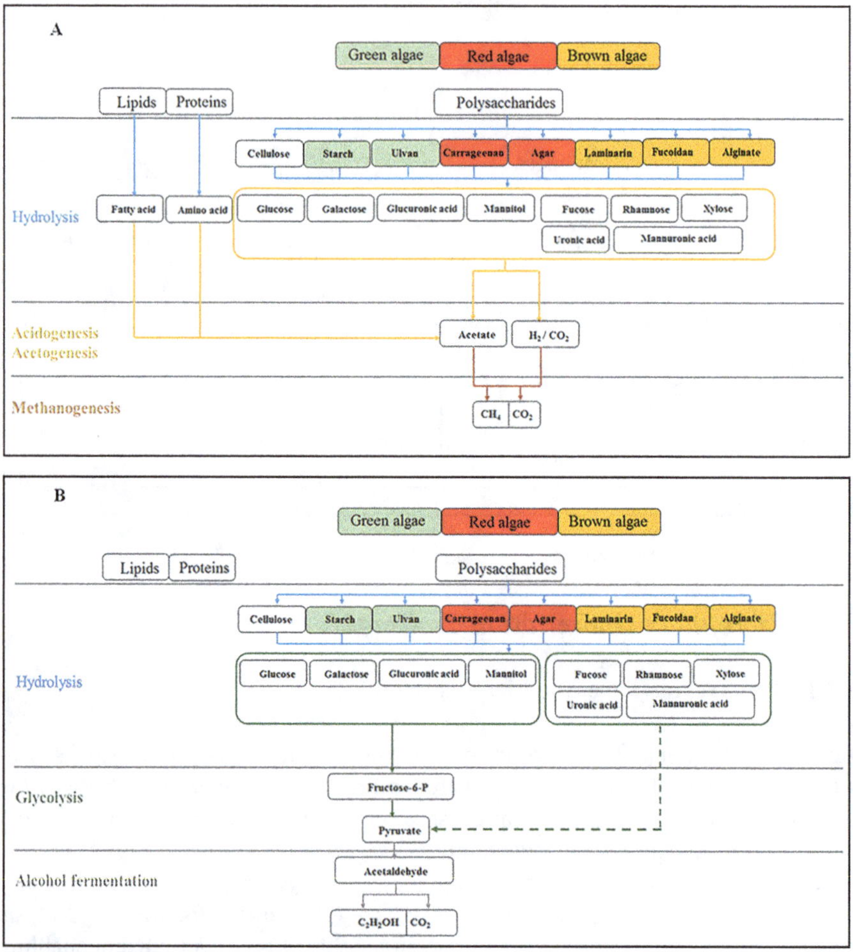

Fig. 12.3 Major steps for biogas (**a**) and ethanol (**b**) products from seaweeds (Maneein et al. 2018). Red, green and brown colours are attributed to each group or components of red, green and brown macroalgae, respectively

5.3.2 Prioritizing Bioactive Components Extraction (PBCE)

Álvarez-Viñas et al. (2019), in their review, gave a well described conventional and alternative techniques, applied for some red algae, in order to have zero wastes. The strategy consists on the extraction of the products with highest value before the other products. Many schemes and examples where all wastes are used two alternatives for the solid waste were described depending on the objective. It can be processed to monomeric sugars, for rare sugars production (Bayu and Handayani 2018), or simply, the use of sugars as yeasts carbon source for the production of biofuels or its catalytic conversion into valuable chemicals (Cesário et al. 2018). In the same way (Kostas et al. 2016b) in his study gave a micro array technique to screen 24 strains of yeasts for their ability to metabolize macroalgae monosaccharides because they contain a diverse array of monomeric sugars than lignocellulosic feedstocks. The PBCE was also described by other researchers (Balina et al. 2017) who presented a conceptual model for valuable product production along with biofuel production (Fig. 12.4).

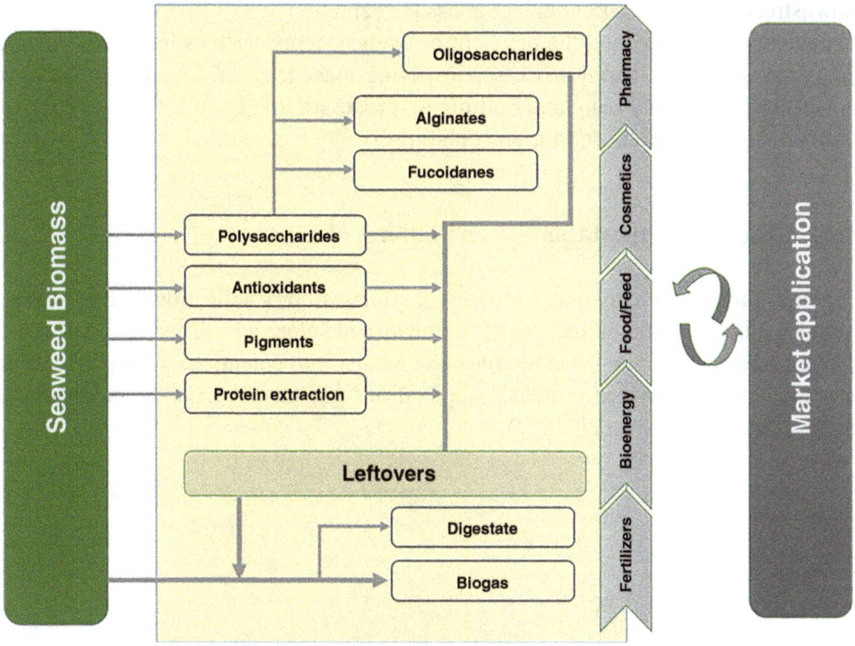

Fig. 12.4 Conceptual model for seaweed biorefinery (Balina et al. 2017)

5.3.3 Biochemical Conversion: Anaerobic Digestion (AD) and Fermentation

Milledge et al. (2019) gave a brief review about AD of algae for bioenergy production. This technique would be the closest to industrial exploitation because it could reduce cost related to the step of drying wet biomass (Milledge and Heaven 2014, 2017). New processing methods are needed to reduce costs and increase the net energy balance because the practical yields of biogas production from AD is under the theoretical maximum. The microbial communities involved in the fermentation and AD processes will also increase biofuel yields (McKennedy and Sherlock 2015). Ramirez (2017) in his thesis gives a review about the optimization of AD of seaweeds.

With respect to fermentation process, Brethauer and Wyman (2010) gave a review about the continuous hydrolysis and fermentation for cellulosic ethanol production. They stated that nowadays, the continuous fermentation system could enable both higher yields and productivities than in batch fermentations.

5.3.4 Open-System Thermodynamics (OST)

Gravitis et al. (2008) reported that this technique, well described by Offei 2021, should be applied in order to strengthen zero emissions and biorefinery concepts. This author reported that the thermochemical processes could be considered as downstream processes after algal biomass fermentation. OST provides a framework to analyze and understand the behavior of open systems, such as living organisms, ecosystems, and industrial processes involving mass transfer. Besides, provides a powerful tool for analyzing and optimizing processes that involve the exchange of both energy and matter with the surroundings.

5.3.5 Thermochemical Conversion Platform (TCP)

Through three thermochemical processes; combustion, gasification and pyrolysis, biomass can be converted into fuels, chemicals, and electricity. Thermal conversion system uses heat to change the biomass chemically into potential energy products or intermediates for further synthesis applications via the precited thermochemical processing routes (Ng et al. 2017).

6 Strategies for Ecofriendly Extraction of Seaweed Bioproducts

The extraction of seaweed bioproducts can be approached in an eco-friendly manner to minimize environmental impact and promote sustainable practices, contributing to the conservation of marine ecosystems while harnessing the potential of seaweed for various applications. It is very important to conduct a comprehensive life cycle assessment (LCA) of the entire extraction process to identify potential environmental impacts at each stage. This assessment can help optimize the process and identify areas where improvements can be made to reduce the overall environmental footprint. Enzyme-assisted extraction (EAE), microwave-assisted extraction (MAE), ultrasound-assisted extraction (UAE), supercritical fluid extraction (SFE), and pressurized liquid extraction (PLE) are new extraction technologies explored in terms of their efficiency and ecofriendly approaches compared to older ones and their potential prospective for extracting marine algae bioproducts.

6.1 *Enzyme-Assisted Extraction (EAE)*

Seaweeds are sources of complex chemicals and heterogeneous biomolecules. The sulfated and branched polysaccharides in seaweeds are associated with proteins and various bound ions, including calcium and potassium (Ahn et al. 2004; Liang et al. 2014). It is necessary to break down these complex molecules to extract marine algal bioactive compounds. Applications of cell wall degrading enzymes such as carbohydrases and proteases to marine algae at optimal temperature and pH conditions have been found to break down the cell wall and release the desired bioactive compounds. Some of the commonly used enzymes are viscozyme, cellucast, termamyl, ultraflo, carrageenase, agarase, xylanase, kojizyme, neutrase, alcalase, and umamizyme (Herrero et al. 2006; Wijesinghe and Jeon 2012). Commonly used approaches for marine algal bioactive compound extraction are listed in Table 12.1.

Enzyme-assisted extraction (EAE) is the environmentally friendly technology as it alleviates the use of solvents in the process. This technique facilitates a high yield of bioactive compounds where unnecessary components from cell walls are removed and desired bioactive compounds are extracted into the medium. The use of common food grade enzymes such as cellulase, a-amylase, pepsin, etc. makes EAE a relatively low-cost technique for extraction purposes (Jeon et al. 2011).

EAE have high catalytic efficiency. They also preserve the original efficacy of the compounds to a high degree (Hong et al. 2014).

Siriwardhana and his collaborators have found positive effect of pH on solubilization yield of Hizikia fusiformis using carbohydrases and proteases mixtures (Siriwardhana et al. 2008). The highest effect of enzymatic treatments were reached at 9 h incubation for Amylase, Celluclast and Protamex, while for Alcalase and the

Table 12.1 Type of enzymes employed for marine algae (Herrero et al. 2006; Wijesinghe and Jeon 2012; Jeon et al. 2011)

Enzyme	Temperature (°C)	pH	Enzyme composition
Viscozyme	50	4.5	Arabanase, cellulase, b-glucanase, hemi-cellulase, and xylanase
Cellucast	50	4.5	Group of enzymes, catalyzing the breakdown of cellulose into glucose, cellobiose, and higher glucose polymer
Termamyl	60	6.0	Heat-stable a-amylase
Ultraflo	60	7.0	Heat-stable multiactive b-glucanase
Neutrase	50	6.0	Endoprotease
Flavourzyme	50	7.0	Endoprotease and exopeptidase activities
Alcalase	50	8.0	a-Endoprotease

protease mixture the maximum effect on solubilisation was reached at 6 h (Park et al. 2004).

Many studies have been carried out to determine the antioxidant activity of seaweed extracts. Enzyme assisted extracts have a higher antioxidant activity as compared to extracts obtained using conventional extraction methods (Siriwardhana et al. 2008). Cian et al. (2012) carried out proteolysis of phycobiliproteins from Pyropiacolumbina to obtain a peptidic fraction rich in low molecular weight peptides that exhibit immunosuppressive, antihypertensive, and antioxidant properties (Cian et al. 2012).

The antioxidant activity of seaweed extracts has been investigated in several studies. When compared to Sargassum horneri extracts prepared using traditional extraction procedures, EAE have higher antioxidant activity (Matanjun et al. 2008). Moreover, it has been found in a previous study that the carbohydrases enhanced the release of the fucoidan fraction, whereas the proteolytic systems favored the extraction of phenolics, and these are more potent antioxidants. Therefore, depending on the target products, different enzymatic cocktails could be recommended (Casas et al. 2019).

6.2 *Ultrasound-Assisted Extraction (UAE)*

Ultrasound waves are high-frequency sound waves above human hearing capacity, that is, above 20 kHz. Thus, it can pass through solid, gas, and liquid media by rarefactions and compression. If the pressure exceeds the tensile strength of the liquid, then the formation of vapor bubbles occurs. These vapor bubbles undergo an implosive collapse in strong ultrasound fields, which is known as bubble cavitation (Luque-Garcia and De Castro 2003). The implosion of cavitation bubbles generates macroturbulence, high-velocity interparticle collisions, and perturbation in microporous particles of the biomass (Shirsath et al. 2012). Cavitation near liquid–solid interfaces direct a fast-moving stream of liquid through the cavity at the surface.

Impingement by these microjets results in surface peeling, erosion, and particle breakdown, facilitating the release of bioactive compounds from the biological matrix. This effect increases the efficiency of extraction by increasing mass transfer by eddy and internal diffusion mechanisms (Shirsath et al. 2012). There are two main types of ultrasound equipment that can be employed for extraction purposes: an ultrasonic water bath and an ultrasonic probe system fitted with horn transducers (Vilkhu et al. 2011). Ultrasonic water baths are relatively inexpensive and are commonly used to sonicate samples. The more powerful ultrasonic probe system with horn transducers introduces vibrations directly into the sample and may be used in batch or continuous mode. Moreover, Ultrasound has been used as pretreatment for many extraction and analytical techniques. UAE has been successfully used to obtain higher yields of bioactive compounds such as laminarin, phenolic compounds, fucose, and uronic acids from *Ascophyllum nodosum* (Jung et al. 2013b).

6.3 Microwave-Assisted Extraction (MAE)

MAE is another novel technique that has been extensively utilized to obtain high-value biologically active compounds from algae and plants Microwaves are non-ionizing electromagnetic radiation with a frequency ranging from 300 MHz to 300 GHz (Matos et al. 2021). MAE transfers energy by the twin mechanisms of dipole rotation and ionic conduction. The radiation frequency corresponds to the rotational motion of molecules; in condensed matter, energy absorption immediately causes energy redistribution between molecules and homogeneous heating of the medium (Álvarez-Viñas et al. 2019). MAE causes disruptions of hydrogen bonds and migration of dissolved ions, resulting in increased penetration of solvent into the matrix, which facilitates the extraction of target compounds. Due to the significant pressure developed inside the matrix, there is an increase in the porosity of the biological matrix, resulting in a higher penetration of solvent into the matrix. Thus, it has been reported to be more efficient as compared to other conventional extraction methods such as Soxhlet extraction, heat reflux extraction, UAE, and maceration (Kubrakova and Toropchenova 2008).

Microwave assisted aqueous two-phase extraction for the simultaneous recovery and separation of polysaccharides from Sargassum pallidum. The results displayed a maximum polysaccharide yield of 0.75 ± 0.04% of top phase and 6.81 ± 0.33% of bottom phase at 15 min, 95 °C, microwave power 830 W, 22.0% ammonium sulphate (w/w), 21.0% ethanol (w/w) and the ratio of material to liquid 1:60 (g/ml) (Routray and Orsat 2012).

However, MAE is not suitable for use with heat sensitive bioactive compounds. Recently, MAE has been employed to extract fucoidans, carotenoids, and minerals from marine macro and microalgae. MAE has also been applied for extraction of sulfated polysaccharides (fucoidan) from brown seaweed Fucus vesiculosus (Rodriguez-Jasso et al. 2011). Additionally, microwave assisted extraction using deep eutectic solvent was employed for the extraction of polysaccharides from

Fucale. A maximum of 116.33 mg/g of polysaccharide was procured in 35 min at 168 °C with excellent antioxidant and anticancer activities in vitro.

6.4 Supercritical Fluid Extraction (SFE)

Supercritical fluid extraction is a method in which the separation of the components takes place from the matrix by using supercritical fluids as an extracting solvent (Rai et al. 2016). Different parameters involved in the Supercritical fluid extraction process such as pressure, temperature, co-solvents or solvent flow rate have been optimized in order to improve extraction performance and the selectivity of the recovered compounds. Kurt Zosel, first used the SFE technique in 1969 for extracting valuable compounds at an industrial scale (Michalak 2018). Since then, many compounds like lipids, fatty acids, antioxidants have been extracted using this supercritical fluid extraction method (Abhari and Khaneghah 2020).

Some researchers studied the effect of extraction conditions to obtain fatty acids from Hypneacharoides algae using supercritical CO_2. Temperature values from 40 °C to 50 °C and pressure values from 24.1 MPa to 37.9 MPa were investigated (Abhari and Khaneghah 2020). Lipid recovery as well as the ratio of unsaturated fatty acids were found to increase with extraction pressure and temperature. Algal extracts from SFE demonstrate antimicrobial activity due to the presence of an indolic derivative (Cheung 1999).

6.5 Ultrahigh Pressure Extraction (UPE) and Pressurized Fluid Extraction (PFE)

The use of ultrahigh pressure (more than 1000 bar) extraction has been claimed to increase the yield of recovery of certain compounds with very low solubility at lower extraction pressures. It provided a better recovery of organic compounds from a natural matrix, comparable to those methods that use organic solvents, the use of modifiers increases the range of materials that can be extracted. Food grade modifiers such as ethanol can often be used and help in the collection of the extracted material (Mendiola et al. 2008).

Pressurized liquid extraction (PLE), also called accelerated solvent extraction, has been recognized as a promising technology for the extraction of a wide range of biologically active compounds from different natural sources. The PLE applies high temperatures (up to 200 °C) and pressures (up to 200 bar) using low solvent volumes, which favors rapid extraction of the desired compounds. Plaza et al. (2010) reported that PLE was a suitable technique to produce extracts with antioxidant and antimicrobial activities from *Himanthalia elongatae*.

The various benefits associated with pressurized liquid extraction are; it is faster, easier to use, automated, requires minimal solvent, saves time and money, increases productivity, improves reproducibility and significantly minimizes exposure to solvents (Abbott et al. 2007). These advantages completely satisfy the agenda of achieving a green and sustainable environment. Saravana et al. (2016) utilized pressurized liquid extraction to extract fucoidan from brown seaweed Saccharina japonica (Saravana et al. 2016). 8.23% of crude fucoidan was extracted with a molecular weight ranging between 83.39–183.32 kDa. The obtained fucoidan displayed excellent antioxidant activities. The PLE technique was utilized as an alternative green approach by Santoyo and collaborators to extract polysaccharide fractions from the edible brown seaweed *Himanthalia elongatae* (Santoyo et al. 2011).

6.6 The Use of Deep Eutectic Solvents (DESs)

Polluting, hazardous and non-renewable solvents tend to be less used due to the reinforcement of environmental and safety regulations (Chemat et al. 2012). Thus, Investigations for green solvents has been a key step toward green processing, accordingly. Deep eutectic solvents (DESs) are arrangements made from a eutectic mixture of Brønsted–Lewis bases and acids, which has various types of cationic or anionic groups. DESs are categorized as a type of ionic liquids because of their specific properties. They can integrate with single or several complexes in a mixed form to form a eutectic function with a melting point lower than that of a particular compound (Singh et al. 2013). Polyols, choline chloride, urea, sugars, and organic acids are extensively used to form DESs (Zainal-Abidin et al. 2017). DESs are advantageous as they contain non-toxic quaternary ammonium salts and are biodegradable and inexpensive. In addition, they have several properties such as low volatility, low melting point, and high thermal stability. Recently, DESs have been used in several applications such as organic synthesis, catalysis, extraction media, enzyme reaction, electrode position, and chromatography (Dai et al. 2013).

It has been confirmed that the extraction yield of hydrophilic ascorbic acid and phlorotannins significantly raised as the water content in the DES increased. The compounds rich in hydroxyl groups like ascorbic acid and total phlorotannins are good hydrogen donors and preferably form bonds with hydrogen bond acceptors like choline chloride (Obluchinskaya et al. 2019). Recently, Roy et al. (2021) reported on the extraction of another carotenoid astaxanthin from a marine species with DES. Green solvents for green processing have also been reported by Zollmann et al. (2019).

7 Future Considerations

Seaweed biorefinery is a rapidly developing field with several emerging trends that are likely to shape its future. Potential trends and considerations in relation with seaweed biorefinery:

- *Increased utilization of seaweed biomass*: Seaweeds are abundant and fast-growing biomass resources that offer numerous applications. In the future, there is likely to be an increased focus on utilizing different types of seaweed species and optimizing their cultivation methods to maximize biomass production for biorefinery processes.
- *Integrated biorefinery systems*: Future seaweed biorefineries may adopt integrated systems that utilize various components of seaweed biomass to produce a wide range of valuable products. This could involve extracting bioactive compounds for pharmaceuticals, food additives, and functional ingredients, while also utilizing the remaining biomass for bioenergy production, bioplastics, and other high-value applications.
- *Sustainable cultivation methods*: Sustainable and environmentally friendly cultivation methods, such as seaweed farming in open ocean environments or integrated multi-trophic aquaculture (IMTA), are likely to gain more prominence. Combine seaweed cultivation with other aquaculture practices, such as fish or shellfish farming, to create an ecological balance it is of great interest. The seaweed absorbs excess nutrients produced by the other organisms, reducing the risk of eutrophication and improving water quality. These methods minimize the environmental impact and allow for large-scale cultivation of seaweeds without competing with traditional agriculture for land and freshwater resources.
- *Development of novel extraction techniques*: Efficient extraction of desired compounds from seaweed biomass is essential for biorefinery processes. Future trends may involve the development of innovative and more sustainable extraction techniques, such as green solvents, supercritical fluid extraction, or enzymatic processes, which can enhance the extraction efficiency and minimize the use of harmful chemicals.
- *Valorization of waste streams*: Seaweed biorefineries can generate various waste streams during the production process. Future trends may involve finding innovative ways to valorize these waste streams, such as using them for the production of biofertilizers, animal feed, or converting them into value-added products through biotechnological processes.
- *Technological advancements*: The field of seaweed biorefinery is likely to benefit from ongoing technological advancements. This includes developments in analytical techniques, genetic engineering, fermentation processes, and bioprocessing technologies, which can improve the efficiency and cost-effectiveness of seaweed biorefinery operations.
- *Commercialization and market expansion*: As the industry matures and technology advances, there is likely to be an increased commercialization of seaweed biorefinery products. This may lead to a broader range of seaweed-derived

products entering the market, including sustainable alternatives to conventional materials, novel food products, and innovative functional ingredients.

It's important to note that these trends are speculative and based on the current trajectory of seaweed biorefinery research and development. The actual future trends may vary depending on various factors, including technological breakthroughs, market demands, regulatory frameworks, and sustainability considerations.

8 Conclusions

Seaweed biorefineries have the potential to play a significant role in the future of sustainable bio-based industries. Some aspects that hold promise for the future of seaweed biorefineries can be briefly summarize:

- Biofuel production: Seaweeds, such as macroalgae, can be used as a feedstock for biofuel production. They are rich in carbohydrates, which can be converted into biofuels like ethanol and butanol. As the world seeks alternative energy sources to reduce dependence on fossil fuels, seaweed-based biofuels offer a renewable and environmentally friendly option.
- Bioplastics and biomaterials: Seaweeds contain polysaccharides like alginate and carrageenan, which can be extracted and used to produce biodegradable bioplastics. These bioplastics have the potential to replace conventional plastics derived from non-renewable resources, reducing environmental pollution. Additionally, seaweed-based biomaterials can find applications in various industries, including packaging, textiles, and construction.
- Food and feed production: Seaweeds have long been consumed as food in many cultures, and their popularity is growing globally due to their nutritional value and health benefits. Seaweed biorefineries can optimize the processing of seaweed for food applications, including the extraction of proteins, lipids, and other valuable compounds. These components can be used to develop innovative plant-based food products and supplements. Seaweed can also be used as feed for livestock, contributing to sustainable and nutritious animal diets.
- Pharmaceuticals and nutraceuticals: Seaweeds contain a wide range of bioactive compounds with potential pharmaceutical and nutraceutical applications. These compounds, such as antioxidants, antimicrobials, and anti-inflammatory agents, can be extracted and used in the development of novel drugs and health supplements. Seaweed biorefineries can focus on the isolation and purification of these bioactive compounds to meet the increasing demand for natural and sustainable healthcare products.
- Waste treatment and bioremediation: Seaweeds have the ability to absorb and accumulate heavy metals and other pollutants from the surrounding water. Seaweed biorefineries can be utilized for bioremediation purposes, where seaweeds are cultivated in polluted waters to remove contaminants. Additionally, seaweed biomass can be used as a potential substrate for the production of bio-

char or as a component in wastewater treatment systems, providing an eco-friendly approach to waste management.
- Carbon sequestration and climate change mitigation: Seaweeds are highly efficient at absorbing and storing carbon dioxide (CO_2) from the atmosphere. Large-scale cultivation of seaweeds can contribute to carbon sequestration, helping to mitigate climate change. Moreover, seaweed biorefineries can harness the carbon captured during cultivation and processing to produce carbon-neutral or even carbon-negative products, further reducing the carbon footprint of various industries.

While seaweed biorefineries offer numerous potential benefits, there are still challenges to address, such as scaling up production, optimizing cultivation techniques, and developing cost-effective processing methods. However, with ongoing research and advancements in technology, seaweed biorefineries have a promising future in contributing to sustainable industries and addressing environmental and societal challenges. The global seaweed market is influenced by several factors, including increasing consumer awareness of the nutritional and health benefits of seaweeds, growing demand for natural and sustainable products, and advancements in seaweed cultivation techniques. However, market dynamics can vary across regions and specific seaweed types.

Acknowledgments ALEHOOP project under grant agreement No 887259, PROMISEANG under grant agreement 101036768, InnoProtein project under grant agreement 101112072, SYLPLANT under grant agreement 101112555, have all received funding from the Bio Based Industries Joint Undertaking (JU). The JU-CBE receives support from the European Union's Horizon 2020 research and innovation programme and the Bio Based Industries Consortium. LIKE-A-PRO project received funding from European Union's Horizon Europe (HORIZON) under Grant Agreement No. 101083961.

References

Abbott AP, Barron JC, Ryder KS, Wilson D (2007) Eutectic-based ionic liquids with metal-containing anions and cations. Chem Eur J 13:6495–6501

Abenavoli LM, Cuzzupoli F, Chiaravalloti V, Proto AR (2016) Traceability system of olive oil: a case study based on the performance of a new software cloud. Agron Res 14:1247–1256

Abhari K, Khaneghah AM (2020) Alternative extraction techniques to obtain isolate and purify proteins and bioactive from aquaculture and by-products. In: Advances in food and nutrition research. Elsevier, pp 35–52

Aden A, Ruth M, Ibsen K, Jechura J, Neeves K, Sheehan J, Wallace B, Montague L, Slayton A, Lukas J (2002) Lignocellulosic biomass to ethanol process design and economics utilizing co-current dilute acid prehydrolysis and enzymatic hydrolysis for corn Stover. National Renewable Energy Lab, Golden

Ahn C-B, Jeon Y-J, Kang D-S, Shin T-S, Jung B-M (2004) Free radical scavenging activity of enzymatic extracts from a brown seaweed *Scytosiphon lomentaria* by electron spin resonance spectrometry. Food Res Int 37:253–258

Alba K, Kontogiorgos V (2018) Seaweed polysaccharides (agar, alginate carrageenan)

Aresta M, Dibenedetto A, Barberio G (2005) Utilization of macro-algae for enhanced CO2 fixation and biofuels production: development of a computing software for an LCA study. Fuel Process Technol 86:1679–1693
Azizi S, Ahmad MB, Namvar F, Mohamad R (2014) Green biosynthesis and characterization of zinc oxide nanoparticles using brown marine macroalga *Sargassum muticum* aqueous extract. Mater Lett 116:275–277
Álvarez-Viñas M, Flórez-Fernández N, Torres MD, Domínguez H (2019) Successful approaches for a red seaweed biorefinery. Mar Drugs 17:620
Baghel RS, Suthar P, Gajaria TK, Bhattacharya S, Anil A, Reddy CRK (2020) Seaweed biorefinery: a sustainable process for valorising the biomass of brown seaweed. J Clean Prod 263:121359
Balina K, Romagnoli F, Blumberga D (2017) Seaweed biorefinery concept for sustainable use of marine resources. Energy Procedia 128:504–511
Bayu A, Handayani T (2018) High-value chemicals from marine macroalgae: opportunities and challenges for marine-based bioenergy development. In: IOP conference series: earth and environmental science. IOP Publishing, p 12046
Bhateria R, Dhaka R (2014) Algae as biofuel. Biofuels 5:607–631
Billakanti JM, Catchpole OJ, Fenton TA, Mitchell KA, MacKenzie AD (2013) Enzyme-assisted extraction of fucoxanthin and lipids containing polyunsaturated fatty acids from *Undaria pinnatifida* using dimethyl ether and ethanol. Process Biochem 48:1999–2008
Bioenergy IEA (2009) Bioenergy – a sustainable and reliable energy source. International Energy Agency, Paris
Bocanegra A, Bastida S, Benedi J, Rodenas S, Sanchez-Muniz FJ (2009) Characteristics and nutritional and cardiovascular-health properties of seaweeds. J Med Food 12:236–258
Brethauer S, Wyman CE (2010) Continuous hydrolysis and fermentation for cellulosic ethanol production. Bioresour Technol 101:4862–4874
Casas MP, Conde E, Domínguez H, Moure A (2019) Ecofriendly extraction of bioactive fractions from *Sargassum muticum*. Process Biochem 79:166–173
Cesário MT, da Fonseca MMR, Marques MM, de Almeida MCMD (2018) Marine algal carbohydrates as carbon sources for the production of biochemicals and biomaterials. Biotechnol Adv 36:798–817
Chang C-J, Yang Y-H, Liang Y-C, Chiu C-J, Chu K-H, Chou H-N, Chiang B-L (2011) A novel phycobiliprotein alleviates allergic airway inflammation by modulating immune responses. Am J Respir Crit Care Med 183:15–25
Chemat F, Vian MA, Cravotto G (2012) Green extraction of natural products: concept and principles. Int J Mol Sci 13:8615–8627
Chen H, Qiu T, Rong J, He C, Wang Q (2015) Microalgal biofuel revisited: an informatics-based analysis of developments to date and future prospects. Appl Energy 155:585–598
Cheung PCK (1999) Temperature and pressure effects on supercritical carbon dioxide extraction of n-3 fatty acids from red seaweed. Food Chem 65:399–403
Chisti Y (2007) Biodiesel from microalgae. Biotechnol Adv 25:294–306
Chisti Y (2013) Constraints to commercialization of algal fuels. J Biotechnol 167:201–214
Cho M, You S (2015) Sulfated polysaccharides from green seaweeds. In: Springer handbook of marine biotechnology. Springer, pp 941–953
Cian RE, Martínez-Augustin O, Drago SR (2012) Bioactive properties of peptides obtained by enzymatic hydrolysis from protein byproducts of *Porphyra columbina*. Food Res Int 49:364–372
Cotas J, Gomes L, Pacheco D, Pereira L (2023) Ecosystem services provided by seaweeds. Hydrobiology 2:75–96
Dai Y, van Spronsen J, Witkamp G-J, Verpoorte R, Choi YH (2013) Natural deep eutectic solvents as new potential media for green technology. Anal Chim Acta 766:61–68
Davis TA, Volesky B, Mucci A (2003) A review of the biochemistry of heavy metal biosorption by brown algae. Water Res 37:4311–4330

De Jong E, Jungmeier G (2015) Biorefinery concepts in comparison to petrochemical refineries. In: Industrial biorefineries & white biotechnology. Elsevier, pp 3–33

Doh H, Lee MH, Whiteside WS (2020) Physicochemical characteristics of cellulose nanocrystals isolated from seaweed biomass. Food Hydrocoll 102:105542

Duan H, Takaishi Y, Momota H, Ohmoto Y, Taki T (2002) Immunosuppressive constituents from *Saussurea medusa*. Phytochemistry 59:85–90

El Bakali M, Aba M (2023) Aquaponics as a sustainable food production system with promising development perspectives in Morocco. In: Gabriel NN, Omoregie E, Abasubong KP (eds) Emerging sustainable aquaculture innovations in Africa. Sustainability sciences in Asia and Africa. Springer, Singapore. https://doi.org/10.1007/978-981-19-7451-9_16

Farghali M, Mohamed IMA, Osman AI et al (2023) Seaweed for climate mitigation, wastewater treatment, bioenergy, bioplastic, biochar, food, pharmaceuticals, and cosmetics: a review. Environ Chem Lett 21:97–152

Fleurence J (1999) Seaweed proteins: biochemical, nutritional aspects and potential uses. Trends Food Sci Technol 10:25–28

Fung A, Hamid N, Lu J (2013) Fucoxanthin content and antioxidant properties of *Undaria pinnatifida*. Food Chem 136:1055–1062

Gomez-Zavaglia A, Prieto Lage MA, Jimenez-Lopez C, Mejuto JC, Simal-Gandara J (2019) The potential of seaweeds as a source of functional ingredients of prebiotic and antioxidant value. Antioxidants 8:406

González-López N, Moure A, Domínguez H (2012) Hydrothermal fractionation of *Sargassum muticum* biomass. J Appl Phycol 24:1569–1578

Gravitis J, Abolins J, Kokorevics A (2008) Integration of biorefinery clusters towards zero emissions. Environ Eng Manag J 7:569–577

Grosse-Damhues J, Glombitza K-W (1984) Isofuhalols, a type of phlorotannin from the brown alga Chorda filum. Phytochemistry 23:2639–2642

Gao Y, Fangel JU, Willats WGT, Vivier MA, Moore JP (2015) Dissecting the polysaccharide-rich grape cell wall changes during winemaking using combined high-throughput and fractionation methods. Carbohydr Polym 133:567–577. https://doi.org/10.1016/J.CARBPOL.2015.07.026

Heaven S, Milledge J, Zhang Y (2011) Comments on 'anaerobic digestion of microalgae as a necessary step to make microalgal biodiesel sustainable'. Biotechnol Adv 29:164–167

Heo S-J, Park E-J, Lee K-W, Jeon Y-J (2005) Antioxidant activities of enzymatic extracts from brown seaweeds. Bioresour Technol 96:1613–1623

Heo S-J, Hwang J-Y, Choi J-I, Han J-S, Kim H-J, Jeon Y-J (2009) Diphlorethohydroxycarmalol isolated from Ishige okamurae, a brown algae, a potent α-glucosidase and α-amylase inhibitor, alleviates postprandial hyperglycemia in diabetic mice. Eur J Pharmacol 615:252–256

Herrero M, Cifuentes A, Ibañez E (2006) Sub-and supercritical fluid extraction of functional ingredients from different natural sources: plants, food-by-products, algae and microalgae: a review. Food Chem 98:136–148

Hoek C, Mann D, Jahns HM, Jahns M (1995) Algae: an introduction to phycology. Cambridge University Press

Holdt SL, Kraan S (2011) Bioactive compounds in seaweed: functional food applications and legislation. J Appl Phycol 23:543–597

Hong IK, Jeon H, Lee SB (2014) Comparison of red, brown and green seaweeds on enzymatic saccharification process. J Ind Eng Chem 20:2687–2691

Horikawa M, Noro T, Kamei Y (1999) In vitro anti-methicillin-resistant *Staphylococcus aureus* activity found in extracts of marine algae indigenous to the coastline of Japan. J Antibiot (Tokyo) 52:186–189

Horn SJ, Aasen IM, Østgaard K (2000) Ethanol production from seaweed extract. J Ind Microbiol Biotechnol 25:249–254

Hosikian A, Lim S, Halim R, Danquah MK (2010) Chlorophyll extraction from microalgae: a review on the process engineering aspects. Int J Chem, Eng, p 2010

Humphrey AM (2004) Chlorophyll as a color and functional ingredient. J Food Sci 69:C422–C425

Hwang H, Chen T, Nines RG, Shin H, Stoner GD (2006) Photochemoprevention of UVB-induced skin carcinogenesis in SKH-1 mice by brown algae polyphenols. Int J Cancer 119:2742–2749
Indergaard M (1991) Animal and human nutrition. Seaweed Resour Eur Use Potential:21–64
Isaza Martínez JH, Torres Castañeda HG (2013) Preparation and chromatographic analysis of phlorotannins. J Chromatogr Sci 51:825–838
Jeon Y, Wijesinghe WAJP, Kim S (2011) Enzyme-assisted extraction and recovery of bioactive components from seaweeds. Handb Mar Macroalgae Biotechnol Appl Phycol:221–228
Jespersen L, Strømdahl LD, Olsen K, Skibsted LH (2005) Heat and light stability of three natural blue colorants for use in confectionery and beverages. Eur Food Res Technol 220:261–266
Jung KA, Lim S-R, Kim Y, Park JM (2013a) Potentials of macroalgae as feedstocks for biorefinery. Bioresour Technol 135:182–190
Jung J, Vermerris W, Gallo M, Fedenko J, Erickson J, Altpeter F (2013b) RNA interference suppression of lignin biosynthesis increasesfermentable sugar yields for biofuel production from field-grownsugarcane. Biotechnol J 6:709–716
Kadam SU, Tiwari BK, O'Donnell CP (2013) Application of novel extraction technologies for bioactives from marine algae. J Agric Food Chem 61:4667–4675
Kamm B, Kamm M (2004) Principles of biorefineries. Appl Microbiol Biotechnol 64:137–145
Kamm B, Kamm M, Soyez K (1998) The green biorefinery, concept of technology. In: First international symposium on green biorefinery. Neuruppin, Society of Ecological Technology and System Analysis, Berlin
Kerrison PD, Stanley MS, Edwards MD, Black KD, Hughes AD (2015) The cultivation of European kelp for bioenergy: site and species selection. Biomass Bioenergy 80:229–242
Kim SM, Shang YF, Um B (2011) A preparative method for isolation of fucoxanthin from Eiseniabicyclis by centrifugal partition chromatography. Phytochem Anal 22:322–329
Kim D-H, Eom S-H, Kim TH, Kim B-Y, Kim Y-M, Kim S-B (2013) Deodorizing effects of phlorotannins from edible brown alga Eiseniabicyclis on methyl mercaptan. J Agric Sci 5:95
Kostas E, Stuart W, Daniel W, David C (2016a) Optimization of a total acid hydrolysis based protocol for the quantification of carbohydrate in macroalgae. J Algal Biomass Util 7:21–36
Kostas E, White DA, Du C, Cook DJ (2016b) Selection of yeast strains for bioethanol production from UK seaweeds. J Appl Phycol 28:1427–1441
Kraan S (2013) Mass-cultivation of carbohydrate rich macroalgae, a possible solution for sustainable biofuel production. Mitig Adapt Strateg Glob Chang 18:27–46
Kubrakova IV, Toropchenova ES (2008) Microwave heating for enhancing efficiency of analytical operations. Inorg Mater 44:1509–1519
Liang W, Mao X, Peng X, Tang S (2014) Effects of sulfate group in red seaweed polysaccharides on anticoagulant activity and cytotoxicity. Carbohydr Polym 101:776–785
Lim SN, Cheung PCK, Ooi VEC, Ang PO (2002) Evaluation of antioxidative activity of extracts from a brown seaweed, *Sargassum siliquastrum*. J Agric Food Chem 50:3862–3866
Liu H, Gu L (2012) Phlorotannins from brown algae (*Fucus vesiculosus*) inhibited the formation of advanced glycation endproducts by scavenging reactive carbonyls. J Agric Food Chem 60:1326–1334
Lundquist TJ, Woertz IC, Quinn NWT, Benemann JR (2010) A realistic technology and engineering assessment of algae biofuel production. Energy Biosci Inst:1
Luque-Garcia JL, De Castro MDL (2003) Ultrasound: a powerful tool for leaching. TrAC Trends Anal Chem 22:41–47
Mæhre HK, Malde MK, Eilertsen K, Elvevoll EO (2014) Characterization of protein, lipid and mineral contents in common Norwegian seaweeds and evaluation of their potential as food and feed. J Sci Food Agric 94:3281–3290
Maneein S, Milledge JJ, Nielsen BV, Harvey PJ (2018) A review of seaweed pre-treatment methods for enhanced biofuel production by anaerobic digestion or fermentation. Fermentation 4:100
Matanjun P, Mohamed S, Mustapha NM, Muhammad K, Ming CH (2008) Antioxidant activities and phenolics content of eight species of seaweeds from North Borneo. J Appl Phycol 20:367–373

Matos GS et al (2021) Advances in extraction methods to recover added-value compounds from seaweeds: sustainability and functionality. Food Secur 10(3):516

McKennedy J, Sherlock O (2015) Anaerobic digestion of marine macroalgae: a review. Renew Sust Energ Rev 52:1781–1790

Mendiola JA, Santoyo S, Cifuentes A, Reglero G, Ibanez E, Señoráns FJ (2008) Antimicrobial activity of sub-and supercritical CO2 extracts of the green algae *Dunaliella salina*. J Food Prot 71:2138–2143

Menetrez MY (2012) An overview of algae biofuel production and potential environmental impact. Environ Sci Technol 46:7073–7085

Michalak I (2018) Experimental processing of seaweeds for biofuels. Wiley Interdiscip Rev Energy Environ 7:e288

Milledge JJ (2012) Microalgae-commercial potential for fuel, food and feed. Food Sci Technol 26:28–30

Milledge JJ, Harvey PJ (2016a) Potential process 'hurdles' in the use of macroalgae as feedstock for biofuel production in the British Isles. J Chem Technol Biotechnol 91:2221–2234

Milledge JJ, Harvey PJ (2016b) Ensilage and anaerobic digestion of *Sargassum muticum*. J Appl Phycol 28:3021–3030

Milledge JJ, Heaven S (2014) Methods of energy extraction from microalgal biomass: a review. Rev Environ Sci Bio/Technol 13:301–320

Milledge JJ, Heaven S (2017) Energy balance of biogas production from microalgae: effect of harvesting method, multiple raceways, scale of plant and combined heat and power generation. J Mar Sci Eng 5:9

Milledge JJ, Nielsen BV, Maneein S, Harvey PJ (2019) A brief review of anaerobic digestion of algae for bioenergy. Energies 12:1166

Murphy F, Devlin G, Deverell R, McDonnell K (2013) Biofuel production in Ireland – an approach to 2020 targets with a focus on algal biomass. Energies 6:6391–6412

Nakamura T, Nagayama K, Uchida K, Tanaka R (1996) Antioxidant activity of phlorotannins isolated from the brown alga Eisenia bicyclis. Fish Sci 62:923–926

Namvar F, Rahman HS, Mohamad R, Baharara J, Mahdavi M, Amini E, Chartrand MS, Yeap SK (2014) Cytotoxic effect of magnetic iron oxide nanoparticles synthesized via seaweed aqueous extract. Int J Nanomedicine 9:2479

Ng DKS, Ng KS, Ng RTL (2017) Integrated biorefineries. Encycl Sustain Technol:299–314. https://doi.org/10.1016/B978-0-12-409548-9.10138-1

Niu J-F, Wang G-C, Tseng C-K (2006) Method for large-scale isolation and purification of R-phycoerythrin from red alga *Polysiphonia urceolata* Grev. Protein Expr Purif 49:23–31

Nomura M, Kamogawa H, Susanto E, Kawagoe C, Yasui H, Saga N, Hosokawa M, Miyashita K (2013) Seasonal variations of total lipids, fatty acid composition, and fucoxanthin contents of *Sargassum horneri* (Turner) and *Cystoseira hakodatensis* (Yendo) from the northern seashore of Japan. J Appl Phycol 25:1159–1169

Obluchinskaya ED, Daurtseva AV, Pozharitskaya ON, Flisyuk EV, Shikov AN (2019) Natural deep eutectic solvents as alternatives for extracting phlorotannins from brown algae. Pharm Chem J 53:243–247

Orejuela-Escobar L, Gualle A, Ochoa-Herrera V, Philippidis GP (2021) Prospects of microalgae for biomaterial production and environmental applications at biorefineries. Sustain For 13:3063

Øverland M, Mydland LT, Skrede A (2019) Marine macroalgae as sources of protein and bioactive compounds in feed for monogastric animals. J Sci Food Agric 99:13–24

Pande M, Bhaskarwar AN (2012) Biomass conversion to energy. In: Biomass conversion. Springer, pp 1–90

Park P-J, Shahidi F, Jeon Y-J (2004) Antioxidant activities of enzymatic extracts from an edible seaweed *Sargassum horneri* using ESR spectrometry. J Food Lipids 11:15–27

Pasquet V, Chérouvrier J-R, Farhat F, Thiéry V, Piot J-M, Bérard J-B, Kaas R, Serive B, Patrice T, Cadoret J-P (2011) Study on the microalgal pigments extraction process: performance of microwave assisted extraction. Process Biochem 46:59–67

Pavia H, Toth GB (2000) Influence of light and nitrogen on the phlorotannin content of the brown seaweeds *Ascophyllum nodosum* and *Fucus vesiculosus*. Hydrobiologia 440:299–305

Pérez MJ, Falqué E, Domínguez H (2016) Antimicrobial action of compounds from marine seaweed. Mar Drugs 14:52

Phillips GO, Williams PA (2009) Handbook of hydrocolloids. Elsevier

Plaza M, Amigo-Benavent M, Del Castillo MD, Ibáñez E, Herrero M (2010) Facts about the formation of new antioxidants in natural samples after subcritical water extraction. Food Res Int 43:2341–2348

Prabhu MS, Israel A, Palatnik RR, Zilberman D, Golberg A (2020) Integrated biorefinery process for sustainable fractionation of *Ulva ohnoi* (Chlorophyta): process optimization and revenue analysis. J Appl Phycol 32:2271–2282

Ragan MA, Glombitza K-W (1986) Phlorotannins, brown algal polyphenols. In: Round FE, Chapman DJ (eds) Progress in phycological research. Biopress Ltd., Bristol, pp 129–241

Rahhou A, Layachi M, Akodad M, El Ouamari N, Rezzoum NE, Skalli A, Oudra B, El Bakali M, Kolar M, Imperl J, Petrova P, Moumen A, Baghour M (2023) The bioremediation potential of Ulva lactuca (Chlorophyta) causing green tide in Marchica lagoon (NE Morocco, Mediterranean Sea): biomass, heavy metals, and health risk assessment. Water 15:1310. https://doi.org/10.3390/w15071310

Rai A, Mohanty B, Bhargava R (2016) Fitting of broken and intact cell model to supercritical fluid extraction (SFE) of sunflower oil. Innov Food Sci Emerg Technol 38:32–40

Ramírez CHV (2015) Biogas production from seaweed biomass: a biorefinery approach.

Ramirez CHV (2017) Biogas production from seaweed biomass: a biorefinery approach

Rioux L-E, Turgeon SL, Beaulieu M (2007) Characterization of polysaccharides extracted from brown seaweeds. Carbohydr Polym 69:530–537

Rodriguez-Jasso RM, Mussatto SI, Pastrana L, Aguilar CN, Teixeira JA (2011) Microwave-assisted extraction of sulfated polysaccharides (fucoidan) from brown seaweed. Carbohydr Polym 86:1137–1144

Routray W, Orsat V (2012) Microwave-assisted extraction of flavonoids: a review. Food Bioprocess Technol 5:409–424

Roy VC, Ho TC, Lee H-J, Park J-S, Nam SY, Lee H, Getachew AT, Chun B-S (2021) Extraction of astaxanthin using ultrasound-assisted natural deep eutectic solvents from shrimp wastes and its application in bioactive films. J Clean Prod 284:125417

Rudolph B (2000) Seaweed products: red algae of economic significance. Mar Freshw Prod Handb:515–529

Rupérez P (2002) Mineral content of edible marine seaweeds. Food Chem 79:23–26

Sachindra NM, Sato E, Maeda H, Hosokawa M, Niwano Y, Kohno M, Miyashita K (2007) Radical scavenging and singlet oxygen quenching activity of marine carotenoid fucoxanthin and its metabolites. J Agric Food Chem 55:8516–8522

Santoyo S, Plaza M, Jaime L, Ibañez E, Reglero G, Señorans J (2011) Pressurized liquids as an alternative green process to extract antiviral agents from the edible seaweed *Himanthalia elongata*. J Appl Phycol 23:909–917

Saravana PS, Cho Y-J, Park Y-B, Woo H-C, Chun B-S (2016) Structural, antioxidant, and emulsifying activities of fucoidan from *Saccharina japonica* using pressurized liquid extraction. Carbohydr Polym 153:518–525

Sekar S, Chandramohan M (2008) Phycobiliproteins as a commodity: trends in applied research, patents and commercialization. J Appl Phycol 20:113–136

Shirsath SR, Sonawane SH, Gogate PR (2012) Intensification of extraction of natural products using ultrasonic irradiations – a review of current status. Chem Eng Process Process Intensif 53:10–23

Singh BS, Lobo HR, Pinjari DV, Jarag KJ, Pandit AB, Shankarling GS (2013) Ultrasound and deep eutectic solvent (DES): a novel blend of techniques for rapid and energy efficient synthesis of oxazoles. Ultrason Sonochem 20:287–293

Siriwardhana N, Kim K, Lee K, Kim S, Ha J, Song CB, Lee J, Jeon Y (2008) Optimisation of hydrophilic antioxidant extraction from Hizikia fusiformis by integrating treatments of enzymes, heat and pH control. Int J Food Sci Technol 43:587–596

Smit AJ (2004) Medicinal and pharmaceutical uses of seaweed natural products: a review. J Appl Phycol 16:245–262

Tabassum MR, Xia A, Murphy JD (2018) Biomethane production from various segments of brown seaweed. Energy Convers Manag 174:855–862

Targett NM, Arnold TM (1998) Minireview—predicting the effects of brown algal phlorotannins on marine herbivores in tropical and temperate oceans. J Phycol 34:195–205

Terasaki M, Hirose A, Narayan B, Baba Y, Kawagoe C, Yasui H, Saga N, Hosokawa M, Miyashita K (2009) Evaluation of recoverable functional lipid components of several brown seaweeds (Phaeophyta) from Japan with special reference to fucoxanthin and fucosterol contents 1. J Phycol 45:974–980

Torres MD, Kraan S, Domínguez H (2019) Seaweed biorefinery. Rev Environ Sci Bio/Technol 18:335–388

Vera J, Castro J, Gonzalez A, Moenne A (2011) Seaweed polysaccharides and derived oligosaccharides stimulate defense responses and protection against pathogens in plants. Mar Drugs 9:2514–2525

Vilkhu K, Manasseh R, Mawson R, Ashokkumar M (2011) Ultrasonic recovery and modification of food ingredients. In: Ultrasound technologies for food and bioprocessing. Springer, pp 345–368

Waterman PG, Mole S (1994) Analysis of phenolic plant metabolites. Blackwell Scientific

Wei N, Quarterman J, Jin Y-S (2013) Marine macroalgae: an untapped resource for producing fuels and chemicals. Trends Biotechnol 31:70–77

Weiland P (2010) Biogas production: current state and perspectives. Appl Microbiol Biotechnol 85:849–860

Wijesinghe W, Jeon Y-J (2012) Enzyme-assistant extraction (EAE) of bioactive components: a useful approach for recovery of industrially important metabolites from seaweeds: a review. Fitoterapia 83:6–12

Zainal-Abidin MH, Hayyan M, Hayyan A, Jayakumar NS (2017) New horizons in the extraction of bioactive compounds using deep eutectic solvents: a review. Anal Chim Acta 979:1–23

Zamalloa C, Vulsteke E, Albrecht J, Verstraete W (2011) The techno-economic potential of renewable energy through the anaerobic digestion of microalgae. Bioresour Technol 102:1149–1158

Zollmann M, Robin A, Prabhu M, Polikovsky M, Gillis A, Greiserman S, Golberg A (2019) Green technology in green macroalgal biorefineries. Phycologia 58:516–534

Chapter 13
Seaweeds in Integrated Multi-Trophic Aquaculture: Environmental Benefits and Bioactive Compounds Production

Eleonora Curcuraci, Claire Hellio, Concetta Maria Messina, and Andrea Santulli

1 Introduction

1.1 Integrated Multi-Trophic Aquaculture (IMTA)

Over the last decades aquaculture has expanded rapidly ensuring the global supply of protein from fish, since the global consumption of aquatic foods has increased and fishery resources have continued to decline (FAO 2022) (Fig. 13.1).

According to FAO, in 2022 aquaculture industrial sector reached 122.6 million tonnes, which was estimated to be worth USD 281.5 billion (FAO 2022). Since over the years, a large part of consumers have become very aware and demand increasingly higher guarantees in terms of quality, traceability and production conditions (Bottema et al. 2021), in the last decades the aquaculture sector is committed to ensuring responsible farming practices that provide food products while maintaining the integrity of aquatic ecosystems and sustainability of the aquaculture industry (FAO 2020). To achieve these goals, it is necessary to establish management practices that are efficient, diversified in terms of products and sustainable from both an environmental and a social point of view (Troell et al. 2009).

An innovative and promising way of meeting the need for fish products in a sustainable basis is represented by Integrated Multi-trophic Aquaculture (IMTA)

E. Curcuraci · C. M. Messina (✉)
Department of Earth and Marine Sciences DiSTeM, University of Palermo, Trapani, Italy
e-mail: concetta.messina@unipa.it

C. Hellio
LEMAR, IRD, CNRS, Ifremer, Université de Brest, Brest, France

A. Santulli
Department of Earth and Marine Sciences DiSTeM, University of Palermo, Trapani, Italy

Istituto di Biologia Marina, Consorzio Universitario della Provincia di Trapani, Trapani, Italy

F. Ozogul et al. (eds.), *Seaweeds and Seaweed-Derived Compounds*,
https://doi.org/10.1007/978-3-031-65529-6_13

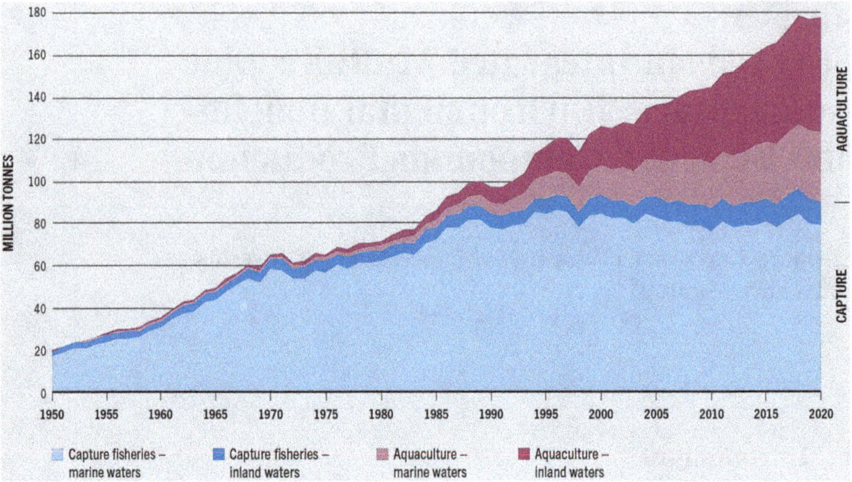

Fig. 13.1 World capture fisheries and aquaculture production. (Source: FAO 2022)

systems (FAO 2022), a polyculture system that, recreating a simplified natural ecosystem, maximises the economic and productive yields and minimises the environmental impacts (Knowler et al. 2020; Thomas et al. 2021; Mansour et al. 2022).

This ideal system, that has its origins in Asia (Hughes and Black 2016; FAO 2013), is based on the co-culture of pellet-farmed species with organic (e.g., fish, echinoderms, molluscs) and inorganic (e.g., micro/macro-algae, macrophytes) extractive species (Troell et al. 2009), recreating the natural food chain through the exploitation of species belonging to different trophic levels and having complementary functions (Knowler et al. 2020; Thomas et al. 2021; Mansour et al. 2022). In this way, faces naturally produced by the fish, as well as unconsumed feed are reused as nutrients by the co-cultured species in the aquaculture facility (Thomas et al. 2021; Mansour et al. 2022; Tacon 2020). As in a natural food chain, at every level, in aquatic cultivated organisms there is a loss of energy, uneaten feed, nutrient excretions and faeces. An ecologically engineer balanced IMTA system includes feeding species, inorganic and organic extractive species such as microalgae, seaweeds and plants, deposit-feeders such as polychaetes, sea cucumbers and sea urchins (Thomas et al. 2021; Mansour et al. 2022; Cutajar et al. 2022; Fraga-Corral et al. 2022; Albrektsen et al. 2022; Hargrave et al. 2022) and filter feeders (mussels) that recover nutrients in the form of particulate organic matter (POM) (Hargrave et al. 2022). IMTA extractive species, acting as living filters, ensures nutrient recycling improving the environmental performance of aquaculture sites (Fraga-Corral et al. 2022; Campanati et al. 2021), biomass production as food, feed and as source of high value bioactive molecules (Mansour et al. 2022; Fraga-Corral et al. 2022; Chopin 2013). Indeed extracted and isolated bioactive compounds and natural pigments of the co-cultivated species, are recognised for their use as functional foods in nutraceuticals and in cosmeceuticals and pharmaceuticals industries, due to their antioxidant and antimicrobial properties which are important for the welfare of farmed species and human health (Mansour et al. 2022; Fraga-Corral et al. 2022). This, as well as having a role in biomitigation of the impact of aquaculture activity, leads to an increase in the commercial role and important economic value of co-cultures, as well as sustainability of the activity, product diversification with a guarantee of higher economic stability and higher consumer acceptability due to the application of sustainable management practices (Troell et al. 2009; Knowler et al. 2020; Barrington et al. 2010). Therefore, IMTA increases yields through a more efficient use of resources developing new value chains (Fraga-Corral et al. 2022).

In Asia, extractive organisms such as molluscs and seaweeds are widely utilized in IMTA's industrial sectors and their high profitability stimulates their production (Thomas et al. 2021; Hughes and Black 2016; FAO 2013). Only recently IMTA farming has found a place and attracted interest among producers in Western countries (Hough 2022), where aquaculture is mainly dominated by monospecific cultivation of fish, which has led to a low reuse of the waste produced by the aquaculture sector (FAO 2018, 2020; Thomas et al. 2021). This slow adoption of technologies to support IMTA can be attributed to poor legislative, regulatory and social progress and support, consumer's misinformation, economic reasons such as lack of direct financial benefits to the entrepreneur and lack of foresight in responsible management of coastal waters to ensure sustainability and profitability of the adoption of IMTA systems (Knowler et al. 2020; Thomas et al. 2021; Barrington et al. 2010; Hough 2022). The legislative framework for IMTA systems is complex and non-harmonised, often focused on monoculture and characterised by low consumer awareness. These conditions could be fixed through closer collaboration between the research sector, through the development of pilot-scale IMTA facilities and subsequent scale-up, through improved regulatory frameworks by legislators and through consumer empowerment for the development of the bioeconomy (Knowler et al. 2020; Thomas et al. 2021; Fraga-Corral et al. 2022; Buck et al. 2018; Wang et al. 2020).

1.2 Seaweeds Role in IMTA Systems

1.2.1 Environmental Benefits

Important benefits related to the co-cultivation of seaweeds in IMTAs are connected to the capacity of the co-cultivated extractive species to uptake and/or utilize nutrients, establishing new value chains and more efficient and cost-effective inorganic and organic pollutants removal methods, compared to the conventional treatment techniques (Knowler et al. 2020; Thomas et al. 2021; Mansour et al. 2022; Pilone 2021).

In accordance with the principles promoted by the European Commission and by FAO at Rio+20 conference fostering the "Blue Growth" concept, which refers to a green economy and sustainable development of the maritime and coastal sectors (Eikeset et al. 2018), it is necessary to implement the concept of reuse of resources by recognising the potentialities of by-products produced within aquaculture facilities (Campanati et al. 2021). Undigested feed and faeces are some of the main concerns related to intensive aquaculture farming, in fact, particulate and dissolved organic matter exceeding from this activity, may lead to eutrophication condition (Campanati et al. 2021; Gao et al. 2022), although the concentrations of dissolved nutrients and solids released in the wastewater depend on the farming practice: cages, ponds, raceways or recirculating aquaculture systems (RAS) (Campanati et al. 2021). Among dissolved nutrients of aquaculture wastewaters, the most significant are nitrogen (N) and phosphorus (P) (Campanati et al. 2021). Nitrogen (N) is one of the main causes of eutrophication and represents the 60–90% of dissolved waste produced by aquaculture activities (Campanati et al. 2021; Van Rijn 2013). Dissolved inorganic nitrogen (DIN) is released as toxic (NH_3) and non-toxic ammonia (NH_4^+). Ammonia can be nitrified to nitrate (NO_3^-) not well tolerated by aquatic species, or reduced to nitrite (NO_2^-), a highly reactive oxidant and toxic, by biological activities (Campanati et al. 2021; Van Rijn 2013). Phosphorus (P), that is considered a limiting element in the marine environment, is present in wastewaters as orthophosphate, polyphosphate, pyrophosphate and metaphosphate and lost such as feed uneaten and mainly realised as solid with faeces (Campanati et al. 2021).

Nitrogen and phosphorous uptake and recycling from wastewater effluents are at the base of bioremediation, defined as "a sustainable strategy focused on alleviating the negative effects of aquaculture effluents" (Martinez-Porchas et al. 2014) (Fig. 13.2).

More specifically, "phycoremediation" refers to the use of algae for wastewater remediation (Koul et al. 2022). Since algae assimilate nitrogen either as NH_4^+or NO_3^-, and store phosphorus, as essential elements for their growth, bioremediation studies have been conducted by using seaweeds due to their capacity to reduce chemical and biological oxygen demand and to remove nutrients. This ability to biofilter water recycling nutrients and by-products in aquaculture, makes the co-cultivation of seaweeds in IMTAs a key element in increasing the sustainability of the aquaculture sector (Machado et al. 2020, 2022; Hargrave et al. 2022; Barrington et al. 2010; Gao et al. 2022; Sickander and Filgueira 2022; Massocato et al. 2022; Aquilino et al. 2020; Khanjani et al. 2022; Azevedo et al. 2022; Biris-Dorhoi et al.

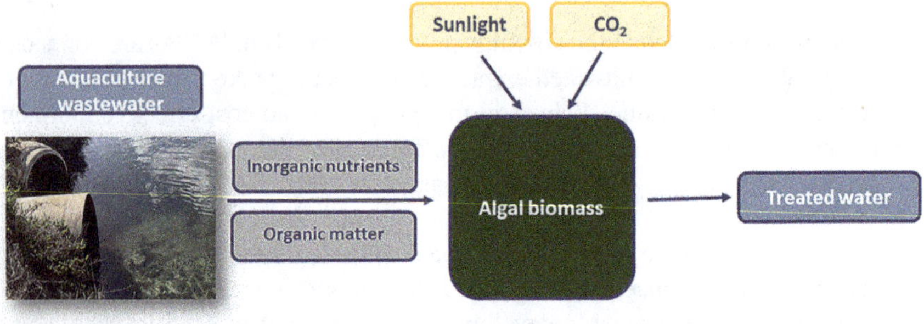

Fig. 13.2 Integrated Multi-trophic Aquaculture (IMTA) bioremediation

2020; Tziveleka et al. 2021; da Silva et al. 2022; Zhou et al. 2006; Yang et al. 2006; Spanò et al. 2022; Minicante et al. 2022; Lubsch and Timmermans 2019; Barbosa et al. 2020; ; Pimentel et al. 2020; Campos et al. 2022).

Another crucial environmental benefit related to the co-cultivation of extractive species in IMTA is linked to "Global warming" mitigation, a major consequence of the increasing concentration of CO_2 in the atmosphere, which is leading to rising temperatures on a global scale (Gao et al. 2022; Duarte et al. 2017; Mashoreng et al. 2019; Ould and Caldwell 2022; Trevathan-Tackett et al. 2015). This is driving the scientific community and industry sectors to investigate mitigation strategies able to decrease the concentration of atmospheric CO_2 and to contribute to countering climate change (Gao et al. 2022; Duarte et al. 2017; Mashoreng et al. 2019; Ould and Caldwell 2022; Trevathan-Tackett et al. 2015). Numerous studies have highlighted the ability of algae and marine plants in CO_2 sequestration, introducing the concept of "Blue Carbon" (Duarte et al. 2017; Macreadie et al. 2019). The ability of algae to mitigate climate change has been quantified as greater than that of terrestrial plants, due to their high productivity rate (Gao et al. 2022; Mashoreng et al. 2019). On this basis, it is possible to imagine an improvement in the efficiency of "Carbon sequestration" by maximising algal cultivation on an industrial scale through the establishment of economic incentives for companies wishing to adopt technologies aimed at optimising production yields to mitigate eutrophication and climate change (Gao et al. 2022; Duarte et al. 2017; Mashoreng et al. 2019; Ould and Caldwell 2022; Trevathan-Tackett et al. 2015).

1.2.2 Bioactive Compounds

Since aquatic organisms are an important source of valuable metabolites whit proved bioactivity, over the past 20 years, interest in marine drug discovery (Fig. 13.3) and macroalgae applications have increased (Biris-Dorhoi et al. 2020; Van Den Burg et al. 2021; Savio et al. 2021), as well as the interest by industries and consumers in functional foods, obtained from traditional foods enriched with nutraceuticals, bioactive compounds extracted from natural plants or animal sources (Biris-Dorhoi et al. 2020; Siró et al. 2008; Freitas et al. 2012).

According to FAO (2022), in the aquaculture industry, the cultivation of algae, which in 2020 reached half a million tonnes (+1.4% compared to 2019), plays an important role in non-food aquaculture production, as extractive species (57.4% of total global aquaculture production) (FAO 2020) (Fig. 13.4).

On this basis, the co-cultivation of algae in IMTAs ensures a higher profitability and productivity, since seaweeds, thanks to their texturising agents (phycocolloids: carrageenan, alginates, agars) and the production and accumulation of protective substances, known as secondary metabolites, with bioactive properties, represent a viable response to the growing demand for biologically active compounds by the pharmacological, chemical and cosmeceutical industries, the agricultural and aquaculture sectors, for the production of functional feeds and foods (Mansour et al. 2022; Buck et al. 2018; Azevedo et al. 2022; Biris-Dorhoi et al. 2020; Tziveleka

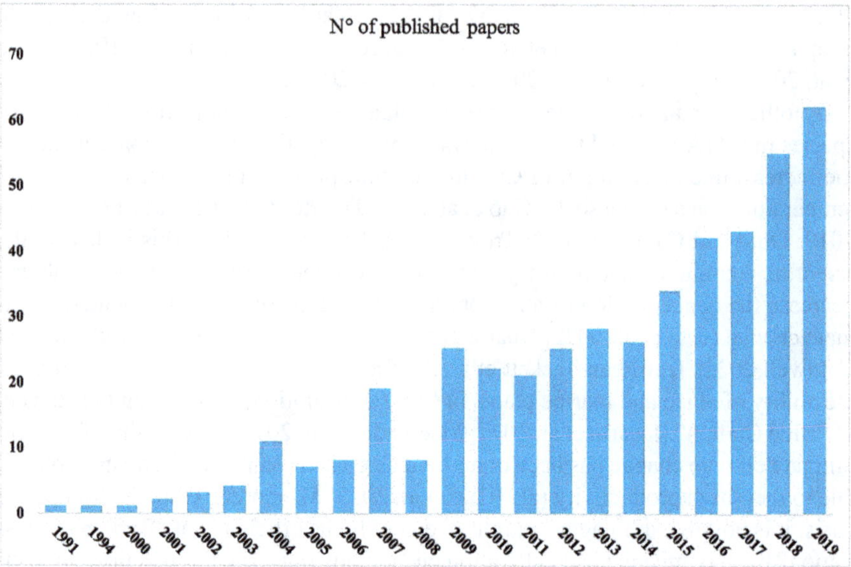

Fig. 13.3 Number of publications per year on the subject of "marine drug discovery", listed in the source index of the Web of Science (Clarivate Analytics) from 1991 to 2019. (Source: Savio et al. 2021)

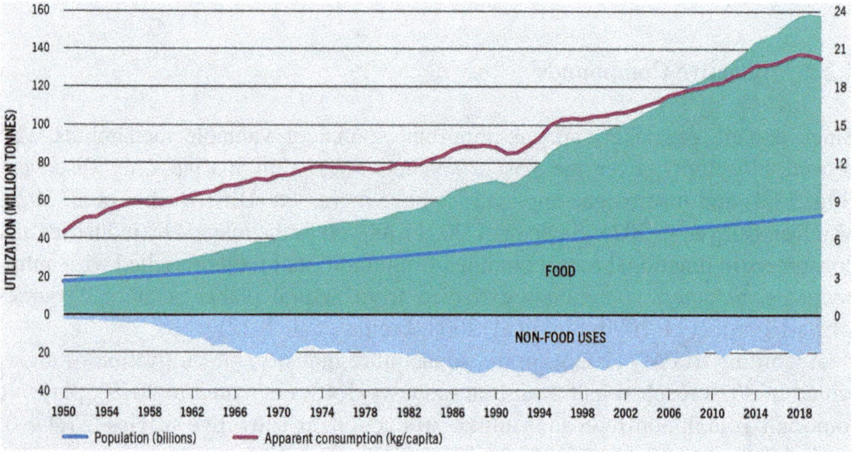

Fig. 13.4 World fisheries and aquaculture production: utilization and apparent consumption. (Source: FAO 2020)

et al. 2021; Minicante et al. 2022; Lubsch and Timmermans 2019; Machado et al. 2020; Pimentel et al. 2020; Campos et al. 2022; Edreva et al. 2008; De Almeida et al. 2011; Dubber and Harder 2008; Kiron 2012; Pradhan et al. 2022; Lomartire and Gonçalves 2022; Afonso et al. 2021a; Wang et al. 2018; Stabili et al. 2019a; Freile-Pelegrín et al. 2020; Grote 2019; Lopes et al. 2019; Passos et al. 2021).

Valuable metabolites produced by algae can be divided into primary metabolites and complex secondary metabolites. Primary metabolites are produced converting CO_2 in lipids, carbohydrates and proteins during photosynthesis and are essential to algal growth, survival, and proliferation (Salehi et al. 2019). Secondary metabolites are produced for protection and survival (Salehi et al. 2019; Karthikeyan et al. 2022; Hulkko et al. 2022) and are widely recognised to have high antiviral, antibacterial and cytotoxic activity and to act as natural antioxidants (Pradhan et al. 2022; Lomartire and Gonçalves 2022; Salehi et al. 2019; Pradhan et al. 2020). Among them, photosynthetic pigments (carotenoids, xanthophylls and chlorophylls), rich in health-beneficial, are produced for the adaptation to extreme environmental conditions (Hulkko et al. 2022), countering the oxidative and photo-oxidative stress that lead to reactive oxygen species (ROS) generation damaging the photosynthetic apparatus (Hulkko et al. 2022), are widely used in the food, nutraceutical and pharmacological industries (Arena et al. 2021). Phenolic compounds, which include phenolic acids (hydroxybenzoic acids and hydroxycinnamic acids) and flavonoids, have a key role as defense against environmental stress (Hulkko et al. 2022; Ramawat and Mérillon 2013) and are recognized as important healthy natural antioxidant phytochemicals that, inhibiting macromolecular oxidation, have anticancer, anti-microbial, anti-inflammatory, anti-diabetic and antiviral properties (Tziveleka et al. 2021; Salehi et al. 2019; Hulkko et al. 2022). The bioavailability and potential beneficial effects of algal metabolites have shown promising results in both *in vivo* and *in vitro* tests (Salehi et al. 2019; Pradhan et al. 2020).

1.3 Seaweeds in IMTA Systems

The international scientific community and industrial sectors are focusing on promoting the co-cultivation of extractive species, such as seaweeds, IMTA systems to mitigate the environmental impact of aquaculture by improving its efficiency and sustainability (Bottema et al. 2021; Knowler et al. 2020; Thomas et al. 2021; Biswas et al. 2020; García-Poza et al. 2020). Several seaweeds species have been utilized for the development of IMTA systems. Their selection for cultivation is strictly linked to their growth rate, nitrogen availability, their resistance to epiphytes and pathogens, their biofiltering capacity, their bioactive compounds production and their profitability (Khanjani et al. 2022; García-Poza et al. 2020; Torres et al. 2019; Soto 2009). Thanks to their ability to act as biofilters, to produce biomass, to uptake nutrients and to produce bioactive valuable compounds, the most commonly cultivated species in IMTA systems belong to the following genera (Khanjani et al. 2022; Minicante et al. 2022; Campos et al. 2022; García-Poza et al. 2020; Soto 2009):

- *Chaetomorpha* is a filamentous green algae with a rapid growth rate, which has proven to be highly suitable for its inclusion in ponds with high concentrations of nutrients such as nitrogen and phosphorus (Aquilino et al. 2020). *Chaetomorpha* has also been shown to be rich in valuable compounds with beneficial properties for both fish and human health (Freile-Pelegrín et al. 2020; Stabili et al. 2019b).

- *Chondrus* is a red seaweed intensively cultivated for the extraction of carrageenan, a polysaccharide widely exploited by several industries and the extraction of lipids. Both extracts find applications in food, feed and pharmaceutical sectors (Azevedo et al. 2022; Campos et al. 2022; Grote 2019; Lopes et al. 2019; Passos et al. 2021). Furthermore, this macroalga its been utilized for biomass growth and nutrients removal in IMTAs (Khanjani et al. 2022; Azevedo et al. 2022; Campos et al. 2022).
- *Gracilaria* is a red seaweed extensively cultivated, especially in East Asia (FAO 2020), industrially important for agar extraction (De Almeida et al. 2011), for the production of antimicrobial compounds such as acrylic acid and eicosanoids, which are derived from the oxidative metabolism of polyunsaturated fatty acids (PUFA) such as eicosanoic acid and arachidonic acid (De Almeida et al. 2011). Furthermore, due to the remarkable ability to accumulate and utilize dissolved nutrients in water, *Gracilaria* can be cultivated for bioremediation (Zhou et al. 2006; Yang et al. 2006; Spanò et al. 2022). *Gracilaria* has demonstrated to be a good candidate for its cultivation in IMTAs, for both the reduction of organic loads in pond effluents and for the simultaneous production of bioactive compounds (Khanjani et al. 2022; Minicante et al. 2022; Campos et al. 2022; García-Poza et al. 2020).
- *Kappaphycus* is a red seaweed rich in polysaccharides, carrageenan essential for the hydrocolloids industry and metabolites with antioxidant activity. It is intensively cultivated in IMTAs (especially in Asia) and has proved high biomass production and biofiltration capacity (Azevedo et al. 2022; Biris-Dorhoi et al. 2020; Tzivelека et al. 2021; da Silva et al. 2022; Campos et al. 2022; Grote 2019).
- *Laminaria* is a brown seaweed with large amounts of carbohydrates, proteins and secondary metabolites used in food, feed and medicine fields due to their proved antifouling and antibacterial activity (Lubsch and Timmermans 2019; Dubber and Harder 2008). Their co-cultivation in IMTA systems has revealed their proficiency in producing biomass rich in bioactive compounds, as well as their nutrient uptake capacity for biomitigation (Hargrave et al. 2022; Barrington et al. 2010; Minicante et al. 2022; Lubsch and Timmermans 2019; Barbosa et al. 2020).
- *Macrocystis* is a brown macroalga largely utilized as biostimulant and biofertilizer (Minicante et al. 2022; Pradhan et al. 2022), it is used for the extraction of alginates (Buck et al. 2018; Kiron 2012; Pradhan et al. 2022) and has also found applications in the formulation of animal feed (Buck et al. 2018; Kiron 2012). Pre-clinical studies have demonstrated the potential pharmacological antidiabetic activity of its extracted bioactive compounds (Lomartire and Gonçalves 2022) and a higher rate of survival on farmed fish fed with *Macrocystis* alginates (Kiron 2012).
- *Palmaria* is an edible red algae with high protein content (Biris-Dorhoi et al. 2020; Campos et al. 2022; Grote 2019), which production has increased considerably in recent years due to the increasing consumer demand (Lopes et al.

2019). *Palmaria* is rich in antioxidants with health-promoting properties (Grote 2019; Lopes et al. 2019). Therefore, its cultivation, in addition to providing an important source of bioactive compounds, has shown significant nutrient removal in IMTA systems due to its ability to act as a biofilter (Khanjani et al. 2022; Grote 2019).

- *Porphyra*, that is one of the most cultivated genus in IMTAs, is a commercially valuable seaweed, due to its high protein content and its amino acid profile, bio-filtration capacity and due to its rapid capability to assimilate nutrients that make this genus highly recommended for bioremediation approaches (Sickander and Filgueira 2022; Machado et al. 2020; Pimentel et al. 2020; Campos et al. 2022).
- *Saccharina* is a brown macroalga with a high carbohydrates content (Biris-Dorhoi et al. 2020), used as dried feed (Mansour et al. 2022), in nutraceutical and cosmeceutical fields (Minicante et al. 2022), due to the antioxidant capacity and prebiotic effects of its bioactive molecules (Afonso et al. 2021b), as biofilter due to their high capacity to uptake nutrients (Hargrave et al. 2022; Gao et al. 2022; Lubsch and Timmermans 2019), and for their rapid growth rate in aquaculture systems (Hargrave et al. 2022).
- *Ulva* has a fast growth rate (Machado et al. 2020, 2022), its largely utilized in IMTAs due to its high rate of nutrients removal (Mansour et al. 2022; Massocato et al. 2022; Minicante et al. 2022) and its cultivation is also linked to the production of nutraceuticals, cosmetics, food, hydrocolloids, paper and bioplastics (Biris-Dorhoi et al. 2020; Minicante et al. 2022).
- *Undaria* is an edible seaweed, mainly produced as human food (FAO 2020). It is a valuable source of bioactive secondary metabolites able to prevent cancer, its progression and healing due to their antioxidant, antiinflammatory and antiproliferative ability. In addition, antiviral and antibacterial activities have been reported for this genus (Biris-Dorhoi et al. 2020; Wang et al. 2018). Other applications include their use in cosmeceutical fields (Minicante et al. 2022).

2 Conclusions

Many studies, in recent years, have highlighted the most important aspects for selecting seaweed species for their co-cultivation in IMTA systems: their ecological role, growth rate, nutrient uptake capacity, production of valuable secondary metabolites and their economic value.

The production of seaweeds in well designed and managed IMTA systems might therefore provide socio-economic benefits and environmental sustainability of the aquaculture sector by nutrients removal, climate mitigation and biomass production as a source of valuable bioactive compounds for commercial sectors such as nutraceuticals, cosmeceuticals and pharmaceuticals.

Conflicts of Interest The authors declare no conflict of interest.

Funding This research received no external funding.

References

Afonso C, Correia AP, Freitas MV, Mouga T, Baptista T (2021a) In vitro evaluation of the antibacterial and antioxidant activities of extracts of gracilaria gracilis with a view into its potential use as an additive in fish feed. Appl Sci 11:6642

Afonso C, Matos J, Guarda I, Gomes-Bispo A, Gomes R, Cardoso C, Gueifão S, Delgado I, Coelho I, Castanheira I et al (2021b) Bioactive and nutritional potential of Alaria esculenta and Saccharina latissima. J Appl Phycol 33:501–513

Albrektsen S, Kortet R, Skov PV, Ytteborg E, Gitlesen S, Kleinegris D, Mydland L-T, Hansen JØ, Lock E-J, Mørkøre T et al (2022) Future feed resources in sustainable salmonid production: a review. Rev Aquac 14:1–23

Aquilino F, Paradiso A, Trani R, Longo C, Pierri C, Corriero G, de Pinto MC (2020) Chaetomorpha linum in the bioremediation of aquaculture wastewater: optimization of nutrient removal efficiency at the laboratory scale. Aquaculture 523:735133

Arena R, Lima S, Villanova V, Moukri N, Curcuraci E, Messina C, Santulli A, Scargiali F (2021) Cultivation and biochemical characterization of isolated Sicilian microalgal species in salt and temperature stress conditions. Algal Res 59:102430

Azevedo G, Torres MD, Almeida PL, Hilliou L (2022) Exploring relationships between seaweeds carrageenan contents and extracted hybrid carrageenan properties in wild and cultivated Mastocarpus stellatus, Chondrus crispus and Ahnfeltiopsis devoniensis. Algal Res 67:102840

Barbosa M, Fernandes F, Pereira DM, Azevedo IC, Sousa-Pinto I, Andrade PB, Valentão P (2020) Fatty acid patterns of the kelps Saccharina latissima, Saccorhiza polyschides and Laminaria ochroleuca: influence of changing environmental conditions. Arab J Chem 13:45–58

Barrington K, Ridler N, Chopin T, Robinson S, Robinson B (2010) Social aspects of the sustainability of integrated multi-trophic aquaculture. Aquac Int 18:201–211

Biris-Dorhoi ES, Michiu D, Pop CR, Rotar AM, Tofana M, Pop OL, Socaci SA, Farcas AC (2020) Macroalgae—a sustainable source of chemical compounds with biological activities. Nutrients 12:1–23

Biswas G, Kumar P, Ghoshal TK, Kailasam M, De D, Bera A, Mandal B, Sukumaran K, Vijayan KK (2020) Integrated multi-trophic aquaculture (IMTA) outperforms conventional polyculture with respect to environmental remediation, productivity and economic return in brackishwater ponds. Aquaculture 516:734626

Bottema MJM, Bush SR, Oosterveer P (2021) Assuring aquaculture sustainability beyond the farm. Mar Policy 132:104658

Buck BH, Troell MF, Krause G, Angel DL, Grote B, Chopin T (2018) State of the art and challenges for offshore integrated multi-trophic aquaculture (IMTA). Front Mar Sci 5:165

Campanati C, Willer D, Schubert J, Aldridge DC (2021) Sustainable intensification of aquaculture through nutrient recycling and circular economies: more fish, less waste, Blue Growth. Rev Fish Sci Aquac 30:143–169

Campos BM, Ramalho E, Marmelo I, Noronha JP, Malfeito-Ferreira M, Mata P, Diniz MS (2022) Proximate composition, physicochemical and microbiological characterization of edible seaweeds available in the Portuguese market. Front Biosci 14:26

Chopin T (2013) Aquaculture aquaculture, integrated multi-trophic (IMTA) aquaculture integrated multi-trophic (IMTA). In: Sustainable food production. Springer, New York, pp 184–205

Cutajar K, Falconer L, Massa-Gallucci A, Cox RE, Schenke L, Bardócz T, Sharman A, Deguara S, Telfer TC (2022) Culturing the sea cucumber Holothuria poli in open-water integrated multi-trophic aquaculture at a coastal Mediterranean fish farm. Aquaculture 550:1–11

da Silva EG, Castilho-Barros L, Henriques MB (2022) Economic feasibility of integrated multi-trophic aquaculture (mussel Perna perna, scallop Nodipecten nodosus and seaweed Kappaphycus alvarezii) in Southeast Brazil: a small-scale aquaculture farm model. Aquaculture 552:738031

De Almeida CLF, Falcão HDS, Lima GRDM, Montenegro CDA, Lira NS, de Athayde-Filho PF, Rodrigues LC, De Souza MDFV, Barbosa-Filho JM, Batista LM (2011) Bioactivities from marine algae of the genus Gracilaria. Int J Mol Sci 12:4550–4573
Duarte CM, Wu J, Xiao X, Bruhn A, Krause-Jensen D (2017) Can seaweed farming play a role in climate change mitigation and adaptation? Front Mar Sci 4:100
Dubber D, Harder T (2008) Extracts of Ceramium rubrum, Mastocarpus stellatus and Laminaria digitata inhibit growth of marine and fish pathogenic bacteria at ecologically realistic concentrations. Aquaculture 274:196–200
Edreva A, Velikova V, Tsonev T, Dagnon S, Gesheva E (2008) Stress-protective role of secondary metabolites: diversity of functions and mechanisms. Gen Appl Plant Physiol 34:67–78
Eikeset AM, Mazzarella AB, Davíðsdóttir B, Klinger DH, Levin SA, Rovenskaya E, Stenseth NC (2018) What is blue growth? The semantics of "sustainable development" of marine environments. Mar Policy 87:177–179
FAO (2013) The state of world fisheries and aquaculture, 2012. Choice Rev 50:50-5350
FAO (2018) The state of world fisheries and aquaculture 2018. FAO. ISBN 9789251305621
FAO (2020) The state of world fisheries and aquaculture 2020. Sustainability in action. FAO, Rome. ISBN 9789251326923
FAO (2022) The state of the world fisheries and aquaculture. Sofia
Fraga-Corral M, Ronza P, Garcia-Oliveira P, Pereira AG, Losada AP, Prieto MA, Quiroga MI, Simal-Gandara J (2022) Aquaculture as a circular bio-economy model with Galicia as a study case: how to transform waste into revalorized by-products. Trends Food Sci Technol 119:23–35
Freile-Pelegrín Y, Chávez-Quintal C, Caamal-Fuentes E, Vázquez-Delfín E, Madera-Santana T, Robledo D (2020) Valorization of the filamentous seaweed Chaetomorpha gracilis (Cladophoraceae, Chlorophyta) from an IMTA system. J Appl Phycol 32:2295–2306
Freitas AC, Rodrigues D, Rocha-Santos TAP, Gomes AMP, Duarte AC (2012) Marine biotechnology advances towards applications in new functional foods. Biotechnol Adv 30:1506–1515
Gao G, Gao L, Jiang M, Jian A, He L (2022) The potential of seaweed cultivation to achieve carbon neutrality and mitigate deoxygenation and eutrophication. Environ Res Lett 17:014018
García-Poza S, Leandro A, Cotas C, Cotas J, Marques JC, Pereira L, Gonçalves AMM (2020) The evolution road of seaweed aquaculture: cultivation technologies and the industry 4.0. Int J Environ Res Public Health 17:6528. ISBN 3512392407
Grote B (2019) Recent developments in aquaculture of Palmaria palmata (Linnaeus) (Weber & Mohr 1805): cultivation and uses. Rev Aquac 11:25–41
Hargrave MS, Nylund GM, Enge S, Pavia H (2022) Co-cultivation with blue mussels increases yield and biomass quality of kelp. Aquaculture 550:737832
Hough C (2022) Regional review on status and trends in aquaculture development in Europe – 2020. FAO Fish Aquac Circ No 1232/1
Hughes AD, Black KD (2016) Going beyond the search for solutions: understanding trade-offs in European integrated multi-trophic aquaculture development. Aquac Environ Interact 8:191–199
Hulkko LSS, Chaturvedi T, Thomsen MH (2022) Extraction and quantification of chlorophylls, carotenoids, phenolic compounds, and vitamins from halophyte biomasses. Appl Sci 12:840
Karthikeyan A, Joseph A, Nair BG (2022) Promising bioactive compounds from the marine environment and their potential effects on various diseases. J Genet Eng Biotechnol 20:14
Khanjani MH, Zahedi S, Mohammadi A (2022) Integrated multitrophic aquaculture (IMTA) as an environmentally friendly system for sustainable aquaculture: functionality, species, and application of biofloc technology (BFT). Environ Sci Pollut Res 1:3
Kiron V (2012) Fish immune system and its nutritional modulation for preventive health care. Anim Feed Sci Technol 173:111–133
Knowler D, Chopin T, Martínez-Espiñeira R, Neori A, Nobre A, Noce A, Reid G (2020) The economics of integrated multi-trophic aquaculture: where are we now and where do we need to go? Rev Aquac 12:1579–1594
Koul B, Sharma K, Shah MP (2022) Phycoremediation: a sustainable alternative in wastewater treatment (WWT) regime. Environ Technol Innov 25:102040

Lomartire S, Gonçalves AMM (2022) An overview of potential seaweed-derived bioactive compounds for pharmaceutical applications. Mar Drugs 20:1–32

Lopes D, Melo T, Meneses J, Abreu MH, Pereira R, Domingues P, Lillebø AI, Calado R, Rosário Domingues M (2019) A new look for the red macroalga Palmaria palmata: a seafood with polar lipids rich in EPA and with antioxidant properties. Mar Drugs 17:533

Lubsch A, Timmermans KR (2019) Uptake kinetics and storage capacity of dissolved inorganic phosphorus and corresponding dissolved inorganic nitrate uptake in Saccharina latissima and Laminaria digitata (Phaeophyceae). J Phycol 55:637–650

Machado M, Machado S, Pimentel FB, Freitas V, Alves RC, Oliveira MBPP (2020) Amino acid profile and protein quality assessment of macroalgae produced in an integrated multi-trophic aquaculture system. Food Secur 9:1382

Macreadie PI, Anton A, Raven JA, Beaumont N, Connolly RM, Friess DA, Kelleway JJ, Kennedy H, Kuwae T, Lavery PS et al (2019) The future of blue carbon science. Nat Commun 10:1–13

Mansour AT, Ashour M, Alprol AE, Alsaqufi AS (2022) Aquatic plants and aquatic animals in the context of sustainability: cultivation techniques, integration, and blue revolution. Sustain For 14:3257

Martinez-Porchas M, Martinez-Cordova LR, Lopez-Elias JA, Porchas-Cornejo MA (2014) Bioremediation of aquaculture effluents. In: Microbial biodegradation and bioremediation. s, pp 542–555

Mashoreng S, La Nafie YA, Isyrini R (2019) Cultivated seaweed carbon sequestration capacity. In: IOP conference series: earth and environmental science, vol 370. IOP Publishing

Massocato T, Robles-carnero V, Bastos E, Avilés A (2022) Short-term nutrient removal efficiency by Ulva pseudorotundata (Chlorophyta): potential use for integrated multi-trophic aquaculture (IMTA). J Appl Phycol 35:233–250

Minicante SA, Bongiorni L, De Lazzari A (2022) Bio-based products from Mediterranean seaweeds: Italian opportunities and challenges for a sustainable blue economy. Sustain For 14:5634

Ould E, Caldwell GS (2022) The potential of seaweed for carbon capture. CAB Rev Perspect Agric Vet Sci Nutr Nat Resour 17:1–9

Passos R, Correia AP, Ferreira I, Pires P, Pires D, Gomes E, do Carmo B, Santos P, Simões M, Afonso C et al (2021) Effect on health status and pathogen resistance of gilthead seabream (Sparus aurata) fed with diets supplemented with Gracilaria gracilis. Aquaculture 531:735888

Pilone FR (2021) The waste management of large-scale recirculating aquaculture systems and potential value-added products from the waste stream. DUNE Digit. 67.

Pimentel FB, Cermeño M, Kleekayai T, Harnedy PA, Fitzgerald RJ, Alves RC, Oliveira MBPP (2020) Effect of in vitro simulated gastrointestinal digestion on the antioxidant activity of the red seaweed Porphyra dioica. Food Res Int 136:109309

Pradhan B, Nayak R, Patra S, Jit BP, Ragusa A, Jena M (2020) Molecules bioactive metabolites from marine algae as potent pharmacophores against oxidative stress-associated human diseases: a comprehensive review. Molecules 26:1–25

Pradhan B, Bhuyan PP, Patra S, Nayak R, Behera PK, Behera C, Behera AK, Ki J-S, Jena M (2022) Beneficial effects of seaweeds and seaweed-derived bioactive compounds: current evidence and future prospective. Biocatal Agric Biotechnol 39:102242

Ramawat KG, Mérillon JM (2013) Phenolic compounds: introduction. In: *Natural products*, pp 1544–1580. ISBN 9783642221446

Salehi B, Sharifi-rad J, Seca AML, Pinto DCGA (2019) Current trends on seaweeds: looking at chemical. Molecules 24:4182

Savio S, Congestri R, Rodolfo C (2021) Are we out of the infancy of microalgae-based drug discovery? Algal Res 54:102173

Sickander O, Filgueira R (2022) Factors affecting IMTA (integrated multi-trophic aquaculture) implementation on Atlantic Salmon (Salmo salar) farms. Aquaculture 561. https://doi.org/10.1016/j.aquaculture.2022.738716

Siró I, Kápolna E, Kápolna B, Lugasi A, Functional food. (2008) Product development, marketing and consumer acceptance-a review. Appetite 51:456–467

Soto D (2009) Integrated mariculture: a global review (No. 529). Food Agric Organ United Nations (FAO) Fish Aquac Tech Pap 529:183

Spanò N, Di Paola D, Albano M, Manganaro A, Sanfilippo M, D'Iglio C, Capillo G, Savoca S (2022) Growth performance and bioremediation potential of Gracilaria gracilis (Steentoft, L.M. Irvine & Farnham, 1995). Int J Environ Stud 79(4):748–760

Stabili L, Cecere E, Licciano M, Petrocelli A, Sicuro B, Giangrande A (2019a) Integrated multi-trophic aquaculture by-products with added value: the Polychaete Sabella spallanzanii and the seaweed Chaetomorpha linum as potential dietary ingredients. Mar Drugs 17:677

Stabili L, Acquaviva MI, Angilè F, Cavallo RA, Cecere E, Del Coco L, Fanizzi FP, Gerardi C, Narracci M, Petrocelli A (2019b) Screening of chaetomorpha linum lipidic extract as a new potential source of bioactive compounds. Mar Drugs 17:1–20

Tacon AGJ (2020) Trends in global aquaculture and aquafeed production: 2000–2017. Rev Fish Sci Aquac 28:43–56

Thomas M, Pasquet A, Aubin J, Nahon S, Lecocq T (2021) When more is more: taking advantage of species diversity to move towards sustainable aquaculture. Biol Rev 96:767–784

Torres P, Santos JP, Chow F, dos Santos DYAC (2019) A comprehensive review of traditional uses, bioactivity potential, and chemical diversity of the genus Gracilaria (Gracilariales, Rhodophyta). Algal Res 37:288–306

Trevathan-Tackett SM, Kelleway J, Macreadie PI, Beardall J, Ralph P, Bellgrove A (2015) Comparison of marine macrophytes for their contributions to blue carbon sequestration. Ecology 96:3043–3057

Troell M, Joyce A, Chopin T, Neori A, Buschmann AH, Fang JG (2009) Ecological engineering in aquaculture – potential for integrated multi-trophic aquaculture (IMTA) in marine offshore systems. Aquaculture 297:1–9

Tziveleka LA, Tammam MA, Tzakou O, Roussis V, Ioannou E (2021) Metabolites with antioxidant activity from marine macroalgae. Antioxidants 10:1431

Van Den Burg SWK, Dagevos H, Helmes RJK (2021) Towards sustainable European seaweed value chains: a triple P perspective. ICES J Mar Sci 78:443–450

Van Rijn J (2013) Waste treatment in recirculating aquaculture systems. Aquac Eng 53:49–56

Wang L, Park YJ, Jeon YJ, Ryu BM (2018) Bioactivities of the edible brown seaweed, Undaria pinnatifida: a review. Aquaculture 495:873–880

Wang X, Cuthbertson A, Gualtieri C, Shao D (2020) A review on mariculture effluent: characterization and management tools. Water (Switzerland) 12:1–24

Yang YF, Fei XG, Song JM, Hu HY, Wang GC, Chung IK (2006) Growth of Gracilaria lemaneiformis under different cultivation conditions and its effects on nutrient removal in Chinese coastal waters. Aquaculture 254:248–255

Zhou Y, Yang H, Hu H, Liu Y, Mao Y, Zhou H, Xu X, Zhang F (2006) Bioremediation potential of the macroalga Gracilaria lemaneiformis (Rhodophyta) integrated into fed fish culture in coastal waters of North China. Aquaculture 252:264–276

Chapter 14
The Primary Bioactive Compounds of Seaweeds

Sevim Polat and Yeşim Ozogul

1 Introduction

Marine macroalgae, known as seaweeds, is a large group of aquatic plants. They are photosynthetic multicellular organisms generally living in shallow waters of marine ecosystems. Their size can vary from a few millimeters to many meters in length. Seaweeds are divided into three main groups, depending on their pigmentation, storage products, cell wall contents and morphological properties: red algae (Rhodophyta), brown (Phaeophyceae) and green algae (Chlorophyta) (Choudhary et al. 2021; Gomes et al. 2022). There are about 10,000 species of seaweed found in different habitats and climatic zones (Mouritsen 2013). Seaweeds have been used by humans since ancient times. They are rich in proteins, minerals, vitamins, pigments and polysaccharides (Holdt and Kraan 2011; Hentati et al. 2020). The chemical composition of seaweeds may vary seasonally, as well as depending on environmental conditions, location and physiological status of the species (Leandro et al. 2020). These organisms can live in extreme habitats where environmental conditions change rapidly and produce some metabolites to adapt to their environment. They can synthesize secondary metabolites which include phenolic compounds, peptides, vitamins, sterols and some other species-specific bioactive compounds that cannot be produced by terrestrial plants in response to environmental conditions (Charoensiddhi et al. 2017). These compounds show many biological effects such as antimicrobial, anticoagulant, anticancer, antioxidant, antitumor, anti-obesity, anti-inflammatory activities (Hentati et al. 2020). About 221 algae species

S. Polat (✉)
Department of Marine Biology, Faculty of Fisheries, Cukurova University, Adana, Turkey
e-mail: sevcan@cu.edu.tr

Y. Ozogul
Department of Seafood Processing Technology, Faculty of Fisheries, Cukurova University, Adana, Turkey

F. Ozogul et al. (eds.), *Seaweeds and Seaweed-Derived Compounds*,
https://doi.org/10.1007/978-3-031-65529-6_14

including green, brown and red seaweeds are used commercially in the world. Of these, approximately 145 species are employed in food and 101 species are used in the phycocolloid industry (Zemke-White and Ohno 1999). Phycocolloids such as carrageenan, agar agar and alginic acid are the most important commercial products of seaweeds. Of these, agar and carrageenan are obtained from red algae, alginic acid and fucoidan from brown algae, and ulvan from green algae. These are mainly used as emulsifiers, stabilizers and gelling agents in food and pharmaceutical industry (Cotas et al. 2020). In addition, with the understanding of the bioactive properties of secondary metabolites such as phenolic compounds, sterols and peptides, the number of studies on their use in human health has increased. The seaweed market, which was over 11.7 billion US$ in 2016, is estimated to reach 22.13 billion US$ by 2024. In this manner, the seaweeds market is expected to grow 8.9% per year until 2050 (Leandro et al. 2020).

This chapter focuses on the structural and biological properties of primary bioactive compounds of seaweeds as well as their potential to use for health benefits due to their various functional properties.

2 The Primary Bioactive Compounds of Seaweeds

2.1 Pigments

Seaweeds are divided into three main classes in terms of their color: green, brown and red seaweeds (Silva et al. 2020). Each group is characterized by specific combinations of photosynthetic pigments. Algae pigments can be categorized into three major groups: carotenoids, chlorophylls and phycobiliproteins (Aryee et al. 2018; Hentati et al. 2020). All autotrophic seaweeds have chlorophyll *a* and the accessory pigment β-carotene (Gallardo 2015). However, the presence of other pigments varies with taxonomic groups. The color of marine algae typically means that they contain phycobilin for red, chlorophyll for green, and fucoxanthin for brown algae (Khalid et al. 2018; Yalçın et al. 2021). Green seaweeds have high chlorophyll a and b content, lower amounts of xanthophylls and carotene. Brown seaweeds contain high levels of fucoxanthin, xanthophyll, and low amounts of other xanthophylls, carotene, chlorophyll a and c. Red seaweeds contain high amounts of phycocyanin and phycoerythrin low amounts of carotene, chlorophyll a, d and xanthophylls (Gallardo 2015; Freitas et al. 2022). Phycobilin pigments found in red seaweeds are considered an adaptation to live in deep waters (Cotas et al. 2020; Freitas et al. 2022). These pigments have numerous health benefits attracted attention in the food industry as well as in cosmetics and pharmaceutical products (Aryee et al. 2018). Therefore, algal pigments are important for the treatment and prevention of chronic degenerative diseases.

2.1.1 Chlorophyll

Chlorophyll (Chl) is the main pigment found in all photosynthetic algae groups (Lourenço-Lopes et al. 2020). It has a great importance in photosynthesis and many other biological functions. Many forms of Chl are found in nature, including chlorophyll a, b, c, d, e and f. Different forms of chlorophyll show structural differences, resulting in differences in their absorption properties (Aryee et al. 2018). Chlorophyll a is the primary molecule responsible for photosynthesis and absorbs light from the red and blue portions of the electromagnetic spectrum. All seaweeds contain chlorophyll *a*, converting light energy into chemical energy and that gives green color to algae. Chl b is one of the major pigments and found in green seaweeds. It is involved in photosynthesis and absorbs primarily blue light. Chlorophyll b, as a complementary pigment, serves to harvest light in a ratio of 1:3 (to Chl a) (Rebeiz 2014). Chl c present in brown seaweeds and differ from other chlorophylls in that they are consist of Mg-phytoporphyrins rather than Mg-chlorins (Zapata et al. 2006). Chlorophyll d is found in red seaweeds and plays a role in capturing the red part of light. It is also different from the other forms of chlorophyll with regards to structure. It has been suggested that chlorophyll d could use infrared light (700–750 nm) that can not be absorbed by chlorophyll a (Chen and Blankenship 2011). It is reported that the differences in absorption wavelength of chlorophyll d compared to chlorophyll a may make it a potential component for use in biorefinery (Queiroz et al. 2017).

Chlorophyll a and its derivatives show antioxidant properties, and chlorophyll extract has a strong peroxyl radical scavenging capacity (Hong et al. 2020). Chlorophyll is converted to pheophytin, pyropheofitin and pheophorbide after human consumption of processed plant foods (Choudhary et al. 2021). These products have an antimutagenic impact and could have a significant role in prevention of cancer (Holdt and Kraan 2011). Besides these, Lee et al. (2021) reported that the chlorophyll derivative of *Grateloupia elliptica,* a red seaweed significantly suppressed the intracellular lipid accumulation and the chlorophyll obtained from this species could be used to prevent obesity (Table 14.1).

2.1.2 Carotenoids

Carotenoids are one of the most common pigments and present in seaweed, higher plants and photosynthetic bacteria (Holdt and Kraan 2011; Maoka 2020). They are photosynthetic pigments in red, yellow or orange colors. Carotenoids are linear polyenes that are catagorized as carotenes (lycopene, α, y, β-carotene) and xanthophylls (violaxanthin, fucoxanthin, lutein, neoxanthin, zeaxanthin) (Silva et al. 2020). Carotenes are compounds with long hydrocarbon chains and do not contain oxygen, while xanthophylls are known as compounds with oxygen atoms in their structure (Pereira et al. 2021). It has been found that there are approximately 50 types of carotene and ~800 types of xanthophylls (Gupta et al. 2021). Red seaweeds have α- and β-caroten, lutein and zeaxanthin, while brown seaweeds contain violaxanthin, fucoxanthin and β-carotene. Green seaweeds also contain β-carotene,

Table 14.1 Bioactive properties and potential applications of seaweed pigments

Substance	Seaweed species	Properties and potential applications	References
Total chlorophyll	*Spyridia filamentosa*, *Caulerpa sertularioides*, *Rhizoclonium riparium*	Antioxidant, antimutagenic and antiproliferative activity	Osuna-Ruiz et al. (2016)
Chlorophyll	*Grateloupia elliptica*	Antiobesity	Lee et al. (2021)
Chlorophyll *a*	*Enteromorpha prolifera*	Antioxidant activity	Cho et al. (2011)
Pheophytin *a*	*Sargassum fulvellum*	Neuroprotective	Ina et al. (2007)
β-caroten	*Porphyra* sp., *Palmaria palmata*	Antioxidant, cancer prevention	Holdt and Kraan (2011)
Fucoxanthin	*Himanthalia elongata*	Antioxidant and antimicrobial activity	Rajauria and Abu-Ghannam (2013)
Fucoxanthin	*Sargassum siliquastrum*	Photoprotection activity, decreased production of intracellular ROS induced by UV exposure	Heo and Jeon (2009)
Fucoxanthin	*Sargassum siliquastrum*	Cryptoprotective, inhibition of DNA damage and intracellular ROS formation	Heo et al. (2008)
Lutein	*Porphyra tenera*	Antimutagenic	Okai et al. (1996)
Phycobilin	*Phorphyridium cruentum*, *Phormidium valderianum*	Antioxidant, anti inflammatory, antitumour, liver protecting, fabric dye and cosmetic uses	Sekar and Chandramohan (2008)
Phycoerythrin	*Gracilaria corticata* var. *corticata*	Natural colorant, antioxidant property, can be used in beverages and other nutraceutical benefits	Sudhakar et al. (2014)
Phycoerythrin	*Kappaphycus alvarezii*	Antioxidant activity	Uju et al. (2020)

violaxanthin, lutein, zeaxanthin and neoxanthin. Of these, β-carotene is one of the most abundant carotenes in algae groups (Freitas et al. 2022). It was reported that β-carotene content was the highest (4500 mg kg^{-1}) in a red seaweed, *Porphyra* (Holdt and Kraan 2011). There are many functions of β-carotene such as anti-inflammatory, antioxidant, retinoprotective and dermoprotector (Chidambara Murthy et al. 2005). Moreover, β-carotene is a source of retinol (vitamin A) and is converted to vitamin A in human body. This pigment is also commonly employed by the food and cosmetic industries (Gupta et al. 2021).

Xanthophylls are the second type of carotenoids that give yellow to brownish color to the seaweeds. Fucoxanthin is a type of xanthophyll and is one of the most common carotenoids (Aslanbay Güler and İmamoğlu 2021). It is found in brown seaweeds with high concentrations (Choudhary et al. 2021; Aslanbay Güler and İmamoğlu 2021; Pereira et al. 2021). It has been reported that the amount of fucoxanthin in brown seaweeds ranges from 172 to 720 mg kg dry weight and it was maximal in *Fucus serratus* (Holdt and Kraan 2011). Fucoxanthin is the most widely used pigment of brown seaweeds due to its properties such as antioxidant, anti-inflammatory and anti-cancer activities (Pereira et al. 2021) (Table 14.1). Food supplements containing fucoxanthin, which contribute to weight loss and contribute improvement of brain, eye, liver and joint health, are found under the trade names Fucovital and ThinOgen® (Pereira et al. 2021). Lutein and zeaxanthin found in red and green seaweeds was reported to decrease the risk of cataracts and macular degeneration (Ravikrishnan et al. 2011). Carotenoids such as lutein, zeaxanthin and carotene may also reduce cancer risk, boost immunity, improve skin health, and help prevent certain diseases such as cardiovascular and diabetes (Maoka 2020). It has been also found that lutein showed antiviral activity as an anti-HBV agent (Gomes et al. 2022).

2.1.3 Phycobilins

Phycobilins are open chain tetrapyrroles groups present as photosynthetic accessory pigment in cyanobacteria and some eucaryotic algae (Ismail and Osman 2016; Mysliwa-Kurdziel and Solymosi 2017). Phycobilins that are covalently bonded to proteins are referred to phycobiliproteins (Aryee et al. 2018). As a difference to chlorophyll and carotenoids, phycobiliproteins are water soluble (Holdt and Kraan 2011). In terms of characteristics of light absorption and kinds of bilins, phycobiliproteins can be catagorized into four classes: phycoerythrins, phycocyanins, phycoerythrocyanins and allophycocyanins (Cotas et al. 2020). Allophycocyanins appear bluish green, phycocyanins are dark blue, and phycoerythrins are purple, which show different absorption maxima (Mysliwa-Kurdziel and Solymosi 2017; Lage-Yusty et al. 2013). Phycocyanin is a blue, light-harvesting pigment in cyanobacteria, Rhodophyta and Cryptophyta. This pigment gives a distinct bluish green colour of many cyanobacteria (Elumalai et al. 2014). Phycoerythrin is an important accessory pigment in red seaweeds. Francavilla et al. (2013) reported that *Gracilaria gracilis* seems a promising source of phycoerythrin, since its concentration is in the range of 3.6 and 7 mg/g dw, depending on season. In addition to use as a colorant, phycoerythrin is also used in fluorescent applications due to its role in fluorescent detection systems (Sekar and Chandramohan 2008). Allophycocyanin is found in cyanobacteria and red seaweeds (Sekar and Chandramohan 2008). This blue-green pigment is located in the core of phycobilisomes (Freitas et al. 2022). Cotas et al. (2020) stated that allophycocyanin is a minor component of phycobiliproteins, its extraction has low yields, and high purity is difficult to achieve due to the presence of other co-extractable proteins. Moreover, Cherdkiatikul and Suwanwong (2014) reported that allophycocyanin showed higher peroxyl radical scavenging activity than c-phycocyanin, while the similar activity of c-phycocyanin for hydroxyl radicals was higher than allophycocyanin. Phycobiliproteins have been generally applied as natural colourants for cosmetic and food industries (Table 14.1). Phycobiliproteins have also been indicated to have antiviral, antioxidant, neuroprotective, anti-inflammatory, anti-tumour, hepatoprotective, lipase inhibition activities and are regarded as therapeutic agent in the prevent or treatment of health problems (Holdt and Kraan 2011; Lage-Yusty et al. 2013). Thus, phycocyanin and other phycobiliproteins are used in many fields such as food, cosmetics, biotechnology and medicine (Eriksen 2008).

2.2 *Polysaccharides*

Polysaccharides are biopolymers composed of simple chain monosaccarides or simple sugars linked by glycosidic bonds. They are catagorized as mucopolysaccharides, storage polysaccharides and structural polysaccharides (Pal et al. 2014; Rengasamy et al. 2020). Seaweeds are main source of polysaccharides, which contain lots of cell wall polysaccharides as well as storage polysaccharides. Therefore, polysaccharides are very important to algae both as a structural element and as a storage material (Wang et al. 2017). The total polysaccharide levels in the seaweed species can vary significantly and range from 4% to 76% of dry weight (Holdt and Kraan 2011). Polysaccharides can hold water up to 20 times their weight and form hydrogel, which characterize them to be called as hydrocolloids or phycocolloids. Hydrogels can be formed by heating or cooling of polysaccharides (Venugopal 2019). Functional properties of polysaccharides make them precious food additives for many commercial applications such as thickeners, stabilizers, emulsifiers, water retention compounds, foam stabilizers and regulating of viscosity (Tseng 2001; Venugopal 2019). Polysaccharides have many health benefits containing antioxidant, anticarcinogenic, anti inflammatory and antiviral activities (de Jesus Raposo et al. 2015; Bilal and Iqbal 2019). The main polysaccharides in seaweeds include agar, carrageenean, alginate, fucoidan, laminaran, mannitol and ulvans.

2.2.1 Agar

Agar, which is one of the cell wall contitutes of some red seaweeds is a mixture of polysaccharides consisting of agaropectin and agarose (Holdt and Kraan 2011; Cotas et al. 2020). Agar is a sulfated polysaccharide and mainly obtained from red seaweeds, *Gelidium* and *Gracilaria* (Pal et al. 2014; Leandro et al. 2020). It was reported that the agar concentration of *Gracilaria* can reach 31% in dry weight (Holdt and Kraan 2011). The agar quality in these two genera differs and the agar of *Gelidium* is of high quality and has been reported to be suitable for use in the pharmaceutical industry. However, agar extracted from *Gracilaria* is widely used in the food industry (Leandro et al. 2020; Din et al. 2019). Agar is soluble in boiling water, so it is in gel form at 32 to 40 °C, and become liquid if heated above 80 °C (Bilal and Iqbal 2019). Due to its gel forming ability, agar (about 90% of agar) is commonly applied in food industry as thickener, stabilizer and emulsifier (McHugh 2003). In addition to the widespread use of agar in the food industry, there have been many studies on its therapeutic use in medical field in recent years (Pereira 2018). It is employed in microbiology, chromatography, gel electrophoresis, and biotechnology (Venugopal 2019). It was reported that agar-agar reduce blood glucose level and has an anti-aggregation impact on red blood cells (Holdt and Kraan 2011). Agar has also been used as a laxative because it passes through the intestines without being digested, thus increases the fecal volume and promotes peristalsis (Ferrara 2020). Agaro-oligosaccharides obtained from agar hydrolytically was reported to display antioxidant activity by Chen et al. (2005). They found that agaro-oligosaccharides can show in vivo and in vitro hepatoprotective impact by scavenging oxidative damage caused by reactive oxygen species. Furthermore, agaro-oligosaccharides was reported to suppress an enzyme related to proinflammatory cytokine and nitric oxide production (Enoki et al. 2010).

2.2.2 Carrageenan

Carrageenans are anionic sulfated polysaccharides composed of linear polymers of galactans (Holdt and Kraan 2011; Pal et al. 2014). According to their chemical structure, carrageenans are categorized as iota (i), kappa (κ) and lambda (λ) carrageenans (McKim et al. 2016). The different types of carrageenan are associated with presence of 3, 6-anhydro-D-galactose as well as the number and position of sulfate groups (Chauhan and Saxena 2016; Venugopal 2019). Carrageenan gels are considered as thermally reversible, as with agar, as they melt on heating and gel again on cooling (Venugopal 2009).

Although carrageenan was first obtained from *Chondrus crispus*, the most of carrageenan is now generated from *Euchema denticulatum* and *Kappaphycus alvarezii* species (McHugh 2003; Leandro et al. 2020). A various type of carrageenans are acquired from different red seaweeds. κ-carrageenan, i-carrageenan and λ-carrageenan are obtained from *Kappaphycus alvarezii, Eucheuma denticulatum* and from various species of the *Chondrus* and *Gigartina*, respectively (Cotas et al. 2020).

Carrageenans are employed in the food, chemical and pharmaceutical applicatons (Rafiquzzaman et al. 2016). The usage of carrageenan in food dates back to ancient times (about 600 years ago) (Kraan 2012). Carrageenans are used as thickeners and stabilizers in the food industry owing to their functional properties such as emulsification, gelling and complexing with polyelectrolytes (Venugopal 2019) (Table 14.2). Food applications using carrageenan (E407) include canned human and pet foods, salad dressings, sweet mousses, ice cream, baking fillings, and instant desserts (Kraan 2012). When examined in terms of human health, it has been reported that carrageenan has antioxidant, anticarcinogenic, anti-tumor, antiviral, antifungal properties, and is also used in the treatment of stomach and intestinal disorders (Zhou et al. 2006; Skoler-Karpoff et al. 2008; Soares et al. 2016; Rafiquzzaman et al. 2016). Carrageenan has no nutritional value, and it passes through the digestive system without alteration because it is resistant to microbial and enzymatic degradation, thus it has a similar effect to dietary fiber (McKim et al. 2016). It has been reported that carrageenan is effective in reducing oxidative damage by scavenging free radicals generated in the human body (Sokolova et al. 2011; Pereira 2018). Rafiquzzaman et al. (2016) investigated the antioxidant properties of

Table 14.2 Bioactive properties and potential applications of seaweed polysaccharides

Substance	Seaweed species	Properties and potential applications	References
Agar	*Gracilariopsis chorda*	Neuroprotective	Mohibbullah et al. (2015)
Agar	*Gelidium* spp., *Gracilaria* spp., *Hypnea* spp.	Antiinflammatory, anti HIV, laxative	Venugopal (2019)
Carrageenan	*Eucheuma*, *Gigartina*, *Hypnea*, *Soliera*, *Chondrus crispus*	Tissue engineering, drug delivery, wound healing, biosensors	Venkatesan et al. (2015)
Carrageenan	*Chondrus armatus*, *C. pinnulatus*, *Tichocarpus crinitus*	Antioxidant activity	Sokolova et al. (2011)
Carrageenan	*Kappaphycus alvarezii*, *Ulva fasciata*	Food applications, edible films for food packaging	Ganesan et al. (2018)
Carrageenan	*Corallina*	Anticoagulant and antibacterial activity	Sebaaly et al. (2014)
Mannitol	*Ascophyllum nodosum*, *Laminaria digitata*	Bioethanol production	Chades et al. (2018)
Mannitol	*Turbinaria ornata*, *Sargassum mangravense*	Antioxidant, antibacterial effect	Zubia et al. (2008)
Alginate	*Laminaria digitata*	Production of micro and nanoparticles for drug delivery	Fertah et al. (2017)
Alginate	*Laminaria japonica*, *Lessonia flavicans*, *L. nigrescens*, *Macrocystis pyrifera*, *Ecklonia maxima*	Food applications, pharmaceuticals, textile dyeing, cosmetics	Zhang (2018)
Ulvan	*Ulva lactuca*	Tissue engineering, drug delivery	Venkatesan et al. (2015)
Ulvan	*Ulva lactuca*	Nanogels to carry and deliver water insoluble bioactive compounds	Bang et al. (2019)
Ulvan	*Ulva fasciata*	Food packaging	Ganesan et al. (2018)
Fucoidan	*Dictyota mertensii*	Antioxidant activity, protect bone tissue from oxidative stress, treatment of bone formation disorders	Fidelis et al. (2019)
Fucoidan	*Nemacystus decipiens*	Antithrombotic activity	Cui et al. (2018)
Fucoidan	*Undaria pinnatifida*	Antitumour activity against prostate cancer, cervical cancer, alveolar carcinoma, and hepatocellular carcinoma	Synytsya et al. (2010)
Fucoidan	*Lessonia vadosa*	Anticoagulant activity	Chandía and Matsuhiro (2008)

(continued)

Table 14.2 (continued)

Substance	Seaweed species	Properties and potential applications	References
Laminarin	*Laminaria digitata*	Anticancer effects on human colon cancer	Park et al. (2013)
Laminarin	*Eisenia, Laminaria, Saccharina*	Anticancer, antioxidant, antiinflammatory and tissue engineering	Zargarzadeh et al. (2020)

carrageenan extract of *Hypnea musciformis*. They found that carrageenan showed antioxidant activity, and the degree of this activity varies with structural features and analysis method. Besides, it has been reported that sulfated carrageenans have inflammatory property and thus can activate monocytes and macrophages for the treatment of tuberculosis (Venugopal 2019). Another medical use of carrageenan is related to its anticoagulant activity. Sebaaly et al. (2014) found that the carrageen extract of a red seaweed, *Corallina* showed potent anticoagulant effect than the sulfated galactan extract. Additionally, carrageenans are employed as suspending agents and stabilizers in other medicines, creams and lotions (Holdt and Kraan 2011). According to recent market research, it has been determined that the global carrageenan market will rise up to 1 Billion US dollars by 2024 (Market Research Future 2021).

2.2.3 Alginate

Alginate is a natural anionic polysaccharide occuring in the cell wall of brown algae (Fertah et al. 2017; Pereira 2018). This term is used for sodium, potasium, calcium and magnesium salts of alginic acid (Venugopal 2019). Their presence in different salt forms in the cell wall of the brown seaweeds makes the tissues stronger and flexible (Hentati et al. 2020). Alginic acid is not soluble in water and its extraction is based on the conversion of insoluble alginic acid salts in the cell wall of seaweed to soluble alginate suitable for water extraction. The main purpose of the extraction process is to generate dry and powdered sodium alginate (Pereira 2018). Alginates are groups of more than 200 compounds with similar physicochemical properties but slight differences (Namvar et al. 2013). In chemical structure, they are linear copolymers of β-(1–4)-linked D-mannuronic acid (M) and α-(1–4)-linked L-guluronic acid (G) residues (Venugopal 2019). They are arranged as homopolymers poly-M (MM) or poly-G (GG), or heteropolymers (MG or GM) (Hentati et al. 2020; Zhang et al. 2021). Different M and G contents affect the physical and biochemical properties of the alginate and determine its industrial use (Jahandideh et al. 2021; Venugopal 2019). Alginates with more G-blocks form stronger, mechanically stable gels, while those with more M-blocks form elastic gels (Namvar et al. 2013). It has been stated that alginate containing more G units is suitable for industrial applications, while alginate with more M units is suitable for pharmaceutical and environmental applications (Osman et al. 2020; Silva et al. 2020). The safety of alginate as one of the safest food additives has been approved by FAO/WHO (Vijayalakshmi et al. 2017). Alginates are mainly used as gelling agents and viscosifiers (Pereira 2018). They are important for the pharmaceutical industry due to their high water retention capacity (Solanki and Solanki 2012). Alginates have bioactive properties such as antibacterial, antihypertensive, anti-obesity, anti-ulcer, anti-inflammatory and anticancer activities (Holdt and Kraan 2011; Venugopal 2019). Since sodium alginate has a high antibacterial effect and protective properties on the wound, it is used in the treatment of wounds (Venkatesan et al. 2015; Hentati et al. 2020). Alginates and its salts could not be digested by intestinal enzymes (Holdt and Kraan 2011; Pereira 2018). It has been shown that alginates control digestive enzymes in vitro, and might be applied in weight loss treatment and are therefore classified as a dietary fibre (Houghton et al. 2015). In addition, alginates have been used in the adsorption of heavy metals and toxic chemicals in industries such as paper, textile and food packaging (Park et al. 2007; Parreidt et al. 2018).

Alginate production from brown seaweed, most of which is collected from wild stocks, is approximately 26,500 tonnes globally and has an annual market value of $318 million. The brown seaweeds, *Ascophyllum*, *Laminaria* and *Macrocystis* are commercially important for alginate production (Kraan 2012).

2.2.4 Fucoidan

Fucoidans, also known as fucans are sulfated polysaccharides commonly found in the cell wall of brown seaweeds (Kraan 2012; Pereira 2018). They contain L-fucose as major sugar and sulfate ester groups with other monosaccharides such as arabinose, xylose, glucose and mannose (Holdt and Kraan 2011; Bilal and Iqbal 2019). The amount of fucoidan in seaweed is about 10–20% of the total dry weight. The structural composition of fucoidans may vary depending on the species of seaweed, season, location and extraction methods (Yang and Lim 2021). Fucoidan is thought to strengthen the cell wall and is associated with protection against drying out when seaweed is exposed to low tide (Holdt and Kraan 2011).

Brown seaweeds such as *Fucus vesiculosus*, *Ascophyllum nodosum*, *Saccharina latissima*, *Undaria pinnatifida*, *Saccharina japonica* and *Cladosiphon okamuranus tokida* are main producers of fucoidan. However, it has been reported that EFSA and FDA only allow extracts of fucoidan with a specific chemical characteristic and obtained from *U. pinnatifida* and *F. vesiculosus* (Leandro et al. 2020).

Fucoidans displays numerous bioactive activities, containing anticoagulant, antioxidant, antimicrobial, anti-mutagenic, anti-tumor, and anti-inflammatory (Table 14.2). Moreover, the bioactive properties of fucoidans are related to their molecular weight and sulfate level as well as the position of sulfate groups, glucuronic acid and fucose content (Li et al. 2008; Holdt and Kraan 2011). A higher content of sulfate groups in natural fucoidans generally indicates higher anticoagulant activity (Li et al. 2008). It was reported that the anticoagulant and antithrombotic activity of *A. nodosum* fucoidan increase with an increase in molecular weight and sulfate content (Berteau and Mulloy 2003). Chandía and Matsuhiro (2008) found that fucoidan of the brown seaweed (*Lessonia vadosa*) with 320 kDa molecular weight exhibited better anticoagulant activity while that of 32 kDa showed weak anticoagulant property.

Some research has been conducted on the anticancer and antitumor potential of fucoidan (Moghadamtousi et al. 2014). The anticancer impact of natural and modified fucoidan fractions of *Padina boryana* was investigated in vitro on colorectal carcinoma cells DLD-1 and HCT-116. At the end of the study, it was found that all fucoidans did not have cytotoxicity below 400 μg/mL and inhibited the colony formation of cancer cells at the level of 200 μg/mL (Usoltseva et al. 2018). In another study, the inhibitory effect of fucoidan on the growth of B16 melanoma and Lewis lung carcinoma in mice was investigated by Koyanagi et al. (2003). The results showed that the increased number of sulfate groups in the fucoidan contributed to its antitumor and antiangiogenic properties. Fernando et al. (2018) found that fucoidans of *Chnoospora minima* and *Sargassum polycystum* can suppress collagenase and elastase activity and fucoidans of these seaweeds have antioxidant, whitening anti-inflammatory, and anti-wrinkle properties in vitro. Fidelis et al. (2019) obtained six fucoidans (F0.3, F0.5, F0.7, F1.0, F1.5, F2.1) from *Dictyota mertensii*. They found that fucoidans (F0.7, F1.5, and F2.1) of *D. mertensii* prevented bone tissue from oxidative stress and may represent possible adjuvants for the treatment of bone formation disorders. Fucoidan is a non-toxic compound and could be used as a food supplement (Leandro et al. 2020). The recommended amount for fucoidan in the prevention of diseases is 1–2 g/day (Florez-Mendez and González 2019). Fucoidans obtained from *Fucus vesiculosus, Undaria pinnatifida* and *Macrocystis pyrifera* have been available commercially.

2.2.5 Laminarin

Laminarin is one of the most abundant polysaccharides stored in brown seaweeds (Chojnacka et al. 2012). It is found in *Laminaria* and *Saccharina*, but less in *Fucus*, *Ascophyllum* and *Undaria* (Kraan, 2012). The content changes with season and habitat and can reach 32% of dry weight. The chemical structure of laminarian is made up a β(1,3)-glucose in the main chain and random β(1,6)-linked side-chains (Choudhary et al. 2021). The ratio of β(1,3)-glucose to β(1,6)-glucose varies significantly among different algae (Bilal and Iqbal 2019). Laminarin does not gel or form a viscous solution, thus its potential for use is in medical and pharmaceutical applications (Holdt and Kraan 2011). Isolation of laminarin can be accomplished by hot water extraction, ultrafiltration and gel chromatography (Venugopal 2009). Laminarian is a source of dietary fiber and acts as a prebiotic. Since laminarins are resistant to hydrolysis in the upper gastrointestinal tract, they regulate intestinal metabolism by arranging mucus composition, short-chain fatty acid production and intestinal pH (Namvar et al. 2013). It has also antimicrobial and antioxidant activity and laminarin β-glucans have a protective effect against bacterial infections, as well as the negative effects of ionizing radiation (Venugopal 2019). With appropriate chemical modifications, the bioactivity of laminarin can be modified. For example, laminarin exhibits anticoagulant activity, following structural modifications (such as sulphation, reduction or oxidation) (Holdt and Kraan 2011). Products comprising 1 → 3:1 → 6-β-D-glucans, laminarin and fucoidan are produced by the health sector owing to their beneficial effects on the immune system (Holdt and Kraan 2011). Laminarin is also a potential anticancer agent, and many studies have been conducted on its use in cancer treatment (Venugopal 2019). Ji and Ji (2014) investigated the anticancer impact of commercial laminarin on human colorectal adenocarcinoma (LoVo) cells and found that it induced apoptosis in LoVo cells via mitochondrial and DR pathways. They concluded that laminarin can be used for the treatment and prevention of certain types of digestive system cancers. The use of these polysaccharides as food supplements or nutraceuticals is promising in reducing rate of cancer and the side impacts of anticancer drugs (Sanjeewa et al. 2017).

2.2.6 Mannitol

Mannitol is an important monomeric sugar alcohol found in brown algae (Kraan 2012; Shrivastava et al. 2021). This compound is extracted mainly from *Laminaria*, *Ecklonia*, *Fucus* and *Sargassum* species (Sartal et al. 2012). Although its amount varies between species, the level of mannitol in brown algae can reach 30% in dry weight (Iwamoto and Shiraiwa 2005; Zubia et al. 2008). However, seaweeds, such as *A. nodosum* and *Fucus* species also produce less than 16% dw of mannitol (Stiger-Pouvreau et al. 2016). It is stated that mannitol has important physiological functions due to its high solubility and high compatibility with organic macromolecules (Iwamoto and Shiraiwa 2005). Since mannitol has a sweetness of approximately 50% of that of sucrose and a low glycemic index, it is suitable for use in many food products (Shrivastava et al. 2021). In this respect, mannitol has the potential to be used in the production of food products for diabetics (Table 14.2). Moreover, it can replace sucrose to make sugar-free compound coatings (Holdt and Kraan 2011). Mannitol is a food additive known as code E421 (Stiger-Pouvreau et al. 2016). It is also used in the manufacture of chewable tablets due to its pleasant cooling sensation in the mouth (Holdt and Kraan 2011; Shrivastava et al. 2021). The hygroscopic property of mannitol is very low and it does not absorb water from the air until the humidity level of the environment reaches 98%. This property makes mannitol a suitable as coating material for products such as hard candy, dried fruit and chewing gum (Qin 2018). Mannitol has many applications in health-related fields. It is a powerful antioxidant and therefore scavenges free hydroxyl radicals to reduce the damage of neurological disorders (Shrivastava et al. 2021). Mannitol cannot be absorbed in the stomach, however, it can penetrate the bloodstream to draw water into the capillaries, causing the capillaries to dilate and blood pressure to reduce (Qin 2018).

The moisturizing and antioxidant properties of mannitol are important for cosmetic and pharmaceutical applications. Zubia et al. (2008) investigated the alginate, mannitol and phenolic contents of brown seaweed *Sargassum mangarevense* and *Turbinaria ornata,* and their antioxidant and antimicrobial activities. They concluded that *S. mangarevense* may be the target seaweed species used in cosmetic products due to its high mannitol and phenolic content (Table 14.2). Furthermore, in a study conducted by Lötze and Hoffman (2016), the impact of three different commercial products (Kelpak®, Afrikelp® and Basfoliar®) obtained from *Ecklonia maxima* were studied on biological activity of mung bean root growth. The root growth was higher in Kelpak with high mannitol level compared to other stimulants and they concluded that higher mannitol content in seaweed products may be beneficial under stress conditions and may contribute to better root survival.

2.2.7 Ulvan

Ulvan is a sulfated polysaccharide found in the cell wall of green seaweeds belonging to Ulvales such as *Ulva conglobate*, *U. lactuca*, *U. rigida*, *U. gigantea*, *U. olivascens*, and *U. clathrata* (Pankiewicz et al. 2016). It is a water-soluble polysaccharide and can rise up to 29% of the dry weight (Venugopal 2019). It is mostly composed of disaccharide repeating sequences consisted of sulfated rhamnose and uronic acids (iduronic acid, glucuronic acid), or xylose. The main repeating disaccharides are aldobiuronic acids (Pereira 2018; Kidgell et al. 2019). The structure and extraction yield of Ulvan are affected by some factors such as temperature and ionic conditions such as acidic or alkaline (Zhong et al. 2020; Kidgell et al. 2019). The chemical structure of ulvan is similar to mammalian glycosaminoglycans and therefore has potential for use in functions performed by mammalian polysaccharides (Kidgell et al. 2019). Ulvan has great potential in food, medicine and agricultural applications due to its physicochemical and biological properties. Studies have revealed that ulvan exhibits a variety of therapeutic properties such as anti-inflammatory, antioxidant, immunostimulating, anti-tumor, anti-hyperlipidemia, antibacterial, antiviral and anticoagulant activities (Venkatesan et al. 2015; Zhong et al. 2020). The antioxidant activity of Ulvan varies depending on its molecular weight (Venugopal 2019). Kidgell et al. (2019) reported that structural features such as the degree of sulfation, constituent sugars, bonds, and branching also influence the bioactivity of the ulvan.

Ulvan polysaccharides are used in the production of some biomaterials. Films, based on polymers such as ulvan polysaccharides have been reported to be non-toxic and environmentally friendly (Lakshmi et al. 2020). Ganesan et al. (2018) prepared edible films from ulvan of *Ulva fasciata* for use in food applications (Table 14.2). The results showed that the films prepared with the combination of ulvan and glycerol greatly improved the physico-chemical and mechanical properties of the film, which are vital for food packaging. Sulastri et al. (2021) reported that ulvan hydrogel films can decrease hydroxyl radicals and control the growth of gram negative and positive bacteria (*Pseudomonas aeruginosa*, *Staphylococcus aureus*, *Streptococcus epidermidis* and *Escherichia coli*). Therefore, many studies have been conducted on the use of ulvan-based membranes and films in wound dressing and medicine (Lakshmi et al. 2020).

There are many studies on the use of ulvan in cancer treatment. Thanh et al. (2016) found that Ulvan exhibited significant cytotoxic activity against hepatocellular carcinoma, breast cancer and cervical cancer. Ulvans are also used in the synthesis of some chemicals. A component of ulvan, rhamnose is used in the synthesis of aromatic compounds and production of rhamnose from *Monostroma* was reported to be patented (Holdt and Kraan 2011).

2.3 Proteins

The protein content of different taxonomic groups of seaweeds have been determined in previous studies (Dawczynski et al. 2007; Pirian et al. 2018; Biancarosa et al. 2017; Naseri et al. 2020). Protein level was in the range of 5–47% of dry weight basis. It was indicated that seaweed protein level relies on the species, on habitat, collection time and, and some environmental dynamics such as intensity of light, temperature, and nutrient levels in the water (Makkar et al. 2016; Lorenzo et al. 2017). Protein extraction method also affects the protein content of algae. It is well established that brown seaweed contains lower protein level than those of red and green seaweed (Angell et al. 2016; Biancarosa et al. 2017). The protein digestibility of seaweed is reported to inferior to animal protein. This is owing to their complex structure of polysaccharide that could obstruct protein accessibility by the gastrointestinal enzymes and also their contents of antinutritional compounds for example; phytic acids, phenolic compounds, and protease inhibitors (Maehre et al. 2016).

2.3.1 Peptides

Seaweeds are well- known as a potential resource of bioactive peptides. Protein extraction is significant for detection of algal-derived bioactive peptides. A numerous techniques have been applied to obtain high protein extraction yields, including chemical, enzymatic, ultrasound/sonication, high hydrostatic pressure, microwave, pulsed electric field, sub- and supercritical fluid extraction methods. The advantages and disadvantages of these methods were given in a review study by Cermeño et al. (2020). Protein fractions are placed in disparate fragments of seaweeds such as their cell wall, organelles and cytoplasm (Sudhakar et al. 2019). Protein extraction of seaweed generally includes interruption of cell walls and extract of other non-protein nitrogen composites while the determination of algal-derived bioactive peptides involves the extraction, fraction, isolation, and intensity of the proteins (Cermeño et al. 2020).

Bioactive peptides are generally sequences 2 to 30 amino acids in length, exhibiting hormone-like useful properties when they are removed from their parent protein (Cian et al. 2012; Lafarga et al. 2020). In other words, they are short protein fragments that contain bioactive properties with physiological and pharmacological effects such as antioxidant, antithrombotic, antihypertensive, antimicrobial, anti-inflammatory and antidiabetic, varying on amino acid composition and their sequence (Capriotti et al. 2015; Bleakley and Hayes 2017; Alboofetileh et al. 2021; Echave et al. 2021). Currently, products with bioactive peptides exist in the market. The biological impact of marine bioactive peptides depends on the chemical structure of these molecules and also biological factors such as location and season (Hayes and Tiwari 2015). In addition, structural alteration of these molecules during extraction and purification processes also affects their bioactive properties (Garcia-Vaquero et al. 2017). Thus, it is essential to determine the relation between their chemical structure and biological activities of these compounds.

It was well established that algal derived peptides and their protein hydrolysates have bioactive properties, which can be used in different industries. For example; Indumathi and Mehta (2016) enzymatically isolated a novel anticoagulant peptide from seaweed *Porphyra yezoensis,* commercially known as Nori. Antioxidative and antihypertensive activities of peptides were also found in Nori by Bhatia et al. (2011) and Qu et al. (2010), respectively. Inhibition of α-amylase enzyme is a important way to reduce level of glucose in the blood for diabetes mellitus. Admassu et al. (2018) detected α-amylase inhibitory peptides from red seaweed (*Porphyra* species). In addition, dipeptidyl peptidase IV inhibitory peptide for controlling type 2 diabetes mellitus was derived from green seaweed *Ulva* spp. hydrolysates (Cian et al. 2022). It was also indicated that the peptides obtained from seaweed *Laminaria japonica* had great inhibitory impact on liver cancer (Chen et al. 2021).

2.3.2 Free Amino Acids

The AA (amino acid) profile of seaweed is crucial for determining the nutritional value of their proteins. Seaweed proteins have considerable levels of all of essential AAs. Essential AAs (histidine, arginine, isoleucine, leucine, methionine, threonine, lysine, phenylalanine, valine and tryptophan) cannot be produced by mammals, thus they should be obtained from the diet. Seaweed protein involves all AA, especially alanine, glycine, proline, arginine, aspartic and glutamic acids. As for non essential AA, brown, green and red seaweeds have the similar levels (Černá 2011). The AA content of 34 edible seaweeds, including *Laminaria* sp., *Undaria pinnatifida, Hizikia fusiforme,* and *Porphyra* sp. strains obtained from German food stores were determined by Dawczynski et al. (2007). They reported that glutamic and aspartic acid were the most common AAs in seaweed species and seaweeds are also rich in threonine, valine, leucine, lysine, glycine, alanine. In addition, Astorga-España et al. (2016) investigated AA score and AA profile in 73 green, brown and red seaweeds from Chile. Differences were detected between seaweed colour and the genera in terms of non-essential and essential AA. AA score and essential AA index exhibited that brown seaweeds had higher protein quality than green and red seaweeds. In red and green seaweeds sulfur AA were the limiting AA whereas in brown seaweed leucine was the limiting AA. They concluded that AA content depend on the seaweed genus and colour.

Gaillard et al. (2018) investigated AA contents of four brown (*Pelvetia canaliculata*, *Laminaria digitata*, *Alaria esculenta* and *Saccharina latissima*), three red (*Palmaria palmata*, *Porphyra* sp. and *Mastocarpus stellatus*), and two green (*Ulva* sp. and *Cladophora rupestris*) seaweed species from Northern Norway. They reported that the seaweeds were rich in serine, threonine, valine, lysine, glycine, arginine and leucine and low in histidine, cysteine, and methionine. Glutamine was the most common AA, followed by aspartic acid and alanine. Mouritsen et al. (2019) also determined free amino acid levels of some brown seaweeds, such as genera *Nereocystis*, *Laminaria*, *Macrocystis*, *Saccharina*, *Alaria*, *Undaria*, *Himanthalia*, *Postelsia*, *Sargassum*, *Ecklonia*, *Corda* and *Fucus*. The levels of all amino acids apart from cysteine changed significantly among 37 samples. The levels of alanine and glutamine for most of species were considerably higher than other free amino acids.

Pirian et al. (2020) evaluated some of macroalgae species, *Sirophysalis trinodis*, *Sargassum boveanum*, *Hypnea caroides*, *Galaxaura rugosa*, *Palisda perforata*, *Caulerpa sertularioides*, *Bryopsis corticolans*, and *Caulerpa racemose* from the Persian Gulf. They found that the levels of amino acids changed with species, and all species had all AA (EAA). The red alga *Galaxaura rugosa* had the greatest amino acid level and phenyl alanine and leucine were the most found EAAs.

Mohammed et al. (2021) assessed AA content of Irish edible red seaweeds (*Porphyra umbilicalis*, *Palmaria palmata*) and brown (*Alaria esculenta*, *Himanthalia elongata*). Red seaweed species had higher content of each amino acid (except methionine, histidine, cystine, tyrosine and phenylalanine) than those of brown seaweeds. This results from higher protein contents of red seaweed species. Both aspartic and glutamic acids were the most abundant non essential AAs in seaweeds.

Umami is known as fifth basic taste along with sweet, salty, sour, and bitter. Umami was first proposed by Japanese scientist, Kikunae Ikeda, discovering that brown seaweed kombu (*Saccharina japonica*) contained high level of free glutamate. Umami is monosodium glutamate–like free amino acids mainly glutamic and aspartic acids, which can be found in seaweeds, giving characteristic taste and aroma of the seaweeds. Milinovic et al. (2020) determined umami amino acids in 12 seaweeds, four brown (*Saccorhiza polyschides*, *Bifurcaria bifurcata*, *Undaria pinnatifida*, *Fucus vesiculosus*) and two green (*Ulva rigida*, *Codium tomentosum*), six red (*Chondrus crispus*, *Chondracanthus teedei* var. *lusitanicus*, *Nemalion helminthoides*, *Grateloupia turuturu*, *Gracilaria gracilis*, *Osmundea pinnatifida*), from the Portuguese seashore. Red seaweeds exhibited the maximum umami level, especially, *Gracilaria gracilis*, *Osmundea pinnatifida* and *Chondrus crispus*.

It is apparent that the seaweeds contain all the essential AAs and also non essential AAs in different proportions. It was also well established that AA content of the seaweeds varies with species, harvesting time, season, location, processing and extraction methods (Sánchez-Machado et al. 2003; Ortiz et al. 2006; Dawczynski et al. 2007; Kwon et al. 2010; Tabarsa et al. 2012; Mouritsen et al. 2019; Pirian et al. 2020; Vilcanqui et al. 2021; Norakma et al. 2022).

2.4 Lipids

The lipid content of seaweeds is usually low (1 to 5% of dry weight-DM), but they contain high omega 3 fatty acids, which inhibit cardiovascular diseases. Pirian et al. (2020) determined lipid contents of some seaweed species from the Persian Gulf. They reported that the total lipid level ranged from 1.27% to 9.13% (DW), relying on the algal species. Higher lipid content was found in green algal species (*Caulerpa sertularioides*, *Caulerpa racemose* and *Bryopsis corticolans*). Jeliani et al. (2021) also studied 10 species of seaweeds from the Persian Gulf, indicating that the lipid level was recorded between 1.5 and 4% DW with higher lipid level in red seaweeds (*Champia globulifera*, *Hypnea pannosa*, and *Centroceras clavulatum*). Mohammed et al. (2021) studied nutriotional value of brown (*Alaria esculenta*, *Himanthalia elongata*) and red (*Porphyra umbilicalis*, *Palmaria palmata*) seaweeds from the coast of Ireland. They found that seaweed lipid levels ranged from 1 to 2% DW. The seaweeds composition is greatly variable, depending on the species, habitat, season, and environmental conditions such as light intensity, water temperature and nutrient levels (Makkar et al. 2016; Gaillard et al. 2018).

2.4.1 Fatty Acids

A knowledge of the fatty acid profile of seaweeds can provide functional food and nutraceutical applications. Fish are considered as the main sources of n-3 PUFA in the human diet with beneficial impacts on human health (Brouwer et al. 2006). Due o the high demand for n-3 PUFA, seaweeds are also regarded as alternative source of these fatty acids. Although seaweeds contain low level of lipid, their the essential PUFAs level is high (Biris-Dorhoi et al. 2020; Bruni et al. 2020). Seaweeds have ability for synthesis of essential fatty acids; linoleic acid (C18:2n-6) and alpha-linolenic (C18:3 n-3,) which cannot be produced by human body and should be provided in the diet. Moreover, seaweeds compared to land plants, can synthesize the long-chain PUFA for example, arachidonic (ARA, C20:4n-6), eicosapentaenoic (EPA, C20:5n-3) and DHA (C22:6n-3) (Biris-Dorhoi et al. 2020). Although EPA and DHA are not classified as essential, they have important role for protection against some diseases, including cardiovascular diseases, mental disorders, and inflammation (Narayan et al. 2006). It was also reported that seaweeds usually contain an optimum ratio of n-3 PUFA/n-6 PUFA (approximately 1) (Wells et al. 2017).

The fatty acid profiles of red, green, and brown seaweeds from the Persian Gulf was determined by Rohani-Ghadikolaei et al. (2012). Among the SFA, myristic acid (C14:0) and palmitic acid (C16:0) were the main fatty acids. As for MUFA, oleic acid (C18:1n-9) and palmitoleic acid (C16:1n-7) were the most abundant fatty acids. Linolenic acid (C18:3n-3) was the most common PUFA. Similar results were also reported in previous studies (Matanjun et al. 2009; Polat and Ozogul 2009, 2013).

Schmid et al. (2018) investigated seaweeds from southern-Australian as a promising source of fatty acids (FA), including 17 Phaeophyceae (Ochrophyta 11), Chlorophyta, and 33 Rhodophyta species. They found that total fatty acid (TFA) levels ranged from 0.6 to 7.8 in % of DW, with the highest level in the Phaeophyceae. In addition, most of species had high levels of PUFA and a low ratio of n-6/n-3 PUFA, indicating the potential use of animal feed and nutraceutical and food industries. A n-6/n-3 fatty acid ratio of 1 is regarded as optimum for diets of human to avoid diseases. WHO (World Health Organization) suggests a n-6/n-3 ratio lower than 10 in diets to decrease inflammatory, neurological and cardiovascular disorders (FAO/WHO 2008). However, the European Nutritional Societies recommend an n-6/n-3 ratio of 5 (Alles et al. 2014). FDA (Food and Drug Administration) has recommended for the consumption of EPA and DHA a daily intake of at least 0.8 g of EPA and DHA, providing a useful effect on health.

Rocha et al. (2021) assessed essential FA of 5 red seaweeds (*Gracilaria gracilis*, *Calliblepharis jubata*, *Asparagospis armata*, *Grateloupia turuturu* and *Chondracanthus teedei* var. *lusitanicus*) and 3 brown seaweeds (*Undaria pinnatifida*, *Colpomenia peregrina* and *Sargassum muticum*) from Portugal for dietary supplements. *Undaria pinnatifida* and *Calliblepharis jubata* were found to be the most suitable for industrial application They concluded that these two seaweeds contained high levels of PUFA+HUFA, with n-3 FA being the most prominent, thus highest nutraceutical potential.

Fatty acids levels and profile of seaweeds varies with biotic and abiotic parameters, and also genetic characteristics of seaweeds (Cotas et al. 2020). For example; it was found that seaweed species in the warm water had higher SFAs, oleic acid and lower PUFA levels than species in the cold water (Khotimchenko and Gusarova 2004). It was also indicated that the life cycle of seaweed can affect the FA content (Pereira and Silva 2021).

2.4.2 Phospholipids

Fatty acids are generally found in polar lipids, which include phospholipids, glycolipids, and betaine lipids, plastid membranes and structures of cell in seaweeds (Harwood and Guschina 2009). Polar lipids (phospholipids and glycolipids) from seaweeds have been paid attention since they are important carriers of PUFAs and a wide range of bioactive properties (Lordan et al. 2011; Plouguerné et al. 2014). Phospholipids have some important functions, including emulsifying properties, supplement of n-3 fatty acids with high bioavailability of PUFA and useful nutritional effects (Lu et al. 2017). Therefore, phospholipids provide useful nutritional effects by means of the transportation of n-3 fatty acids, which are important for healthy human brain or also for prevention of diseases for instance cardiovascular risks, cancer or neurological disorders (Küllenberg et al. 2012; Cornish et al. 2017).

Rey et al. (2019) characterized the polar lipid profile of *Saccharina latissima.* They identified 197 molecular species of polar lipids, containing phospholipids, glycolipids and betaine lipids. Some molecular species determined are carriers of

PUFAs. They indicated that *S. latissima* is regarded as a potential source of natural bioactive lipid composites with potential therapeutic and health benefits. da Costa et al. (2021) identified polar lipids composition from *Grateloupia turuturu*, including 205 lipid species, which were distributed in betaine lipids, phospholipids, glycolipids and phosphosphingolipids. Some of the lipid species had biological activity since extracts of polar lipid exhibited antioxidant activity and anti-inflammatory activity. They concluded that *G. turuturu* can be used as a source of nutritive biomass and bioactive compounds for pharmaceutical, nutraceutical and cosmeceutical applications. da Costa (2018) also investigated the lipidome of *Porphyra dioica* conchocelis and blade life stages. 100 and 110 lipid species were determined in the lipidome of conchocelis and the blade, respectively. They are found in phospholipids, glycolipids, inositephosphoceramides, and betaine lipids. It was reported that stages of life cycle contained a similar composition of glycolipids. However, there was a difference in phospholipids profile between the two life stages. These differences were attributed to abundance and number of molecular species, indicating moves in the lipids of extraplastidial membranes instead of in plastidial membranes. This finding suggests likely use of these life stages for human diet to avoid chronic diseases.

2.4.3 Glycolipids

Glycolipids have health benefits and are utilised as food or feed and also for biotechnological purposes (Plouguerné et al. 2014). Glycolipids are mostly found in photosynthetic membranes and are very important for providing energy. In addition, they have function in membranes for preventing cells towards chemical tension, stabilizing bilayers of membrane, as well as performing as markers for cellular detection (Boudière et al. 2014). Biological activities of glycolipids of seaweeds include antibacterial, antiviral, and antitumor properties, allowing the pharmacological applications of these compounds (Kendel et al. 2015). Zhang et al. (2014) reported that these bioactivities of glycoglycerolipids result from the sugar moiety, the place of the glycerol link to the sugar, the size and site of the acyl chain, and the anomeric configuration of the sugar.

The glycolipids and phospholipids classes of some seaweeds such as *Grateloupia turuturu*, (Kendel et al. 2015; da Costa et al. 2021), *Saccharina latissima* (Rey et al. 2019), *Palmaria palmata* (Banskota et al. 2014), *Osmundaria obtusiloba* (De Souza et al. 2012), *Bifurcaria bifurcata* and *Sargassum muticum* (Santos et al. 2020) were identified. Therefore, lipidomics can be used to identify the best profile of polar lipids with higher PUFA content, thus selecting the seaweed species for industrial application.

2.4.4 Sterols

Plant sterols are an important constituent of the membranes of all eukaryotic organisms, controling membrane fluidity and permeability. Plant sterols, knows as phytosterols, were reported to contain more than 250 diverse sterols and linked composites in numerous plant and marine substances. The most common sterols are sitosterol, stigmasterol and campesterol (Piironen et al. 2000). Bakar et al. (2019) determined sterols contents of *Dictyota dichotoma* and *Sargassum granuliferum* from Malaysia. It was determined that both *S. granuliferum* and *D. dichotoma* contained mostly stigmasterol, campesterol and β-sitosterol. These phytosterols are the most abundant in plant, maintaining the physiology and structure of cell membranes. However, Kendel et al. (2015) reported that the major constituent of sterol fraction from *Ulva armoricana* and *Solieria chordalis* was cholesterol (35%, 43%, respectively) whereas these seaweeds contained other sterols such as cholest-4-en-3-one, regarded as derivative of cholesterol found in animal and plant tissues. It may produced from the biosynthesis or the autoxidation of cholesterol. Kendel et al. (2013) also indicated that total sterol composition in red alga *Grateloupia turuturu* varied during season. In all seasons cholesterol (38%) was the main sterol with a lesser level in both summer and spring.

Osuna-Ruiz et al. (2019) investigated sterol composition in green (*Caulerpa sertularioides*, *Ulva expansa*, *Codium isabelae*, *Rhizoclonium riparium*), red (*Gracilaria vermiculophylla*, *Spyridia filamentosa*), and brown (*Padina durvillaei*) seaweeds from Mexico. The highest levels of cholesterol+dehydrocholesterol (>90%) were determined in *G. vermiculophylla* and S. *filamentosa.* However, β-sitosterol was the main sterol (71–77%) in *C. sertularioides*, *C. isabelae* and *R. riparium.* Stigmasterol was the highest in *R. riparium* (14%). The main sterols were fucosterol+isofucosterol (79%) in *U. expansa* and *P. durvillaei.*

2.5 *Polyphenols*

Plants generate polyphenols as secondary metabolites and play important roles in many activities such as growth, pigmentation, reproduction and fighting against environmental stresses (Duthie et al. 2003). These secondary metabolites exhibit a various phenolic structures and vary structurally from simple molecules to high molecular polymers. Thus, the polyphenol can be categorised according to chemical structure, source of origin, and function. For example; according to chemical structure of aglycones (the composite staying after the glycosyl group on a glycoside is replaced by a hydrogen atom), polyphenols can be divided into flavonoids, phenolic acids, lignans, stilbenes, etc. Among them, flavonoids are extensively found and separated into six subclasses; flavanols, flavonols, flavones, flavanones, anthocyanins and isoflavones (Valdes et al. 2015). Seaweeds are regarded as a source of polyphenols, including flavonoids, phlorotannins, phenolic terpenoids, bromophenols, and mycosporine-like amino acids (Wells et al. 2017). Phlorotannins are the main polyphenolic compound present only in the brown seaweeds whereas the greatest amount of phenolic compounds found in red and green seaweeds are flavonoids, bromophenols and phenolic acids (Cotas et al. 2020a).

Plant polyphenols are regarded as strong antioxidants against dangerous harmful free radicals or reactive oxygen species (ROS) due to their high redox capacity, allowing them to perform as reducing agents, singlet oxygen quenchers, hydrogen donors, and metal chelators (Fraga et al. 2010). The potential useful properties of polyphenols contain antimicrobial, antioxidant, anticancer, antiviral, anti-inflammatory, anti-photoaging, anti-allergic, and antidiabetic activities (Gómez-Guzmán et al. 2018; Wekre et al. 2019; Cotas et al. 2020a; Ummat et al. 2020). Nutritional, epidemiological and clinical research showed that polyphenols have health benefits associated with some chronic diseases such as cardiovascular diseases, cancer, diabetes and obesity (Déléris et al. 2016). Due to their nutritious and well-being benefits, polyphenols derived in seaweeds are being gradually explored for potential use in nutraceuticals, foods, cosmetic, and pharmaceutical industries.

It is challenging to standardize polyphenol analyses in seaweeds due to their variable level of polyphenol and also a high dissolved salt concentration. Rajauria (2018) developed a method with RP–HPLC (reverse phase-high performance liquid chromatography) coupled to DAD (a diode array detector) and ESI–MS (negative ion electrospray mass spectrometer) in order to identify and quantify phenolic compounds in brown seaweed, *H. elongata*. Seven phenolic compounds were identified and their levels were found as phloroglucinol (394.1 μg/g), gallic acid (96.3 μg/g), caffeic acid (44.4 μg/g), chlorogenic acid (38.8 μg/g), myricetin (8.6 ± 0.85 μg/g), ferulic acid (17.6 μg/g), and quercetin (4.2 ± 0.15 μg/g). It was indicated that the developed method was accurate, sensitive, and suitable for routine polyphenols analysis in seaweed. Wekre et al. (2019) also determined polyphenols in a green macroalga, *Ulva intestinalis*, from the western coast of Norway by using quantitative qNMR (quantitative nuclear magnetic resonance), HPLC-DAD (high performance liquid chromatography (HPLC) with wavelength detector), and the Folin-Ciocalteu assay (TPC). The qNMR method gave 5.5% (DW) polyphenols, while HPLC-DAD and TPC gave 1.1% (DW) and 0.4% (DW) respectively in the crude extract. They concluded that more research is required for the optimization of processes such as extraction, identification, and quantification for polyphenols in seaweeds.

Ummat et al. (2020) investigated UAE (ultrasound assisted extraction) as a novel extraction technology in order to increase quality and yield of the phenolic compounds in 11 brown seaweed species. They optimised UAE method and improved the extraction yield in comparison to solvent extraction. The highest yield of total phenolics (572.3 mg gallic acid equivalent/g), total flavonoids (281.0 mg quercetin equivalent/g) and total phlorotannins (476.3 phloroglucinol equivalent/g) was found in *Fucus vesiculosus*. They concluded that UAE was an effective, and green method to obtain polyphenols from 11 brown seaweeds.

Standardization and optimization of extraction and isolation techniques of the polyphenols derived from seaweeds still need to be developed in order to obtain better quality, purity, and quantity of these compounds as well as considering more ecological and sustainability aspects.

References

Admassu H, Gasmalla MA, Yang R, Zhao W (2018) Identification of bioactive peptides with α-amylase inhibitory potential from enzymatic protein hydrolysates of red seaweed (Porphyra spp). J Agric Food Chem 66(19):4872–4882. https://doi.org/10.1021/acs.jafc.8b00960

Alboofetileh M, Hamzeh A, Abdollahi M (2021) Seaweed proteins as a source of bioactive peptides. Curr Pharm Des 27(11):1342–1352. https://doi.org/10.2174/1381612827666210208153249

Alles MS, Eussen SR, Van Der Beek EM (2014) Nutritional challenges and opportunities during the weaning period and in young childhood. Ann Nutr Metab 64(3–4):284–293. https://doi.org/10.1159/000365036

Angell AR, Mata L, de Nys R, Paul NA (2016) The protein content of seaweeds: a universal nitrogen-to-protein conversion factor of five. J Appl Phycol 28(1):511–524. https://doi.org/10.1007/s10811-015-0650-1

Aryee A, Agyei D, Akanbi T (2018) Recovery and utilization of seaweed pigments in food processing. Curr Opin Food Sci 19:113–119. https://doi.org/10.1016/j.cofs.2018.03.013

Aslanbay Güler B, İmamoğlu E (2021) Trends in a natural product fucoxanthin. Ege J Fish Aquat Sci 38(1):117–124. https://doi.org/10.12714/egejfas.38.1.15

Astorga-España MS, Rodríguez-Galdón B, Rodríguez-Rodríguez EM, Díaz-Romero C (2016) Amino acid content in seaweeds from the Magellan Straits (Chile). J Food Compost Anal 53:77–84. https://doi.org/10.1016/j.jfca.2016.09.004

Bakar K, Mohamad H, Tan HS, Latip J (2019) Sterols compositions, antibacterial, and antifouling properties from two Malaysian seaweeds: Dictyota dichotoma and Sargassum granuliferum. J App Pharm Sci 9:47–53. https://doi.org/10.7324/JAPS.2019.91006

Bang TH, Van TTT, Hung LX, Mınh Ly B, Nhut ND, Thuy TTT, Huy BT (2019) Nanogels of acetylated ulvan enhance the solubility of hydrophobic drug curcumin. Bull Mater Sci 42:1. https://doi.org/10.1007/S12034-018-1682-3

Banskota AH, Stefanova R, Sperker S, Lall SP, Craigie JS, Hafting JT, Critchley AT (2014) Polar lipids from the marine macroalga Palmaria palmata inhibit lipopolysaccharide-induced nitric oxide production in RAW264.7 macrophage cells. Phytochemistry 101:101–108. https://doi.org/10.1016/j.phytochem.2014.02.004

Berteau O, Mulloy B (2003) Sulfated fucans, fresh perspectives: structures, functions, and biological properties of sulfated fucans and an overview of enzymes active toward this class of polysaccharide. Glycobiology 13(6):29R–40R. https://doi.org/10.1093/glycob/cwg058

Bhatia S, Sharma K, Sharma A, Namdeo AG, Chaugule BB (2011) Antioxidant potential of Indian porphyra. Pharmacologyonline 1:248–257

Biancarosa I, Espe M, Bruckner CG, Heesch S, Liland N, Waagbø R, Torstensen B, Lock EJ (2017) Amino acid composition, protein content, and nitrogen-to-protein conversion factors of 21 seaweed species from Norwegian waters. J Appl Phycol 29(2):1001–1009. https://doi.org/10.1007/Fs10811-016-0984-3

Bilal M, Iqbal H (2019) Marine seaweed polysaccharides-based engineered cues for the modern biomedical sector. Mar Drugs 18(1):7. https://doi.org/10.3390/md18010007

Biris-Dorhoi ES, Michiu D, Pop CR, Rotar AM, Tofana M, Pop OL, Socaci SA, Farcas AC (2020) Macroalgae – a sustainable source of chemical compounds with biological activities. Nutrients 12(10):3085. https://doi.org/10.3390/nu12103085

Bleakley S, Hayes M (2017) Algal proteins: extraction, application, and challenges concerning production. Food Secur 6:33. https://doi.org/10.3390/foods6050033

Boudière L, Michaud M, Petroutsos D, Rébeillé F, Falconet D, Bastien O, Roy S, Finazzi G, Rolland N, Jouhet J, Block MA, Maréchal E (2014) Glycerolipids in photosynthesis: composition, synthesis and trafficking. Biochim Biophys Acta 1837(4):470–480. https://doi.org/10.1016/j.bbabio.2013.09.007

Brouwer IA, Geelen A, Katan MB (2006) n− 3 Fatty acids, cardiac arrhythmia and fatal coronary heart disease. Progress Lipid Res 45(4):357–367. https://doi.org/10.1016/j.plipres.2006.02.004

Bruni L, Secci G, Mancini S, Faccenda F, Parisi G (2020) A commercial macroalgae extract in a plant-protein rich diet diminished saturated fatty acids of Oncorhynchus mykiss walbaum fillets. Ital J Anim Sci 19(1):373–382. https://doi.org/10.1080/1828051X.2020.1745097

Capriotti AL, Caruso G, Cavaliere C, Samperi R, Ventura S, Zenezini Chiozzi R, Lagana A (2015) Identification of potential bioactive peptides generated by simulated gastrointestinal digestion of soybean seeds and soy milk proteins. J Food Comp Anal 44:205–213. https://doi.org/10.1016/j.jfca.2015.08.007

Cermeño M, Kleekayai T, Amigo-Benavent M, Harnedy-Rothwell P, FitzGerald RJ (2020) Current knowledge on the extraction, purification, identification, and validation of bioactive peptides from seaweed. Electrophoresis 41(20):1694–1717. https://doi.org/10.1002/elps.202000153

Chades T, Scully SM, Ingvadottir EM, Orlygsson J (2018) Fermentation of mannitol extracts from brown macro algae by thermophilic *Clostridia*. Front Microbiol 9:1931. https://doi.org/10.3389/fmicb.2018.01931

Chandía NP, Matsuhiro B (2008) Characterization of a fucoidan from Lessonia vadosa (Phaeophyta) and its anticoagulant and elicitor properties. Int J Biol Macromol 42(3):235–240. https://doi.org/10.1016/j.ijbiomac.2007.10.023

Charoensiddhi S, Conlon MA, Franco CMM, Zhang W (2017) The development of seaweed-derived bioactive compounds for use as prebiotics and nutraceuticals using enzyme technologies. Trends Food Sci Technol 70:20–33. https://doi.org/10.1016/j.tifs.2017.10.002

Chauhan PS, Saxena A (2016) Bacterial carrageenases: an overview of production and biotechnological applications. 3 Biotech 6(2):146. https://doi.org/10.1007/s13205-016-0461-3

Chen M, Blankenship RE (2011) Expanding the solar spectrum used by photosynthesis. Trends Plant Sci 16:427–431. https://doi.org/10.1016/j.tplants.2011.03.011

Chen H-M, Zheng L, Yan X-J (2005) The preparation and bioactivity research of Agaro-oligosaccharides. Food Technol Biotechnol 43(1):29–36

Chen H, Wu Y, Chen Y, Li Y, McGowan E, Lin Y (2021) P-268 seaweed laminaria japonica peptides possess strong anti-liver cancer effects. Ann Oncol 32:S189. https://doi.org/10.1016/j.annonc.2021.05.322

Cherdkiatikul T, Suwanwong Y (2014) Production of the α and β subunits of *Spirulina* allophycocyanin and C-phycocyanin in *Escherichia coli*: a comparative study of their antioxidant activities. J Biomol Screen 19(6):959–965. https://doi.org/10.1177/1087057113520565

Chidambara Murthy KN, Vanitha A, Rajesha J, Mahadeva Swamy M, Sowmya PR, Ravishankar GA (2005) In vivo antioxidant activity of carotenoids from Dunaliella salina green microalga. Life Sci 76:1381–1390. https://doi.org/10.1016/j.lfs.2004.10.015

Cho ML, Lee HS, Kang IJ, Won MH, You SG (2011) Antioxidant properties of extract and fractions from Enteromorpha prolifera, a type of green seaweed. Food Chem 127:999–1006. https://doi.org/10.1016/j.foodchem.2011.01.072

Chojnacka KJ, Saeid A, Witkowska Z, Tuhy L (2012) Biologically active compounds in seaweed extracts – the prospects for the application. Open Conf Proc J 3:20–28. https://doi.org/10.2174/1876326X01203020020

Choudhary B, Chauhan OP, Mishra A (2021) Edible seaweeds: a potential novel source of bioactive metabolites and nutraceuticals with human health benefits. Front Mar Sci 8:740054. https://doi.org/10.3389/fmars.2021.740054

Cian RE, Martínez-Augustin O, Drago SR (2012) Bioactive properties of peptides obtained by enzymatic hydrolysis from protein byproducts of Porphyra columbina. Food Res Int 49:364–372. https://doi.org/10.1016/j.foodres.2012.07.003

Cian RE, Nardo AE, Garzón AG, Añon MC, Drago SR (2022) Identification and in silico study of a novel dipeptidyl peptidase IV inhibitory peptide derived from green seaweed Ulva spp. hydrolysates. LWT 154:112738. https://doi.org/10.1016/j.lwt.2021.112738

Cornish ML, Critchley AT, Mouritsen OG (2017) Consumption of seaweeds and the human brain. J Appl Phycol 29:2377–2398. https://doi.org/10.1007/s10811-016-1049-3

Cotas J, Leandro A, Pacheco D, Gonçalves AMM, Pereira L (2020) A comprehensive review of the nutraceutical and therapeutic applications of red seaweeds (Rhodophyta). Life 10:19. https://doi.org/10.3390/life10030019

Cotas J, Leandro A, Monteiro P, Pacheco D, Figueirinha A, Gonçalves AM, da Silva DJ, Pereira L (2020a) Seaweed phenolics: from extraction to applications. Mar Drugs 18(8):384. https://doi.org/10.3390/md18080384

Cui K, Tai W, Shan X, Hao J, Li G, Yu G (2018) Structural characterization and anti-thrombotic properties of fucoidan from Nemacystus decipiens. Int J Biol Macromol 120:1817–1822. https://doi.org/10.1016/j.jphotobiol.2008.11.011

Černá M (2011) Seaweed proteins and amino acids as nutraceuticals. Adv Food Nutr Res 64:297–312. https://doi.org/10.1016/b978-0-12-387669-0.00024-7

da Costa E, Azevedo V, Melo T, Rego AM, Evtuguin DV, Domingues P, Calado R, Pereira R, Abreu MH, Domingues MR (2018) High-resolution lipidomics of the early life stages of the red seaweed *Porphyra dioica*. Molecules 23(1):187. https://doi.org/10.3390/molecules23010187

da Costa E, Melo T, Reis M, Domingues P, Calado R, Abreu MH, Domingues MR (2021) Polar lipids composition, antioxidant and anti-inflammatory activities of the Atlantic red seaweed Grateloupia turuturu. Mar Drugs 19(8):414. https://doi.org/10.3390/md19080414

de Jesus Raposo M, de Morais A, de Morais R (2015) Marine polysaccharides from algae with potential biomedical applications. Mar Drugs 13(5):2967–3028. https://doi.org/10.3390/md13052967

Dawczynski C, Schubert R, Jahreis G (2007) Amino acids, fatty acids, and dietary fibre in edible seaweed products. Food Chem 103(3):891–899. https://doi.org/10.1016/j.foodchem.2006.09.041

De Souza LM, Sassaki GL, Romanos MTV, Barreto-Bergter E (2012) Structural characterization and anti-HSV-1 and HSV-2 activity of glycolipids from the marine algae Osmundaria obtusiloba isolated from Southeastern Brazilian coast. Mar Drugs 10(4):918–931. https://doi.org/10.3390/md10040918

Déléris P, Nazih H, Bard JM (2016) Seaweeds in human health. In: Fleurence J, Levine I (eds) Seaweed in health and disease prevention, 1st edn. Elsevier, London, pp 319–367

Din SS, Chew KW, Chang YK, Show PL, Phang SM, Juan JC (2019) Extraction of agar from Eucheuma cottonii and Gelidium amansii seaweeds with sonication pretreatment using autoclaving method. J Oceanol Limnol 37:871–880. https://doi.org/10.1007/s00343-019-8145-6

Duthie GG, Gardner PT, Kyle JA (2003) Plant polyphenols: are they the new magic bullet? Proc Nutr Soc 62:599–603. https://doi.org/10.1079/pns2003275

Echave J, Fraga-Corral M, Garcia-Perez P, Popović-Djordjević J, Avdović EH, Radulović M, Xiao J, Prieto MA, Simal-Gandara J (2021) Seaweed protein hydrolysates and bioactive peptides: extraction. Purif Appl Mar Drugs 19(9):500. https://doi.org/10.3390/md19090500

Elumalai S, Gopal R, Jegan Sangeetha T, Roopsingh D (2014) Extraction of phycocyanin an important pharmaceutical phycobiliproteins from Cyanobacteria. Int J Pharm Res Dev 6(4):67–74

Enoki T, Okuda S, Kudo Y, Takashima F, Sagawa H, Kato I (2010) Oligosaccharides from agar inhibit pro-inflammatory mediator release by inducing heme oxygenase 1. Biosci Biotechnol Biochem 74(4):766–770. https://doi.org/10.1271/bbb.90803

Eriksen NT (2008) Production of phycocyanin-a pigment with applications in biology, biotechnology, foods and medicine. Appl Microbiol Biotechnol 80(1):1–14. https://doi.org/10.1007/s00253-008-1542-y

FAO/WHO (2008) Expert consultation. Fats and fatty acids in human nutrition. Ann Nutr Metab 55

Fernando S, Sanjeewa A, Samarakoon K, Kim H-S, Gunasekara U, Park Y-J, Abeytunga T, Lee W, Jeon Y-J (2018) The potential of fucoidans from Chnoospora minima and Sargassum polycystum in cosmetics: antioxidant, anti-inflammatory, skin-whitening, and antiwrinkle activities. J Appl Phycol 30:3223–3232. https://doi.org/10.1007/s10811-018-1415-4

Ferrara L (2020) Seaweeds: a food for our future. J Food Chem Nanotechnol 6(2):56–64. https://doi.org/10.17756/jfcn.2020-084

Fertah M, Belfkira A, Dahmane EM, Taourirte M, Brouillette F (2017) Extraction and characterization of sodium alginate from Moroccan Laminaria digitata brown seaweed. Arab J Chem 10:S3707–S3714. https://doi.org/10.1016/j.arabjc.2014.05.003

Fidelis GP, Silva CHF, Nobre LTDB, Medeiros VP, Rocha HAO, Costa LS (2019) Antioxidant fucoidans obtained from tropical seaweed protect pre-osteoblastic cells from hydrogen peroxide-induced damage. Mar Drugs 17(9):506. https://doi.org/10.3390/md17090506

Florez-Mendez J, González L (2019) Role of the consumption of fucoidans and beta-glucans on human health: an update of the literature. Revi Chil Nutr 46:768–775. https://doi.org/10.4067/S0717-75182019000600768

Fraga CG, Galleano M, Verstraeten SV, Oteiza PI (2010) Basic biochemical mechanisms behind the health benefits of polyphenols. Mol Aspects of Med 31(6):435–445. https://doi.org/10.1016/j.mam.2010.09.006

Francavilla M, Franchi M, Monteleone M, Caroppo C (2013) The red seaweed Gracilaria gracilis as a multi products source. Mar Drugs 11(10):3754–3776. https://doi.org/10.3390/md11103754

Freitas MV, Pacheco D, Cotas J, Mouga T, Afonso C, Pereira L (2022) Red seaweed pigments from a biotechnological perspective. Phycology 2:1–29. https://doi.org/10.3390/phycology2010001

Gaillard C, Bhatti HS, Novoa-Garrido M, Lind V, Roleda MY, Weisbjerg MR (2018) Amino acid profiles of nine seaweed species and their in situ degradability in dairy cows. Anim Feed Sci Technol 241:210–222. https://doi.org/10.1016/j.anifeedsci.2018.05.003

Ganesan RA, Shanmugan M, Bhat R (2018) Producing novel edible films from semi refined carrageenan (SRC) and ulvan polysaccharides for potential food applications. Int J Biol Macromol 112:1164–1170. https://doi.org/10.1016/j.ijbiomac.2018.02.089

Garcia-Vaquero M, Lopez-Alonso M, Hayes M (2017) Assessment of the functional properties of protein extracted from the brown seaweed *Himanthalia elongata* (Linnaeus) SF gray. Food Res Int 99:971–978. https://doi.org/10.1016/j.foodres.2016.06.023

Gomes L, Monteiro P, Cotas J, Gonçalves AMM, Fernandes C, Gonçalves T, Pereira L (2022) Seaweeds' pigments and phenolic compounds with antimicrobial potential. Biomol Concepts 13(1):89–102. https://doi.org/10.1515/bmc-2022-0003

Gómez-Guzmán M, Rodríguez-Nogales A, Algieri F, Gálvez J (2018) Potential role of seaweed polyphenols in cardiovascular-associated disorders. Mar Drugs 16(8):250. https://doi.org/10.3390/md16080250

Gupta AK, Seth K, Maheshwari K, Baroliya PK, Meena M, Kumar A, Vinayak V, Harish (2021) Biosynthesis and extraction of high-value carotenoid from algae. Front Biosci (Landmark Ed) 26(6):171–190. https://doi.org/10.52586/4932

Gallardo, T (2015) Marine algae: general aspects (biology, systematics, field and laboratory techniques). In: Pereira L, Neto JM, (eds) Marine algae: biodiversity, taxonomy, environmental assessment, and biotechnology. CRC Press, Boca Raton, pp 1–67

Harwood JL, Guschina IA (2009) The versatility of algae and their lipid metabolism. Biochimie 91:679–684. https://doi.org/10.1016/j.biochi.2008.11.004

Hayes M, Tiwari B (2015) Bioactive carbohydrates and peptides in foods: an overview of sources, downstream processing steps and associated bioactivities. Int J Mol Sci 16:22485–22508. https://doi.org/10.3390/ijms160922485

Hentati F, Tounsi L, Djomdi D, Pierre G, Delattre C, Ursu AV, Fendri I, Abdelkafi S, Michaud P (2020) Bioactive polysaccharides from seaweeds. Molecules 25(14):3152. https://doi.org/10.3390/molecules25143152

Heo S-J, Jeon YJ (2009) Protective effect of fucoxanthin isolated from Sargassum siliquastrum on UV-B induced cell damage. J Photochem Photobiol B 95(2):101–107. https://doi.org/10.1016/j.jphotobiol.2008.11.011

Heo S-J, Ko S-C, Kang S-M, Kang H-S, Kim J-P, Kim S-H, Lee K-W, Cho MG, Jeon Y-J (2008) Cytoprotective effect of fucoxanthin isolated from brown algae Sargassum siliquastrum against H2O2-induced cell damage. Eur Food Res Technol. 228:145–151. https://doi.org/10.1007/s00217-008-0918-7

Holdt SL, Kraan S (2011) Bioactive compounds in seaweed: functional food applications and legislation. J Appl Phycol 23:543–597. https://doi.org/10.1007/s10811-010-9632-5

Hong JE, Lim JH, Kim TY, Jang HY, Oh HB, Chung BG, Lee SY (2020) Photo-oxidative protection of chlorophyll *a* in c-phycocyanin aqueous medium. Antioxidants (Basel) 9(12):1235. https://doi.org/10.3390/antiox9121235

Houghton D, Wilcox MD, Chater PI, Brownlee IA, Seal CJ, Pearson JP (2015) Biological activity of alginate and its effect on pancreatic lipase inhibition as a potential treatment for obesity. Food Hydrocoll 49:18–24. https://doi.org/10.1016/j.foodhyd.2015.02.019

Ina A, Hayashi K, Nozaki H, Kamei Y (2007) Pheophytin a, a low molecular weight compound found in the marine brownalga Sargassum fulvellum, promotes the differentiation of PC12 cells. Int J Dev Neurosci 25:63–68. https://doi.org/10.1016/j.ijdevneu.2006.09.323

Indumathi P, Mehta A (2016) A novel anticoagulant peptide from the Nori hydrolysate. J Funct Foods 20:606–617. https://doi.org/10.1016/j.jff.2015.11.016

Ismail MM, Osman MEH (2016) Seasonal fluctuation of photosynthetic pigments of most common red seaweeds species collected from Abu Qir, Alexandria. Egypt Rev Biol Mar Oceanogr 51(3):515–525. https://doi.org/10.4067/S0718-19572016000300004

Iwamoto K, Shiraiwa Y (2005) Salt-regulated mannitol metabolism in algae. Mar Biotechnol 7(5):407–415. https://doi.org/10.1007/s10126-005-0029-4

Jahandideh A, Ashkani M, Moini N (2021) Biopolymers in textile industries. In: Thomas S, Gopi S, Amalraj A (eds) Biopolymers and their industrial applications. Elsevier, Cambridge, pp 193–218. https://doi.org/10.1016/B978-0-12-819240-5.00008-0

Jeliani ZZ, Pirian K, Sohrabipour J, Sorahinobar M, Soltani M, Sourinejad I, Yousefzadi M (2021) Assessment of fatty acid and amino acid composition of macroalgae from the Persian Gulf to characterize their suitability for nutritional supplements. Biol Bull 48(6):752–762. https://doi.org/10.1134/S1062359021130033

Ji CF, Ji YB (2014) Laminarin-induced apoptosis in human colon cancer LoVo cells. Oncol Lett 7(5):1728–1732. https://doi.org/10.3892/ol.2014.1952

Kendel M, Couzinet-Mossion A, Viau M, Fleurence J, Barnathan G, Wielgosz-Collin G (2013) Seasonal composition of lipids, fatty acids, and sterols in the edible red alga Grateloupia turuturu. J Appl Phycol 25:425–432. https://doi.org/10.1007/s10811-012-9876-3

Kendel M, Wielgosz-Collin G, Bertrand S, Roussakis C, Bourgougnon N, Bedoux G (2015) Lipid composition, fatty acids and sterols in the seaweeds Ulva armoricana, and Solieria chordalis from Brittany (France): an analysis from nutritional, chemotaxonomic, and antiproliferative activity perspectives. Mar Drugs 13:5606–5628. https://doi.org/10.3390/md13095606

Khalid S, Abbas M, Saeed F, Bader-Ul-Ain H, Suleria HAR (2018) Therapeutic potential of seaweed bioactive compounds. In: Maiti S (ed) Seaweed biomaterials. IntechOpen. https://doi.org/10.5772/intechopen.74060

Khotimchenko SV, Gusarova IS (2004) Red algae of peter the great bay as a source of arachidonic and eicosapentaenoic acids. Russ J Mar Biol 30(3):183–187. https://doi.org/10.1023/B:RUMB.0000033953.67105.6b

Kidgell JT, Magnusson M, de Nys R, Glasson CRK (2019) Ulvan: a systematic review of extraction, composition and function. Algal Res 39:101422. https://doi.org/10.1016/j.algal.2019.101422

Koyanagi S, Tanigawa N, Nakagawa H, Soeda S, Shimeno H (2003) Oversulfation of fucoidan enhances its anti-angiogenic and antitumor activities. Biochem Pharmacol 65(2):173–179. https://doi.org/10.1016/s0006-2952(02)01478-8

Kraan S (2012) Algal polysaccharides, novel applications and outlook. In: Chang C-F (ed) Carbohydrates; comprehensive studies on glycobiology and glycotechnology, 1st edn. Intech, pp 489–532

Küllenberg D, Taylor LA, Schneider M, Massing U (2012) Health effects of dietary phospholipids. Lipids Health Dis 11(1):1–16. https://doi.org/10.1186/1476-511X-11-3

Kwon KT, Jung GW, Chun BS (2010) Amino acid recovery from Brown seaweed (Undaria pinnatifida) using subcritical water hydrolysis. Korean Chem Eng Res 48(6):747–751

Lafarga T, Acién-Fernández FG, Garcia-Vaquero M (2020) Bioactive peptides and carbohydrates from seaweed for food applications: natural occurrence, isolation, purification, and identification. Algal Res 48:101909. https://doi.org/10.1016/j.algal.2020.101909

Lage-Yusty M-A, Caramés-Adán P, López-Hernández J (2013) Determination of phycobiliproteins by constant-wavelength synchronous spectrofluorimetry method in red algae. CyTA-J Food 11(3):243–247. https://doi.org/10.1080/19476337.2012.728629

Lakshmi DS, Sankaranarayanan S, Gajaria TK, Li G, Kujawski W, Kujawa J, Navia R (2020) A short review on the valorization of green seaweeds and ulvan: feedstock for chemicals and biomaterials. Biomolecules 10(7):991. https://doi.org/10.3390/biom10070991

Leandro A, Pacheco D, Cotas J, Marques JC, Pereira L, Gonçalves AMM (2020) Seaweed's bioactive candidate compounds to food industry and global food security. Life (Basel) 10(8):140. https://doi.org/10.3390/life10080140

Lee HG, Lu YA, Je JG, Jayawardena TU, Kang MC, Lee SH, Kim TH, Lee DS, Lee JM, Yim MJ, Kim H-S, Jeon Y-J (2021) Effects of ethanol extracts from Grateloupia elliptica, a red seaweed, and its chlorophyll derivative on 3T3-L1 adipocytes: suppression of lipid accumulation through downregulation of adipogenic protein expression. Mar Drugs 19:91. https://doi.org/10.3390/md19020091

Li B, Lu F, Wei X, Zhao R (2008) Fucoidan: structure and bioactivity. Molecules 13(8):1671–1695. https://doi.org/10.3390/molecules13081671

Lordan S, Ross RP, Stanton C (2011) Marine bioactives as functional food ingredients: potential to reduce the incidence of chronic diseases. Mar Drugs 9:1056–1100. https://doi.org/10.3390/md9061056

Lorenzo JM, Agregán R, Munekata PES, Franco D, Carballo J, Sahin S, Lacomba R, Barba FJ (2017) Proximate composition and nutritional value of three macroalgae: Ascophyllum nodosum, Fucus vesiculosus and Bifurcaria bifurcata. Mar Drugs 15(11):360. https://doi.org/10.3390/md15110360

Lötze E, Hoffman EW (2016) Nutrient composition and content of various biological active compounds of three South African-based commercial seaweed biostimulants. J Appl Phycol 28:1379–1386. https://doi.org/10.1007/s10811-015-0644-z

Lourenço-Lopes C, Fraga-Corral M, Jimenez-Lopez C, Pereira AG, Garcia-Oliveira P, Carpena M, Prieto MA, Simal-Gandara J (2020) Metabolites from macroalgae and its applications in the cosmetic industry: a circular economy approach. Resources 9:101. https://doi.org/10.3390/resources9090101

Lu FSH, Nielsen NS, Baron CP, Jacobsen C (2017) Marine phospholipids: the current understanding of their oxidation mechanisms and potential uses for food fortification. Crit Rev Food Sci Nutr 57:2057–2070. https://doi.org/10.1080/10408398.2014.925422

Maehre HK, Edvinsen GK, Eilertsen KE, Elvevoll EO (2016) Heat treatment increases the protein bioaccessibility in the red seaweed dulse (Palmaria palmata), but not in the brown seaweed winged kelp (Alaria esculenta). J Appl Phycol 28(1):581–590. https://doi.org/10.1007/s10811-015-0587-4

Makkar HP, Tran G, Heuzé V, Giger-Reverdin S, Lessire M, Lebas F, Ankers P (2016) Seaweeds for livestock diets: a review. Anim Feed Sci Technol 212:1–17. https://doi.org/10.1016/j.anifeedsci.2015.09.018

Maoka T (2020) Carotenoids as natural functional pigments. J Nat Med 74:1–16. https://doi.org/10.1007/s11418-019-01364-x

Market Research Future (2021) Global Carrageenan Market. https://www.marketresearchfuture.com/reports/carrageenan. January 2022

Matanjun P, Mohamed S, Mustapha NM, Muhammad K (2009) Nutrient content of tropical edible seaweeds, Eucheuma cottonii, Caulerpa lentillifera and Sargassum polycystum. J Appl Phycol 21(1):75–80. https://doi.org/10.1007/s10811-008-9326-4

McHugh DJ (2003) A guide to the seaweed industry, FAO Fisheries Technical Paper No. 441. Food and Agriculture Organization of the United Nations, Rome, p 105

McKim JM, Baas H, Rice GP, Willoughby JA, Weiner ML, Blakemore W (2016) Effects of carrageenan on cell permeability, cytotoxicity, and cytokine gene expression in human intestinal and hepatic cell lines. Food Chem Toxicol 96:1–10. https://doi.org/10.1016/j.fct.2016.07.006

Milinovic J, Campos B, Mata P, Diniz M, Noronha JP (2020) Umami free amino acids in edible green, red, and brown seaweeds from the Portuguese seashore. J Appl Phycol 32(5):3331–3339. https://doi.org/10.1007/s10811-020-02169-2

Moghadamtousi SZ, Karimian H, Khanabdali R, Razavi M, Firoozinia M, Zandi K, Abdul Kadir H (2014) Anticancer and antitumor potential of fucoidan and fucoxanthin, two main metabolites isolated from brown algae. Sci World J 768323. https://doi.org/10.1155/2014/768323

Mohammed HO, O'Grady MN, O'Sullivan MG, Hamill RM, Kilcawley KN, Kerry JP (2021) An assessment of selected nutritional, bioactive, thermal and technological properties of brown and red Irish seaweed species. Food Secur 10(11):2784. https://doi.org/10.3390/foods10112784

Mohibbullah M, Hannan MA, Choi JY, Bhuiyan MM, Hong YK, Choi JS, Choi IS, Moon IS (2015) The edible marine alga Gracilariopsis chorda alleviates hypoxia/reoxygenation-induced oxidative stress in cultured hippocampal neurons. J Med Food 18(9):960–971. https://doi.org/10.1089/jmf.2014.3369

Mouritsen OG (2013) Seaweeds: edible, available, and sustainable. University of Chicago Press, Chicago

Mouritsen OG, Duelund L, Petersen MA, Hartmann AL, Frøst MB (2019) Umami taste, free amino acid composition, and volatile compounds of brown seaweeds. J Appl Phycol 31(2):1213–1232. https://doi.org/10.1007/s10811-018-1632

Mysliwa-Kurdziel B, Solymosi K (2017) Phycobilins and phycobiliproteins used in food industry and medicine. Mini Rev Med Chem 17(13):1173–1193. https://doi.org/10.2174/1389557516666160912180155

Namvar FM, Tahir P, Mohamad R, Mahdavi M, Abedi PFN, Tahereh Rahman H, Jawaid M (2013) Biomedical properties of edible seaweed in cancer therapy and chemoprevention trials: a review. Nat Prod Commun 8(12):1811–1820. https://doi.org/10.1177/F1934578X1300801237

Narayan B, Miyashita K, Hosakawa M (2006) Physiological effects of eicosapentaenoic acid (EPA) and docosahexaenoic acid (DHA) – a review. Food Rev Int 22:291–307. https://doi.org/10.1080/87559120600694622

Naseri A, Marinho GS, Holdt SL, Bartela JM, Jacobsen C (2020) Enzyme-assisted extraction and characterization of protein from red seaweed *Palmaria palmata*. Algal Res 47:101849. https://doi.org/10.1016/j.algal.2020.101849

Norakma MN, Zaibunnisa AH, Razarinah WW (2022) The changes of phenolics profiles, amino acids and volatile compounds of fermented seaweed extracts obtained through microbial fermentation. Mater Today Proc 48:815–821. https://doi.org/10.1016/J.MATPR.2021.02.366

Okai Y, Higashi-Okai K, Yano Y, Otani S (1996) Identification of antimutagenic substances in an extract of edible red alga, *Porphyra tenera* (Asadusa-nori). Cancer Lett 100:235–240

Ortiz J, Romero N, Robert P, Araya J, Lopez-Hernández J, Bozzo C, Navarrete E, Osorio A, Rios A (2006) Dietary fiber, amino acid, fatty acid and tocopherol contents of the edible seaweeds Ulva lactuca and Durvillaea Antarctica. Food Chem 99(1):98–104. https://doi.org/10.1016/j.foodchem.2005.07.027

Osman N, Suliman T, Osman K (2020) Characterization of native alginates of common Alginophytes from the Red Sea Coast of Sudan. Int J Second Metab 7(4):266–274. https://doi.org/10.21448/ijsm.685864

Osuna-Ruiz I, López-Saiz CM, Burgos-Hernández A, Velázquez C, Nieves-Soto M, Hurtado-Oliva MA (2016) Antioxidant, antimutagenic and antiproliferative activities in selected seaweed species from Sinaloa, Mexico. Pharm Biol 54(10):2196–2210. https://doi.org/10.3109/13880209.2016.1150305

Osuna-Ruiz I, Nieves-Soto M, Manzano-Sarabia MM, Hernández-Garibay E, Lizardi-Mendoza J, Burgos-Hernández A, Hurtado-Oliva MÁ (2019) Gross chemical composition, fatty acids, sterols, and pigments in tropical seaweed species off Sinaloa, Mexico. Ciencias Marinas 45(3):101–120. https://doi.org/10.7773/cm.v45i3.2974

Pal A, Kamthania M, Kumar A (2014) Bioactive compounds and properties of seaweeds-a review. Open Access Library J 1:1–17. https://doi.org/10.4236/oalib.1100752

Pankiewicz R, Łęska B, Messyasz B, Fabrowska J, Sołoducha M, Pikosz M (2016) First isolation of polysaccharidic ulvans from the cell walls of freshwater algae. Algal Res 19:348–354. https://doi.org/10.1016/j.algal.2016.02.025

Park HG, Kim TW, Chae MY, Yoo IK (2007) Activated carbon-containing alginate adsorbent for the simultaneous removal of heavy metals and toxic organics. Process Biochem 42(10):1371–1377. https://doi.org/10.1016/j.procbio.2007.06.016

Park HK, Kim IH, Kim J, Nam T-J (2013) Induction of apoptosis and the regulation of ErbB signaling by laminarin in HT29 human colon cancer cells. Int J Mol Med 32:291–295. https://doi.org/10.3892/ijmm.2013.1409

Parreidt TS, Müller K, Schmid M (2018) Alginate-based edible films and coatings for food packaging applications. Food Secur 7(10):170. https://doi.org/10.3390/foods7100170

Pereira L (2018) Biological and therapeutic properties of the seaweed polysaccharides. Int Biol Rev 2(2):1–50. https://doi.org/10.18103/ibr.v2i2.1762

Pereira L, Silva P (2021) A concise review of the red macroalgae Chondracanthus teedei (Mertens ex Roth) Kützing and Chondracanthus teedei var. lusitanicus (JE De Mesquita Rodrigues) Bárbara & Cremades. J Appl Phycol 33(1):111–131. https://doi.org/10.1007/s10811-020-02243-9

Pereira AG, Otero P, Echave J, Carreira-Casais A, Chamorro F, Collazo N, Jaboui A, Lourenço-Lopes C, Simal-Gandara J, Prieto MA (2021) Xanthophylls from the sea: algae as source of bioactive carotenoids. Mar Drugs 19:188. https://doi.org/10.3390/md19040188

Piironen V, Lindsay DG, Miettinen TA, Toivo J, Lampi AM (2000) Plant sterols: biosynthesis, biological function, and importance to human nutrition. J Sci Food Agric 80:939–966

Pirian K, Jeliani ZZ, Sohrabipour J, Arman M, Faghihi MM, Yousefzadi M (2018) Nutritional and bioactivity evaluation of common seaweed species from the Persian Gulf. Iran J Sci Technol Trans A: Sci 42(4):1795–1804. https://doi.org/10.1007/s40995-017-0383

Pirian K, Jeliani ZZ, Arman M, Sohrabipour J, Yousefzadi M (2020) Proximate analysis of selected macroalgal species from the Persian Gulf as a nutritional resource. Trop Life Sci Res 31(1):1–17. https://doi.org/10.21315/tlsr2020.31.1.1

Plouguerné E, da Gama BAP, Pereira RC, Barreto-Bergter E (2014) Glycolipids from seaweeds and their potential biotechnological applications. Front Cell Infect Microbiol 4:174. https://doi.org/10.3389/fcimb.2014.00174

Polat S, Ozogul Y (2009) Fatty acid, mineral and proximate composition of some seaweeds from the northeastern Mediterranean coast. Ital J Food Sci 21:317–324

Polat S, Ozogul Y (2013) Seasonal proximate and fatty acid variations of some seaweeds from the northeastern Mediterranean coast. Oceanologia 55(2):375–391. https://doi.org/10.5697/oc.55-2.375

Qin Y, (2018) Applications of bioactive seaweed substances in functional food products. In: Qin Y, (ed) Bioactive seaweeds for food applications. Academic, London, pp 111–134

Qu W, Ma H, Pan Z, Luo L, Wang Z, He R (2010) Preparation and antihypertensive activity of peptides from Porphyra yezoensis. Food Chem 123(1):14–20. https://doi.org/10.1016/j.foodchem.2010.03.091

Queiroz MI, Fernandes AS, Deprá M, Jacob-Lopes E, Zepka LQ (2017) Introductory chapter: chlorophyll molecules and their technological relevance. In: Jacob-Lopes LQE, Zepka Queiroz MI (eds) Chlorophyll. IntechOpen. https://doi.org/10.5772/67953

Rafiquzzaman M, Raju A, Lee J, Gyuyou N, Jo G, Kong I-S (2016) Improved methods for isolation of carrageenan from *Hypnea musciformis* and its antioxidant activity. J Appl Phycol 28:1265–1274. https://doi.org/10.1007/s10811-015-0605-6

Rajauria G (2018) Optimization and validation of reverse phase HPLC method for qualitative and quantitative assessment of polyphenols in seaweed. J Pharm Biomed Anal 148:230–237. https://doi.org/10.1016/j.jpba.2017.10.002

Rajauria G, Abu-Ghannam N (2013) Isolation and partial characterization of bioactive fucoxanthin from Himanthalia elongata brown seaweed: a TLC-based approach. Int J Anal Chem 802573. https://doi.org/10.1155/2013/802573

Ravikrishnan R, Rusia S, Ilamurugan G, Salunkhe U, Deshpande J, Shankaranarayanan J, Shankaranarayana ML, Soni MG (2011) Safety assessment of lutein and zeaxanthin (LutemaxTM 2020): subchronic toxicity and mutagenicity studies. Food Chem Toxicol 49(11):2841–2848. https://doi.org/10.1016/j.fct.2011.08.011

Rebeiz CA (2014) Chlorophyll biosynthesis and technological applications. Springer, Dordrecht

Rengasamy KR, Mahomoodally MF, Aumeeruddy MZ, Zengin G, Xiao J, Kim DH (2020) Bioactive compounds in seaweeds: an overview of their biological properties and safety. Food Chem Toxicol 135:111013. https://doi.org/10.1016/j.fct.2019.111013

Rey F, Lopes D, Maciel E, Monteiro J, Skjermo J, Funderud J, Raposo D, Domingues P, Calado R, Domingues MR (2019) Polar lipid profile of Saccharina latissima, a functional food from the sea. Algal Res 39:101473. https://doi.org/10.1016/j.algal.2019.101473

Rocha CP, Pacheco D, Cotas J, Marques JC, Pereira L, Gonçalves AM (2021) Seaweeds as valuable sources of essential fatty acids for human nutrition. Int J Environ Res Public Health 18(9):4968. https://doi.org/10.3390/ijerph18094968

Rohani-Ghadikolaei K, Abdulalian E, Ng WK (2012) Evaluation of the proximate, fatty acid and mineral composition of representative green, brown and red seaweeds from the Persian Gulf of Iran as potential food and feed resources. J Food Sci Technol 49:774–780. https://doi.org/10.1007/s13197-010-0220-0

Stiger-Pouvreau V, Bourgougnon N, Deslandes E (2016) Carbohydrates from seaweeds. In: Fleurence J, Levine I (eds) Seaweed in health and disease prevention, 1st edn. Elsevier, London, pp 223–274

Sánchez-Machado DI, López-Cervantes J, López-Hernández J, Paseiro-Losada P, Simal-Lozano J (2003) High-performance liquid chromatographic analysis of amino acids in edible seaweeds after derivatization with phenyl isothiocyanate. Chromatographia 58(3):159–163

Sanjeewa KKA, Lee JS, Kim WS, Jeon YJ (2017) The potential of brown-algae polysaccharides for the development of anticancer agents: an update on anticancer effects reported for fucoidan and laminaran. Carbohydr Polym 177451-459:451–459. https://doi.org/10.1016/j.carbpol.2017.09.005

Santos F, Monteiro JP, Duarte D, Melo T, Lopes D, da Costa E, Domingues MR (2020) Unraveling the lipidome and antioxidant activity of native *Bifurcaria bifurcata* and ınvasive *Sargassum muticum* seaweeds: a lipid perspective on how systemic intrusion may present an opportunity. Antioxidants 9(7):642. https://doi.org/10.3390/antiox9070642

Schmid M, Kraft LG, van der Loos LM, Kraft GT, Virtue P, Nichols PD, Hurd CL (2018) Southern Australian seaweeds: a promising resource for omega-3 fatty acids. Food Chem 265:70–77. https://doi.org/10.1016/j.foodchem.2018.05.060

Sebaaly C, Kassem S, Grishina E, Kanaan H, Sweidan A, Chmit MS, Kanaan HM (2014) Anticoagulant and antibacterial activities of polysaccharides of red algae Corallina collected from lebanese coast. J Appl Pharma Sci 4(4):30–37. https://doi.org/10.7324/JAPS.2014.40406

Sekar S, Chandramohan M (2008) Phycobiliproteins as a commodity: trends in applied research, patents and commercialization. J Appl Phycol 20:113–136. https://doi.org/10.1007/s10811-007-9188-1

Sartal, GC, Alonso MCB, Barrera PB (2012) Application of seaweeds in the food industry. In: Kim SK, (ed) Handbook of marine macroalgae: biotechnology and applied phycology. Blackwell, Oxford, pp 522–531

Shrivastava A, Sharma S, Kaurav M, Sharma A (2021) Characteristics and analytical methods of mannitol: an update. Int J Appl Pharm 13(5):20–32. https://doi.org/10.22159/ijap.2021v13i5.42068

Silva A, Silva SA, Carpena M, Garcia-Oliveira P, Gullón P, Barroso MF, Prieto MA, Simal-Gandara J (2020) Macroalgae as a source of valuable antimicrobial compounds: e-extraction and applications. Antibiotics 9(10):642. https://doi.org/10.3390/antibiotics9100642

Skoler-Karpoff S, Ramjee G, Ahmed K, Altini L, Plagianos MG, Friedland B, Govender S, De Kock A, Cassim N, Palanee T, Dozier G, Maguire R, Lahteenmaki P (2008) Efficacy of Carraguard for prevention of HIV infection in women in South Africa: a randomised, double-blind, placebo-controlled trial. Lancet 372:1977–1987. https://doi.org/10.1016/s0140-6736(08)61842-5

Soares F, Fernandes C, Silva P, Pereira L, Gonçalves T (2016) Antifungal activity of carrageenan extracts from the red alga *Chondracanthus teedei* var. *lusitanicus*. J Appl Phycol 28(5):2991–2998. https://doi.org/10.1007/s10811-016-0849-9

Sokolova E, Barabanova A, Bogdanovich R, Khomenko V, Soloveva T, Yermak I (2011) In vitro antioxidant properties of red algal polysaccharides. Biomed Prev Nutr 1(3):161–167. https://doi.org/10.1016/j.bionut.2011.06.011

Solanki G, Solanki R (2012) Alginate dressings: an overview. Int J Biomed Res 3(1):24–28. https://doi.org/10.7439/ijbr.v3i1.322

Sudhakar MP, Saraswathi M, Nair BB (2014) Extraction, purification and application study of R-Phycoerythrin from Gracilaria corticata (J. Agardh) J. Agardh var. corticata. Indian J Nat Prod Resour 5(4):371–374

Sudhakar MP, Kumar BR, Mathimani T, Arunkumar K (2019) A review on bioenergy and bioactive compounds from microalgae and macroalgae-sustainable energy perspective. J Clean Prod 228:1320–1333. https://doi.org/10.1016/j.jclepro.2019.04.287

Sulastri E, Zubair MS, Lesmana R, Mohammed AFA, Wathoni N (2021) Development and characterization of ulvan polysaccharides-based hydrogel films for potential wound dressing applications. Drug Des Devel Ther 15:4213–4226. https://doi.org/10.2147/DDDT.S331120

Synytsya A, Kim WJ, Kim SM, Pohl R, Synytsya A, Kvasnicka F, Copíková J, Park YI (2010) Structure and antitumour activity of fucoidan isolated from sporophyll of Korean brown seaweed Undaria pinnatifida. Carbohydr Polym 81:41–48. https://doi.org/10.1016/j.carbpol.2010.01.052

Tabarsa M, Rezaei M, Ramezanpour Z, Robert Waaland J, Rabiei R (2012) Fatty acids, amino acids, mineral contents, and proximate composition of some brown seaweeds 1. J Phycol 48(2):285–292. https://doi.org/10.1111/j.1529-8817.2012.01122

Thanh TTT, Quach TMT, Nguyen TN, Luong DV, Bui ML, Tran TTV (2016) Structure and cytotoxic activity of ulvan extracted from green seaweed Ulva lactuca. Int J Biol Macromol 93:695–702. https://doi.org/10.1016/j.ijbiomac.2016.09.040

Tseng CK (2001) Algal biotechnology industries and research activities in China. J Appl Phycol 13:37–380

Uju S, Dewi NPSUK, Santoso J, Setyaningsih I, Hardingtyas SD, Yopi Y (2020) Extraction of phycoerythrin from Kappaphycus alvarezii seaweed using ultrasonication. IOP Conf Ser Earth Environ Sci 414:012028. IOP Publishing. https://doi.org/10.1088/1755-1315/414/1/012028

Ummat V, Tiwari BK, Jaiswal AK, Condon K, Garcia-Vaquero M, O'Doherty J, O'Donnell C, Rajauria G (2020) Optimisation of ultrasound frequency, extraction time and solvent for the recovery of polyphenols, phlorotannins and associated antioxidant activity from brown seaweeds. Mar Drugs 18(5):250. https://doi.org/10.3390/Fmd18050250

Usoltseva RV, Anastyuk SD, Ishina IA, Isakov VV, Zvyagintseva TN, Thinh PD, Zadorozhny PA, Dmitrenok PS, Ermakova SP (2018) Structural characteristics and anticancer activity in vitro of fucoidan from brown alga Padina boryana. Carbohydr Polym 184:260–268. https://doi.org/10.1016/j.carbpol.2017.12.071

Valdes L, Cuervo A, Salazar N, Ruas-Madiedo P, Gueimonde M, Gonzalez S (2015) The relationship between phenolic compounds from diet and microbiota: impact on human health. Food Funct 6:2424–2439. https://doi.org/10.1039/c5fo00322

Venkatesan J, Lowe B, Anil S, Manivasagan P, Kheraif AA, Kang K-H, Kim S-K (2015) Seaweed polysaccharides and their potential biomedical applications. Starch/Stärke 67:381–390. https://doi.org/10.1002/star.201400127

Venugopal V (2009) Marine products for healthcare functional and bioactive nutraceutical compounds from the ocean. CRC Press, Boca Raton

Venugopal V (2019) Sulfated and non-sulfated polysaccharides from seaweeds and their uses: an overviev. Ecronicon. EC Nutr, pp 126–141

Vijayalakshmi K, Latha S, Rose MH, Sudha PN (2017) Industrial applications of alginate. In: Sudha PN (ed) Industrial applications of marine biopolymers. CRC Press, pp 545–575

Vilcanqui Y, Mamani-Apaza LO, Flores M, Ortiz-Viedma J, Romero N, Mariotti-Celis MS, Huamán-Castilla NL (2021) Chemical characterization of brown and red seaweed from southern Peru, a sustainable source of bioactive and nutraceutical compounds. Agronomy 11(8):1669. https://doi.org/10.3390/agronomy11081669

Wang HD, Li XC, Lee DJ, Chang JS (2017) Potential biomedical applications of marine algae. Bioresour Technol 244:1407–1415. https://doi.org/10.1016/j.biortech.2017.05.198

Wekre ME, Kåsin K, Underhaug J, Holmelid B, Jordheim M (2019) Quantification of polyphenols in seaweeds: a case study of Ulva intestinalis. Antioxidants 8(12):612. https://doi.org/10.3390/antiox8120612

Wells ML, Potin P, Craigie JS, Raven JA, Merchant SS, Helliwell KE, Smith AG, Camire ME, Brawley SH (2017) Algae as nutritional and functional food sources: revisiting our understanding. J Appl Phycol 29(2):949–982. https://doi.org/10.1007/s10811-016-0974-5

Yalçın S, Uzun M, Karakaş Ö, Başkan SK, Okudan EŞ, Apak MR (2021) Determination of total antioxidant capacities of algal pigments in seaweed by the combination of high-performance liquid chromatography (HPLC) with a cupric reducing antioxidant capacity (CUPRAC) assay. Anal Lett 54(14):2239–2258. https://doi.org/10.1080/00032719.2020.1855439

Yang JY, Lim SY (2021) Fucoidans and bowel health. Mar Drugs 19(8):436. https://doi.org/10.3390/md19080436

Zapata M, Garrido JL, Jeffrey SW (2006) Chlorophyll c pigments: current status. In: Grim B, Porra RJ, Rüdiger W, Scheer H (eds) Chlorophylls and bacteriochlorophylls, biochemistry, biophysics, functions and applications. Springer, pp 38–53

Zargarzadeh M, Amaral AJR, Custódio CA, Mano JF (2020) Biomedical applications of laminarin. Carbohydr Polym 232:115774. https://doi.org/10.1016/j.carbpol.2019.115774

Zemke-White WL, Ohno M (1999) World seaweed utilisation: an end-of-century summary. J Appl Phycol 11:369–376

Zhang J (2018) Seaweed industry in China. Innovation Norway, Beijing, p 31

Zhang J, Li C, Yu G, Guan H (2014) Total synthesis and structure-activity relationship of glycoglycerolipids from marine organisms. Mar Drugs 12:3634–3659. https://doi.org/10.3390/md12063634

Zhang H, Cheng J, Ao Q (2021) Preparation of alginate-based biomaterials and their applications in biomedicine. Mar Drugs 19:264. https://doi.org/10.3390/md19050264

Zhong H, Gao X, Cheng C, Liu C, Wang Q, Han X (2020) The structural characteristics of seaweed polysaccharides and their application in gel drug delivery systems. Mar Drugs 18(12):658. https://doi.org/10.3390/md18120658

Zhou G, Sheng W, Yao W, Wang C (2006) Effect of low molecular lambdacarrageenan from Chondrus ocellatus on antitumor H-22 activity of 5- Fu. Pharmacol Res 53(2):129–134. https://doi.org/10.1016/j.phrs.2005.09.009

Zubia M, Payri C, Deslandes E (2008) Alginate, mannitol, phenolic compounds and biological activities of two range extending brown algae, Sargassum mangarevense and Turbinaria ornata (Phaeophyta: Fucales), from Tahiti (French Polynesia). J Appl Phycol 20:1033–1043. https://doi.org/10.1007/s10811-007-9303-3

Chapter 15
Seaweeds as Growth Promoter and Crop Protectant: Modern Agriculture Application

Johnson Marimuthu Alias Antonysamy, Vidyarani George, Silvia Juliet Iruthayamani, and Shivananthini Balasundaram

1 Introduction

Plant growth regulators are organic compounds that modify physiological processes of plants other than the nutrients. They are said to act inside the cells to help regulate plants metabolism or inhibit specific enzyme. In 1930s, the significance of plant growth regulators (PGRs) was recognized. And today, specific plant growth regulators are used to modify the growth rate of crop and their pattern during the developmental stages (URL 2022). Algae help in reducing the soil erosion by regulating the water flow in soils. Soil fertility, soil reclamation, bio-controlling of agricultural pests, and formation of microbiological crust, agricultural wastewater treatment and recycling of treated water are also regulated through the usage of algae. They also help to improve the carbon content, aeration and nitrogen fixation (Abdel-Raouf et al. 2012).

One of the eight major classes of bio-stimulants is seaweed extracts. The seaweed extracts contain additional sugar alcohols, vitamins, phenolic compounds along with mineral nutrients as well as natural hormones such as abscisic acid, cytokinins and auxins, etc. The seaweed extracts are widely used in the food, chemical, medicine and agriculture industries. As a bio-stimulant it not only improve the soil, growth and development of plants it also helps in improving the pest resistance of crops, diseases, drought and various stresses. At present various research on seaweed extract application on crops are conducted worldwide (Chen et al. 2019). Various phytohormones such as Auxins, Gibberellins, Cytokinins, Abscisic acid, Ethylene, Betaine and Polyamines are fund in different concentrations in seaweed concentrates and marine macroalgal extracts. These extracts help to stimulate the growth of plant when they are applied exogenously. Not only phytohormones but

J. M. Alias Antonysamy (✉) · V. George · S. J. Iruthayamani · S. Balasundaram
Centre for Plant Biotechnology, Department of Botany, St. Xavier's College (Autonomous), Tirunelveli, Tamil Nadu, India

F. Ozogul et al. (eds.), *Seaweeds and Seaweed-Derived Compounds*,
https://doi.org/10.1007/978-3-031-65529-6_15

also vitamins namely, C, B, (thiamine), B_2 (riboflavin), B_{12}, D_3, E, K, niacin, pantothenic, folic and folinic acids are present in algae (Panda et al. 2012).

Abdel-Raouf et al. (2012), studied discussed the significant aspects of seaweeds as well as their uses in agriculture. Algae when they're used as biofertilizers and soil stabilizers play a crucial role in agriculture. Seaweeds are used as fertilizers, resulting in less nitrogen and phosphorous runoff than the one from the use of livestock manure. These organisms are cultivated round the world and used as human food supplements. Seaweeds are an important source of iodine. Iodine levels in milk depend upon what the cow producing the milk has been fed with. Feeding milk cattle with seaweeds can increase the number of iodine in milk.

Chen et al. (2019) carried out a field study to understand the possible effect of seaweed application in sugarcane variety (YT03-373) regarding the phenology, yield and quality attributions. Seaweed extract were applied on the seedling and tillering stages. Traits analyzed for the phenological stages showed an increase rate of 10.83–12.87% enhanced production. And for other traits the sugarcane production increased by 15.09% and net income improved up to 32.64%. And so the foliar application of seaweed helped in growth of the plant, cold and pest resistance and also gave a positive affirmation for economic returns.

John (2017), studied the effect of locally produced seaweed on two crops i.e. Potato and grape. The seaweed extract was applied in various concentration (0, 2, 3, 4, 5, 6, and 7 L/ha) as soil drench as well as foliar sprays for two cropping seasons for 2010 and 2011 for potato; 2011 and 2012 for grapes. There was no difference in the leaf nutrient value whereas the mineral elements concentration enhanced with increased concentration of seaweed extract. The antioxidant values and the oxygen radical absorbance capacity of polyphenols were higher in lower seaweed extract treatment plants. Crude fibre and proteins value increased in the foliar treatment of the seaweeds in the crops. Potato leaf chlorophyll and the growth parameters also increased with increased concentration of seaweed extracts. And therefore, 7 L/ha which was the highest concentration showed significant increased in growth and yield for both potato and grape crops.

Hamidreza and Sergei (2019), evaluated the effect of humic acid, plant growth promoting rhizobacteria and seaweed extracts for various traits in sweet basil. The final result showed highest value for humic acid whereas for seaweed extract only few traits expressed high values. Plant growth promoting rhizobacteria showed no significant effect except for essential oil. Finally humic acid showed the best result compared to both Plant growth promoting rhizobacteria and seaweed extracts.

Nabti et al. (2016) reviewed the impact of seaweeds used as biofertilizers in the agricultural crop production. The role of seaweed as biostimulants is emphasized in the agriculture. Organic farming trends has increased the demand for biofertilizers and also created opportunities to investigate the potential of seaweed application. The seaweeds are rich in secondary metabolites and this makes them resistant to various climatic and environmental conditions. This further improves the fertility of the soil and induces the plant growth. The whole review contained the information and scientific progress of the seaweeds made in the field of crop production with further investigations and their applications.

Panda et al. (2012), reviewed the papers evaluating the use of seaweed extracts used as plant growth regulators in the agriculture. The seaweed extracts are found to contain many secondary metabolites which help in crop growth, yield. They further help the plants from biotic as well as abotic stress and these further increase their filed application. The authors reviewed the research papers which are indeed supporting his work and giving the required progress. Finally the literature proves as the best evidence for seaweed extracts to be their growth promoter which helps in organic agriculture.

Zodape (2001) explained that dried and fresh seaweeds liquid extracts were use as biofertilizers. Algae extracts were now commercially available as maxicrop, SM3, kelpak and cytokin. Seaweed extracts are used as chemical constituents, and it has different types of effects on seed germination and seedling growth, some species of seaweeds extracts used as foliar spray on cultivated race before harvest, seaweed spray used as shelf-life and storage purpose in peach trees and soil improver, resistance of plants from plant pathogen like mildew.

Ashour et al. (2021) compared seaweed bio-fertilizers performance and fertilizer in vegetative growth, yield, nutrients content and bioactive compounds. True Algae Max (TAM) is the SLF (Seaweed Liquid Fertilizer) extract that commercially available as a plant growth stimulant on nutritional and antioxidant activity of *Capsicum annuum*. TAM is rich in phytochemical compounds such as ascorbic acid, phenolics and flavonoids that showed good antioxidant activity and DPPH inhibition of 70.33%. The yield of C. *annuum* improved in all TAM treatments especially in concentration 0.5% resulted in maximum yield and significant amount of profuse biological molecules like chlorophyll, ascorbic, phenolic compounds, flavonoids and total nutrients. Compared to NPK treatments total antioxidant activity of C. *annuum* treated with TAM 0.5% improved from 162.16 to 190.95 mg $g^{-1.}$

Deshmukh and Phonde (2013) studied the effect of seaweed extract on growth, yield and quality of sugarcane. NPK (Nitrogen, Phosphorus and Potassium) fertilizers in seven different doses were applied. The recommended dose of fertilizers (250:115:115 kg N, P and K ha^{-1}) and other packages of practices for sugarcane were imposed uniformly for all the treatments including control. The field observations of biometric, growth and yield parameters were taken at different stages of the crop and at harvest. Sugarcane juice quality for brix and pol percentage was analyzed by ICUMS method. The results shows the highest cane yield 89.23 t ha^{-1} was obtained at 1500 g ha^{-1} seaweed extract was applied in soil along with RDF at 45 and 120 days after planting and that was significantly higher than all other treatments either by soil or foliar application. The highest commercial was found in 1500 g ha^{-1} treatment at 14.96 t ha^{-1} followed by 14.66 t ha^{-1}. Foliar treatment did not show any significant response and it indicates that soil application of 1000 g ha^{-1} extract was beneficial to increase the cane and sugar yield. In ICYMS analysis the result shows that the juice quality was not affected either by soil or foliar application of seaweed extract.

Ravneet et al. (2018) determined the effect of Stimplex on the yield of six organic leafy vegetables. Organic seeds of six leafy vegetables were taken and each plant seeds were seeded. Total of 15 plants for each leafy green were assigned to untreated and Stimplex treated plants. An equal dose of 3.2 mL/L of Stimplex was applied to all treated plants by foliar spray method and untreated plants were sprayed with water. The foliar spray application was applied 14 day intervals for 6 weeks. Both treated and untreated plants were harvested manually at the end of the crop season. The result showed all the crop yields were responded positively to the Stimpex treatment. Lettuce plants treated with Stimpex had more weight than untreated plant with average per plant weight of 262.27 and 343.37 g. In mustard also the treated plants weighted more (500.77 g) when compared to the untreated (368.63 g) plant. Collard, Swiss chard and Kale plants treated with Stimplex showed a significant increase in weight compared to untreated plants. Weight of the untreated collard, Swiss chard and kale plants were 354.43 g, 401.33 g and 388.40 g, while Stimplex treated collard, Swiss chard and kale plants weighed 423.00 g, 445.70, and 456.20 g respectively. The Stimplex treated amaranth plant did not showed any significant difference in weight.

Anisimov and Chaikina (2014) investigated the effect of aqueous extract of red algae *Neorhodomela larix*, *Tichocarpus crinitus*, *Saccharina japonica*, *Sargassum pallidum*, *Ulva fenestrata* and *Codium fragile* on the length of seedling roots of soybean. The results revealed, the extracts in different degrees have a positive effect on the length of the roots of soybean seedlings. The maximum stimulatory effect showed extract of *Codium fragile* at concentration at 10^{-5} gSW mL^{-1} and the roots on 18.0% were longer than the control roots.

The comparableness of the performance of TAM as a foliar spray on cucumber with regard to conventional NPK foliar spray fertilizing routines was studied by Shimaa et al. (2021). Three seaweed species *Ulva lactuca*, *Jania rubens* and *Pterocladia capillacea* were collected and TAM was prepared. From crude TAM extract phytochemical, physical, chemical and biochemical analysis were conducted and estimated. The experiment was carried out in the greenhouse condition. Before the experiments the soil samples were collected and analyzed. The fertilizer treatments were 100% NPK mineral as control (C_0), 25% TAM + 75% NPK (C_{25}), 50% NPK mineral (C_{50})+ 50% TAM, 75% TAM + 25% NPK and 100% TAM (C_{100}). The result showed the height trait was recorded 193.7 cm under C_0 treatment and it reached 209.7 cm in C_{100} treatment. The leaf area parameter was also influenced by the use of TAM was 306 plant/cm^{-2}. TAM has different dose influence on the fruit of cucumber. The total yield was highly affected by C_{50} and C_{75} treatment reaching 4.07 kg/m^2 for both concentrations and it reduced as 2.47 kg/m^2 without treatment.

Israel et al. (2020) studied the biomass of *Padina durvillaei* and *Ulva lactuca* due to their plant growth enhancing properties. The PGRs and antioxidant properties of the two seaweeds were identified in this specific study to know their potential in improving the growth and crop yield. Phytochemical analysis showed the presence of sulfates, flavonoids and phenolics content in *P. durvillaei* which is more compared to *Ulva sps*. The identification showed the presence of PGRs such as GA1, GA4, ABA, IAA, tZ, IP and DHZ. And the contents of GA4, tZ and DHZ where more in *P. durvillaei* compared to *U. lactuna*. And finally, these results reveal that both the species are suitable candidates for biofertilizers application.

Sathya et al. (2019), studied the effect of liquid fertilizer made of three seaweeds with different concentration on *Cajanus cajan*. Growth parameters showed best result for *Chaetomorpha linum* whereas maximum biochemical parameter and photosynthetic effect was seen for *Sargassum wightii*. Finally the best results were seen for Sargassum wightii and Caetomorpha linun at low concentration.

Izabela et al. (2016), researched to screen the biostimulant properties of supercritical CO_2 macroalgal extracts for species of *Polysiphonia, Ulva and Cladophora.* The secondary metabolites were determined using inducticely coupled plasma atomic emission spectroscopy (inorganic compounds) and high-performance liquid chromatography and spectrophotometry (organic compounds). The extracts were tested on garden cress and wheat. Final observations showed that the algal extracts gave best results for stimulating growth for the cultivated plants.

Chitra and Sreeja (2013) study the effect of Seaweed liquid fertilizer prepared from Algae on *Vigna radiata*. The seaweeds *Gracilaria corticata* and *Caulerpa peltata* were collected and the extracts were prepared using distilled water. *Vigna radiata* seeds were soaked in different concentration of extracts for 12 hrs. Then the seeds were observed for 3 days. Germination was recorded maximum at lowest concentration of both algae treatments. Maximum radical and hypocotyl length was observed at the higher concentration of G. *corticata*. Chitra and Sreeja (2013) explained that phytotoxicity was found to be decrease at higher concentration of *Gracilaria*. *Caulerpa* treated plants the growth index and vigour index were founds to decrease from lower to higher concentration.

The effect of seaweed liquid fertilizer (SLF) of algae *Codium tomentosum* and *Sargassum vulgare* on growth, germination and photosynthetic pigment of wheat plant were investigated (El-Din 2015). Various concentrations 10%, 20%, 30%, 40%, and 50% of SLFs were prepared. In each concentration of SLF, 60 seeds were soaked for 12 h and the seeds soaked in water were used as control. The seed germination for *Sargassum vulgare* was 98% while for *Codium tomentosum* is 97% in the 20% concentration. Likewise, the length of shoot and root, fresh and dry eight of the seedlings were also increased in a significant level in the 20% concentration when compared to the control. The chlorophyll content in the leaves also showed an enhancement with the application of SLF. The results for the brown algae were better to the green algae.

The authors used the seaweed aqueous extracts of *Corallina elongata, Corallina officinalis, Jania rubens* and *Ulva fasciata* to study the on germination and growth of maize (*Zea mays* L.) (Fayzi et al. 2020). The study was experimented using 10% seaweed extracts, a commercial biostimulant and distilled water as control. All three extracts were applied for 10 days through in vitro or by foliar spray in the greenhouse conditions. The highest results were observed for seeds soaked in *Jania rubens* extract with a result more than 50% to the control. In the in vitro application, the seaweed extracts of *J. rubens* and *C. elongata* showed greater result with enhanced growth and improved chlorophyll content. There was a increase of 10% to 53% in the fresh and dry weight, leaf, root and shoot length. And hence, the seaweed extracts of *Jania rubens* showed best results when compared to other three species as well as the commercial biostimulants.

Hussein et al. (2021) investigated the impact of seed priming with seaweed liquid extracts (SLEs) prepared from three Egyptian seaweeds *Ulva fasciata*, *Cystoseira compressa*, and *Laurencia obtusa* in a concentration of 20 g L^{-1} on seed germination and seedling growth of *Vigna sinensis* and *Zea mays* and the effectiveness of these extracts in ameliorating salinity stress. Laurancia extract (LLE) induced maximum positive response in germination as well as producing the maximum increases in all seedling morphological parameters with percent of response 87.14% for hypocotyl growth of V. *sinensis* and 85.71% coleoptile elongation of Z. *mays*. The maximum amylase and protease activities were recorded with LLE priming. *Ulva*, *Cystoseira*, *Laurencia* liquid extract consortium (UCLLE) induced the highest promotion effect for both plants giving significant increments for all growth criteria, dry biomass 0.301 and 0.438 g seedling-1, Chl a 18.98 and 13.79 mg g^{-1} FW, total carbohydrates 460.88 and 518.14 g seedling-1, protein 219.72 and 207.49 g seedling-1 respectively.

2 Green Seaweeds

The effect of two seaweed extracts on tomato germination and seedling growth were observed by Reis et al. (2020). *Ulva flexuosa* and *Ulva lactuca* were collected, washed and dried in an oven at 50 °C. The extraction sample was obtained from the slurry of 10 g of crushed seaweed in 100 mL of water and autoclaved for 40 min and centrifuged. Tomato seeds were purchased from a commercial supplier for germination bioassay. The germinated seeds were immersed 10 min in the extract solution at 0.2 gL^{-1} and 0.4 gL^{-1} concentrations. Germination tests were carried out in an incubator under 20–30 °C and at 16-h light/8-h dark. The liquid extract of U. *lactuca* significantly increased the germination percentages of tomato seeds and the highest total germination percentage was observes with the U. *lactuca* flour treatment at 2 gL^{-1}.

The proximate composition of green algae *Ulva rigida* was determined and the evaluation of potential natural fertilizer on wheat plant was studied by Salma et al. (2021). The widely distributed and edible seaweed *U. rigida* was collected and the aqueous extract was prepared in different concentrations. From the leaf tissues of the microalgae the chlorophyll and carotenoid amount was estimated and the result shows the presence of chlorophyll a (13%), chlorophyll b (7.5%) and carotenoid (4.5%). Protein and lipid content was also determines by AOAC method and it responed that the level of protein was relatively higher than the lipid content. By using HPLC method the amino acid content was observed. The observation revealed two types of amino acids. The essential amino acids are methionine, leucine, isoleucine, lysine, phenylaline, tyrosine, arginine, threonine, valine and tryptophan. The non-essential amino acids are histidine, aspartic acid, glutamic acid, serine, proline, glycine and alanine. Also the seaweed showed significant different between total phenolic and α tocopherol content. The plant treated with the algae extract results in the length of aerial part with significantly greater compared with untreated plants.

Simon et al. (2021) examined the efficacy of Plant Growth Promoter (PGP) on early seedling growth related to nutrient content and composition of the growth substrate. Totally 16 plant species were used in this trails. First the seeds were sown in trays containing a low nutrient potting mix. After the sufficient growth the seedlings were transferred to separate pots and each plant was treated with products solutions. Sixteen plant promoting products were tested. The growth of the plants treated with PGP was compared with untreated plants. Nitrogen concentration of each PGP was determined by Dumas Style elemental analyzer. Totally 148 PGP plant combinations were used to compare plant growth in treated plants with the plants observed in water as control. The REML mixed model analysis indicated there were significant differences in the effect of different PGP classes. All the general class of PGP produced a significant positive effect on the shoot growth. The PGP based on animal waste shows the highest average effect and the results for individual products were also highly variable. The growing media have significant influence on the response of the plants to the PGP and its average effect was seen when using the high nutrient potting mix were lower compared with those obtained when using the low nutrient potting mix and organic soil. Most of the seaweed based PGP produced negligible and negative effects on plant growth, so the growing media and class of PGP interaction were also statistically significant. On seedling growth the seaweed based product dose only have significant differences occurring at recommended rate and above. The trail results confirmed that the beneficial effects of PGP on plant are highly variable when comparing with individual products, classes of products and testing the same product under different conditions.

The seaweed liquid fertilizer of marine algae (*Enteromorpha intestinalis*) were studied to know there potential on seed germination, yield, biochemical parameters and pigment characteristics in *Glycine max* (Chetna et al. 2015). The morphological and bio-chemical parameters showed highest value in 60% application of SLF whereas phenol content showed best value for 40% application. Notable benzoic compounds were found during GCMS analysis which was involved in PGRs. And so SLF were effective and alternative options for chemical fertilizers as they showed significant effect in the studied crop.

Chanthini et al. (2019) evaluated the liquid seaweed extract of green seaweed *Chaetomorpha antennina* (CA-LSE) for their biostimulant activity on seed germination and growth of tomato. The CA-LSEs were able to stimulate early emergence of tomato seeds and in addition to increasing their germination percentage and energy. They also exerted a positive influence on the vegetative growth, resulting in increased plant height, leaf-branch number and yield. The CA-LSEs were influenced the plant's biochemical profile and displayed a linear increase in pigment contents like chlorophyll a & b and carotenoids, total soluble solids, phenols as well as ascorbic acid contents. The bio-stimulant potential was attributed to the elements present in CA-LSEs, elucidated by EDX analysis that revealed the presence of six elements (O, Na, Mg, S, Cl and Ca) and they were essential plant nutrients. These results support the bio-stimulant potential of *C. antennina*, which can be applied as a bio-fertilizer that is economic, renewable, efficient and eco-friendly, and also can be regarded as a potential catalyst for sustainable agricultural food production.

3 Red Seaweeds

Diana et al. (2021) investigated the use of principal carbohydrates of 5 different seaweeds namely, *Gracilaria gracilis* (Slender wart weed), *Asparagospis armata* (Harpoon weed), *Calliblepharis jubata* (False eyelash weed), *Chondracanthus teedei* var. *lusitanicus* and *Grateloupia turuturu* (Devil's tongue weed); and three brown seaweeds *Colpomenia peregrina* (Oyster thief), *Sargassum muticum* (Wireweed) and *Undaria pinnatifida* (Wakame) on kale growth. The result showed positive impact for three polymers tested i.e. iota-carrageenan, kappa/iota-carrageenan and agar on the growth of kale. And the results were further confirmed using the quantification methods.

Pramanick et al. (2013) conducted an experiment to study the seaweed saps of *Kappaphycus*, *Gracilaria* on growth, yield and quality improvement of green gram in the alluvial soil of West Bengal. The foliar spray was applied twice in each different concentration. The traits enhanced with the highest concentration of seaweed extract. The results showed the improved quality of crop and nutrient uptake such as nitrogen, phosphorous and potassium.

4 Brown Seaweeds

Abdel-Mawgoud et al. (2010) studied the effect of foliar application of seaweed extract on hybrids variety of three watermelons namely, Gizal (red-color flesh), Envy (Seedless, red-color flesh) and Yellow Crimson watermelon hybrid (yellow color flesh hybrid) for two successive cultivation seasons of 2008 and 2009. The watermelons were sprayed twice after 5 and 9 weeks with *Ascophyllum nodosum* at four different concentrations (0, 1, 2, and 3 g/L). Gizal hybrid had higher vegetative growth while yellow hybrid had higher fruit weight. Seaweed application was observed to be positive and correlated with the concentration applied. Seaweed extract application increased the growth parameters and yield response.

Seaweeds which are naturally derived products are used as biostimulents. They are said to be organic products whose growth promoting properties are not recognized in farming. In accordance with maize bioassay, different doses of filtrate seaweed extract from *Ascophyllum nodosum* where applied to organic lettuce to study the biostimulant effect on growth and nutrient uptake of vegetables. The dry weight and uptake of nutrient of lettuce shoot increased with increased dilution rate. And so filtrate seaweed extract are good stimulants for the vegetables in early stage development in nursery production in organic farming (Alessandra et al. 2014).

Sumangala et al. (2019) studied the effect of seaweed liquid extract of *Sargassum crassifolium* on roses for revealing its growth and flowering effects. Seven different treatments T1 (10% once a week), T2 (10% twice a week), T3 (20% once a week), T4 (20% twice a week), T5 (30% once a week), T6 (30% twice a week) and T7—control (distilled water) where applied at random level with ten replications. The potential for increase in the height, leaf area, biomass and flower number and dry weight were observed in samples of 20% seaweed application. The nutrients and growth promoter presence in *S. crassifoliam* with an optimum concentration could be the reason. And so 20% application of the seaweed application is more suitable for flowering and growth of roses.

John and Yuvaraj (2014) evaluated the effect of seaweed liquid fertilizer on seed germination, shoot and root length, biochemical and pigment content of *Vigna radiata* using *Colpomenia sinuosa*. The SLF was applied in different concentration. The highest concentration gave the best and maximum potential result.

The effect of seaweed liquid fertilizer of *Dictyota dichotoma* studied on growth and yield parameters of *Abelmoschus esculantus* using different concentrations (Sasikumar et al. 2011). The results showed that the use of seaweed extracts as SLF enhanced the parameters such as the biomass, growth of roots and shoots, leaves, flowers and even the fruits. And the final results prove as evidence that seaweed liquid fertilizers are best organic or biofertilizers.

Thirumaran et al. (2009), studies the effect of seaweed liquid fertilizer of *Rosenvingea intricata* on *Abelmoschus exculentus*. The final result showed that the various parameters observed gave an enhanced potential when applied with seaweed extracts. 20% SLF with or without the chemical fertilizer gave a higher growth, best yield, pigment value and the soil profile was also improved tremendously.

Seaweed extracts of four species were used to study the effect on seed germination and plant growth of *Capsicum annum* (Jayasinghe et al. 2016). The Seaweed Liquid Fertilizer (SLF) was treated with and without the chemical fertilizers. Randomized complete block design was used for with 9 concentrations and 4 replications for the experiment. Each concentration with Recommended rate of Chemical Fertilizer (RCF) was applied for 1 week and the result was observed after 2 weeks. The best results were met for 75% SLF plus RCF for *Sargassum wightii* when compared with other three species. The combination of SLF and RCF were proposed to have enhanced the growth, quality and yield of *Capsicum annum*.

Alam et al. (2012) examined the plant growth and associated soil microbials in several strawberry cultivars following the extracts of *Ascophyllum nodosum*. Greenhouse and field experiments were establishes over plots of Albion, Camarosa, Chandler and Festival strawberries. Soluble A. *nodosum* extract powder was applied once or twice per week for over 8 weeks and roots were examined weekly application of SAEP. The results indicate maximum productivities had found at 1 and 2 g SAEP L^{-1} in both field and greenhouse. Chandler was the most responsive cultivar and Albion was the least responsive to SAEP application. Soluble extract powder increased colony counts in greenhouse and field soil samples with maximum counts at 4 g L^{-1} SAEP in green house and 1 and 2 g L^{-1} SAEP. Metabolic activities of soil microbes were found using SAEP application. The maximum AWCD, substrate diversity, evenness and richness were founds at 4 g L^{-1} and 2 g L^{-1} SAEP. The highest respiration rate showed between 0.10 and 0.40 g per week. SAEP applications increased root and shoot growth in strawberry, berry yield and rhizophere microbial diversity and physiological activity.

Fatimah et al. (2018) conducted the study to determine the effect of different *Sargassum* species concentration on *in vitro* seed germination of two plants. *Sargassum* species extracts were prepared using four different solvents which are methanol, hexane, dichloromethane and water. The crude extract was separated using rotary evaporator. Phytochemical screening was conducted to test the presence of alkaloids, terpenoids, tannins, saponins and flavonoids. The phytochemical screening of *Sargassum* species showed the occurrence of bioactive components in various extract. It exhibited the presence of alkaloids, terpenoids, tannins and saponins. Terpenoids and alkaloids were the dominant components in all extract. For in vitro plant growth the media was prepared by adding macronutrients and micronutrients, iron and vitamins with all different extract solution with different concentration. *Capsicum annum* and *Lycopersicon esculentum* seeds were taken and surface of the seeds were sterilized with sodium hypochloride. Then the sterilized seeds were inoculated on the MS medium with 0.50, 0.100. 2.50 and 5.00 mg/L of extract concentration. The culture was observed alternately for 4 weeks and the root lengths of the studied species were measured. *Lycopersicon esculentum* seedling development was classified at optimum concentration at 2.50 mg/L. Most of the crude solvents showed the positive development in 2.50 mg/L compared with MSO.

Tensingh et al. (2017) revealed the effect of SLF (Seaweed Liquid Fertilizer) of *Sargassum wightii* in the yield of Okra and the population dynamics of general micro flora and nutrient content of SLF treated and untreated soil. The seaweed was collected and shade dried to prepare the SLF in different concentrations. With this extract the nursery experiment in okra, soil nutrient content and soil microorganism were also studied. The treatment was given to 10 days old seedlings. The plants treated with 5.0% of SLF showed the maximum number of flowers and fruits and maximum harvest index with significant increase in the length and weight of the fruits. The soil microbial dynamics showed that the population of bacteria and fungi were increased with increasing of SLF concentrations. Also the macro and micro nutrients like N, P, K, Fe, Mn, Zn and Cu were also increased with SLF concentrations.

Seaweed extracts were one of the inexpensive sources of naturally occurring plant growth biostimulants in agriculture. Hidangmayum and Sharma (2017) studied the effect of seaweed extracts of *Ascophyllum nodosum* as a plant biostimulant on the growth of Onion. Six treatments were conducted randomly with four replications like T0- Control (0.00%), T1- (0.35%), T2- (0.45%), T3- (0.55%), T4- (0.65%) and T5- (0.75%). On the basis, they concluded that treatment receiving 0.55% was found to be the best treatment in terms of leaf number (9.08/plant), plant height (55.20 cm/plant), Crop growth rate (33.65 g/m^2/day), fresh bulb diameter (5.13 cm/plant), bulb fresh weight (120.21 g/plant), harvest index (77.44%), chlorophyll 'a' (0.81 mg/g), chlorophyll 'b' (0.58 mg/g), carotenoid content (0.61 mg/g), bulb sulphur content (1.80 ppm), bulb protein content (1.19 mg/g) and leaf protein content (0.46 mg/g). While the higher concentration of the extract shows decreasing trend. Our study provides important information on optimization of seaweed liquid extracts on onion crop.

The seaweed concentrate Kelpak made from the brown kelp *Ecklonia maxima* is used worldwide as a biostimulant for lot of agricultural crops. Kannan et al. (2014) investigated the plant growth stimulating effects of phloroglucinol and eckol that were isolated from E. maxima. The isolated eckol activity was compared with commercial products Kelpak, phloroglucinol and IBA. The isolated eckol stimulated maize growth with shoot and root elongation and number of seminal roots also exhibited the improvement of biochemical activities of alpha amylase compared to commercial products.

5 Biopesticides from Seaweeds

Suganya et al. (2018) investigated two seaweeds, *Sargassum wightii* and *Halimeda gracilis* to develop new insecticides. The five different extracts namely acetone, chloroform, methanol, ethanol and water with varied polarity were obtained. The late third instar larvae of *Anopheles stephensi*, *Aedes aegypti*, and *Culex tritaeniorhynchus* are used to assess the insecticidal potential. *S. wightii* showed the LC50 value lower than 50 ppm for all the tested mosquito species in its ethanol extracts. The study revealed that the seaweed have a possibility to develop as a novel insecticides.

Seaweeds fertilizing industry has raised popularly with the increased of organic farming. Seaweeds enable the plants to withstand drought, disease and frost as they act as natural plant growth stimulants. The major constraints in tomato and sunflower production are root rotting fungi species. The ethanol and water extracts of three seaweeds *Spatoglossum variabile*, *Melanothamnus afaqhusainii* and *Halimeda tuna* were compared with nematicide carbofuran and fungicide Topsin –M in the screen house as well as under the field conditions. Individual application of seaweeds and fungicides showed the similar effect of suppressing of fungi growth in the plants but when both the seaweeds and the fungicide were applied together, the suppressing effect of disease in the plants was in greater level and it even produced high fresh shoot weight, root length and maximum yield of tomato in the field conditions (Table 15.1 and Fig. 15.1).

6 Conclusion

Seaweed extracts used as biostimulants and pesticides have a positive impact on the crop production and it also increases the quality of the crop. When compared with commercial biostimulants, the natural seaweed extracts produce better results. And moreover the increased in organic farming is directly proportional to the increase in demand for biofertilizers or biostimulants. The marine algae are rich in secondary metabolites. This makes the seaweed extracts natural sources for compounds which are resistant to various climatic and environmental conditions. In this review a total

Table 15.1 Seaweeds as growth promoters

S. No	Plants name	Seaweed name	Algae type	Category	Reference
1.	Soybean	*Neorhodomela larix*	Red	Biostimulant	Anisimov and Chaikina (2014)
		Tichocarpus crinitus	Red		
		Saccharina japonica	Brown		
		Sargassum pallidum	Brown		
		Ulva fenestrata	Green		
		Codium fragile	Green		
2.	Strawberry	*Ascophyllum nodosum*	Brown	Seaweed liquid fertilizer	Alam et al. (2012)
3.	*Vigna radiata*	*Gracilaria corticata*	Red	Seaweed liquid fertilizer	Chitra and Sreeja (2013)
		Caulerpa peltata	Green		
4.	Tomato	*Ulva flexuosa*	Green	Seaweed liquid fertilizer	Reis et al. (2020)
		Ulva lactuca	Green		
5.	Wheat	*Ulva rigida*	Green	Biostimulant	Salma et al. (2021)
6.	Cucumber	*Ulva lactuca*	Green	Biostimulant	Shimaa et al. (2021)
		Jania rubens	Red		
		Pterocladia capillacea	Red		
7.	*Capsicum annum*	*Sargassum spp*	Brown	Biostimulant	Fatimah et al. (2018)
8.	*Lycopersicon esculentum*	*Sargassum* spp	Brown		
9.	Okra	*Sargassum wightii*	Brown	Seaweed liquid fertilizer	Tensingh et al. (2017)
10.	watermelon	*Ascophyllum nodosum*	Brown	Biostimulant	Abdel-Mawgoud et al. (2010)
11.	lettuce	*Ascophyllum nodosum*	Brown	Biostimulant	Alessandra et al. (2014)
12.	Rose	*Sargassum crassifolium*	Brown	Seaweed liquid fertilizer	Sumangala et al. (2019)
13.	*Vigna radiata*	*Colpomenia sinuosa*	Brown	Seaweed liquid fertilizer	John and Yuvaraj (2014
14.	*Abelmoschus esculantus*	*Dictyota dichotoma*	Brown	Seaweed liquid fertilizer	Sasikumar et al. (2011)
15.	*Abelmoschus exculentus*	*Rosenvingea intricata*	Brown	Seaweed liquid fertilizer	Thirumaran et al. (2009)
16.	*Capsicum annum*	*Sargassum wightii*	Brown	Seaweed liquid fertilizer	Jayasinghe et al. (2016)

(continued)

Table 15.1 (continued)

S. No	Plants name	Seaweed name	Algae type	Category	Reference
17.	Kale	*Gracilaria gracilis*	Red	Biostimulant	Diana et al. (2021)
		Asparagospis armata	Red		
		Calliblepharis jubata	Red		
		Chondracanthus teedei var. *lusitanicus*	Red		
		Grateloupia turuturu	Red		
		Colpomenia peregrine	Brown		
		Sargassum muticum	Brown		
		Undaria pinnatifida	Brown		
18.	Green gram	*Kappaphycus*	Red	Seaweed liquid fertilizer	Pramanick et al. (2013)
		Gracilaria	Red		
19.	*Cajanus cajan*	*Chaetomorpha linum*	Green	Seaweed liquid fertilizer	Sathya et al. (2019)
		Sargassum wightii	Brown		
20.	Garden cress	*Polysiphonia*	Red	Biostimulant	Izabela et al. (2016)
		Ulva	Green		
		Cladophora	Green		
21.	Wheat	*Polysiphonia*	Red		
		Ulva	Green		
		Cladophora	Green		
22.	Tomato	*Chaetomorpha antennina*	Green	Biostimulant	Chanthini et al. (2019)
23.	*Vigna sinensis*	*Ulva fasciata*	Green	Biostimulant	Hussein et al. (2021)
		Cystoseira compressa	Brown		
		Laurencia obtusa	Brown		
24	*Zea mays*	*Ulva fasciata*	Green	Biostimulant	Hussein et al. (2021)
		Cystoseira compressa	Brown		
		Laurencia obtusa	Brown		
25	Onion	*Ascophyllum nodosum*	Brown	Biostimulant	Hidangmayum and Sharma (2017)
26	Maize	*Ecklonia maxima*	Brown	Biostimulant	Kannan et al. (2014)
27	Tomato	*Spatoglossum variabile*	Brown	Biopesticide	Sultana et al. (2011)
		Melanothamnus afaqhusainii	Red		
		Halimeda tuna	Green		
28	Sunflower	*Spatoglossum variabile*	Brown	Biopesticide	Sultana et al. (2011)
		Melanothamnus afaqhusainii	red		
		Halimeda tuna	Green		

(continued)

Table 15.1 (continued)

S. No	Plants name	Seaweed name	Algae type	Category	Reference
29	Maize	*Corallina elongata,*	Red	Biostimulant	Fayzi et al. (2020)
		Corallina officinalis	Red		
		Jania rubens	Red		
		Ulva fasciata	Green		
30	Wheat	*Codium tomentosum*	Green	Seaweed liquid fertilizer	El-Din (2015)
		Sargassum vulgare	Brown		

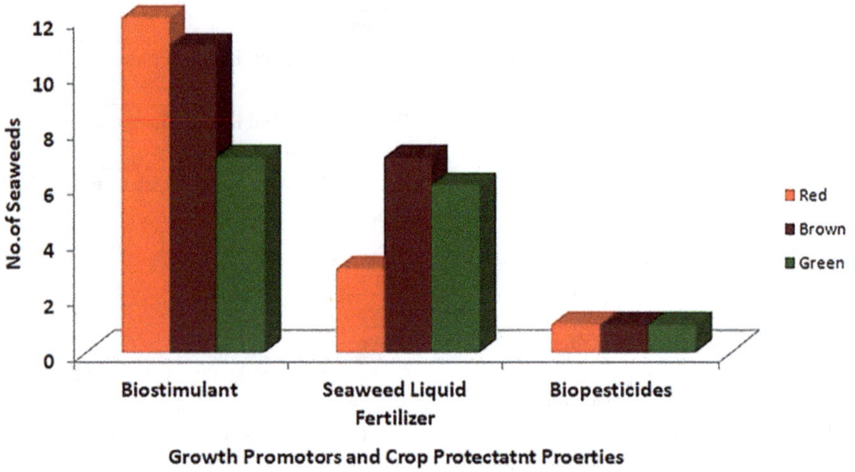

Fig. 15.1 Growth promoters and crop protectant properties of seaweeds

of 44 seaweeds possessed growth promoting properties and 3 seaweeds contains crop protecting properties are compiled. Further more research on commonly available seaweed and its function should be studied to make it more effective to apply on the crops for enhanced growth and better quality and quantity.

References

Abdel-Mawgoud AMR, Tantaway AS, Hafez MM et al (2010) Seaweed extract improves growth, yield and quality of different watermelon hybrids. Res J Agric Biol Sci 6(2):161–168

Abdel-Raouf N, Al-Homaidan AA, Ibraheem IBM (2012) Agricultural importance of algae. Afr J Biotechnol 11(54):11648–11658

Alam MZ, Braun G, Norrie J et al (2012) Effect of *Ascophyllum* extract application on plant growth, fruit yield and soil microbial communities of Strawberry. Can J Plant Sci 93:23–36

Alessandra T, Andrea M, Marco R et al (2014) Filtrate seaweed extract as biostimulant in nursery organic horticulture. In: Proceedings of the 4th ISOFAR scientific conference. Building organic bridges at the organic world congress, pp 13–15

Anisimov MM, Chaikina EL (2014) Effect of seaweed extracts on the growth of seedling roots of soybean (Glycine max (L.) Merr.) seasonal changes in the activity. Int J Curr Res Acad Rev 2(3):19–23

Ashour M, Hassan SM, Elshobary ME et al (2021) Impact of commercial seaweed liquid extract (TAM) biostimulant and its bioactive molecules on growth and antioxidant activities of hot pepper (*Capsicum annuum*). Plan Theory 10:1045. https://doi.org/10.3390/plants10061045

Chanthini KMP, Sengattayan SN, Vethamonickam SR, Annamalai T, Sengodan K, Haridoss S, Narayanan SS, Palanikani R, Soranam R (2019) *Chaetomorpha antennina* (Bory) Kützing derived seaweed liquid fertilizers as prospective bio-stimulant for *Lycopersicon esculentum* (Mill). Biocatal Agric Biotechnol 20:101190

Chen D, Huang Y, Shen D et al (2019) Effect of seaweed extracts on promoting growth and improving stress resistance in sugarcane. Agric Res 11(4):69–72

Chetna M, Saumya R, Nikhil S et al (2015) *Enteromorpha intestinalis* derived seaweed liquid fertilizers as prospective biostimulant for *Glycine max*. Braz Arch Biol Technol 58(6):813–820

Chitra G, Sreeja PS (2013) A comparative study on the effect of seaweed liquid fertilizers on the growth and yield of *Vigna radiata* (L.). Nat Environ Pollut Technol 12(2):359–362

Deshmukh PS, Phonde DB (2013) Effect of seaweed extract on growth, yield and quality of Sugarcane. Int J Agric Sci 9(2):750–753

Diana P, Joao C, Carolina PR et al (2021) Seaweeds carbohydrates polymers as plant growth promoters. Carbohydr Polym Technol Appl 2:1–13

El-Din SMM (2015) Utilization of seaweed extracts as bio-fertilizers to stimulate the growth of wheat seedlings. Egypt J Exp Biol 11(1):31–39

Fatimah S, Alimon A, Daud N (2018) The effect of seaweed extract (Sargassum sp) used as fertilizer on plant growth of *Capsicum Annum* (Chilli) and *Lycopersicon Esculentum* (Tomato). Indones J Sci Technol 3(2):115–123

Fayzi L, Dayan M, Cherifi O, Boufous EH, Cherifi K (2020) Biostimulant effect of four Moroccan seaweed extracts applied as Seed treatment and foliar spray on maize. Asian J Plant Sci 19(4):419–428

Hamidreza B, Sergei B (2019) The effect of humic acid, plant growth promoting rhizobacteria and seaweed on growth parameters, essential oil and chlorophyll content in sweet basil (Ocimum basilicum L.). Glob Sci J 7(7):19–32

Hidangmayum A, Sharma R (2017) Effect of different concentration of commercial seaweed liquid extract of *Ascophylum nodosum* on germination of onion (*Allium cepa* L.). Int J Sci Res 6(7):1488–1491

Hussein MH, Eladl E, Bakry AFA, Nesrein E, Maha ME (2021) Seaweed extracts as prospective plant growth bio-stimulant and salinity stress alleviator for *Vigna sinensis* and *Zea mays*. J Appl Phycol 33:1273–1291

Israel BG, Ana KDL, Emmanuel MM et al (2020) Identification and quantification of plant growth regulators and antioxidant compounds in aqueous extracts of *Padina durvillaei* and *Ulva lactuca*. Agronomy 10:866. https://doi.org/10.3390/agronomy10060866

Izabela M, Boguslawa G, Piotr PW et al (2016) Supercritical fluid extraction of algae enhances levels of biologically active compounds promoting plant growth. Eur J Phycol 51:243–252

Jayasinghe PS, Pahalawattaarachchi V, Ranaweera KKDS (2016) Effect of seaweed liquid fertilizer on plant growth of Capsicum annum. Discover 22(244):723–734

John RVO (2017) Seaweed extract effects on potato (*Solanum Tuberosum* 'BP1') and grape (*Vitis Vinifera* var. Sultana) production. Thesis submitted in the Department of Biodiversity and Conservation Biology (BCB), University of the Western Cape

John PPJ, Yuvaraj P (2014) Effect of seaweed liquid fertilizer of *Colpomenia sinuosa* (Mert. ex Roth) Derbes & Solier (Brown Seaweed) on *Vigna radiata* (L.) R. Wilczek. In Koothankuzhi, Tirunelveli district, Tamil Nadu, India. Int J Pure App Biosci 2(3):17–184

Kannan RRR, Manoj GK, Wendy AS, Johannes VS (2014) Eckol – a new growth stimulant from the brown seaweed Eckolonia maxima. J Appl Phycol. https://doi.org/10.1007/s10811-014-0337-z

Nabti E, Jha B, Hartmann A (2016) Impact of seaweeds on agricultural crop production as biofertilizer. Int J Environ Sci Technol. https://doi.org/10.1007/s13762-016-1202-1

Panda D, Pramanik K, Nayak BR (2012) Seaweed extracts as plant growth regulators for sustainable agriculture. In. J Bio-Resource Environ Agric Sci 3(3):404–411

Pramanick B, Brahmachari K, Ghosh A (2013) Effect of seaweed saps on growth and yield improvement of green gram. Afr J Agric Res 8(13):1180–1186

Ravneet K, Sandhu ND et al (2018) Assessing seaweed extract as a biostimulant on the yield of organic leafy greens in Tennessee. J Agric Univ P R 102(1–2):53–64

Reis RP, Antônio CSA, Ana CC et al (2020) Effects of extracts of two Ulva spp. seaweeds on tomato germination and seedling growth. Res Soc Dev 9(11):e61691110174. http://dx.doi.org/10.33448/rsd-v9i11.10174

Salma L, Mrid RB, Kabach I et al (2021) The effect of foliar application of *Ulva rigida* extract on the growth and biochemical parameters of wheat plants. E3S Web Conf 234:00103

Sasikumar K, Govindan T, Anuradha C (2011) Effect of seaweed liquid fertilizer of Dictyaota dichotoma on growth and yield of Abelmoschus esculantus L. Eur J Exp Biol 1(3):223–227

Sathya B, Indu H, Seenivasan R et al (2019) Influence of Seaweed Liquid fertilizer on the Growth and Biochemical composition of Legume crop, Cajanus cajan (L.) Mill sp. J Phytol 2(5):50–63

Shimaa MH, Ashour M, Sakai N et al (2021) Impact of seaweed liduid extract biostimulant on growth, yield and chemical composition of cucumber (*Cucumis sativus*). Agriculture 11:320

Simon H, Charles NM, Wendy YYL et al (2021) Seedling responses to organically-derived plant growth promoters: an effects-based approach. Plan Theory 10:660

Suganya S, Ishwarya R, Jayakumar R, Govindarajan M, Alharbi NS, Kadaikunnan S, Khaled JM, Al-anbr MN, Vaseeharan B (2018) New insecticides and antimicrobials derived from Sargassum wightii and Halimeda gracillis seaweeds: toxicity against mosquito vectors and antibiofilm activity against microbial pathogens. S Afr J Bot 125:466–480

Sultana V, Baloch GN, Ara J, Ehteshamul-Haque S, Tariq RM, Athar M (2011) Seaweeds as an alternative to chemical pesticides for the management of root diseases of sunflower and tomato. J Appl Bot Food Qual 84:162–168

Sumangala K, Srikrishnah S, Sutharsan S (2019) Roses growth and flowering responding to concentration and frequency of seaweed (*Sargassum crassifolium* L.) liquid extract application. Curr Agric Res J 7(2):236–244

Tensingh NB, Priyatharsini SL, Sheeba PC et al (2017) Effect of seaweed liquid fertilizer of Sargassum wightii on the yield characters of *Abelmoschus esculentus* (L.) Moench. Int J Adv Sci Eng Technol 4(9):4511–4518

Thirumaran G, Arumugam M, Arumugam R et al (2009) Effect of seaweed liquid fertilizer on growth and pigment concentration of Abelmoschus esculentus (l) medikus. Am Eurasian J Agron 2(2):57–66

URL (2022). http://extension.ag.ron.iastate.edu/compendium/compendiumpdfs/plant%20growth%20regulations.pdf

Zodape ST (2001) Seaweed as a biofertilizer. J Sci Ind Res 60:378–382

Chapter 16
Sustainable Encapsulation Materials Derived from Seaweed

Nikola Nowak, Wiktoria Grzebieniarz, Ewelina Jamróz, and Fatih Ozogul

1 Encapsulation-Methods, Encapsulating Agents, Controlled Release, Release Mechanisms and Encapsulation Capacity

Encapsulation was introduced by Bungenberg de Jong in 1932 in the Netherlands (Campos et al. 2013). This is a technique of incorporating solids, liquids or gases into coatings, carriers or embedding them in matrices in order to obtain closed capsules with controlled release properties at a strictly defined place and time and at a concrete speed (Escobar-Puentes et al. 2022). Apart from achieving such effects, encapsulation provides protection for the encapsulated substance against external factors (i.e. light, gases, temperature, humidity) (Comunian and Favaro-Trindade 2016). The produced structures are characterised by a strong, thin, semi-permeable and spherical membrane, their size ranging from a few nanometres to a few millimetres. This method allows to improve production efficiency, which is why it has been introduced into various areas of life, such as the food industry, pharmacy, cosmetics, chemistry or even printing (Naveena and Nagaraju 2020).

In order to select the appropriate encapsulation method, certain criteria should be taken into account (Dias et al. 2017). The end-use as well as the final particle size, release mechanisms and type of encapsulant, must be defined. When selecting an encapsulation technology for food applications, the type of food product and potential adaptation to industrial production should also be considered (Premjit et al. 2022).

N. Nowak (✉) · W. Grzebieniarz · E. Jamróz
Department of Chemistry, University of Agriculture, Cracow, Poland
e-mail: nikola.nowak@urk.edu.pl

F. Ozogul
Department of Seafood Processing Technology, Faculty of Fisheries, Cukurova University, Adana, Turkey

F. Ozogul et al. (eds.), *Seaweeds and Seaweed-Derived Compounds*,
https://doi.org/10.1007/978-3-031-65529-6_16

For many years, research has been carried out on encapsulation techniques and their application in industrial production. Some of them are widely-known and described as spray-drying, but there are also some that require extensive research. In Table 16.1, examples of encapsulation techniques are presented.

In accordance with the definition provided in the European Directive, release control is the transport of bioactive compounds, within a specific time frame, in which a given stimulus takes place (Katouzian et al. 2017). The encapsulation of bioactive ingredients makes it possible to target their release at a specific destination. This mechanism allows to take advantage of the full benefits concerning active compounds (McClements 2018). The release may be under full control of speed, time and place, and may be influenced by endogenous or exogenous stimuli. By using controlled release processes, it is possible to obtain a system that can be implemented in the storage of food products with the potential to gradually activate during food storage (Wang and Chen 2014).

There are many existing release mechanisms that have already been developed. One of the basic and simple mechanisms is the diffusion system, which relies on the release of the active compound from the inside of the particle to the external environment. Other methods used are swelling, fragmentation, degradation, surface erosion (heterogenic), surface erosion (homogenic), bulk erosion and dissolution (Boostani and Jafari 2020). In addition to these mechanisms, controlled release also has profiles that depend on selected rate, time and destination. Release profiles include burst, controlled, delayed and sustained release types (Goonoo et al. 2014).

Decisive factors influencing the release of bioactive ingredients include type of active ingredient, the matrix, the selected encapsulation system and the external environment (Katouzian et al. 2017). The structure, porosity as well as the composition of the matrix used can directly influence the release of the encapsulated compounds (Mahfoudhi et al. 2016). The hydrophobicity of the polysaccharide reduces the rate of release, while its molecular weight affects its profile (Sabliov and Astete 2008). Another factor influencing release rate is the size of the encapsulated particles. The produced small (micro-) capsules have a preference for quick release in the initial phase followed by a slower discharge. Nanoparticles are released slowly, which is due to their smaller surface area (Akbarbaglu et al. 2019).

2 Encapsulated Substances

The advantages of using bioactive compounds, i.e. vitamins, minerals, enzymes, antimicrobial and antioxidant substances, plant extracts, micro- and macro-elements in food, are undeniable (Devi et al. 2017). That is why it is so important to provide them with protection against the negative influence of the external environment, due to the common problem of their stability, sensitivity to temperature, pH, light and oxidation. The effects of these factors contributes to shortening shelf-life and even degradation, leading from poor sensory perception to health effects (Liu et al. 2022a). Most frequently, the active compound is very quickly neutralised or

Table 16.1 Examples of encapsulation methods for various ingredients

Encapsulation technique	Description
Spray-drying	The most frequently applied, simple and economical method in the food industry allowing to obtain a large amount of materials. It is widely-used for encapsulating vitamins, minerals, dyes, oils, flavourings and active ingredients (Alvim et al. 2016). With its use, we can obtain particles of high quality and size in the order of 10–50 μm. The process itself begins with the preparation of an emulsion consisting of a hydrophobic component and a wall material. The sample prepared in this way is sprayed under the influence of hot, compressed air in a drying chamber, during which capsules are produced by thermodynamic phenomena. The capsules are collected in a cyclone and the air is eliminated in conditions of lower humidity and temperature. The disadvantage of the method regards significant discrepancies in the shapes and sizes of the produced capsules, as well as high temperature, which may affect degradation of the active ingredient (Naveena and Nagaraju 2020).
Spray-cooling/ spray-chilling	Spray-cooling, compared to spray-drying, is a method that utilises a temperature below the melting point of the lipid, causing the capsules to solidify when the solution is sprayed, resulting in the encapsulation of the component (Alvim et al. 2016). The sprayed material is an active substance enclosed in a lipid core. The lipids used in this method are most often of hydrophobic nature. The method is widely-used to deliver encapsulated bioactive ingredients to the gastrointestinal tract (Paucar et al. 2016).
Freeze-drying	This process includes three phases, i.e. freezing, sublimation and desorption. It allows to preserve the nutritional properties of encapsulated substances as it does not expose them to high temperatures (Wongsasulak et al. 2014).
Extrusion	A simple, economical and mild method often used to encapsulate various types of bacteria. It consists in preparing an aqueous suspension of the polymer with an active substance, and then, it is extruded into a hardening substance through a hole with a selected diameter. This allows the generation of capsules having relatively uniform size and shape compared to spray-drying. There are several extrusion types, i.e. hot-melt extrusion, melt injection, centrifugal/co-extrusion, electrostatic/electro-spinning. Using the former, it is possible to obtain the most uniform capsules with the highest release control (Hegde and Chandra 2005). This method can apply pulsations or vibrations, as well as an electrostatic field to create nanocapsules.
Coacervation method/phase separation technique	This is first encapsulation method developed, in which the process begins with the preparation of a polysaccharide solution, protein or mixtures thereof. Then, it is incorporated into the carrier material, and the entire suspension is subjected to a change in pH and/or temperature, or to the action of electrolytic components to precipitate (Xiao et al. 2014). Implementing this method, it becomes possible obtain particles with a size of 2–1200 μm. The disadvantage of the method is the difficulty in encapsulating hydrophilic compounds, requiring the use of a double emulsion at the initial stage of the process.

(continued)

Table 16.1 (continued)

Encapsulation technique	Description
Ionic gelation	A process that uses low temperatures, thus, preventing any possible oxidation of the product. Sodium alginate is the polysaccharide most commonly-used in this method due to its mild nature, easy gelation and safety (Leong et al. 2016). In this method, sodium alginate is applied as an encapsulating material that forms a capsule upon contact with a $CaCl_2$ solution. The method enables the formation of alginate capsules with a size smaller than 40 μm.
Emulsification	A method used to make an aqueous solution mixture of a polysaccharide and oil, then, homogenising it to produce an emulsion. The polymer capsules are collected via filtration and their size is estimated to be 25 μm to 2 mm. Emulsifying agents are used to obtain smaller capsule sizes.

deconcentrated in the food product. Therefore, it is of great significance to ensure the stability of active ingredients introduced into food during their entire processing and storage, so that they do not lead to a loss in functionality. The challenge for scientists is to overcome problems such as avoiding degradation, reacting upon contact with food products and losing active properties. Functional compounds can also contribute to changing the taste, smell or texture of packaged products, which further leads to negative perception and lack of product approval by the consumer, as it is very often unfavourable to the senses (Keawchaoon and Yoksan 2011).

One of the methods aimed at avoiding these problems is to encapsulate the extracted active ingredients by incorporating them into a polysaccharide matrix. This technology allows for stabilising these compounds and controlling their release, while not affecting taste, smell or texture. The application of functional compound encapsulation reduces the emission of after-tastes by minerals or vitamins.

Some greatly important factors for encapsulation in food is flavour and aroma (Esfanjani et al. 2018). This process is expected to be fully-preserved until the food product is prepared. Encapsulating these elements is considered to be one of the leading methods of their stabilisation, concealment or preservation (Smaoui et al. 2021). Alkhatib et al. (2020) formed alginate beads to encapsulate black cumin oil for stability and to mask its flavour. The encapsulation of flavours and aromas is not limited to the food industry. Almurisi et al. (2020) masked the taste of paracetamol by enclosing it in alginate beads coated with chitosan.

Vitamins and minerals are ingredients that are mostly synthesised in the human body, but some of them are necessary to be supplied in the daily diet. The possibility of their encapsulation makes it possible to ensure the stability of these compounds, as well as to eradicate the problem of their solubility and improving bioavailability (Gupta et al. 2015).

Essential oils are volatile compounds with high sensitivity to oxidation, humidity, temperature and light, and also have antioxidant as well as antimicrobial properties. They are increasingly used in the food industry, mainly in meat, as they belong to the Generally Recognized as Safe (GRAS) group, which ensures their safety as ingredients for use in food (Mishra et al. 2020; Hassoun and Çoban 2017). Encapsulating essential oils allows for an increased inhibitory effect on the growth of microorganisms (Wu et al. 2015), and in the form of capsules, they are characterised by high efficiency (Benavides et al. 2016). Mukurumbira et al. (2022) created a work on the encapsulation of essential oils and their potential use in antimicrobial active packaging. A similar work was also presented by Bakry et al. (2016), who considered the process of microencapsulation of oils in terms of benefits, techniques, but also applications other than only active packaging.

One of the type of live bioactive ingredients used for encapsulation are probiotics. They include microorganisms that have only benefits for their intended use (Reid 2016). The most commonly-used microbes in the encapsulation method are *Bifidobacterium, Saccharomyces* and *Lactobacillus* (Coghetto et al. 2016). Their encapsulation provides protection against bacteriophages and external factors, which allows them to extend their viability and increase fermentation efficiency (De Prisco and Mauriello 2016).

3 Marine Polysaccharides Extracted from Seaweed

Seaweed consists of various active ingredients, however, as much as 76% are polysaccharides having many functional properties, including those: antioxidant, anti-inflammatory, antiproliferative and anticancerous (Shofia et al. 2018). The rapidly developing field of carrier development means that polysaccharides obtained from seaweed are increasingly used in drug delivery systems due to their ability to form hydrogels at a specific pH, and high compatibility with tissues due to their hydrophilic nature (Nakamura et al. 2008).

Microalgae are a multicellular species, including *Rhodophyta* (red), *Heterocontophyta* (green) and *Chlorophyte* (blue) (Stirk and van Staden 2022). The size of the algae market is estimated by the Food and Agriculture Organization of the United Nations to be at over USD 5.65 billion, of which the main consumers are China, Japan and South Korea (Casoni et al. 2020). These plants are a rich source of active ingredients (Lourenço-Lopes et al. 2020), i.e. polysaccharides, polyphenols (Cotas et al. 2020), as well as natural dyes (Lourenço-Lopes et al. 2020). They have antibacterial, antifungal, antiviral, anti-inflammatory and anticancer properties (Dewi et al. 2018a). Due to such a plethora of properties, they have a very wide range of applications, i.e. the cosmetics, fuel, food, pharmaceutical industries, as well as on the fertiliser market (Cassani et al. 2022).

Polysaccharides of marine origin are known to mankind and have been used since time immemorial. Agar was discovered in Japan in 1658, while carrageenan was used as a food additive as early as in the fifteenth century (Khalil et al. 2017), when it began to be considered as an element of food (Craigie 2011), and the first symposium on their the topic took place in 1950 (Dillehay et al. 2008). However, the use and production of seaweed on an industrial scale and its broad application did not occur until the twenty-first century. The amount of polysaccharides extracted from marine algae is closely dependent on several biological, physical and environmental factors. The main dependencies are the species of algae and the period of their harvest, which later directly affects structural properties (molecular weight, nature, number and sulphate group positions and the type of glycosidic bond) (Melo et al. 2002).

Polymers obtained from various types of seaweed, i.e. agar, alginate, furcellaran, fucoidin, agarose, ulvana and carrageenan, are good substrates for the creation of modern materials due to their low oxygen permeability and full insulation for fats and oils (Khalil et al. 2017). The majority of polysaccharides are obtained from the algae cell wall, but some may be found in plastids (Qiu et al. 2022; Lim et al. 2021).

Alginate is obtained from the cell walls of brown marine algae (*Phaeophyceae*). It is of linear structure, comprising two acids (β-D-mannuronic and α-L-glucuronic) covalently linked in different sequences (Norcino et al. 2022). It is insoluble in water, but has the ability to swell when exposed to this element. Alginate can be extracted in two different ways, by adding acid, resulting in the formation of alginic acid, and also with a calcium salt (Fertahi et al. 2021). They are used in medicine for the controlled delivery of drugs (Agarwal et al. 2015), as a tableting agent (Sanchez-Ballester et al. 2021), wound dressings (Valor et al. 2021), impression materials in dentistry (Demajo et al. 2016) and in tissue engineering (Aguero et al. 2021). In the packaging industry, it is a compound commonly-used in the production of films (Choi et al. 2022; Riahi et al. 2022) as well as a carrier material for incorporating active ingredients (Toprakçı and Şahin 2022; Yuan et al. 2022).

Agar is an agarose polysaccharide that is obtained from the cell walls of red algae *Rhodophyceae* (Schmidt et al. 2010). It is most commonly obtained from the genera *Gelidium* and *Gracilaria* (McHugh 2003). Its extraction is possible by acidic or alkaline treatment of the algae cell wall (Arvizu-Higuera et al. 2007). It is fully soluble in water >85 °C and forms a gel when cooled. In the food industry, it is utilised as a gelling and/or stabilising agent (Wang et al. 2013).

Carrageenan is a linear sulphate polysaccharide obtained from red algae of the *Rrhodophyceae* class. It comprises three leading classes: kappa, iota and lambda, each having different properties and applications. They are obtained by alcohol precipitation or alkaline treatment (Filipović-Vinceković et al. 2005). In the food industry, they are used as gelling (Prajapati et al. 2014), as well as stabilising and thickening agents (Saha 2010).

Furcellaran is an anionic polysaccharide obtained from the red algae *Furcellaria lumbricalis*. It is of linear structure, similar to carrageenan. It consists of D-galactose, 3,6-anyhydro-D-galactose and D-galactose-4-sulfates. Jamróz et al. (2014) applied furcellaran as an encapsulation core to form a suspension that was stable for at least 10 days. It is also commonly-used to create films (Jamróz et al. 2021a, 2022) and coatings (Kulawik et al. 2022).

Fucoidan comprises sulphite polysaccharides isolated from brown algae. It has a great variety and a broad spectrum of activity (Zvyagintseva et al. 2021). It is characterised by bioactive, anticoagulant, anti-cancerous, antiviral and immune-modulating properties (Shen et al. 2018). It is a practically non-toxic compound with a very wide range of applications in dietary supplements, cosmetics and on pharmacological markets (Xue et al. 2021).

Ulvan is a sulphated polysaccharide derived from *Heterocontophyta* and it has various antioxidant properties. It is widely-used in drug delivery systems (Don et al. 2022), as wound dressings (Ren et al. 2022), and also in the production of films (Don et al. 2021).

Natural polymers have a broad application in the industry. They are present in various forms on the pharmacy, biomedicine and tissue engineering market, but also in the field of cosmetology. Their main advantages include wide and easy availability, economic considerations, being renewable, full naturalness and biodegradability, biocompatibility, gelling ability and encapsulation efficiency (Khalil et al. 2017). The human body is able to fully metabolise and remove them from its pathways. In the food industry, they are used as gelling, thickening and stabilising ingredients. However, in recent years, they have become the leading materials for food coating and encapsulation of biologically-active ingredients. Edible polysaccharides and, in particular, those derived from the extraction of marine algae, have excellent translocating properties for proteins, enzymes, vitamins, probiotics, compounds with antioxidant and antimicrobial activity as well as pharmaceuticals (Karbowiak et al. 2009).

The food industry is considered to be mainstream in the development of polysaccharides. Due to the increasing demand for healthier foods, there is a need to extend the quality of packaged food. In order to meet the expectations of consumers, entrepreneurs are implementing innovative methods to extend the shelf-life of fresh products, but also processed foods. One of them is the introduction of modern packaging or coatings with intelligent and active properties. For over a dozen years, this field has been enjoying considerable development and interest, also from a scientific point of view. Various types of functional compounds are added to polysaccharide-based materials to improve taste, colour and texture. Nonetheless, there is a trend towards displacing synthetic compounds in favour of biologically-active compounds with potential health benefits. The active compounds are most often obtained from plants or animals and processed into extracts or granules.

Their disadvantage primarily regards poor mechanical properties compared to their synthetic counterparts (Saurabh et al. 2013). In order to overcome this limitation, at least to some extent, modifications of such materials are carried out with nano- or micro-sized structures. The inclusion of such substances in polymer chains increases their mechanical strength (Rhim 2011; Martins et al. 2013), and sometimes even makes it possible to inhibit the development of specific pathogens (Kanmani and Rhim 2014).

In following sections, an is made attempt to characterise coatings, films, micro- and nano-capsules using polysaccharides derived from seaweed (Fig. 16.1).

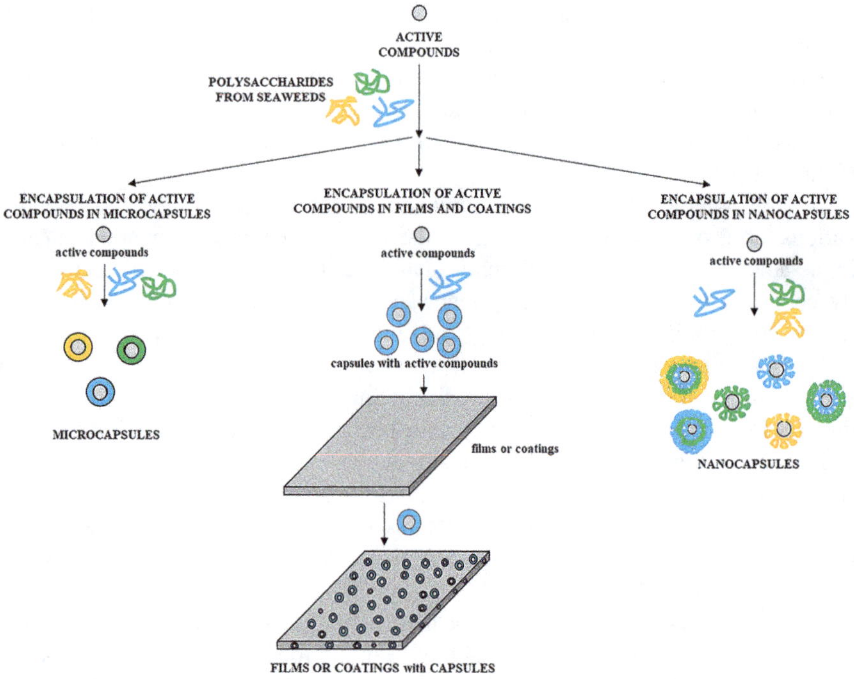

Fig. 16.1 Potential use of polysaccharides derived from sea- algae in the encapsulation of active ingredients

4 Polysaccharides Derived from Seaweed as Films and Coatings with Encapsulated Active Ingredients

The polysaccharide coating is a fully-edible material, applied in a liquid form to food products, most often in order to increase quality (Kang et al. 2013). Films made of polysaccharides do not have to be edible materials, but they are intended to protect the packaged product against environmental factors. They are produced in the form of various shapes of sheets and then applied to a food product (Jamróz et al. 2021b). Films and coatings can also be used as matrices for encapsulating functional ingredients, thus, it is likely to achieve stabilisation of a given compound and its controlled release for a desired purpose (Table 16.2).

Kurczewska et al. (2021) developed alginate films with halloysite in order to encapsulate salicylic acid into antimicrobial materials. This experiment allowed to show that these films released the encapsulated ingredient more slowly and in a more controlled manner compared to analogous pectin-based materials. These films also demonstrated inhibitory activity against the growth of *P. aeruginosa, S. aureus, E. coli* and *S. Typhimurium* bacteria.

Navarro et al. (2016) investigated the effect of the encapsulant on the properties of alginate films. As encapsulation compounds, β-cyclodextrin, trehalose and

Table 16.2 Examples of using films as methods for various ingredients as encapsulation carries for active ingredients

Film	Encapsulating agent	Encapsulated substance	References
Alginate	Halloysite	Salicylic acid	Kurczewska et al. (2021)
Alginate	β-cyclodextrin, trehalose, Tween-20	Thyme oil	Navarro et al. (2016)
Alginate	–	Lemongrass oil	Riquelme et al. (2017)
Alginate, chitosan	–	Yerba mate	Anbinder et al. (2011)
Alginate/polyvinyl alcohol	–	Limonene	Levi et al. (2011)
Alginate	–	N-hexanal	Hambleton et al. (2009a)
Alginate, chitosan	–	Indomethacin (IDM)	Ye et al. (2005)
Alginate	–	NLC (nano-structured lipid carriers)	Khorrami et al. (2021)
Sodium alginate	B-cyclodextrin	Carvacrol (CAR)	Cheng et al. (2019)
Iota carrageenan	–	Aroma	Hambleton et al. (2009b)

Tween-20 were used, while the encapsulated substance was thyme oil. Depending on the applied compound, differences in rheology and particle size were observed. It was also indicated that alginate films with β-cyclodextrin had the best encapsulation potential and, moreover, showed antimicrobial activity.

Levi et al. (2011) created films based on sodium alginate and polyvinyl alcohol with encapsulated flavouring, i.e. d-limonene. This combination proved to be a good matrix for aroma encapsulation. A limonene concentration up to 1% w/v was the most suitable as no significant release occurred, while, in turn, amounts above 5% w/v turned out to be unfavourable due to excessive release. Research on the production of films with encapsulated flavouring was also conducted by Hambleton et al. (2009a), who made composites based on alginate with encapsulated n-hexanal. The analyses allowed to show changes in alginate structures due to encapsulation of n-hexanal in them. The structure of the film was disturbed while improving its homogeneity and stability, while the alginate itself proved to be an efficient matrix for n-hexanal encapsulation.

Multilayer films are also starting to become increasingly used in the encapsulation technique. Ye et al. (2005) produced multilayer films based on alginate and chitosan using the LBL method. Khorrami et al. (2021) created alginate-based films with generated NLC capsules to functionalise the polymer. These films met the acceptable standards in terms of mechanical and optical properties, and were also characterised by the ability to absorb UV radiation and high thermal stability. Cheng et al. (2019) designed films on the basis of sodium alginate, with micro-encapsulated carvacrol. This matrix proved to be stable and effective for controlled release of the encapsulated compound. The produced βCD-CARM/SA films showed antimicrobial activity for the *Trichoderma sp.*, while and extending the shelf-life of the mushrooms stored.

Hambleton et al. (2009b) proved that the iota-carrageenan matrix is good for the encapsulation of aromatic compounds and has odour-barrier properties. Research on the encapsulation of aromas in terms of their release was carried out by the same researchers in 2008, showing that capsules diffuse slowly, which confirms that carrageenan is a good matrix for encapsulating bioactive compounds (Hambleton et al. 2008). Research on the iota-carrageenan matrix was carried out by Fabra et al. (2012), producing films with n-hexanal and d-limonene capsules. Analyses concerning evaluation of temperature and solvent influence on the release of encapsulated substances showed their slow release, which makes carrageenan a good matrix for the encapsulation process.

5 Polysaccharides Derived from Seaweed as Microencapsulation Components

There are existing studies in which it is indicated that seaweed polysaccharides have been used to develop a controlled release system via microencapsulation methods. In contrast, seaweed proteins have not been extensively studied with regard to controlled release (Rawiwan et al. 2022).

The susceptibility of ingredients to oxidation, instability to external and internal factors such as pH, heat or light, limits their potential use. To eradicate these disadvantages, the active ingredients can be enclosed in capsules. Zhang et al. (2016) proved that the encapsulation of β-galactosidase in carrageenan hydrogel beads caused the enzyme activity to increase regardless of pH and thermal conditions. Bioactive tea extracts (TE) are sensitive to external factors, therefore, Baltrusch et al. (2022) enclosed TE in microcapsules with thin films based on alginate, carrageenan and starch via the spray-drying method. The highest microencapsulation efficiencies (ME) were obtained for microcapsules based on carrageenan (92.72 ± 1.8%), and the lowest for starch microcapsules (60.25 ± 7.14%). Noor et al. (2022) received microspheres of Piper beetle polyphenols via the extrusion method using alginate. The alginate microspheres had a higher encapsulation efficiency (~84%) and the polyphenols in the microspheres were also resistant to a wide range of temperatures, pHs and durations. The simulated digestion study indicated that the encapsulated polyphenols are released from the small intestine and the resulting capsule protects the active ingredients in the stomach. Moreover, encapsulating them in an alginate capsule improved the stability as the polyphenols and the antioxidant were protected for a period of 24 months. By contrast, Gu et al. (2021) proved that combinations of carrageenan with sodium alginate gave better encapsulation of egg yolk immunoglobulin Y (IgY) than in the case of alginate beads alone. The research carried out on digestion allowed to indicate that the composite beads do not swell during the gastric phase, but rather at the time of the intestinal phase. In addition, IgY digestion was slower in the beads based on alginate-carrageenan than in the alginate beads.

The microencapsulation of the *Laurus nobilis L.* extract was successfully performed via the spray-drying method with the use of modified chitosan, sodium alginate and arabic gum as encapsulating agents. Rapid release was observed in deionised water, with the *L. nobilis L.* extract completely released after 6 min in acacia gum microcapsules, or after 9 and 15 min for modified chitosan and sodium alginate particles, respectively (Chaumun et al. 2020). Kavoosi et al. (2018) examined the encapsulation efficiency of zataria essential oil into seaweed polymers including agar, alginate and carrageenan. They considered that the results obtained clearly indicate the seaweed polymer may be promising pharmaceutical excipients for the encapsulation of essential oils, however, of the three proposed, carrageenan demonstrates the most potential. Sun et al. (2019) enclosed carvacrol in a pectin-alginate matrix using the spray-drying method. The microcapsule did not adversely affect the antimicrobial or antioxidant activity of the sensitive phenolic compound.

El-Deeb et al. (2022) analysed the effect of extracted novel *Agaricus bisporus* MH751906 polysaccharides and their formulas on alginate/kappa carrageenan microcapsules in order to exert an immunotherapeutic effect on the activation of intestinal resident NK cells (Natural Killer Cells) against colon cancer. Due to the oral delivery of the presented microcapsules, human NK cells were activated, exhibiting 74.09% cytotoxic activity against Caco-2 human colon cancer cells, in which the majority of the cancer cell population was arrested in the G0/G1 phase, leading to apoptosis. Abraham et al. (2021) checked how the type of alginate influences the release profile of active ingredients (seaweed and spirulina powder) in the model gastrointestinal tract, using the Gastrointestinal Severity Index (GSI). The medium viscosity alginate beads exhibited better retention capacity for the active ingredients than the high- and low-viscosity one. Moreover, compared to the wet beads, those freeze-dried demonstrated a slow and controlled release of protein, phlorotannin and antioxidants. *In vitro* digestion has shown that alginate beads can be used as carriers in the intestinal digestive delivery system as they show storage stability and the ability to effectively perform in the GSI model. Toprakçı and Şahin (2022) checked how the concentrations of calcium chloride (2–15%, w/v), sodium alginate (1–2%, w/v) and the hardening time (15–45 min) affected the encapsulation of leaf extract olive. The best parameters of the encapsulation efficiency for the active ingredient were noted at a 2.34% concentration of calcium chloride and 2% concentration of sodium alginate for 26 min of hardening duration. Bacteriophages were enclosed in pure alginate beads and a mixture of alginate with carrageenan, chitosan or whey protein. Phages were viable in the whey protein-alginate beads after exposure to a pH of 2.5 for 2 h (Silva Batalha et al. 2021). The betacyanins of red dragon fruit (*Hylocereus polyrhizus L.*) were enclosed with alginate microspheres. Half-life ($T_{1/2}$) and total betacyanin retention (%) in the microspheres during storage improved in comparison to the uncoated extract and commercial betanin solution (control) (Fathordoobady et al. 2021). Sarıyer et al. (2020) attempted to deliver bovine serum albumin (BSA) protein to the intestines in the form of alginate/kappa(κ)-carrageenan double network hydrogel beads at different pHs and $CaCl_2$/KCl ratios. The presence of alginate improved the mechanical properties of the capsule, while the proportion of carrageenan caused improvement in the release of protein in the simulated intestinal fluid (SIF). In this study, it was noted that the highest encapsulation efficiency (83–89%) concerned hydrogel beads prepared at a pH lower than the BSA isoelectric point, then, 67–72% of them were released in the SIF within 4–5 h, and, therefore, the target delivery of proteins was achieved for the intestines.

Hard vegetarian capsules contain hydroxypropyl methylcellulose in their composition, however, their production volume is low. An interesting alternative to gelatin hard capsules is carrageenan reinforced with cellulose nanocrystals. The optimal composition of these components leads to the production of mechanically strong capsules that may assume a protective role for these sensitive ingredients (Hamdan et al. 2021). The research team comprising He et al. (2017) also obtained hard capsules based on carrageenan and locust bean gum, and the obtained material was characterised by good mechanical properties and storage stability. In literature items on the subject, there are many examples of using marine algae polysaccharides as microcapsule coating materials, and there is also information on the encapsulation of ingredients or extracts from algae in various types of microcapsules (Table 16.3).

The addition of bioactive compounds derived from algae to food products requires the development of encapsulation systems to ensure the stability of very sensitive and perishable ingredients of this type. Kaushalya and Gunathilake (2022) isolated the phlorotannins from *Sargassum ilicifolum*, collected from the southern coastal region of Sri Lanka, and encapsulated it in a cross-linked chitosan tripolyphosphate (TPP) coating. Encapsulation did not improve the storage stability or maintenance of the anti-diabetic potential of florotannins. A very important criterion for enclosing the active ingredients derived from algae is the choice of the coating material. Nkurunziza et al. (2021) developed an encapsulation method using coating materials (dextrin, maltodextrin, lactose, acacia, whey protein, gelatin and sodium caseinate) with spray-drying phenolic compounds from brown seaweed (*Saccharina japonica*) obtained as a result of subcritical water extraction. The authors report that the best coating materials are gelatin and whey protein, which retain more than 86% of phenolic compounds in aqueous extracts, while in the case of polysaccharides, an efficiency of less than 40% was observed. Shofia et al. (2018) compared the encapsulation of exopolysaccharides extracted from brown seaweed (*Sargassum longifolium*) in an orange oil nanoemulsion (NE) prepared by ultrasonication and nanostructured lipid carrier (NLC), further prepared by hot solvent diffusion. Although both carriers showed good anti-tumour activity, in research, it has been showed that the polysaccharide is better encapsulated in NLC than NE.

Seaweed beds are marine ecosystems that provide a habitat and foraging for fish as well as crustaceans. Their area is being diminished by environmental changes and pollution. There are many methods of restoring seaweed resources, but these efforts are inadequate. Jung et al. (2020) developed the encapsulation method of seaweed zygotes with Ca-alginate to improve their attachment using the brown alga

Table 16.3 Examples of using components from seaweed as care and wall materials in the encapsulation process

Wall material	Core material	Method	Functional properties	References
Polysaccharides from seaweed as coating material				
Alginate	Jabuticaba peel extract Propolis extracts	Ionotropic gelation by dispensing it as liquid droplets	Encapsulation efficiency reached 89.6% for the total concentration of phenols and 98.1% for the total concentration of monomeric anthocyanins.	Dalponte Dallabona et al. (2020)
Carrageenan	Ethanol extract from sepal flower (*Hibiscus sabdariffa Linn*)	Freeze-drying	The size of the microcapsules totalled 14.91 μm. The best efficiency was achieved with the ratio of extract to carrageenan equalling 1:7 and this totalled 86.01%.	Sumarni et al. (2019)
<u>1st layer</u> Alginate microcapsules coated with composite from egg whites (EA) and stearic acid (SA) <u>2nd layer</u> Cassava starch granules	*Lactobacillus acidophilus* probiotic	Technique combining electro-spraying and fluidised bed-coating	Encapsulation efficiency was above 90%. High degree of protection when exposed to humid heat.	Pitigraisorn et al. (2017)
Seaweed as core material				
Maltodextrin (10, 20, and 30% (w/v))	*Sargassum aquifolium* extract (brown seaweed)	Spray-drying	The best umami flavour enhancing microcapsules from seaweed extract made by adding 30% MD as a coating material. Microcapsules have low moisture content and very good solubility, which further has a good effect on product stability.	Pranata et al. (2022)

(continued)

Table 16.3 (continued)

Wall material	Core material	Method	Functional properties	References
Palm stearin as the solid lipid core	Fucoxanthin from *Sargassum thunbergii*	–	The size of the microcapsules was 19.19 μm. Encapsulation efficiency and fucoxanthin microcapsule load capacity were 98.3% and 0.04%, respectively. Fucoxanthin in microcapsules showed higher stability than free fucoxanthin to light, moisture and temperature. The cumulative amount of fucoxanthin released from the microcapsules was 22.92% in simulated gastric fluid (SGF) and 56.55% in simulated intestinal fluid (SIF).	Wang et al. (2017)
Maltodextrin (CM), maltodextrin-alginate (CMA) and maltodextrin-fish gelatin (CMG)	Chlorophyll from sea grape (*Caulerpa racemosa*)	Freeze-drying	CM and CMA appeared more fragile than CMG as shown in the SEM photos. The size of the microcapsules was: CM—9.061–469.9 nm, CMA—9.707–363.5 nm, and CMG—11.49–433.2 nm. Chlorophyll release time was longer for the CMG microcapsules than for CM and CMA.	Kurniasih et al. (2018)
Maltodextrin (MD) and carrageenan (C)	Phycocyanin from *Spirulina sp.*	Spray-dry method	The best parameters were obtained for microcapsules at the concentration of maltodextrin in carrageenan 1.0% (w/v): encapsulation efficiency—approx. 12.89%, moisture approx. 8.36%, phycocyanin content approx. 2.83% and antioxidant activity approx. 49.05%.	Dewi et al. (2016)

(continued)

Table 16.3 (continued)

Wall material	Core material	Method	Functional properties	References
Maltodextrin and Na-alginate	Phycocyanin from *Spirulina sp.*	Spray-drying method	The obtained microcapsules were added to jelly candy in various concentrations. The best results for the colour, pH and moisture of the candy were achieved with the addition of 3% microcapsules.	Dewi et al. (2018b)

Sargassum fulvellum. In the experiment, PVC panels and concrete bricks were covered with closed and non-encapsulated zygotes. It was observed that the density and growth rate of encapsulated zygotes was 4 and 7 times higher than that of those non-encapsulated. Moreover, the density and growth rate were higher on concrete bricks. The method developed in this way allows to increase the adhesion of seaweed spores in the marine environment.

6 Polysaccharides Derived from Seaweed as Nanocapsules

The use of polymers as materials for the preparation of nanocapsules has been raising more and more interest due to their biodegradability, biocompatibility, non-toxicity and a very wide range of potential applications. Recently, hydrogel nanoparticles, also known as nanogels, have been an area of intense research. These are water-swollen polymer networks that are highly analogous to the extracellular matrix of cells. Due to the high water content, they are characterised by a porous structure with soft and flexible mechanical properties. Their size is in the range of 1–1000 nm and they are formed by physically or chemically cross-linked polymer networks. They further enable the incorporation of DNA, proteins and drugs. Due to their polymeric origin, they can absorb compounds through salt bridges, hydrogen bonds or the hydrophilic-hydrophobic forces of polymer chains (Bardajee et al. 2020; Van Vlierberghe et al. 2011; Karbarz et al. 2017; Qureshi and Khatoon 2019). Of particular interest are hydrogel nanoparticles derived from polymers sensitive to changes in environmental conditions, such as pH, temperature, the influence of ionic or electric field strength, type of solvent or magnetic field (Sivakumaran et al. 2013; Echeverria et al. 2018). Changes in external stimuli can cause changes in particle size, polymer structure or the arrangement of the gel network, which allows for the controlled release of drugs (Caldorera-Moore and Peppas 2009; Motornov et al. 2010). In recent years, significant progress has been made with regard to the distribution and delivery of drugs in the human body. The latest lines of action focus on improving the pharmacokinetics of a drug through the development of an intelligent carrier, directional delivery or slow release. Hydrogel nanoparticles are an area of vast research, mainly delivery systems or carriers of active ingredients. Thanks to their polymeric origin, they are characterised by high hydrophilicity and biocompatibility, and their sizes at a nanometric scale allow for increasing bioavailability and achieving a controlled delivery of drugs or active ingredients (Daniel-da-Silva et al. 2011; Dalby et al. 2013; Janes et al. 2001). In Table 16.4, examples are given of nanocapsules coated with various seaweed polysaccharides.

Encapsulation of polymers derived from marine algae usually takes place with the use of an encapsulating or cross-linking agent, such as calcium chloride and Tween-80, or a suitable method—nano spray-drying or layer-by-layer method based on the zeta potential values of polymers. The encapsulation of polysaccharides, as well as the encapsulation of active compounds therein without the addition of an encapsulating substance, is possible, but it is not as effective as with the use of a surfactant. De and Robinson (2003) described the formation of chitosan/alginate and chitosan/poly-1-lysine nanospheres with the use of calcium chloride. They showed that to produce a negatively charged pre-gel of calcium alginate, the concentration of calcium chloride must be less than 0.2 of the mass ratio of alginate. The ratio of positively charged polymers to alginate is also significant. They proved that the optimal ratio for chitosan and poly-l-lysine to sodium alginate is 0.1.R2 = 0.98. Exceeding the mass ratio of 0.25 for PLL-alginate and 0.33 for chitosan-alginate causes formation of microspheres and their precipitation due to excessive aggregation. The use of calcium chloride as a cross-linking agent compared to the use of proteins (haemoglobin and myoglobin) was also analysed by Wei et al. (2018), who showed that oxidised sodium alginate in the presence of Ca^{2+} ions can form stable nanogels, and the addition of myoglobin and haemoglobin allows the polymer to cross-link. In the second attempt, they did not use calcium ions, but only the protein itself. As a consequence, it was not possible to obtain an alginate nanogel, which proves the high importance of the use of an encapsulating agent. In the cross-linking of alginate particles with calcium chloride, the mixing process is a fundamental factor, otherwise aggregation of the nanoparticles may occur. Mokhtari et al. (2017) attempted to add $CaCl_2$ solution in a spray form to maximise the collision of calcium ions with sodium alginate hydroxyl groups. As a consequence, they managed to obtain sodium alginate nanocapsules containing 785 nm peppermint extract, but also showed that the most important independent variable for particle size is the concentration of the polymer itself. Alaysuy et al. (2022) encapsulated red cabbage extract into calcium alginate in the presence of alum-potassium mortar, which was then applied to cotton gauze to obtain a smart, therapeutic wound-dressing that would respond to changes in pH. The obtained nanocapsules had a size of 50–175 nm and a colourimetric change of the dressing was observed from purple at 579 nm to pink at 437 nm. A homogeneous emulsion using mechanical agitation was achieved by Basha et al. (2021) using an emulsifier such as castor oil. High-pressure homogenisation allowed to reduce the droplet size and, at the same time, obtain nanoemulsions. The cross-link and zeta potential of the mixture components were used. The negatively charged sodium alginate was combined with a mixture of aloe vera gel and insulin, which was positively charged. The highest encapsulation efficiency (26.9%) was obtained when the ratio of alginate to aloe-vera gel was 3:3,

Table 16.4 Examples of nanocapsules based on polysaccharides from seaweed

Polymer	Encapsulating substance/method	Active compounds	Properties	References
Sodium alginate	$CaCl_2$, castor oil	Aloe gel	Oral administration of insulin.	Basha et al. (2021)
Sodium alginate	Cystamine dihydrochloride	Graphene oxide, doxorubicin hydrochloride	Delivering an anti-cancer drug (doxorubicin).	Xu et al. (2018)
Sodium alginate and gelatin	Technique of inverted mini-emulsion	Curcumin	Anti-tumour activity against MC-7 cells.	Sarika et al. (2016)
Sodium alginate	$CaCl_2$	*Lactobacillus rhamnosus GG*	Protection of probiotics by delaying the penetration of gastric fluid into the nanocomposite. Controlled release of the probiotic in the intestine.	Kim et al. (2021)
Sodium alginate	Cystamine dihydrochloride	Doxorubicin hydrochloride	Diagnosis and chemotherapy of neoplasms. Directional delivery of anti-cancer drugs.	Podgórna et al. (2017)
Sodium alginate	$CaCl_2$	Valnemulin hydrochloride	Prolonged and controlled drug release against *Staphylococcus aureus*.	Liu et al. (2022b)
Sodium alginate, chitosan	EDC/NH, $CaCl_2$	*Lactobacillus reuteri*	Directional, enteric drug delivery—probiotic.	Ding et al. (2022)
Sodium alginate	$CaCl_2$, Tween-80, Span-80	Quercetin, glycyrrhizin	Directional drug delivery—quercitin. Drug carrier against acute liver damage.	Zhao et al. (2021)

(continued)

Table 16.4 (continued)

Polymer	Encapsulating substance/method	Active compounds	Properties	References
Sodium alginate	$CaCl_2$	Transforming growth factor Beta 3 (TGF-ß3)	Controlled release of bioactive ingredients during chondrogenic differentiation of mesenchymal stem cells (MSCs). Reconstruction of damaged articular cartilage.	Mahmoudi et al. (2020)
Sodium alginate, chitosan	Tripolyphosphate	Silver sulfadiazine	Controlled drug delivery Treatment of burn wounds	El-Feky et al. (2017)
Carrageenan	Technique of inverted micro-emulsion	Methylene blue	Directional therapies. As a thermo-sensitive drug carrier.	Daniel-da-Silva et al. (2011)
Carrageenan	$FeCl_3$	Levodopa	Delayed release of levodopa.	Bardajee et al. (2020)
Carrageenan and chitosan	N, N′- Methylene bisacrylamide (MBA)	Rivastigmine	Directional drug delivery.	Rahmani et al. (2021)
Carrageenan, alginate, pectin, acacia, carboxymethylcellulose	Nano spray-drying	Low-density lipoprotein, curcumin	Oral delivery of lipophilic nutrients.	Zhou et al. (2016)
Furcellaran, chitosan	Layer-by-layer	Doxorubicin	Directional delivery of anti-cancer drugs.	Milosavljevic et al. (2020)

while the zeta potential of the nanoemulsion was then −40 mV ± 0.71. Insulin enclosed in a highly-charged emulsion retained its structure stability and biological activity. The insulin-containing nanogel showed promising results for *in vitro* research and indicates its potential use as a carrier for the oral delivery of inulin. The biomedical application of alginate nanogel was also investigated by Xu et al. (2018) who attempted to produce an alginate nanogel enriched with graphene oxide and carrying an anti-cancer drug. The encapsulation efficiency of doxorubicin (DOX) was very high and remained at the level of 97.2 ± 1.2%. Such high efficiency was achieved, among others, due to the strong electrostatic interactions of the substance with the negatively charged nanogel. Additionally, the hydrophilic nanogel prevented the formation of aggregates concerning graphene nanoparticles loaded with DOX (the size remaining at 230 ± 19 nm), which alone, tended to form micro-sized aggregates (approximately 1025 nm). Pei et al. (2018) also enclosed a DOX alginate matrix in order to design a pro-drug gel for cancer diagnosis and therapy. The alginate nanogel turned out to be a good matrix for drug encapsulation, and its modifications made it possible to obtain stability under physiological pH conditions and specific, pH-dependent drug release. Additionally, the nanogel exhibited strong fluorescence in an acidic environment, characteristic for the micro-environment of cancer cells. As a result, the obtained nanocapsules can be successfully used as a non-invasive method for tracking the location in cancer cells in real time. DOX encapsulation is the subject of various studies due to its anti-cancer properties. In the polymer matrix (sodium alginate/collagen) with the addition of protein (keratin), DOX nanocapsules were also synthesised by Sun et al. (2017). They used a simple, 2-step method of nanogel synthesis in a neutral aqueous solution. The keratin/alginate solution was formed by the oxidative cross-linking of alginate chains interspersed with cross-linked keratin, the cross-linking agent being hydrogen peroxide. The size of the nanocapsules obtained in this way was uniform and remained within the range of 60–90 nm, which correlated with the alginate content in the mixture (alginate: keratin mass ratio 1:1). In the nanogel prepared this way, the DOX solution (1 mg/ml) was placed in a nitrogen atmosphere. The method of synthesis developed in this manners allows to optimise conditions while minimising cytotoxicity. Due to the interphase disulfide cross-linking and macromolecular hydrogen bonds resulting from cross-linking, it was possible to obtain high stability of the encapsulated drug and its controlled release. The disintegration of the capsule at the target site was due to the reduction and action of enzymes, which resulted in a rapid and efficient drug release compared to standard therapy. In contrast to the free drug, it was also characterised by a longer retention time due to easier accumulation in tumours, thanks to which better anti-tumour activity was demonstrated.

As reported in the literature, the appropriate size of a nanoparticle for drug delivery to neoplastic cells should range from 100 to 600 nm, due to the vascularisation of tumours (Adibkia et al. 2017). Such a size of the nanoparticles was achieved by Sarika et al. (2016), who encapsulated curcumin as a potential anti-cancer drug in a sodium alginate gelatin nanogel. In their experiment, they used the negative zeta potential of alginate/gelatin nanogels, the charge of which was −36.9 mV. Negatively surface-charged nanoparticles made of hydrophilic polymers are considered suitable for drug delivery to cancer cells because the hydrophilic surface allows the nanoparticle to escape from macrophage uptake (Xi et al. 2012; Schmitt et al. 2010). Curcumin was also used as an encapsulated active ingredient by Zhou et al. (2016) by enclosing it together with LDL from egg yolk in a matrix comprising five different polymers, comparing the properties of the capsules obtained in this way. Curcumin, as a model, lipophilic bioactive compound, was a specific identifier of the encapsulation potential regarding the analysed polymers. Comparing gum arabic, sodium alginate, carrageenan, carboxymethylcellulose and pectin, it was noticed that each of the polymers had its own critical curcumin loading factor, and the amount of added curcumin correlated with the size of the obtained nanoparticles. The carrageenan- and alginate-based nanogels showed significantly higher PDI values (>0.4) and higher zeta potential values (>44 mV) compared to the remaining polymers (>30 mV). Similarly, in terms of the size of the obtained nanoparticles, these had the largest particle size range (10–1000 nm), but the average value was approx. 100 nm. Such a wide range of sizes may suggest destruction of the nanogel structure during complexation. The very high zeta potential of both polysaccharides (−70 mV at a pH of 7) induced excessive electrostatic interactions between the LDL fraction of the chicken egg and the polymers, which, in turn, disrupted the protein-phospholipid connection. As a result of these interactions, LDL proteins aggregated and gelled during mixing and heating. However, the average size of the nanocapsules of these polymers (100 nm) indicates that the disturbance occurred only to a small extent.

Another biomedical application of alginate nanogels is enteral drug delivery. In their research, Kim et al. (2021) developed a nanogel based on sodium alginate and bentonite as encapsulating matrices for *Lactobacillus rhamnosus GG* probiotic bacteria. In the conducted experiment, it was possible to obtain nanocapsules resistant to gastric acids and allow for a controlled release of capsules in the intestine. The combination of the polymer with bentonite allowed to overcome the limitations of using nanocapsules of the polymer itself. Bentonite is characterised by water impermeability and buffering properties. Exfoliated bentonite created a nanocomposite network, and, as a consequence of vigorous mixing, the alginate molecules migrated to the bentonite inter-layer spaces. Ultimately, the bentonite-alginate matrix delayed gastric juice penetration of the encapsulated probiotics and helped to maintain the micro-environmental pH.

The encapsulation of the probiotic in the nanogel was also performed by Ding et al. (2022), creating a double W/O/W emulsion using calcium alginate hydrogel beads. The encapsulated bacterial strain was *Lactobacillus reuteri*. As in the above-mentioned work, Kim et al. (2021) the development of nanocapsules allowed for the controlled, sustained release of the drug at the target site. The use of the double W/O/W emulsion sealed with beads of alginate hydrogel also improved the viability of *Lactobacillus reuteri* cells during long-term storage and freezing. The concentration of sodium alginate influenced the swelling and the structure of the hydrogel beads. Probiotic cells were surrounded by CMKGM-CS (carboxymethyl konjac glucomannan), which was connected to the polymer by hydrogen bonds. The concentration of the polymer influenced the amount of these bonds, thus mediating the controlled release of the drug.

Directional drug delivery by encapsulation in a polymer matrix has also been studied by Zhao et al. (2021). They enclosed quericitin in the alginate nanogel matrix, increasing its solubility, and consequently managed to improve its antioxidant activity and bioavailability. The antioxidant activity of querecitin was increased 81-fold, which made it possible to reduce the level of toxicity through the possibility of using a lower dose. The nanogel was also enriched with glycyrrhizin, which not only increased the effectiveness of targeted therapy, but also had an anti-inflammatory function. Ultimately, the alginate nanocapsules coating these active ingredients showed excellent liver protection by reducing AST, ALT and TBIL levels, reducing oxidative stress, and suppressing inflammatory mRNA expression with a consequent reduction in acute liver damage.

Encapsulation of valnemulin hydrochloride in an alginate nanogel (Liu et al. 2022b) for the antibacterial activity against *Staphylococcus aureus* allowed to obtain excellent mechanical properties and stability of nanoparticles and high uniformity of sizes (229.4 ± 1.5 nm). Enclosing the drug in a nanocapsule allowed for prolonged release and directional delivery of the drug to the site infected with bacteria. Analysis of prolonged and directional drug release under *in vitro* conditions was based on the determination of the relationship between the release and the pH of the environment. Based on the knowledge so far, it was assumed that the pH of the site infected with SCV is about 5.5, which is the cause of anaerobic fermentation and, consequently, inflammation. Comparison of the release rate of valnemulin hydrochloride from nanogels in a pH 5.5 environment with that of a pH 7.4 environment showed that about 61% ± 1.8% was released in an acidic environment after 48 h, compared to 56% of the drug released in a neutral pH environment. In addition, nanogels were characterised by a strong antibacterial effect and high therapeutic effectiveness.

The inverse microemulsion technique has been successfully used to prepare carrageenan-based gels. Daniel-da-Silva et al. (2011) combined the water-in-oil microemulsion method with thermally induced gelling. Thanks to this method, it is possible to change the size of nanoparticles by manipulating the concentration of water, surfactant or polymer. However, the use of a high surfactant concentration is often undesirable, especially in potential biomedical applications. In this study, they managed to produce stable nanoparticles with a size of about 100 nm at a constant concentration of the surfactant by selecting the appropriate concentration of the polymer (1% and 4% by weight). The use of a thermosensitive polymer such as carrageenan, also allowed to obtain thermosensitive nanospheres in physiological conditions (37–45 °C) and undergoing reversible changes in response to thermal stimuli.

The phenomenon of reducing the size of polymer nanoparticles as a result of increasing the concentration of the surfactant was used by Rodriguez et al. (2020). They used the sonication assisted microemulsion method and consequently obtained carrageenan nanoparticles with a size of 400–600 nm. In turn, Rahmani et al. (2021) synthesised nanogels based on carrageenan and chitosan as a carrier for the drug—rivastagmine. For the synthesis of polymer nanogels, they used the method of one-way copolymerisation with free radical precipitation. Thanks to the use of these methods, the nanocapsules were characterised by a narrow size range (18–22 nm) and sensitivity to changes in pH conditions, thanks to which the drug release rate could be controlled by changing the pH of the environment and accelerated when changing the pH from 1.2 to 7.4.

Another polymer derived from seaweed used to create nanocapsules is furcellaran. It was applied to trap the drug doxorubicin (Milosavljevic et al. 2020), and also as a core for the formation of nanocapsules in a complex with a protein—bovine albumin (Jamróz et al. 2014). In this research, it was shown, as in the case of alginate and carrageenan, that the average size of polymer nanocapsules depends on its concentration. Furcellaran is a very good polymer for the formation of protein complexes due to its anionic nature (membrane potential −48 mV to −27 mV at ionic strength of 0.05 mol/dm^3 and 0.15 mol/dm^3, respectively). Complexes are formed as a result of electrostatic interactions between a negatively charged polymer and a cationic protein. Jamróz et al. (2014) used the anionic nature of the polysaccharide not only to form a complex, but mainly as a core for the formation of nanocapsules using the layer-by-layer method, developing an innovative method of creating polymer-protein capsules. The negatively charged core was covered with polydiallyl dimethyl ammonium chloride—a cationic polyelectrolyte, achieving a zeta potential of +50 mV. Then, poly(sodium-4-styrenesulfonate) was applied to this layer, and the procedure was repeated in order to obtain the desired number of layers. As a consequence, it was possible to obtain nanocapsules of the polymer-protein complex stable for at least 10 days with a size range of 60–80 nm. The same method was used by Milosavljevic et al. (2020) to enclose an anticancer drug in a polymer capsule of the furcellaran-chitosan complex. The synthesised capsules showed excellent compatibility and non-toxicity to eukaryotic cells. Against neoplastic cells, they demonstrated great internal penetration ability and, consequently, induced apoptase. The use of polymers for encapsulation allowed to obtain a non-toxic biomaterial that is a drug carrier, enabling directional action and controlled release. A very high efficiency of encapsulation of the active ingredient (97%) and target sensitivity to changes in the pH of the environment were obtained, thus, creating an alternative, simple method for the directional, efficient and selective delivery of bioactive compounds.

7 Conclusions and Future Trends

Due to its unique properties, such as biodegradability, non-toxicity, biocompatibility, hydrophilic nature or the ability to control properties by changing environmental conditions, nano- and micro-capsules of polysaccharides derived from marine algae are becoming more and more popular. The encapsulation of biologically-active agents or drugs inside capsules allows for the creation of innovative, safe and directional methods of treatment or for obtaining a complete modification of the properties of compounds known to us so far. The sensitivity of these polysaccharides to changes in environmental conditions initiates many new possibilities and potential applications that will certainly be a further area of intensive research.

However, there are still a few steps that need to be taken to get the most out of algae and their polysaccharides in the encapsulation process. First of all, the process of extracting polysaccharides from marine algae should be optimised, whereby, the amount of chemicals used in this process will be minimised. Achieving the best use of chemicals while reducing waste-water production is currently the greatest challenge when extracting polysaccharides from seaweed. The next step will be the commercial use of marine algae polysaccharides as different applications can be found for each type of extracted sugar. The encapsulation process is one issue, while the other is the impact of nanocapsules on human health and the environment. Detailed research is needed to determine the safe use of this type of material before allowing its commercial use.

Acknowledgements This work was supported by the National Centre for Research and Development, Poland [Grant No.: LIDER/6/0016/L-11/19/NCBR/2020]. This work was also supported by the Scientific Research Project Office of Cukurova University under contract no: FBA-2022-15153.

References

Abraham RE et al (2021) Release of encapsulated bioactives influenced by alginate viscosity under in-vitro gastrointestinal model. Int J Biol Macromol 170:540–548

Adibkia K, Yaqoubi S, Dizaj SM (2017) Pharmaceutical and medical applications of nanofibers. In: Keservani RK, Sharma AK, Kesharwani RK (eds) Novel approaches for drug delivery. IGI Global, Hershey, pp 338–363

Agarwal T et al (2015) Calcium alginate-carboxymethyl cellulose beads for colon-targeted drug delivery. Int J Biol Macromol 75:409–417

Aguero L et al (2021) Functional role of crosslinking in alginate scaffold for drug delivery and tissue engineering: a review. Eur Polym J 160:110807

Akbarbaglu Z et al (2019) Influence of spray drying encapsulation on the retention of antioxidant properties and microstructure of flaxseed protein hydrolysates. Colloids Surf B: Biointerfaces 178:421–429

Alaysuy O et al (2022) Development of green and sustainable smart biochromic and therapeutic bandage using red cabbage (Brassica oleracea L. Var. capitata) extract encapsulated into alginate nanoparticles. Int J Biol Macromol 211:390–399

Alkhatib H, Mohamed F, Doolaanea A (2020) Document details. J Drug Deliv Sci Technol 60:102030. ISSN 1773-2247. https://doi.org/10.1016/j.jddst.2020.102030

Almurisi SH et al (2020) Taste masking of paracetamol encapsulated in chitosan-coated alginate beads. J Drug Deliv Sci Technol 56:101520

Alvim ID et al (2016) Comparison between the spray drying and spray chilling microparticles contain ascorbic acid in a baked product application. LWT-Food Sci Technol 65:689–694

Anbinder PS et al (2011) Yerba mate extract encapsulation with alginate and chitosan systems: interactions between active compound encapsulation polymers. J Encapsul Adsorpt Sci 1:80–87

Arvizu-Higuera DL et al (2007) Effect of alkali treatment time and extraction time on agar from Gracilaria vermiculophylla. In: Nineteenth international seaweed symposium. Springer

Bakry AM et al (2016) Microencapsulation of oils: a comprehensive review of benefits, techniques, and applications. Compr Rev Food Sci Food Saf 15(1):143–182

Baltrusch KL et al (2022) Spray-drying microencapsulation of tea extracts using green starch, alginate or carrageenan as carrier materials. Int J Biol Macromol 203:417–429

Bardajee GR et al (2020) Multi-stimuli responsive nanogel/hydrogel nanocomposites based on κ-carrageenan for prolonged release of levodopa as model drug. Int J Biol Macromol 153:180–189

Basha SK et al (2021) Development of nanoemulsion of Alginate/Aloe vera for oral delivery of insulin. Mater Today: Proc 36:357–363

Benavides S et al (2016) Development of alginate microspheres containing thyme essential oil using ionic gelation. Food Chem 204:77–83

Boostani S, Jafari SM (2020) Controlled release of nanoencapsulated food ingredients. In: Release and bioavailability of nanoencapsulated food ingredients. Elsevier, pp 27–78

Caldorera-Moore M, Peppas NA (2009) Micro-and nanotechnologies for intelligent and responsive biomaterial-based medical systems. Adv Drug Deliv Rev 61(15):1391–1401

Campos E et al (2013) Designing polymeric microparticles for biomedical and industrial applications. Eur Polym J 49(8):2005–2021

Casoni AI et al (2020) Sustainable and economic analysis of marine macroalgae based chemicals production-process design and optimization. J Clean Prod 276:122792

Cassani L et al (2022) Thermochemical characterization of eight seaweed species and evaluation of their potential use as an alternative for biofuel production and source of bioactive compounds. Int J Mol Sci 23(4):2355

Chaumun M et al (2020) In vitro evaluation of microparticles with Laurus nobilis L. extract prepared by spray-drying for application in food and pharmaceutical products. Food Bioprod Process 122:124–135

Cheng M et al (2019) Characterization and application of the microencapsulated carvacrol/sodium alginate films as food packaging materials. Int J Biol Macromol 141:259–267

Choi I et al (2022) Characterization of ionically crosslinked alginate films: effect of different anion-based metal cations on the improvement of water-resistant properties. Food Hydrocoll 131:107785

Coghetto CC et al (2016) Viability and alternative uses of a dried powder, microencapsulated Lactobacillus plantarum without the use of cold chain or dairy products. LWT-Food Sci Technol 71:54–59

Comunian TA, Favaro-Trindade CS (2016) Microencapsulation using biopolymers as an alternative to produce food enhanced with phytosterols and omega-3 fatty acids: a review. Food Hydrocoll 61:442–457

Cotas J et al (2020) Seaweed phenolics: from extraction to applications. Mar Drugs 18(8):384

Craigie JS (2011) Seaweed extract stimuli in plant science and agriculture. J Appl Phycol 23(3):371–393

Dalby MJ et al (2013) Hydrogel nanoparticles for drug delivery. Nanomedicine (Lond) 8(11):1744–1745

Dalponte Dallabona I et al (2020) Development of alginate beads with encapsulated jabuticaba peel and propolis extracts to achieve a new natural colorant antioxidant additive. Int J Biol Macromol 163:1421–1432

Daniel-da-Silva AL et al (2011) Synthesis and swelling behavior of temperature responsive κ-carrageenan nanogels. J Colloid Interface Sci 355(2):512–517

De S, Robinson D (2003) Polymer relationships during preparation of chitosan–alginate and poly-l-lysine–alginate nanospheres. J Control Release 89(1):101–112

De Prisco A, Mauriello G (2016) Probiotication of foods: a focus on microencapsulation tool. Trends Food Sci Technol 48:27–39

Demajo JK et al (2016) Effectiveness of disinfectants on antimicrobial and physical properties of dental impression materials. Int J Prosthodont 29(1):63–67

Devi N et al (2017) Encapsulation of active ingredients in polysaccharide–protein complex coacervates. Adv Colloid Interface Sci 239:136–145

Dewi EN, Purnamayati L, Kurniasih RA (2016) Antioxidant activities of phycocyanin microcapsules using maltodextrin and carrageenan as coating materials. J Teknol 78(4–2). https://doi.org/10.11113/jt.v78.8151

Dewi IC et al (2018a) Anticancer, antiviral, antibacterial, and antifungal properties in microalgae. In: Microalgae in health and disease prevention. Elsevier, pp 235–261

Dewi E, Kurniasih R, Purnamayati L (2018b) The application of microencapsulated phycocyanin as a blue natural colorant to the quality of jelly candy. IOP Conf Ser: Earth Environ Sci 116:012047

Dias DR et al (2017) Encapsulation as a tool for bioprocessing of functional foods. Curr Opin Food Sci 13:31–37

Dillehay TD et al (2008) Monte Verde: seaweed, food, medicine, and the peopling of South America. Science 320(5877):784–786
Ding X et al (2022) Carboxymethyl konjac glucomannan-chitosan complex nanogels stabilized double emulsions incorporated into alginate hydrogel beads for the encapsulation, protection and delivery of probiotics. Carbohydr Polym 289:119438
Don T-M et al (2021) Crosslinked complex films based on chitosan and ulvan with antioxidant and whitening activities. Algal Res 58:102423
Don T-M et al (2022) Preparation and characterization of fast dissolving ulvan microneedles for transdermal drug delivery system. Int J Biol Macromol 207:90–99
Echeverria C et al (2018) Functional stimuli-responsive gels: hydrogels and microgels. Gels 4(2):54
El-Deeb NM et al (2022) Alginate/κ-carrageenan oral microcapsules loaded with Agaricus bisporus polysaccharides MH751906 for natural killer cells mediated colon cancer immunotherapy. Int J Biol Macromol 205:385–395
El-Feky GS et al (2017) Alginate coated chitosan nanogel for the controlled topical delivery of Silver sulfadiazine. Carbohydr Polym 177:194–202
Escobar-Puentes AA et al (2022) Encapsulation of probiotics. In: Probiotics. Elsevier, pp 185–208
Esfanjani AF, Assadpour E, Jafari SM (2018) Improving the bioavailability of phenolic compounds by loading them within lipid-based nanocarriers. Trends Food Sci Technol 76:56–66
Fabra MJ et al (2012) Influence of temperature and NaCl on the release in aqueous liquid media of aroma compounds encapsulated in edible films. J Food Eng 108(1):30–36
Fathordoobady F et al (2021) Encapsulation of betacyanins from the peel of red dragon fruit (Hylocereus polyrhizus L.) in alginate microbeads. Food Hydrocoll 113:106535
Fertahi S et al (2021) Recent trends in organic coating based on biopolymers and biomass for controlled and slow release fertilizers. J Control Release 330:341–361
Filipović-Vinceković N et al (2005) Phase behavior in mixtures of cationic surfactant and anionic polyelectrolytes. Colloids Surf A Physicochem Eng Asp 255(1):181–191
Goonoo N et al (2014) Naltrexone: a review of existing sustained drug delivery systems and emerging nano-based systems. J Control Release 183:154–166
Gu L et al (2021) Formulation of alginate/carrageenan microgels to encapsulate, protect and release immunoglobulins: egg yolk IgY. Food Hydrocoll 112:106349
Gupta C et al (2015) Iron microencapsulation with blend of gum arabic, maltodextrin and modified starch using modified solvent evaporation method–Milk fortification. Food Hydrocoll 43:622–628
Hambleton A et al (2008) Protection of active aroma compound against moisture and oxygen by encapsulation in biopolymeric emulsion-based edible films. Biomacromolecules 9(3):1058–1063
Hambleton A et al (2009a) Influence of alginate emulsion-based films structure on its barrier properties and on the protection of microencapsulated aroma compound. Food Hydrocoll 23(8):2116–2124
Hambleton A et al (2009b) Interface and aroma barrier properties of iota-carrageenan emulsion–based films used for encapsulation of active food compounds. J Food Eng 93(1):80–88
Hamdan MA et al (2021) Tuning mechanical properties of seaweeds for hard capsules: a step forward for a sustainable drug delivery medium. Food Hydrocoll Health 1:100023
Hassoun A, Çoban ÖE (2017) Essential oils for antimicrobial and antioxidant applications in fish and other seafood products. Trends Food Sci Technol 68:26–36
He H et al (2017) κ-Carrageenan/locust bean gum as hard capsule gelling agents. Carbohydr Polym 175:417–424
Hegde PS, Chandra T (2005) ESR spectroscopic study reveals higher free radical quenching potential in kodo millet (Paspalum scrobiculatum) compared to other millets. Food Chem 92(1):177–182
Jamróz E et al (2014) Albumin–furcellaran complexes as cores for nanoencapsulation. Colloids Surf A Physicochem Eng Asp 441:880–884

Jamróz E et al (2021a) The influence of lingonberry extract on the properties of novel, double-layered biopolymer films based on furcellaran, CMC and a gelatin hydrolysate. Food Hydrocoll 124:107334

Jamróz E et al (2021b) The effects of active double-layered furcellaran/gelatin hydrolysate film system with Ala-Tyr peptide on fresh Atlantic mackerel stored at −18 °C. Food Chem 338:127867

Jamróz E et al (2022) Utilisation of soybean post-production waste in single-and double-layered films based on furcellaran to obtain packaging materials for food products prone to oxidation. Food Chem 387:132883

Janes K, Calvo P, Alonso M (2001) Polysaccharide colloidal particles as delivery systems for macromolecules. Adv Drug Deliv Rev 47(1):83–97

Jung SM et al (2020) A new approach to the restoration of seaweed beds using Sargassum fulvellum. J Appl Phycol 32(4):2575–2581

Kang H-J et al (2013) Inhibitory effect of soy protein coating formulations on walnut (Juglans regia L.) kernels against lipid oxidation. LWT-Food Sci Technol 51(1):393–396

Kanmani P, Rhim J-W (2014) Development and characterization of carrageenan/grapefruit seed extract composite films for active packaging. Int J Biol Macromol 68:258–266

Karbarz M et al (2017) Recent developments in design and functionalization of micro-and nano-structural environmentally-sensitive hydrogels based on N-isopropylacrylamide. Appl Mater Today 9:516–532

Karbowiak T et al (2009) From macroscopic to molecular scale investigations of mass transfer of small molecules through edible packaging applied at interfaces of multiphase food products. Innov Food Sci Emerg Technol 10(1):116–127

Katouzian I et al (2017) Formulation and application of a new generation of lipid nano-carriers for the food bioactive ingredients. Trends Food Sci Technol 68:14–25

Kaushalya KGD, Gunathilake KDPP (2022) Encapsulation of phlorotannins from edible brown seaweed in chitosan: effect of fortification on bioactivity and stability in functional foods. Food Chem 377:132012

Kavoosi G et al (2018) Microencapsulation of zataria essential oil in agar, alginate and carrageenan. Innov Food Sci Emerg Technol 45:418–425

Keawchaoon L, Yoksan R (2011) Preparation, characterization and in vitro release study of carvacrol-loaded chitosan nanoparticles. Colloids Surf B: Biointerfaces 84(1):163–171

Khalil HA et al (2017) Seaweed based sustainable films and composites for food and pharmaceutical applications: a review. Renew Sustain Energy Rev 77:353–362

Khorrami NK et al (2021) Fabrication and characterization of alginate-based films functionalized with nanostructured lipid carriers. Int J Biol Macromol 182:373–384

Kim J et al (2021) Exfoliated bentonite/alginate nanocomposite hydrogel enhances intestinal delivery of probiotics by resistance to gastric pH and on-demand disintegration. Carbohydr Polym 272:118462

Kulawik P et al (2022) Biological activity of biopolymer edible furcellaran-chitosan coatings enhanced with bioactive peptides. Food Control 137:108933

Kurczewska J, Ratajczak M, Gajecka M (2021) Alginate and pectin films covering halloysite with encapsulated salicylic acid as food packaging components. Appl Clay Sci 214:106270

Kurniasih R, Dewi E, Purnamayati L (2018) Effect of different coating materials on the characteristics of chlorophyll microcapsules from Caulerpa racemosa. IOP Conf Ser: Earth Environ Sci 116:012030. IOP Publishing

Leong J-Y et al (2016) Advances in fabricating spherical alginate hydrogels with controlled particle designs by ionotropic gelation as encapsulation systems. Particuology 24:44–60

Levi S et al (2011) Limonene encapsulation in alginate/poly (vinyl alcohol). Proc Food Sci 1:1816–1820

Lim C et al (2021) Bioplastic made from seaweed polysaccharides with green production methods. J Environ Chem Eng 9(5):105895

Liu K et al (2022a) Co-encapsulation systems for delivery of bioactive ingredients. Food Res Int 155:111073

Liu J et al (2022b) Composite inclusion complexes containing sodium alginate composite nanogels for pH-responsive valnemulin hydrochloride release. J Mol Struct 1263:133054
Lourenço-Lopes C et al (2020) Metabolites from macroalgae and its applications in the cosmetic industry: a circular economy approach. Resources 9(9):101
Mahfoudhi N, Ksouri R, Hamdi S (2016) Nanoemulsions as potential delivery systems for bioactive compounds in food systems: preparation, characterization, and applications in food industry. In: Emulsions. Elsevier, pp 365–403
Mahmoudi Z et al (2020) Promoted chondrogenesis of hMCSs with controlled release of TGF-β3 via microfluidics synthesized alginate nanogels. Carbohydr Polym 229:115551
Martins JT et al (2013) Biocomposite films based on κ-carrageenan/locust bean gum blends and clays: physical and antimicrobial properties. Food Bioproc Tech 6(8):2081–2092
McClements DJ (2018) Encapsulation, protection, and delivery of bioactive proteins and peptides using nanoparticle and microparticle systems: a review. Adv Colloid Interface Sci 253:1–22
McHugh DJ (2003) A guide to the seaweed industry, vol 441. Food and Agriculture Organization of the United Nations Rome
Melo M et al (2002) Isolation and characterization of soluble sulfated polysaccharide from the red seaweed Gracilaria cornea. Carbohydr Polym 49(4):491–498
Milosavljevic V et al (2020) Encapsulation of doxorubicin in furcellaran/chitosan nanocapsules by layer-by-layer technique for selectively controlled drug delivery. Biomacromolecules 21(2):418–434
Mishra AP et al (2020) Combination of essential oils in dairy products: a review of their functions and potential benefits. LWT 133:110116
Mokhtari S, Jafari SM, Assadpour E (2017) Development of a nutraceutical nano-delivery system through emulsification/internal gelation of alginate. Food Chem 229:286–295
Motornov M et al (2010) Stimuli-responsive nanoparticles, nanogels and capsules for integrated multifunctional intelligent systems. Prog Polym Sci 35(1–2):174–211
Mukurumbira A et al (2022) Encapsulation of essential oils and their application in antimicrobial active packaging. Food Control 136:108883
Nakamura S et al (2008) Effect of controlled release of fibroblast growth factor-2 from chitosan/fucoidan micro complex-hydrogel on in vitro and in vivo vascularization. J Biomed Mater Res A 85(3):619–627
Navarro R et al (2016) Effect of type of encapsulating agent on physical properties of edible films based on alginate and thyme oil. Food Bioprod Process 97:63–75
Naveena B, Nagaraju M (2020) Microencapsulation techniques and its application in food industry. Int J Chem Stud 8(1):2560–2563
Nkurunziza D et al (2021) Effect of wall materials on the spray drying encapsulation of brown seaweed bioactive compounds obtained by subcritical water extraction. Algal Res 58:102381
Noor A et al (2022) Alginate based encapsulation of polyphenols of Piper betel leaves: development, stability, bio-accessibility and biological activities. Food Biosci 47:101715
Norcino LB et al (2022) Development of alginate/pectin microcapsules by a dual process combining emulsification and ultrasonic gelation for encapsulation and controlled release of anthocyanins from grapes (Vitis labrusca L.). Food Chem 391:133256
Paucar OC et al (2016) Production by spray chilling and characterization of solid lipid microparticles loaded with vitamin D3. Food Bioprod Process 100:344–350
Pei M et al (2018) Alginate-based cancer-associated, stimuli-driven and turn-on theranostic prodrug nanogel for cancer detection and treatment. Carbohydr Polym 183:131–139
Pitigraisorn P et al (2017) Encapsulation of Lactobacillus acidophilus in moist-heat-resistant multilayered microcapsules. J Food Eng 192:11–18
Podgórna K et al (2017) Gadolinium alginate nanogels for theranostic applications. Colloids Surf B: Biointerfaces 153:183–189
Prajapati VD et al (2014) RETRACTED: carrageenan: a natural seaweed polysaccharide and its applications. Carbohydr Polym 105:97–112

Pranata B et al (2022) Microencapsulation of umami flavor enhancer from Indonesian waters brown seaweed. Curr Res Nutr Food Sci J 10(1):349–359

Premjit Y et al (2022) Current trends in flavor encapsulation: a comprehensive review of emerging encapsulation techniques, flavour release, and mathematical modelling. Food Res Int 151:110879

Qiu S-M et al (2022) Bioactive polysaccharides from red seaweed as potent food supplements: a systematic review of their extraction, purification, and biological activities. Carbohydr Polym 275:118696

Qureshi MA, Khatoon F (2019) Different types of smart nanogel for targeted delivery. J Sci Adv Mater Devices 4(2):201–212

Rahmani Z, Ghaemy M, Olad A (2021) Preparation of nanogels based on kappa-carrageenan/chitosan and N-doped carbon dots: study of drug delivery behavior. Polym Bull 78(5):2709–2726

Rawiwan P et al (2022) Red seaweed: a promising alternative protein source for global food sustainability. Trends Food Sci Technol 123:37–56

Reid G (2016) Probiotics: definition, scope and mechanisms of action. Best Pract Res Clin Gastroenterol 30(1):17–25

Ren Y et al (2022) hUC-MSCs lyophilized powder loaded polysaccharide ulvan driven functional hydrogel for chronic diabetic wound healing. Carbohydr Polym 288:119404

Rhim J-W (2011) Effect of clay contents on mechanical and water vapor barrier properties of agar-based nanocomposite films. Carbohydr Polym 86(2):691–699

Riahi Z et al (2022) Alginate-based multifunctional films incorporated with sulfur quantum dots for active packaging applications. Colloids Surf B: Biointerfaces 215:112519

Riquelme N, Herrera ML, Matiacevich S (2017) Active films based on alginate containing lemongrass essential oil encapsulated: effect of process and storage conditions. Food Bioprod Process 104:94–103

Rodriguez S et al (2020) Synthesis of highly stable κ/ι-hybrid carrageenan micro-and nanogels via a sonication-assisted microemulsion route. Polym Renew Resour 11(3–4):69–82

Sabliov C, Astete C (2008) Encapsulation and controlled release of antioxidants and vitamins. In: Delivery and controlled release of bioactives in foods and nutraceuticals. Elsevier, pp 297–330

Saha D (2010) Hydrocolloids as thickening and gelling agents in food: a critical review. J Food Sci Tech Mys 47:587–597. https://doi.org/10.1007/s13197-010-0162-6

Sanchez-Ballester NM, Bataille B, Soulairol I (2021) Sodium alginate and alginic acid as pharmaceutical excipients for tablet formulation: structure-function relationship. Carbohydr Polym 270:118399

Sarika P, James NR, Raj DK (2016) Preparation, characterization and biological evaluation of curcumin loaded alginate aldehyde–gelatin nanogels. Mater Sci Eng C 68:251–257

Sarıyer S et al (2020) pH-responsive double network alginate/kappa-carrageenan hydrogel beads for controlled protein release: effect of pH and crosslinking agent. J Drug Deliv Sci Technol 56:101551

Saurabh CK et al (2013) Radiation dose dependent change in physiochemical, mechanical and barrier properties of guar gum based films. Carbohydr Polym 98(2):1610–1617

Schmidt ÉC et al (2010) Effects of UVB radiation on the agarophyte Gracilaria domingensis (Rhodophyta, Gracilariales): changes in cell organization, growth and photosynthetic performance. Micron 41(8):919–930

Schmitt F et al (2010) Chitosan-based nanogels for selective delivery of photosensitizers to macrophages and improved retention in and therapy of articular joints. J Control Release 144(2):242–250

Shen P et al (2018) Bioactive seaweeds for food applications. Academic Press, Cambridge

Shofia SI et al (2018) Efficiency of brown seaweed (Sargassum longifolium) polysaccharides encapsulated in nanoemulsion and nanostructured lipid carrier against colon cancer cell lines HCT 116. RSC Adv 8(29):15973–15984

Silva Batalha L et al (2021) Encapsulation in alginate-polymers improves stability and allows controlled release of the UFV-AREG1 bacteriophage. Food Res Int 139:109947

Sivakumaran D et al (2013) Tuning drug release from smart microgel–hydrogel composites via cross-linking. J Colloid Interface Sci 392:422–430
Smaoui S et al (2021) Recent advancements in encapsulation of bioactive compounds as a promising technique for meat preservation. Meat Sci 181:108585
Stirk WA, van Staden J (2022) Bioprospecting for bioactive compounds in microalgae: antimicrobial compounds. Biotechnol Adv 59:107977
Sumarni N et al (2019) Microcapsule efficiency of ethanol extract of rosella petal flower (hibiscus sabdariffa linn) coated crude carrageenan (Eucheuma cottony). J Phys: Conf Ser 1280(2):022074. IOP Publishing
Sun Z et al (2017) Bio-responsive alginate-keratin composite nanogels with enhanced drug loading efficiency for cancer therapy. Carbohydr Polym 175:159–169
Sun X, Cameron RG, Bai J (2019) Microencapsulation and antimicrobial activity of carvacrol in a pectin-alginate matrix. Food Hydrocoll 92:69–73
Toprakçı İ, Şahin S (2022) Encapsulation of olive leaf antioxidants in microbeads: application of alginate and chitosan as wall materials. Sustain Chem Pharm 27:100707
Valor D et al (2021) Supercritical solvent impregnation of alginate wound dressings with mango leaves extract. J Supercrit Fluids 178:105357
Van Vlierberghe S, Dubruel P, Schacht E (2011) Biopolymer-based hydrogels as scaffolds for tissue engineering applications: a review. Biomacromolecules 12(5):1387–1408
Wang Y, Chen L (2014) Cellulose nanowhiskers and fiber alignment greatly improve mechanical properties of electrospun prolamin protein fibers. ACS Appl Mater Interfaces 6(3):1709–1718
Wang Y-Z, Zhang X-H, Zhang J-X (2013) New insight into the kinetic behavior of the structural formation process in agar gelation. Rheol Acta 52(1):39–48
Wang X et al (2017) Isolation of fucoxanthin from Sargassum thunbergii and preparation of microcapsules based on palm stearin solid lipid core. Front Mater Sci 11(1):66–74
Wei X et al (2018) Ion-assisted fabrication of neutral protein crosslinked sodium alginate nanogels. Carbohydr Polym 186:45–53
Wongsasulak S, Pathumban S, Yoovidhya T (2014) Effect of entrapped α-tocopherol on mucoadhesivity and evaluation of the release, degradation, and swelling characteristics of zein–chitosan composite electrospun fibers. J Food Eng 120:110–117
Wu H et al (2015) Microcapsule preparation of allyl isothiocyanate and its application on mature green tomato preservation. Food Chem 175:344–349
Xi J, Zhou L, Dai H (2012) Drug-loaded chondroitin sulfate-based nanogels: preparation and characterization. Colloids Surf B: Biointerfaces 100:107–115
Xiao Z et al (2014) A review of the preparation and application of flavour and essential oils microcapsules based on complex coacervation technology. J Sci Food Agric 94(8):1482–1494
Xu X et al (2018) Formation of graphene oxide-hybridized nanogels for combinative anticancer therapy. Nanomedicine 14(7):2387–2395
Xue M et al (2021) Neuroprotective effect of fucoidan by regulating gut-microbiota-brain axis in alcohol withdrawal mice. J Funct Foods 86:104726
Ye S et al (2005) Deposition temperature effect on release rate of indomethacin microcrystals from microcapsules of layer-by-layer assembled chitosan and alginate multilayer films. J Control Release 106(3):319–328
Yuan Y et al (2022) Effect of calcium ions on the freeze-drying survival of probiotic encapsulated in sodium alginate. Food Hydrocoll 130:107668
Zhang Z et al (2016) Encapsulation of lactase (β-galactosidase) into κ-carrageenan-based hydrogel beads: impact of environmental conditions on enzyme activity. Food Chem 200:69–75
Zhao F-Q et al (2021) Glycyrrhizin mediated liver-targeted alginate nanogels delivers quercetin to relieve acute liver failure. Int J Biol Macromol 168:93–104
Zhou M et al (2016) Effects of different polysaccharides on the formation of egg yolk LDL complex nanogels for nutrient delivery. Carbohydr Polym 153:336–344
Zvyagintseva TN et al (2021) Structural diversity of fucoidans and their radioprotective effect. Carbohydr Polym 273:118551

Chapter 17
State of the World's Commercially Seaweeds Genetic Resources for Food and Feeds

Stefanie Verstringe, Robin Vandercruyssen, Hannes Carmans, Monica Trif, Geert Bruggeman, and Alexandru Vasile Rusu

1 Introduction

Seaweeds, also known as marine macroalgae, have gained significant attention as a valuable genetic resource for food and other applications. They are a diverse group of photosynthetic organisms found in marine environments worldwide and have been used for centuries in various cultures as a food source and for medicinal purposes. In recent years, there has been a growing interest in the commercial cultivation and utilization of seaweeds due to their nutritional value, environmental benefits, and potential for industrial applications (Leandro et al. 2020; Cai et al. 2021).

Genetic resources refer to the heritable materials, such as genes and genetic information, found in plants, animals, and microorganisms. In the context of seaweeds, genetic resources include the genetic diversity within different species, strains, and populations of seaweeds. This genetic diversity is important because it provides a wide range of traits and characteristics that can be used for various purposes, including food production (Lourenço-Lopes et al. 2020).

Commercially, seaweeds are primarily cultivated for their use as food and food ingredients. They are a rich source of essential nutrients, such as vitamins, minerals, and dietary fiber. Seaweeds also contain unique compounds, including

S. Verstringe (✉) · R. Vandercruyssen · H. Carmans · G. Bruggeman
Nutrition Sciences N.V., Drongen, Belgium
e-mail: stefanie.verstringe@nusciencegroup.com

M. Trif
Food Research Department, Centre for Innovative Process Engineering GmbH, Stuhr, Germany

A. V. Rusu
Strategic Research Department, Biozoon Food Innovations GmbH, Bremerhaven, Germany

F. Ozogul et al. (eds.), *Seaweeds and Seaweed-Derived Compounds*,
https://doi.org/10.1007/978-3-031-65529-6_17

polysaccharides, proteins, and bioactive compounds, which have potential health benefits and functional properties (Choudhary et al. 2021; Polat et al. 2021).

By exploring the genetic diversity of seaweeds, researchers and breeders can identify strains or varieties with desirable traits, such as improved nutritional composition, taste, texture, and growth characteristics. These genetic resources can be used to develop new seaweed cultivars through selective breeding or genetic engineering techniques. For example, breeders may aim to develop seaweeds with higher protein content, reduced saltiness, or improved growth rates to meet specific market demands (Barbier et al. 2020; Morais et al. 2020; Verstringe et al. 2023).

In addition to food production, seaweed genetic resources have applications in various other sectors. They can be used for the production of biofuels, bioplastics, and pharmaceuticals (Polat et al. 2021). Seaweeds also have the potential to be used in environmental management, such as bio-remediation and carbon sequestration, due to their ability to absorb and utilize nutrients from their surroundings.

To ensure the sustainable and responsible use of seaweed genetic resources, it is important to have proper management and conservation strategies in place. This includes maintaining biodiversity, preventing the loss of unique genetic material, and respecting the rights of local communities who rely on seaweeds for their livelihoods. International agreements and conventions, such as the Nagoya Protocol on Access to Genetic Resources and the Fair and Equitable Sharing of Benefits Arising from their Utilization, provide a framework for the fair and equitable use of genetic resources.

Seaweeds offer immense potential as a nutritious food source and sustainable feed ingredient. Ongoing research and development in this field aim to explore new applications, improve cultivation techniques, and promote the integration of seaweeds into our diets and agricultural systems. Seaweeds are highly nutritious and rich in essential minerals, vitamins, dietary fibers, and bioactive compounds. They are particularly abundant in iodine, calcium, iron, and vitamin K. Additionally, they are low in calories and fat, making them a healthy food choice. Their high mineral content promotes bone health and helps prevent iron deficiency. They also contain antioxidants and anti-inflammatory compounds, which may have positive effects on cardiovascular health, immune function, and cancer prevention. Some seaweed species, like kelp, are a good source of dietary fiber, aiding digestion.

The current use of macroalgae is mainly for human consumption products (82%) followed by pharmaceuticals and cosmetics (12.2%). Animal feed accounts only for 2.9% of the current use (Ferdouse et al. 2017). However, global demand for seaweed is on the rise as its applications in the feed and food industry show great potential. Seaweed sales are rising each year and are predicted to rise to around 10 billion US dollars by 2024 (McCullough 2019). Although Asia produces the vast majority of the total seaweed volume, most companies producing commercial feed products are located in the Western world (Ferdouse et al. 2017).

Seaweeds are also used as a feed ingredient for livestock, aquaculture, and pets. Seaweed or marine macroalgae are considered as promising sustainable alternatives to conventional feed resources since they have a high growth rate and could be cultivated in saltwater, without the need of arable land (Øverland et al. 2019). Seaweeds can enhance the nutritional profile of animal feed, improve gut health, and reduce the environmental impact of livestock and aquaculture production. Macroalgae are an excellent source of vitamins and nutritionally important minerals, including I, Zn, Na, Ca, Mn and Fe, due to their ability to absorb inorganic compounds (Øverland et al. 2019). Seaweed have also a high fibre and antioxidant content, which, together with essential amino acids and omega-3-fatty acids, make the biomass a suitable food and feed source (Miyashita et al. 2013). The content of protein is highly variable, but various seaweed species contain a sufficient protein level for human and animal consumption (Morais et al. 2020). Seaweeds are rich in polysaccharides and complex carbohydrates, such as laminarin, fucoidans, carrageenans and ulvans, which are associated with the prebiotic actions. These bioactive components have been reported to possess a variety of biological and health promoting properties (Evans and Critchley 2014; Øverland et al. 2019; Miyashita et al. 2013). Their antimicrobial and immunomodulating properties make them ideal candidates that could substitute the in-feed antibiotics (Morais et al. 2020; Øverland et al. 2019). However, more studies are required to assess the beneficial effects of bioactive compounds and may facilitate increased commercial use of macroalgal products as feed ingredients (Makkar et al. 2015; Øverland et al. 2019).

The composition of proteins, minerals, lipids and fibers differs greatly and depends on a number of factors such as the species, harvest time, geographical location and on external conditions such as water, temperature, light intensity, salinity and nutrient concentration in water (Øverland et al. 2019; Morais et al. 2020). Over 50% of the most common seaweed species have a higher proportion of total amino acids than maize or rice (Mæhre et al. 2014). However, the main limitation for the use of seaweed in animal feed is the unfavorable essential amino acid profile (EAAs), which makes seaweed less suitable as a protein source, especially for monogastric animals (Mæhre et al. 2014). When consumed alone, macroalgae can have a detrimental impact on animals and are thus foremost used as additives and alternative ingredient in feed (Morais et al. 2020). Protein quality analysis and EAA concentration measurements are essential for determining the nutritional value of the algae biomass. This analysis is necessary for the next processing steps as seaweed can be used fresh, dried, liquefied or cooked (Morais et al. 2020; Mæhre et al. 2014).

Macroalgae are a diverse group of marine organisms and could be classified into brown (Phaeophyta), red (Rhodophyta) and green (Chlorophyta) algae. Brown seaweed species are of lesser nutritional value than green or red seaweed due to their low protein and high mineral content. However, brown algae have a higher diversified content on bioactive molecules with high commercial interest (Morais et al. 2020). Most commercial feed products are derived of green seaweed, given that green seaweed species contain a good protein content and some important bioactive compounds. Currently, red seaweed species have gained a renewed interest since recent studies have demonstrated that some red seaweed species are capable of reducing enteric methane production of ruminants (Kandasamy et al. 2012; Kinley et al. 2020).

2 Tradition of Seaweeds Being Used as Food and Feed

The tradition of using seaweeds as food dates back thousands of years and is prevalent in many coastal cultures around the world (Araújo et al. 2021). They come in various colors, shapes, and sizes and offer a wide range of nutritional and culinary benefits.

Seaweeds are used in various cuisines for centuries, particularly in East Asian countries like Japan, China, and Korea. The consumption of seaweeds is associated with several health benefits. Some examples of cultures with a tradition of using seaweeds as food:

- East Asia: Seaweeds have been a staple in the diets of several East Asian countries, including Japan, China, and Korea, for centuries. In Japan, seaweeds such as nori (used in sushi), wakame (used in miso soup), and kombu (used in broths and stews) are widely consumed. These seaweeds are not only rich in minerals and vitamins but also add unique flavors and textures to dishes.
- North Atlantic: In countries like Ireland, Scotland, and Iceland, there is a long-standing tradition of using seaweeds in traditional dishes. In Ireland, for instance, dulse (*Palmaria palmata*) has been harvested and eaten for centuries, either dried or as an ingredient in soups, stews, and salads. Carrageen moss (*Chondrus crispus*) is also used to make a traditional Irish dessert known as "carrageen pudding."
- Polynesia: In Polynesian cultures, seaweeds, known as limu, have played an important role as a food source. In Hawaii, limu is used in various dishes, including poke (a raw fish salad), laulau (a traditional dish wrapped in ti leaves), and lomi-lomi salmon (a side dish with tomatoes and onions).
- Scandinavia: In coastal regions of Scandinavia, seaweeds have been historically used as a food source. For example, in Norway, kelp (*Laminaria digitata*) has been harvested and consumed for centuries, often boiled and served as a side dish with fish or as an ingredient in traditional stews.
- Indigenous cultures: Many indigenous communities around the world, particularly those living near coastal areas, have traditionally harvested and consumed seaweeds. For instance, the Maori people of New Zealand have a long-standing tradition of using seaweeds in their cuisine, including the use of karengo (a type of red seaweed) as a seasoning or ingredient in traditional dishes.

Seaweeds are increasingly being incorporated into modern cuisine, innovative food products, and even sustainable farming practices. They can be used fresh, dried, or processed into various forms such as sheets, flakes, powders, or extracts. Seaweeds are commonly used in soups, salads, sushi, wraps, and as a seasoning or garnish.

Seaweeds or marine macroalgae have a long tradition of being used as feed or feed supplements for animals. Ancient Greek sagas writings from 45 BC report that when feed was scarce, seaweed from the shore would be collected and washed where after it was fed to the cattle (Heuzé et al. 2017; Morais et al. 2020; Evans and Critchley 2014). The use of seaweeds for human applications goes back even further than this since supplementation of macroalgae in animal feed is associated with the domestication of animals (Evans and Critchley 2014). Whenever accessible, seaweed species were undoubtedly used to supplement both human and their domesticated animal's nutritional sources (Abowei and Tawari 2011).

The tradition of seaweed as food or feed source could also be traced by the origin of the term "kelp". This term would refer to any large brown seaweed and originated from the fourteenth century (Evans and Critchley 2014). The term was also used to refer to the burnt ash of various brown seaweeds, primarily *Laminaria* and *Ascophyllum,* and was used as a source of soda and iodine. Kelp, today, generally refers to dried marine plants of any species and "kelp meal" is used in the animal feed industry, denoting the dried, ground seaweed meal of the mixed or separated brown seaweeds (Evans and Critchley 2014).

In the nineteenth and early twentieth century there were numerous reports of seaweeds or kelp being part of the livestock diet. Macroalgae were commonly used to feed cattle during hard winter times due to limited feed supplies (Chapman and Chapman 1980; Heuzé et al. 2017). In the province of Prince Edward Island, Canada, horses would be well fed during winter, while for livestock it would be a matter of survival. In Iceland, seaweeds were grazed by sheep near beaches and would also be fed to horses, cattle and sheep during long periods of fodder scarcity. According to numerous reports, seaweed would be collected, dried and stored in barns so it could be used ad feedstuff for winter. In other regions in Europe (Scotland, France, Scandinavia), seaweeds were often, occasionally or systematically, fed to livestock, especially ruminants and pigs (Evans and Critchley 2014; Chapman and Chapman 1980). Due to the lack of knowledge about protein requirements and nutrition, these practices were common during the nineteenth century. Early scientific studies in the twentieth century, however, showed that the nutritive value of seaweeds was too poor for animal feed, especially at high inclusion rates (Evans and Critchley 2014; Heuzé et al. 2017). Even though scientific and nutritional reports in the twentieth century were conflicting, animal breeders maintained confidence in the inclusion of kelp meal in their animal rations. This was most likely because they observed benefits despite the academic trials being unable to measure significant differences (Evans and Critchley 2014).

The misconception that seaweed had no significant feed value for animals, except as a source of iodine or other trace minerals, was debunked in the 1970s, when it was discovered that chelated micromineral sources were more efficient for the delivery of microelements than conventional inorganic sources (Evans and Critchley 2014). In addition, recent research has demonstrated that supplementation of macroalgae is connected to a potent prebiotic action of its complex carbohydrates. This led to a whole range of benefits to animal health as well as productivity, and will undoubtedly renew considerable interest in the use of macroalgae in commercial animal agriculture (Evans and Critchley 2014).

3 Seaweed Exploitation for Food and Feed

3.1 In Europe

Although macroalgae aquaculture (land-based and at sea) is developing in several countries in Europe, seaweed used for animal feeding appears to result mostly from harvesting from wild stocks (Heuzé et al. 2017). Today, 68% of the macroalgae producing units comes from harvesting from wild stocks with Spain, France and Ireland being the countries with the highest number of seaweed harvesting companies (Dos Santos Fernandes De Araujo et al. 2021). The processing industries are always located close to production areas and are a major driver in the persistence of seaweed production (netalgae 2012). Although Europe has a long tradition of excellent research in phycology, it has little experience in commercial seaweed cultivation, representing only 32% of the macroalgae production units. The reason being that the cost of commercial cultivation in the western world exceeds the sales price making a profitable operation a challenge (Fernand et al. 2017; Dos Santos Fernandes De Araujo et al. 2021).

Commercial harvesting of seaweeds for animal feed applications began in Europe at the beginning of the twentieth century (Evans and Critchley 2014). The most common species used in animal feed applications is undoubtedly *Ascophyllum nodosum* because of its large size, abundance, and easy accessibility (Evans and Critchley 2014; Øverland et al. 2019). It is commonly known as “Rockweed”, and is found in the cold waters of the North Atlantic Ocean (Mohi El-Din et al. 2008).

Laminaria hyperborea, Laminaria digitata are other brown seaweed species that are now used in animal rations (Øverland et al. 2019; Heuzé et al. 2017; Makkar et al. 2015). These brown seaweed species often form forests in their habitat and are considered among the most ecologically dynamic and biologically diverse habitats on the planet (netalgae 2012). Depending on the location, stringent regulations are applied on the harvest in order to protect the marine eco-environment and it is widely recognized that a transition from wild stock seaweed harvesting to aquaculture is needed to avoid the overexploitation of wild seaweed resources. Seaweed aquaculture companies in Europe have already been established at a fully commercial scale, with Norway the country with the highest number of seaweed aquaculture companies. In Spain, the development of commercial-scale aquaculture of edible kelp species has been receiving increasing interest since the beginning of the 2000s. Other countries like France, Denmark and Portugal have already several sea-based (coastal) or land-based production facilities (Dos Santos Fernandes De Araujo et al. 2021).

Currently, algae production in Europe remains limited by a series of technological, regulatory and market-related barriers. In addition, significant knowledge gaps still exist regarding the dimension, capability, organization and structure of the algae production in Europe (Dos Santos Fernandes De Araujo et al. 2021). However, the European Union recognizes the potential of the algae biomass value chain and aims to support the sustainable growth and development of the algae production as soon as the acknowledged economic, social and environmental challenges are addressed (Dos Santos Fernandes De Araujo et al. 2021).

3.2 In Asia

Around 96% of the global macroalgal production today is cultivated, while only 4% is wild harvested. In Asia, seaweeds are usually cultivated, which results in far greater seawead production than from collection (Heuzé et al. 2017; Makkar et al. 2015; Ferdouse et al. 2017). The largest seaweed-producing countries around the world are China, Indonesia, the Philippines, and Korea and account in total for more than 95% of the total seaweed production volume (Ferdouse et al. 2017).

The kelp industrial processing industry in China started from 1950s mainly for being consumed as food and producing iodine. However, due to water pollution and environmental regulation many industries shut down, with only some of them surviving. Today, most seaweed is exploited for their high alginate content and more and more cultivated kelp gradually ends up as feed, mostly for abalone (*Haliotis* sp.) and sea cucumber aquaculture (Zhang 2018). The Chinese coastal provinces Shandong and Fujian offer ideal off-shore locations for seaweed farming. Abalone is farmed in kelp farms as a polyculture. The cultivated kelp is then harvested during well-defined months, depending on the climate of the region. *Saccharina* species are primarily harvested as food, but are also regularly used as feed for abalones in Asia. Driven by increasing demand for kelp feed from abalone farmers, the area used for farming *Saccharina japonica* increased significantly the last decennia (Hwang and Park 2020). Other seaweeds such as Wakame (*Undaria* spp), Hijiki (*Sargassumm* spp), JiangLi (*Gracilaria* spp), etc., are also used due to shortage of kelp supply (Zhang 2018). Among the *Undaria* spp is the Asian laminarian kelp (*Undaria pinnatifida*), which is native to Japan, Korea, and China, with a mild flavor and is commonly used in Japanese cuisine. It is rich in minerals, vitamins, and dietary fiber and is also known for its potential health benefits. The latter species is an important edible seaweed crop and is also cultivated extensively to feed juvenile abalone (*Haliotis* sp.). However, the natural growth period of this species is not long enough to cover the period for which fresh seaweed is required for feeding abalone. *Ecklonia stolonifera* and *E. cava* are perennial algae and continue to grow and produce valuable biomass. In Korea, they are used for abalone feed, especially during summer months, when kelp is not available (Hwang and Park 2020).

Despite having commercial applications in the food and feed industry, *Undaria pinnatifida* is seen as highly invasive. With the increase of marine farming in the 1970s, this species has been spread all over the world, having a negative effect on native marine flora and fauna (Lee 2001; Hwang and Park 2020).

Nonetheless, mariculture is regarded as the only option to supply the increasing demands for seaweeds in a sustainable manner. Since the 1970s, technologies for culturing a range of seaweed species have been developed successively in Asia. Today, the aquaculture production in Asia, and especially China, is state-of-the-art, which should inspire the future farming initiatives in the western world (Ferdouse et al. 2017).

3.3 The Rest of the World

South-Africa is believed to have a commercial seaweed industry that dates back to the 1950s, today however, their seaweed resources are considered underexploited. The country mostly collects seaweeds from wild stocks rather than harvesting from aquaculture. However, commercial seaweed species such as *Ulva* and *Gracilaria* are grown in land ponds, raceways or along the coast line, and are predominantly used as feed source for the abalone (*Haliotis midae* L.) farming (Amosu et al. 2013; Ferdouse et al. 2017). *Ecklonia maxima*, a genus of kelp, is harvested from wild in South-Africa and would also serve as abalone feed (Amosu et al. 2013; Ferdouse et al. 2017). Tanzania and Morocco are important seaweed producers on the African continent and lots of their seaweed products are exported, however, almost none have feed applications (Ferdouse et al. 2017).

In South-America, Chile has been the top producer of algal commodities. Nowadays, *Gracilaria chilensis* is the most important cultivated seaweed in Chile, mainly for its food applications. Gracilaria spp. are widely cultivated for their high agar content. Agar is a gelatinous substance extracted from seaweeds and is used in various industries, including food, pharmaceuticals, and cosmetics. *Macrocystis pyrifera* (Giant kelp) is one of the largest seaweeds and forms dense underwater forests along temperate coastlines, which could be used for abalone foraging, is another commercially important species and has picked up the interest of some Chilean companies (Ferdouse et al. 2017). This species is distributed mainly in the southern hemisphere (South America, Southern Africa, Australia and New Zealand) and in the North-East Pacific coast (Cruz-Suárez et al. 2000). It is commercially important for its biomass production and is used as a source of alginates, biofuels, and also harvested off the coasts of California (USA) for feeding abalone (Rodríguez-Montesinos and Hernández-Carmonal 1991; Guiry 2021). North-America is one of the main destinations for the seaweed trade, mainly for food and non-food products. Most companies manufacturing seaweed feed are located in Canada and mainly manufacture products derived from *Ascophylum nodosum* or *Laminaria* species. However, in recent years, seaweed farming has also taken off in the United States of America, with mariculture farms in New England, the Pacific Northwest, and Alaska (www.fisheries.noaa.gov/).

Australia has a renewed interest towards feed applications of seaweed since scientists discovered an *Asparagopsis* species, native to Australia, which significantly reduces the enteric methane emissions and has potential to increase livestock productivity (Kinley et al. 2020). An Australian startup (FutureFeed) is currently developing a value chain which aims to manufacture and supply feed to livestock producers.

4 Commercially Available Seaweed Products

The growing population urges the animal production industry to follow, leading to a greater demand for feed. This makes many look for feed alternatives that are sustainable, safe and affordable (Morais et al. 2020; Cherry et al. 2019). Marine macroalgae have a high potential in feed applications and some seaweed-based commercial feed products are already available.

Seaweeds or macroalgae are a diverse group of multicellular organisms and are broadly classified into brown (*Phaeophyta*), green (*Chlorophyta*) and red (*Rhodophyta*) algae (Morais et al. 2020). Like terrestrial plants, the chemical composition of macroalgae varies considerably between species and with season of harvest, growth habitat, and environmental conditions. Even within a small geographic area, growth rate and chemical composition depends on numerous factors, e.g. harvest season, sunlight, salinity, depth in the sea or local water currents (Øverland et al. 2019; Morais et al. 2020). Aside from their basic nutritional value, seaweeds contain a number of pigments, defensive and storage compounds, and secondary metabolites that could have beneficial effects on farmed fish (Wan et al. 2019).

Seaweed are rich in minerals and it has been suggested that bioactive components from macroalgae trigger a range of health-promoting effects (Øverland et al. 2019). Most of the bioactive compounds are complex polysaccharides such as ulvans and mannans from green, fucoidans and laminarans from brown and carrageenans from red seaweeds (Kulshreshtha et al. 2020). These polysaccharides have been associated with prebiotic, anti-bacterial, anti-inflammatory and antioxidant functionalities and are neither digested nor absorbed by the host, but serve as a substrate for bacterial fermentation in the colon (Kulshreshtha et al. 2020). Dietary supplementation of seaweeds to the diet of various livestock species, including pigs, cattle, poultry and marine animals, has shown beneficial effects on the immune status and intestinal health, often decreasing pathogens, often conferring an increase of animal performance (Øverland et al. 2019; Sobolewska et al. 2017).

An overview of some commercial feed products derived from seaweed is given in Table 17.1.

4.1 *Brown Seaweed Species*

Although having lower crude protein level in comparison to green and red seaweed, most commercial algae products in Europe originate from brown seaweed species (Table 17.1). This is probably due to their high content in bioactive compounds which have beneficial effects on animal health as well as performance (Makkar et al. 2015; Del Tuffo et al. 2019). Besides a high mineral content (Table 17.2), brown algae have the highest concentration iodine (around 1.2% of seaweed dry biomass) among the three seaweed classes and algae derived iodine has been used

Table 17.1 Commercial feed products derived from seaweed and their benefits

Company name	Commercial product	Product form/use	Seaweed species	Targeted animal(s)	Benefits	Reference(s)
Acadian Seaplants (Canada)	Tasco®	Air-dried supplement (seaweed meal)	*Ascophyllum nodosum* (Brown)	Livestock, horses, pets	Prebiotic effect Immune function support Stress reduction	Kandasamy et al. (2012) Evans and Critchley (2014)
AddiCan (Canada)	AquaArom®	Seaweed meal	*Laminaria* sp., kelp (Brown)	Fish	Increases palatability Prebiotic effect Improve growth performance	Kamunde et al. (2019)
Algea, The Arctic Company (Norway)	Algeafeed	Dried and grinded (seaweed meal)	*Ascophyllum nodosum* (Brown)	Cattle, fish, horses, poultry, pigs, pets	Healthy and balanced growth Prebiotic effect Improve performance	Novoa-Garrido et al. (2014, 2016)
Agri-Growth International Inc. (Canada)	NA	Dried and sized	*Ascophyllum nodosum* (Brown)	Beef Cattle, Horses, Swine, Sheep, Goats, Poultry	Improves feed utilization Improves production Support immune function Antibiotic alternative Prevents wool shedding	NA
BioAtlantis (Ireland)	LactoShield®	Extract	*Laminaria* sp. (Brown)	Young pigs	Health and Performance improvement	Draper et al. (2016)
BioAtlantis (Ireland)	NeoShield®	Extract	*Laminaria* sp. (Brown)	Pigs	Antibiotic effect Support immune function	McDonnell et al. (2016)

Company name	Commercial product	Product form/use	Seaweed species	Targeted animal(s)	Benefits	Reference(s)
BioAtlantis (Ireland)	DiNovo®	Extract	*Laminaria* sp. (Brown)	Chickens	Prebiotic effect	Sobolewska et al. (2017)
CH4 Global (New Zealand)	Under development	NA	*Asparagopsis taxiformis* (Red)	Ruminant livestock	Prevents enteric methane production	McCauley et al. (2020).
FutureFeed (Australia)	Under development	NA	*Asparagopsis* sp. (Red)	Ruminant livestock	Prevents enteric methane production	McCauley et al. (2020)
Greener Grazing (Vietnam)	Under development	NA	*Asparagopsis taxiformis* (Red)	Ruminant livestock	Prevents enteric methane production	McCauley et al. (2020)
Hebridean Seaweed Company (Great Britain)	Seaquim®	Green or Brown Powder	*Ascophyllum nodosum* (Brown)	Ruminant livestock	Support immune function Improve animal health and performance	NA
Marifeed (South-Africa)	Abfeed®	Formulation/pellet form	*Ulva* sp. (Green)	Abalone	Improve growth performance	Dlaza et al. (2008) Amosu (2016)
Ocean Harvest Technology (Ireland)	Oceanfeed® Swine	Powder	Blend of Red, Brown and Green seaweeds	Pigs	Support immune function Prebiotic effect Improve performance	Del Tuffo et al. (2019) Ruiz et al. (2018)
Ocean Harvest Technology (Ireland)	OceanFeed® Pet	Powder	Blend of Red, Brown and Green seaweeds	Dogs	Support immune function Prebiotic effect	NA
Olmix (France)	MFeed+®	NA	*Ulva* sp. (Green) and *Solieria chordalis*	Swine, poultry and fish	Improve animal performance Prebiotic effect	Kulshreshtha et al. (2020)

(continued)

Table 17.1 (continued)

Company name	Commercial product	Product form/use	Seaweed species	Targeted animal(s)	Benefits	Reference(s)
PacificBio (Australia)	NA	(free choice)	*Ulva* sp. (Green)	Marine animals, chickens	Support immune function Improve performance	Li et al. (2018)
PacificBio (Australia)	NA	Dried and milled	*Oedogonium intermedium* (Green)	Cattle	Support immune function	NA
Sea6 Energy Pvt. Ltd. (India)	Eqqua Royyal	Processed macroalgal extract	NA	Shrimp	Support immune function Lower stress Possible prebiotic effect	NA
Sea6 Energy Pvt. Ltd. (India)	Poultry Feed Additive	NA	NA	Broiler and layer chicken	Improving health Boost immunity	NA
Symbrosia (US)	Under development	NA	*Asparagopsis taxiformis* (Red)	Ruminant livestock	Prevents enteric methane production	McCauley et al. (2020)
The Seaweed Company (the Netherlands)	TopHealth Under development	NA	NA	Poultry/Equine/Swine/Pet/Aqua/Dairy	Antibiotic alternative Support immune function Improve animal health and performance	NA
Volta Greentech (Sweden)	Under development	NA	*Asparagopsis taxiformis* (Red)	Ruminant livestock	Prevents enteric methane production	McCauley et al. (2020)

NA not available

Table 17.2 Chemical (based on dry matter) and essential amino acid (EAA) composition of commercial seaweed species

Analysis	*Ascophyllum nodosum*	*Macrocystis pyrifera*	*Laminaria* and *Saccharina* sp.	*Sargassum* sp.	*Palmaria palmata*	*Aragopsis taxiformis*	*Ulva* sp.
Crude protein (%)	8 ± 2.7 (5)	10.1 ± 2.3 (7)	9.8 ± 2.2 (49)	8.5 ± 1.8 (10)	19.1 ± 6.1 (12)	18.2	18.6 ± 7.3 (14)
Crude fibre (%)	5.5 (4.1–6.8)	8 ± 2.5 (5)	6.6 (5.5–7.7)	10.1 ± 2.4 (9)	1.5	NA	9.6 ± 4.1 (7)
Gross energy (MJ/kg)	14.6 (14.5–14.7)	9 ± 0.4 (5)	NA	9.1 (8.9–9.2)	NA	NA	14.7 ± 2.6 (4)
Ca (g/kg)	20	14.1 ± 1.5 (3)	8.8	3.8 ± 2.6 (4)	NA	1.98	29.2 ± 28.9 (3)
Na (g/kg)	NA	36.9 ± 9.9 (3)	25.3	NA	3.3 ± 0.3 (51)	12.3	(11.0–29.3)
K (g/kg)	24	67.5 ± 22.4 (3)	59.5	46.2 (15.9–76.6)		2.02	22.1 (15.1–29.0)
Mg (g/kg)	8	39 ± 22.8 (3)	5.5	7.7 (7.5–7.9)	NA	1.35	16.7 ± 3.2 (3)
Mn (mg/kg)	12 ± 3 (8)	11	6 ± 2 (12)	214 ± 106 (4)	11	26.0	101
Zn (mg/kg)	181 ± 114 (8)	12	111 ± 70 (12)	214 ± 151 (5)	143	37.0	45 (28–61)
Cu (mg/kg)	28 ± 16 (1)	2	14 ± 9 (12)	7 ± 4 (5)	24	6.00	12 (7–17)
Fe (mg/kg)	134 ± 36 (8)	117	233 ± 233 (12)	7291 ± 6327 (5)	153	939	1246 (1052–1440)
EAAs (g/16gN)							
Methionine	1.3(0.7–1.9)	1.9 ± 0.4(5)	1.7(0.9–2.4)*	NA	NA	NA	1.6(1.3–1.9)
Cysteine	NA	2.6 ± 0.9(4)	2.2(1.2–3.2)*	NA	NA	NA	5.9
Lysine	4.6 (4.3–4.9)	4.7 ± 0.7 (5)	5.8 (3.9–7.7) *	NA	NA	NA	3.8 (3.7–3.9)

Adapted from Makkar et al. (2015); values for *Aragopsis taxiformis* come from Brooke et al. (2020)

Ascophyllum nodosum, Macrocystis pyrifera, Sargassum sp., *Laminaria* and *Saccharina* sp. are brown seaweed species; *Palmaria palmata* and *Aragopsis taxiformis* are red seaweed species; *Ulva* sp. are green seaweed species

Values are mean ± SD (n); other values are either single or average of two values, each value given in bracket; * denotes values for *Saccharina japonica* only

NA not available

for the nutritional enrichment of the meat of several fish species (Morais et al. 2020; Øverland et al. 2019).

Brown seaweeds mainly contain several functional polysaccharides including laminarin, alginic acid and fucoidan (McDonnell et al. 2016; Makkar et al. 2015; Morais et al. 2020). These bioactive polysaccharides possess various antioxidant, anti-inflammatory and immuno- and bacterial modulation properties (Makkar et al. 2015; Del Tuffo et al. 2019; McDonnell et al. 2016; Sobolewska et al. 2017).

Brown seaweeds constitute the majority of seaweed used in terrestrial animal feed, however, they are less investigated for application in aquafeeds compared to other seaweed classes (Kamunde et al. 2019). Seaweed meal of *A. nodosum* is reported to be beneficial to health of terrestrial animals by affecting the gastrointestinal tract microflora, thereby increasing stress resistance through activation of the immune system (Kandasamy et al. 2012). In addition, supplementation of *A. nodosum* extracts to animal feed has an inhibitory effect on the growth of pathogenic bacteria, including *Escherichia coli, Pseudomonas, Salmonella, Streptococcus* and *Enterococcus* sp., in a range of animals (Braden et al. 2004; Kandasamy et al. 2012; Novoa-Garrido et al. 2014; Braden et al. 2004). Addition of *A. nodosum* seaweed meal is also shown to increase animal performance by improving nutrient uptake, and antioxidant activity (Kandasamy et al. 2012). For cattle, nutrients digestion and rumen activity were enhanced, which reflected in higher weight gain and feed conversion (Mohi El-Din et al. 2008).

Laminaria digitata is frequently used as feed for shellfish, however, much remains unknown about the effect of seaweeds in nutrition of aquatic animals (Kamunde et al. 2019; Zhang 2018). It is suggested that seaweed meal from *Laminaria* sp. can be mixed with fish food up to 10% to increase food consumption and enhance growth performance, as well as to improve antioxidant capacity (Kamunde et al. 2019). Extracts derived from *Laminaria* are already commercialized and could be added to feed for animals, particularly pigs and chickens (Table 17.2).

Laminaria spp. commercially cultivated for their high alginate content. Alginate is a polysaccharide used as a gelling and thickening agent in the food, pharmaceutical, and textile industries. *Laminaria* extracts containing laminarin and fucoidan are reported to improve the intestinal health in weanling pigs, simultaneously enhancing growth performance and animal health (McDonnell et al. 2016). These two bioactive compounds were also found to be beneficial to pigs challenged by Salmonella typhimurium and Porcine Parvovirus (PPV), by decreasing the bacterial and viral load (Draper et al. 2016).

Kelp meal or kelp extracts, derived from *Laminaria* sp. and *A. nodosum,* could thus be used for a multiple animal species, including pigs, broiler chicken, cattle and fish (Sobolewska et al. 2017; McDonnell et al. 2016; Morais et al. 2020; Draper et al. 2016). However, the use of intact *Laminaria* sp. and *A. nodosum* as replacements for major protein sources in formulated feed for monogastric animals is prohibited since these brown seaweeds contain low levels of metabolizable energy, a low concentration of essential amino acids (EAAs) and a high mineral content (Øverland et al. 2019; Mæhre et al. 2014).

Fresh and intact seaweed is mostly used for shellfish feed as fishmeal replacement (Zhang 2018; O'Mahoney et al. 2014). While *Macrocystis pyrifera* is used for shrimp feeding, *Sargassum* is often used as feedstock for abalone, especially when other kelp species are scarce (Zhang 2018).

Besides the high content in the nutritionally important minerals such as Na, K, Ca and Mg (Table 17.2), *Sargassum* sp. also contains good amounts of polyphenols, which have strong antioxidant effects (Morais et al. 2020). Mixing species of this genus with diverse seaweeds, such as *Ulva* sp., *Porphyra* sp., *Gracilaria* sp., and other brown seaweed species, led to encouraging results for the use of seaweed as fish and shellfish feed and could offer a cost-effective way of harnessing the beneficial effects of seaweeds in animal production (O'Mahoney et al. 2014; Morais et al. 2020).

Porphyra spp. (Nori): *Nori* is a popular edible seaweed used mainly in sushi rolls. Porphyra species are cultivated for their thin, papery texture and high protein content.

Worldwide, several seaweed companies have been commercially producing high value seaweed-based commercial feed products for animal nutrition (Table 17.1).

4.2 Green Seaweed Species

Ulva sp., known as sea lettuce, has a good protein content and low energy content (Table 17.2; Makkar et al. 2015). Most commercial feed products are derived from *Ulva* since this is a one of the most common genera of green seaweeds (Table 17.1).

Although the amino acid profiles of macroalgae in general have been perceived as limiting, the amino acid profile in *Ulva* species is characterized by high lysine and sulfur containing amino acids, with up to 75% of the crude protein is in the form of true protein (Table 17.2; Machado et al. 2015). *Ulva* has the potential as an alternative source of proteins for animal feeding, having higher protein content than other green seaweeds (Makkar et al. 2015).

Ulva is filled with vitamins and contains high levels of micro-elements such as Cu, Zn, Mn and other nutritionally important minerals such as Ca and Mg (Table 17.2; Morais et al. 2020). *Ulva* contains high insoluble dietary fibre, such as glucans, and soluble fibre such ulvan (Table 17.2; Makkar et al. 2015; Kandasamy et al. 2012). The latter is a sulphated polysaccharide present in the cell wall and represents up to 36% of *Ulva* sp. dry biomass. It regulates immune functions and act as an antioxidant and antibiotic. A high level of ulvan in *Ulva* sp. is believed to have anticoagulant, antiviral, anti-inflammatory, immunomodulatory and anticancer activities (Morais et al. 2020).

Some *Ulva* species are used as livestock feed, mostly for marine animals and poultry (Table 17.1). Adding *Ulva* to diets in powder form can improve meat quality and amylase activity in the duodenal content of chicken (Øverland et al. 2019). Studies on laying chickens showed that the addition of *Ulva* up to 1% of feed rations improves egg production, increases egg weight, improves shell strength and brightens yolk colour (Li et al. 2018). It is also demonstrated that the Newcastle disease virus, a disease affecting poultry and other bird species, is inhibited by ulvan (Li et al. 2018). For marine animals such as fish it was discovered that using seaweed meals as supplement enhance growth, lipid metabolism, physiological activity, stress response, disease resistance and carcass quality of various fish species (Morais et al. 2020).

Ulva sp. are also much used for feeding juvenile abalone. Feeding trials with incorporated cultured *Ulva* show that abalone growth improved, most likely result of the higher protein content of the cultivated *Ulva* (Amosu 2016). Trials with Abfeed®, a seaweed product for abalone, led to improved growth over other feeds, which was even greater when fresh seaweed was incorporated into the diet of *H. midae* (Dlaza et al. 2008).

Despite the benefits on animal health and performance, the efficient use of *Ulva* as a single feed for animals is generally prevented due to the limited information on the bioavailability of the polysaccharides (Øverland et al. 2019).

Oedogonium intermedium is used as a freshwater green seaweed and could be used as a supplement to low quality forage diets for ruminant livestock (Neveux et al. 2020; Machado et al. 2015).

4.3 Red Seaweed Species

Red algae (*Rhodophyta*) have an interesting nutritional profile since they contain higher protein levels than green and brown seaweed, reaching 47% (*Neopyropia tenera*) of a dry matter (Makkar et al. 2015; Morais et al. 2020). Next to the presence of numerous minerals (Table 17.2), red algae contain a high content of soluble fibers, including sulphated galactans (agars and carrageenans), xylans and floridean starch (Kandasamy et al. 2012; Morais et al. 2020). The iodine content in red algae is high, especially in *Gracilaria* sp. (reaching 0.4% of seaweed dry biomass) which is often used as feed for shellfish (Zhang 2018).

Eucheuma spp. are red seaweeds cultivated for their carrageenan content. Carrageenan is a natural thickener and stabilizer used in the food and pharmaceutical industries.

The dried form of red seaweed species such as *Neopyropia tenera, Gracilaria* sp. and *Palmaria palmata* (Table 17.1), may be used as protein sources in formulated feed for animals, especially fish. Inclusion of dried and milled *Palmaria palmata* (150 g kg^{-1}) in the diet of Atlantic salmon, showed no adverse effects on growth rate or feed conversion ratio and resulted in enhanced coloration through deposition of red algal pigments (Øverland et al. 2019). Although protein digestibility for the intact algae may be low, most studies show the great potential of several red macroalgae as feed ingredients for fish (Øverland et al. 2019).

In recent years, the interest towards red seaweed species has increased significantly since several species have shown to lower enteric methane production of ruminants. Enteric methane is a natural by-product of microbial fermentation of nutrients in the digestive tract of animals and accounts for approximately 15% of the world's entire total of greenhouse gas emissions (Brooke et al. 2020). Recent studies have shown that supplementation of the red seaweed *Asparagopsis* significantly reduces enteric methane emissions (Kandasamy et al. 2012; Kinley et al. 2020).

The effectiveness of macroalgae in reducing enteric methane during rumen incubation has been linked to the concentration of halogenated bioactives such as bromoform and di-bromochloromethane (Brooke et al. 2020). These bioactive compounds are abundantly present in marine algae and especially in *Asparagopsis* species, and prevent the formation of methane by inhibiting a specific enzyme in the gut during the digestion of feed (Brooke et al. 2020).

It is recently explored that the supplementation of *Asparagopsis* seaweed to cattle not only drastically reduces the greenhouse gas contribution, but also improves weight gain of cattle (Machado et al. 2015; Kinley et al. 2020). Thus, *Asparagopsis* supplementation to feed may be an effective strategy to improve profitability as well as sustainability of beef and dairy operations. Currently, the bottleneck in the implementation of an *A. taxiformis*-based methane mitigation strategy is the availability of this bioactive seaweed (Brooke et al. 2020). The growth in controlled conditions has not yet been achieved at a large scale and has led to the exploration of alternative seaweed species and strains (McCauley et al. 2020; Brooke et al. 2020).

Nowadays, several projects are set up to further explore the potential of *Asparagopsis* sp. as alternatives to conventional ruminant feed. In Spain and Norway, grant funding was allocated to investigate the nutritive value of various seaweed species. Commercially, various companies based in countries such as Sweden, USA, New Zealand, Australia and Vietnam, aim to scale up the production of selected red macroalgae as feed additives (McCauley et al. 2020).

Seaweeds provide a unique taste profile and can be a healthy and flavorful addition to various dishes. Seaweeds can be a great option to replace high-sodium flavorings *(*Milinovic et al. 2021*)*. Species such *Ascophyllum nodosum, Saccharina latissima* and *Fucus vesiculosus* have been studied for taste enhancement and salt replacement (TASTE EU project n.d.), and such flavoring components might eventually take the place of processed food's high-sodium flavoring, which would lower heart disease and advance the seaweed processing sector. Seaweeds offer a natural and healthy alternative while adding a unique umami taste to your food. Few types of seaweed well-known for such properties: *Nori* has a mild and slightly sweet flavor. Nori sheets can be toasted and crumbled afterwards to use them as a seasoning or sprinkle them on salads, rice, or other dishes; Kombu can simmer in water to create a stock or added to soups, stews, and braises for a rich umami taste. It can also be used as a natural tenderizer for beans and legumes; Wakame has a delicate flavor and is often used in seaweed salads and miso soup, to stir-fries, noodle dishes, or soups to enhance the taste and texture; Dulse has a slightly smoky and salty taste. It can be eaten raw, roasted, or added to various dishes such as salads, soups, and stir-fries. Dulse flakes can also be used as a seasoning or sprinkled over popcorn for a unique flavor; Arame is a mild and sweet seaweed, and has a tender texture and can be soaked and added to salads, stir-fries, or sautéed with vegetables; Hijiki has a slightly sweet and nutty flavor. It is often used in Japanese cuisine and can be added to salads, stir-fries, and rice dishes. It is also a good source of minerals and is considered nutritious.

When using seaweed as a replacement for high-sodium flavorings, it's important to consider the overall salt content in the dish. While seaweed naturally contains some sodium, it is generally lower in sodium than many commercial seasonings.

5 Future Perspective

Overall, seaweed genetic resources hold great potential for the development of innovative food products and other applications. Their unique nutritional composition, versatility, and sustainable cultivation make them an attractive option for the future of food production and sustainability.

Commercial seaweeds refer to various species of marine algae that are cultivated and harvested for economic purposes. These seaweeds are valuable resources with numerous applications in various industries, including food, pharmaceuticals, cosmetics, agriculture, and biofuels. The genetic resources of commercially important seaweeds play a crucial role in their cultivation, improvement, and development of new varieties with desirable traits.

Seaweed genetic resources encompass the genetic diversity found within different species of seaweeds and the variations within each species. They include the naturally occurring genetic variations as well as the genetic modifications achieved through breeding or genetic engineering techniques. The genetic resources of commercially important seaweeds are typically conserved in gene banks and research institutions for future use.

The exploitation of seaweed genetic resources allows for the selection and cultivation of seaweed strains with desirable traits such as high growth rates, increased biomass production, improved nutritional profiles, enhanced disease resistance, and tolerance to environmental stressors. These traits are important for optimizing seaweed cultivation practices and ensuring sustainable production for various applications (Cotas et al. 2023).

The genetic resources of commercially important seaweeds are studied and utilized by researchers and scientists to understand the genetic basis of important traits, develop breeding programs for seaweed improvement, and develop new varieties through selective breeding or genetic engineering techniques. For example, researchers may identify specific genes or genetic markers associated with desirable traits and use them in breeding programs to develop improved seaweed varieties.

Furthermore, the genetic resources of commercially important seaweeds are also explored for the discovery of novel bioactive compounds with potential applications in pharmaceuticals, nutraceuticals, and cosmeceuticals. Seaweeds are known to produce a wide range of bioactive compounds such as antioxidants, antimicrobial agents, anti-inflammatory substances, and polysaccharides with various biological activities. Understanding the genetic basis of these compounds can facilitate their production and utilization in different industries.

Overall, the genetic resources of commercially important seaweeds are valuable assets that contribute to the development of sustainable seaweed cultivation practices, the improvement of seaweed varieties, and the discovery of novel bioactive compounds. The conservation, study, and utilization of these genetic resources are essential for the continued growth and innovation in the seaweed industry. The examples represent only a fraction of the commercially important seaweed genetic resources, as there are numerous other species with potential applications in various industries, including food, agriculture, biotechnology, and environmental sustainability.

The commercial cultivation of seaweeds, known as seaweed farming or mariculture, is gaining momentum globally. Various species of seaweeds, such as kelp, nori, and wakame, are cultivated in coastal areas using floating or submerged systems. Seaweeds grow rapidly, often reaching harvestable size within a few months, making them an attractive crop for cultivation. While seaweed cultivation is generally sustainable, it is important to ensure that it is carried out in an environmentally responsible manner. Proper management is necessary to prevent the introduction of non-native species, protect marine ecosystems, and minimize the potential for habitat alteration.

Funding This work was supported from the Bio-Based Industries Joint Undertaking under the European Union's Horizon 2020 research and innovation program, under grant agreement No. 887259 (ALEHOOP) and, under grant agreement 101036768 (PROMISEANG), under grant agreement 101112072 (InnoProtein), under grant agreement 101112555 (SYLPLANT), and from the European Union's Horizon 2020 research and innovation program under grant agreement No. 862704 (NextGenProteins). LIKE-A-PRO project received funding from European Union's Horizon Europe (HORIZON) under Grant Agreement No. 101083961.

References

Abowei JFN, Tawari CC (2011) A review of the biology, culture, exploitation and utilization potentials seaweed resources: case study in Nigeria. Res J Appl Sci Eng Technol 3:290–303

Amosu A (2016) Production and uses of *Ulva armoricana*: the South African perspective. J Environ Sci 5:17–22

Amosu A, Robertson-Andersson D, Maneveldt G, Anderson R, Bolton J (2013) South African seaweed aquaculture: a sustainable development example for other African coastal countries. Afr J Agric Res 8:5260–5271

Araújo R, Vázquez Calderón F, Sánchez López J, Azevedo IC, Bruhn A, Fluch S, Garcia Tasende M, Ghaderiardakani F, Ilmjärv T, Laurans M, Mac Monagail M, Mangini S, Peteiro C, Rebours C, Stefansson T, Ullmann J (2021) Current status of the algae production industry in Europe: an emerging sector of the blue bioeconomy. Front Mar Sci 7:626389

Barbier M, Araújo R, Rebours C, Jacquemin B, Holdt S, Charrier B (2020) Development and objectives of the PHYCOMORPH European guidelines for the sustainable aquaculture of seaweeds (PEGASUS). Bot Mar 63(1):5–16

Braden KW, Blanton JR, Allen VG, Pond KR, Miller MF (2004) *Ascophyllum nodosum* supplementation: a pre-harvest intervention for reducing *Escherichia coli* O157:H7 and *Salmonella* sp. in feedlot steers. J Food Prot 67:1824–1828

Brooke C, Roque B, Najafi N, Gonzalez M, Pfefferlen A, DeAnda V, Ginsburg D, Harden M, Nuzhdin S, Salwen J, Kebreab E, Hess M (2020) Methane reduction potential of two Pacific coast macroalgae during in-vitro ruminant fermentation. Front Mar Sci 7. https://doi.org/10.3389/fmars.2020.00561

Cai J, Lovatelli A, Aguilar-Manjarrez J, Cornish L, Dabbadie L, Desrochers A, Diffey S, Garrido Gamarro E, Geehan J, Hurtado A, Lucente D, Mair G, Miao W, Potin P, Przybyla C, Reantaso M, Roubach R, Tauati M, Yuan X (2021) Seaweeds and microalgae: an overview for unlocking their potential in global aquaculture development. FAO Fisheries and Aquaculture Circular No. 1229. FAO, Rome

Chapman VJ, Chapman DJ (1980) Seaweed as animal fodder, manure and for energy. In: Seaweeds and their uses. Springer, Dordrecht

Cherry P, O'Hara C, Magee PJ, McSorley EM, Allsopp PJ (2019) Risks and benefits of consuming edible seaweeds. Nutr Rev 77(5):307–329

Choudhary B, Chauhan OP, Mishra A (2021) Edible seaweeds: a potential novel source of bioactive metabolites and nutraceuticals with human health benefits. Front Mar Sci 8:740054

Cotas J, Gomes L, Pacheco D, Pereira L (2023) Ecosystem services provided by seaweeds. Hydrobiology 2:75–96

Cruz-Suárez LE, Ricque-Marie D, Tapia-Salazar M, Guajardo-Barbosa C (2000) Utilization of kelp (*Macrocystis pyrifera*) in shrimp feeding. Avances en Nutrición Acuícola V. Memorias del V Simp. Int. de Nutrición Acuícola, Mérida

Del Tuffo L, Laskoski F, Vier CM, Tokach MD, Dritz SS, Woodworth JC, DeRouchey JM, Goodband RD, Constance LA, Niederwerder M, Arkfeldt E (2019) Effects of oceanfeed swine feed additive on performance of sows and their offspring. Kansas Agric Exp Stn Res Rep 5(8). https://doi.org/10.4148/2378-5977.7834

Dlaza TS, Maneveldt G, Viljoen C (2008) Growth of post-weaning abalone Haliotis midae fed commercially available formulated feeds supplemented with fresh wild seaweed. Afr J Mar Sci 30:199–203

Dos Santos Fernandes De Araujo R, Vazquez Calderon F, Sanchez Lopez J, Azevedo I, Bruhn A, Fluch S, Garcia Tasende M, Ghaderiardakani F, Ilmjarv T, Laurans M, Mac Monagail M, Mangini S, Peteiro C, Rebours C, Stefansson T, Ullmann J (2021) Current status of the algae production industry in Europe: an emerging sector of the Blue Bioeconomy. Front Mar Sci 7:626389, JRC122250, ISSN 2296-7745

Draper J, Walsh AM, McDonnell M, O'Doherty J (2016) Maternally offered seaweed extracts improves the performance and health status of the postweaned pig. J Anim Sci 94:391–394

Evans FD, Critchley AT (2014) Seaweeds for animal production use. J Appl Phycol 26:891–899

Ferdouse F, Lovstad Holdt S, Smith R, Murúa P, Yang L (2017) The global status of seaweed production, trade and utilization. FAO Globefish Res Program 124:1–57

Fernand F, Israel A, Skjermo J, Wichard T, Timmermans KR, Golberg A (2017) Offshore macroalgae biomass for bioenergy production: environmental aspects, technological achievements and challenges. Renew Sust Energ Rev 75:35–45

Guiry MD (2021) Phaeophyceae: Brown Algae. https://www.seaweed.ie/algae/phaeophyta.php. Accessed on 6 Nov 2021

Heuzé V, Tran G, Giger-Reverdin S, Lessire M, Lebas F (2017) Seaweeds (marine macroalgae). Feedipedia, a programme by INRA, CIRAD, AFZ and FAO

Hwang EK, Park CS (2020) Seaweed cultivation and utilization of Korea. Algae 35(2):107–121

Kamunde C, Sappal R, Melegy TM (2019) Brown seaweed (AquaArom) supplementation increases food intake and improves growth, antioxidant status and resistance to temperature stress in Atlantic salmon, Salmo salar. PLoS One 14(7):e0219792

Kandasamy S, Khan W, Evans F, Critchley AT, Prithiviraj B (2012) Tasco®: a product of *Ascophyllum nodosum* enhances immune response of *Caenorhabditis elegans* against *Pseudomonas aeruginosa* infection. Mar Drugs 10:84–105

Kinley RD, Martinez-Fernandez G, Matthews M-K, de Nys R, Magnusson M, Tomkins NW (2020) Mitigating the carbon footprint and improving productivity of ruminant livestock agriculture using a red seaweed. J Clean Prod 259:120836

Kulshreshtha G, Hincke MT, Prithiviraj B, Critchley AT (2020) A review of the varied uses of macroalgae as dietary supplements in selected poultry with special reference to laying hen and broiler chickens. J Mar Sci Eng 8:536

Leandro A, Pereira L, Gonçalves AMM (2020) Diverse applications of marine macroalgae. Mar Drugs 18:17

Lee WG (2001) Introduced plants, negative effects of. In: Encyclopedia of biodiversity, 2nd edn. Academic Press, pp 345–356

Li Q, Luo J, Wang C, Tai W, Wang H, Zhang X, Liu K, Jia Y, Lyv X, Wang L et al (2018) Ulvan extracted from green seaweeds as new natural additives in diets for laying hens. J Appl Phycol 30:2017–2027

Lourenço-Lopes C, Fraga-Corral M, Jimenez-Lopez C, Pereira AG, Garcia-Oliveira P, Carpena M, Prieto MA, Simal-Gandara J (2020) Metabolites from macroalgae and its applications in the cosmetic industry: a circular economy approach. Resources 9:101

Machado L, Kinley RD, Magnusson M et al (2015) The potential of macroalgae for beef production systems in Northern Australia. J Appl Phycol 27:2001–2005

Mæhre HK, Malde MK, Eilertsen K-E, Elvevoll EO (2014) Characterization of protein, lipid and mineral contents in common Norwegian seaweeds and evaluation of their potential as food and feed. J Sci Food Agric 94:3281–3290

Makkar H, Tran G, Heuzé V, Giger-Reverdin S, Lessire M, Lebas F, Ankers P (2015) Seaweeds for livestock diets: a review. Anim Feed Sci Technol 212. https://doi.org/10.1016/j.anifeedsci.2015.09.018

McCauley JI, Labeeuw L, Jaramillo-Madrid AC et al (2020) Management of enteric methanogenesis in ruminants by algal-derived feed additives. Curr Pollut Rep 6:188–205

McCullough C (2019) Harvesting seaweed for cattle feed. https://www.allaboutfeed.net/all-about/new-proteins/harvesting-seaweed-for-cattle-feed/. Accessed on 6 Nov 2021

McDonnell M, Bouwhuis M, Sweeney T, O'Shea C, O'Doherty J (2016) Effects of dietary supplementation of galactooligosaccharides and seaweed-derived polysaccharides on an experimental *Salmonella Typhimurium* challenge in pigs. J Anim Sci 94:153–156

Milinovic J, Mata P, Diniz M, Noronha JP (2021) Umami taste in edible seaweeds: the current comprehension and perception. Int J Gastron Food Sci 23:100301

Miyashita K, Mikami N, Hosokawa M (2013) Chemical and nutritional characteristics of brown seaweed lipids: A review. J Funct Foods 5(4):1507–1517. https://doi.org/10.1016/j.jff.2013.09.019

Mohi El-Din AMA, Gaafar HMA, El-Nahas HM, Ragheb EE, Mehrez AF (2008) Effect of natural feed additives on performance of growing friesian calves. Egypt J Anim Prod 45:401–413

Morais T et al (2020) Seaweed potential in the animal feed: a review. J Mar Sci Eng 8:559

netalgae (2012) Seaweed industry in Europe. https://www.seaweed.ie/irish_seaweed_contacts/. Accessed on 6 Nov 2021

Neveux N, Nugroho A, Roberts D, Vucko M, de Nys R (2020) Selecting extraction conditions for the production of liquid biostimulants from the freshwater macroalga *Oedogonium intermedium*. J Appl Phycol 32:539–551

Novoa-Garrido M, Aanensen L, Lind V, Larsen HJS, Jensen SK, Govasmark E, Steinshamn H (2014) Immunological effects of feeding macroalgae and various vitamin E supplements in Norwegian white sheep-ewes and their offspring. Livest Sci 167:126–136

Novoa-Garrido M, Rebours C, Aanensen L, Torp T, Lind V, Steinshamn H (2016) Effect of seaweed on gastrointestinal microbiota isolated from Norwegian white sheep. Acta Agric Scand A Anim Sci 66(3):152–160

O'Mahoney M, Rice O, Mouzakitis G, Burnell G (2014) Towards sustainable feeds for abalone culture: evaluating the use of mixed species seaweed meal in formulated feeds for the Japanese abalone, *Haliotis discus hannai*. Aquaculture 430:9–16

Øverland M, Mydland LT, Skrede A (2019) Marine macroalgae as sources of protein and bioactive compounds in feed for monogastric animals. J Sci Food Agric 99(1):13–24

Polat S, Trif M, Rusu A, Šimat V, Čagalj M, Alak G, Meral R, Özogul Y, Polat A, Özogul F (2021) Recent advances in industrial applications of seaweeds. Crit Rev Food Sci Nutr 8:1–30

Rodríguez-Montesinos YE, Hernández-Carmonal G (1991) Seasonal and geographic variations of *Macrocystis pyrifera* chemical composition at the western coast of Baja California. Cienc Mar 17:91–107

Ruiz ÁR, Gadicke P, Andrades SM, Cubillos R (2018) Supplementing nursery pig feed with seaweed extracts increases final body weight of pigs. Austral J Vet Sci 50(2):83–87

Sobolewska A, Elminowska-Wenda G, Bogucka J, Dankowiakowska A, Kułakowska A, Szczerba A, Stadnicka K, Szpinda M, Bednarczyk M (2017) The influence of in ovo injection with the prebiotic DiNovo® on the development of histomorphological parameters of the duodenum, body mass and productivity in large-scale poultry production conditions. J Anim Sci Biotechnol 8:45

TASTE EU project (n.d.). https://cordis.europa.eu/project/id/315170

Verstringe S, Vandercruyssen R, Carmans H, Rusu AV, Bruggeman G, Trif M (2023) Alternative proteins for food and feed. In: Galanakis CM (ed) Biodiversity, functional ecosystems and sustainable food production. Springer, Cham

Wan AH, Davies SJ, Soler-vila A, Fitzgerald RD, Johnson MP (2019) Macroalgae as a sustainable aquafeed ingredient. Rev Aquac 11:458–492

Zhang J (2018) Seaweed industry in China. Innovation Norway China. www.fisheries.noaa.gov. Accessed on 6 Nov 2021

Chapter 18
The Legal Status and Compliance of Seaweed and Seaweed-Derived Compounds

Lubna Ahmed and Catherine Barry-Ryan

1 Introduction

Seaweed or marine macroalgae are plant-like organisms that normally live attached to rocks or other hard substances in coastal areas. They can also be found in the freshwater ecosystems ie lakes and rivers. Macroalgae belong to three different groups on the basis of their thallus colour: brown, red and green algae (Wang et al. 2015).

The global market value of macroalgae production is more than 8 billion USD (FAO 2021). The majority of the global production is cultivated now-a-days with a smaller portion of naturally growing macroalgae harvested. Currently, the global cultivation of macroalgae is approximately 32 million tons fresh weight (FW), and harvesting of natural macroalgae is approximately 1 million tons FW annually (FAO 2021).

Asian countries, specifically China and Indonesia are leaders for macroalgae cultivation (>80%), whereas Chile and Norway being big producers along with China and Japan for harvesting of wild macroalgae (Araujo et al. 2021). The most cultivated species are the brown algae *Saccharina japonica* (Japanese kelp), the green algae *Monostroma* spp. and the red algae *Porphyra* spp. (Japanese nori).

In Europe, Norway is the largest collector of wild seaweed (*Laminaria digitata*, *Laminaria hyperborea*) in Europe, followed by Scotland, France, Denmark and Sweden harvesting brown algae *Ascophyllum nodosum* (rockweed) and *Saccharina latissimi* (sugar kelp) (Weinberger et al. 2020).

L. Ahmed (✉)
Department of Agriculture, Food and Animal Health, School of Health and Science, Dundalk Institute of Technology, Dunkdal, Ireland
e-mail: lubna.ahmed@dkit.ie

C. Barry-Ryan
Food Innovation Lab, School of Food Science and Environmental Health, Technological University Dublin, Dublin, Ireland

F. Ozogul et al. (eds.), *Seaweeds and Seaweed-Derived Compounds*,
https://doi.org/10.1007/978-3-031-65529-6_18

2 Legislation and Licencing on Macroalgae Production

There is no specific regulatory framework on the cultivation or harvesting available for most countries. However, all the general environmental regulations are applicable. Below is a list of main regulations in the EU:

1. The Habitats Directive 92/43/EEC,
2. The Marine Strategy Framework Directive 2008/56/EC,
3. The Maritime Spatial Planning Directive 2014/89/EU,
4. The Alien Species Regulations 1143/2014/EU and 708/2007/EC, and
5. The Environmental Impact Assessment Directive 2011/92/EU.
6. The Water Framework Directive 2000/60/EC,
7. The Organic Food Regulation 2018/848/EU has a section specified for the cultivation of organic algae (Part III: 1. Production rules for algae and aquaculture animals, 2. Requirements for algae).

One of the limiting factors that the macroalgae industry often faces for its development is the legal barriers, specially the substantive procedural requirements for obtaining a license to cultivate or harvest the seaweed (Araujo et al. 2021). The environmental licensing process is time-consuming, it depends on the sociopolitical context and the bureaucratic fragmentation of political authority. The knowledge and resources of the applicant and the permitting agency and the novelty or complexity of the technology that are used for the production of the seaweed can also be lingering this process (Barbier et al. 2019).

3 The Uses of Macroalgae

For centuries seaweed have been used as food, feed and fertiliser. In the late seventeenth century the ash of the macroalgae was used in glass production. In the nineteenth century macroalgae was used for extracting iodine and in the twentieth century the production of alginate started from seaweed, which still is one of the key macroalgal food products.

Macroalgae is a rich source of bioactive products which can have potentially high market value (European Commission 2017; Garcia-Vaquero and Hayes 2016). Now-a-days huge interest has grown in macroalgae as raw material by the trend of using natural source in foods (including food use as such, food ingredients, food supplements, and food additives), feed and feed additives (as a source of potash, iodine, and algal polysaccharides) and nutraceuticals and cosmetics (Pimentel et al. 2018), using bio-based materials for plastic replacement (Lakshmi et al. 2020), and using biofuels for fossil fuels replacement, fertilisers and agricultural biostimulants (Joshi et al. 2018).

The benefit of macroalgae production is that they do not need land, fertilisation, or freshwater (FAO 2021; Wan et al. 2018). The growing demand of seaweed in different industries has led to more interest in macroalgae cultivation in Europe (European Commission 2017; Barbier et al. 2019), Start-up companies as well as the traditional aquaculture and fish industry are getting interested in the potential of native European macroalgae species.

4 Legislation on Specific Uses of Macroalgae

To bring the seaweed derived products into the EU Market the legislation that applies to them is an important aspect. Regulations implies on these products impact on all the stakeholders ie., the macroalgae cultivators or harvesters, manufacturers of the ingredients and the end-users who would incorporate them in final products. Regulations and legal principles, as well as procedures for pre-market authorisation influence market access. In the end these factors influence the scope and direction of macroalgae innovation research.

5 Food and Food Additives

Due to high nutritional value, bioactive properties and good safety records worldwide, macroalgae species are considered as potential raw materials for food (Cherry et al. 2019). Macroalgal species that have not been used as food and food additives in EU prior to 15th May, 1997 must conform to the Novel Food Regulation 2015/2283/EU in order to be used as food and food additives. However, this regulation contains a provision for easy entry of the species which have been considered as safe for consumption for at least 25 years in a non-EU country and were not consumed in EU before the stated date. Following rigorous safety assessments or historical data, EU commission has maintained a catalogue of approved novel foods as authorised by the Regulation (EU) 2017/2470. This catalogue includes 22 species of seaweed of which only *Sphaerotrichia divaricata*, underwent safety assessments and authorisation. The most common macroalgae in this authorised list are European red (e.g., *Gracilaria verrucosa*, *Porphyra tenera*, *Palmaria palmata*), brown (e.g., *Alaria esculenta*, *Fucus vesiculosus*, *Saccharina latissima*), and green (e.g., *Enteromorpha* sp., *Ulva lactuca*, *Monostroma nitidum*) algae species. A number of *Ulva* and *Monostroma* species have great potential as food due to their high content on proteins and polyunsaturated fatty acids (Garcia-Vaquero and Hayes 2016). In addition, two macroalgae products such as fucoidan extracts from *Fucus vesiculosus* and *Undaria pinnatifida*, have been included in the novel foods list for use in foods and food supplements. Phlorotannins extracts from brown alga *Ecklonia cava* have been authorised as a novel food for use in food supplements only (Commission Regulation (EU) 2017/2470). A separate legislation (2018/464/EU) deals with the safety concerns arising from the metal and iodine content in seaweed, and products based on seaweed. The European Commission has directed the member countries to monitor cadmium, lead, mercury, arsenic, and iodine contents in macroalgae based food and food additives through this legislation. However, there is no maximum levels in EU for cadmium, mercury or inorganic arsenic in macroalgae (Biancarosa et al. 2017). Therefore, a default value of 0.01 mg/kg has been set as maximum level for mercury in algae according to the Regulation (EC) No 396/2005 on maximum residue levels of pesticides in food.

6 Animal Feed and Feed Additives

Macroalgae have been used for animal and aqua-feed for their high content of protein and minerals, mainly iodine (Karatzia et al. 2012). Regulation 68/2013 of the European Commission defines algae meals, algae oil, and algae extracts as acceptable feed materials. A maximum level of toxic pollutants in feed is defined in Directive 2002/32/EC. According to Directive 2002/32/EC, the maximum level for arsenic in macroalgae feed materials has been set at 40 mg/kg for macroalgae meal and at 10 mg/kg for complete and supplemental feed for domestic animals. Furthermore, the feed operator is required to provide proof that the inorganic arsenic level is less than 2 mg/kg upon request. According to Directive 2002/32/EC, the maximum amounts of lead, cadmium, and mercury in feed materials are 10 mg/kg, 1 mg/kg, and 0.1 mg/kg, respectively.

Many European companies offer products made from *Ascophyllum nodosum* (Norwegian kelp), which can be used as an animal feed for fish, horses, pigs, and ruminants for the high content of polyphenols, peptides, fatty acids and polysaccharides (Karatzia et al. 2012). Furthermore, it has been noted that feeding ruminants macroalgae reduces their methane emissions (Maia et al. 2016). The mineral status of organic dairy calves was enhanced by using a combination of the three macroalgal species (*Ulva rigida*, *Sargassum muticum*, and *Saccorhiza polyschides*) as a mineral supplement in their feed (Rey-Crespo et al. 2014). This included higher levels of iodine and selenium in the milk. *Ulva rigida* and *Cystoseira barbata*, for example, have been demonstrated to be suitable as feed ingredients in studies on the utilisation of macroalgae as aqua-feed (Ergün et al. 2009).

7 Cosmetics

Ingredients from macroalgae could be used in cosmetics to address consumer demand for natural products and sustainability (Couteau and Coiffard 2016; Wang et al. 2015). The fundamental EU regulations for cosmetics are found in Regulation 1223/2009/EC. Each cosmetic product that is released onto the EU market must follow a centralized notification process. The EU site is used by importers and manufacturers to report products (Cosmetic Products Notification Portal, CPNP). Assessing the product's safety, maintaining a product information file (PIF), and disclosing any potentially major adverse effects are all legal obligations. Ingredients that are permitted, limited, and unacceptable are listed in legally enforceable lists. The legal prerequisites and limitations for each item are listed in CosIng, a database of cosmetic ingredients maintained by the EU (https://ec.europa.eu/growth/tools-databases/cosing/index.cfm). For instance, fucoidan is a widely utilized component for skin conditioning and antioxidant purposes. The free circulation of goods between EU member states combined with a centralised, scientific approach produces a clear regulatory framework for the cosmetics industry (Zakaria 2015).

Bioactive substances found in macroalgae include polysaccharides, proteins, peptides, amino acids, lipids, vitamins, minerals, pigments, and phenolic compounds. Macroalgae could be included into a variety of cosmetic products, including deodorants, make-up, and items for the skin, hair, and mouth (Pimentel et al. 2018). Moisturising and antiaging products are the product categories where macroalgae are most frequently used, according to Couteau and Coiffard (2016). *Alaria esculenta* (winged kelp, brown algae), *Chondrus crispus* (Irish moss), *Ulva lactuca* (sea lettuce), *Postelsia palmaeformis* (sea palm), red algae *Furcellaria lumbricalis* and *Fucus vesiculosus* are among the macroalgal species that are utilised in cosmetics (Joshi et al. 2018).

8 Biofuel

The carbohydrate-rich macroalgae are a possible source of third-generation biofuels, which do not require agricultural land for production and do not compete with food crops for land or fresh water resources. As biogas does not require drying or lipid extraction, it is currently a more energy-efficient and financially viable way to produce biogas. The demand for bioethanol and biogas has increased as a result of the EU Member States' regulations and the updated Renewable Energy Directive (EU 2018/2001). The remaining biomass from anaerobic digestion may be used as organic fertiliser if the conditions outlined in EU Regulation 2019/1009 are satisfied. The Commission believes that renewable biofuels can help reduce carbon emissions in transportation. The Delegated Regulation on Indirect Land-Use Change (EU) 2019/807 on the sustainability standards for biofuels was published by the EU in 2019. Identification of people who pose a high risk of indirect land use change and certification of those who pose a low risk are the objectives. There are numerous barriers to the production of biogas from macroalgal biomass. The collection and preparation of the biomass consume a lot of energy, and the high concentrations of phenols, heavy metals, sulfides, salts, and volatile acids may prevent biomethanation. Seasonality also causes an irregular supply of biomass in terms of quantity and quality. Hence, it is not yet possible to produce economical biogas only from macroalgae. Co-digestion with other feedstocks, and most crucially, an integrated strategy for algae growing and co-production of numerous high-value products (i.e., biorefinery) or a mix of these are more environmentally friendly and economically viable techniques (Balina et al. 2017; Torres et al. 2019).

9 Fertilisers

The practice of harvesting and using macroalgae as fertiliser is centuries old. The west coast of Ireland, for instance, has long been fertilised with *Palmaria palmata* and *Laminaria digitata* in Europe. For the manufacture of high-quality compost for agricultural crop production, the green macroalga Ulva sp. is a possible contender (Akila et al. 2019; Torres et al. 2019). As a soil conditioner to increase crop output, macroalgae and its extracts have a long history of use. Brown macroalgae such *Ascophyllum nodosum*, *Ecklonia maxima*, *Fucus serratus*, *Laminaria digitata*, *L. hyperborea*, *Macrocystis pyrifera*, and *Sargassum* spp. are frequently utilised for the synthesis of biostimulants (Sharma et al. 2014). They are abundant in polysaccharides like alginates, fucoidans, and laminarin that improve the defense mechanisms of plants.

Macroalgae may accumulate significant amounts of heavy metals, which can limit their usage in agriculture, depending on the water quality at their growth site. The Regulation EU 2019/1009 specifies the upper limits for a number of heavy metals, including cadmium, hexavalent chromium, mercury, nickel, lead, and inorganic arsenic 2 in fertilisers. The present allowable amount of cadmium in sewage sludge biofertilizers in Sweden is 2 mg/kg (Regulation 1998: 944), although the Swedish Environmental Protection Agency has proposed a limit of 0.8 mg/kg for the year 2030 (Swedish-EPA 2013). Moreover, Regulation EU 2019/1009 now sets a limit cadmium content in biofertilizers at 1.5 mg/kg. In order to prevent hazardous concentrations of heavy metals from limiting the use of fertilisers in agriculture, the collecting site and species composition must be considered even if macroalgae are a promising optional raw material for the manufacturing of bio-fertilisers. A method to lessen the amount of hazardous pollutants could be to combine the macroalgae biomass with other bio-fertilisers.

Due to inconsistent laws and a vague definition of what constitutes a bio-stimulant, the European market for these products has suffered. The adoption, marketing, and use of bio-stimulants in agriculture have been governed by the laws of each individual EU member state up until quite recently. As a result, expanding their business into other EU nations has become challenging and time-consuming for the producers. The Regulation EU 2019/1009, which establishes a standard definition and regulations for bio-stimulants, is timely in light of the market for them that is expanding quickly. Both producers of bio-stimulants and farmers now face fewer obstacles because of the uniform EU law. In this rule, the maximum limits for heavy metals are specified, and they are the same as the limits established for biofertilisers. The labels of bio-stimulants may only list qualities that have been scientifically proven. A few general guidelines for defending plant bio-stimulant claims are put forth by Ricchi et al. (2019), including the number of field trials necessary for each crop. The European Council for Standardisation (CEN) is establishing standards to facilitate the implementation of the Regulation.

10 Herbal Medicines

Because of certain properties such as polysaccharides and fibres like alginates and carrageenan, macroalgae can have therapeutic effects. *Fucus vesiculosus*, has a dry weight that is 50% fibre (fucoidans, which are complex polysaccharides). The insoluble fiber in *Fucus* contributes to increased colonic fermentation and improved gut flora, which also has an anti-obesity effect (Beaulieu 2019).

In the EU, pharmaceuticals made containing macroalgal components must adhere to the EU's drug laws. The EU Directive 2001/83/EC outlines the regulations for human medications. 1 A pharmaceutical product must meet the following criteria: Any substance or combination of substances that is "(a) presented as having properties for treating or preventing disease in human beings; or (b) any substance or combination of substances which may be used in or administered to human beings with a view to restoring, correcting or modifying physiological functions by exerting a pharmacological, immunological or metabolic action, or with a view to making a medical diagnosis" (Directive 2001/83/EC, Title I, Article 1).

The Directive 2011/83/EC establishes additional regulatory processes for (a) herbal medical products and (b) traditional herbal medicinal products in addition to the "regular" regulatory process for synthetic medicines. Algae fall under the category of herbal substances, and both the ingredients and preparations made from algae are considered herbal medicines. The scientific literature must demonstrate established medical usage, acknowledged efficacy, and an acceptable level of safety before herbal medicinal products can be authorised. The marketer must show only traditional use in order to get a product registered as a traditional herbal medicine; no claims of therapeutic efficacy are permitted. Conventional use refers to a period of at least 30 years, including at least 15 of those years spent in the EU. An illustration of a traditional herbal medicine product is a powder tablet that is used in conjunction with a weight loss regimen for overweight individuals. In all of Europe, there are commercially accessible tablets containing 130 mg of *Fucus vesiculosus* (bladderwrack). As a conventional herbal medication, the item was authorised in 2019. A product with a macroalgae-derived substance that improves health (such a prebiotic or a bioactive peptide) may also be marketed as a dietary supplement rather than as a herbal remedy (Beaulieu 2019).

11 Conclusion

Macroalgae are intriguing for the development of novel goods due to a number of their characteristics. Product regulations should safeguard consumer rights and public health while not placing an undue burden on businesses in Europe or other countries. EU directives establish common goals for all member states, but governments are free to choose how to get there. Discrepancies may result from directives to some extent. Hence, they do not need to be implemented at the level of the member states; in turn, EU legislation are directly applicable in member states. Harmonized EU regulations make the working environment clear to all parties. The creation of methodology for safety evaluation at the EU level, along with centralised processes for novel foods, food additives, feed, cosmetics, and fertilizers, aid in the realisation of the single market for goods.

Acknowledgments ALEHOOP project has received funding from the Bio Based Industries Joint Undertaking (JU) under grant agreement No 887259. The JU-CBE receives support from the European Union's Horizon 2020 research and innovation programme and the Bio Based Industries Consortium.

References

Akila V, Arjunan M, Sahaya Suheektha D, Balakhrisnan S, Munusamy Ayyasamy P, Rajakumar S (2019) Biogas and biofertiliser production of marine macroalgae: an effective anaerobic digestion of *Ulva* sp. Biocatal Agric Biotechnol 18:101035

Araujo R, Sanchez Lopez J, Landa L, Lusser M (2021) Current status of the algae production industry in Europe: an emerging sector of the blue bioeconomy. Front Mar Sci 7:626389

Balina K, Romagnoli F, Blumberga D (2017) Seaweed biorefinery concept for sustainable use of marine resources. Energy Procedia 128:504–511

Barbier M, Charrier B, Araujo R, Holdt S, Jacquemin B, Rebours C (2019) PEGASUS – PHYCOMORPH European guidelines for a sustainable aquaculture of seaweeds. In: Barbier M, Charrier B (eds) COST action FA1406, Roscoff

Beaulieu L (2019) Insights into the regulation of algal proteins and bioactive peptides using proteomic and transcriptomic approaches. Molecules 24(9):1708

Biancarosa I, Espe M, Bruckner CG, Heesch S, Liland N, Waagbø R, Torstensen B, Lock EJ (2017) Amino acid composition, protein content, and nitrogen-to-protein conversion factors of 21 seaweed species from Norwegian waters. J Appl Phycol 29:1001–1009

Cherry P, Yadav S, Strain C, Allsopp P, McSorley E, Ross R, Stanton C (2019) Prebiotics from seaweeds: an ocean of opportunity? Mar Drugs 17(6):327

Couteau C, Coiffard L (2016) Seaweed applications in cosmetics. In: Fleurence J, Levine I (eds) Seaweed in health and disease prevention. Academic Press, London

Ergün S, Soyutürk M, Güroy B, Güroy D, Merrifield D (2009) Influence of Ulva meal on growth, feed utilisation, and body composition of juvenile Nile tilapia (*Oreochromis niloticus*) at two levels of dietary lipid. Aquac Int 17:355–361

European Commission (2017) Food from the ocean. Scientific Opinion No. 3/2017 Publications Office of the European Union, Luxembourg. Available at: https://ec.europa.eu/research/sam/pdf/sam_food-from-oceans_report.pdf

FAO (2021) Seaweeds and microalgae: an overview for unlocking their potential in global aquaculture development. https://openknowledge.fao.org/server/api/core/bitstreams/43da43ad-be75-4aaa-ba4a-e375b1c65cfc/content

Garcia-Vaquero M, Hayes M (2016) Red and green macroalgae for fish, animal feed and human functional food development. Food Rev Intl 32(1):15–45

Joshi S, Kumari R, Upasani VN (2018) Applications of algae in cosmetics: an overview. Int J Innov Res Sci Eng Technol 7(2):1269–1278

Karatzia M, Christaki E, Bonos E, Karatzias C, Florou-Paneri P (2012) The influence of dietary Ascophyllum nodosum on haematologic parameters of dairy cows. Ital J Anim Sci 11(2):e31

Lakshmi S, Sankaranarayanan S, Galaria T (2020) A short review on the valorisation of green seaweeds and Ulvan: FEEDSTOCK for chemicals and biomaterials. Biomol Ther 10(7):991

Maia M, Fonseca A, Oliveira H, Mendonç C, Cabrita A (2016) The potential role of seaweeds in the natural manipulation of rumen fermentation and methane production. Nat Sci Rep 6:32321
Pimentel FB, Alves RC, Rodrigues F, Beatriz M, Oliveira P (2018) Macroalgae-derived ingredients for cosmetic industry—an update. Cosmetics 5(1):2
Rey-Crespo F, Lopez-Alonso M, Miranda M (2014) The use of seaweed from the Galician coast as a mineral supplement in organic dairy cattle. Animal 8(4):580–586
Ricchi M, Tilbury L, Daridon B, Sukalac K (2019) General principles to justify plant biostimulant claims. Front Plant Sci 10:494
Sharma S, Fleming C, Selby C, Rao J, Martin T (2014) Plant biostimulants: a review on the processing of macroalgae and use of extracts for crop management to reduce abiotic and biotic stresses. J Appl Phycol 26:465–490
Swedish-EPA (2013) Hållbar Återföring Av Fosfor (sustainable phosphorous recycling, in Swedish with English summary), report 6580. Stockholm, Sweden
Torres MD, Kraan S, Domínguez H (2019) Seaweed biorefinery. Rev Environ Sci Biotechnol 18(2):335–388
Wan AHL, Davies SJ, Soler-Vila A, Fitzgerald R, Johnson MP (2018) Macroalgae as a sustainable aquafeed ingredient. Rev Aquac 11(3):458–492
Wang HM, Chen CC, Huynh P, Chang JS (2015) Exploring the potential of using algae in cosmetics. Bioresour Technol 184:255–362
Weinberger F, Paalme T, Wikstrcim SA (2020) Seaweed resources of the Baltic Sea, Kattegat and German and Danish North Sea coasts. Bot Mar 63(1):61–72
Zakaria Z (2015) Regulation of cosmetics: what has Malaysia learnt from the European system? J Consum Policy 38:39–59

Chapter 19
Global Initiatives, Future Challenges, and Trends for the Wider Use of Seaweed

Pranav Nakhate and Yvonne van der Meer

1 Introduction

According to the European Commission report on the common fishery policy, the aquaculture sector has a high potential for sustainable and inclusive growth. It is evidenced by the fact that various EU countries are investigating seaweed-based cultivation, biorefineries, and products (European Commission 2015). The estimated market value of seaweed as a biobased raw material for various products is nearly €10 billion, with an expected growth of 6% annually (Cottier-Cook et al. 2016). According to the estimates, 70% more food will be required worldwide by 2050 compared to the present consumption, wherein the abundant availability of seaweed will surely help in addressing this challenge (Searchinger and Heimlich 2015). Seaweed-based products are widely accepted in the European market for food, cosmetics, and other biomolecules like carrageenan. Traditionally, seaweed is the most underutilized biomass, which constitutes only 2% of the overall sea-based food, despite occupying nearly 50% of the overall sea-based biomass (O'Shea et al. 2019; Searchinger and Heimlich 2015). Nearly 100,000 seaweed species are identified in various parts of the world, and few of them are already used for food, biofuel, nutraceutical, cosmetic and feed applications. Seaweed contains carbohydrates like glucose, galactose, alginate, xylose, and mannitol that can be processed into economically essential products (Guiry et al. 2014). Seaweed is one of the fast-growing species in the world, exceeding any terrestrial plants. Wild Kelp, for example, can grow faster than bamboo, with a growth rate of 7–14 cm per day. It was reported that the sugar kelp grows nearly 1.1 cm/day, reaching around 2.25 m in a year. In terms of productivity, seaweed outcasts the land's most productive crop, sugarcane, with

P. Nakhate · Y. van der Meer (✉)
Aachen-Maastricht Institute for Biobased Materials (AMIBM), Maastricht University, Geleen, the Netherlands
e-mail: yvonne.vandermeer@maastrichtuniversity.nl

F. Ozogul et al. (eds.), *Seaweeds and Seaweed-Derived Compounds*,
https://doi.org/10.1007/978-3-031-65529-6_19

an average production of 3–13 kg biomass per m^2 (Agostinia et al. 2015). The rapid growth rate of seaweed also signifies the CO_2 absorption capacity of the plant. It is estimated that 50–100 tons of CO_2 are absorbed per hectare of its production, meaning that the emissions released from the processing of seaweed-based products are quickly reabsorbed (Asmus and de Jonge 2021). Seaweed cultivation is considered the most gentle type among other aquaculture activities, as it does not require vast quantities of nutrients and fertilizers (Cottier-Cook et al. 2016).

According to the latest publications and EU conventions, Europe is already overutilizing the land-based biomass for the energy sector without considering any environmental safeguards; seaweed serves as a golden opportunity to establish a high-potential biobased industry from scratch. Various commercial products, apart from the conventional ones, are highly researched (Camia et al. 2018), and the technology is also sufficiently running at Technology Readiness Level 4, wherein the scale-up is excessively being researched. Cultivation and associated infrastructure have reportedly benefitted the environment; therefore, governments promote it globally (Cottier-Cook et al. 2016). For example, the Dutch climate agreement discusses 14,000 km^2 of seaweed cultivation (Van Den Burg et al. 2021). Moreover, with all the benefits, one may ask, is seaweed a potential solution to the present sustainability issues? For enhancing and supporting sustainable production at a global level, seaweed aquaculture needs additional research, investment, and directive legislation. Despite the current establishments, the growth in this sector has historically been linear for decades due to the lack of effective implementation of policies. The EU had set some common legislations for farm management, which affects the sustainable governing of seaweed cultivation. Such common legislation includes;

- Minimize the damage to the environment,
- Maintain the genetic diversity of the wild stocks,
- Avoid cultivation of non-native/alien species (subjected to change within the EU), and
- Take biosecurity measures for disease/parasite control.

It has been reported that the present legislative framework for biosecurity measures includes a variety of inconsistencies in cultivation-related terminologies, implementation of responsibilities, alignments with biosecurity and hazards, and evidence-based particulars (Asmus and de Jonge 2021). For seaweed farming, such legislations are assumed to improve the environmental impacts. For example, there is a negative emission potential in combining seaweed cultivation & harvesting with carbon capture and storage. This may help reduce excess CO_2 from the atmosphere and help in other subsequently allied policies (Asmus and de Jonge 2021; Cottier-Cook et al. 2021; Van Den Burg et al. 2021). However, the legislations have traditionally discussed the biosecurity measures and risks, as those are urgently required to limit the environmental impacts. Policymakers in the EU have established that the wide acceptance of seaweed or seaweed-based products is slightly hindered by technical aspects and largely hindered by commercial and legislative aspects (Searchinger and Heimlich 2015; Van Den Burg et al. 2021). The present literature also suggested

Fig. 19.1 Priority action areas for the broader use of seaweed

few public involvements, minor industry involvements, fewer long-term plans, and a lack of climate progression in the present form of the legislation.

Interestingly, seaweed-producing Asian countries have taken the initiative to boost the wide acceptance of seaweed by developing a policy-based framework associated with aquaculture. Such frameworks discuss the industry's involvement and cover the social, economic, and environmental aspects. This chapter walks you through the current global initiatives and trends in seaweed-based aquaculture and future challenges and strategies (Fig. 19.1).

With all the seaweed's potential, it is essential to find the right pathways to unlock this potential. Based on the importance of seaweed, breadths of actions are needed to tackle the present issues. In order to embrace a pragmatic approach, these priority actions are divided into six major areas that need urgent attention. These six areas of protection involve optimized seaweed farming, global mapping initiatives, creating a public-private partnership, developing seaweed policies, tackling the future challenges, and setting up the future trend. While the priority actions are highlighted in these six areas, the present book chapter endeavored to articulate how global initiatives, policy development, and tackling future trends and challenges are linked to sustainable seaweed farming for its wider acceptance.

2 Global Initiatives and Trends of Seaweed

Seaweed is the most promising and rapidly developing industry in the aquaculture sector. Seaweed cultivation restores and enhances biodiversity by giving habitat to aquatic species. According to the 2019 estimates, this industry has supported nearly six million small to medium scale farmers, especially from the low-income Asian and African countries, with a 15 billion market value (Wegeberg et al. 2013). Seaweed cultivation has received enormous interest from developing or developed countries due to its functionality and applicability in various sectors, helping them achieve the UN's Sustainable Development Goals. Since the expansion in the last few decades, various initiatives have boosted its global outreach. These initiatives have helped society's social, economic, and environmental aspects.

Phillippines is the third-largest seaweed producer globally after China and Indonesia, with nearly 200,000 people directly or indirectly associated with seaweed farming (Mateo et al. 2020). *Kappaphycus* is one of the economically significant red seaweeds, primarily cultivated in tropical waters. Colloids and carrageenans extracted from the *Kappaphycus* species have a considerable market demand for their applicability in the food, nutraceutical, and pharmaceutical industries. The successful commercial cultivation of *Kappaphycus* was firstly recorded in southern parts of the Phillippines in the 1960s with traditional cultivation techniques. Since then, the exponential growth of *Kappaphycus* has been observed in various parts of the Philippines in the last decade. The selling price of the dried *Kappaphycus* was around 1.9–2.4 USD per kg of dry seaweed, making it one of the leading sources of income in the period 2005–2010 (Bureau & Department of Environment and Natural Resources 2016). The *Kappaphycus* production declined in the Philippines, whereas Indonesia became a leading supplier of *Kappaphycus* from 2008 onwards, leading to the destruction of the economic cushion in the Phillippines. After that, the Phillippines government joined hands with the public-private partners to establish the national seaweed R&D program to resolve the existing issue and achieve maximum production of *Kappaphycus.* The program explored various aspects, like molecular taxonomy, tissue culture, sporulation factors, strain selection, and disease control for maximum production. The joint program identified that the root cause of the declined production was repetitive plantation leading to the limited growth and consistent occurrence of ice-ice disease (Hurtado et al. 2015). The program identified that the extract of *Ascophyllum nodosum*, a brown seaweed, can be used as a growth hormone to regenerate the growth of plantlets, which was supported by various other publications. Such studies have reported that *Ascophyllum nodosum* extract can improve the chlorophyll content, the number of leaves, the flavonoid content, vegetative propagation, and the tolerance toward marine abiotic stress (Kumar and Sahoo 2011; Kumari et al. 2011; Rayorath et al. 2008). This program also focused on improving cultivation management. The seaweed farmers from the Phillippines traditionally preferred the cultivation according to their economic needs, which significantly affected the quality and quantity of the crop (Villanueva et al. 2011). The program also observed that the long-line cultivation of a minimum of 30 days with approximately 500 g m^{-1} line could improve the carrageenan content and its molecular weight. The national seaweed R&D program also trained the seaweed farmers with the latest cultivation technologies, basic seaweed biology, and seasonal variation of the crop (Hurtado et al. 2013). The program also initiated interesting activities associated with the world bank to assist the local fishing activities. This initiative supports the livelihood of the local seaweed farming community by giving an additional source of income, hence increasing the per capita income and empowering the local women through their direct involvement in farming (Kunjuraman et al. 2019).

Indonesia is the second-largest seaweed producer, with nearly 200 tons of seaweed exported in 2019, estimated at around $280 million (Mariño et al. 2019).

Moreover, Indonesia is the largest carrageenan producer from dry seaweed, with a net export worth $162 million in 2017 (United Nations 2021). Seaweed farming also supports nearly one million directly linked households in Indonesia. However, similar to the Philippines, Indonesia also experienced challenges in the sustainable growth of seaweed in spreading diseases like ice-ice and other pests, dysfunctional value chains, and volatile prices. Such factors limited the growth potential and the quality of the carrageenan (Food and Agriculture Organization of the United Nations 2016). The lack of crop variation was also linked with the disease vulnerability. Taking this into account, the national government of Indonesia started working on re-engineering the present value chain. The government was partnering with the private sector in a consortium and advised farmers on the short time cultivation cycle and use of genetically varied seedlings for high production (Food and Agriculture Organization of the United Nations 2016; Kambey et al. 2020). This consortium also conducted workshops on sustainable crop production to help the local communities manage the technologies, unskilled labor, and capital investment. A biosecurity policy was also implemented wherein the regulations were made on strain selection and coastal geography improvement, ensuring the pest and disease-free cultivation of seaweed. This consortium also made financial arrangements, including a tax waiver on equipment and seedlings import. A transition to contract-based farming was also suggested for the stable and consistent income of the farmers. Digital incorporation was also suggested for better coherence and transparency within the supply chain (Kambey et al. 2020). The effective initiatives by the consortium resulted in improved family income with extended cultivation practices, as 31 out of 34 provinces are involved in this activity as of 2017 (Kambey et al. 2020, 2021).

The successful implementation of seaweed farming in the neighboring countries accumulated greater interest in Malaysia. The Malaysian government agency has taken many initiatives in the recent decade to help promote seaweed farming. In accordance with this, the Minimum Estate System (MES) and the Cluster System (CS) were developed to improve seaweed cultivation and supply chain scientifically (Wegeberg et al. 2013). With such initiatives, the government trained the local community in using modern technology, crop variations, skill improvement, and supply chain management. The MES and CS activities significantly reduced the labor force engaged with conventional farming and diverted the majority to modern farming techniques, resulting in flexible, cost-effective, and sustainable cultivation (Marine Scotland 2022). The MEC and CS initiatives also provided a significant opportunity for women in coastal areas resulting in improved family income. Apart from this, the Malaysian government is also giving attention to developing seaweed farming as community-based tourism (CBT) to enhance the region's social and economic aspects (Hussin et al. 2013).

Seaweed farming on Faroe Island was at a peak in the 1980s due to the natural habitat of wild seaweed. The alginate was extracted from the wild brown seaweed that still has a tremendous commercial value. However, the alginate concentration was eventually reduced due to a seaweed quality issue, and hence the farming was affected severely (Wegeberg et al. 2013). In the last decade, few private sector companies rejuvenated the seaweed farming sector on Faroe Island with improved quality seaweed and exported it to countries like Germany, the UK, The Netherlands, Denmark, Estonia, and North America (Christensen 2020). These companies used innovative offshore cultivation techniques like Integrated Multi-Trophic Aquaculture (IMTA) along with wind farms. IMTA is already in use in some parts of Asian farming systems; however, it has been recently scaled up in Norway and Canada (Barrington et al. 2010; Kleitou et al. 2018). This technique includes ring-shaped nutrient flow amongst various species on the trophic level. This bioremediation process via IMTA helps in reducing the harvest waste and clean-up for the new cultivation, which otherwise would cause marine eutrophication problems (Barrington et al. 2010; Kleitou et al. 2018). The wastes from the fish aquaculture, including fecal matter, drop down through the water channels and become food for the Bivalvia, whereas the dissolved waste is used as a nutrient for the seaweed cultivation. The seaweed and bivalve harvest remaining then become food for the fish aquaculture and close the loop (Barrington et al. 2010). It was estimated that if this integrated farming business model were implemented at a larger scale, this would help clean the ocean and add financial benefits to the coastal farmers (Christensen 2020).

In various South Asian islands, seaweed cultivation has been rapidly adopted, leading to the change in the island's demography. These islands' local cultivators and residents claimed to have a considerable return on investment since the adoption of seaweed farming (Steenbergen et al. 2017). Apart from the cultural rope, the island provides all the necessary materials required for cultivation. Moreover, the cultivation sites are nearby coastal areas, which provide easy access for the locals to their job. Seaweed cultivation is often considered a family business where men, women, and youngsters can work together fulfilling various tasks. The local governments encourage women from the coastal community to participate in cultivation activities. Women are welcomed and cognized in the agriculture sector, as according to the estimates, women's involvement has increased not less than 6% in the agricultural sector to nearly 45% (Ghosh and Ghosh 2014). Besides, the fifth goal of the United Nations Sustainable Development Goals (SDGs) also recognizes women's involvement in development projects, ensuring their recognition and development on par with men ("The Sustainable Development Goals Report," 2016), wherein seaweed cultivations are promoted as a focal point in achieving this. The demographic study conducted by Kronen et al. on the Soloman island in 2013 revealed that women actively participated in all the seaweed related activities, including cultivation (33%), harvesting (34%), maintenance (34%), drying (32%), packaging (24%), and selling (24%) (Ponia et al. 2013). The study concluded that the initiative of women's involvement has a positive chain effect leading to a better knowledge of strain and cultivation, enhanced monthly income, increased business opportunity, and better management (Kunjuraman et al. 2019).

According to the Food and Agricultural Organisation of United Nations (FAO), the world population will reach 9 billion by 2050, which would require a massive amount of nutritious food. Seaweed, as it contains a large number of proteins and vitamins, can be a nutritious and sustainable alternative to current food sources

(Bouga and Combet 2015). Seaweed consumption has various health benefits like lowering blood pressure, preventing stroke, etc., as reported in multiple reports (Al-Thawadi 2018). Considering these health benefits, many European countries are developing initiatives in gastronomy to estimate the new scope of seaweed consumption. One study surveyed the Italian population for their acceptance of seaweed as a part of their diet. Out of 257 people, 57% agreed and showed interest in seaweed inclusion as a part of a daily diet. It was reported that the seaweed characteristics and its health benefit were the most influential factors in the public acceptance (Palmieri and Forleo 2020). These surveys indicate the global trend in indicating the willingness of people to accept seaweed-based products in their daily routine. The seaweed-based vegetation market in European countries has significantly increased in the past few years. The reported market in western Europe by 2013 was €34 million and expected to grow 5% every year till 2040 (Marine Scotland 2022). Similar initiatives are being carried out in Canada, as Vancouver Island started the Seaweed day festival in 2021, where seaweed-based food products were highly promoted.

FAO has taken the initiative in supporting numerous projects to promote seaweed-based development in the past few decades. It is working on establishing a collaborative management framework for risk-based farming and an information system to secure the genetic data (Resources & Food 2019). Some of the FAO's initiatives include developing a practical manual, hands-on training, developing a technical platform, and facilitating collaboration amongst different parties. Few of the FAO global initiatives are grouped in Table 19.1 below.

3 Seaweed Policies

As discussed in the global initiatives section, seaweed is used in various countries for food, feed, pharmaceutical, agricultural, and other applications. Seaweed farming also reportedly restores and increases biodiversity and helps mitigate climate change by carbon sequestration. The consistent expansion of the seaweed-related activities benefits the development of a new ocean economy and conserves the marine ecosystem (Capuzzo and McKie 2016). However, this potentially dynamic industry lacks the infrastructure for commercial farming, as there are issues with supply chain management, seasonal variation, and high cost. Moreover, inconsistent yield, ecological effects, unregularly practices, and lack of information on the operational costs hinder the overall development (Capuzzo and McKie 2016). Experts emphasize that these challenges need to be addressed with community building, incentive programs, and policy-making. However, the absence of policy support is one of the most discussed underlying reasons reported by many authors, as there are only a few regulations on seaweed farming across the globe. In many countries, seaweed policies are majorly sub-part of the overall aquaculture policy due to a lack of information, leading to uncertainties in the seaweed farming sector (Barbier et al. 2020). Until the last decade, the inclusion of seaweed in biosecurity policies was

Table 19.1 FAO's initiatives to promote Seaweed (Cai et al. 2021)

Project Title	Country	Nature of Project	Timeframe
TCP/URT/3601/C1: Support to Seaweed Diseases and Die-off Understanding and Eradication in Zanzibar	Zanzibar, United Republic of Tanzania	Biosecurity	2016–17
FMM/GLO/112/MUL: Baby 4—Blue Growth Initiative in Support of Food Nutrition Security, Poverty Alleviation, and Healthy Oceans	Kenya	Policy planning & business planning in the public-private sector	2015–17
TCP/BGD/3704 (19/I/BGD/238): Support to Seaweed Cultivation, Processing, and Marketing Through Assessment and Capacity Development	Bangladesh	Capacity building	2019–21
TCP/INS/3502: (Indonesia)—Decent Work for Food Security and Sustainable Rural Development.	Indonesia	Managing socio-economic impacts	2015–17
GCP/CHI/039/GFF: Strengthening the Adaptive Capacity to Climate Change in the Fisheries and Aquaculture Sector of Chile	Chile	Climate change	2017–21
GCP/RLA/230/IFA: Farmers' Organizations for Africa, Caribbean, and Pacific Countries Program in the Caribbean Region	Grenada and Saint Lucia	Poverty alleviation and food security	2020–23
GCP/SLC/202/SCF: Climate Change for the Eastern Caribbean Fisheries Sector	Dominica, Grenada, Saint Lucia, Trinidad and Tobago, Antigua and Barbuda, Saint Kitts and Nevis, Saint Vincent and the Grenadines	Adaption to the climate change	2007–21

debatable. It was proposed as an aquatic animal health holder under the sanitary trading standards of the World Organization for animal health and a plant holder in the International Plant Protection Convention (IPPC). IPPC is a multilateral treaty managed by United Nations FAO, responsible for the coordinated efforts in disease & pest control and safeguarding natural flora. It suggested that the cultivated aquatic plants should be a part of their sanitary standards to ensure their risk analysis (Bolton 2020; Food and Agriculture Organiization of the United Nations 2017). As a result, IPPC is one of the few organizations that define "aquatic plants" in their International Sanitary and Phytosanitary Measures (ISPMs) policy framework (Campbell et al. 2020). Nevertheless, the inclusive representation of entire seaweed species and their genetic interpretation is still missing.

4 Why is Policy Important?

The experts believe that developing a seaweed-related policy guide can bring uniformity to co-operate the farming operations, reduce the risk of uncertainty, and serve as a one-stop guide for the seaweed farmers and investors. Establishing seaweed policies can lead to process regulations and licensing, which can help farmers determine suitable farming locations and avoid conflicts with other marine activities. The policy establishment can help track down the potential farming and harvesting impacts, which may lead to marine ecology conservation. Similarly, the introduction of seaweed-related policies can help scale up the activities leading to higher productivity and yield. The information on algal strains, production skills, and technical know-how can be spread uniformly with the policy intact (Valderrama et al. 2013). It can also help acquire capital investments and establish the infrastructure, including storage, facilities, and distribution. The lack of knowledge of the size and location of the market is often an obstacle to any relatively new product. In the case of seaweed, the market is still non-static, as few products have global demand while many others have local demand. The policies and subsequent regulations can help seaweed explore the vast market for products, by-products, and co-products. Biosecurity is the most discussed and promoted policy in the aquaculture sector, which deals with natural risks and liabilities associated with the biological product. This policy ensures consistent productivity and yield by identifying pests and diseases, reporting outbreaks, and developing seed banks & nurseries (Kimathi et al. 2018). Such policies in the seaweed sector can help avoid the natural casualties caused by "ice-ice" disease predominantly affecting the seaweed crop in Asian countries. The seasonal variations and price volatility are the significant factors that affect the seaweed farmers' livelihood and impact the entire value chain. The seaweed-related policies and the subsequent regulations can assure consistent supply by promoting dedicated storage facilities (Valderrama et al. 2013). As there is little mention of seaweed in the existing policies, it is essential to make an effort to develop an international policy that can serve as a solution for the investor and farmers (Fig. 19.2).

5 Global Initiatives in Making Seaweed-Related Policies

As discussed earlier, there is a lack of seaweed specifications in the existing aquaculture policies. However, a specific international policy is required to manage the sustainable growth of seaweed and flourish its products. IPPC has taken a direct initiative to include seaweed as an aquatic plant in its ISPMs policy framework. Moreover, other international bodies have also included seaweed in their policy framework. The country-specific policies are gathered from the literature and presented in Table 19.2.

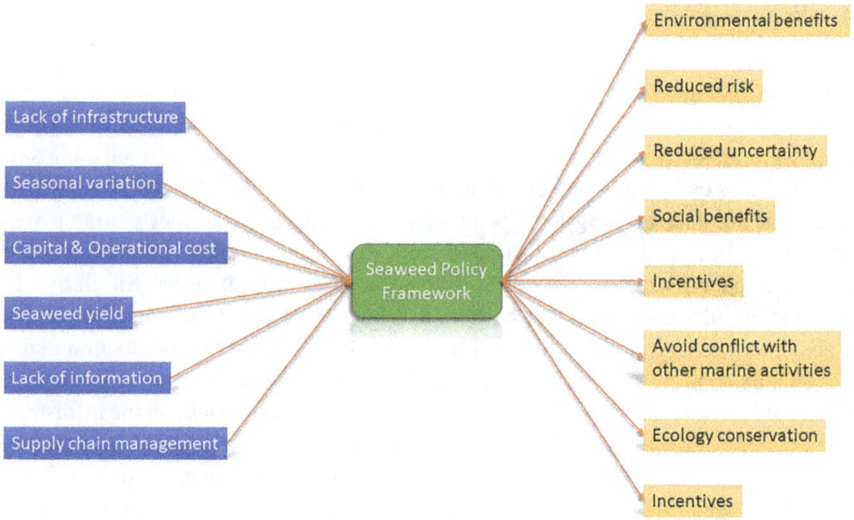

Fig. 19.2 The importance of policy frameworks for the broad acceptance of seaweed

In the Union Republic of Tanzania, the biosecurity policy is operated by two government bodies. Both government bodies have started explicitly collecting the field data to include seaweed in their biosecurity policy. It was observed that the seaweed aquaculture was mentioned in 69% of the biosecurity policy, focusing on four main biosecurity components: seaweed transportation, quarantine measures, the introduction of alien seaweed species, and import regulations. The prime objective of including seaweed in the biosecurity policy is to directly line up the seaweed sector to the Tanzanian Poverty Alleviation strategy and United Nations millennium SDGs (Ministry of Communication, Science, and Technology, Government of Tanzania 2010; United Republic of Tanzania 2015). In Indonesia, the government is proactive in developing and promoting the policy framework for the sustainable growth of aquaculture. The recent regulations focus 27% on seaweed farming, concentrating on the empowerment of farmers, national competency for seaweed processing, tax-free import of equipment, and seedling distribution (Kambey et al. 2020).

The main associated policy framework in the EU for seaweed includes Water Framework Directives (WFD) 2000/60/EC, The Habitats Directive 92/43/EEC, Alien Species Regulations (ASR) 2014/1143/EU, and Maritime Spatial Planning Directive (MSPD) 2014/89/EU. The WFD was established to protect eutrophication and ecology, wherein the ASR was introduced to protect and manage invasive alien species in the aquatic habitat (Schratter-Sehn et al. 2000). The goal of the Habitat Directives is to conserve the wild flora and fauna to promote a sustainable ecosystem in the EU. The MSPD framework was developed to maintain the marine environmental status and estimate the effects on the biodiversity, ecosystem, and aquatic species (The European Parliment and the Council of the European Union 2014).

Table 19.2 Global initiatives in making seaweed-related policies

Country	Policy	Policy focus	Seaweed related focus	References
Tanzania	Tanzania Aquaculture Development Strategy	Capture fisheries and finfish aquaculture	plant quarantine and phytosanitary measures	Ministry of livestock and fisheries, Tanzania (2018)
Zanzibar island	Zanzibar Research Agenda (2015–2020)	Education, health, fisheries, and aquaculture	plant quarantine and phytosanitary measures	United Republic of Tanzania (2015)
Indonesia	Aquaculture farm management regulations (PPRI No. 28/2017)	Fish cultivation & utilization, Infrastructure support, Quality control, environmental impact management	Ecology interaction, Incentivization	Kambey et al. (2020)
Indonesia	Fish quarantine requirements (MMAF-PER No. 10/MEN/2012)	Fish quarantine protocol, Pests & disease control	Pests and pathogen control	Kambey et al. (2020)
Philippines	Code of Good Aquaculture Practices (GAqP) for Seaweed (PNS/BAFS 208:2017)	Guidance on good aquaculture practices, risk and hazard management	Guidance on site selection, harvesting, transportation, and sanitization	Mateo et al. (2020)
Malaysia	the Malaysian Good Agricultural Practices in 2014 (MyGAP)	Management of food aquaculture practices	Crop certification & Crops standardization	Kambey et al. (2021)
Finland	Maritime Spatial Planning Directive (MSPD) 2014/89/EU, Finnish Bioeconomy Strategy (2014), Aquaculture Strategy 2022	ecosystem approach to promoting sustainable economic development and ecological protection	Good management of space use, Maximum production with minimum environmental impacts	Barbier et al. (2020)
Sweden	The Marine Strategy Framework Directive (MSFD) 2008/56/EC, Swedish Environmental Code Ds 2000:61, and the Plan and Building Act (2010:900).	to achieve or maintain Good Environmental Status of the marine areas	Aquaculture activities to get proteins, oil, vitamins, and produce ingredients for medicine, feed, fuel, and food.	Barbier et al. (2020), Camarena Gómez and Lähteenmäki-Uutela (2021)

(continued)

Table 19.2 (continued)

Country	Policy	Policy focus	Seaweed related focus	References
Estonia	Estonia 35	To develop aquaculture, fisheries and maintain the aquatic ecosystem	Mentioned as a potential aquaculture activity.	Camarena Gómez and Lähteenmäki-Uutela (2021)
Germany	National Bioeconomy Strategy 2030	Aquaculture improvement	National Bioeconomy Strategy 2030	Camarena Gómez and Lähteenmäki-Uutela (2021)
UK	Marine and Coastal access act and Environmental Impact Assessment Directive (97/11/EC)	Inshore and Offshore marine management	Commercial activities are mentioned in Annex II, but the circumstances are not clear	Capuzzo and McKie (2016)

Many European countries have adopted these frameworks wherein attempts have been made to include seaweed within such frameworks. Associated with this, the Finnish government adopted the MSPD framework of the EU and developed the MSP2030, where the goal is to incorporate the innovations within different sectors to produce new products from the marine biomass (European Commission 2012). The Swedish government is also adopting the MSPD framework from the EU to specify the marine activities, including energy extraction, sand extraction, maritime and commercial shipping, energy transmission, etc. However, seaweed is not explicitly included in such frameworks; the extracted value-added products of seaweed like protein, vitamins, oil, food ingredients, medicine, and fuel are already included (Governmnet Offices of Sweden 2015). The same story followed in the other EU countries like Germany, Denmark, Poland, Estonia, Latvia, and other countries like the UK and Russia, as they have either adapted to the EU policy or are on the verge of establishing their policies. The bottom line is that seaweed has gained much interest in the EU countries, and these countries are preparing norms to support the seaweed for its wider acceptance. The UK adopted the Marine and Coastal Access Act (MCAA), which deals with marine licensing for England, Welsh, Northern Ireland, and Scotland. Seaweed farming and related activities are not covered in this act; however, the Environmental Impact Assessment Directive (97/11/EC) mentions the Environmental Impact Assessment (EIA) requirement for seaweed farming (Capuzzo and McKie 2016).

6 Policy-Based Recommendations for the Global Outreach of Seaweed

Policy-making and its implementation can help in sustainable product making, distribution, and extended outreach. Several seaweed policies and legislative frameworks are analyzed in the policy section of this chapter, and it is observed that there are no specific policies dedicated to seaweed farming, production, or its supply chain. The deficiencies in the present framework involve inconsistent definitions, unclear responsibilities, insufficient and evidence less data, and limited alignment with the biosecurity framework. Based on the interpretation of the present policies, various experts have made policy recommendations that may help seaweed achieve the extended outreach. The positive aspect is that government organizations across the globe have started to consider seaweed in their policy-making, which surely can boost the seaweed sector. Establishing the equilibrium between ocean health, economic growth, and social justice is essential. The experts recommend developing the research centers to investigate the new indigenous species for the disease and pests control and higher yield. Pathogen profiling may help in also help in understanding the interactions and genetic variations in a particular geography. Such research establishments can also set the industry-oriented standards for the pathogen-resistant seaweed, which may help small to large-scale industries. The experts also recommended nurturing the genetic diversity in the wild seaweed stock by blocking the use of non-indigenous species for commercial cultivation (Cottier-Cook et al. 2016). Establishing a seaweed seed bank is also the recommendation by the experts to maintain the disease and pest-resistant seeds to be used in case of a disease outbreak. Experts strongly believe that priority has to be given to improvising the biosecurity program. The Biosecurity frameworks are adopted in many countries, especially in the food production and processing sectors (Rodgers et al. 2015; Stentiford et al. 2017; Subasinghe et al. 2012). However, the inclusion of seaweed has lagged, making it more vulnerable to diseases, especially in lower-income countries. It is essential to include the seaweed in the biosecurity policy and communicate this to the farmers for its implementation at the farm level, making them well informed about the good farm practices, quarantine procedures, and incentivizing the diagnosis (Kambey et al. 2020; Subasinghe et al. 2012). Experts insist on applying the social welfare systems to seaweed farming. Though seaweed activities and their monetary contribution to the overall aquaculture sector are limited, it is the prime source of income for an entire family in many underdeveloped countries. Moreover, coastal and rural women get associated with seaweed farming as a family business, making it an effective tool for women's empowerment. The association of social welfare systems with seaweed farming can help in improving the socio-economic status and achieving the UN's SDG 5, 8, and 13 related to gender equality, economic growth, and climate actions respectively (Kambey et al. 2020; Subasinghe et al. 2012). Long-term Incentivization or innovative financial schemes that safeguard the business, especially in the case of disease spread, is also recommended by the experts, which can also attract investors. Experts suggested the integration of seaweed farming with the IMTA, which can reduce the ecological impacts and give economic diversification. Developing a feedstock and market assessment tool for facilitating the licensing, future investment, and promotion, establishing a seaweed-specific regulation, and developing a competent authority at the decision-making level will boost the seaweed products and, hence, their application in various fields (Campbell et al. 2020).

7 Global Trend

The growth in the seaweed industry is impressive and expected to increase further due to effective market growth and enhanced activities. Seaweed consistently finds its space in various food cuisines and dishes due to its texture and flavor. A variety of benefits are offered by seaweed as it contains essential health nutrients responsible for the sound functioning of the thyroid, insulin production, and oral care. Seaweed, however, has been traditionally considered for food applications in Asian countries, as described earlier (Blikra et al. 2021). The Nordic food trend has started including locally sourced plants, especially seaweed, in their cuisine in the last few decades, as a few companies are working on developing a salmon substitute for vegan people. Dry seaweed flakes are used as a salt replacement, whereas some seaweed like *Polysiphonia lanosa or Palmaria palmate is used as a flavoring agent. It has also been used as a bulking agent to bulk out unnecessary ingredients. It is also categorized as a superfood thanks to the abundant presence of proteins and dietary supplements. Seaweed powder is also reportedly used as a coloring agent and texture improver in dairy, beverage, fish, or meat products* (Mouritsen 2012).

Apart from its traditional usage, attention is now given to its biochemical composition as it contains a variety of phytochemicals, nutrients, minerals, dietary fibers, and minerals (Holdt and Kraan 2011; Maehre et al. 2016; Stévant et al. 2017). Algotherapy is discussed frequently, wherein seaweed extract is used for skincare routine and skin therapy. So, algotherapy has become a common practice in seaside resorts. The demand for seaweed in nutricosmetics, cosmetics, and cosmeceuticals is increasing, being a natural ingredient (Lopez-Hortas et al. 2021). Seaweed contains hydrocolloids that reportedly offer many features to cosmetics products, such as thickening, emulsifying, moistening, texturizing, and gelling agents (Pereira 2018). The extract of some seaweed species, like *Codium tomentosum*, is used as moisturizer, whereas the extract of other species, like *Tetraselmis suecica*, is used in antiaging formulations due to the presence of phenolic compounds (Lopez-Hortas et al. 2021; Pereira 2018). Ulvan species are well known for their antioxidant properties, and hence their extract is used in body lotions, gels, and moisturizers (Goldberg 1943; Wang et al. 2013) (Fig. 19.3).

Seaweed has found a way in the pharmaceutical and nutraceutical sector as seaweed-based oil is now approaching the market with a higher content of omega-3. The comprehensive knowledge of and ever-expanding trends in seaweed-based products indicate the enormous commercial scope of seaweed extracts (Ganesan

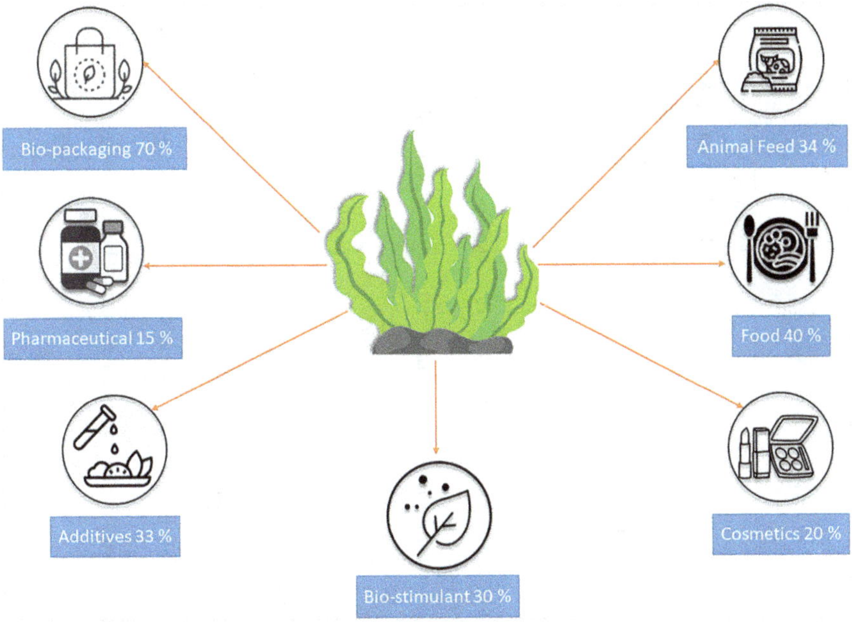

Fig. 19.3 Potential market share by 2030 (Vincent et al. 2020)

et al. 2019). The fucoxanthin and fucoidan extracted from seaweed species like Ecklonia stolonifera, Eisenia bicyclis, and Sargassum horneri, have been reportedly used to lower the blood glucose level (Ma et al. 2014; Moon et al. 2011). Seaweed is also reported to increase the immunological response, making it suitable for immunotherapy, antibiotics, or vaccines. *The seaweed-derived phlorotannins and phloroglucinols have been studied as chemo-preventing, antiaging, and antiallergic agents* (Martins et al. 2014). *Marinomed Biotech AG also introduces an effective antiviral nasal spray made out of carrageens into the market.*

In addition, seaweed is extensively researched in bioplastic applications where many companies are working on developing biobased cutleries, glasses, or packaging materials. The use of seaweed residues in chemicals and fuel-making is already widely discussed in different forums (Hayashi et al. 2020). *Various active seaweed-based products are being introduced to multiple emerging markets across the globe and provide the broader spectrum of seaweed and its emerging trend; seaweed can become a catalyst for the next generation of the green industrial revolution. Many companies are putting their efforts into making value-added products from seaweed, and a few of them are listed below in* Table 19.3.

Apart from the companies mentioned in Table 19.3, *various start-ups, NGOs, and non-profit organizations are working on the maximum utilization of seaweed to tackle various social challenges that arise today. However, the success of these seaweed-based companies depends on various factors like market creation,*

Table 19.3 Companies involved in the production of seaweed-based products

Company	Country	Segment	Benefit	References
Notpla (fka Skipping Rocks Lab)	*UK*	*Sustainable packaging*	*Create plastic packaging with low environmental impacts*	https://www.notpla.com/
Algiknit	*USA*	*Textile*	*Developing durable textile yarn from seaweed*	http://algiknit.com
Neutrogena Corporation	*USA*	*Cosmetics*	*A hydrogel* Neutrogena MaskiD™ face mask	https://www.neutrogena.com/
Blue Evolution	*USA*	*Food*	*Plant-based meat, soup, and sausage.* *Unique marketable products like Alaskan Kombu, Alaskan Wakame, etc.*	https://www.blueevolution.com/
Centre for Process Innovation	*UK*	*Biofuel*	*Assessing the feasibility of bioenergy production from sugary seaweed via anaerobic digestion*	http://www.uk-cpi.com
Symbrosia, Inc.	*USA*	*Feed*	*Developed seaweed-based commercial product* SeaGraze™ for animal feed	https://symbrosia.co/
Seaweed & Co Ltd	*Germany*	*Nutraceuticals*	*Developed a wide range of PureSea® nutraceutical products*	https://www.seaweedandco.com/
Sway	*USA*	*Bioplastic*	*Compostable replacement of single-use plastic*	https://swaythefuture.com/
Acadian Seaplants	*Canada*	*Nutraceutical and Cosmetics*	*Wide range of products in the* Acadian SeaPlus™ series	http://www.acadianseaplants.com
Sea6energy	*India*	*Seaweed farming*	*Automated farming and marine infrastructure solutions*	http://www.sea6energy.com/
Volta Greentech	*Sweden*	*Feed*	*Developed seaweed-based commercial product* VoltaSeaweed for animal feed	https://voltagreentech.com/
Marinova Pty Ltd	*Australia*	*Food and nutraceuticals*	*Çommercial production of* Maritech®organic fucoidan extracts	http://www.marinova.com.au/

people's perception, efficient supply chains, and investors' willingness to provide capital.

8 Tackling Future Challenges

According to the estimates, seaweed has the potential to achieve 30% of the market share in most of the segments mentioned in Table 19.3 (Vincent et al. 2020). Competitive dynamics suggest a vast market potential for seaweed and seaweed-based products across the globe. However, the present ecosystem imposes structural constraints that need urgent attention. Moreover, many system-level improvements are needed by 2030, including expansion of the farming capacity, regulations reforms, and innovative biorefineries, to name a few (Vincent et al. 2020).

One of the first issues to tackle is maintaining the continuous demand and quality. Seaweed needs to grow consistently with specific biochemical structures and compositions. In order to achieve this, the hatchery phase needs to be appropriately in line with the physical and chemical properties of the surrounding environment and genetic selection (Charrier et al. 2017). The traditional vegetative propagation method is prone to pests and disease, reducing productivity and concentration of bioactive compounds. Therefore, research institutes need to develop new micro-propagation and tissue culture techniques that lead to pest and disease-resistant plants (Jong et al. 2015). The innovations and market demand should drive the seaweed manufacturing process and extraction processes of bioactive compounds. Most western consumers consider seaweed as traditional Asian cuisine and are unaware of its existence in other products. The investors and producers need to promote awareness on that front by providing specific knowledge tools like media promotions, workshops, conferences, incentives, and documentaries. In the case of food applications, the explicit seaweed demand is not recorded, making it difficult for consumers to locate the seaweed products. Consumers, especially from the western world, have less knowledge about seaweed-based product recipes, hence, the variation in taste judgment. Innovative marketing strategies can guide the consumers about the seaweed and promote the seaweed-based cooked products to the households.

As discussed earlier, government organizations can promote seaweed-based knowledge through public procurement and policy-making. In addition, government organizations can arrange public funds, scholarships, and incubation hubs, thereby creating a tangible environment to promote seaweed-based start-ups. While using seaweed as an integral part of food, pharmaceuticals, nutraceuticals, or supplementary materials, certifications and safety standards are required. Most of the time, the standards are not clearly communicated, or there is a lack of standardization associated with seaweed. As the seaweed can absorb various nutrients and heavy metals from the sea, the constituents, the area of production, and health safety issues should be published alongside the product (Barbier et al. 2020). This activity requires research organizations, industry stakeholders, and government organizations to work in a coalition.

The inappropriate grouping, complexity in the application, and higher processing timeline affect the potential investor's interest in seaweed-based activities. As discussed earlier, seaweed is not explicitly defined in the aquaculture policies, making it difficult to process the seaweed-based licensing. In many cases, licensers and applicators do not know where and how to apply for the seaweed licensing. The potential solution to this dilemma is to involve seaweed in national aquaculture policies or allow the social license to operate (SLO). Seaweed farming can be performed alongside other aquatic activities like fishery, finfish, or oyster farming, thereby obtaining symbiotic association. Integrating such activities can help farmers earn extra income, boost the seaweed farming capacity, and reduce the licensing complexity (Steenbergen et al. 2017). Such multi-trophic farming can become an initial phase in attracting the potential investors and linking the small-scale farmers with the commercial activities (Billing et al. 2021).

Though seaweed farming is considered an excellent strategy to mitigate climate change, this may not translate to local coastal acceptance. The primary concerns of the local coastal residents are the foul smell, emission perception, traffic congestion, conflicts with other marine activities, visual impacts, marine life, health concerns, and external interventions on local properties. The consumer's perception or insecurities might affect the potential expansion of seaweed. The government should take the initiative to mitigate the doubts of the local community by arranging a knowledge drive, thereby promoting the socio-economic and environmental benefits (Firestone et al. 2012; Jensen et al. 2018). Though societal impacts are not calculated in the seaweed sector yet, the government should advertise the social equity associated with seaweed farming as the vast number of women participate in this activity.

Another essential activity is to train the workforce for sustainable seaweed production. Thousands of workforces are required in terms of hatchery specialists, nursery experts, onshore cultivators, harvesters, extractors, transporters, and other personnel associated with processing of seaweed. As most cultivators are unfamiliar with modern technologies, government organizations and NGOs should arrange the knowledge drive and training workshops. The coastal cultivators can apply modern knowledge to their farms and get most of the benefits. Maintaining the seaweed production cost is another crucial segment that needs to be prioritised in the future. Unlike Asian cultivation, Western seaweed cultivation is not considered an economically feasible activity. It is reported that current selling prices per kg of seaweed are higher than the import cost, making it challenging to attract new potential investors. To reduce the production cost, various actors play a significant role. To achieve economic feasibility, Western cultivators need to learn from the Asian seaweed farms in terms of techniques and adapt the new methodologies. Newer technologies in the cultivation or hatchery process and adapting seasonal variation can yield higher than traditional seedling twine (Kimathi et al. 2018). Remote sensing and automation at every stage may help achieve higher yields and better processing. AtSea Nova, a Belgium-based company, has provided automated mechanical solutions and floating machines for seaweed farming, significantly reducing losses and yield (Vincent et al. 2020). Identifying new revenue streams via biorefineries is another solution to reduce the per kg cost of seaweed. Various activities like CO_2 sequestration, nutrient uptake, and eco-tourism can help in generating alternative revenues and expanding the existing farms. Additionally, the valorization of seaweed, prioritizing the highest valorized product value in a cascaded approach, and sharing the facilities may help reduce the seaweed production costs.

In general, seaweed farming is a frontrunner in aquaculture with benefits like climate change mitigation and societal and economic benefits, and so on. However, the hurdles associated with seaweed farming are the lack of consent, legitimacy, consumer perception and trust, and compatible costing. Nevertheless, adopting the latest technologies, networking, advertising, promoting, policy-making, and communicating the importance of seaweed farming can help expand seaweed farms in Western countries.

9 Conclusions

From a sustainability point of view, seaweed should be considered an integral part of the Western food, feed, pharmaceutical, nutraceutical, and cosmetics market instead of a newly added product. Though the seaweed market is still taking shape, the prerequisites for creating a value chain are still not developed enough to support the growth. The production cost is still high, the identification of the market-specific products is still lacking, the environmental benefits are not quantified enough, automation in cultivation and harvesting is still lacking, policies are not defined well, and the consumer's perception is still not tackled enough. With all these challenges, the seaweed sector should focus on a qualitative assessment and rethinking its role in each sector with acquiring potential environmental benefits. All the stakeholders should understand their role in upbringing the seaweed sector. For researchers, the focus should be on understanding the potential environmental benefits, thereby communicating it to the policymakers. Researchers can also benefit the seaweed sector by developing genetically modified diseases and pests resistant crops that can generate a high yield. It is essential for the seaweed producers to understand the integration of seaweed farming with the other cost-benefit activities, whereas implementing the seaweed cascading model can reduce production costs. Policymakers must develop seaweed-specific aquaculture policies to help acquire licenses, integrate with other aquaculture activities, and develop an ecosystem for potential investors. They should also focus on understanding consumer behavior and develop incentive systems to boost the overall production. Seaweed farming justifies the environmental benefits and advocates the socio-economic behaviors of coastal rural community development and women empowerment. Therefore, by assigning the frameworks in the ecological interaction, social, and economic segments, the nature-inclusive seaweed farming sector can extend to receive multiple benefits concerning the people, planet, and profit perspective.

References

Agostinia S, Chiavaccib E, Matteuccia M, Torellic M, Pittob L, Lionettia V (2015) Barley beta-glucan promotes MnSOD expression and enhancesangiogenesis under oxidative microenvironment. J Cell Mol Med 19(1):227–238

Al-Thawadi S (2018) Public perception of algal consumption as an alternative food in the Kingdom of Bahrain. Arab J Basic Appl Sci 25(1):1–12. https://doi.org/10.1080/25765299.2018.1449344

Asmus R, de Jonge VN (2021) Use of coastal and estuarine food web models in politics and management: The need for an entire ecosystem approach. Ocean Coast Manag 207(105607):1–2. https://doi.org/10.1016/j.ocecoaman.2021.105607

Barbier M, Araújo R, Rebours C, Jacquemin B, Holdt SL, Charrier B (2020) Development and objectives of the PHYCOMORPH European Guidelines for the Sustainable Aquaculture of Seaweeds (PEGASUS). Bot Mar 63(1):5–16. https://doi.org/10.1515/bot-2019-0051

Barrington K, Ridler N, Chopin T, Robinson S, Robinson B (2010) Social aspects of the sustainability of integrated multi-trophic aquaculture. Aquac Int 18(2):201–211. https://doi.org/10.1007/s10499-008-9236-0

Billing SL, Rostan J, Tett P, Macleod A (2021) Is social license to operate relevant for seaweed cultivation in Europe? Aquaculture 534(736203):1–10. https://doi.org/10.1016/j.aquaculture.2020.736203

Blikra MJ, Altintzoglou T, Løvdal T, Rognså G, Skipnes D, Skåra T, Sivertsvik M, Noriega Fernández E (2021) Seaweed products for the future: using current tools to develop a sustainable food industry. Trends Food Sci Technol 118:765–776. https://doi.org/10.1016/j.tifs.2021.11.002

Bolton JJ (2020) The problem of naming commercial seaweeds. J Appl Phycol 32(2):751–758. https://doi.org/10.1007/s10811-019-01928-0

Bouga M, Combet E (2015) Emergence of seaweed and seaweed-containing foods in the UK: focus on labeling, iodine content, toxicity and nutrition. Food Secur 4(2):240–253. https://doi.org/10.3390/foods4020240

Bureau BM, Department of Environment and Natural Resources (2016) Philippine biodiversity strategy and action plan (2015-2028): bringing resilience to Filipino Communities. In: Cabrido C (ed) Quezon City, Philippines: BMB-DENR, United Nations Development Programme – Global Environment Facility. Foundation for the Philippine Environment

Cai J, Lovatelli A, Aguilar-Manjarrez J, Cornish L, Dabbadie L, Desrochers A, Diffey S, Garrido Gamarro E, Geehan J, Hurtado A, Lucente D, Mair G, Miao W, Potin P, Przybyla C, Reantaso M, Roubach R, Tauati M, Yuan X (2021) Fisheries and seaweeds and microalgae: an overview for unlocking. In: FAO Fisheries and Aquaculture Circular, vol 1229

Camarena Gómez T, Lähteenmäki-Uutela A (2021) European and National Regulations on Seaweed Cultivation and Harvesting. https://www.submariner-network.eu/images/grass/FINAL-GRASS_GoA_3.2._SYKE_regulation_report.pdf

Camia A, Rober N, Jonsson R, Pilli R, Garcia-Condado S, Lopez-Lozano R, van der Velde M, Ronzon T, Gurria P, M'Barek R, Tamosiunas S, Fiore G, Araujo R, Hoepffner N, Marelli L, Giuntoli J (2018) Biomass production, supply, uses and flows in the European Union. First results from an integrated assessment. In: EUR 28993 EN. Publications Office of the European Union, Luxembourg., ISBN 978-92-79-77237-5, JRC109869, https://doi.org/10.2760/181536. https://doi.org/10.2760/539520

Campbell I, Kambey CSB, Mateo JP, Rusekwa SB, Hurtado, Anicia Q, Msuya FE, Stentiford, Grant D, Cottier-Cook EJ (2020) Biosecurity policy and legislation of the seaweed aquaculture industry in Tanzania. J Appl Phycol 32(6):4411–4422. https://doi.org/10.1007/s10811-020-02194-1

Capuzzo E, McKie T (2016) Seaweed in the UK and abroad – status, products, limitations, gaps and Cefas role. Centre for Environment, Fisheries & Aquaculture Science. https://assets.publishing.service.gov.uk/government/uploads/system/uploads/attachment_data/file/546679/FC002I__Cefas_Seaweed_industry_report_2016_Capuzzo_and_McKie.pdf

Charrier B, Abreu MH, Araujo R, Bruhn A, Coates JC, De Clerck O, Katsaros C, Robaina RR, Wichard T (2017) Furthering knowledge of seaweed growth and development to facilitate sustainable aquaculture. New Phytol 216(4):967–975. https://doi.org/10.1111/nph.14728
Christensen LD (2020) Seaweed cultivation in The Faroe Islands: analyzing the potential for forward and fiscal linkages. Mar Policy 119(104015):1–11. https://doi.org/10.1016/j.marpol.2020.104015
Cottier-Cook EJ, Nagabhatla N, Badis Y, Campbell ML, Chopin T, Dai W, Fang J, He P, Hewitt CL, Kim GH, Huo Y, Jiang Z, Kema G, Li X, Liu F, Liu H, Liu Y, Lu Q, Luo Q et al (2016) Safeguarding the future of the global seaweed aquaculture industry. United Nations University (INWEH) and Scottish Association for Marine Science. https://inweh.unu.edu/safeguarding-the-future-of-the-global-seaweed-aquaculture-industry/
Cottier-Cook J, Nagabhatla N, Asri A, Beveridge M, Bianchi P, Bolton J, Bondad-Reantaso M, Brodie J, Buschmann A, Cabarubias J, Campbell I, Chopin T, Critchley A, Lombaerde P, Doumeizel V, Gachon C, Hayashi L, Hewitt C, Huang J, et al (2021) Ensuring the sustainable future of the rapidly expanding global seaweed aquaculture industry -a vision (Issue 06). https://www.researchgate.net/publication/356104942_Ensuring_the_Sustainable_Future_of_the_Rapidly_Expanding_Global_Seaweed_Aquaculture_Industry_-A_Vision
European Commission (2012) COM(2012) 494 final - Blue Growth: opportunities for marine and maritime sustainable growth
European Commission (2015) The new Common Fisheries Policy: sustainability in depth
Firestone J, Kempton W, Lilley MB, Samoteskul K (2012) Public acceptance of offshore wind power across regions and through time. J Environ Plan Manag 55(10):1369–1386. https://doi.org/10.1080/09640568.2012.682782
Food and Agriculture Organiization of the United Nations, F (2017) Recommendation on: IPPC coverage of aquatic plants. In: International Plant Protection Convention
Food and Agriculture Organization of the United Nations, F (2016) The State of World Fisheries and Aquaculture. In: The State of World Fisheries and Aquaculture 2016. Contributing to food security and nutrition for all, Rome
Ganesan AR, Tiwari U, Rajauria G (2019) Seaweed nutraceuticals and their therapeutic role in disease prevention. Food Sci Human Wellness 8(3):252–263. https://doi.org/10.1016/j.fshw.2019.08.001
Ghosh MM, Ghosh A (2014) Financial inclusion strategies of banks: study of Indian States. Int J Appl Financ Manag Perspect © Pezzottaite Journals 3(2):990–996
Goldberg SL (1943) The use of water soluble chlorophyll in oral sepsis. An experimental study of 300 cases. Am J Surg 62(1):117–123. https://doi.org/10.1016/S0002-9610(43)90301-0
Governmnet Offices of Sweden (2015) A Swedish maritime strategy – for people , jobs and the environment: strategy for the development of the maritime industries. Ministry of Enterprise and Innovation
Guiry MD, Guiry GM, Morrison L, Miranda SV, Mathieson AC, Parker BC, Langangen A, John DM, Bárbara I, Carter CF, Garbary DJ (2014) Algae base: an on-line resource for algae. Cryptogam Algol 35(2):105–115. https://doi.org/10.7872/crya.v35.iss2.2014.105
Hayashi L, de Cantarino J, Critchley AT (2020) Challenges to the future domestication of seaweeds as cultivated species: understanding their physiological processes for large-scale production. In: Advances in botanical research, vol 95. Elsevier Ltd. https://doi.org/10.1016/bs.abr.2019.11.010
Holdt SL, Kraan S (2011) Bioactive compounds in seaweed: functional food applications and legislation. J Appl Phycol 23(3):543–597. https://doi.org/10.1007/s10811-010-9632-5
Hurtado AQ, Montaño MNE, Martinez-Goss MR (2013) Commercial production of carrageenophytes in The Philippines: ensuring long-term sustainability for the industry. J Appl Phycol 25(3):733–742. https://doi.org/10.1007/s10811-012-9945-7
Hurtado AQ, Neish IC, Critchley AT (2015) Developments in production technology of Kappaphycus in The Philippines: more than four decades of farming. J Appl Phycol 27(5):1945–1961. https://doi.org/10.1007/s10811-014-0510-4

Hussin R, Yasir SM, Kunjuraman V (2013) Potential of seaweed cultivation as a community-based rural tourism product: a stakeholders' perspectives. Adv Environ Biol 9(5):154–156

Jensen CU, Panduro TE, Lundhede TH, Nielsen ASE, Dalsgaard M, Thorsen BJ (2018) The impact of on-shore and off-shore wind turbine farms on property prices. Energy Policy 116(May 2017):50–59. https://doi.org/10.1016/j.enpol.2018.01.046

Jong LW, Thien VY, Yong YS, Rodrigues KF, Yong WTL (2015) Micropropagation and protein profile analysis by SDS-PAGE of Gracilaria changii (Rhodophyta, Solieriaceae). Aquacult Rep 1:10–14. https://doi.org/10.1016/j.aqrep.2015.03.002

Kambey CSB, Campbell I, Sondak CFA, Nor ARM, Lim PE, Cottier-Cook EJ (2020) An analysis of the current status and future of biosecurity frameworks for the Indonesian seaweed industry. J Appl Phycol 32(4):2147–2160. https://doi.org/10.1007/s10811-019-02020-3

Kambey CSB, Campbell I, Cottier-Cook EJ, Nor ARM, Kassim A, Sade A, Lim PE (2021) Evaluating biosecurity policy implementation in the seaweed aquaculture industry of Malaysia, using the quantitative knowledge, attitude, and practices (KAP) survey technique. Mar Policy 134:104800. https://doi.org/10.1016/j.marpol.2021.104800

Kimathi A, Mirera HOD, Mwaluma J, Wainana M, Ntabo J, Wairimu ME (2018) Seaweed farming policy brief - 1 (Issue 001)

Kleitou P, Kletou D, David J (2018) Is Europe ready for integrated multi-trophic aquaculture? A survey on the perspectives of European farmers and scientists with IMTA experience. Aquaculture 490:136–148. https://doi.org/10.1016/j.aquaculture.2018.02.035

Kumar G, Sahoo D (2011) Effect of seaweed liquid extract on growth and yield of Triticum aestivum var. Pusa Gold. J Appl Phycol 23(2):251–255. https://doi.org/10.1007/s10811-011-9660-9

Kumari R, Kaur I, Bhatnagar AK (2011) Effect of aqueous extract of Sargassum johnstonii Setchell & Gardner on growth, yield and quality of Lycopersicon esculentum Mill. J Appl Phycol 23(3):623–633. https://doi.org/10.1007/s10811-011-9651-x

Kunjuraman V, Hossin A, Hussin R (2019) Women in Malaysian seaweed industry: motivations and impacts. Kajian Malaysia 37(2):49–74. https://doi.org/10.21315/km2019.37.2.3

Lopez-Hortas L, Florez-Fernandez N, Torres MD, Ferreira-Anta T, Casas MP, Balboa EM, Falque E, Domínguez H (2021) Applying seaweed compounds in cosmetics, cosmeceuticals and nutricosmetics. Mar Drugs 19(10):1–30. https://doi.org/10.3390/md19100552

Ma A-C, Chen Z, Wang T, Song N, Yan Q, Fang Y-C, Guan H-S, Liu H-B (2014) Isolation of the molecular species of Monogalactosyldiacylglycerolsfrom Brown Edible SeaweedSargassum horneriand their Inhibitory effects on Triglyceride accumulation in 3T3-L1 Adipocytes. J Agric Food Chem 62:11157–11116

Maehre HK, Edvinsen GK, Eilertsen KE, Elvevoll EO (2016) Heat treatment increases the protein bioaccessibility in the red seaweed dulse (Palmaria palmata), but not in the brown seaweed winged kelp (Alaria esculenta). J Appl Phycol 28(1):581–590. https://doi.org/10.1007/s10811-015-0587-4

Marine Scotland SG (2022) Understanding the potential scale for seaweed-based industries in Scotland. For Marine Scotland

Mariño M, Breckwoldt A, Teichberg M, Kase A, Reuter H (2019) Livelihood aspects of seaweed farming in Rote Island, Indonesia. Marine Policy 107(103600):1–9. https://doi.org/10.1016/j.marpol.2019.103600

Martins A, Vieira H, Gaspar H, Santos S (2014) Marketed marine natural products in the pharmaceutical and cosmeceutical industries: tips for success. Mar Drugs 12:1066–1101

Mateo JP, Campbell I, Cottier-Cook EJ, Luhan MRJ, Ferriols VMEN, Hurtado AQ (2020) Analysis of biosecurity-related policies governing the seaweed industry of The Philippines. J Appl Phycol 32(3):2009–2022. https://doi.org/10.1007/s10811-020-02083-7

Ministry of Communicaation, Science, and Technology, Government of Tanzania, G (2010) The national research and development policy. Government of Tanzania, GoT

Ministry of livestock and fisheries, Tanzania, T. U. R. of T (2018) Tanzania Aquaculture Development Strategy (In Kiswahili)

Moon HE, Islam MN, Ahn BR, Chowdhury SS, Sohn HS, Jung HA, Choi JS (2011) Protein tyrosine phosphatase 1B and α-glucosidase inhibitory phlorotannins from edible brown algae, Ecklonia stolonifera and Eisenia bicyclis. Biosci Biotechnol Biochem 75(8):1472–1480. https://doi.org/10.1271/bbb.110137

Mouritsen OG (2012) The emerging science of gastrophysics and its application to the algal cuisine. Flavour 1(6):1–9. https://doi.org/10.1186/2044-7248-1-6

O'Shea T, Jones R, Markham A, Norell E, Scott J, Theuerkauf SA, Waters T (2019) Towards a Blue Revolution: catalyzing private investment in sustainable aquaculture production systems. The Nature Conservancy and Encourage Capital, Arlington

Palmieri N, Forleo MB (2020) The potential of edible seaweed within the western diet. A segmentation of Italian consumers. Int J Gastron Food Sci 20(100202):1–9. https://doi.org/10.1016/j.ijgfs.2020.100202

Pereira L (2018) Seaweeds as source of bioactive substances and skin care therapy-Cosmeceuticals, algotheraphy, and thalassotherapy. Cosmetics 5(68). https://doi.org/10.3390/cosmetics5040068

Ponia B, Pickering T, Teitelbaum A, Meloti A, Kama J, Diaake S, Mgwaerobo J (2013) Social and economic dimensions of carrageenan seaweed farming in Solomon Islands. In: Social and economic dimensions of carrageenan seaweed farming, vol 17. Royal Society of Chemistry, pp 147–161. https://doi.org/10.1039/c4gc02169j

Rayorath P, Jithesh MN, Farid A, Khan W, Palanisamy R, Hankins SD, Critchley AT, Prithiviraj B (2008) Rapid bioassays to evaluate the plant growth promoting activity of Ascophyllum nodosum (L.) Le Jol. using a model plant, Arabidopsis thaliana (L.) Heynh. J Appl Phycol 20(4):423–429. https://doi.org/10.1007/s10811-007-9280-6

Resources AG, Food FOR (2019) The state of the world's aquatic genetic resources for food and agriculture. FAO Commission on Genetic Resources for Food and Agriculture assessments, Rome. https://doi.org/10.4060/ca5256en

Rodgers CJ, Carnegie RB, Chávez-Sánchez MC, Martínez-Chávez CC, Furones Nozal MD, Hine PM (2015) Legislative and regulatory aspects of molluscan health management. J Invertebr Pathol 131:242–255. https://doi.org/10.1016/j.jip.2015.06.008

Schratter-Sehn AU, Schmidt WFO, Kielhauser R, Langer H, Karcher KH (2000) Establishing a framework for Community action in the field of water policy. Off J Eur Commun 168(1):1–72

Searchinger T, Heimlich R (2015) Avoiding bioenergy competition for food and land. World Resources Institute: 1–44

Steenbergen DJ, Marlessy C, Holle E (2017) Effects of rapid livelihood transitions: examining local co-developed change following a seaweed farming boom. Mar Policy 82:216–223. https://doi.org/10.1016/j.marpol.2017.03.026

Stentiford GD, Sritunyalucksana K, Flegel TW, Williams BAP, Withyachumnarnkul B, Itsathitphaisarn O, Bass D (2017) New paradigms to help solve the global aquaculture disease crisis. PLoS Pathog 13(2):1–6. https://doi.org/10.1371/journal.ppat.1006160

Stévant P, Rebours C, Chapman A (2017) Seaweed aquaculture in Norway: recent industrial developments and future perspectives. Aquac Int 25(4):1373–1390. https://doi.org/10.1007/s10499-017-0120-7

Subasinghe RP, Arthur JR, Bartley DM, De Silva SS, Halwart M, Hishamunda N, Mohan CV, Sorgeloos P (2012) Farming the Waters for People and Food. Proc Global Conf Aquacult 2010:1–910

The European Parliment and the Council of the European Union (2014) Directive 2014/89/EU of the European Parliment and of the Council of 23 July 2014 establishing a framework for maritime spatial planning. Off J Eur Union

The Sustainable Development Goals Report (2016) United Nations. https://doi.org/10.1177/000331979004100307

United Nations (2021) International trade statistics yearbook, vol I

United Republic of Tanzania (2015) Natinal Fisheries Policy. Ministry of Livestock and Fisheries Development, pp 1–58

Valderrama D, Cai J, Hishamunda N, Ridler N (2013) Social and economic dimensions of carrageenan seaweed farming: a global synthesis. In: FAO fisheries and aquaculture technical paper
Van Den Burg SWK, Dagevos H, Helmes RJK (2021) Towards sustainable European seaweed value chains: a triple P perspective. ICES J Mar Sci 78(1):443–450. https://doi.org/10.1093/icesjms/fsz183
Villanueva RD, Romero JB, Montaño MNE, de la Peña PO (2011) Harvest optimization of four Kappaphycus species from The Philippines. Biomass Bioenergy 35(3):1311–1316. https://doi.org/10.1016/j.biombioe.2010.12.044
Vincent A, Stanley A, Ring J (2020) Hidden champion of the ocean: Seaweed as a growth engine for a sustainable Eropean future
Wang J, Jina W, Hou Y, Niu X, Zhanga H, Zhanga Q (2013) Chemical composition and moisture-absorption/retention ability of polysaccharides extracted from five algae. Int J Biol Macromol 57:26–29
Wegeberg S, Mols-Mortensen A, Engell-Sørensen K (2013) Integreret akvakultur i Grønland og på Færøerne. Undersøgelse af potentialet for dyrkning af tang og muligt grønlandsk fiskeopdræt. Aarhus Universitet, DCE – Nationalt Center for Miljø og Energi, 48 s. - Videnskabelig rapport fra DCE - Nationalt Center fo

Index

F. Ozogul et al. (eds.), *Seaweeds and Seaweed-Derived Compounds*,
https://doi.org/10.1007/978-3-031-65529-6

V

Printed in the USA
CPSIA information can be obtained
at www.ICGtesting.com
CBHW050447281024
16499CB00003B/58